# Algebra 1

Boswell
Kanold
Stiff

## Applications • Equations • Graphs

# Worked-Out Solution Key

The Solution Key provides step-by-step solutions
for all the exercises in the Student Edition.

**McDougal Littell**
A HOUGHTON MIFFLIN COMPANY
Evanston, Illinois • Boston • Dallas

ISBN-13: 978-0-618-02052-2  ISBN-10: 0-618-02052-7

12 13 14 15 16 -CKI- 11 10 09 08 07

# Contents

# CHAPTER 1

## Think and Discuss (p. 1)

1. 17.8 pounds per square inch

2. $4.45 \times 10$ or 44.5 pounds per square inch

## Skill Review (p. 2)

1. $60\% = 0.6, \frac{3}{5}$

2. $33\% = 0.33, \frac{33}{100}$

3. $0.08\% = 0.0008, \frac{1}{1250}$

4. $150\% = 1.5, 1\frac{1}{2}$

5. $899 < 901$

6. $64.1 > 64.03$

7. $2050 > 2005$

8. $0.099 > 0.01$

9. $A = (3.5)(3.5)$
   $= 12.25 \text{ cm}^2$
   $P = 4(3.5)$
   $= 14 \text{ cm}$

10. $A = \frac{1}{2}(3)(4)$
    $= 6 \text{ ft}^2$
    $P = 5 + 3 + 4$
    $= 12 \text{ ft}$

11. $A = (5.2)(1.6)$
    $= 8.32 \text{ mi}^2$
    $P = 2(5.2) + 2(1.6)$
    $= 13.6 \text{ mi}$

## Lesson 1.1

### 1.1 Guided Practice (p. 6)

1. To evaluate a variable expression is to replace each letter that represents one or more numbers in a collection of these letters, numbers, and operations with a number.

2. a. multiplication   b. division
   c. addition   d. subtraction

3. $\frac{5}{r}$ or $5 \div r$

4. By using unit analysis the solution will have the proper units.

5. $5y = 5(6) = 30$

6. $\frac{24}{y} = \frac{24}{6} = 4$

7. $y + 19 = 6 + 19 = 25$

8. $y - 2 = 6 - 2 = 4$

9. $y \div 3 = 6 \div 3 = 2$

10. $27 - y = 27 - 6 = 21$

11. $2.5y = 2.5(6) = 15$

12. $3.2 + y = 3.2 + 6 = 9.2$

13. $\text{time} = \frac{15 \text{ mi}}{5 \text{ mi/h}} = \frac{15}{5} \text{ mi} \cdot \frac{h}{mi} = 3 \text{ h}$

14. $\text{perimeter} = 3 \text{ cm} + 4 \text{ cm} + 5 \text{ cm} = 12 \text{ cm}$

15. $\text{distance} = (60 \text{ mi/h})(2.3 \text{ h}) = 60(2.3) \frac{mi}{h} \cdot h = 138 \text{ mi}$

16. $\text{simple interest} = (\$100)\left(\frac{0.05}{\text{year}}\right)(2.5 \text{ years}) =$
    $(\$100)(0.05)(2.5) \cdot \frac{1}{\text{year}} \cdot \text{year} = \$12.50$

17. $\text{time} = \frac{\text{distance}}{\text{rate}}$
    $\text{time} = t \text{ h}$
    $\text{distance} = 10 \text{ mi}$
    $\text{rate} = 1.25 \text{ mi/h}$
    $t = \frac{10}{1.25}$
    $t = 8 \text{ h}$

18. $\text{time} = \frac{mi}{mi/h} = mi \cdot \frac{h}{mi} = h$

### 1.1 Practice and Applications (pp. 6–8)

19. $12x = 12(5) = 60$

20. $b - 7 = 24 - 7 = 17$

21. $0.5d = 0.5(0.5) = 0.25$

22. $9 + p = 9 + 11 = 20$

23. $r(10) = 8.2(10) = 82$

24. $3.67a = 3.67(2) = 7.34$

25. $\frac{6.3}{x} = \frac{6.3}{3} = 2.1$

26. $\frac{2}{3} \cdot x = \frac{2}{3} \cdot \frac{1}{3} = \frac{2}{9}$

27. $\frac{d}{12} = \frac{60}{12} = 5$

28. $\frac{18}{t} = \frac{18}{3} = 6$

29. $\frac{5}{8} - p = \frac{5}{8} - \frac{3}{16} = \frac{7}{16}$

30. $\frac{1}{2} + t = \frac{1}{2} + \frac{1}{2} = 1$

31. $\text{Simple interest} = Prt = (80)(0.02)(1.5) = 2.4$

    After one and a half years, you will have $2.40 of simple interest.

32. $\text{Perimeter} = 4s = 4(7) = 28$

    The square has a perimeter of 28 feet.

    $\text{Perimeter} = 4s = \text{ft} + \text{ft} + \text{ft} + \text{ft} = \text{ft}$

33. She incorrectly represented 2.5% numerically as 2.5. The value should have been written as 0.025.

34. $\text{Simple interest} = (\$300)\left(\frac{0.025}{\text{year}}\right)(0.75 \text{ year}) = \$5.63$

35. $\text{Average speed} = \frac{d}{t} = \frac{75 \text{ mi}}{55 \text{ min}} = 1\frac{4}{11} \frac{mi}{min}$

    The average speed was $1\frac{4}{11}$ miles per minute.

36. $\text{Average speed} = \frac{d}{t} = \frac{40 \text{ ft}}{5 \text{ sec}} = 8 \frac{ft}{sec}$

    The average speed was 8 feet per second.

37. $\text{Average speed} = \frac{d}{t} = \frac{4 \text{ km}}{30 \text{ min}} = \frac{2 \text{ km}}{15 \text{ min}}$

    The average speed was $\frac{2}{15}$ kilometer per minute.

38. $\text{Average speed} = \frac{d}{t} = \frac{24{,}986.73 \text{ mi}}{216.06 \text{ h}} \approx 115.65 \frac{mi}{h}$

    The average speed was 115.65 miles per hour.

**39.** time $= \dfrac{\text{distance}}{\text{rate}} = \dfrac{288 \text{ km}}{96 \frac{\text{km}}{\text{h}}} = 3$ hours

It should take you 3 hours.

**40.** Calories burned $= 2.7 \dfrac{\text{Cal}}{\text{min}} (30 \text{ min}) = 81$ Calories

**41.** rate $= \dfrac{387 \text{ Cal}}{90 \text{ min}} = 4.3 \dfrac{\text{Cal}}{\text{min}}$

**42.** in-line skating

**43.**

| Mountain, Continent | h | $\dfrac{h}{0.3048}$ |
|---|---|---|
| Mt. McKinley, North America | 6194 | 20,322 |
| Mt. Elbrus, Europe | 5633 | 18,481 |
| Mt. Kilimanjaro, Africa | 5963 | 19,564 |
| Mt. Aconcagua, South America | 6959 | 22,831 |

| Mountain, Continent | $29,029 - \dfrac{h}{0.3048}$ |
|---|---|
| Mt. McKinley, North America | 8707 |
| Mt. Elbrus, Europe | 10,548 |
| Mt. Kilimanjaro, Africa | 9465 |
| Mt. Aconcagua, South America | 6198 |

**44.** batting average $= \dfrac{\text{hits}}{\text{at bats}}$      **45.** B

batting average $= a$

hits $= 213$

at bats $= 686$

$a = \dfrac{213}{686}$

$a \approx 0.310$

Alex Rodriguez's batting average for 1998 was 0.310.

**46.** simple interest $= (\$300)\left(\dfrac{0.045}{\text{year}}\right)(10 \text{ years}) = \$135$

C

**47.** perimeter $= 2(28 \text{ cm}) + 2(21 \text{ cm}) = 56 \text{ cm} + 42 \text{ cm} =$

98 cm

B

**48.** time $= \dfrac{\text{distance}}{\text{rate}} = \dfrac{60.5 \text{ ft}}{90 \frac{\text{mi}}{\text{h}}}$

$= \dfrac{60.5 \text{ ft}}{90 \frac{\text{mi}}{\text{h}}}\left(\dfrac{1 \text{ mi}}{5280 \text{ ft}}\right)\left(\dfrac{3600 \text{ sec}}{1 \text{ h}}\right)$

$= \dfrac{60.5 \text{ ft}}{90 \frac{\text{mi}}{\text{h}}}\left(\dfrac{1 \text{ mi}}{5280 \text{ ft}}\right)\left(\dfrac{3600 \text{ sec}}{1 \text{ h}}\right) = \dfrac{21,7800}{47,5200} \text{ sec}$

$\approx 0.46$ seconds

**49.** time $= \dfrac{\text{distance}}{\text{rate}} = \dfrac{17 \text{in.}}{90 \frac{\text{mi}}{\text{h}}}$

$= \dfrac{17 \text{ in.}}{90 \frac{\text{mi}}{\text{h}}}\left(\dfrac{1 \text{ mi}}{63,360 \text{ in.}}\right)\left(\dfrac{3600 \text{ sec}}{1 \text{ h}}\right)$

$= \dfrac{17 \text{ in.}}{90 \frac{\text{mi}}{\text{h}}}\left(\dfrac{1 \text{ mi}}{63,360 \text{ in.}}\right)\left(\dfrac{3600 \text{ sec}}{1 \text{ h}}\right) = \dfrac{61,200}{5,702,400} \text{sec}$

$\approx 0.01$ second

### 1.1 Mixed Review (p. 8)

**50.** $1\frac{7}{8} + \frac{3}{4} = 1\frac{7}{8} + \frac{6}{8} = 2\frac{5}{8}$     **51.** $\frac{2}{5} - \frac{1}{10} = \frac{4}{10} - \frac{1}{10} = \frac{3}{10}$

**52.** $5\frac{3}{4} \cdot \frac{1}{3} = \frac{23}{4} \cdot \frac{1}{3} = \frac{23}{12} = 1\frac{11}{12}$

**53.** $2\frac{1}{5} \div \frac{4}{5} = \frac{11}{5} \div \frac{4}{5} = \frac{11}{5} \cdot \frac{5}{4} = \frac{11}{4} = 2\frac{3}{4}$

**54.** $3\frac{2}{3} \cdot 3 = \frac{11}{3} \cdot \frac{3}{1} = \frac{33}{3} = 11$

**55.** $\frac{6}{5} \div \frac{3}{10} = \frac{6}{5} \cdot \frac{10}{3} = 2 \cdot 2 = 4$

**56.** $4\frac{5}{9} - \frac{1}{5} = \frac{41}{9} - \frac{1}{5} = \frac{205}{45} - \frac{9}{45} = \frac{196}{45} = 4\frac{16}{45}$

**57.** $8\frac{9}{10} + 1\frac{1}{5} = \frac{89}{10} + \frac{6}{5} = \frac{89}{10} + \frac{12}{10} = \frac{101}{10} = 10\frac{1}{10}$

**58.** area $=$ length $\cdot$ width

area $= 4 \text{ cm} \cdot 8 \text{ cm} = 32 \text{ cm}^2$

**59.** area $=$ length $\cdot$ width

area $= 10 \text{ in.} \cdot 7 \text{ in.} = 70 \text{ in.}^2$

**60.** area $=$ length $\cdot$ width

area $= 6 \text{ m} \cdot 28 \text{ m} = 168 \text{ m}^2$

**61.** $2x = 2(2) = 4$      **62.** $(x)(x) = (2)(2) = 4$

**63.** $3x = 3(2) = 6$      **64.** $(x)(x)(x) = (2)(2)(2) = 8$

**65.** $4x = 4(2) = 8$      **66.** $(x)(x)(x)(x) = (2)(2)(2)(2) = 16$

**67.** $5x = 5(2) = 10$

**68.** $(x)(x)(x)(x)(x) = (2)(2)(2)(2)(2) = 32$

## Lesson 1.2

### 1.2 Guided Practice (p. 12)

**1.** base, exponent, power

**2.** The base for $3x^2$ is $x$. The base for $(3x)^2$ is $3x$.

**3.** $3x^2 = 3(4)^2 = 3(16) = 48$

$(3x)^2 = (3 \cdot 4)^2 = 12^2 = 144$

**4.** C     **5.** D     **6.** B     **7.** A     **8.** $x^2 = 3^2 = 9$

**9.** $(x + 1)^3 = (3 + 1)^3 = 4^3 = 64$

**10.** $2x^2 = 2(3)^2 = 2(9) = 18$

**11.** $(2x)^3 = (2 \cdot 3)^3 = 6^3 = 216$

**12.** $(x - 1)^4 = (3 - 1)^4 = 2^4 = 16$

**13.** $5^x = 5^3 = 125$     **14.** $(3x)^4 = (3 \cdot 3)^4 = 9^4 = 6561$

**15.** $10^x = 10^3 = 1000$

**16.** $6s^2 = 6(35)^2 = 6(1225) = 7350$

The surface area of the stereo speaker is 7350 cm$^2$.

# Chapter 1 continued

*1.2 Practice and Applications (pp. 12–14)*

**17.** $2^3$    **18.** $p^2$    **19.** $9^y$    **20.** $b^8$    **21.** $3^4y$

**22.** $t^2$    **23.** $c^6$    **24.** $5x^5$    **25.** $(4x)^3$

**26.** $10^2 = 10 \cdot 10 = 100$    **27.** $5^2 = 5 \cdot 5 = 25$

**28.** $8^2 = 8 \cdot 8 = 64$    **29.** $6^4 = 6 \cdot 6 \cdot 6 \cdot 6 = 1296$

**30.** $10^5 = 10 \cdot 10 \cdot 10 \cdot 10 \cdot 10 = 100{,}000$

**31.** $7^4 = 7 \cdot 7 \cdot 7 \cdot 7 = 2401$

**32.** $4^6 = 4 \cdot 4 \cdot 4 \cdot 4 \cdot 4 \cdot 4 = 4096$

**33.** $9^3 = 9 \cdot 9 \cdot 9 = 729$    **34.** $2^5 = 2 \cdot 2 \cdot 2 \cdot 2 \cdot 2 = 32$

**35.** $4^n = 4^5 = 1024$    **36.** $b^4 = 9^4 = 6561$

**37.** $x^6 = 10^6 = 1{,}000{,}000$    **38.** $c^6 = 2^6 = 64$

**39.** $w^3 = 13^3 = 2197$    **40.** $p^2 = 2.5^2 = 6.25$

**41.** $(x + y)^2 = (5 + 3)^2 = 8^2 = 64$

**42.** $m - n^2 = 25 - 4^2 = 25 - 16 = 9$

**43.** $(a - b)^4 = (4 - 2)^4 = 2^4 = 16$

**44.** $c^3 + d = 4^3 + 16 = 64 + 16 = 80$

**45.** $(d - 3)^2 = (13 - 3)^2 = 10^2 = 100$

**46.** $16 + x^3 = 16 + 2^3 = 16 + 8 = 24$

**47.** $9^5 = 59{,}049$    **48.** $2^{10} = 1024$

**49.** $5^9 = 1{,}953{,}125$    **50.** $3^{11} = 177{,}147$

**51.** $8^6 = 262{,}144$    **52.** $12^7 = 35{,}831{,}808$

**53.** $6^8 = 1{,}679{,}616$    **54.** $13^5 = 371{,}293$

**55.** $(5w)^3 = (5 \cdot 5)^3 = 25^3 = 15{,}625$

**56.** $6t^4 = 6(3)^4 = 6(81) = 486$

**57.** $7b^2 = 7(7)^2 = 7(49) = 343$

**58.** $2x^2 = 2(15)^2 = 2(225) = 450$

**59.** $(8x)^3 = (8 \cdot 2)^3 = 16^3 = 4096$

**60.** $5y^5 = 5(2)^5 = 5(32) = 160$

**61.**

| Side length, $s$ | 1 | 2 | 3 | 4 | 5 |
|---|---|---|---|---|---|
| Area, $s^2$ | 1 | 4 | 9 | 16 | 25 |

**62.**

| Power | $10^2$ | $100^2$ | $1000^2$ | $10{,}000^2$ |
|---|---|---|---|---|
| Evaluate | 100 | 10,000 | 1,000,000 | 100,000,000 |

The number of zeros increases by two for each additional zero in the base of the power.

**63.** $A = s^2$

$A = 12.2^2$

$A = 148.84 \text{ ft}^2$

Jean needs 148.84 square feet of carpet.

**64.** $V = s^3$

$V = 9.5^3$

$V = 857.375 \text{ in.}^3$

Kuhn needed 857.38 cubic inches of liquid glass.

**65.** $V = s^3 = 12^3 = 1728 \text{ in.}^3$

The volume of the storage space is 1728 cubic inches.

**66.** volume $=$ length $\cdot$ width $\cdot$ height $= 50 \cdot 19.5 \cdot 3 = 2925 \text{ m}^3$

The swimming pool can hold 2925 cubic meters of water.

**67.** volume $= \frac{1}{3} \cdot$ height $\cdot$ area of base $= \frac{1}{3} \cdot 100 \cdot (200)^2 \approx 1{,}333{,}333.33 \text{ ft}^3$

The volume of space is about 1,333,333.33 cubic feet.

**68. a.** $V = \frac{1}{2}s^3 = \frac{1}{2}(6)^3 = 108 \text{ in.}^3$

**b.** A

**c.** Answers may vary.

**69.**

| $9^1$ | $9^2$ | $9^3$ | $9^4$ | $9^5$ | $9^6$ |
|---|---|---|---|---|---|
| 9 | 81 | 729 | 6561 | 59,049 | 531,441 |

| $9^7$ | $9^8$ |
|---|---|
| 4,782,969 | 43,046,721 |

The pattern for the last digit is "9, 1, 9, 1, . . . ."

**70.**

| $8^1$ | $8^2$ | $8^3$ | $8^4$ | $8^5$ |
|---|---|---|---|---|
| 8 | 64 | 512 | 4096 | 32,768 |

| $8^6$ | $8^7$ | $8^8$ |
|---|---|---|
| 262,144 | 2,097,152 | 16,777,216 |

The pattern for the last digit is "8, 4, 2, 6, 8, 4, 2, 6, . . . ."

**71.**

| $7^1$ | $7^2$ | $7^3$ | $7^4$ | $7^5$ |
|---|---|---|---|---|
| 7 | 49 | 343 | 2401 | 16,807 |

| $7^6$ | $7^7$ | $7^8$ |
|---|---|---|
| 117,649 | 823,543 | 5,764,801 |

The pattern for the last digit is "7, 9, 3, 1, 7, 9, 3, 1, . . . ."

*1.2 Mixed Review (p. 14)*

**72.** perimeter $= 4x = 4(1.7) = 6.8$

**73.** perimeter $= 6x = 6(1.7) = 10.2$

**74.** perimeter $= 2(2x) + x = 2(2 \cdot 1.7) + 1.7 = 8.5$

**75.** $\frac{5}{8} = 0.625 = 62.5\%$    **76.** $\frac{3}{4} = 0.75 = 75\%$

**77.** $\frac{11}{20} = 0.55 = 55\%$    **78.** $\frac{4}{25} = 0.16 = 16\%$

**79.** $7x = 7(3) = 21$    **80.** $y + 2 = 10 + 2 = 12$

**81.** $\frac{a}{2} = \frac{8}{2} = 4$    **82.** $m - 5 = 17 - 5 = 12$

**83.** $\frac{9}{b} = \frac{9}{4} = 2\frac{1}{4}$    **84.** $9b = 9(4) = 36$

# Chapter 1 *continued*

## Technology Activity 1.2 (p. 15)

**1.** $2^{25} \approx 3.36\text{E}7$

$2^{50} \approx 1.1\text{E}15$

$2^{75} \approx 3.8\text{E}22$

$2^{100} \approx 1.3\text{E}30$

**2.** $3^5 = 243$

$3^7 = 2187$

$3^9 = 19{,}683$

$3^{11} = 177{,}147$

$3^{13} \approx 1.59\text{E}6$

**3.** $4^2 = 16$

$4^3 = 64$

$4^4 = 256$

$4^5 = 1024$

$4^6 = 4096$

**4.** $5^4 = 625$

$5^8 = 390{,}625$

$5^{12} \approx 2.44\text{E}8$

$5^{16} \approx 1.5\text{E}11$

**5.** $6^2 = 36$

$6^4 = 1296$

$6^6 = 46{,}656$

$6^8 \approx 1.68\text{E}6$

$6^{10} \approx 6.05\text{E}7$

**6.** $10^4 = 10{,}000$

$10^7 = 1\text{E}7$

$10^{10} = 1\text{E}10$

$10^{13} = 1\text{E}13$

$10^{16} = 1\text{E}16$

## Lesson 1.3

### Activity (p. 16)

**1.** $(3 \cdot 4)^2 + (8 \div 4)$

**2.** $3 \cdot (4^2 + 8 \div 4)$

**3.** $((3 \cdot 4)^2 + 8) \div 4$

### 1.3 Guided Practice (p. 19)

**1.** Begin with operations within grouping symbols. Next, evaluate powers. Then do multiplications and divisions from left to right. Last, do additions and subtractions from left to right.

**2.** the exponent

**3.** the operation that is leftmost

**4.** $x^4 - 3 = 2^4 - 3 = 16 - 3 = 13$

**5.** $5 \cdot 6y = 5 \cdot 6(5) = 5 \cdot 30 = 150$

**6.** $a^3 + 10a = 3^3 + 10(3) = 27 + 10(3) = 27 + 30 = 57$

**7.** $\dfrac{16}{x} - 2 = \dfrac{16}{4} - 2 = 4 - 2 = 2$

**8.** $\dfrac{22}{x} \div 2 + 16 = \left(\dfrac{22}{11} \div 2\right) + 16 = (2 \div 2) + 16 = 1 + 16 = 17$

**9.** $\dfrac{16}{n} + 2^3 - 10 = \dfrac{16}{8} + 2^3 - 10 = \dfrac{16}{8} + 8 - 10 = 2 + 8 - 10 = 0$

**10.** $(x + 5) \div 4 = (9 + 5) \div 4 = 14 \div 4 = 3\frac{1}{2}$

**11.** $b + 6 \div 4 = 1.5 + (6 \div 4) = 1.5 + 1.5 = 3$

**12.** No. He can use grouping symbols around $12 \div 4$.

### 1.3 Practice and Applications (pp. 19–21)

**13.** $3 + 2x^3 = 3 + 2(2)^3 = 3 + 2(8) = 3 + 16 = 19$

**14.** $y^4 \div 8 = (4)^4 \div 8 = 256 \div 8 = 32$

**15.** $6 \cdot 2p^2 = 6 \cdot 2(5)^2 = 6 \cdot 2(25) = 12(25) = 300$

**16.** $t^5 - 10t = 3^5 - 10(3) = 243 - 10(3) = 243 - 30 = 213$

**17.** $13 + 3b = 13 + 3(7) = 13 + 21 = 34$

**18.** $3r^2 - 17 = 3(6)^2 - 17 = 3(36) - 17 = 108 - 17 = 91$

**19.** $\dfrac{x}{7} + 16 = \dfrac{14}{7} + 16 = 2 + 16 = 18$

**20.** $27 - \dfrac{24}{b} = 27 - \dfrac{24}{8} = 27 - 3 = 24$

**21.** $\dfrac{4}{5} \div n + 13 = \left(\dfrac{4}{5} \div \dfrac{1}{5}\right) + 13 = 4 + 13 = 17$

**22.** $\dfrac{9}{10} \cdot y - \dfrac{3}{10} = \left(\dfrac{9}{10} \cdot \dfrac{1}{2}\right) - \dfrac{3}{10} = \dfrac{9}{20} - \dfrac{3}{10} = \dfrac{3}{20}$

**23.** $4 + 9 - 1 = 12$

**24.** $3 \cdot 2 + \dfrac{5}{9} = 6 + \dfrac{5}{9} = 6\dfrac{5}{9}$

**25.** $6 \div 3 + 2 \cdot 7 = (6 \div 3) + (2 \cdot 7) = 2 + 14 = 16$

**26.** $5 + 8 \cdot 2 - 4 = 5 + (8 \cdot 2) - 4 = 5 + 16 - 4 = 17$

**27.** $16 \div 8 \cdot 2^2 = 16 \div 8 \cdot 4 = 2 \cdot 4 = 8$

**28.** $2 \cdot 3^2 \div 7 = 2 \cdot 9 \div 7 = 18 \div 7 = 2\dfrac{4}{7}$

**29.** $10 \div (3 + 2) + 9 = 10 \div 5 + 9 = 2 + 9 = 11$

**30.** $7[(18 - 6) - 6] = 7(12 - 6) = 7(6) = 42$

**31.** $[(7 - 4)^2 + 3] + 15 = (3)^2 + 3 + 15 = 9 + 3 + 15 = 27$

**32.** $3(2.7 \div 0.9) - 5 = 3(3) - 5 = 9 - 5 = 4$

**33.** $6(5 - 3)^2 + 3 = 6(2)^2 + 3 = 6(4) + 3 = 24 + 3 = 27$

**34.** $[10 + (5^2 \cdot 2)] \div 6 = [10 + (25 \cdot 2)] \div 6 = (10 + 50) \div 6 = 60 \div 6 = 10$

**35.** $\dfrac{1}{3}(9 \cdot 3) + 18 = \dfrac{1}{3}(27) + 18 = 9 + 18 = 27$

**36.** $\dfrac{1}{2} \cdot 26 - 3^2 = \dfrac{1}{2} \cdot 26 - 9 = 13 - 9 = 4$

**37.** $2.5 \cdot 0.5^2 \div 5 = 2.5 \cdot 0.25 \div 5 = 0.625 \div 5 = 0.125$

**38.** $\dfrac{9 \cdot 2}{4 + 3^2 - 1} = \dfrac{9 \cdot 2}{4 + 9 - 1} = \dfrac{18}{4 + 9 - 1} = \dfrac{18}{12} = 1\dfrac{1}{2}$

**39.** $\dfrac{13 - 4}{18 - 4^2 + 1} = \dfrac{13 - 4}{18 - 16 + 1} = \dfrac{9}{3} = 3$

**40.** $\dfrac{5^3 \cdot 2}{1 + 6^2 - 8} = \dfrac{125 \cdot 2}{1 + 36 - 8} = \dfrac{250}{1 + 36 - 8} = \dfrac{250}{29} = 8\dfrac{18}{29}$

**41.** The vendor only obtained the sales tax for the $12.48 ring.

**42.** B   To have the correct order of operations, the numerator needs to be grouped.

**43.** Calculator B;   $15 - [(6 \div 3) \times 4]$

**44.** Calculator A;   $15 - (9 \div 3) + 7$

**45.** Calculator A;   $15 + (10 \div 5) + 4$

**46.** Calculator B;   $(4 \times 3) + (6 \div 2)$

**47.** $(2 \times \$49.99) + (3 \times \$44.10)$

**48.** $35(\$230 + \$300 + \$50 + \$25 + \$100 + \$200) - \$3200$

**49.** $35(905) - 3200 = 31{,}675 - 3200 = \$28{,}475$

# Chapter 1 *continued*

**50.** $3(7) + 1(5) + 1(4) + 1(0)$

**51.** $3(7) + 1(5) + 1(4) + 1(0) = 21 + 5 + 4 + 0 = \$30$

**52.** $h = 2$ meters, $b_1 = 6$ meters, $b_2 = 10$ meters

$A = \frac{1}{2}h(b_1 + b_2)$

$A = \frac{1}{2}(2)(6 + 10) = \frac{1}{2}(2)(16) = (1)(16) = 16 \text{ m}^2$

**53.** surface area $= (2\pi r \cdot h) + (2 \cdot \pi r^2)$

**54.** $h = 10.5$ centimeters, $r = 2.5$ centimeters, $\pi = 3.14$

surface area $= (2 \cdot 3.14 \cdot 2.5 \cdot 10.5) + (2 \cdot 3.14 \cdot 2.5^2)$

$= (2 \cdot 3.14 \cdot 2.5 \cdot 10.5) + (2 \cdot 3.14 \cdot 6.25)$

$= (164.85) + (39.25) = 204.1 \text{ cm}^2$

**55. a.** $0.1(5 \cdot 2.5) + 0.2(10 \cdot 1.3)$

  **b.** $0.1(5 \cdot 2.5) + 0.2(10 \cdot 1.3)$

  $= 0.1(12.5) + 0.2(13)$

  $= 1.25 + 2.6 = \$3.85$

  **c.** There will be \$3.35 left over. First, find out how much the supplies would have been without the discount. Subtract the discount from the total without the discount. Finally, subtract that total from \$25.

**56. a.** $2 \cdot (3^3 + 4)$  **b.** $(2 \cdot 3)^3 + 4$

  **c.** $2 \cdot 3^{(3+4)}$  **d.** $(2 \cdot 3)^{(3+4)}$

**57.** Answers may vary.

### 1.3 Mixed Review (p. 22)

**58.** $8a = 8(4) = 32$  **59.** $\frac{24}{x} = \frac{24}{3} = 8$

**60.** $c + 15 = 12.5 + 15 = 27.5$  **61.** $\frac{4}{3} \cdot x = \frac{4}{3} \cdot \frac{1}{6} = \frac{2}{9}$

**62.** $9d = 9(0.5) = 4.5$  **63.** $\frac{5}{16} - p = \frac{5}{16} - \frac{3}{8} = -\frac{1}{16}$

**64.** $(6w)^2 = (6 \cdot 5)^2 = 30^2 = 900$

**65.** $4(t^3) = 4(3^3) = 4(27) = 108$

**66.** $9b^2 = 9(8)^2 = 9(64) = 576$

**67.** $5x^2 = 5(16)^2 = 5(256) = 1280$

**68.** $(7x)^3 = (7 \cdot 2)^3 = 14^3 = 2744$

**69.** $6y^5 = 6(4)^5 = 6(1024) = 6144$

**70.** simple interest $= (120)(0.025)(2) = \$6$

You would earn six dollars.

**71.** time $= \frac{170}{68} = 2\frac{1}{2}$

It would take $2\frac{1}{2}$ hours.

**72.** volume $=$ length $\cdot$ width $\cdot$ height

volume $= 4^3 = 64 \text{ cm}^3$

The volume of the cube is 64 cm³.

area $=$ length $\cdot$ width $= 4 \cdot 4 = 16 \text{ cm}^2$

The area of one side is 16 cm².

### Quiz 1 (p. 22)

**1.** $6x = 6(3) = 18$  **2.** $\frac{36}{x} = \frac{36}{3} = 12$

**3.** $x + 29 = 3 + 29 = 32$  **4.** $x - 2 = 3 - 2 = 1$

**5.** $x \div 3 = 3 \div 3 = 1$  **6.** $21 - x = 21 - 3 = 18$

**7.** $4.5x = 4.5(3) = 13.5$

**8.** $13.7 + x = 13.7 + 3 = 16.7$

**9.** average speed $= \frac{120 \text{ mi}}{3 \text{ h}} = 40 \frac{\text{mi}}{\text{h}}$

**10.** average speed $= \frac{90 \text{ mi}}{1.5 \text{ h}} = 60 \frac{\text{mi}}{\text{h}}$

**11.** average speed $= \frac{360 \text{ mi}}{6 \text{ h}} = 60 \frac{\text{mi}}{\text{h}}$

**12.** $6^3$  **13.** $4^5$  **14.** $(5y)^3$  **15.** $5^2$  **16.** $36 \cdot t$  **17.** $(7x)^2$

**18.** $x^5 = 6^5 = 7776$  **19.** $2x^3 = 2(6)^3 = 2(216) = 432$

**20.** $(2x)^3 = (2 \cdot 6)^3 = 12^3 = 1728$

**21.** $x^2 - 3 = 6^2 - 3 = 36 - 3 = 33$

**22.** $\frac{6 \cdot 3}{7 + (2^3 - 1)} = \frac{6 \cdot 3}{7 + (8 - 1)} = \frac{18}{7 + (8 - 1)} =$

$\frac{18}{14} = 1\frac{2}{7}$

**23.** $\frac{2^5 - 12}{2(5^2 - 5)} = \frac{32 - 12}{2(25 - 5)} = \frac{32 - 12}{2(20)} = \frac{32 - 12}{40} =$

$\frac{20}{40} = \frac{1}{2}$

**24.** $\frac{3^2 - 3}{2 \cdot 9} = \frac{9 - 3}{2 \cdot 9} = \frac{9 - 3}{18} = \frac{6}{18} = \frac{1}{3}$

**25.** $\frac{2(17 + 2 \cdot 4)}{6^2 - 11} = \frac{2(17 + 2 \cdot 4)}{36 - 11} = \frac{2(17 + 8)}{36 - 11} =$

$\frac{2(25)}{36 - 11} = \frac{50}{36 - 11} = \frac{50}{25} = 2$

**26.** volume $= 1.2^3 = 1.728 \text{ ft}^3$

## Lesson 1.4

### Developing Concepts Activity 1.4 (p. 23)
### Exploring the Concept

**1.**

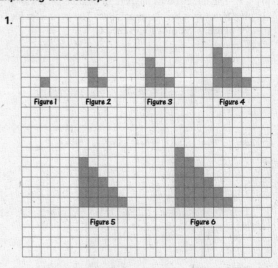

Figure 1  Figure 2  Figure 3  Figure 4

Figure 5  Figure 6

# Chapter 1 *continued*

**2.**

| Figure | 1 | 2 | 3 | 4 | 5 | 6 |
|---|---|---|---|---|---|---|
| Perimeter | 4 | 8 | 12 | 16 | 20 | 24 |
| Pattern | $4 \cdot 1$ | $4 \cdot 2$ | $4 \cdot 3$ | $4 \cdot 4$ | $4 \cdot 5$ | $4 \cdot 6$ |

**3.** $4 \cdot 25 = 100$

**4.** $4n$. The table shows the pattern that the perimeter is four times the figure number. The perimeter for the $N^{th}$ figure is $n$ multiplied by four.

## Drawing Conclusions

**1.**

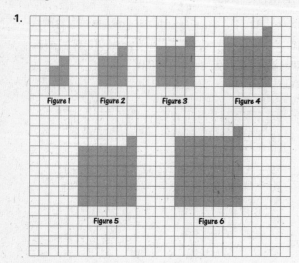

Figure 1  Figure 2  Figure 3  Figure 4

Figure 5  Figure 6

**2.**

| Figure | 1 | 2 | 3 |
|---|---|---|---|
| Perimeter | 10 | 14 | 18 |
| Pattern | $(4 \cdot 1) + 6$ | $(4 \cdot 2) + 6$ | $(4 \cdot 3) + 6$ |

| Figure | 4 | 5 | 6 |
|---|---|---|---|
| Perimeter | 22 | 26 | 30 |
| Pattern | $(4 \cdot 4) + 6$ | $(4 \cdot 5) + 6$ | $(4 \cdot 6) + 6$ |

**3.** $4n + 6$. The table shows the pattern that the perimeter is six plus four times the figure number. The perimeter for the $N^{th}$ figure is six plus four times $n$.

**4.**

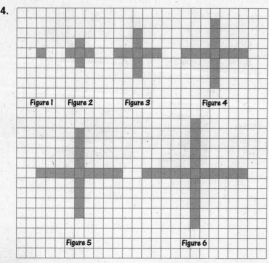

Figure 1  Figure 2  Figure 3  Figure 4

Figure 5  Figure 6

| Figure | 1 | 2 | 3 |
|---|---|---|---|
| Perimeter | 4 | 12 | 20 |
| Pattern | $(8 \cdot 1) - 4$ | $(8 \cdot 2) - 4$ | $(8 \cdot 3) - 4$ |

| Figure | 4 | 5 | 6 |
|---|---|---|---|
| Perimeter | 28 | 36 | 44 |
| Pattern | $(8 \cdot 4) - 4$ | $(8 \cdot 5) - 4$ | $(8 \cdot 6) - 4$ |

$8n - 4$. The table shows the pattern that the perimeter is the product of eight times the figure number subtracted by four. The perimeter is the product of eight times $n$ subtracted by four.

## Activity (p. 25)

**1.** E; 4    **2.** C; 7    **3.** A; 5    **4.** B; 3    **5.** D; 2

## 1.4 Guided Practice (p. 27)

**1.** equation   **2.** expression   **3.** expression

**4.** equation   **5.** inequality   **6.** inequality

**7.** Left side is $8 + 3x$. Right side is $5x - 9$.

**8.** Jan subtracted before she multiplied.

$$3(6) - 4 = 14$$
$$18 - 4 = 14$$
$$14 = 14$$

**9.** B   **10.** D   **11.** C   **12.** A   **13.** B

## 1.4 Practice and Applications (pp. 27–30)

**14.** $3(4) + 1 \overset{?}{=} 13$

$12 + 1 \overset{?}{=} 13$

$13 = 13$

4 is a solution.

**15.** $5 + 3^2 \overset{?}{=} 17$

$5 + 9 \overset{?}{=} 17$

$14 \neq 17$

3 is not a solution.

**16.** $4(2) + 2 \overset{?}{=} 2(2) + 8$

$8 + 2 \overset{?}{=} 4 + 8$

$10 \neq 12$

2 is not a solution.

**17.** $2(1)^3 + 3 \overset{?}{=} 5$

$2(1) + 3 \overset{?}{=} 5$

$2 + 3 \overset{?}{=} 5$

$5 = 5$

1 is a solution.

**18.** $5(5) - 10 \overset{?}{=} 11$

$25 - 10 \overset{?}{=} 11$

$15 \neq 11$

5 is not a solution.

**19.** $4(7) - 4 \overset{?}{=} 30 - 7$

$28 - 4 \overset{?}{=} 30 - 7$

$24 \neq 23$

7 is not a solution.

**20.** $6(5) - 5 \overset{?}{=} 20$

$30 - 5 \overset{?}{=} 20$

$25 \neq 20$

5 is not a solution.

**21.** $\frac{36}{4} - 9 \overset{?}{=} 9$

$9 - 9 \overset{?}{=} 9$

$0 \neq 9$

36 is not a solution.

# Chapter 1 *continued*

**22.** $10 + 4(10) \stackrel{?}{=} 60 - 2(10)$

$10 + 40 \stackrel{?}{=} 60 - 20$

$50 \neq 40$

10 is not a solution.

**23.** $10 + \frac{14}{7} \stackrel{?}{=} 12$

$10 + 2 \stackrel{?}{=} 12$

$12 = 12$

14 is a solution.

**24.** $6^2 - 5 \stackrel{?}{=} 20$

$36 - 5 \stackrel{?}{=} 20$

$31 \neq 20$

6 is not a solution.

**25.** $4(3) - 3 \stackrel{?}{=} \frac{12}{3}$

$12 - 3 \stackrel{?}{=} 4$

$9 \neq 4$

3 is not a solution.

**26.** What number can be added to 3 to obtain 8?

Answer: 5

**27.** What number can be added to 6 to obtain 11?

Answer: 5

**28.** What number can 11 be subtracted from to obtain 20?

Answer: 31

**29.** What number can be multiplied by 3 to obtain 12?

Answer: 4

**30.** What number can be divided by 4 to obtain 5?

Answer: 20

**31.** What number can be multiplied by 4 to obtain 36?

Answer: 9

**32.** What number can be multiplied by 4 and have 1 subtracted from the result to obtain 11?

Answer: 3

**33.** What number can be multiplied by 2 and have 1 subtracted from the result to obtain 9?

Answer: 5

**34.** What number can be squared to obtain 144?

Answer: 12

**35.** What number can be divided by 7 to obtain 3?

Answer: 21

**36.** What number can be multiplied by 5 and have 2 subtracted from the result to obtain 3?

Answer: 1

**37.** What number can be cubed to obtain 125?

Answer: 5

**38.** $3 - 2 \stackrel{?}{<} 6$

$1 < 6$

3 is a solution.

**39.** $5 + 4 \stackrel{?}{>} 8$

$9 > 8$

4 is a solution.

**40.** $5 + 5(1) \stackrel{?}{\geq} 10$

$5 + 5 \stackrel{?}{\geq} 10$

$10 \geq 10$

1 is a solution.

**41.** $4(2) - 1 \stackrel{?}{\geq} 8$

$8 - 1 \stackrel{?}{\geq} 8$

$7 \not\geq 8$

2 is not a solution.

**42.** $3(5) - 15 \stackrel{?}{<} 0$

$15 - 15 \stackrel{?}{<} 0$

$0 \not< 0$

5 is not a solution.

**43.** $11(9) \stackrel{?}{\leq} 9 - 7$

$99 \stackrel{?}{\leq} 9 - 7$

$99 \not\leq 2$

9 is not a solution.

**44.** $6 + 3 \stackrel{?}{\leq} 8$

$9 \not\leq 8$

3 is not a solution.

**45.** $29 - 4(7) \stackrel{?}{>} 5$

$29 - 28 \stackrel{?}{>} 5$

$1 \not> 5$

7 is not a solution.

**46.** $6^2 + 6 \stackrel{?}{>} 40$

$36 + 6 \stackrel{?}{>} 40$

$42 > 40$

6 is a solution.

**47.** $22 - 7 \stackrel{?}{\geq} 15$

$15 \geq 15$

22 is a solution.

**48.** $6(7) - 16 \stackrel{?}{<} 20$

$42 - 16 \stackrel{?}{<} 20$

$26 \not< 20$

7 is not a solution.

**49.** $2^3 - 2 \stackrel{?}{\leq} 8$

$8 - 2 \stackrel{?}{\leq} 8$

$6 \leq 8$

2 is a solution.

**50.** $9 + 2(9) \stackrel{?}{<} 30$

$9 + 18 \stackrel{?}{<} 30$

$27 < 30$

9 is a solution.

**51.** $4(3 \cdot 4 + 2) \stackrel{?}{>} 50$

$4(12 + 2) \stackrel{?}{>} 50$

$4(14) \stackrel{?}{>} 50$

$56 > 50$

4 is a solution.

**52.** $\frac{3 + 5}{3} \stackrel{?}{\leq} 4$

$\frac{8}{3} \stackrel{?}{\leq} 4$

$2\frac{2}{3} \leq 4$

3 is a solution.

**53.** $\frac{25 - 5}{5} \stackrel{?}{\geq} 4$

$\frac{20}{5} \stackrel{?}{\geq} 4$

$4 \geq 4$

5 is a solution.

**54.** $6^2 - 10 \stackrel{?}{>} 16$

$36 - 10 \stackrel{?}{>} 16$

$26 > 16$

6 is a solution.

**55.** $8(21 - 8) \stackrel{?}{<} 100$

$8(13) \stackrel{?}{<} 100$

$104 \not< 100$

8 is not a solution.

**56.** C **57.** G **58.** F **59.** H **60.** I **61.** B **62.** D

**63.** A **64.** E

**65.** 20 represents the number of square feet needed for each station.

*x* represents the number of computer stations.

400 represents the area available for computer stations.

$20x = 400$

$x = 20$ computer stations

$20(20) = 400$

# Chapter 1 *continued*

**66.** 8 represents the number of complete screens needed to receive a bonus.

$x$ represents the number of bonuses.

96 represents the number of screens completed.

$$8x = 96$$
$$x = 12 \text{ bonuses}$$
$$8(12) = 96$$

**67.** 15 represents the cost of filling the tank one time.

$x$ represents the number of times you can completely fill the tank.

65 represents the total amount of money you have available for gas.

$$15(4) \overset{?}{\le} 65$$
$$60 \le 65$$

Yes, you can completely fill the tank four times.

**68.** 5 represents the money you save each week.

$n$ represents the number of weeks.

250 represents the cost of the violin and bow.

$$5(52) \overset{?}{\ge} 250$$
$$260 \ge 250$$

Yes, you will have enough money in a year.

**69.**
$$d = rt$$
$$10{,}000 = r(25)$$
$$r = 400 \,\tfrac{\text{meters}}{\text{minute}}$$

**70.**
$$I = Prt$$
$$300 = 1000(0.03)t$$
$$t = 10 \text{ years}$$

**71.**
$$d = rt$$
$$300 = 60t$$
$$t = 5 \text{ hours}$$

**72.**
$$A = lw$$
$$\frac{800}{20} = \frac{15}{3}l$$
$$l = 8 \text{ yards} = 24 \text{ feet}$$

**73.**
$$C = \tfrac{5}{9}(F - 32)$$
$$= \tfrac{5}{9}(32 - 32)$$
$$= \tfrac{5}{9}(0)$$
$$= 0°C$$

**74.**
$$V = s^3$$
$$27 = s^3$$
$$s = 3 \text{ inches}$$

**75.**
$$A = lw$$
$$80 = 10w$$
$$w = 8 \text{ feet}$$

**76.**
$$S = 6s^2$$
$$96 = 6s^2$$
$$16 = s^2$$
$$s = 4 \text{ feet}$$

**77. a.** The passenger jet's length is 1.212 times the wingspan.

**b.** The passenger jet is longer than it is wide.

**78.** Note: Speeds are in miles per hour.

| Airplane type | X-15A-2 | Supersonic transport | Commercial jet |
|---|---|---|---|
| **Mach number, m** | 6.7 | 2.2 | 0.9 |
| **Speed, v** | 4422 | 1452 | 594 |

**79.** B   $75p \ge 2600$   **80.** B   $100x \le 13{,}000$

**81.** A   $250 = 238 + 12$, so, $250 \ge x + 12$

**82.** $1.79x \ge 1400 + 50(4)$

$$1.79x \ge 1400 + 200$$
$$1.79x \ge 1600$$
$$x \ge 893.85$$

You must sell at least 894 greeting cards.

**83.** As the price per card increases, the number of cards you must sell decreases.

If the price is too high, people may not buy the cards. A price that is too low may not generate enough money to cover costs.

## 1.4 Mixed Review (p. 30)

**84.** $1.2n = 1.2(4.8) = 5.76$   **85.** $\frac{1}{12} + x = \frac{1}{12} + \frac{1}{6} = \frac{3}{12} = \frac{1}{4}$

**86.** $b - 12 = 43 - 12 = 31$   **87.** $\frac{4}{5} \cdot y = \frac{4}{5} \cdot \frac{1}{5} = \frac{4}{25}$

**88.** $3^3$   **89.** $y^2$   **90.** $6^5$   **91.** $c^4$   **92.** $5^4$   **93.** $(5y)^4$

**94.** $3^2$   **95.** $(9a)^6$   **96.** $7^2$   **97.** $1^3$   **98.** $6 \cdot x^6$   **99.** $t^6$

**100.**
$$C = \tfrac{5}{9}(F - 32)$$
$$C = \tfrac{5}{9}(78 - 32)$$
$$C = \tfrac{5}{9}(46)$$
$$C = 25\tfrac{5}{9}°C$$

The temperature of the water is $25\tfrac{5}{9}°C$.

## Lesson 1.5

### Developing Concepts Activity 1.5 (p. 31)
### Exploring the Concept

**1.** the lowest score you can get on the final test to get an A

**2.** 5 100-point tests given

One final 200-point test

Previous scores of 88, 92, 87, 98, and 81

Need an average of 90 points to get an A

**3.** The lowest score you can get on the final test to get an A is 184.

### Drawing Conclusions

**1.** First find the labor charge per hour, then find the basic service fee.

Labor charge per hour $= x$

Basic service fee $= y$

$$4x - 2x = 180 - 120 \qquad 2x + y = 120$$
$$2x = 60 \qquad\qquad 2(30) + y = 120$$
$$x = 30 \qquad\qquad 60 + y = 120$$
$$\qquad\qquad\qquad y = 120 - 60$$
$$\qquad\qquad\qquad y = 60$$

# Chapter 1 *continued*

The basic service fee is $60.

First, a verbal model was written. Next, labels were assigned. Then an algebraic model was written. The model was solved and this information was used to write another model. The second model was solved and the answer was checked.

**2.** $\dfrac{80 + 75 + 79 + 86 + ?}{6} = 80$

$\dfrac{80 + 75 + 79 + 86 + 160}{6} = 80$

The lowest score Jerry can get is 160.

The strategy used was Guess, Check, and Revise.

**3.** $40x + 5x \overset{?}{<} 8x$

$40 + 5(14) \overset{?}{<} 8(14)$

$40 + 70 \overset{?}{<} 112$

$110 \overset{?}{<} 112$

The member must visit 14 times.

Mental math was used to find the number of visits.

## 1.5 Guided Practice (p. 35)

**1.** subtraction  **2.** $7 - n$  **3.** yes

**4.** You can solve a verbal model by translating it into a mathematical expression or an equation or an inequality. Then you can solve using algebra or a problem-solving strategy.

**5.** C  **6.** D  **7.** B  **8.** A  **9.** $x + 10 = 24$

**10.** $7y = 42$  **11.** $\dfrac{20}{n} \le 2$  **12.** $x + 10 > 14$

## 1.5 Practice and Applications (pp. 35–38)

**13.** $x + 9$  **14.** $\frac{1}{2}x$  **15.** $\frac{1}{2}x + 3$  **16.** $x + 7$  **17.** $\dfrac{x}{0.2}$

**18.** $4x$  **19.** $\dfrac{2^3}{x}$  **20.** $10 - x$  **21.** $5^2 - x$  **22.** $29 - x$

**23.** $9 > 3s$  **24.** $25 = \dfrac{y}{3.5}$  **25.** $14x = 1$

**26.** $10d - 9 = 11$  **27.** $3(x - 2) = 10$  **28.** $5 - 8 = 4y$

**29.** $(38 - n) - 23 < 8$  **30.** $t + (7 + s^2) = 10$

**31.** $(20 - x) - 5 \le 10$  **32.** $14 + 12y \le 50$

**33.** $9 + \dfrac{b}{10} \ge 11$  **34.** $\dfrac{70}{7p} = 1$  **35.** $q \ge 100$

**36.** $x^2 + 44 = 3x^4$  **37.** $\dfrac{35}{t} \le 7$  **38.** $50\left(\dfrac{20}{n}\right) \ge 250$

**39.** $b = e - 1.50$  **40.** $s = c + \dfrac{1}{5}$  **41.** $c = 3r + 229$

**42.** $\dfrac{C}{d} = \pi$  **43.** $V \le 30 - 3$ or $s^3 \le 30 - 3$

**44.** $25m \ge 500$  **45.** $P = 4(s - 2)$  **46.** $300 \cdot \dfrac{x}{100} \cdot t \le 72$

**47.** $A = \frac{1}{2}(7 + 9)(h + 7)$  **48.** $c^2 = 4^2 + 3^2$

**49.** $\boxed{\text{Number of subscriptions}} \cdot \boxed{\text{Cost of each subscription}} =$

$\boxed{\text{Amount of money the club needs to raise}}$

**50.** cost of subscription $= \$15$

number of subscriptions $= n$

amount of money the club needs to raise $= \$315$

$15n = 315$

**51.** $15n = 315$

$n = 21$

**52.** The club needs to sell 21 subscriptions.

**53.** It is reasonable that the science club can sell 21 subscriptions.

**54.** fine per mi/h over speed limit $= 20$

miles per hour over speed limit $= m$

amount of ticket $= 260$

**55.** $20m = 260$

**56.** $20m = 260$

$m = 13$

The solution represents the number of mi/h over the speed limit Jeff was driving.

**57.** $\dfrac{\$}{\left(\frac{\text{mi}}{\text{h}}\right)}\left(\dfrac{\text{mi}}{\text{h}}\right) = \$$

$\left(\dfrac{\text{mi}}{\text{h}}\right) = \$ \div \dfrac{\$}{\left(\frac{\text{mi}}{\text{h}}\right)}$

$\dfrac{\text{mi}}{\text{h}} = \dfrac{\text{mi}}{\text{h}}$

**58. a.** $95 + x > 120 + (45 - x)$

**b.** $95 + x > 120 + 45 - x$

$95 + x > 165 - x$

$2x > 70$

$x > 35$

36 is the smallest value that is a solution.

**59.** weeks working summer job $= 12$

extra money needed to save each week $= 20 + x$

cost of model with aluminum rims $= 480$

$12(20 + x) = 480$

**60.** $12(20 + x) = 480$

$20 + x = 40$

$x = \$20$

This is the extra money needed to be saved each week.

**61.** $x \ge 18\left(\dfrac{544}{17} + 5\right)$  **62.** A  **63.** B  **64.** A

**65.** $300 + x = 4(135)$

$300 + x = 540$

$x = 240$

$240 extra needs to be raised.

**66.** $1.25x = 240$

$x = 192$

192 students must contribute.

### 1.5 Mixed Review (p. 38)

**67.** = **68.** > **69.** > **70.** > **71.** < **72.** <

**73.**

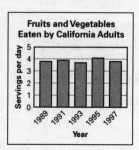

**Fruits and Vegetables Eaten by California Adults**

**74.** $x = 1(31)$

$x = 31$ feet

**75.** $4 + 3x = 4 + 3(2) = 4 + 6 = 10$

**76.** $y \div 8 = 32 \div 8 = 4$

**77.** $5 \cdot 2p^2 = 5 \cdot 2(6)^2 = 5 \cdot 2(36) = 5 \cdot 72 = 360$

**78.** $t^4 - t = 7^4 - 7 = 2401 - 7 = 2394$

### Quiz 2 (p. 39)

**1.** $12 \div (7 - 3)^2 + 2 = 12 \div (4)^2 + 2 = 12 \div 16 + 2 =$ $\frac{3}{4} + 2 = 2\frac{3}{4}$

**2.** $32 - 5 \cdot (2 + 1) + 4 = 32 - 5 \cdot 3 + 4 =$ $32 - 15 + 4 = 21$

**3.** $x^2 + 4 - x = 6^2 + 4 - 6 = 36 + 4 - 6 = 34$

**4.** $y \div 3 + 2 = 30 \div 3 + 2 = 10 + 2 = 12$

**5.** $\frac{r}{s} \cdot 7 = \frac{30}{5} \cdot 7 = 6 \cdot 7 = 42$

**6.** $5x^2 - y = 5(4)^2 - 26 = 5(16) - 26 = 80 - 26 = 54$

**7.** $2(9) + 6 \overset{?}{=} 18$

$18 + 6 \overset{?}{=} 18$

$24 \neq 18$

9 is not a solution.

**8.** $13 - 3(2) \overset{?}{=} 7$

$13 - 6 \overset{?}{=} 7$

$7 = 7$

2 is a solution.

**9.** $4(2) + 7 \overset{?}{=} 5 + 5(2)$

$8 + 7 \overset{?}{=} 5 + 10$

$15 = 15$

2 is a solution.

**10.** $2(6) + 6 \overset{?}{=} 4(6)$

$12 + 6 \overset{?}{=} 24$

$18 \neq 24$

6 is not a solution.

**11.** $3(2) - 4 \overset{?}{>} 0$

$6 - 4 \overset{?}{>} 0$

$2 > 0$

2 is a solution.

**12.** $8 - 2(3) \overset{?}{>} 4$

$8 - 6 \overset{?}{>} 4$

$2 \not> 4$

3 is not a solution.

**13.** $\frac{x}{4} = 2.65$

$x = 4(2.65)$

$x = \$10.60$

The pizza cost $10.60.

### Math & History (p. 39)

**1.** Answers may vary.

**2.** Order of operations, $(2 + 7)^2 - 6 \div 3 = 9^2 - 2 =$ $81 - 2 = 79$; change a mixed number to an improper fraction, $1\frac{3}{7} = 1 \times 7 + 3 = 10 \rightarrow \frac{10}{7}$

## Lesson 1.6

### 1.6 Guided Practice (p. 43)

**1.** Data are information, facts, or numbers that describe something. *Sample answer:* the heights of all the students in your class

**2.** line graph

**3.**

| Event \ Year | 1992 | 1996 | 2000 |
|---|---|---|---|
| 100 m | | | |
| 200 m | | | |
| 400 m | | | |
| 800 m | | | |
| 1500 m | | | |
| 3000 m | | | |

**4.** false **5.** true **6.** false **7.** false

### 1.6 Practice and Applications (pp.43–45)

**8.** true **9.** false **10.** true

**11.**

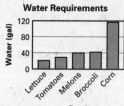

**Water Requirements**

**12.** paper; about 80 million tons

**13.** paper; about 33 million tons **14.** plastics; about 1 million tons **15.** 6 years **16.** $4.25 **17.** 1991

# Chapter 1 continued

**18.**

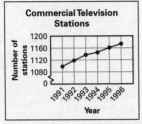

The line graph shows an increase in the number of commercial television stations.

**19.**

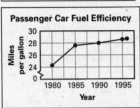

**20. a.** CDs; cassettes; albums

**b.** 1995 and 1996

**c.** The standard bar graph is useful in comparing the individual values of data. The stacked bar graph is useful in comparing the total value for a group of data.

**21.** The line graph on the left shows the data with a small increase between 1990 and 1996. The line graph on the right shows the data with a big increase between 1990 and 1996.

The graph on the right could be misleading. Because of the jump in the vertical axis, the data looks like it is sharply increasing.

### 1.6 Mixed Review (p. 45)

**22.** $5(3) + 2 \stackrel{?}{=} 17$

$15 + 2 \stackrel{?}{=} 17$

$17 = 17$

3 is a solution.

**23.** $12 - 2(4) \stackrel{?}{=} 6$

$12 - 8 \stackrel{?}{=} 6$

$4 \neq 6$

4 is not a solution.

**24.** $3(2) - 4 \stackrel{?}{=} 12 - 5(2)$

$6 - 4 \stackrel{?}{=} 12 - 10$

$2 = 2$

2 is a solution.

**25.** $2(5) + 8 \stackrel{?}{=} 4(5) - 2$

$10 + 8 \stackrel{?}{=} 20 - 2$

$18 = 18$

5 is a solution.

**26.** $x + 3 = 5$  **27.** $\frac{x}{3} = 12$  **28.** A

## Lesson 1.7

### 1.7 Guided Practice (p. 49)

1. input, output  **2.** domain, range

3. words, equation, input-output table, graph

4. The table is a function since there is one output for each input.

5. The table is a function since there is one output for each input.

**6.** The table is not a function since there is more than one output for one of the input values.

**7.**

| Time (hours) | 0.25 | 0.50 | 0.75 |
|---|---|---|---|
| Distance traveled (miles) | 44.5 | 89.0 | 133.5 |

| Time (hours) | 1.00 | 1.25 | 1.50 |
|---|---|---|---|
| Distance traveled (miles) | 178.0 | 222.5 | 267.0 |

**8.** The domain is 0.25, 0.50, 0.75, 1.00, 1.25, 1.50.

The range is 44.5, 89, 133.5, 178, 222.5, 267.

**9.**

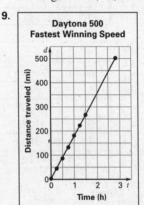

### 1.7 Practice and Applications (pp. 49–51)

**10.** The table represents a function since there is one output for each input.

**11.** The table represents a function since there is one output for each input.

**12.** The table does not represent a function since there is more than one output for one of the input values.

**13.**

| Input, x | Output, y |
|---|---|
| 0 | 2 |
| 1 | 5 |
| 2 | 8 |
| 3 | 11 |

**14.**

| Input, x | Output, y |
|---|---|
| 0 | 21 |
| 1 | 19 |
| 2 | 17 |
| 3 | 15 |

**15.**

| Input, x | Output, y |
|---|---|
| 0 | 0 |
| 1 | 5 |
| 2 | 10 |
| 3 | 15 |

**16.**

| Input, x | Output, y |
|---|---|
| 0 | 1 |
| 1 | 7 |
| 2 | 13 |
| 3 | 19 |

**17.**

| Input, x | Output, y |
|---|---|
| 0 | 1 |
| 1 | 3 |
| 2 | 5 |
| 3 | 7 |

**18.**

| Input, x | Output, y |
|---|---|
| 0 | 4 |
| 1 | 5 |
| 2 | 6 |
| 3 | 7 |

**19.**

| Input, x | Output, y |
|---|---|
| 1 | 6.5 |
| 1.5 | 8.5 |
| 3 | 14.5 |
| 4.5 | 20.5 |
| 6 | 26.5 |

**20.**

| Input, x | Output, y |
|---|---|
| 1 | 29 |
| 1.5 | 27.5 |
| 3 | 23 |
| 4.5 | 18.5 |
| 6 | 14 |

**21.**

| Input, x | Output, y |
|---|---|
| 1 | 19 |
| 1.5 | 16 |
| 3 | 13 |
| 4.5 | 12 |
| 6 | 11.5 |

**22.**

| Input, x | Output, y |
|---|---|
| 1 | 4 |
| 1.5 | 5 |
| 3 | 8 |
| 4.5 | 11 |
| 6 | 14 |

**23.**

| Input, x | Output, y |
|---|---|
| 1 | 0.5 |
| 1.5 | 1.75 |
| 3 | 8.5 |
| 4.5 | 19.75 |
| 6 | 35.5 |

**24.**

| Input, x | Output, y |
|---|---|
| 1 | 2.5 |
| 1.5 | 3.75 |
| 3 | 10.5 |
| 4.5 | 21.75 |
| 6 | 37.5 |

**25.**

| Input | Function | Output |
|---|---|---|
| $d = 1$ | $W = 360 \cdot 1$ | $W = 360$ |
| $d = 2$ | $W = 360 \cdot 2$ | $W = 720$ |
| $d = 3$ | $W = 360 \cdot 3$ | $W = 1080$ |
| $d = 4$ | $W = 360 \cdot 4$ | $W = 1440$ |
| $d = 5$ | $W = 360 \cdot 5$ | $W = 1800$ |

**26.**

| Input C | 0 | 5 | 10 | 15 | 20 | 25 | 30 | 35 | 40 |
|---|---|---|---|---|---|---|---|---|---|
| Output F | 32 | 41 | 50 | 59 | 68 | 77 | 86 | 95 | 104 |

**27.**

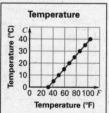

**28.** The equation $F = \frac{9}{5}C + 32$ is a function because for each input of $C$ there is exactly one output of $F$.

**29.** The domain is $0°, 5°, 10°, 15°, 20°, 25°, 30°, 35°, 40°$.

The range is $32°, 41°, 50°, 59°, 68°, 77°, 86°, 95°, 104°$.

**30.**

| Input, n | Output, C |
|---|---|
| 1 | 6.25 |
| 2 | 7.5 |
| 3 | 8.75 |
| 4 | 10 |
| 5 | 11.25 |
| 6 | 12.5 |
| 7 | 13.75 |

**31.** The domain is 1, 2, 3, 4, 5, 6, 7.

The range is 6.25, 7.5, 8.75, 10, 11.25, 12.5, 13.75.

**32.**

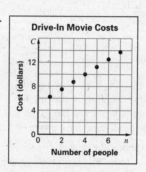

**33.** $C = 0.75\left(\frac{n}{2}\right)$

or $C = 0.375n$

**34.**

| Input, n | Output, C |
|---|---|
| 6 | 2.25 |
| 8 | 3 |
| 10 | 3.75 |
| 12 | 4.5 |

**35.** The input is the number of signs and the output is the total cost of the poster board.

**36.** $C = 0.75\left(\frac{14}{2}\right)$

$C = 0.75(7)$

$C = 5.25$

No. You will need $.25 more.

**37.** number of days $= n$

amount of money you earn $= a$

$a = 2n + 5$

**38.** $a = 2(7) + 5 = 14 + 5 = 19$

You will earn $19.

**39.** profit $= p$

number of sandwich plates $= n$

$p = (2 - 0.85)n$

$p = 1.15n$

**40.** $p = 1.15n - 50$

**41.** $p = 1.15(75) - 50 = 36.25$

The profit would be $36.25.

# Chapter 1 *continued*

**42.** The increase would reduce the profit by $.05.

$$p = (1.15 - 0.05)n - 50$$

$$p = 1.1n - 50$$

**43.**
$$100 = 1.1x - 50$$

$$100 + 50 = 1.1x$$

$$\frac{150}{1.1} = x$$

$$136.36 = x$$

You would have to sell 137 sandwich plates.

**44.** In 2000, $543.8 billion will be spent.

In 2005, $686.3 billion will be spent.

## 1.7 Mixed Review (p. 52)

**45.** $x^6$  **46.** $9^3$  **47.** $y^7$  **48.** $15^4$

**49.** $\dfrac{3(1) + 3(9)}{5} = \dfrac{3 + 27}{5} = \dfrac{30}{5} = 6$

**50.** $6 + \dfrac{7 + 2}{3} = 6 + \dfrac{9}{3} = 6 + 3 = 9$

**51.** $3(8)^2 + 27 = 3(64) + 27 = 192 + 27 = 219$

**52.** $2(4)^3 - 3(6) = 2(64) - 3(6) = 128 - 18 = 110$

**53.** $7(2) + 2 \overset{?}{=} 4(2) + 8$

$14 + 2 \overset{?}{=} 8 + 8$

$16 = 16$

2 is a solution.

**54.** $5(4) - 1 \overset{?}{=} 3(4) + 2$

$20 - 1 \overset{?}{=} 12 + 2$

$19 \neq 14$

4 is not a solution.

**55.** $32 - 5(3) \overset{?}{>} 11$

$32 - 15 \overset{?}{>} 11$

$17 > 11$

3 is a solution.

**56.** $7(1) + 2 \overset{?}{<} 12$

$7 + 2 \overset{?}{<} 12$

$9 < 12$

1 is a solution.

**57.** $9 - 7 \overset{?}{\geq} 12 - 9$

$2 \not\geq 3$

9 is not a solution.

**58.** $4 + 2 \overset{?}{\leq} 2(4) - 2$

$4 + 2 \overset{?}{\leq} 8 - 2$

$6 \leq 6$

4 is a solution.

**59.** $3750 - 3500 = 250$

The company's sales decreased by $250 million.

## Quiz 3 (p. 52)

**1.**

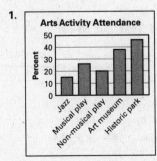

**2.** *Sample answer:* The activity with the highest percentage was historic parks. Jazz had the lowest attendance percentage.

---

**3.** B

**4.** Let $t$ = time (min) and $a$ = altitude (ft)

$$a = 25t + 200$$

**5.**

| Input, $t$ | Output, $a$ |
|---|---|
| 0 | 200 |
| 1 | 225 |
| 2 | 250 |
| 3 | 275 |
| 4 | 300 |
| 5 | 325 |
| 6 | 350 |

## Chapter 1 Review (pp. 54–56)

**1.** $a + 14 = 23 + 14 = 37$  **2.** $1.8x = 1.8(10) = 18$

**3.** $\dfrac{m}{1.5} = \dfrac{15}{1.5} = 10$  **4.** $\dfrac{15}{y} = \dfrac{15}{7.5} = 2$

**5.** $p - 12 = 22 - 12 = 10$  **6.** $b(0.5) = 9(0.5) = 4.5$

**7.** $t = \dfrac{d}{r}$

$t = \dfrac{6}{3}$

$t = 2$

It will take 2 hours.

**8.** $8^4 = 4096$  **9.** $(2 + 3)^5 = 5^5 = 3125$

**10.** $s^2 = 1.5^2 = 2.25$

**11.** $6 + (b^3) = 6 + (3^3) = 6 + 27 = 33$

**12.** $2x^4 = 2(2)^4 = 2(16) = 32$

**13.** $(5x)^3 = (5 \cdot 5)^3 = 25^3 = 15{,}625$

**14.** $4 + 21 \div 3 - 3^2 = 4 + 21 \div 3 - 9 =$

$4 + (21 \div 3) - 9 = 4 + 7 - 9 = 2$

**15.** $(14 \div 7)^2 + 5 = 2^2 + 5 = 4 + 5 = 9$

**16.** $\dfrac{6 + 2^2}{17 - 6 \cdot 2} = \dfrac{6 + 4}{17 - 6 \cdot 2} = \dfrac{6 + 4}{17 - 12} = \dfrac{10}{5} = 2$

**17.** $\dfrac{x - 3y}{6} = \dfrac{15 - 3(2)}{6} = \dfrac{15 - 6}{6} = \dfrac{9}{6} = 1\dfrac{1}{2}$

**18.** $2(4) - 3 \overset{?}{=} 2$

$8 - 3 \overset{?}{=} 2$

$5 \neq 2$

4 is not a solution.

**19.** $2^2 - 2 \overset{?}{=} 2$

$4 - 2 \overset{?}{=} 2$

$2 = 2$

2 is a solution.

**20.** $9(3) - 3 \overset{?}{>} 24$

$27 - 3 \overset{?}{>} 24$

$24 \not> 24$

3 is not a solution.

**21.** $5(5) + 2 \overset{?}{\leq} 27$

$25 + 2 \overset{?}{\leq} 27$

$27 \leq 27$

5 is a solution.

# Chapter 1 *continued*

**22.** number of bottles · price per bottle = money given

number of bottles of juice = $N$

money given = 75

price per bottle = 0.75

$0.75N = 75$

$N = 100$

You can buy 100 bottles of juice.

**23.** More women have won the Wimbledon title than men.

**24.** The men have won 54 titles.

The women have won 80 titles.

The women have won more titles than the men.

**25.**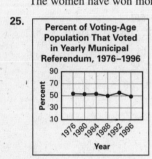

Percent of Voting-Age Population That Voted in Yearly Municipal Referendum, 1976–1996

**26.** The percent voting has decreased in the years 1980, 1988, and 1996. The percent voting has increased in the years 1984 and 1992.

**27.** perimeter of a rectangle = (2 · length of one side) + (2 · length of other side)

perimeter = $P$

one side = $2w$

other side = $3w$

$P = 2(2w) + 2(3w)$ or $P = 10w$

**28.**

| Input, $w$ | Output, $P$ |
|------------|-------------|
| 1 | 10 |
| 2 | 20 |
| 3 | 30 |
| 4 | 40 |
| 5 | 50 |

**29.** The domain is 1, 2, 3, 4, 5.

The range is 10, 20, 30, 40, 50.

## Chapter 1 Test *(p. 57)*

**1.** $5y + x^2 = 5(3) + 5^2 = 5(3) + 25 = 15 + 25 = 40$

**2.** $\dfrac{24}{y} - x = \dfrac{24}{3} - 5 = 8 - 5 = 3$

**3.** $2y + 9x - 7 = 2(3) + 9(5) - 7 = 6 + 45 - 7 = 44$

**4.** $(5y + x) \div 4 = (5 \cdot 3 + 5) \div 4 = (15 + 5) \div 4 = 20 \div 4 = 5$

**5.** $2x^3 + 4y = 2(5)^3 + 4(3) = 2(125) + 4(3) = 250 + 12 = 262$

**6.** $8(x^2) \div 25 = 8(5^2) \div 25 = 8(25) \div 25 = 200 \div 25 = 8$

**7.** $(x - y)^3 = (5 - 3)^3 = 2^3 = 8$

**8.** $x^4 + 4(y - 2) = 5^4 + 4(3 - 2) = 625 + 4(3 - 2) = 625 + 4(1) = 625 + 4 = 629$

**9.** $(5y)^4$ **10.** $9^3$ **11.** $6^n$ **12.** $5 \cdot (4 + 6) \div 2$

**13.** $2\frac{1}{2}$ hours is enough time.

$35\left(2\frac{1}{2}\right) \overset{?}{\geq} 85$

$87.5 \geq 85$

**14.** $7n$ **15.** $x \geq 90$ **16.** $\dfrac{m}{2}$ or $m \div 2$

**17.** $(2 \cdot 3)^2 \overset{?}{=} 2 \cdot 3^2$

$6^2 \overset{?}{=} 2 \cdot 9$

$36 \neq 18$

false

**18.** $3 \div 12 \overset{?}{=} 4$

$\frac{1}{4} \neq 4$

false

**19.** $8 - 6 \overset{?}{=} 6 - 8$

$2 \neq -2$

false

**20.** $0.10(38) \overset{?}{=} 0.38$

$3.8 \neq 0.38$

false

**21.** $8 \overset{?}{\leq} (3)^2 + 3$

$8 \overset{?}{\leq} 9 + 3$

$8 \leq 12$

true

**22.** $9(3) \overset{?}{>} 3^3$

$9(3) \overset{?}{>} 27$

$27 \not> 27$

false

**23.** $35x = 3920$

$x = 112$

112 seniors have paid.

**24.** $y = 1.75[(1.2)(0.5)x]$

or $y = 1.05x$

**25.** $y = 1.75[(1.2)(0.5)(20)] = 21$

It will cost $21.

**26.** about 38 million students

**27.** The K–8 category includes nine grades while the 9–12 and College categories only include four grades.

**28.** the ninth through twelfth graders

**29.** Each of the bars is longer for 2000 than for 1995, so the number of students is higher in 2000. This can be checked by adding the estimated values for the different years.

## Chapter 1 Standardized Test *(pp. 58–59)*

**1.** $r = \dfrac{d}{t}$

$r = \dfrac{209.2}{2}$

$r = 104.6$ B 104.6 km/h

**2.** perimeter = $2(6.4) + 2(3) = 12.8 + 6 = 18.8$

A 18.8

# Chapter 1 *continued*

**3.** $[(5 \cdot 9) \div x] + 6 = [(5 \cdot 9) \div 3] + 6 =$
$[45 \div 3] + 6 = 15 + 6 = 21$

    D  21

**4.** $4t + 5 = 21$

    $4t = 21 - 5$

    $4t = 16$

    $t = 4$

  $t^2 - 3 = (4)^2 - 3 = 16 - 3 = 13$

    B  13

**5.** $7b - 2 = 47$

    $7b = 47 + 2$

    $7b = 49$

    $b = 7$

    B  7

**6.** $6 \cdot (15 + 8) - [(2 \cdot 7) - 4]^2 = 6 \cdot (23) - [14 - 4]^2 =$
$6 \cdot 23 - [10]^2 = 6 \cdot 23 - 100 = 138 - 100 = 38$

    B  38

**7.** $3 \cdot 5 - 4 \ ? \ 3 \cdot (5 - 4)$

    $15 - 4 \ ? \ 3 \cdot 1$

    $11 > 3$

    A  The number in column A is greater.

**8.** $2x \div 5 + 7 \ ? \ 2x \div (5 + 7)$

    D  The relationship cannot be determined from the information given.

**9.** $42 - (5^2 + 2) \ ? \ 42 - (5^2) + 2$

    $42 - (25 + 2) \ ? \ 42 - 25 + 2$

    $42 - 27 \ ? \ 42 - 25 + 2$

    $15 < 19$

    B  The number in column B is greater.

**10.** $A = 6.3^2$

    $A = 39.69$

    E  39.69 cm$^2$

**11.** $50 - x^2 = 1$

    $-x^2 = 1 - 50$

    $-x^2 = -49$

    $\dfrac{-x^2}{(-1)} = \dfrac{-49}{(-1)}$

    $x^2 = 49$

    $x = 7$

    C  7

**12.** $20 - x \geq x + 2$

    $20 - 2 \geq x + x$

    $18 \geq 2x$

    $9 \geq x$

    A  9

**13.** $3.4(11) - 2.3(12) = 37.4 - 27.6 = 9.8$

    D  9.8

**14.** B  $150 - 6n$   **15.** A  $5(8 - x)$   **16.** E  $2 + 3y$

**17.** D  $y = x^2 + 3$   **18.** B  I and II

**19.** about nine pounds per person more

**20.** about $5\frac{1}{2}$ pounds per person more

**21.** The differences between the countries looks greater because of the jump in the axis.

**22.**

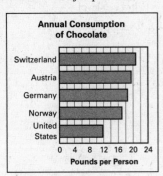

Annual Consumption of Chocolate

**23.** $y = 200 + x$

**24.**

| Input | Output |
|-------|--------|
| 0 | 200 |
| 1 | 201 |
| 2 | 202 |
| 3 | 203 |
| 4 | 204 |
| 5 | 205 |
| 6 | 206 |
| 7 | 207 |
| 8 | 208 |
| 9 | 209 |
| 10 | 210 |

**25.** The domain is 0, 1, 2, 3, 4, 5, 6, 7, 8, 9, 10.

The range is 200, 201, 202, 203, 204, 205, 206, 207, 208, 209, 210.

**26.**

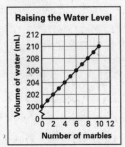

Raising the Water Level

# CHAPTER 2

## Think & Discuss (p. 61)

**1.** *Sample answers:* Below-zero temperature readings, golf scores, debt, elevation below sea level

**2.** 8 ft/sec downward; $10\frac{2}{3}$ ft/sec upward

## Skill Review (p. 62)

**1.** 4.07, 4.6, 5.06, 5.46, 5.5, 6.1

**2.** $\frac{25}{11}, 2\frac{3}{5}, 2\frac{5}{8}, 2\frac{2}{3}, \frac{11}{4}, 2\frac{5}{6}$

**3.** $\frac{9}{14} - \frac{3}{7} = \frac{9}{14} - \frac{6}{14}$
$= \frac{3}{14}$

**4.** $\frac{5}{6} + \frac{2}{3} = \frac{5}{6} + \frac{4}{6}$
$= \frac{9}{6}$
$= 1\frac{1}{2}$

**5.** $1\frac{3}{4} + 2\frac{1}{8} = 1\frac{6}{8} + 2\frac{1}{8}$
$= 3\frac{7}{8}$

**6.** $8\frac{7}{15} - 2\frac{4}{5} = 8\frac{7}{15} - 2\frac{12}{15}$
$= 5\frac{10}{15}$
$= 5\frac{2}{3}$

**7.** $\frac{5}{9} \div \frac{2}{9} = \frac{5}{\cancel{9}} \cdot \frac{\cancel{9}^{1}}{2}$
$= \frac{5}{2}$
$= 2\frac{1}{2}$

**8.** $\frac{^{1}7}{3\cancel{9}} \cdot \frac{\cancel{3}^{1}}{14_{2}} = \frac{1}{6}$

**9.** $6\frac{6}{7} \cdot 4\frac{1}{12} = \frac{48}{7} \cdot \frac{49}{12}$
$= \frac{^{4}\cancel{48}}{_{1}\cancel{7}} \cdot \frac{\cancel{49}^{7}}{\cancel{12}_{1}}$
$= \frac{28}{1}$
$= 28$

**10.** $9\frac{1}{6} \div 1\frac{3}{8} = \frac{55}{6} \div \frac{11}{8}$
$= \frac{^{5}\cancel{55}}{_{3}\cancel{6}} \cdot \frac{\cancel{8}^{4}}{\cancel{11}_{1}}$
$= \frac{20}{3}$
$= 6\frac{2}{3}$

**11.** $4 + 7 - 3 \cdot 2 = 4 + 7 - 6 = 11 - 6 = 5$

**12.** $64 \div 16 + 5 - 7 = 4 + 5 - 7 = 2$

**13.** $9 - 8 \div 4 + 12 = 9 - 2 + 12 = 7 + 12 = 19$

**14.** $16 - 9 \cdot 5 \div 3 = 16 - 45 \div 3 = 16 - 15 = 1$

**15.** $8 \cdot 3 - 6 \cdot 4 = 24 - 24 = 0$

**16.** $12.8 - 7.5 \div 1.5 + 3.2 = 12.8 - 5 + 3.2 = 11$

## Lesson 2.1

### 2.1 Guided Practice (p. 67)

**1.** $3 > -5$

**2.**
$-2.5 > -3.5$ because $-2.5$ is to the right of $-3.5$.

**3.** Answers may vary. Ex: $-5$. The opposite of $-5$ is 5.

**4.** $>$    **5.** $<$    **6.** $<$    **7.** $<$

**8.**
$-5, 0, 3$

**9.**
$-4, 2, 3$

**10.**
$-\frac{3}{4}, -\frac{1}{2}, \frac{1}{4}$

**11.**
$-1.1, -1, -0.1$

**12.** 12   **13.** 4.1   **14.** $\frac{1}{5}$   **15.** 103   **16.** positive

**17.** negative

### 2.1 Practice and Applications (p. 67–70)

**18.**
$-6 < 4, \ 4 > -6$

**19.**
$-6.4 < -6.3$,
$-6.3 > -6.4$

**20.**
$-7 < 2, \ 2 > -7$

**21.**
$-0.1 > -0.11$,
$-0.11 < -0.1$

**22.**
$5.7 > -4.2, \ -4.2 < 5.7$

**23.**
$-2.8 < 3.7, \ 3.7 > -2.8$

**24.**
$-2.7 < \frac{3}{4}, \ \frac{3}{4} > -2.7$

**25.**
$-0.5 < -\frac{1}{3}$,
$-\frac{1}{3} > -0.5$

**26.**
$-1\frac{5}{6} < -1\frac{7}{9}, \ -1\frac{7}{9} > -1\frac{5}{6}$

**27.** $-1.8, -0.66, 0.7, 3, 4.6, 4.66$

**28.** $-0.2, -0.03, -0.02, 0, 0.2, 2.0$

**29.** $-3, -2.6, -\frac{1}{2}, 0, \frac{1}{2}, 4.8$

**30.** $-5.1, -5, -4\frac{1}{2}, 3.4, 3\frac{1}{2}, 4.1$

**31.** $-5.8, -\frac{3}{4}, -\frac{1}{2}, \frac{1}{3}, 2.4, 7$

**32.** $-6.11, -6.1, -6.08, -6.02, 6.03, 6.07$

# Chapter 2 *continued*

**33.** $-8$ **34.** $3$ **35.** $-3.8$ **36.** $2.5$ **37.** $\frac{5}{6}$ **38.** $-3\frac{4}{5}$

**39.** $2.01$ **40.** $\frac{1}{9}$ **41.** $7$ **42.** $4$ **43.** $4.5$ **44.** $\frac{2}{3}$ **45.** $\frac{4}{5}$

**46.** $2$ **47.** $4.3$ **48.** $-\frac{8}{9}$ **49.** $0.09$ **50.** $-4, 4$ **51.** $0$

**52.** $3.8$ **53.** $-1, 1$ **54.** False; $|-4| = 4$

**55.** False; $a = -1 \quad -(-1) = 1$

opposite of 1 is $-1$

**56.** True **57.** False; $|-4| = 4$ **58.** $<$ **59.** $\geq$

**60.** $\geq$ **61.** $\leq$ **62.** $12{,}799$ **63.** $-8$ **64.** $-282$

**65.** $0$ **66.** Sirius; $-1.46$ **67.** Regulus; $1.35$

**68.** Spica, Pollux, Regulus, Deneb

**69.** Canopus, Vega, Sirius, Arcturus

**70.** Regulus, Deneb, Pollux, Spica, Altair, Procyon, Vega, Arcturus, Canopus, Sirius

**71.** Hiromi Kobayashi **72.** Annika Sörenstam

**73.** Luciana Bemvenuti, Liz Early, Hiromi Kobayashi, Nancy Scranton

**74.** Michele Estill, Jenny Lidback, Joan Pitcock, Annika Sörenstam

**75.** Annika Sörenstam **76.** 6 ft/sec, $-6$ ft/sec

**77.** 429 ft/min, 429 ft/min **78.** 3 m/sec, $-3$ m/sec

**79.** 10°F, 25 mi/h at $-29$°F **80.** 5°F, 15 mi/h at $-25$°F

**81.** 10°F, 20 mi/h at $-24$°F **82.** Answers may vary.

**83.** B **84.** C **85.** D

**86.** Yes; $-|x| = |-x|$ is true when, and only when, $x = 0$.

**87.** always

### 2.1 Mixed Review (p. 70)

**88.** $\frac{2}{4} + \frac{3}{8} = \frac{4}{8} + \frac{3}{8} = \frac{7}{8}$ **89.** $\frac{2}{3} + \frac{1}{6} = \frac{4}{6} + \frac{1}{6} = \frac{5}{6}$

**90.** $\frac{2}{5} + \frac{1}{4} = \frac{8}{20} + \frac{5}{20} = \frac{13}{20}$ **91.** $\frac{5}{8} + \frac{1}{3} = \frac{15}{24} + \frac{8}{24} = \frac{23}{24}$

**92.** $3\frac{2}{7} + 4\frac{1}{2} = 3\frac{4}{14} + 4\frac{7}{14} = 7\frac{11}{14}$

**93.** $1\frac{7}{9} + 4\frac{3}{7} = 1\frac{49}{63} + 4\frac{27}{63} = 5\frac{76}{63} = 6\frac{13}{63}$

**94.** $5x + 3 = 5(2) + 3 = 10 + 3 = 13$

**95.** $2a - 7 = 2(6) - 7 = 12 - 7 = 5$

**96.** $3y + 12 = 3(0) + 12 = 0 + 12 = 12$

**97.** $6b - 39 + c = 6(15) - 39 + 2 = 90 - 39 + 2 = 53$

**98.** $\frac{5}{6}a - b = \frac{5}{6}(6) - 5 = 5 - 5 = 0$

**99.** $\frac{x}{5} - 2y = \frac{12}{5} - 2\left(\frac{4}{5}\right) = \frac{12}{5} - \frac{8}{5} = \frac{4}{5}$

**100.** $z - 5 = 8$ **101.** $r + 8 = 17$ **102.** $\frac{3}{4}y + 9 < 6w$

## Lesson 2.2

### Developing Concepts Activity 2.2 (p. 71)
### Drawing Conclusions

**1.** Model 4 and 5.

Combine the tiles.
There are nine positive tiles.

$4 + 5 = 9$

**2.** Model 3 and 3.

Combine the tiles.
There are six positive tiles.

$3 + 3 = 6$

**3.** Model $-4$ and $-2$.

Combine the tiles.
There are six negative tiles.

$-4 + -2 = -6$

**4.** Model $-1$ and $-7$.

Combine the tiles.
There are eight negative tiles.

$-1 + -7 = -8$

**5.** Model $-3$ and 2.

Group pairs of positive and negative tiles. There is one negative tile remaining.

$-3 + 2 = -1$

**6.** Model 5 and $-2$.

Group pairs of positive and negative tiles. There are three positive tiles remaining.

$5 + -2 = 3$

**7.**

Model −6 and 6.

Group pairs of positive and negative tiles. There are no tiles remaining.

$-6 + 6 = 0$

**8.**

Model 2 and −2.

Group pairs of positive and negative tiles. There are no tiles remaining.

$2 + -2 = 0$

**9.**

Model −6 and −3.

Combine the tiles. There are nine negative tiles.

$-6 + -3 = -9$

**10.** true; *Sample answer:* $5 + 4 = 9$

**11.** false; *Sample answer:* $-4 + -4 = -8$

**12.** true; *Sample answer:* $5 + -8 = -3$

**13.** true; *Sample answer:* $-4 + -1 = -5$

## 2.2 Guided Practice (p. 75)

**1.** losses

**2.** no

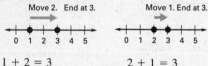

$1 + 2 = 3$    $2 + 1 = 3$

commutative property of addition

**3.**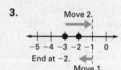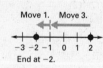

$-3 + 2 + -1 = -2$    $2 + -3 + -1 = -2$

**4.** $-5 + 9 = 4$   **5.** $-2$   **6.** $1$   **7.** $-5$   **8.** $0$   **9.** $5$

**10.** $-3\frac{1}{2}$   **11.** $100 + 18 = 118°F$

## 2.2 Practice and Applications (pp. 75–77)

**12.**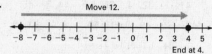

$-8 + 12 = 4$

**13.**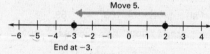

$2 + (-5) = -3$

**14.**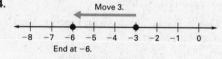

$-3 + -3 = -6$

**15.**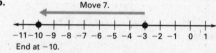

$-3 + (-7) = -10$

**16.**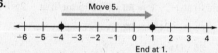

$-4 + 5 = 1$

**17.**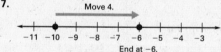

$-10 + 4 = -6$

**18.**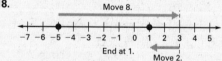

$-5 + 8 + (-2) = 1$

**19.**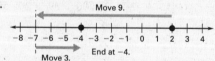

$2 + (-9) + 3 = -4$

**20.**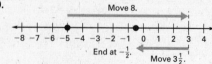

$-5 + 8 + \left(-3\frac{1}{2}\right) = -\frac{1}{2}$

**Algebra 1**
Chapter 2 Worked-out Solution Key

# Chapter 2 *continued*

**21.** 2  **22.** 52  **23.** 19  **24.** $-5$  **25.** $-19$  **26.** 3

**27.** 2  **28.** 0  **29.** $-8.2$  **30.** 1.4  **31.** 6  **32.** $-3.7$

**33.** identity property of addition

**34.** commutative property of addition

**35.** property of zero  **36.** associative property of addition

**37.** $-2.95 + 5.76 + (-88.6) = 2.81 + (-88.6) = -85.79$

**38.** $10.97 + (-51.14) + (-40.97) = -40.17 + (-40.97)$
$$= -81.14$$

**39.** $20.37 + 190.8 + (-85.13) = 211.17 + (-85.13)$
$$= 126.04$$

**40.** $300.3 + (-22.24) + 78.713 = 278.06 + 78.713$
$$= 356.773$$

**41.** $-1.567 + (-2.645) + 5308.34 = -4.212 + 5308.34$
$$= 5304.128$$

**42.** $-7344.28 + 2997.65 + (-255.11)$
$$= -4346.63 + (-255.11) = -4601.74$$

**43.** $5 + x + (-8) = 5 + 2 + -8 = -1$

**44.** $4 + x + 10 + (-10) = 4 + 3 + 10 + (-10) = 7$

**45.** $-24 + 6 + x = -24 + 6 + 8 = -10$

**46.** $-6 + x + 4 = -6 + (-3) + 4 = -5$

**47.** $2 + (-5) + x + 14 = 2 + (-5) + (-8) + 14 = 3$

**48.** $-11 + (-2) + 11 + x = -11 + (-2) + 11 + (-10)$
$$= -12$$

**49.** $x + (-6) + (-11) = -7 + (-6) + (-11) = -24$

**50.** $9 + x + (-8) + (-3) = 9 + (-12) + (-8) + (-3)$
$$= -14$$

**51.**

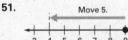
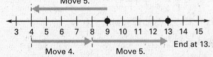

No, they only gained 13 yards.

**52.** Na atom 1: 1, ion
Na atom 2: 0, not an ion
Na atom 1 is Na$^+$ symbol

**53.** F atom 1: 0, not an ion
F atom 2: $-1$, ion
F atom 2 has F$^-$ symbol.

**54.** Because neutrons have no charge, they neither add nor take away from the atom's charge. Identity Property

**55.** $3514.65 + 5674.25 = \$9188.90$
$-8992.88 + -1207.03 = -10,199.91$
No, the losses were larger by $1011.01.

**56.** 6:00 A.M.  **57.** 10:00 A.M.  **58.** C; $318.84

**59.** D; $-\$322.15$  **60.** B; $16.69

**61.** $-(a + b) = -a + (-b)$ True; *Sample answers:*
$-(-1 + -4)$ and $-(-1) + -(-4)$ are both equal to 5;
$-(-3 + -5)$ and $-(-3) + -(-5)$ are both equal to 8.

---

*2.2 Mixed Review (p. 77)*

**62.** $\frac{4}{5} - \frac{2}{5} = \frac{2}{5}$  **63.** $\frac{8}{9} - \frac{2}{3} = \frac{8}{9} - \frac{6}{9} = \frac{2}{9}$

**64.** $\frac{3}{4} - \frac{5}{12} = \frac{9}{12} - \frac{5}{12} = \frac{4}{12} = \frac{1}{3}$  **65.** $\frac{7}{8} - \frac{1}{4} = \frac{7}{8} - \frac{2}{8} = \frac{5}{8}$

**66.** $4\frac{2}{3} - 2\frac{1}{5} = 4\frac{10}{15} - 2\frac{3}{15} = 2\frac{7}{15}$

**67.** $\frac{5}{6} - \frac{1}{9} = \frac{15}{18} - \frac{2}{18} = \frac{13}{18}$

**68.** $7\frac{9}{10} - 5\frac{3}{7} = 7\frac{63}{70} - 5\frac{30}{70} = 2\frac{33}{70}$

**69.** $\frac{5}{12} - \frac{3}{16} = \frac{20}{48} - \frac{9}{48} = \frac{11}{48}$

**70.** $a^4 + 8 = 10^4 + 8 = 10,000 + 8 = 10,008$

**71.** $79 - v^3 = 79 - 4^3 = 79 - 64 = 15$

**72.** $t^2 - 7t + 12 = 8^2 - 7(8) + 12 = 64 - 56 + 12 = 20$

**73.** $2(x^2) + 8(x) - 5 = 2(3^2) + 8(3) - 5$
$$= 2(9) + 24 - 5$$
$$= 18 + 24 - 5$$
$$= 37$$

**74.** $7 + 5 \overset{?}{=} 11$
$12 \neq 11$
7 is not a solution.

**75.** $12 - 2(4) \overset{?}{=} 18$
$12 - 8 \overset{?}{=} 18$
$4 \neq 18$
4 is not a solution.

**76.** $7(3) - 15 \overset{?}{=} 6$
$21 - 15 \overset{?}{=} 6$
$6 = 6$
3 is a solution.

**77.** $3 + 2(6) \overset{?}{=} 9 + 6$
$3 + 12 \overset{?}{=} 15$
$15 = 15$
6 is a solution.

**78.** $3(5) - 7 \overset{?}{=} 5 + 1$
$15 - 7 \overset{?}{=} 6$
$8 \neq 6$
5 is not a solution.

**79.** $6(8.5) + 5 \overset{?}{=} 8(8.5) - 12$
$51 + 5 \overset{?}{=} 68 - 12$
$56 = 56$
8.5 is a solution.

**80.** $C = 6 + 0.85n$

| Input | Output |
|-------|--------|
| 0 | 6.00 |
| 1 | 6.85 |
| 2 | 7.70 |
| 3 | 8.55 |
| 4 | 9.40 |
| 5 | 10.25 |

**81.** Domain = 0, 1, 2, 3, 4, 5
Range = 6.00. 6.85,
7.70, 8.55,
9.40, 10.25

**82.**

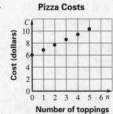

Pizza Costs

---

# Chapter 2 *continued*

## Lesson 2.3

**Developing Concepts Activity 2.3** *(p. 78)*
**Drawing Conclusions**

1. ⊞⊞⊞⊞⊞⊞⊞   Model of seven

⊞⊞⊞⊞⊞⟨⊞⊞⟩   Subtract 2 by taking away 2 1-tiles.

$7 - 2 = 5$

2. ⊟⊟⊟⊟⊟⊟⊟   Model of $-7$

⊟⊟⊟⊟⟨⊟⊟⊟⟩   Subtract $(-3)$ by taking away 3 $-1$-tiles.

$-7 - (-3) = -4$

3. ⊟⊟⊟⊟⊟   Model of $-5$

⊟⊟⊟⊟⟨⊟⟩   Subtract $(-1)$ by taking away 1 $-1$-tile.

$-5 - (-1) = -4$

4. ⊟⊟⊟⊟⊟⊟   Model of $-6$

⟨⊟⊟⊟⊟⊟⊟⟩   Subtract $(-6)$ by taking away 6 $-1$-tiles.

$-6 - (-6) = 0$

5.    Model of two

Add 1 "zero pair."

⊞⊞⊞
⊟   Subtract 3 by taking away 3 1-tiles.

$2 - 3 = -1$

6. ⊞⊞⊞⊞⟨⊞⊞⊞⟩   Model of four

Add 3 "zero pairs."

⟨⊞⊞⊞⊞⊞⊞⊞⟩
⊟⊟⊟   Subtract 7 by taking away 7 1-tiles.

$4 - 7 = -3$

7.    Model of three

Add 2 "zero pairs."

⟨⊞⊞⊞⊞⊞⟩
⊟⊟   Subtract 5 by taking away 5 1-tiles.

$3 - 5 = -2$

8. ⊞⊞⊞⊞⊞⟨⊞⊞⊞⟩   Model of five

Add 3 "zero pairs."

⊟⊟⊟   Subtract 8 by taking away 8 1-tiles.

$5 - 8 = -3$

9.    Model of zero

Add 5 "zero pairs."

⊞⊞⊞⊞⊞
⟨⊟⊟⊟⊟⊟⟩   Subtract $-5$ by taking away 5 $-1$-tiles.

$0 - (-5) = 5$

10. True
Ex. $2 - 1 = 1$

⊞⟨⊞⟩

$2 + (-1) = 1$

⊞⊞
⊟ } $= 0$

11. True
Ex. $2 - (-1) = 3$

⊞⊞⊞
⟨⊟⟩

$2 + 1 = 3$

⊞⊞⊞

### 2.3 Guided Practice *(p. 82)*

1. No. Terms are the parts being added. Therefore, $-7x$ is a term.

2. $-7$

3. Solve left to right

$5 - 7 - (-4) = -2 - (-4)$

Add the opposite of $-4$.

$-2 - (-4) = -2 + 4 = 2$

4. $4 + (-5) = -1$   **5.** $3 + 8 = 11$

6. $0 + (-7) = -7$

7. $2 + 3 + (-6) = 5 + (-6) = -1$

8. $-2.4 + (-3) = -5.4$   **9.** $-3.6 + 6 = 2.4$

10. $\frac{1}{2} + \left(\frac{-1}{4}\right) = \frac{2}{4} + \left(\frac{-1}{4}\right) = \frac{1}{4}$

11. $\frac{2}{3} + \frac{1}{6} + \left(-\frac{1}{3}\right) = \frac{4}{6} + \frac{1}{6} + \left(-\frac{2}{6}\right) = \frac{3}{6} = \frac{1}{2}$

12. 12 and $-5x$   **13.** $-7y^2$, $12y$, and $-6$

14. $23$, $-5w$, and $-8y$

15.

| Input | Function | Output |
|-------|----------|--------|
| $-2$ | $y = -(-2) - 3$ | $-1$ |
| $-1$ | $y = -(-1) - 3$ | $-2$ |
| $0$ | $y = -0 - 3$ | $-3$ |
| $1$ | $y = -1 - 3$ | $-4$ |
| $2$ | $y = -2 - 3$ | $-5$ |

As $x$ increases by 1, $y$ decreases by 1.

# Chapter 2 *continued*

**2.3 Practice and Applications** *(pp. 82–84)*

**16.** −5  **17.** −7  **18.** 13  **19.** 5  **20.** −19  **21.** −15

**22.** −218  **23.** −6  **24.** 2.7  **25.** −4.9  **26.** 9.2

**27.** 9.5  **28.** −1  **29.** 3  **30.** $-1\frac{3}{8}$  **31.** $-1\frac{3}{20}$  **32.** 4

**33.** 9  **34.** −2.9  **35.** −23.1

**36.** $2 - (-4) - 7 = 2 + 4 + (-7) = 6 + (-7) = -1$

**37.** $8 - 11 - (-6) = 8 + (-11) + 6 = -3 + 6 = 3$

**38.** $4 + (-3) - (-5) = 4 + (-3) + 5 = 1 + 5 = 6$

**39.** $3 - (-8) + (-9) = 3 + 8 + (-9) = 11 + (-9) = 2$

**40.** $14 + (-7) - 12 = 14 + (-7) + (-12)$
$= 7 + (-12) = -5$

**41.** $-7 + 42 - 63 = -7 + 42 + (-63)$
$= 35 + (-63) = -28$

**42.** $-8 - (-12) + 3 = -8 + 12 + 3 = 4 + 3 = 7$

**43.** $6 - 1 + 10 - (-8) = 6 + (-1) + 10 + 8$
$= 5 + 10 + 8 = 23$

**44.** $14 - 8 + 17 - (-23) = 14 + (-8) + 17 + 23$
$= 6 + 17 + 23 = 46$

**45.** $2.3 + (-9.1) - 1.2 = 2.3 + (-9.1) + (-1.2)$
$= -6.8 + (-1.2) = -8$

**46.** $1.3 + (-1.3) - 4.2 = 1.3 + (-1.3) + (-4.2)$
$= 0 + (-4.2) = -4.2$

**47.** $-8.5 - 3.9 + (-16.2) = -8.5 + (-3.9) + (-16.2)$
$= -12.4 + (-16.2) = -28.6$

**48.** $8.4 - 5.2 - (-4.7) = 8.4 + (-5.2) + 4.7$
$= 3.2 + 4.7 = 7.9$

**49.** $-\frac{4}{9} - \frac{2}{3} + \left(-\frac{5}{6}\right) = \frac{-8}{18} + \left(-\frac{12}{18}\right) + \left(-\frac{15}{18}\right)$
$= \frac{-20}{18} + \left(-\frac{15}{18}\right) = \frac{-35}{18} = -1\frac{17}{18}$

**50.** $\frac{7}{12} - \left(-\frac{3}{4}\right) + \left(-\frac{1}{8}\right) = \frac{14}{24} + \frac{18}{24} + \left(-\frac{3}{24}\right)$
$= \frac{32}{24} + \left(-\frac{3}{24}\right) = \frac{29}{24} = 1\frac{5}{24}$

**51.** −4 and −$y$  **52.** −$x$ and −7  **53.** −3$x$ and 6

**54.** 9 and −28$x$  **55.** −9 and 4$b$  **56.** $a$, 3$b$, and −5

**57.** $x$, −$y$, and −7  **58.** −3$x$, 5, and −8$y$

**59.** $y = x - 8$

| Input | Function | Output |
|---|---|---|
| −2 | $y = -2 - 8$ | −10 |
| −1 | $y = -1 - 8$ | −9 |
| 0 | $y = 0 - 8$ | −8 |
| 1 | $y = 1 - 8$ | −7 |

**60.** $y = 12 - x$

| Input | Function | Output |
|---|---|---|
| −2 | $y = 12 - (-2)$ | 14 |
| −1 | $y = 12 - (-1)$ | 13 |
| 0 | $y = 12 - 0$ | 12 |
| 1 | $y = 12 - 1$ | 11 |

**61.** $y = -x + 12.1$

| Input | Function | Output |
|---|---|---|
| −2 | $y = -(-2) + 12.1$ | 14.1 |
| −1 | $y = -(-1) + 12.1$ | 13.1 |
| 0 | $y = -(0) + 12.1$ | 12.1 |
| 1 | $y = -1 + 12.1$ | 11.1 |

**62.** $y = -8.5 - (-x)$

| Input | Function | Output |
|---|---|---|
| −2 | $y = -8.5 - [-(-2)]$ | −10.5 |
| −1 | $y = -8.5 - [-(-1)]$ | −9.5 |
| 0 | $y = -8.5 - (-0)$ | −8.5 |
| 1 | $y = -8.5 - (-1)$ | −7.5 |

**63.** $y = 27 + x$

| Input | Function | Output |
|---|---|---|
| −2 | $y = 27 + (-2)$ | 25 |
| −1 | $y = 27 + (-1)$ | 26 |
| 0 | $y = 27 + 0$ | 27 |
| 1 | $y = 27 + 1$ | 28 |

**64.** $y = -x + 13 - x$

| Input | Function | Output |
|---|---|---|
| −2 | $y = -(-2) + 13 - (-2)$ | 17 |
| −1 | $y = -(-1) + 13 - (-1)$ | 15 |
| 0 | $y = -0 + 13 - 0$ | 13 |
| 1 | $y = -1 + 13 - 1$ | 11 |

**65.** $5.3 - (-2.5) - 4.7 = 5.3 + 2.5 + (-4.7)$
$= 7.8 + (-4.7) = 3.1$

**66.** $8.9 - (-2.1) - 7.3 = 8.9 + 2.1 + (-7.3)$
$= 11 + (-7.3) = 3.7$

**67.** $-4.89 + 2.69 - (-3.74) = -4.89 + 2.69 + 3.74$
$= -2.2 + 3.74 = 1.54$

**68.** $-7.85 + 5.96 - (-2.49) = -7.85 + 5.96 + 2.49$
$= -1.89 + 2.49 = 0.6$

**69.** $-13.87 - (-13.87) + 5.8 = -13.87 + 13.87 + 5.8$
$= 0 + 5.8 = 5.8$

# Chapter 2 continued

**70.** $-15.7 + 0.01 + (-34.44)$
$= -15.7 + 0.01 + (-34.44)$
$= -15.69 + (-34.44)$
$= -50.13$

**71.** $725 - 450 = 275$ ft.; up

**72.** $1.19 - 1.13 = -\$.06/\text{gallon}$

**73.** $286.90 - 287.56 = -\$.66/\text{oz}$

**74.** $287.37 - 286.90 = \$.47/\text{oz}$

**75.** $287.37 - 287.56 = -\$.19/\text{oz}$

**76.** $1100 - 690 = 410$ ft

**77.** Start to Cave: 410   Cave to Oak Ridge: 1040 ft
Oak Ridge to Trail Junction: $-235$ ft
Trail Junction to Mt. Parker: 1110 ft

**78.** $410 + 1040 + (-235) + 1110 = 2325$ ft

**79.** $3741 - 11{,}042 = -7301$
$3079 - 3741 = -662$
$1196 - 3079 = -1883$
$1273 - 1196 = 77$
$(-38) - 1273 = -1311$
$7983 - (-38) = 8021$

**80.** $11{,}042 - 7301 - 662 - 1883 + 77 - 1311 +$
$8021 = 7983$

**81. a.**

| Month | Change |
|-------|--------|
| Jan. | — |
| Feb. | $-5{,}000$ |
| Mar. | $5{,}000$ |
| Apr. | $1{,}000$ |
| May | $5{,}000$ |
| Jun. | $13{,}000$ |
| Jul. | $3{,}000$ |
| Aug. | $-5{,}000$ |
| Sep. | $-7{,}000$ |
| Oct. | $-7{,}000$ |
| Nov. | $10{,}000$ |
| Dec. | $-12{,}000$ |

**b.** Negative—decrease in number of visitors
Positive—increase in number of visitors

**c.** Answers may vary.

**82.** True. When you subtract a negative number, you add its opposite, which is positive. The sum of 2 positive numbers is positive.

**83.** True. When you subtract a positive number, you add its opposite, which is negative. The sum of 2 negative numbers is negative.

## 2.3 Mixed Review (p. 85)

**84.** $89 - 8 \cdot 5 - 27 = 89 - 40 - 27 = 49 - 27 = 22$

**85.** $\frac{10}{3} - \frac{2}{3} \cdot 4 + 5 = \frac{10}{3} - \frac{8}{3} + 5 = \frac{10}{3} - \frac{8}{3} + \frac{15}{3} = \frac{17}{3} = 5\frac{2}{3}$

**86.** $12 \cdot 9 \div 6 - 13.5 = 108 \div 6 - 13.5 = 18 - 13.5 = 4.5$

**87.** $17 + 100 \div 25 - 5 = 17 + 4 - 5 = 21 - 5 = 16$

**88.** $5 \cdot \frac{8}{9} - \frac{6}{9} + 51 \div 3 = \frac{40}{9} - \frac{6}{9} + 17 = \frac{40}{9} - \frac{6}{9} + \frac{153}{9}$
$= \frac{187}{9} = 20\frac{7}{9}$

**89.** $13 + 11 \cdot 7 - 6 \div 3 = 13 + 77 - 2 = 90 - 2 = 88$

**90.** $25 - \left[\frac{3}{10}(6 \cdot 5) - 2\right] = 25 - \left[\frac{3}{10}(30) - 2\right]$
$= 25 - [9 - 2] = 25 - 7 = 18$

**91.** $(27 \div 9) \div (7 - 5) = (3) \div (2) = 1\frac{1}{2}$

**92.** $[(12 \cdot 9) \div 6] - 13.5 = [108 \div 6] - 13.5$
$= 18 - 13.5 = 4.5$

**93.** False; 5.78 million

**94.** False; male—up 0.17 million; female—0

**95.** Yes; $2943.21 + 4988.97 = 7932.18$ profits
$5126.55 + 1807.81 = 6934.36$ losses
$7932.18 - 6934.36 = 997.82$ profit

## Quiz 1 (p. 85)

**1.** $-6.2, -5.32, 5.04, 5.3, 5.31, 6.3$

**2.** $-1.6, -1.07, -0.28, 0.18, 1.06, 1.16$

**3.** $-7.5, -7\frac{1}{3}, 7.3, 7\frac{1}{3}, 7.5, 7\frac{2}{3}$

**4.** $-\frac{33}{5}, -6\frac{2}{5}, -6.3, 6.05, 6.42, \frac{33}{5}$

**5.** 3.76   **6.** 75   **7.** $-345$   **8.** 27.5   **9.** 7   **10.** 67.4

**11.** $-11 + 35 = 24$

**12.** $29 - 501 = -472$

**13.** $-17 - (-14) = -17 + 14 = -3$

**14.** $-8 + 12 + (-5) = 4 + (-5) = -1$

**15.** $-32 - (-27) - 9 = -32 + 27 + (-9)$
$= -5 + (-9) = -14$

**16.** $35 - 0 - (-19) = 35 - 0 + 19 = 54$

**17.** $12 + (-1.2) + (-7) = 10.8 + (-7) = 3.8$

**18.** $-143 - (-60) - 8 = -143 + 60 + (-8)$
$= -83 + -8 = -91$

**19.** $14.1 + (-75.2) + 60.7 = -61.1 + 60.7 = -0.4$

**20.** $47.3 + (-2.1) + (-11.3) + 12.9$
$= 45.2 + (-11.3) + 12.9$
$= 33.9 + 12.9$
$= 46.8$ in.

## Lesson 2.4

### 2.4 Guided Practice (p. 89)

**1.** 2 rows; 3 columns   **2.** $2 \times 3$ matrix

**3.** 2; the number of Republican members of the House of Representatives from Arkansas

# Chapter 2 *continued*

**4.**

|        | New | Regular |
|--------|-----|---------|
| Comedy | 25  | 215     |
| Drama  | 30  | 350     |
| Horror | 26  | 180     |

**5.** $\begin{bmatrix} -3 & 0 \\ -6 & 4 \\ 1 & -4 \end{bmatrix} + \begin{bmatrix} 2 & -4 \\ 1 & -3 \\ -1 & 9 \end{bmatrix} = \begin{bmatrix} -3 + 2 & 0 + (-4) \\ -6 + 1 & 4 + (-3) \\ 1 + (-1) & -4 + 9 \end{bmatrix}$

$= \begin{bmatrix} -1 & -4 \\ -5 & 1 \\ 0 & 5 \end{bmatrix}$

$\begin{bmatrix} -3 & 0 \\ -6 & 4 \\ 1 & -4 \end{bmatrix} - \begin{bmatrix} 2 & -4 \\ 1 & -3 \\ -1 & 9 \end{bmatrix} =$

$\begin{bmatrix} -3 - 2 & 0 - (-4) \\ -6 - 1 & 4 - (-3) \\ 1 - (-1) & -4 - 9 \end{bmatrix} = \begin{bmatrix} -5 & 4 \\ -7 & 7 \\ 2 & -13 \end{bmatrix}$

**6.** $\begin{bmatrix} 1 & 8 & -2 \\ -4 & -5 & 6 \end{bmatrix} + \begin{bmatrix} -1 & 9 & 2 \\ 3 & 3 & -5 \end{bmatrix} =$

$\begin{bmatrix} 1 + (-1) & 8 + 9 & -2 + 2 \\ -4 + 3 & -5 + 3 & 6 + (-5) \end{bmatrix} = \begin{bmatrix} 0 & 17 & 0 \\ -1 & -2 & 1 \end{bmatrix}$

$\begin{bmatrix} 1 & 8 & -2 \\ -4 & -5 & 6 \end{bmatrix} - \begin{bmatrix} -1 & 9 & 2 \\ 3 & 3 & -5 \end{bmatrix} =$

$\begin{bmatrix} 1 - (-1) & 8 - 9 & -2 - 2 \\ -4 - 3 & -5 - 3 & 6 - (-5) \end{bmatrix} = \begin{bmatrix} 2 & -1 & -4 \\ -7 & -8 & 11 \end{bmatrix}$

## 2.4 Practice and Applications (pp. 89–91)

**7.** yes  **8.** no  **9.** no  **10.** yes

**11.** $\begin{bmatrix} 3 & -2 \\ 5 & 1 \end{bmatrix} + \begin{bmatrix} 4 & -3 \\ -8 & -2 \end{bmatrix} = \begin{bmatrix} 3 + 4 & -2 + -3 \\ 5 + -8 & 1 + -2 \end{bmatrix}$

$= \begin{bmatrix} 7 & -5 \\ -3 & -1 \end{bmatrix}$

**12.** $\begin{bmatrix} 4 & -1 \\ -5 & -9 \end{bmatrix} + \begin{bmatrix} -6 & -3 \\ 2 & -3 \end{bmatrix} = \begin{bmatrix} 4 + -6 & -1 + -3 \\ -5 + 2 & -9 + -3 \end{bmatrix}$

$= \begin{bmatrix} -2 & -4 \\ -3 & -12 \end{bmatrix}$

**13.** $\begin{bmatrix} 1 & -2 & 2 \\ 0 & -3 & 4 \end{bmatrix} + \begin{bmatrix} 3 & -4 & 5 \\ -8 & 1 & 6 \end{bmatrix} =$

$\begin{bmatrix} 1 + 3 & -2 + -4 & 2 + 5 \\ 0 + -8 & -3 + 1 & 4 + 6 \end{bmatrix} = \begin{bmatrix} 4 & -6 & 7 \\ -8 & -2 & 10 \end{bmatrix}$

**14.** $\begin{bmatrix} -2.4 & 1.6 & -7.8 \\ 14.3 & 1.1 & -3.9 \end{bmatrix} + \begin{bmatrix} -2.8 & 5.4 & 2.3 \\ -1.7 & 4.2 & 5.6 \end{bmatrix}$

$= \begin{bmatrix} -2.4 + -2.8 & 1.6 + 5.4 & -7.8 + 2.3 \\ 14.3 + -1.7 & 1.1 + 4.2 & -3.9 + 5.6 \end{bmatrix}$

$= \begin{bmatrix} -5.2 & 7.0 & -5.5 \\ 12.6 & 5.3 & 1.7 \end{bmatrix}$

**15.** $\begin{bmatrix} 6.2 & -1.2 \\ -2.5 & -4.4 \\ 3.4 & -5.8 \end{bmatrix} + \begin{bmatrix} 1.5 & 9.2 \\ 6.6 & -2.2 \\ 5.7 & -7.1 \end{bmatrix} =$

$\begin{bmatrix} 6.2 + 1.5 & -1.2 + 9.2 \\ -2.5 + 6.6 & -4.4 + -2.2 \\ 3.4 + 5.7 & -5.8 + -7.1 \end{bmatrix} = \begin{bmatrix} 7.7 & 8 \\ 4.1 & -6.6 \\ 9.1 & -12.9 \end{bmatrix}$

**16.** $\begin{bmatrix} 2 & 9 & -3 \\ 1 & 8 & -2 \\ -3 & -1 & -7 \end{bmatrix} + \begin{bmatrix} -2 & -6 & 4 \\ -1 & -2 & 5 \\ 2 & 0 & 8 \end{bmatrix} =$

$\begin{bmatrix} 2 + -2 & 9 + -6 & -3 + 4 \\ 1 + -1 & 8 + -2 & -2 + 5 \\ -3 + 2 & -1 + 0 & -7 + 8 \end{bmatrix} = \begin{bmatrix} 0 & 3 & 1 \\ 0 & 6 & 3 \\ -1 & -1 & 1 \end{bmatrix}$

**17.** $\begin{bmatrix} 8 & -3 \\ 4 & -1 \end{bmatrix} - \begin{bmatrix} 7 & 7 \\ -2 & -5 \end{bmatrix} = \begin{bmatrix} 8 - 7 & -3 - 7 \\ 4 - (-2) & -1 - (-5) \end{bmatrix}$

$= \begin{bmatrix} 1 & -10 \\ 6 & 4 \end{bmatrix}$

**18.** $\begin{bmatrix} 4 & 3 \\ -12 & -10 \end{bmatrix} - \begin{bmatrix} -6 & 1 \\ -4 & 2 \end{bmatrix} =$

$\begin{bmatrix} 4 - (-6) & 3 - 1 \\ -12 - (-4) & -10 - 2 \end{bmatrix} = \begin{bmatrix} 10 & 2 \\ -8 & -12 \end{bmatrix}$

**19.** $\begin{bmatrix} -4 & 1 \\ 0 & -13 \\ 2 & -8 \end{bmatrix} - \begin{bmatrix} -6 & 3 \\ -5 & 8 \\ 2 & -7 \end{bmatrix} =$

$\begin{bmatrix} -4 - (-6) & 1 - 3 \\ 0 - (-5) & -13 - 8 \\ 2 - 2 & -8 - (-7) \end{bmatrix} = \begin{bmatrix} 2 & -2 \\ 5 & -21 \\ 0 & -1 \end{bmatrix}$

**20.** $\begin{bmatrix} -5 & 11 & -2 \\ -10 & 4 & 6 \end{bmatrix} - \begin{bmatrix} -3 & 0 & 2 \\ 8 & -5 & -1 \end{bmatrix}$

$= \begin{bmatrix} -5 - (-3) & 11 - 0 & -2 - 2 \\ -10 - 8 & 4 - (-5) & 6 - (-1) \end{bmatrix}$

$= \begin{bmatrix} -2 & 11 & -4 \\ -18 & 9 & 7 \end{bmatrix}$

**21.** $\begin{bmatrix} 3a & 5b \\ c - 6 & d \end{bmatrix} = \begin{bmatrix} -12 & -5 \\ 1 & -3 \end{bmatrix}$

$3a = -12 \quad 5b = -5 \quad c - 6 = 1 \quad d = -3$

$a = -4 \quad\quad b = -1 \quad\quad c = 7 \quad\quad d = -3$

**22.** $\begin{bmatrix} 4a & b + 3 \\ c & d - 3 \end{bmatrix} = \begin{bmatrix} 8 & -1 \\ 0 & -6 \end{bmatrix}$

$4a = 8 \quad b + 3 = -1 \quad c = 0 \quad d - 3 = -6$

$a = 2 \quad\quad b = -4 \quad\quad c = 0 \quad\quad d = -3$

**23.**

|       | Sale | Regular |
|-------|------|---------|
| CDs   | 52   | 3300    |
| tapes | 28   | 1600    |

**24.**

|        | S | M | L  | XL |
|--------|---|---|----|----|
| Shirts | 3 | 7 | 10 | 5  |
| Shorts | 7 | 4 | 2  | 2  |

**25.** $\begin{bmatrix} 36 & 28 \\ 24 & 26 \\ 51 & 32 \end{bmatrix} + \begin{bmatrix} 40 & 31 \\ 26 & 20 \\ 46 & 34 \end{bmatrix} = \begin{bmatrix} 36 + 40 & 28 + 31 \\ 24 + 26 & 26 + 20 \\ 51 + 46 & 32 + 34 \end{bmatrix}$

$= \begin{bmatrix} 76 & 59 \\ 50 & 46 \\ 97 & 66 \end{bmatrix}$

**26.** Over a 2 year period of time, the kennel cared for a total of 76 male beagles, 59 female beagles, 50 male dalmations, 46 female dalmations, 97 male bulldogs, and 66 female bulldogs.

# Chapter 2 *continued*

**27.** $\begin{bmatrix} 0.20 & 0.20 & 0.20 \\ 0.35 & 0.35 & 0.35 \\ 0.45 & 0.45 & 0.45 \end{bmatrix}$

**28.** $\begin{bmatrix} 5.50 & 6.50 & 7.00 \\ 6.50 & 8.00 & 9.50 \\ 7.50 & 8.75 & 11.00 \end{bmatrix} + \begin{bmatrix} 0.20 & 0.20 & 0.20 \\ 0.35 & 0.35 & 0.35 \\ 0.45 & 0.45 & 0.45 \end{bmatrix}$

$= \begin{bmatrix} 5.50 + 0.20 & 6.50 + 0.20 & 7.00 + 0.20 \\ 6.50 + 0.35 & 8.00 + 0.35 & 9.50 + 0.35 \\ 7.50 + 0.45 & 8.75 + 0.45 & 11.00 + 0.45 \end{bmatrix}$

$= \begin{bmatrix} 5.70 & 6.70 & 7.20 \\ 6.85 & 8.35 & 9.85 \\ 7.95 & 9.20 & 11.45 \end{bmatrix}$

**29. a.**

|  | May | June | July |
|---|---|---|---|
| Atlanta, GA | 79.6 | 85.8 | 88.0 |
| San Francisco, CA | 66.5 | 70.3 | 71.6 |
| Anchorage, AK | 54.4 | 61.6 | 65.2 |

**b.**

|  | May | June | July |
|---|---|---|---|
| Atlanta, GA | 58.7 | 66.2 | 69.5 |
| San Francisco, CA | 49.7 | 52.6 | 53.9 |
| Anchorage, AK | 38.8 | 47.2 | 51.7 |

**c.** $\begin{bmatrix} 79.6 - 58.7 & 85.8 - 66.2 & 88.0 - 69.5 \\ 66.5 - 49.7 & 70.3 - 52.6 & 71.6 - 53.9 \\ 54.4 - 38.8 & 61.6 - 47.2 & 65.2 - 51.7 \end{bmatrix} =$

$\begin{bmatrix} 20.9 & 19.6 & 18.5 \\ 16.8 & 17.7 & 17.7 \\ 15.6 & 14.4 & 13.5 \end{bmatrix}$

To find the difference of average temperatures during May, June and July in each city, set up the 2 matrices and subtract.

**30. a.** $\begin{bmatrix} 5 + 2 & -9 + 8 \\ -4 + 1 & 1 + 9 \\ -1 + -3 & 4 + -1 \end{bmatrix} = \begin{bmatrix} 2 + 5 & 8 + -9 \\ 1 + -4 & 9 + 1 \\ -3 + -1 & -1 + 4 \end{bmatrix}$

$\begin{bmatrix} 7 & -1 \\ -3 & 10 \\ -4 & 3 \end{bmatrix} = \begin{bmatrix} 7 & -1 \\ -3 & 10 \\ -4 & 3 \end{bmatrix}$

Yes; each separate addition is commutative.

**b.** yes

**31.** yes; *Sample answer:*

$\left( \begin{bmatrix} 1 & 2 \\ 3 & 4 \end{bmatrix} + \begin{bmatrix} 5 & 6 \\ 7 & 8 \end{bmatrix} \right) + \begin{bmatrix} 0 & 1 \\ 1 & 0 \end{bmatrix} = \begin{bmatrix} 6 & 9 \\ 11 & 12 \end{bmatrix}$

$\begin{bmatrix} 1 & 2 \\ 3 & 4 \end{bmatrix} + \left( \begin{bmatrix} 5 & 6 \\ 7 & 8 \end{bmatrix} + \begin{bmatrix} 0 & 1 \\ 1 & 0 \end{bmatrix} \right) = \begin{bmatrix} 6 & 9 \\ 11 & 12 \end{bmatrix}$

### 2.4 Mixed Review *(p. 91)*

**32.** $(2y)^3$  **33.** $5^2$  **34.** $4^6$  **35.** $x^2 y^3$  **36.** $2^x$  **37.** $3t^4$

**38.** 82  **39.** $-43.7$  **40.** $-4.5$  **41.** 36  **42.** $-0.9$  **43.** 15

**44.** $-25$  **45.** $-21$  **46.** $-32$  **47.** $-3$  **48.** $-11$

**49.** $-14$  **50.** $-24$  **51.** $-3$  **52.** $-12$

### Technology Activity 2.4 *(p. 92)*

**1.** $\begin{bmatrix} 7 & -5 \\ 5 & -3 \\ -9 & -8 \end{bmatrix} + \begin{bmatrix} 0 & -1 \\ 1 & 0 \\ -1 & -6 \end{bmatrix} = \begin{bmatrix} 7 & -6 \\ 6 & -3 \\ -10 & -14 \end{bmatrix}$

$\begin{bmatrix} 7 & -5 \\ 5 & -3 \\ -9 & -8 \end{bmatrix} - \begin{bmatrix} 0 & -1 \\ 1 & 0 \\ -1 & -6 \end{bmatrix} = \begin{bmatrix} 7 & -4 \\ 4 & -3 \\ -8 & -2 \end{bmatrix}$

**2.** $\begin{bmatrix} 52 & 79 & -61 \\ -84 & 21 & -76 \end{bmatrix} + \begin{bmatrix} 10 & 45 & -62 \\ -16 & 98 & -28 \end{bmatrix} =$

$\begin{bmatrix} 62 & 124 & -123 \\ -100 & 119 & -104 \end{bmatrix}$

$\begin{bmatrix} 52 & 79 & -61 \\ -84 & 21 & -76 \end{bmatrix} - \begin{bmatrix} 10 & 45 & -62 \\ -16 & 98 & -28 \end{bmatrix} =$

$\begin{bmatrix} 42 & 34 & 1 \\ -68 & -77 & -48 \end{bmatrix}$

**3.** $\begin{bmatrix} 7.4 & -5.1 \\ -0.6 & -4.6 \\ 9.9 & -9.4 \\ 2.1 & -3.6 \\ -0.5 & 7.1 \end{bmatrix} + \begin{bmatrix} 3.3 & 3.2 \\ -3.2 & 2.7 \\ -8.2 & -1.7 \\ 5.1 & -4.8 \\ -3.9 & 6.1 \end{bmatrix} = \begin{bmatrix} 10.7 & -1.9 \\ -3.8 & -1.9 \\ 1.7 & -11.1 \\ 7.2 & -8.4 \\ -4.4 & 13.2 \end{bmatrix}$

$\begin{bmatrix} 7.4 & -5.1 \\ -0.6 & -4.6 \\ 9.9 & -9.4 \\ 2.1 & -3.6 \\ -0.5 & 7.1 \end{bmatrix} - \begin{bmatrix} 3.3 & 3.2 \\ -3.2 & 2.7 \\ -8.2 & -1.7 \\ 5.1 & -4.8 \\ -3.9 & 6.1 \end{bmatrix} = \begin{bmatrix} 4.1 & -8.3 \\ 2.6 & -7.3 \\ 18.1 & -7.7 \\ -3 & 1.2 \\ 3.4 & 1 \end{bmatrix}$

**4.** $\begin{bmatrix} 44 & -51 & 17 \\ 76 & 63 & 25 \\ -20 & -60 & 42 \end{bmatrix} + \begin{bmatrix} -53 & -18 & 11 \\ 29 & -27 & -62 \\ 29 & -35 & -28 \end{bmatrix} =$

$\begin{bmatrix} -9 & -69 & 28 \\ 105 & 36 & -37 \\ 9 & -95 & 14 \end{bmatrix}$

$\begin{bmatrix} 44 & -51 & 17 \\ 76 & 63 & 25 \\ -20 & -60 & 42 \end{bmatrix} - \begin{bmatrix} -53 & -18 & 11 \\ 29 & -27 & -62 \\ 29 & -35 & -28 \end{bmatrix} =$

$\begin{bmatrix} 97 & -33 & 6 \\ 47 & 90 & 87 \\ -49 & -25 & 70 \end{bmatrix}$

## Lesson 2.5

### Activity *(p. 93)*

| Factor of $-3$ | Factor of $-2$ | Factor of $-1$ |
|---|---|---|
| $(3)(-3) = -9$ | $(3)(-2) = -6$ | $(3)(-1) = -3$ |
| $(2)(-3) = -6$ | $(2)(-2) = -4$ | $(2)(-1) = -2$ |
| $(1)(-3) = -3$ | $(1)(-2) = -2$ | $(1)(-1) = -1$ |

# Chapter 2 *continued*

| Factor of $-3$ | Factor of $-2$ | Factor of $-1$ |
|---|---|---|
| $(0)(-3) = 0$ | $(0)(-2) = 0$ | $(0)(-1) = 0$ |
| $(-1)(-3) = 3$ | $(-1)(-2) = 2$ | $(-1)(-1) = 1$ |
| $(-2)(-3) = 6$ | $(-2)(-2) = 4$ | $(-2)(-1) = 2$ |

## 2.5 Guided Practice (p. 96)

1. commutative property of multiplication

2. no; only when an odd number of factors are negative

3. no; only when an even number of factors are negative

4. There are 3 negative factors $(-1, -5, -5)$.
   Therefore, the product is also negative.

5. $-8$  6. $0$  7. $35$  8. $35$  9. $72$  10. $-240$

11. $70$  12. $1$  13. $2(-6)(-x) = 2(-6)(-4) = 48$

14. $5(x - 4) = 5(-3 - 4) = 5(-7) = -35$

15. $-10(4.7) = -47$ ft

## 2.5 Practice and Applications (pp. 96–98)

16. $-8(3) = -24$  17. $4(-4) = -16$

18. $20(-65) = -1300$  19. $(-1)(-5) = 5$

20. $-7(-1.2) = 8.4$  21. $(-11)\left(\frac{1}{8}\right) = \frac{-11}{8}$

22. $(-15)\left(\frac{3}{5}\right) = -9$  23. $|(-12)(2)| = 24$

24. $(-3)(-1)(-6) = -18$  25. $(13)(-2)(-3) = 78$

26. $5(-2)(7) = -70$  27. $(-4)(-7)\left(\frac{3}{7}\right) = 12$

28. $(-3)(-1)(4)(-6) = -72$

29. $(-13)(-2)(-2)\left(-\frac{2}{13}\right) = 8$

30. $(-3)(-y) = 3y$  31. $(7)(-x) = -7x$

32. $5(-a)(-a)(-a) = -5a^3$

33. $(-4)(-x)(x)(-x) = -4x^3$  34. $-(-b)^3 = b^3$

35. $-(-4)^2(y) = -16y$  36. $|(8)(-z)(-z)(-z)| = 8|z^3|$

37. $-(y^4)(y) = -y^5$  38. $(-b^2)(-b^3)(-b^4) = -b^9$

39. $-\frac{1}{2}(-2x) = x$  40. $\frac{2}{3}\left(-\frac{3}{2}x\right) = -x$

41. $-\frac{3}{7}(-w^2)(7w) = 3w^3$  42. $-8x = -8(6) = -48$

43. $y^3 - 4 = -2^3 - 4 = -12$

44. $3x^2 - 5x = 3(-2^2) - 5(-2) = 12 + 10 = 22$

45. $4a + a^2 = 4(-7) + (-7)^2 = -28 + 49 = 21$

46. $-4(|y - 12|) = -4(|5 - 12|) = -4(7) = -28$

47. $-2(|x - 5|) = -2(|-5 - 5|) = -2(10) = -20$

48. $-2x^2 + 3x - 7 = -2(4^2) + 3(4) - 7$
    $= -32 + 12 - 7 = -27$

49. $9r^3 - (-2r) = 9(2^3) - [-2(2)] = 9(8) - (-4)$
    $= 72 + 4 = 76$

50. $94.51$  51. $-70.70$  52. $644.80$  53. $1809.86$

54. $x^3 - 8.29 = (-2.47)^3 - 8.29 = -15.07 - 8.29 =$
    $= -23.36$

---

55. $8.3 + y^3 = 8.3 + (-4.6)^3 = 8.3 + (-97.34) = -89.04$

56. $4.7b - (-b^2) = 4.7(1.99) - (-1.99^2)$
    $= 9.35 - (-3.96) = 13.31$

57. $x^2 + x - 27.2 = -7^2 + (-7) - 27.2$
    $= 49 + (-7) - 27.2 = 42 - 27.2 = 14.8$

58. true  59. false; $a$ could be any nonpositive number.

60. false; any number times 0 is 0 and 0 cannot be greater than zero.

61. The total amount of money lost can be found by multiplying the loss per pound by the number of pounds of bananas sold.

62. $2956(0.11) = \$325.16$

63. To find the total amount lost, multiply \$.08 per bottle by 3107 bottles of apple juice for a loss of \$248.56.

64. The total decrease is equal to the daily decrease times the number of days helped.

65. $-25(5) = -\$125$   $175 - 125 = \$50$

66. $x + 3(80,000)$
    $x + 240,000$

67. $700,000 + 240,000 = 940,000$ visitors

68. $-8\begin{bmatrix} -4 & -7 \\ 3 & 3 \end{bmatrix} = \begin{bmatrix} -8(-4) & -8(-7) \\ -8(3) & -8(3) \end{bmatrix}$
    $= \begin{bmatrix} 32 & 56 \\ -24 & -24 \end{bmatrix}$

69. $-7\begin{bmatrix} 6 & -4 & 3 \\ -1 & 2^2 & -9 \end{bmatrix} = \begin{bmatrix} -7(6) & -7(-4) & -7(3) \\ -7(-1) & -7(2^2) & -7(-9) \end{bmatrix}$
    $= \begin{bmatrix} -42 & 28 & -21 \\ 7 & -28 & 63 \end{bmatrix}$

70. $-5x\begin{bmatrix} 2x & -6y \\ 4b & -8a \end{bmatrix} = \begin{bmatrix} -5x(2x) & -5x(-6y) \\ -5x(4b) & -5x(-8a) \end{bmatrix}$
    $= \begin{bmatrix} -10x^2 & 30xy \\ -20xb & 40xa \end{bmatrix}$

71. **a.** Answers may vary. *Sample answer:* $\frac{1}{2}(x)(-60)$

    **b.** Answers may vary. *Sample answer:* $\frac{-1}{2}(x) + .62x$

72. C

73. A

   **a.** $\left[\frac{3}{8}(8 - 6) + \frac{1}{4}\right] \cdot (-12) = \left[\frac{3}{8}(2) + \frac{1}{4}\right] \cdot (-12) =$
       $\left[\frac{6}{8} + \frac{2}{8}\right] \cdot (-12) = [1] \cdot (-12) = -12$

   **b.** $\frac{3}{8} \cdot 8 - 6 + \frac{1}{4} \cdot (-12) = 3 - 6 + (-3) = -6$

   **c.** $-\frac{3}{8} \cdot 8 - 6 + \frac{1}{4} \cdot 12 = -3 - 6 + 3 = -6$

   **d.** $-\frac{3}{8} \cdot \left(8 - 6 + \frac{1}{4}\right) \cdot (-12) = -\frac{3}{8} \cdot \left(2\frac{1}{4}\right) \cdot (-12) =$
       $-\frac{27}{32} \cdot (-12) = \frac{324}{32} = 10\frac{1}{8}$

74. $\frac{3}{4} \cdot [-7 \cdot (-4 - 6) + 30] - 11$
    $= \frac{3}{4} \cdot [-7 \cdot (-10) + 30] - 11 = \frac{3}{4} \cdot [70 + 30] - 11$
    $= \frac{3}{4} \cdot [100] - 11 = 75 - 11 = 64$

# Chapter 2 *continued*

**75.** $-3 \cdot \left[\left(2\frac{9}{14} - 3\frac{3}{7}\right) \cdot \frac{28}{11}\right] + 5\left(-9\frac{1}{5} - 9\right)$

$= -3 \cdot \left[-\frac{11}{14} \cdot \frac{28}{11}\right] + 5\left(-18\frac{1}{5}\right)$

$= -3 \cdot [-2] + (-91)$

$= 6 + (-91) = -85$

## 2.5 Mixed Review (p. 98)

**76.** What number plus 4 equals 9?   5

**77.** Seven taken away from what number is 3?   10

**78.** What number times 6 is 18?   3

**79.** What number divided by 8 equals 4?   32

**80.** Seven minus 1 divided by 2 is what number?   3

**81.** What number times itself is 121?   11, $-11$

**82.** 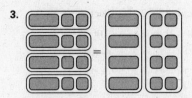   $6 > -3, \quad -3 < 6$

**83.**   $9 > -4, \quad -4 < 9$

**84.**   $-\frac{1}{2} < \frac{1}{3}, \quad \frac{1}{3} > -\frac{1}{2}$

**85.**   $-4.0 < -3.8,$
$-3.8 > -4.0$

**86.**   $0.5 > -2.8, \quad -2.8 < 0.5$

**87.**   $-4.1 < -4.02,$
$-4.02 > -4.1$

**88.** 12 and $-z$   **89.** $-t$ and 5   **90.** $4w$ and $-11$

**91.** 31 and $-15n$   **92.** $-7$ and $4x$

**93.** $m, -2n,$ and $-t^2$   **94.** $c^2, -3c,$ and $-4$

**95.** $y, 6,$ and $-8x$   **96.** $-9a^2, 4,$ and $-2a^3$

**97.** $21.9 - 10 = 11.9$ billion

The deficit went down by 11.9 billion.

## Lesson 2.6

### Developing Concepts Activity 2.6 (p. 99)
### Drawing Conclusions

**1.** $5(x + 2); \ 5x + 10$

**2.**

---

**3.**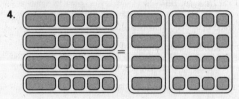

**4.**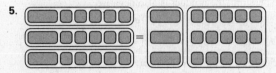

**5.**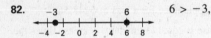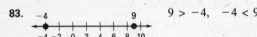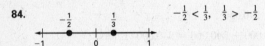

**6.** false; $4(x + 6) = 4x + 24$   **7.** true

**8.** Answers may vary.

$a(b + c) = ab + ac; \ a(b - c) = ab - ac$

## 2.6 Guided Practice (p. 103)

**1.** $4; -7$   **2.** $-6$ and $7, -3x^2$ and $9x^2, 3y$ and $-4y$

**3.** A mistake was made when multiplying the negatives.

$-2(6.5 - 2.1) = -13 + 4.2 = -8.8$

**4.** The distributive property was completed incorrectly.

$5 - 4(3 + x) = 5 - 12 - 4x = -7 - 4x$

**5.** $5w - 40$   **6.** $7y + 133$   **7.** $12y - xy$   **8.** $-4u - 8$

**9.** $-\frac{5}{6}b + 5$   **10.** $-\frac{2}{3}t + 16$   **11.** 6.9   **12.** 2; 0.05; 17.55

**13.** simplified   **14.** $2x^2$   **15.** simplified

**16.** $-4x^2 + 4x + 8$   **17.** $8t^2 + 3t - 4$

**18.** $-9w - 12 + 2x^2$

## 2.6 Practice and Applications (pp. 103–106)

**19.** false; $3(2) + 3(7)$   **20.** true   **21.** true

**22.** false; $9(13) - 2(13)$   **23.** true   **24.** true

**25.** $3x + 12$   **26.** $4w + 24$   **27.** $5y - 10$   **28.** $28 - 4m$

**29.** $-y + 9$   **30.** $-3r - 24$   **31.** $-4t + 32$

**32.** $-2x - 12$   **33.** $x^2 + x$   **34.** $3y - y^2$   **35.** $-r^2 + 9r$

**36.** $-7s - s^2$   **37.** $6x - 2$   **38.** $20 + 15y$   **39.** $-6x + 12$

**40.** $-9a - 54$   **41.** $4x^2 + 32x$   **42.** $-24t + 2t^2$

**43.** $15y^2 - 10y$   **44.** $-2x^2 + 16x$   **45.** $9t + 27$

**46.** $-6w^2 + 3w^3$   **47.** $\frac{5}{2}x - \frac{10}{3}$   **48.** $y^3 - y^2$   **49.** $11x$

**50.** $-7y$   **51.** $-17b$   **52.** $7 - x$   **53.** $y + 4$   **54.** $4 + 2a$

**55.** $-0.8t$   **56.** $\frac{1}{9}w$   **57.** $-101a$   **58.** $5x^2 - 7$

**59.** $5x^3 - 2$   **60.** $5b + 5$

**61.** $(3y + 1)(-2) + y = -6y - 2 + y = -5y - 2$

**62.** $4(2 - a) - a = 8 - 4a - a = 8 - 5a$

# Chapter 2 *continued*

**63.** $12s + (7 - s)2 = 12s + 14 - 2s = 10s + 14$

**64.** $(5 - 2x)(-x) + x^2 = -5x + 2x^2 + x^2 = -5x + 3x^2$

**65.** $7x - 3x(x + 1) = 7x - 3x^2 - 3x = 4x - 3x^2$

**66.** $-4(y + 2) - 6y = -4y - 8 - 6y = -10y - 8$

**67.** $3t(t - 5) + 6t^2 = 3t^2 - 15t + 6t^2 = 9t^2 - 15t$

**68.** $-x^3 + 2x(x - x^2) = -x^3 + 2x^2 - 2x^3 = -3x^3 + 2x^2$

**69.** $4w^2 - w(2w - 3) = 4w^2 - 2w^2 + 3w = 2w^2 + 3w$

**70.** $x + 20 + 0.06(x + 20) \le 58$

$x + 20 + 0.06x + 1.20 \le 58$

$1.06x + 21.2 \le 58$

**71.** no    $1.06(35) + 21.2 \overset{?}{\le} 58$    **72.** A

$37.1 + 21.2 \overset{?}{\le} 58$

$58.3 \nleq 58$

**73.** $97.3(150 - n) + 114n = 97.3(150 - 90) + 114(90) =$

$5838 + 10260 = 16,098$ tons

**74.** $97.3n + 114(150 - n) = 97.3(72) + 114(150 - 72) =$

$7005.6 + 8892 = 15,897.6$ tons

**75.** $T = 5000 + 0.02s + 0.06(5000 - s)$

$= 5000 + 0.02s + 300 - 0.06s = 5300 - 0.04s$

**76.** $T = 5300 - 0.04s = 5300 - 0.04(2000)$

$= 5300 - 80 = 5220$

$T = 5300 - 0.04s = 5300 - 0.04(3000)$

$= 5300 - 120 = 5180$

**77.** $w = 60b + 15(16 - b) = 60b + 240 - 15b$

$= 45b + 240$

**78.**

| Boxes of books | Function | Total W |
|---|---|---|
| 16 | $45(16) + 240$ | 960 |
| 15 | $45(15) + 240$ | 915 |
| 14 | $45(14) + 240$ | 870 |
| 13 | $45(13) + 240$ | 825 |
| 12 | $45(12) + 240$ | 780 |
| 11 | $45(11) + 240$ | 735 |
| 10 | $45(10) + 240$ | 690 |
| 9 | $45(9) + 240$ | 645 |
| 8 | $45(8) + 240$ | 600 |
| 7 | $45(7) + 240$ | 555 |
| 6 | $45(6) + 240$ | 510 |
| 5 | $45(5) + 240$ | 465 |
| 4 | $45(4) + 240$ | 420 |
| 3 | $45(3) + 240$ | 375 |
| 2 | $45(2) + 240$ | 330 |
| 1 | $45(1) + 240$ | 285 |
| 0 | $45(0) + 240$ | 240 |

**79.** $x + (x - 7) + (x - 7) + x$

$= x + x - 7 + x - 7 + x = 4x - 14$

**80.** $(x - 2) + (x + 11) + (2x + 3)$

$= x - 2 + x + 11 + 2x + 3 = 4x + 12$

**81.** $(x + 5)11 = 11x + 55;$   $11x + 11(5) = 11x + 55$

**82.** $9(2x + 4) = 18x + 36;$   $9(2x) + 9(4) = 18x + 36$

**83.** $T = 0.10C + 0.5(0.10C)$

**84.** $0.10c + 0.5(0.10C) = 0.10c + 0.05C$

$T = 0.15C;$   Distributive

**85.** Yes; the two equations are identical.

**86.** $C = 200x(0.06) + (300 - x)(200)(0.05)$

$= 12x + (60,000 - 200x)(0.05)$

$= 12x + 3000 - 10x = 2x + 3000$

**87.** $C = 2x + 3000 = 2(75) + 3000$

$= 150 + 3000 = \$3150$

**88.** $C = 200x(0.05) + (300 - x)(200)(0.06)$

$= 10x + 3600 - 12x = 3600 - 2x$

$= 3600 - 2(125) = 3600 - 250 = \$3350$

**89.** No; $c = 3600 - 2x = 3600 - 2(90)$

$= 3600 - 180 = \$3420$

**90.** Answers may vary.

*Sample answer:*

The area of the rectangle is $(x + 2)4$. The sum of the areas of the small rectangles is $4x + 8$.

**91. a.** A

**b.** $C = 16.99(23 - 8) + 11.99(8) = 254.85 +$

$95.92 = \$350.77$

**c.** $C = 11.99(23 - 13) + 16.99(13) = 119.90 +$

$220.87 = \$340.77$

**d.** $275.77 + 5x \le 300$

$5x \le 24.23$

$x \le 4.85$

The greatest number of people to receive the meadow bouquet is 4.

**92.** Answers may vary.

*Sample answer:*

Suppose $x = 3, y = 4$

$2(xy) = 2x \cdot 2y$

$2(3 \cdot 4) = 2(3) \cdot 2(4)$

$24 = 6 \cdot 8$

$24 \ne 48$

# Chapter 2 *continued*

**93.** $a(b + c) = ab + ac$    distributive property

            $= ba + ca$    commutative property

            $= (b + c)a$    distributive property

## 2.6 Mixed Review (p. 106)

**94.** $\frac{7}{2}$   **95.** $\frac{11}{6}$   **96.** $\frac{1}{7}$   **97.** 1   **98.** 121   **99.** $\frac{1}{435}$   **100.** $\frac{2}{9}$

**101.** $\frac{8}{27}$

**102.** $\dfrac{10 \cdot 8}{4^2 + 4} = \dfrac{80}{16 + 4} = \dfrac{80}{20} = 4$

**103.** $\dfrac{6^2 - 12}{3^2 + 15} = \dfrac{36 - 12}{9 + 15} = \dfrac{24}{24} = 1$

**104.** $\dfrac{75 - 5^2}{11 + (3 \cdot 4)} = \dfrac{75 - 25}{11 + 12} = \dfrac{50}{23} = 2\dfrac{4}{23}$

**105.** $\dfrac{(3 \cdot 7) + 9}{2^3 + 5 - 3} = \dfrac{21 + 9}{8 + 5 - 3} = \dfrac{30}{10} = 3$

**106.** $\dfrac{(2 + 5)^2}{3^2 - 2} = \dfrac{7^2}{9 - 2} = \dfrac{49}{7} = 7$

**107.** $\dfrac{6 + 7^2}{3^3 - 9 - 7} = \dfrac{6 + 49}{27 - 9 - 7} = \dfrac{55}{11} = 5$

**108.** $6 - (-8) - 11 = 6 + 8 + (-11) = 14 + (-11) = 3$

**109.** $4 - 8 - 3 = 4 + (-8) + (-3) = -4 + (-3) = -7$

**110.** $6 + (-13) + (-5) = -7 + (-5) = -12$

**111.** $-7 + 9 - 8 = -7 + 9 + (-8) = 2 + (-8) = -6$

**112.** $20 + (-16) + (-3) = 4 + (-3) = 1$

**113.** $12.4 - 9.7 - (-6.1) = 12.4 - 9.7 + 6.1$

                      $= 2.7 + 6.1 = 8.8$

## Quiz 2 (p. 107)

**1.**
$\begin{bmatrix} 2 & -6 \\ -5 & 5 \\ -7 & 0 \end{bmatrix} + \begin{bmatrix} 3 & -7 \\ 1 & 4 \\ -2 & -8 \end{bmatrix} =$

$\begin{bmatrix} 2+3 & -6+(-7) \\ -5+1 & 5+4 \\ -7+(-2) & 0+(-8) \end{bmatrix} = \begin{bmatrix} 5 & -13 \\ -4 & 9 \\ -9 & -8 \end{bmatrix}$

$\begin{bmatrix} 2 & -6 \\ -5 & 5 \\ -7 & 0 \end{bmatrix} - \begin{bmatrix} 3 & -7 \\ 1 & 4 \\ -2 & -8 \end{bmatrix} =$

$\begin{bmatrix} 2-3 & -6-(-7) \\ -5-1 & 5-4 \\ -7-(-2) & 0-(-8) \end{bmatrix} = \begin{bmatrix} -1 & 1 \\ -6 & 1 \\ -5 & 8 \end{bmatrix}$

**2.** $-63$   **3.** $-21$   **4.** 14   **5.** $-2800$   **6.** $-3$

**7.** 10.8   **8.** 110   **9.** 270   **10.** $7.1t$   **11.** $-13x$

**12.** $-10b^3$   **13.** $-\frac{1}{2}x^4$   **14.** $-43x$   **15.** $2y$   **16.** $-8t$

**17.** $-24 + 3b$   **18.** $20y + 10$

**19.**

| | ft | m |
|---|---|---|
| Sears Tower | 1454 | 443 |
| One World Trade Center | 1368 | 417 |
| Two World Trade Center | 1362 | 415 |

## Math & History (p. 107)

**1.** $628 - (-200) = 828$   years

## Lesson 2.7

### 2.7 Guided Practice (p. 111)

**1.** its reciprocal

**2.** No. Dividing by a number is the same as multiplying by the reciprocal of a number.

**3.** never

**4.** Yes; $a = 6$   $b = 3$; $-\frac{6}{3} = \frac{-6}{3} = \frac{6}{-3}$ are all equal to $-2$.

**5.** The reciprocal of $-\frac{2}{3}$ is $-\frac{3}{2}$.

**6.** $\frac{1}{34}$   **7.** $-\frac{1}{8}$   **8.** $-\frac{4}{3}$   **9.** $-\frac{5}{11}$   **10.** $7 \div \frac{1}{2} = 7\left(\frac{2}{1}\right) = 14$

**11.** $10 \div \left(-\frac{1}{5}\right) = 10\left(-\frac{5}{1}\right) = -50$

**12.** $-4$   **13.** $\dfrac{16}{\frac{-2}{9}} = 16 \div \dfrac{-2}{9} = 16\left(\dfrac{9}{-2}\right) = -72$

**14.** All real numbers except $-2$.   **15.** $\dfrac{-3.06}{18} = -\$.17$

### 2.7 Practice and Applications (p. 111–113)

**16.** 3   **17.** $-5$   **18.** $-6$   **19.** $-7$   **20.** $-8$   **21.** $-11$

**22.** $\frac{1}{2}$   **23.** $-\frac{1}{3}$   **24.** 135   **25.** $-21\frac{7}{9}$   **26.** 52   **27.** $-43\frac{1}{5}$

**28.** $-6$   **29.** $-7$   **30.** $-48$   **31.** 145

**32.** $42y \div \frac{1}{7} = 42y\left(\frac{7}{1}\right) = 294y$

**33.** $6t \div -\frac{1}{2} = 6t\left(-\frac{2}{1}\right) = -12t$

**34.** $58z \div \left(-\frac{2}{5}\right) = 58z\left(-\frac{5}{2}\right) = -145z$

**35.** $-\dfrac{x}{12} \div 3 = -\dfrac{x}{12}\left(\dfrac{1}{3}\right) = -\dfrac{x}{36}$

**36.** $\dfrac{d}{4} \div 6 = \dfrac{d}{4}\left(\dfrac{1}{6}\right) = \dfrac{d}{24}$   **37.** $\dfrac{3y}{4} \div \dfrac{1}{2} = \dfrac{3y}{4}\left(\dfrac{2}{1}\right) = \dfrac{3y}{2}$

**38.** $\dfrac{-2b}{7} \div \dfrac{7}{9} = \dfrac{-2b}{7}\left(\dfrac{9}{7}\right) = \dfrac{-18b}{49}$

**39.** $33x \div \frac{3}{11} = 33x\left(\frac{11}{3}\right) = 121x$

**40.** $49x \div 3\frac{1}{2} = 49x\left(\frac{2}{7}\right) = 14x$

**41.** $8x^2 \div \left(-\frac{4}{5}\right) = 8x^2\left(-\frac{5}{4}\right) = -10x^2$

**42.** $68x \div \left(-\frac{17}{9}\right) = 68x\left(-\frac{9}{17}\right) = -36x$

**43.** $-54x^2 \div \left(-\frac{9}{5}\right) = -54x^2\left(-\frac{5}{9}\right) = 30x^2$

**44.** $\dfrac{42t}{-14z} \div \dfrac{-6}{7t} = \dfrac{42t}{-14z}\left(\dfrac{7t}{-6}\right) = \dfrac{49t^2}{14z} = \dfrac{7t^2}{2z}$

**45.** $8 \cdot \dfrac{x}{8} = x$   **46.** $3\left(-\dfrac{y}{3}\right) = -y$

**47.** $-7 \cdot \left(-\dfrac{2w}{-7}\right) = -2w$

**48.** $\dfrac{18x - 9}{3} = \left(\dfrac{1}{3}\right)18x - 9\left(\dfrac{1}{3}\right) = 6x - 3$

**49.** $\dfrac{22x + 10}{-2} = \left(-\dfrac{1}{2}\right)22x + 10\left(-\dfrac{1}{2}\right) = -11x - 5$

# Chapter 2 *continued*

**50.** $\dfrac{-56 + x}{-8} = \left(-\dfrac{1}{8}\right)-56 + x\left(-\dfrac{1}{8}\right) = 7 - \dfrac{x}{8}$

**51.** $\dfrac{45 - 5x}{5} = \left(\dfrac{1}{5}\right)45 - 5x\left(\dfrac{1}{5}\right) = 9 - x$

**52.** $\dfrac{x - 5}{6} = \dfrac{30 - 5}{6} = \dfrac{25}{6} = 4\dfrac{1}{6}$

**53.** $\dfrac{3r - 7}{11} = \dfrac{3(17) - 7}{11} = \dfrac{44}{11} = 4$

**54.** $\dfrac{3a - b}{a} = \dfrac{3\left(\frac{1}{3}\right) - (-3)}{\frac{1}{3}} = \dfrac{1 + 3}{\frac{1}{3}} = 4\left(\dfrac{3}{1}\right) = 12$

**55.** $\dfrac{15x^2 + 10}{y} = \dfrac{15(-3)^2 + 10}{\frac{2}{3}} = [15(9) + 10]\left(\dfrac{3}{2}\right)$
$\qquad\qquad = 145\left(\dfrac{3}{2}\right) = 217\dfrac{1}{2}$

**56.** $\dfrac{28 - 4x}{y} = \dfrac{28 - 4(2)}{\frac{1}{2}} = (28 - 8)\dfrac{2}{1} = 20\left(\dfrac{2}{1}\right) = 40$

**57.** $\dfrac{3a - 4b}{ab} = \dfrac{3\left(-\frac{1}{3}\right) - 4\left(\frac{1}{4}\right)}{-\frac{1}{3}\left(\frac{1}{4}\right)} = \dfrac{-1 - 1}{\frac{-1}{12}} = -2\left(\dfrac{12}{-1}\right) = 24$

**58.** all real numbers except 0

**59.** all real numbers except 2

**60.** all real numbers except $-2$

**61.** all real numbers except 0

**62.** $82\dfrac{1}{2} \div 13\dfrac{3}{4} = \dfrac{165}{2}\left(\dfrac{4}{55}\right) = 6$ servings

**63.** $-22.5 \div 9 = -2.5$ ft/sec

**64.** $38{,}387 \div 1560 = 24.6$ points/game;
$17{,}440 \div 1560 = 11.2$ rebounds/game

**65.**

| Fraction | $\dfrac{x}{1}$ | $\dfrac{x}{0.1}$ | $\dfrac{x}{0.01}$ | $\dfrac{x}{0.001}$ |
|---|---|---|---|---|
| Product | $1x$ | $10x$ | $100x$ | $1000x$ |

| Fraction | $\dfrac{x}{0.0001}$ | $\dfrac{x}{0.00001}$ |
|---|---|---|
| Product | $10{,}000x$ | $100{,}000x$ |

**66.** The power of ten in the product is the same as the number of digits after the decimal point in the denominator of the fraction.

**67.** $10{,}000{,}000x$ **68.** $550 \div 2.75 = 200$ ft/sec

**69.** $-11$ ft/sec **70.** $550 \div 11 = 50$ sec

**71.** Yes; $a = 2$ $b = 3$
$\dfrac{1}{3} < \dfrac{1}{2}$
By substituting $a = 2$ and $b = 3$ you get $\dfrac{1}{3} < \dfrac{1}{2}$ which is a true statement.

**72.** C $\dfrac{6x - 4}{-\frac{2}{3}} = 6x - 4\left(-\dfrac{3}{2}\right) = -9x + 6$

**73.** D **74. a.** no **b.** yes **c.** yes **d.** no

---

## 2.7 Mixed Review (p. 113)

**75.** 0.75 **76.** 0.6 **77.** 0.1 **78.** 0.28

**79.** 0.17 **80.** 0.29 **81.** 0.75 **82.** 0.45

**83.** $3x^2 = 3(7^2) = 147$ **84.** $4(b^3) = 4\left(\frac{1}{2}^3\right) = 4\left(\frac{1}{8}\right) = \frac{1}{2}$

**85.** $2(y^3) = 2(-5^3) = 2(-125) = -250$

**86.** $5y^3 = 5(4^3) = 5(64) = 320$

**87.** $(6w)^4 = [6(2)]^4 = 12^4 = 20{,}736$

**88.** $12d^2 = 12(9^2) = 972$

**89.** $5x^2 = 5(0.3^2) = 5(0.09) = 0.45$

**90.** $32x^7 = 32(-1)^7 = -32$

**91.** $(7t)^3 = \left[7\left(-\frac{3}{7}\right)\right]^3 = -3^3 = -27$

**92.** $12 - 9 + 7 = 3 + 7 = 10$

**93.** $42 \div 6 + 8 = 7 + 8 = 15$

**94.** $2(11 - 7) \div 3 = 2(4) \div 3 = \frac{8}{3} = 2\frac{2}{3}$

**95.** $8 + (91 \div 13) \cdot \frac{4}{7} = 8 + 7 \cdot \frac{4}{7} = 8 + 4 = 12$

**96.** $\frac{3}{4} \cdot 8 - 6 = 6 - 6 = 0$

**97.** $23 - [(12 \div 3)^2 + 8] = 23 - [4^2 + 8]$
$\qquad\qquad = 23 - [24] = -1$

**98.** $11 \cdot (-5) + 20 = -55 + 20 = -35$

**99.** $-8 \cdot (-9) - 80 = 72 - 80 = -8$

**100.** $-16 + (-6) \cdot (-8) = -16 + 48 = 32$

## Lesson 2.8

### Activity (p. 115)

**1.** Answers may vary. **2.** Answers may vary.

**3.** 0.125; *Sample answer:* The results are about the same.

### 2.8 Guided Practice (p. 117)

**1.** experimental **2.** more likely that it will not occur

**3.** more likely that it will not occur

**4.** Quite likely; the probability is greater than 0.75.

**5.** Unlikely; the probability is 0.25.

**6.** Likely to occur half of the time. The probability is 0.5.

**7.** $\frac{1}{5}$ or 0.2 **8.** 2 to 3

### 2.8 Practice and Applications (p. 117–119)

**9.** $\frac{16}{64} = 0.25$ **10.** $\frac{8}{40} = 0.20$ **11.** $\frac{13}{20} = 0.65$

**12.** $\frac{8}{32} = 0.25$ **13.** 4 to 7 **14.** 3 to 9 = 1 to 3

**15.** 2 to 6 = 1 to 3 **16.** 1 to 7 **17.** 13 to 39 = 1 to 3

**18.** 6 to 30 = 1 to 5 **19.** 0.5 **20.** 1 to 1, or even

**21. a.** $\frac{12}{20}$ or 0.6 **22.** $\frac{13}{90} \approx 0.14$
**b.** 12 to 8 = 3 to 2

# Chapter 2 *continued*

**23.** $\frac{31.2}{69} = 0.45$  **24.** $\frac{62.8}{69} = 0.91$

**25.** 690 to 298, or about 7 to 3

**26.** 0.15  **27.** $\frac{5}{95} = 1$ to 19  **28.** $\frac{129}{3930} = \frac{43}{1310} = 0.03$

**29.** $\frac{759}{3930} = \frac{253}{1310} \approx 0.19$  **30.** 18 to 3912 = 3 to 652

**31.** 4 to 12 = 1 to 3  **32.** Probability $= \frac{5}{20} = 0.25$

Odds $= \frac{5}{15}$ or 1 to 3

Explanations may vary.

**33.** C  $\frac{523}{816} \approx 0.64$  **34.** C  227 to 657

**35.** $A = \pi r^2 = \pi(1^2) = \pi$ square units

**36.** $\frac{25\pi}{81\pi} \approx 0.31$  **37.** $\frac{56\pi}{81\pi} \approx 0.69$  **38.** $\pi$ to $80\pi = 1$ to 80

## 2.8 Mixed Review (p. 120)

**39.** What number plus 17 is 25?; 8

**40.** 5 taken away from what number is 19?; 24

**41.** 2 times what number minus 1 is 10?; $5\frac{1}{2}$

**42.** 11 times what number is 110?; 10

**43.** What number divided by 15 is 8?; 120

**44.** 3 times what number plus 5 is 24.6?; 3.2

**45.** $z - 17 = 9$  **46.** $8 + r < 17$

**47.** $-3 = y + (-6)$  **48.** $-9 = y - 21$

**49.** $-8 + 4 - 9 = -8 + 4 + (-9) = -4 + (-9) = -13$

**50.** $12 - (-8) - 5 = 12 + 8 + (-5) = 20 + (-5) = 15$

**51.** $20 + (-17) - 8 = 20 + (-17) + (-8)$
$$= 3 + (-8) = -5$$

**52.** $-17 + 25 - 34 = -17 + 25 + (-34)$
$$= 8 + (-34) = -26$$

**53.** $-6.3 + 4.1 - 9.5 = -6.3 + 4.1 + (-9.5)$
$$= -2.2 + (-9.5) = -11.7$$

**54.** $2 - 11 + 5 - (-16) = 2 + (-11) + 5 + 16$
$$= -9 + 5 + 16 = -4 + 16 = 12$$

**55.** $-29.4 - (-8) + 4 = -29.4 + 8 + 4$
$$= -21.4 + 4 = -17.4$$

**56.** $\frac{1}{2} + \left(-\frac{4}{5}\right) - \frac{2}{3} = \frac{1}{2} + \left(-\frac{4}{5}\right) + \left(-\frac{2}{3}\right)$
$$= \frac{15}{30} + \left(\frac{-24}{30}\right) + \left(-\frac{20}{30}\right)$$
$$= -\frac{9}{30} + \left(-\frac{20}{30}\right) = \frac{-29}{30}$$

**57.** $-1\frac{3}{8} + 4\frac{3}{4} - 7\frac{1}{2} = -1\frac{3}{8} + 4\frac{6}{8} + \left(-7\frac{4}{8}\right)$
$$= 3\frac{3}{8} + \left(-7\frac{4}{8}\right) = -4\frac{1}{8}$$

## Quiz 3 (p. 120)

**1.** $-28 \div \frac{4}{7} = -28 \cdot \frac{7}{4} = -49$

**2.** $\frac{36}{\frac{2}{3}} = 36 \div \frac{2}{3} = 36\left(\frac{3}{2}\right) = 54$

**3.** $\frac{32}{\frac{1}{4}} = 32 \div \frac{1}{4} = 32\left(\frac{4}{1}\right) = 128$

**4.** $48 \div (-12) = -4$  **5.** $75 \div (-15) = -5$

**6.** $-144 \div 9 = -16$

**7.** $-120 \div \frac{3}{8} = -120\left(\frac{8}{3}\right) = -320$

**8.** $-\frac{13}{27} \div \left(-1\frac{4}{9}\right) = -\frac{13}{27} \div \left(-\frac{13}{9}\right) = -\frac{13}{27}\left(-\frac{9}{13}\right) = \frac{9}{27} = \frac{1}{3}$

**9.** $42x \div (-6) = -7x$  **10.** $-56 \div (-8x) = \frac{7}{x}$

**11.** $9x \div \frac{1}{2} = 9x\left(\frac{2}{1}\right) = 18x$  **12.** $20 \div \frac{4}{x} = 20\left(\frac{x}{4}\right) = 5x$

**13.** $25x \div \frac{5}{7} = 25x\left(\frac{7}{5}\right) = 35x$

**14.** $15t \div \left(-\frac{3}{4}\right) = 15t\left(-\frac{4}{3}\right) = -20t$

**15.** $66y \div \left(-\frac{6}{5}\right) = 66y\left(-\frac{5}{6}\right) = -55y$

**16.** $\frac{-2x}{18} \div (-4) = \frac{-2x}{18}\left(-\frac{1}{4}\right) = \frac{x}{36}$  **17.** $\frac{8}{25} = 0.32$

**18.** Unlikely; the probability is just less than 0.25.

**19.** Certain; a certain event is defined as having a probability of 1.

**20.** Unlikely; although the probability is very small, it is not 0.

**21.** Quite likely; the probability would be 0.70.

## Chapter 2 Review (pp. 122–124)

**1.**
$-6 < 7, \quad 7 > -6$

**2.**
$-4.3 < -3.9, \quad -3.9 > -4.3$

**3.**
$-9 < 8, \quad 8 > -9$

**4.**
$-0.25 < -0.2, \quad -0.2 > -0.25$

**5.**
$-14.9 < 13.9, \quad 13.9 > -14.9$

# Chapter 2 *continued*

**6.**

$$-\left|\tfrac{5}{11}\right| \qquad \left|\tfrac{-4}{7}\right|$$

(number line from $-1$ to $1$ with points marked)

$$-\left|\tfrac{5}{11}\right| < \left|\tfrac{-4}{7}\right|, \quad \left|\tfrac{-4}{7}\right| > -\left|\tfrac{5}{11}\right|$$

**7.** $12 + (-7) = 5$  **8.** $-24 + (-16) = -40$

**9.** $2.4 + (-3.1) = -0.7$

**10.** $9 + (-10) + (-3) = -1 + (-3) = -4$

**11.** $-35 + 41 + (-18) = 6 + (-18) = -12$

**12.** $-2\tfrac{3}{4} + 5\tfrac{3}{8} + \left(-4\tfrac{1}{2}\right) = -2\tfrac{6}{8} + 5\tfrac{3}{8} + \left(-4\tfrac{4}{8}\right)$
$$= 2\tfrac{5}{8} + \left(-4\tfrac{4}{8}\right) = -1\tfrac{7}{8}$$

**13.** $-2 - 7 - (-8) = -2 + (-7) + 8 = -9 + 8 = -1$

**14.** $5 - 11 - (-6) = 5 + (-11) + 6 = -6 + 6 = 0$

**15.** $-18 - 14 - (-15) = -18 + (-14) + 15$
$$= -32 + 15 = -17$$

**16.** $-5.7 + 3.1 - 8.6 = -5.7 + 3.1 + (-8.6)$
$$= -2.6 + (-8.6) = -11.2$$

**17.** $-\tfrac{7}{16} + \left(-\tfrac{3}{8}\right) - \tfrac{13}{4} = -\tfrac{7}{16} + \left(-\tfrac{6}{16}\right) + \left(-\tfrac{52}{16}\right)$
$$= \tfrac{-13}{16} + \left(-\tfrac{52}{16}\right) = \tfrac{-65}{16} = -4\tfrac{1}{16}$$

**18.** $-\tfrac{23}{36} - \left|-\tfrac{4}{9}\right| + \left(-\tfrac{7}{12}\right) = -\tfrac{23}{36} + \left(-\tfrac{16}{36}\right) + \left(-\tfrac{21}{36}\right)$
$$= -\tfrac{39}{36} + \left(-\tfrac{21}{36}\right) = -\tfrac{60}{36} = -1\tfrac{24}{36} = -1\tfrac{2}{3}$$

**19.** $\begin{bmatrix} -3 & -2 \\ 8 & 4 \end{bmatrix} + \begin{bmatrix} 4 & -2 \\ -7 & 5 \end{bmatrix} = \begin{bmatrix} -3+4 & -2+(-2) \\ 8+(-7) & 4+5 \end{bmatrix} =$

$\begin{bmatrix} 1 & -4 \\ 1 & 9 \end{bmatrix}$

$\begin{bmatrix} -3 & -2 \\ 8 & 4 \end{bmatrix} - \begin{bmatrix} 4 & -2 \\ -7 & 5 \end{bmatrix} =$

$\begin{bmatrix} -3-4 & -2-(-2) \\ 8-(-7) & 4-5 \end{bmatrix} = \begin{bmatrix} -7 & 0 \\ 15 & -1 \end{bmatrix}$

**20.** $\begin{bmatrix} -2 & 5 & 9 \\ -3 & 10 & 0 \end{bmatrix} + \begin{bmatrix} -1 & -6 & 11 \\ -2 & -7 & 1 \end{bmatrix}$

$= \begin{bmatrix} -2+(-1) & 5+(-6) & 9+11 \\ -3+(-2) & 10+(-7) & 0+1 \end{bmatrix}$

$= \begin{bmatrix} -3 & -1 & 20 \\ -5 & 3 & 1 \end{bmatrix}$

$\begin{bmatrix} -2 & 5 & 9 \\ -3 & 10 & 0 \end{bmatrix} - \begin{bmatrix} -1 & -6 & 11 \\ -2 & -7 & 1 \end{bmatrix}$

$= \begin{bmatrix} -2-(-1) & 5-(-6) & 9-11 \\ -3-(-2) & 10-(-7) & 0-1 \end{bmatrix}$

$= \begin{bmatrix} -1 & 11 & -2 \\ -1 & 17 & -1 \end{bmatrix}$

**21.** $-36$  **22.** $-40$  **23.** $600$  **24.** $-9$  **25.** $-\tfrac{34}{9} = -3\tfrac{7}{9}$

**26.** $4.2$  **27.** $49.5$  **28.** $64$  **29.** $-84$  **30.** $12$  **31.** $14$

**32.** $-231$  **33.** $5x + 60$  **34.** $9y + 54$  **35.** $5.5b - 55$

**36.** $6.4 - 2w$  **37.** $-3t - 33$  **38.** $-2s - 26$

**39.** $-2.5z + 12.5$  **40.** $-\dfrac{3x}{7} - xy$  **41.** $-4$  **42.** $-17$

**43.** $-3$  **44.** $3$  **45.** $-50$

**46.** $-63 \div 4\tfrac{1}{5} = -63\left(\tfrac{5}{21}\right) = -15$

**47.** $\dfrac{48}{-\tfrac{3}{4}} = 48\left(-\tfrac{4}{3}\right) = -64$  **48.** $-\dfrac{-84}{\tfrac{7}{8}} = -84\left(-\tfrac{8}{7}\right) = 96$

**49.** Probability: $\tfrac{12}{48} = \tfrac{1}{4} = 0.25$
Odds: $\tfrac{12}{36}$ or 1 to 3

**50.** Probability: $\tfrac{9}{81} = \tfrac{1}{9} \approx 0.11$
Odds: $\tfrac{9}{72}$ or 1 to 8

**51.** Probability: $\tfrac{4}{40} = \tfrac{1}{10} = 0.10$
Odds: $\tfrac{4}{36}$ or 1 to 9

**52.** Probability: $\tfrac{51}{68} = \tfrac{3}{4} = 0.75$
Odds: $\tfrac{51}{17}$ or 3 to 1

## Chapter 2 Test *(p. 125)*

**1.** $8$  **2.** $2.7$  **3.** $3.3$  **4.** $4.5$  **5.** $-5$  **6.** $-13.6$

**7.** $-2\tfrac{5}{6} + 3\tfrac{1}{4} = -2\tfrac{10}{12} + 3\tfrac{3}{12} = \tfrac{5}{12}$  **8.** $1$

**9.** $\begin{bmatrix} 3 & -7 \\ \tfrac{1}{2} & 6 \\ 0 & 2 \end{bmatrix} + \begin{bmatrix} 4 & 2 \\ \tfrac{1}{2} & 4 \\ 5 & \tfrac{1}{2} \end{bmatrix} = \begin{bmatrix} 3+4 & -7+2 \\ \tfrac{1}{2}+\tfrac{1}{2} & 6+4 \\ 0+5 & 2+\tfrac{1}{2} \end{bmatrix}$

$= \begin{bmatrix} 7 & -5 \\ 1 & 10 \\ 5 & 2\tfrac{1}{2} \end{bmatrix}$

**10.** $\begin{bmatrix} -5 & -1 \\ 5 & 8 \\ \tfrac{7}{8} & 2\tfrac{1}{2} \end{bmatrix} - \begin{bmatrix} 6 & 3 \\ \tfrac{1}{2} & -2 \\ 0 & 2 \end{bmatrix} = \begin{bmatrix} -5-6 & -1-3 \\ 5-\tfrac{1}{2} & 8-(-2) \\ \tfrac{7}{8}-0 & 2\tfrac{1}{2}-2 \end{bmatrix}$

$= \begin{bmatrix} -11 & -4 \\ 4\tfrac{1}{2} & 10 \\ \tfrac{7}{8} & \tfrac{1}{2} \end{bmatrix}$

**11.** $\begin{bmatrix} 2 & \tfrac{1}{2} & 4 \\ 5 & 16 & -7 \end{bmatrix} + \begin{bmatrix} 5 & \tfrac{1}{2} & -1 \\ -2 & 8 & 4 \end{bmatrix}$

$= \begin{bmatrix} 2+5 & \tfrac{1}{2}+\tfrac{1}{2} & 4+(-1) \\ 5+(-2) & 16+8 & -7+4 \end{bmatrix}$

$= \begin{bmatrix} 7 & 1 & 3 \\ 3 & 24 & -3 \end{bmatrix}$

**12.** $-24$  **13.** $-6$  **14.** $9$  **15.** $840$

**16.** $-56 \div \left(-\tfrac{7}{8}\right) = -56\left(-\tfrac{8}{7}\right) = 64$  **17.** $60$  **18.** $374$

**19.** $-42$  **20.** $-\tfrac{3}{8} \div \tfrac{1}{2} = -\tfrac{3}{8}\left(\tfrac{2}{1}\right) = -\tfrac{3}{4}$

**21.** $\left(1\tfrac{2}{7}\right)\left(1\tfrac{5}{9}\right) = \tfrac{9}{7}\left(\tfrac{14}{9}\right) = 2$  **22.** $\tfrac{1}{8}$

**23.** $-7\tfrac{4}{5} \div \left(-1\tfrac{3}{10}\right) = -\tfrac{39}{5}\left(\tfrac{10}{-13}\right) = 6$

**24.** $5w^3$  **25.** $-64x^3$  **26.** $32 - 8q$  **27.** $-72 - 12y$

**28.** $-14 + 4q$  **29.** $-9y - 99$

**30.** $5(3 - z) - z = 15 - 5z - z = 15 - 6z$

**31.** $14p + 2(5 - p) = 14p + 10 - 2p = 12p + 10$

**32.** $5 - (-8) + x = 5 - (-8) + (-5) = 5 + 8 + (-5) =$
$13 + (-5) = 8$

# Chapter 2 *continued*

**33.** $|-9| - 2x + 5 = |-9| - 2(6) + 5 = 9 - 12 + 5$
$$= 9 + (-12) + 5 = -3 + 5 = 2$$

**34.** $-9x + 12 = -9(-2) + 12 = 18 + 12 = 30$

**35.** $1 - 2x^2 = 1 - 2(-2^2) = 1 - 2(4) = 1 + (-8) = -7$

**36.** $\dfrac{4 - x}{-3} = \dfrac{4 - (-1)}{-3} = \dfrac{5}{-3}$

**37.** $-4x^2 - 8x + 9 = -4(-5^2) - 8(-5) + 9 =$
$$-4(25) + 40 + 9 = -100 + 40 + 9 = -51$$

**38.** $2189.70 + 1527.11 + (-2502.18) + (-266.54) =$
$\$948.09$ profit

**39.** $-8\frac{1}{3}\left(4\frac{1}{2}\right) = -\frac{25}{3}\left(\frac{9}{2}\right) = \frac{-75}{2} = -37\frac{1}{2}$ ft

**40.** $\frac{59}{260} \approx 0.23$

**41.** $\frac{43}{217}$ or 43 to 217

## Chapter 2 Standardized Test *(p. 126–127)*

**1.** A; $-\frac{55}{4} = -13.75$
$$-\frac{37}{3} = -12.\overline{3}$$

**2.** D; $|x| + |y| - 10 = |-9| + |2| - 10 = 9 + 2 - 10 =$
$$11 - 10 = 1$$

**3.** E

**4.** B; $-9 + 3 + (-14) = -6 + (-14) = -20$

**5.** C; $2 - 5 + (-3) = -3 + (-3) = -6$
$$2 + (-5) - 3 = -3 - 3 = -6$$

**6.** B; $-\frac{7}{9} - \frac{1}{3} = -\frac{7}{9} - \frac{3}{9} = \frac{-10}{9}$
$$-\frac{11}{12} - \frac{1}{6} = -\frac{11}{12} - \frac{2}{12} = -\frac{13}{12}$$

**7.** D

**8.** D; $-4 - 6 - (-10) = -4 + (-6) + 10$
$$= -10 + 10 = 0$$

**9.** C; $9 - (-13) + (-17) + (-10)$
$$= 9 + 13 + (-17) + (-10)$$
$$= 22 + (-17) + (-10)$$
$$= 5 + (-10) = -5$$

**10.** C; $\begin{bmatrix} -5 + 8 & -4 + 9 \\ 2 + -6 & -4 + 1 \\ 1 + -1 & -3 + 4 \end{bmatrix} = \begin{bmatrix} 3 & 5 \\ -4 & -3 \\ 0 & 1 \end{bmatrix}$

**11.** E; $\begin{bmatrix} -1 - (-7) & 3 - 3 \\ -6 - 2 & 4 - 6 \\ 2 - 0 & -5 - (-8) \end{bmatrix} = \begin{bmatrix} 6 & 0 \\ -8 & -2 \\ 2 & 3 \end{bmatrix}$

**12.** B; $-\frac{1}{2} \cdot \frac{-2}{3} \cdot \frac{3}{-4} \cdot \frac{4}{5} = -\frac{1}{5}$

**13.** D; $-2m^6 \div 4m^3 = -2(-2^6) \div 4(-2^3)$
$$= -2(64) \div 4(-8)$$
$$= -128 \div -32 = 4$$

**14.** A; $(-10)^4 = 10,000$  **15.** A; $-3 \cdot 24 \div (-9) = 8$
$(-10)^5 = -100,000$     $-12 \cdot 8 \div 6 = -16$

**16.** B; $6(x + 3) - 2(4 - x) = 6x + 18 - 8 + 2x$
$$= 8x + 10$$

**17.** D; $\dfrac{4p + 6pq}{p^2 q} = \dfrac{4(-2) + 6(-2)(-3)}{(-2)^2(-3)}$
$$= \dfrac{-8 + 36}{4(-3)} = \dfrac{28}{-12} = -\dfrac{7}{3}$$

**18.** B; $142.91 - 135.16 = 7.75$ down
$$-7.75 \div 15.5 = -\$.50$$

**19.** C; $\frac{7}{25} = 0.28$  **20.** A; $\frac{5}{20}$ or 1 to 4

**21. a.**

|       | −3.4 | 5.1 | 2 | 0.3 | −1.7 | −2 |
|-------|------|-----|---|-----|------|----|
| −3.4  | A    | B   | A | A   | A    | A  |
| 5.1   | B    | B   | B | B   | B    | B  |
| 2     | A    | B   | B | B   | B    | B  |
| 0.3   | A    | B   | B | B   | A    | A  |
| −1.7  | A    | B   | B | A   | A    | A  |
| −2    | A    | B   | B | A   | A    | A  |

**b.** 17 to 19

**c.** No; the probability that Player A wins is $\frac{17}{36}$, and the probability that Player B wins is $\frac{19}{36}$.

**d.** *Sample answer:* Let the spinner have six sections numbered 1 through 6. The players take turns. On a player's turn, he or she wins if the spinner lands on an odd number. If the spinner lands on an even number, the other player wins. On each turn, each player has an equal probability (0.5) of winning.

# CHAPTER 3

## Think & Discuss (p. 129)

1. $D = 30(1)$
   $= 30$ miles
   $D = 30\left(\frac{1}{2}\right)$
   $= 15$ miles
   $D = 30\left(\frac{1}{12}\right)$
   $= 2\frac{1}{2}$ miles

2. $\dfrac{30}{60}$ mi/min $= \dfrac{1}{2}$ mi/min

## Skill Review (p. 130)

1. $0.11(650) = 71.5$
2. $0.41(71.5) = 29.315$
3. $0.06(250) = 15$
4. $0.042(60) = 2.52$

5. $9 - (-2) \overset{?}{=} 7$
   $11 \neq 7$
   no

6. $-11 \overset{?}{=} -4(3) + 1$
   $-11 \overset{?}{=} -12 + 1$
   $-11 = -11$
   yes

7. $4 - 3(-1) + 5(-1) \overset{?}{=} 2$
   $4 + 3 - 5 \overset{?}{=} 2$
   $2 = 2$
   yes

8. $3(4r + 6) = 12r + 18$

9. $-6(3 - 5z) = -18 + 30z$
10. $(-x + 7)(-3) = 3x - 21$
11. $-9 + 3x$
12. $3a + 2$
13. $1 + 4y$

## Lesson 3.1

### Developing Concepts Activity 3.1 (p. 131)
### Exploring the Concept

1.

2.

3.
   $x = -7$

4.

5.

---

6. One $x$-tile on the left and five $+1$-tiles on the right; 5

### Drawing Conclusions

1. $x + 4 = 6$

   Subtract 4 from each side.

   $x = 2$

2. $x + 5 = 3$

   Subtract 5 from each side.

   $x = -2$

3. $x + 6 = -1$

   Subtract 6 from each side.

   $x = -7$

4. $x + 3 = -3$

   Subtract 3 from each side.

   $x = -6$

5. $x - 1 = 5$

   Add 1 to each side.

   $x = 6$

# Chapter 3 *continued*

**6.** [tiles] $x - 6 = 2$

[tiles] Add 6 to each side.

[tiles] $x = 8$

**7.** [tiles] $x - 7 = -4$

[tiles] Add 7 to each side.

[tiles] $x = 3$

**8.** [tiles] $x - 5 = -5$

[tiles]

Add 5 to each side.

[tiles] $x = 0$

**9.** subtraction; addition

**10.** No; you must subtract 3 from each side to get $x = -7$.

### 3.1 Guided Practice (p. 135)

**1.** equivalent

**2.** $-4 \overset{?}{=} 54 - 58$
$-4 = -4$

**3.** $\quad -3 - x = 1$
$-3 - x + 3 = 1 + 3$    Add 3 to each side.
$\qquad -x = 4$    Simplify.
$\qquad\quad x = -4$    $x$ is the opposite of 4.

**4.** $\quad r + 3 = 2$
$r + 3 - 3 = 2 - 3$
$\qquad r = -1$

**5.** $\quad 9 = x - 4$
$9 + 4 = x - 4 + 4$
$\quad 13 = x$

**6.** $\quad 7 + c = -10$
$7 + c - 7 = -10 - 7$
$\qquad c = -17$

**7.** $\quad -3 = b - 6$
$-3 + 6 = b - 6 + 6$
$\qquad 3 = b$

**8.** $\quad 8 - x = 4$
$8 - x - 8 = 4 - 8$
$\qquad -x = -4$
$\qquad\quad x = 4$

**9.** $\quad r - (-2) = 5$
$\qquad r + 2 = 5$
$r + 2 - 2 = 5 - 2$
$\qquad\quad r = 3$

**10.** 
| Original amount of money | − | Amount spent on lunch | = | Amount left |

Original amount of money $= x$
Amount spent on lunch $= 4.65$
Amount left $= 7.39$
$x - 4.65 = 7.39$

**11.** Add 4.65 to both sides.

**12.** $x - 4.65 + 4.65 = 7.39 + 4.65$
$\qquad\qquad x = 12.04$
You started with $12.04 in your pocket.

### 3.1 Practice and Applications (pp. 135–137)

**13.** Subtract 28.   **14.** Subtract 17.   **15.** Add 15.

**16.** Add 3.

**17.** Subtract $-3$ or add 3.   **18.** Subtract $-12$ or add 12.

**19.** Add $-45$ or subtract 45.   **20.** Add $-2\frac{1}{2}$ or subtract $2\frac{1}{2}$.

**21.** $x = 4 - 7$
$\quad x = -3$

**22.** $\quad x + 5 = 10$
$x + 5 - 5 = 10 - 5$
$\qquad\quad x = 5$

**23.** $\quad t - 2 = 6$
$t - 2 + 2 = 6 + 2$
$\qquad t = 8$

**24.** $\quad 11 = r - 4$
$11 + 4 = r - 4 + 4$
$\quad 15 = r$

**25.** $\quad -9 = 2 + y$
$-9 - 2 = 2 + y - 2$
$\quad -11 = y$

**26.** $\quad n - 5 = -9$
$n - 5 + 5 = -9 + 5$
$\qquad n = -4$

**27.** $\quad -3 + x = 7$
$-3 + x + 3 = 7 + 3$
$\qquad\quad x = 10$

**28.** $\quad \frac{2}{5} = a - \frac{1}{5}$
$\frac{2}{5} + \frac{1}{5} = a - \frac{1}{5} + \frac{1}{5}$
$\qquad \frac{3}{5} = a$

**29.** $\quad r + 3\frac{1}{4} = 2\frac{1}{2}$
$r + 3\frac{1}{4} - 3\frac{1}{4} = 2\frac{1}{2} - 3\frac{1}{4}$
$\qquad\quad r = -\frac{3}{4}$

**30.** $\quad t - (-4) = 4$
$\qquad t + 4 = 4$
$t + 4 - 4 = 4 - 4$
$\qquad\quad t = 0$

**31.** $\quad |-6| + y = 11$
$\qquad 6 + y = 11$
$6 + y - 6 = 11 - 6$
$\qquad\quad y = 5$

**32.** $\quad |-8| + x = -3$
$\qquad 8 + x = -3$
$8 + x - 8 = -3 - 8$
$\qquad\quad x = -11$

**33.** $\quad 19 - (-y) = 25$
$\qquad 19 + y = 25$
$19 + y - 19 = 25 - 19$
$\qquad\quad y = 6$

**34.** $\quad |2| - (-b) = 6$
$\qquad 2 + b = 6$
$2 + b - 2 = 6 - 2$
$\qquad\quad b = 4$

**Algebra 1**
Chapter 3 Worked-out Solution Key

## Chapter 3 *continued*

**35.** $x + 4 - 3 = 6 \cdot 5$
$x + 1 = 30$
$x + 1 - 1 = 30 - 1$
$x = 29$

**36.** $-b = 8$
$b = -8$

**37.** $12 - 6 = -n$
$6 = -n$
$-6 = n$

**38.** $3 - a = 0$
$3 - a - 3 = 0 - 3$
$-a = -3$
$a = 3$

**39.** $4 = -b - 12$
$4 + 12 = -b - 12 + 12$
$16 = -b$
$-16 = b$

**40.** $-3 = -a + (-4)$
$-3 + 4 = -a + (-4) + 4$
$1 = -a$
$-1 = a$

**41.** $-r - (-7) = 16$
$-r + 7 = 16$
$-r + 7 - 7 = 16 - 7$
$-r = 9$
$r = -9$

**42.** C; 22 CDs

**43.** B; 8 members

**44.** A; $-8°F$

**45.** $x + 7 = 24$
$x + 7 - 7 = 24 - 7$
$x = \$17$
The store paid \$17.

**46.** $x - 247 = 7044$
$x - 247 + 247 = 7044 + 247$
$x = 7291$ points
Her 1988 score was 7291 points.

**47.** $43{,}368 + x = 49{,}831$
$43{,}368 + x - 43{,}368 = 49{,}831 - 43{,}368$
$x = 6463$ seats
The number of seats to be added is 6463.

**48.** $x + 9 = 87$
$x + 9 - 9 = 87 - 9$
$x = 78$
Your score would have been 78.

**49.** $x - 732 = 645$
$x - 732 + 732 = 645 + 732$
$x = 1377$ min
The average 18-to-24-year-old spends 1377 minutes.

**50.** $3 + 4 + x = 12$
$7 + x = 12$
$7 + x - 7 = 12 - 7$
$x = 5$ ft

**51.** $8 + 15 + x = 43$
$23 + x = 43$
$23 + x - 23 = 43 - 23$
$x = 20$ cm

**52.** $e + 418 = 4218$
$e + 418 - 418 = 4218 - 418$
$e = 3800$ acres

**53.** $4218 + 3800 + 2764 = c + 248$
$10{,}782 = c + 248$
$10{,}782 - 248 = c + 248 - 248$
$10{,}534$ acres $= c$
Cullen Park is 10,534 acres.

**54. a.** and **b.**

| Month | Cumulative sales | Sales equation | Monthly sales |
|---|---|---|---|
| March | 5,828 | — | 5,828 |
| April | 13,198 | $5{,}828 + x = 13{,}198$ | 7,370 |
| May | 22,254 | $13{,}198 + x = 22{,}254$ | 9,056 |
| June | 31,580 | $22{,}254 + x = 31{,}580$ | 9,326 |
| July | 40,972 | $31{,}580 + x = 40{,}972$ | 9,392 |
| Aug. | 51,401 | $40{,}972 + x = 51{,}401$ | 10,429 |
| Sept. | 68,702 | $51{,}401 + x = 68{,}702$ | 17,301 |
| Oct. | 83,494 | $68{,}702 + x = 83{,}494$ | 14,792 |
| Nov. | 108,897 | $83{,}494 + x = 108{,}897$ | 25,403 |
| Dec. | 158,068 | $108{,}897 + x = 158{,}068$ | 49,171 |

**c.** Answers may vary.

**55.** $x + 4 - 6 + 1 - 8 - 3 + 1 - 6 = 0$
$x - 17 = 0$
$x - 17 + 17 = 0 + 17$
$x = 17$

You started on Floor 17.

### 3.1 Mixed Review  (p. 137)

**56.** $5n = 160$   **57.** $\dfrac{n}{6} = 32$   **58.** $\dfrac{1}{4}n = 36$   **59.** $\dfrac{2}{3}n = 8$

**60.** $8\left(\dfrac{x}{8}\right) = x$   **61.** $\dfrac{1}{8}y \cdot 8 = y$   **62.** $-\dfrac{3}{5}\left(-\dfrac{5}{3}x\right) = x$

**63.** $-4x \div (-4) = x$   **64.** $6\left(-\dfrac{1}{6}x\right) = -x$

**65.** $\dfrac{7}{12}y \cdot \dfrac{12}{7} = y$   **66.** $\dfrac{t}{-4}(-4) = t$   **67.** $-19a \div 19 = -a$

**68.** $4(x + 2) = 4x + 8$   **69.** $7(3 - 2y) = 21 - 14y$

**70.** $-5(-y - 7) = 5y + 35$

**71.** $(3x + 8)(-2) = -6x - 16$

**72.** $-x(x - 6) = -x^2 + 6x$

**73.** $-5x(y + 3) = -5xy - 15x$

**74.** $2y(8 - 7y) = 16y - 14y^2$

**75.** $(-3x - 9y)(-6y) = 18xy + 54y^2$

# Chapter 3 *continued*

## Lesson 3.2

### 3.2 Guided Practice (p. 141)

**1.** addition and subtraction, multiplication and division

**2.** (1) Multiply each side by the same nonzero number.
   (2) Divide each side by the same nonzero number.
   (3) Add the same number to each side.
   (4) Subtract the same number from each side.
   (5) Simplify one or both sides.
   (6) Interchange the two sides.

**3.** (4) Divide each side by 6.
   (5) Multiply each side by 4.
   (6) Multiply each side by $-5$.
   (7) Multiply each side by $\frac{6}{5}$.

**4.** $6x = 18$
$$\frac{6x}{6} = \frac{18}{6}$$
$$x = 3$$

**5.** $\frac{y}{4} = 8$
$$4\left(\frac{y}{4}\right) = 4(8)$$
$$y = 32$$

**6.** $\frac{r}{-5} = 20$
$$-5\left(\frac{r}{-5}\right) = -5(20)$$
$$r = -100$$

**7.** $\frac{5}{6}a = -10$
$$\frac{6}{5}\left(\frac{5}{6}a\right) = \frac{6}{5}(-10)$$
$$a = -\frac{60}{5}$$
$$a = -12$$

**8.** $-7b = -4$
$$\frac{-7b}{-7} = \frac{-4}{-7}$$
$$b = \frac{4}{7}$$

**9.** $-3x = 5$
$$\frac{-3x}{-3} = \frac{5}{-3}$$
$$x = -\frac{5}{3}$$

**10.** $-\frac{3}{8}t = -6$
$$-\frac{8}{3}\left(-\frac{3}{8}t\right) = -\frac{8}{3}(-6)$$
$$t = \frac{48}{3}$$
$$t = 16$$

**11.** $\frac{1}{7}x = \frac{5}{7}$
$$7\left(\frac{1}{7}x\right) = 7\left(\frac{5}{7}\right)$$
$$x = 5$$

**12.** $10\frac{1}{2}x = 630$
$$\frac{21}{2}x = 630$$
$$\frac{2}{21}\left(\frac{21}{2}x\right) = \frac{2}{21}(630)$$
$$x = \frac{1260}{21}$$
$$x = 60 \text{ mi/h}$$

**13.** $\frac{x}{3} = \frac{4}{6}$
$$3\left(\frac{x}{3}\right) = 3\left(\frac{4}{6}\right)$$
$$x = \frac{12}{6}$$
$$x = 2$$

### 3.2 Practice and Applications (pp. 142–144)

**14.** Multiply by 6.

**15.** Divide by $-2$.

**16.** Multiply by $-4$.

**17.** Divide by $\frac{2}{3}$ or multiply by $\frac{3}{2}$.

**18.** Divide by $-\frac{9}{4}$ or multiply by $-\frac{4}{9}$.

**19.** Multiply by $-\frac{4}{3}$.

**20.** $-4x = 44 \qquad x = 11$
$$\frac{-4x}{-4} = \frac{44}{-4}$$
$$x = -11$$
no

**21.** $21x = 7 \qquad x = 3$
$$\frac{21x}{21} = \frac{7}{21}$$
$$x = \frac{1}{3}$$
no

**22.** $\frac{x}{10} = -4 \qquad x = -40$
$$10\left(\frac{x}{10}\right) = 10(-4)$$
$$x = -40$$
yes

**23.** $\frac{2}{3}x = 24 \qquad x = 16$
$$\frac{3}{2}\left(\frac{2}{3}x\right) = \frac{3}{2}(24)$$
$$x = 36$$
no

**24.** $10x = 110$
$$\frac{10x}{10} = \frac{110}{10}$$
$$x = 11$$

**25.** $-21m = 42$
$$\frac{-21m}{-21} = \frac{42}{-21}$$
$$m = -2$$

**26.** $18 = -2a$
$$\frac{18}{-2} = \frac{-2a}{-2}$$
$$-9 = a$$

**27.** $30b = 5$
$$\frac{30b}{30} = \frac{5}{30}$$
$$b = \frac{1}{6}$$

**28.** $-4n = -24$
$$\frac{-4n}{-4} = \frac{-24}{-4}$$
$$n = 6$$

**29.** $288 = 16t$
$$\frac{288}{16} = \frac{16t}{16}$$
$$18 = t$$

**30.** $7r = -56$
$$\frac{7r}{7} = \frac{-56}{7}$$
$$r = -8$$

**31.** $8x = 3$
$$\frac{8x}{8} = \frac{3}{8}$$
$$x = \frac{3}{8}$$

**32.** $-10x = -9$
$$\frac{-10x}{-10} = \frac{-9}{-10}$$
$$x = \frac{9}{10}$$

**33.** $\frac{y}{7} = 12$
$$7\left(\frac{y}{7}\right) = 7(12)$$
$$y = 84$$

# Chapter 3 continued

**34.** $\dfrac{z}{2} = -5$

$2\left(\dfrac{z}{2}\right) = 2(-5)$

$z = -10$

**35.** $\dfrac{1}{2}x = -20$

$2\left(\dfrac{1}{2}\right)x = 2(-20)$

$x = -40$

**36.** $\dfrac{1}{3}y = 82$

$3\left(\dfrac{1}{3}y\right) = 3(82)$

$y = 246$

**37.** $\dfrac{m}{-4} = -\dfrac{3}{4}$

$-4\left(\dfrac{m}{-4}\right) = -4\left(-\dfrac{3}{4}\right)$

$m = 3$

**38.** $0 = \dfrac{4}{5}d$

$\dfrac{5}{4}(0) = \dfrac{5}{4}\left(\dfrac{4}{5}\right)d$

$0 = d$

**39.** $-\dfrac{4}{5}x = 72$

$-\dfrac{5}{4}\left(-\dfrac{4}{5}x\right) = -\dfrac{5}{4}(72)$

$x = -90$

**40.** $-\dfrac{1}{5}y = -6$

$-5\left(-\dfrac{1}{5}y\right) = -5(-6)$

$y = 30$

**41.** $\dfrac{t}{-2} = \dfrac{1}{2}$

$-2\left(\dfrac{t}{-2}\right) = -2\left(\dfrac{1}{2}\right)$

$t = -1$

**42.** $-\dfrac{2}{3}t = -16$

$-\dfrac{3}{2}\left(-\dfrac{2}{3}t\right) = -\dfrac{3}{2}(-16)$

$t = 24$

**43.** $\dfrac{3}{4}z = -5\dfrac{1}{2}$

$\dfrac{4}{3}\left(\dfrac{3}{4}z\right) = \dfrac{4}{3}\left(-5\dfrac{1}{2}\right)$

$z = \dfrac{4}{3}\left(-\dfrac{11}{2}\right)$

$z = -\dfrac{22}{3}$

$z = -7\dfrac{1}{3}$

**44.** $\dfrac{1}{3}y = 5\dfrac{2}{3}$

$3\left(\dfrac{1}{3}y\right) = 3\left(5\dfrac{2}{3}\right)$

$y = 17$

**45.** $\dfrac{3}{4}t = |-15|$

$\dfrac{3}{4}t = 15$

$\dfrac{4}{3}\left(\dfrac{3}{4}t\right) = \dfrac{4}{3}(15)$

$t = 20$

**46.** $-\dfrac{1}{2}b = -|-8|$

$-\dfrac{1}{2}b = -8$

$-2\left(-\dfrac{1}{2}b\right) = -2(-8)$

$b = 16$

**47.** $-6y = -|27|$

$-6y = -27$

$\dfrac{-6y}{-6} = \dfrac{-27}{-6}$

$y = \dfrac{9}{2}$

$y = 4\dfrac{1}{2}$

**48. A** $\dfrac{x}{4} = 37$

$4\left(\dfrac{x}{4}\right) = 4(37)$

$x = 148$ lb

**49.** $52x = 676$

$\dfrac{52x}{52} = \dfrac{676}{52}$

$x = 13$ pieces

**50.** $\dfrac{x}{8} = 50$

$8\left(\dfrac{x}{8}\right) = 8(50)$

$x = 400$ students

**51.** $\dfrac{3}{8}x = 3.3$

$\dfrac{8}{3}\left(\dfrac{3}{8}x\right) = \dfrac{8}{3}(3.3)$

$x = \$8.80$

The whole pizza was $8.80.

**52.** $\dfrac{x}{12} = 45$

$12\left(\dfrac{x}{12}\right) = 12(45)$

$x = 540$ peanuts

**53.** $30{,}000x = 10{,}000$

$\dfrac{30{,}000x}{30{,}000} = \dfrac{10{,}000}{30{,}000}$

$x = \dfrac{1}{3}$ ft$^2$

**54.** $\dfrac{x}{5} = 9$

$5\left(\dfrac{x}{5}\right) = 5(9)$

$x = 45$ seconds

**55. a.** $\dfrac{\text{Total number of frames}}{\text{Number of frames per second}} = \dfrac{\text{Total number}}{\text{of seconds}}$

$\dfrac{x}{24} = 5400$

$24\left(\dfrac{x}{24}\right) = 24(5400)$

$x = 129{,}600$

A 90-minute movie has 129,600 frames.

**b.** $\dfrac{\text{Total number of frames}}{\text{Number of frames restored per hour}} = \dfrac{\text{Number of}}{\text{hours of work}}$

$\dfrac{129{,}600}{8} = y$

$16{,}200 = y$

The number of hours of work needed is 16,200.

**56.** $2(2x) + 2(3x) = 216$

$4x + 6x = 216$

$10x = 216$

$\dfrac{10x}{10} = \dfrac{216}{10}$

$x = 21.6$

The garden's dimensions are 43.2 ft by 64.8 ft.

**57. a.** $30x = 6$

$\dfrac{30x}{30} = \dfrac{6}{30}$

$x = \dfrac{1}{5}$

The least amount of time would be 12 minutes.

**b.** $\dfrac{1}{36}x = 1$

$36\left(\dfrac{1}{36}x\right) = 36(1)$

$x = 36$

The least amount of time would be 36 seconds.

**58.** $\dfrac{6}{10} = \dfrac{x}{5}$

$5\left(\dfrac{6}{10}\right) = 5\left(\dfrac{x}{5}\right)$

$\dfrac{6}{2} = x$

$3$ ft $= x$

**59.** $\dfrac{x}{12} = \dfrac{4}{8}$

$12\left(\dfrac{x}{12}\right) = 12\left(\dfrac{4}{8}\right)$

$x = \dfrac{48}{8}$

$x = 6$ in.

# Chapter 3 *continued*

**60. B** $\dfrac{3}{4} = \dfrac{x}{2}$

$2\left(\dfrac{3}{4}\right) = 2\left(\dfrac{x}{2}\right)$

$\dfrac{3}{2} = x$

You would need $1\frac{1}{2}$ teaspoons of soy sauce.

**61.** $\dfrac{5}{4} = \dfrac{x}{1}$

$\dfrac{5}{4} = x$

You need $1\frac{1}{4}$ cups of chicken.

**62. D**   **63.** $-\dfrac{5}{7}x = -2$

$-\dfrac{7}{5}\left(-\dfrac{5}{7}x\right) = -\dfrac{7}{5}(-2)$

$x = \dfrac{14}{5}$

A

**64. B**

**65.** $\dfrac{150}{100}x = 33{,}000$

$\dfrac{100}{150}\left(\dfrac{150}{100}x\right) = \dfrac{100}{150}(33{,}000)$

$x = 22{,}000$

## 3.2 Mixed Review (p. 144)

**66.** $18 + 5n = 108$

**67.** $9n - 12 = 60$

**68.** $5 + \dfrac{2}{3}n = 11$

**69.** $11 = \dfrac{2}{5}(n - 13)$

**70.** $15 - 8x + 12$

$27 - 8x$

**71.** $4y - 9 + 3y$

$7y - 9$

**72.** $(x + 8)(-2) - 36$

$-2x - 16 - 36$

$-2x - 52$

**73.** $5(y + 3) + 7y$

$5y + 15 + 7y$

$12y + 15$

**74.** $12x - (x - 2)(2)$

$12x - (2x - 4)$

$12x - 2x + 4$

$10x + 4$

**75.** $-25y - 6(-y - 9)$

$-25y + 6y + 54$

$-19y + 54$

**76.** $4 + y = 12$

$4 + y - 4 = 12 - 4$

$y = 8$

**77.** $t - 2 = 1$

$t - 2 + 2 = 1 + 2$

$t = 3$

**78.** $5 - (-t) = 14$

$5 + t = 14$

$5 + t - 5 = 14 - 5$

$t = 9$

**79.** $x - 2 = 28$

$x - 2 + 2 = 28 + 2$

$x = 30$

**80.** $19 - x = 37$

$19 - x - 19 = 37 - 19$

$-x = 18$

$x = -18$

**81.** $-9 - (-a) = -2$

$-9 + a = -2$

$-9 + a + 9 = -2 + 9$

$a = 7$

**82. A** $x + 7 = 24$

$x + 7 - 7 = 24 - 7$

$x = 17$ pictures

Seventeen pictures can be developed.

## Lesson 3.3

### 3.3 Guided Practice (p. 148)

**1.**

| *Solution Steps* | *Explanation* |
|---|---|
| $\dfrac{5x}{2} + 3 = 6$ | Original equation |
| $\dfrac{5x}{2} = 3$ | Subtract 3 from each side. |
| $5x = 6$ | Multiply both sides by 2. |
| $x = \dfrac{6}{5}$ | Divide both sides by 5. |

**2.** The distributive property could be used first in any of Exercises 6–8. However, it would be simpler in Exercise 6 to divide each side by 5 first and in Exercise 7 to multiply each side by $\frac{4}{3}$.

**3.** $4x + 3 = 11$

$4x + 3 - 3 = 11 - 3$

$4x = 8$

$\dfrac{4x}{4} = \dfrac{8}{4}$

$x = 2$

Check:

$4(2) + 3 \overset{?}{=} 11$

$8 + 3 \overset{?}{=} 11$

$11 = 11$

**4.** $\dfrac{1}{2}x - 9 = 11$

$\dfrac{1}{2}x - 9 + 9 = 11 + 9$

$\dfrac{1}{2}x = 20$

$2\left(\dfrac{1}{2}x\right) = 2(20)$

$x = 40$

Check:

$\dfrac{1}{2}(40) - 9 \overset{?}{=} 11$

$20 - 9 \overset{?}{=} 11$

$11 = 11$

**5.** $3x - x + 15 = 41$

$2x + 15 = 41$

$2x + 15 - 15 = 41 - 15$

$2x = 26$

$\dfrac{2x}{2} = \dfrac{26}{2}$

$x = 13$

Check:

$3(13) - 13 + 15 \overset{?}{=} 41$

$39 - 13 + 15 \overset{?}{=} 41$

$26 + 15 \overset{?}{=} 41$

$41 = 41$

**6.** $5(x - 7) = 90$

$\dfrac{5(x - 7)}{5} = \dfrac{90}{5}$

$x - 7 = 18$

$x - 7 + 7 = 18 + 7$

$x = 25$

Check:

$5(25 - 7) \overset{?}{=} 90$

$5(18) \overset{?}{=} 90$

$90 = 90$

**7.** $\dfrac{3}{4}(x + 6) = 12$

$\dfrac{4}{3}\left(\dfrac{3}{4}\right)(x + 6) = \dfrac{4}{3}(12)$

$x + 6 = 16$

$x + 6 - 6 = 16 - 6$

$x = 10$

Check:

$\dfrac{3}{4}(10 + 6) \overset{?}{=} 12$

$\dfrac{3}{4}(16) \overset{?}{=} 12$

$12 = 12$

## Chapter 3 *continued*

**8.** 
$$6x - 4(-3x + 2) = 10$$
$$6x + 12x - 8 = 10$$
$$18x - 8 = 10$$
$$18x - 8 + 8 = 10 + 8$$
$$18x = 18$$
$$\frac{18x}{18} = \frac{18}{18}$$
$$x = 1$$

Check:
$$6(1) - 4[-3(1) + 2] \overset{?}{=} 10$$
$$6 - 4(-3 + 2) \overset{?}{=} 10$$
$$6 - 4(-1) \overset{?}{=} 10$$
$$6 + 4 \overset{?}{=} 10$$
$$10 = 10$$

**9.** Let $x$ = number of errands
$$2x + 5 = 17$$
$$2x + 5 - 5 = 17 - 5$$
$$2x = 12$$
$$\frac{2x}{2} = \frac{12}{2}$$
$$x = 6 \text{ errands}$$
You must run 6 errands.

### 3.3 Practice and Applications (pp. 148–151)

**10.** 
$$9x - 5x - 19 = 21; \; -10$$
$$9(-10) - 5(-10) - 19 \overset{?}{=} 21$$
$$-90 + 50 - 19 \overset{?}{=} 21$$
$$-40 - 19 \overset{?}{=} 21$$
$$-59 \neq 21$$
$-10$ is not a solution.

**11.** 
$$\tfrac{3}{4}x + 1 = -8; \; -12$$
$$\tfrac{3}{4}(-12) + 1 \overset{?}{=} -8$$
$$-9 + 1 \overset{?}{=} -8$$
$$-8 = -8$$
$-12$ is a solution.

**12.** 
$$6x - 4(9 - x) = 106; \; 7$$
$$6(7) - 4(9 - 7) \overset{?}{=} 106$$
$$42 - 4(2) \overset{?}{=} 106$$
$$42 - 8 \overset{?}{=} 106$$
$$34 \neq 106$$
7 is not a solution.

**13.** 
$$7x - 15 = -1; \; -2$$
$$7(-2) - 15 \overset{?}{=} -1$$
$$-14 - 15 \overset{?}{=} -1$$
$$-29 \neq -1$$
$-2$ is not a solution.

**14.** 
$$\tfrac{1}{2}x - 7 = -4; \; 6$$
$$\tfrac{1}{2}(6) - 7 \overset{?}{=} -4$$
$$3 - 7 \overset{?}{=} -4$$
$$-4 = -4$$
6 is a solution.

**15.** 
$$\frac{x}{4} - 7 = 13; \; 24$$
$$\frac{24}{4} - 7 \overset{?}{=} 13$$
$$6 - 7 \overset{?}{=} 13$$
$$-1 \neq 13$$
24 is not a solution.

**16.** 
$$2x + 7 = 15$$
$$2x + 7 - 7 = 15 - 7$$
$$2x = 8$$
$$\frac{2x}{2} = \frac{8}{2}$$
$$x = 4$$

**17.** 
$$3x - 1 = 8$$
$$3x - 1 + 1 = 8 + 1$$
$$3x = 9$$
$$\frac{3x}{3} = \frac{9}{3}$$
$$x = 3$$

**18.** 
$$\frac{x}{3} - 5 = -1$$
$$\frac{x}{3} - 5 + 5 = -1 + 5$$
$$\frac{x}{3} = 4$$
$$3\left(\frac{x}{3}\right) = 3(4)$$
$$x = 12$$

**19.** 
$$\frac{x}{2} + 13 = 20$$
$$\frac{x}{2} + 13 - 13 = 20 - 13$$
$$\frac{x}{2} = 7$$
$$2\left(\frac{x}{2}\right) = 2(7)$$
$$x = 14$$

**20.** 
$$30 = 16 + \tfrac{1}{5}x$$
$$30 - 16 = 16 + \tfrac{1}{5}x - 16$$
$$14 = \tfrac{1}{5}x$$
$$5(14) = 5\left(\tfrac{1}{5}x\right)$$
$$70 = x$$

**21.** 
$$6 = 14 - 2x$$
$$6 - 14 = 14 - 2x - 14$$
$$-8 = -2x$$
$$\frac{-8}{-2} = \frac{-2x}{-2}$$
$$4 = x$$

**22.** 
$$7 + \tfrac{2}{3}x = -1$$
$$7 + \tfrac{2}{3}x - 7 = -1 - 7$$
$$\tfrac{2}{3}x = -8$$
$$\tfrac{3}{2}\left(\tfrac{2}{3}x\right) = \tfrac{3}{2}(-8)$$
$$x = -12$$

**23.** 
$$3 - \tfrac{3}{4}x = -6$$
$$3 - \tfrac{3}{4}x - 3 = -6 - 3$$
$$-\tfrac{3}{4}x = -9$$
$$-\tfrac{4}{3}\left(-\tfrac{3}{4}x\right) = -\tfrac{4}{3}(-9)$$
$$x = 12$$

**24.** 
$$22 = 18 - \tfrac{1}{4}x$$
$$22 - 18 = 18 - \tfrac{1}{4}x - 18$$
$$4 = -\tfrac{1}{4}x$$
$$-4(4) = -4\left(-\tfrac{1}{4}x\right)$$
$$-16 = x$$

**25.** 
$$8x - 3x = 10$$
$$5x = 10$$
$$\frac{5x}{5} = \frac{10}{5}$$
$$x = 2$$

**26.** 
$$-7x + 4x = 9$$
$$-3x = 9$$
$$\frac{-3x}{-3} = \frac{9}{-3}$$
$$x = -3$$

**27.** 
$$x + 5x - 5 = 1$$
$$6x - 5 = 1$$
$$6x - 5 + 5 = 1 + 5$$
$$6x = 6$$
$$\frac{6x}{6} = \frac{6}{6}$$
$$x = 1$$

**28.** 
$$3x - 7 + x = 5$$
$$4x - 7 = 5$$
$$4x - 7 + 7 = 5 + 7$$
$$4x = 12$$
$$\frac{4x}{4} = \frac{12}{4}$$
$$x = 3$$

**29.** 
$$3(x - 2) = 18$$
$$\frac{3(x - 2)}{3} = \frac{18}{3}$$
$$x - 2 = 6$$
$$x - 2 + 2 = 6 + 2$$
$$x = 8$$

# Chapter 3 *continued*

**30.** $12(2 - x) = 6$

$$\frac{12(2 - x)}{12} = \frac{6}{12}$$

$$2 - x = \frac{1}{2}$$

$$2 - x - 2 = \frac{1}{2} - 2$$

$$-x = -1\frac{1}{2}$$

$$x = 1\frac{1}{2}$$

**31.** $\frac{9}{2}(x + 3) = 27$

$$\frac{2}{9}\left(\frac{9}{2}\right)(x + 3) = \frac{2}{9}(27)$$

$$x + 3 = 6$$

$$x + 3 - 3 = 6 - 3$$

$$x = 3$$

**32.** $-\frac{4}{9}(2x - 4) = 48$

$$-\frac{9}{4}\left(-\frac{4}{9}\right)(2x - 4) = -\frac{9}{4}(48)$$

$$2x - 4 = -108$$

$$2x - 4 + 4 = -108 + 4$$

$$2x = -104$$

$$\frac{2x}{2} = \frac{-104}{2}$$

$$x = -52$$

**33.** $17 = 2(3x + 1) - x$

$$17 = 6x + 2 - x$$

$$17 = 5x + 2$$

$$17 - 2 = 5x + 2 - 2$$

$$15 = 5x$$

$$\frac{15}{5} = \frac{5x}{5}$$

$$3 = x$$

**34.** $\frac{4x}{3} + 3 = 23$

$$\frac{4x}{3} + 3 - 3 = 23 - 3$$

$$\frac{4x}{3} = 20$$

$$\frac{3}{4}\left(\frac{4x}{3}\right) = \frac{3}{4}(20)$$

$$x = 15$$

**35.** $-10 = 4 - \frac{7x}{4}$

$$-10 - 4 = 4 - \frac{7x}{4} - 4$$

$$-14 = -\frac{7x}{4}$$

$$-\frac{4}{7}(-14) = -\frac{4}{7}\left(-\frac{7x}{4}\right)$$

$$8 = x$$

**36.** $-10 = \frac{1}{2}x + x$

$$-10 = \frac{3}{2}x$$

$$\frac{2}{3}(-10) = \frac{2}{3}\left(\frac{3}{2}x\right)$$

$$\frac{-20}{3} = x$$

$$-6\frac{2}{3} = x$$

**37.** $5m - (4m - 1) = -12$

$$5m - 4m + 1 = -12$$

$$m + 1 = -12$$

$$m + 1 - 1 = -12 - 1$$

$$m = -13$$

**38.** $55x - 3(9x + 12) = -64$

$$55x - 27x - 36 = -64$$

$$28x - 36 = -64$$

$$28x - 36 + 36 = -64 + 36$$

$$28x = -28$$

$$\frac{28x}{28} = \frac{-28}{28}$$

$$x = -1$$

**39.** $22x + 2(3x + 5) = 66$

$$22x + 6x + 10 = 66$$

$$28x + 10 = 66$$

$$28x + 10 - 10 = 66 - 10$$

$$28x = 56$$

$$\frac{28x}{28} = \frac{56}{28}$$

$$x = 2$$

**40.** $9x - 5(3x - 12) = 30$

$$9x - 15x + 60 = 30$$

$$-6x + 60 = 30$$

$$-6x + 60 - 60 = 30 - 60$$

$$-6x = -30$$

$$\frac{-6x}{-6} = \frac{-30}{-6}$$

$$x = 5$$

**41.** The distributive property was applied incorrectly.

$$2(x - 3) = 5$$

$$2x - 6 = 5$$

$$2x - 6 + 6 = 5 + 6$$

$$2x = 11$$

$$\frac{2x}{2} = \frac{11}{2}$$

$$x = 5\frac{1}{2}$$

**42.** The left side was simplified incorrectly.

$$5 - 3x = 10$$

$$5 - 3x - 5 = 10 - 5$$

$$-3x = 5$$

$$\frac{-3x}{-3} = \frac{5}{-3}$$

$$x = -1\frac{2}{3}$$

**43.** The left side was simplified incorrectly.

$$\frac{1}{4}x - 2 = 7$$

$$\frac{1}{4}x - 2 + 2 = 7 + 2$$

$$\frac{1}{4}x = 9$$

$$4\left(\frac{1}{4}x\right) = 4(9)$$

$$x = 36$$

**44.** $-6x + 3(4x - 1) = 9$     Write the original equation.

$-6x + 12x - 3 = 9$     Apply the distributive property.

$6x - 3 = 9$     Simplify.

$6x = 12$     Add 3 to each side.

$x = 2$     Divide each side by 6.

## Chapter 3 *continued*

**45. a.** $\frac{1}{9}x + 1 = 4$

$x + 9 = 36$

$x = 27$

Preferences may vary.

**b.** $\frac{1}{9}x + 1 = 4$

$\frac{1}{9}x = 3$

$x = 27$

**46. a.** $\frac{1}{8}x - 5 = 3$

$\frac{1}{8}x = 8$

$x = 64$

Preferences may vary.

**b.** $\frac{1}{8}x - 5 = 3$

$x - 40 = 24$

$x = 64$

**47. a.** $\frac{x}{3} + 6 = -2$

$\frac{x}{3} = -8$

$x = -24$

Preferences may vary.

**b.** $\frac{x}{3} + 6 = -2$

$x + 18 = -6$

$x = -24$

**48. a.** $-4x - 2 = 4$

$-4x = 6$

$x = -\frac{6}{4}$

$x = -1\frac{1}{2}$

Preferences may vary.

**b.** $-4x - 2 = 4$

$x + \frac{1}{2} = -1$

$x = -1\frac{1}{2}$

**49. a.** $\frac{2}{3}x + 1 = \frac{1}{3}$

$2x + 3 = 1$

$2x = -2$

$x = -1$

Preferences may vary.

**b.** $\frac{2}{3}x + 1 = \frac{1}{3}$

$\frac{2}{3}x = -\frac{2}{3}$

$x = -1$

**50. a.** If $n$ is an integer, the next two consecutive integers are $n + 1$ and $n + 2$.

**b.** $n + (n + 1) + (n + 2) = 84$

$n + n + 1 + n + 2 = 84$

$3n + 3 = 84$

$3n = 81$

$n = 27$

The three integers are 27, 28, and 29.

**51.** $n + (2n - 9) + (3n + 6) = 123$

$n + 2n - 9 + 3n + 6 = 123$

$6n - 3 = 123$

$6n = 126$

$n = 21$

The three numbers are 21, 33, and 69.

**52.** $90 + 65x = 1000$

$65x = 910$

$x = 14$

The calf will weigh 1000 pounds in 14 months.

**53.** $339 + 34x = 458$

$34x = 119$

$x = 3\frac{1}{2}$ hours

The number of hours of labor was $3\frac{1}{2}$.

**54.** $m\angle A + m\angle B + m\angle C = 180$

$m\angle A = 4x$

$m\angle B = x$

$m\angle C = x - 20$

$4x + x + (x - 20) = 180$

$6x - 20 = 180$

$6x = 200$

$x = 33\frac{1}{3}$

$m\angle A = 133\frac{1}{3}°$; $m\angle B = 33\frac{1}{3}°$; $m\angle C = 13\frac{1}{3}°$

**55.** $4x = 500$

$x = 125$ students

The number of students is 125.

**56.** $18x + 315 = 405$

$18x = 90$

$x = 5$ hours

You worked 5 hours.

**57.** $d = \frac{n}{2} + 26$

$50 = \frac{n}{2} + 26$

$24 = \frac{n}{2}$

$48 = n$

The amount of pressure needed is 48 pounds per square inch.

**58.** $v = -32t + 28$

$0 = -32t + 28$

$-28 = -32t$

$0.875 = t$

It takes 0.875 second.

**59.** Rate of first machine = 1800

Time to complete = $x$

Rate of second machine = 2400

Time to complete = $x$

Total pages = 20(420)

$1800x + 2400x = 20(420)$

$4200x = 8400$

$x = 2$

It would take 2 hours.

**60.** $\frac{\text{pages}}{\text{hour}} \cdot \text{hours} + \frac{\text{pages}}{\text{hour}} \cdot \text{hours} = \text{pages}$

$\text{pages} = \text{pages}$

**61.** $\boxed{\text{Rate of first person}} \cdot \boxed{\text{Time to complete}} + \boxed{\text{Rate of second}} \cdot$

$\boxed{\text{Time to complete}} = \boxed{\text{Total pages}}$

Rate of first person = 15

Time to complete = $x$

Rate of second person = 20

Total pages = 910

$15x + 20x = 910$

$35x = 910$

$x = 26$

It will take 26 hours.

# Chapter 3 *continued*

**62.** There are twice as many nickels as there are quarters. Each nickel is worth 5 cents. Simplify the expression by multiplying: $5(2q) = 10q$.

**63.** $25q + 5(2q) + 10(q - 4) = 500$

$25q + 10q + 10q - 40 = 500$

$45q = 540$

$q = 12$

There are 12 quarters, 24 nickels, and 8 dimes.

**64.** $10d + 5(4d) = 600$

$10d + 20d = 600$

$30d = 600$

$d = 20$

There are 80 nickels.

**65. a.** $\frac{160}{560} = \frac{2}{7}; \frac{2}{7}t$  **b.** $\frac{120}{560}t = \frac{3}{14}t$  **c.** $\frac{2}{7}t + \frac{3}{14}t$

**d.** 1 represents the whole job, so letting $t = $ the time it takes to do the whole job, the fraction of the work Luis can do in that time $\left(\frac{2}{7}t\right)$ plus the fraction of the work Mei can do in that time $\left(\frac{3}{14}t\right)$ equals the whole job.

**e.** $\frac{2}{7}t + \frac{3}{14}t = 1$     Check: $\frac{2}{7}(2) + \frac{3}{14}(2) \overset{?}{=} 1$

$\frac{7}{14}t = 1$                   $\frac{4}{7} + \frac{6}{14} \overset{?}{=} 1$

$t = 2$ hours               $\frac{14}{14} = 1$

**66.** $x + x + 2 + x + 4 = 111$

$3x + 6 = 111$

$3x = 105$

$x = 35$

The integers are 35, 37, and 39; no; if $n$ is the first of the three integers, you could use the equation $3n + 6 = 1111$, which has no integer solutions.

## 3.3 Mixed Review (p. 151)

**67.** $4^2$  **68.** $b^3$  **69.** $a^6$  **70.** $10^1$  **71.** $2^4$  **72.** $(3x)^5$

**73.** $5 + 8 - 3 = 13 - 3 = 10$

**74.** $3^2 \cdot 4 + 8 = 9 \cdot 4 + 8 = 36 + 8 = 44$

**75.** $16.9 - 1.5(1.8 + 0.2) = 16.9 - 1.5(2) = 16.9 - 3 = 13.9$

**76.** $5 \cdot (12 - 4) + 7 = 5(8) + 7 = 40 + 7 = 47$

**77.** $-6 \div 3 - 4 \cdot 5 = -2 - 20 = -22$

**78.** $10 - [4.3 + 2(6.4 \div 8)] = 10 - [4.3 + 2(0.8)] =$
$10 - [4.3 + 1.6] = 10 - [5.9] = 4.1$

**79.** $\dfrac{-5 \cdot 4}{3 - 7^2 + 6} = \dfrac{-20}{3 - 49 + 6} = \dfrac{-20}{-40} = \dfrac{1}{2}$

**80.** $2 - 8 \div \dfrac{-2}{3} = 2 - (-12) = 2 + 12 = 14$

**81.** $\dfrac{(3 - 6)^2 + 6}{-5} = \dfrac{(-3)^2 + 6}{-5} = \dfrac{9 + 6}{-5} = \dfrac{15}{-5} = -3$

**82.** $d = r \cdot t$

$d = 5\left(\dfrac{45}{60}\right) = 3\dfrac{3}{4}$ mi

**83.** hamburgers: \$134 income, \$24 expense;
hot dogs: \$137 income, \$29 expense;
tacos: \$118 income, \$45 expense

**84.** hamburgers

**85.** tacos

## Quiz 1 (p. 152)

**1.** $8 - y = -9$

$-y = -17$

$y = 17$

**2.** $x + \frac{1}{2} = 5$

$x = 4\frac{1}{2}$

**3.** $|-14| + z = 12$

$14 + z = 12$

$z = -2$

**4.** $8b = 5$

$b = \dfrac{5}{8}$

**5.** $\frac{3}{4}q = 24$

$q = 32$

**6.** $\dfrac{n}{-8} = -\dfrac{3}{8}$

$n = 3$

**7.** $\dfrac{x}{5} + 10 = \dfrac{4}{5}$

$x + 50 = 4$

$x = -46$

**8.** $\frac{1}{4}(y + 8) = 5$

$y + 8 = 20$

$y = 12$

**9.** $25x - 4(4x + 6) = -69$

$25x - 16x - 24 = -69$

$9x - 24 = -69$

$9x = -45$

$x = -5$

**10.** $x + 4 = 91$

$x = 87$

You would have had an 87.

**11.** $F = \frac{9}{5}C + 32$

$72 = \frac{9}{5}C + 32$

$40 = \frac{9}{5}C$

$22\frac{2}{9} = C$

The Celsius temperature is $22\frac{2}{9}°$.

## Math & History (p. 152)

**1.** Scale: 1 mm = 1 inch

tread 9 in.

Riser | 8.25 in     traditional stair

tread 11 in.

Riser | 7 in     "7–11" stair

**2.**

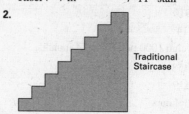

Traditional Staircase

# Chapter 3 *continued*

7-11 Rule Staircase

the traditional staircase; the staircase built using the 7–11 rule

## Developing Concepts Activity (p. 153)
## Drawing Conclusions

**1.**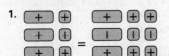

Subtract 3 *x*-tiles from each side.

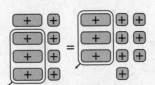

Write new equation.

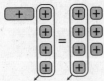

To isolate the *x*-tile, subtract 4 1-tiles from each side.

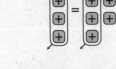

Solution: $x = 3$

**2.**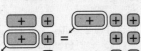

Subtract 1 *x*-tile from each side.

Write new equation.

To isolate the *x*-tile, subtract 3 1-tiles from each side.

Solution: $x = 3$

**3.**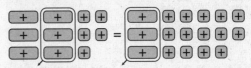

Subtract 3 *x*-tiles from each side.

Write new equation.

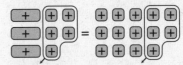

To isolate the *x*-tiles, subtract 5 1-tiles from each side.

Write new equation.

To find the value of *x*, split the tiles on each side into 3 groups to get $x = 3$.

**4.**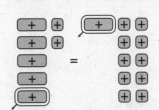

Subtract 1 *x*-tile from each side.

Write new equation.

Subtract 2 1-tiles from each side.

—CONTINUED—

# Chapter 3 *continued*

**4. —CONTINUED—**

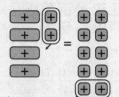

Write new equation.

To find the value of $x$, split the tiles on each side into 4 groups.

Solution: $x = 2$

**5.**

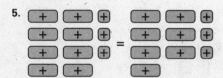

Subtract 7 $x$-tiles from each side.

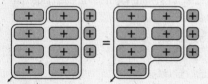

Write new equation.

Subtract 3 1-tiles from each side.

Solution: $x = 0$

**6.**

Subtract 1 $x$-tile from each side.

—CONTINUED—

**6. —CONTINUED—**

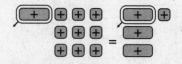

Write new equation.

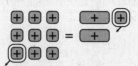

Subtract 1 1-tile from each side.

Write new equation.

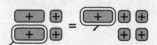

To find $x$, split the tiles on each side of the equation in half.

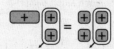

Solution: $x = 4$

**7.** Model of the original equation.

Subtract 1 $x$-tile from each side.

Subtract 2 1-tiles from each side.

**8.** If you subtract 4 $x$-tiles instead of 2 $x$-tiles, you will get a model of the equation $5 = -2x + 9$. Preferences may vary.

## Lesson 3.4

### 3.4 Guided Practice (p. 157)

**1.** $-2(4 - x) = 2x - 8$

$-8 + 2x = 2x - 8$

This is an identity. The given equation is true for all values of $x$ since $2x - 8 = 2x - 8$.

# Chapter 3 *continued*

**2.** $x = 2x$

$0 = x$

true

**3.** $9(9 - x) = 4x - 10$ — Write original equation.

$81 - 9x = 4x - 10$ — Apply distributive property.

$91 - 9x = 4x$ — Add 10 to both sides.

$91 = 13x$ — Add $9x$ to both sides.

$7 = x$ — Divide both sides by 13.

**4.** $2x + 3 = 7x$

$3 = 5x$

$\frac{3}{5} = x$

one solution

**5.** $12 - 2a = -5a - 9$

$12 + 3a = -9$

$3a = -21$

$a = -7$

one solution

**6.** $x - 2x + 3 = 3 - x$

$-x + 3 = 3 - x$

All real numbers are solutions. This is an identity.

**7.** $5x + 24 = 5(x - 5)$

$5x + 24 = 5x - 25$

$5x = 5x - 49$

$0 \neq -49$

no solution

**8.** $\frac{2}{3}(6c + 3) = 6(c - 3)$

$4c + 2 = 6c - 18$

$2 = 2c - 18$

$20 = 2c$

$10 = c$

one solution

**9.** $6y - (3y - 6) = 5y - 4$

$6y - 3y + 6 = 5y - 4$

$3y + 6 = 5y - 4$

$6 = 2y - 4$

$10 = 2y$

$5 = y$

one solution

**10.** B

**11.** $3x + 20 = 5x$

$20 = 2x$

$10 = x$

You must sell 10 pies.

### 3.4 Practice and Application (p. 157–159)

**12.** $7 - 4c = 10c$ — Write original equation.

$7 = 14c$ — Add $4c$ to both sides.

$\frac{1}{2} = c$ — Divide both sides by 14.

**13.** $-8x + 7 = 4x - 5$ — Write original equation.

$7 = 12x - 5$ — Add $8x$ to both sides.

$12 = 12x$ — Add 5 to both sides.

$1 = x$ — Divide both sides by 12.

**14.** $x + 2 = 3x - 1$ — Write original equation.

$2 = 2x - 1$ — Subtract $x$ from both sides.

$3 = 2x$ — Add 1 to both sides.

$1\frac{1}{2} = x$ — Divide both sides by 2.

**15.** $7(1 - y) = -3(y - 2)$ — Write original equation.

$7 - 7y = -3y + 6$ — Apply distributive property.

$7 = 4y + 6$ — Add $7y$ to both sides.

$1 = 4y$ — Subtract 6 from both sides.

$\frac{1}{4} = y$ — Divide both sides by 4.

**16.** $\frac{1}{5}(10a - 15) = 3 - 2a$ — Write original equation.

$2a - 3 = 3 - 2a$ — Apply distributive property.

$4a - 3 = 3$ — Add $2a$ to both sides.

$4a = 6$ — Add 3 to both sides.

$a = \frac{6}{4}$ — Divide both sides by 4.

$a = 1\frac{1}{2}$ — Simplify.

**17.** $5(y - 2) = -2(12 - 9y) + y$ — Write original equation.

$5y - 10 = -24 + 18y + y$ — Apply distributive property.

$5y - 10 = -24 + 19y$ — Add like terms.

$-10 = -24 + 14y$ — Subtract $5y$ from both sides.

$14 = 14y$ — Add 24 to both sides.

$1 = y$ — Divide both sides by 14.

**18.** $4x + 27 = 3x$

$x + 27 = 0$

$x = -27$

**19.** $12y + 21 = 9y$

$3y + 21 = 0$

$3y = -21$

$y = -7$

**20.** $-2m = 16m - 9$

$0 = 18m - 9$

$9 = 18m$

$\frac{1}{2} = m$

**21.** $4n = -28n - 3$

$32n = -3$

$n = -\frac{3}{32}$

**22.** $12c - 4 = 12c$

$-4 \neq 0$

no solution

**23.** $-30d + 12 = 18d$

$12 = 48d$

$\frac{1}{4} = d$

**24.** $6 - (-5r) = 5r - 3$

$6 + 5r = 5r - 3$

$6 \neq -3$

no solution

**25.** $6s - 11 = -2s + 5$

$8s - 11 = 5$

$8s = 16$

$s = 2$

**26.** $12p - 7 = -3p + 8$

$15p - 7 = 8$

$15p = 15$

$p = 1$

**27.** $-12q + 4 = 8q - 6$

$4 = 20q - 6$

$10 = 20q$

$\frac{1}{2} = q$

**28.** $-7 + 4m = 6m - 5$

$-7 = 2m - 5$

$-2 = 2m$

$-1 = m$

**29.** $-7 + 11g = 9 - 5g$

$-7 + 16g = 9$

$16g = 16$

$g = 1$

**30.** $8 - 9t = 21t - 17$

$8 = 30t - 17$

$25 = 30t$

$\frac{5}{6} = t$

**31.** $24 - 6r = 6(4 - r)$

$24 - 6r = 24 - 6r$

all real numbers

**32.** $3(4 + 4x) = 12x + 12$

$12 + 12x = 12x + 12$

all real numbers

**33.** $-4(x - 3) = -x$

$-4x + 12 = -x$

$12 = 3x$

$4 = x$

**34.** $10(-4 + y) = 2y$

$-40 + 10y = 2y$

$-40 = -8y$

$5 = y$

# Chapter 3 *continued*

**35.** $8a - 4(-5a - 2) = 12a$

$\quad 8a + 20a + 8 = 12a$

$\quad\quad 28a + 8 = 12a$

$\quad\quad\quad\quad 8 = -16a$

$\quad\quad\quad -\frac{1}{2} = a$

**36.** $9(b - 4) - 7b = 5(3b - 2)$

$\quad 9b - 36 - 7b = 15b - 10$

$\quad\quad\quad 2b - 36 = 15b - 10$

$\quad\quad\quad\quad -36 = 13b - 10$

$\quad\quad\quad\quad -26 = 13b$

$\quad\quad\quad\quad -2 = b$

**37.** $-2(6 - 10n) = 10(2n - 6)$

$\quad -12 + 20n = 20n - 60$

$\quad\quad\quad -12 \neq -60$

no solution

**38.** $-(8n - 2) = 3 + 10(1 - 3n)$

$\quad -8n + 2 = 3 + 10. - 30n$

$\quad -8n + 2 = 13 - 30n$

$\quad\quad 22n + 2 = 13$

$\quad\quad\quad 22n = 11$

$\quad\quad\quad\quad n = \frac{1}{2}$

**39.** $\frac{1}{2}(12n - 4) = 14 - 10n$

$\quad 6n - 2 = 14 - 10n$

$\quad 16n - 2 = 14$

$\quad\quad 16n = 16$

$\quad\quad\quad n = 1$

**40.** $\frac{1}{4}(60 + 16s) = 15 + 4s$

$\quad 15 + 4s = 15 + 4s$

all real numbers

**41.** $\frac{3}{4}(24 - 8b) = 2(5b + 1)$

$\quad 18 - 6b = 10b + 2$

$\quad\quad 18 = 16b + 2$

$\quad\quad 16 = 16b$

$\quad\quad\quad 1 = b$

**42.** The distributive property is applied incorrectly.

$7(c - 6) = (4 - c)(3)$

$7c - 42 = 12 - 3c$

$\quad c = 5.4$

**43.** Since $-6 = -6$, the equation is true for all real values of $b$. The conclusion $b = -6$ is incorrect.

**44.** The given equation is true for all real numbers.

**45.** $5x = 3x + 25$

$\quad 2x = 25$

$\quad\; x = 12.5$

The membership is worthwhile if you plan to attend 13 or more sessions.

**46.** $450 + 6x = 16x + 8x$

$\quad 450 + 6x = 24x$

$\quad\quad\quad 450 = 18x$

$\quad\quad\quad\; 25 = x$

You must use the gym 25 days.

**47.** The possible number of stories is $3\frac{1}{2}$ more than the actual number, 50.

**48.** Combined height of lowest 7 stories $= 126$ (feet)

Number of other stories $= 43$ (stories)

Average height of other stories $= h$ (feet)

Possible number of stories $= 53\frac{1}{2}$ (stories)

Algebraic model: $126 + 43h = 53\frac{1}{2}h$

**49.** $126 + 43h = 53\frac{1}{2}h$

$\quad\quad 126 = 10\frac{1}{2}h$

$\quad\quad\; 12 = h$

The stories above the 7th floor are about 12 feet high.

**50.** A **51.** C

**52.** $\frac{1}{3}(7x + 5) = 3x - 5$

$\quad 7x + 5 = 9x - 15$

$\quad -2x + 5 = -15$

$\quad\quad -2x = -20$

$\quad\quad\quad x = 10$

C

**53.** Number of subscribers $= x$

Income $\quad\quad\quad\quad = 1.5x$

Expenses $\quad\quad\quad = 100 + x$

$1.5x = 100 + x$

$0.5x = 100$

$\quad x = 200$

The magazine needs 200 subscribers.

**54.** Find the column where income equals expenses. The number of subscribers is 200, the answer to Exercise 53.

### 3.4 Mixed Review (p. 159)

**55.** 0.28 **56.** 0.4 **57.** 0.03 **58.** 0.195

**59.** $0.45 \cdot 84 = 37.8$ **60.** $0.07 \cdot 28.5 = 1.995$

**61.** $0.76 \cdot 540 = 410.4$ **62.** $0.163 \cdot 132 = 21.516$

**63.** $0.08 \cdot 928.5 = 74.28$ **64.** $0.055 \cdot 74 = 4.07$

$\quad\quad \$74.28$ $\quad\quad\quad \$4.07$

**65.** $\dfrac{\text{dollars}}{\text{hour}} \cdot \text{hours} = \text{dollars}$

**66.** $\text{years} \cdot \dfrac{\text{people}}{\text{year}} = \text{people}$

**67.** $\text{miles} \cdot \dfrac{\text{hour}}{\text{miles}} = \text{hour}$

**68.** $\text{meters} \cdot \dfrac{\text{kilometers}}{\text{meter}} = \text{kilometers}$

**69.** $\dfrac{3}{5} = \dfrac{x}{10}$

$\quad 3(10) = 5x$

$\quad\quad 30 = 5x$

$\quad 6 \text{ cm} = x$

**70.** $\dfrac{x}{8} = \dfrac{9}{6}$

$\quad 6x = 9(8)$

$\quad 6x = 72$

$\quad\; x = 12$

# Chapter 3 continued

## Lesson 3.5

### 3.5 Guided Practice (p. 163)

1.

$$2\left(\tfrac{7}{8}\right) + 4\left(\tfrac{3}{16}\right) + 5h = 11$$
$$\tfrac{10}{4} + 5h = 11$$
$$5h = \tfrac{34}{4}$$
$$h = 1.7$$

Each picture is about 1.7 inches high.

2. feet + feet = feet

3. Rate your friend is driving:     52 (mi/h)
   Time after you start driving:     $t$ (h)
   Friend's distance when you leave:   32 (mi)
   Rate you are driving:     60 (mi/h)
   Time after you start driving:     $t$ (h)
   $$52t + 32 = 60t$$

4. $52t + 32 = 60t$
   $$32 = 8t$$
   $$4 = t$$

   Four hours after you leave, you and your friend are side by side. You drive $4(60) = 240$ miles. Your friend drives $4(52) + 32 = 240$ miles.

5.

| Time (in hours) | 1 | 2 | 3 | 4 | |
|---|---|---|---|---|---|
| Your distance | 60 | 120 | 180 | 240 | |
| Friend's distance | 52 | 104 | 156 | 208 | equal |
| Friend's distance + 32 | 84 | 136 | 188 | 240 | |

### 3.5 Practice and Applications (pp. 163–165)

6. B because it shows the length, width, and height of a 3-dimensional box.

7. C   $4x + 36 = 108$
   $$4x = 72$$
   $$x = 18 \text{ in.}$$

8.

| Length | Width | Height | Girth | Combined Length and Girth |
|---|---|---|---|---|
| 36 | 9 | 9 | 36 | 72 |
| 36 | 18 | 18 | 72 | 108 |
| 36 | 25 | 25 | 100 | 136 |

The box with a width and height of 18 inches just meets the regulation.

9. $\dfrac{\text{hours}}{\text{day}} \cdot \text{day} = \text{hours}$

10. $\dfrac{\text{feet}}{\text{minute}} = \text{ft/min}$

11. $\dfrac{\text{inches}}{\frac{\text{inches}}{\text{foot}}} = \text{inches} \cdot \dfrac{\text{foot}}{\text{inches}} = \text{ft}$

12. $\dfrac{\text{miles}}{\text{hour}} \cdot \dfrac{\text{hours}}{\text{minute}} = \text{mi/min}$

13. No; according to the graph the solution is 5 weeks, not 25 weeks.

14.

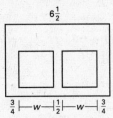

15. Let $w$ = width in inches of photos
    $$2w + 2 = 6\tfrac{1}{2}$$

16. $2w + 2 = 6\tfrac{1}{2}$
    $$2w = 4\tfrac{1}{2}$$
    $$w = 2\tfrac{1}{4}$$

    The photos should be $2\tfrac{1}{4}$ inches wide.

17. Let $x$ be the number of years until the classes are equal.
    $$45 + 3x = 108 - 4x$$
    $$45 + 7x = 108$$
    $$7x = 63$$
    $$x = 9$$

    The classes will be equal in 9 years.

| Time (in years) | 1 | 5 | 9 | |
|---|---|---|---|---|
| Students taking German | 104 | 88 | 72 | equal |
| Students taking Japanese | 48 | 60 | 72 | |

18. $\dfrac{5}{60} = \tfrac{1}{12}$ h
    $$\tfrac{1}{12}(10) = \tfrac{5}{6} \text{ mi}$$

19. $\tfrac{1}{12}(6) = \tfrac{1}{2}$ mi
    total distance = $x + \tfrac{1}{2}$

20. $x + \tfrac{1}{2} = \tfrac{5}{6}$
    $$x = \tfrac{1}{3} \text{ mi}$$

21. Answers may vary.

22. a. Diagrams may vary.
    b. Sally: $21t$; Teresa: $15t$
    c. The sum of the distances they traveled must be 60 miles.
       $$21t + 15t = 60$$
       $$36t = 60$$
       $$t = 1\tfrac{2}{3} \text{ hours}$$
    d. It would be convenient because it would take Sally about 1.6 hours to get there and it would take Teresa about 1.7 hours.

       Data displays may vary.

23. rabbit: $\dfrac{30}{25} = 1.2$ sec
    coyote: $\dfrac{x + 30}{50}$ sec

24. $\dfrac{x + 30}{50} = 1.2$
    $$x + 30 = 60$$
    $$x = 30 \text{ ft}$$

25. If the coyote is 30 feet from the rabbit, both will reach the burrow at the same time. If $x$ is less than 30 ft, the coyote will overtake the rabbit. If $x$ is greater than 30 ft, the rabbit will escape the coyote.

# Chapter 3 *continued*

**26.** Tables and graphs may vary.

**27.** $\frac{3}{11} \approx 0.27$
27%

**28.** $\frac{11}{12} \approx 0.92$
92%

**29.** $\frac{25}{31} \approx 0.81$
81%

**30.** $\frac{100}{201} \approx 0.50$
50%

**31.** $\frac{-6}{2} - 2(11) = -3 - 22 = -25$

**32.** $\frac{3}{5}(-25) - (-10) = -15 + 10 = -5$

**33.** $\frac{-24 + 6}{12} = \frac{-18}{12} = -1.5$

**34.** $\frac{8(7)}{3(8) - 10} = \frac{56}{24 - 10} = \frac{56}{14} = 4$

**35.** $18 + \frac{x}{3} = 9$
$\frac{x}{3} = -9$
$x = -27$

**36.** $8 = -\frac{2}{3}(2x - 6)$
$-12 = 2x - 6$
$-6 = 2x$
$-3 = x$

**37.** $4(2 - n) = 1$
$8 - 4n = 1$
$-4n = -7$
$n = \frac{7}{4}$
$n = 1\frac{3}{4}$

**38.** $-3y + 14 = -5y$
$14 = -2y$
$-7 = y$

**39.** $-4a - 3 = 6a + 2$
$-3 = 10a + 2$
$-5 = 10a$
$-\frac{1}{2} = a$

**40.** $5x - (6 - x) = 2(x - 7)$
$5x - 6 + x = 2x - 14$
$6x - 6 = 2x - 14$
$4x - 6 = -14$
$4x = -8$
$x = -2$

## Lesson 3.6

### Activity (p. 168)

**1.** Student 1: $18.42x - 12.75 = (5.32x - 6.81)3.46$
$\quad 18.42x - 12.75 = 18.41x - 23.56$
$\quad\quad\quad\quad 0.01x = -10.81$
$\quad\quad\quad\quad\quad\quad x = -1081$
Student 2: $18.42x - 12.75 = (5.32x - 6.81)3.46$
$\quad 18.42x - 12.75 = 18.4072x - 23.5626$
$\quad\quad\quad\quad 0.0128x = -10.8126$
$\quad\quad\quad\quad\quad\quad x \approx -844.73$

**2.** Student 1: $-1081$; Student 2: $-844.73$; the second is more accurate because the first involves three round-off errors.

**1.** the symbol "$\approx$" (is approximately equal to)

**2.** 100

**3.** $68x = 442$
$\quad x = 6.5$
They will need 6.5 buses.

**4.** No. It is impossible to rent 0.5 of a bus. They will need a seventh bus.

**5.** 23.4   **6.** 108.2   **7.** $-13.9$   **8.** 63.0

**9.** $2.2x = 15$
$\quad x \approx 6.82$

Check:
$2.2(6.82) \stackrel{?}{=} 15$
$15.004 \approx 15$

**10.** $14 - 9x = 37$
$\quad -9x = 23$
$\quad\quad x \approx -2.56$

Check:
$14 - 9(-2.56) \stackrel{?}{=} 37$
$14 + 23.04 \stackrel{?}{=} 37$
$37.04 \approx 37$

**11.** $2(3b - 14) = -9$
$\quad 6b - 28 = -9$
$\quad\quad 6b = 19$
$\quad\quad\quad b \approx 3.17$

Check:
$2[3(3.17) - 14] \stackrel{?}{=} -9$
$2(-4.49) \stackrel{?}{=} -9$
$-8.98 \approx -9$

**12.** $2.69 - 3.64x = 8.37 + 23.78x$
$\quad 2.69 = 8.37 + 27.42x$
$\quad -5.68 = 27.42x$
$\quad -0.21 \approx x$

Check:
$2.69 - 3.64(-.21) \stackrel{?}{=} 8.37 + 23.78(-.21)$
$2.69 + .7644 \stackrel{?}{=} 8.37 - 4.9938$
$3.4544 \approx 3.3762$

**13.** $1.07x = 35.72$
$\quad x \approx 33.38$
The limit is $33.38.

**14.** $-35.2; -35.19$

**15.** $5.34(6.79) \approx 36.3; 36.26$

**16.** $-7.895 + 4.929 \approx -3.0; -2.97$

**17.** $47.0362 - 39.7204 \approx 7.3; 7.32$

**18.** $5.349 \div 46.597 \approx 0.1; 0.11$

**19.** $-25.349(-1.369) \approx 34.7; 34.70$

**20.** $13x - 7 = 27$
$\quad 13x = 34$
$\quad\quad x = 2.62$

Check:
$13(2.62) - 7 \stackrel{?}{=} 27$
$27.06 \approx 27$

# Chapter 3 *continued*

**21.** $18 - 3y = 5$
$-3y = -13$
$y \approx 4.33$

Check:
$18 - 3(4.33) \stackrel{?}{=} 5$
$5.01 \approx 5$

**22.** $-7n + 17 = -6$
$-7n = -23$
$n \approx 3.29$

Check:
$-7(3.29) + 17 \stackrel{?}{=} -6$
$-6.03 \approx -6$

**23.** $38 = -14 + 9a$
$52 = 9a$
$5.78 \approx a$

Check:
$38 \stackrel{?}{=} -14 + 9(5.78)$
$38 \approx 38.02$

**24.** $47 = 28 - 12x$
$19 = -12x$
$-1.58 \approx x$

Check:
$47 \stackrel{?}{=} 28 - 12(-1.58)$
$47 \approx 46.96$

**25.** $14r + 8 = 32$
$14r = 24$
$r \approx 1.71$

Check:
$14(1.71) + 8 \stackrel{?}{=} 32$
$31.94 \approx 32$

**26.** $358 = 39c - 17$
$375 = 39c$
$9.62 \approx c$

Check:
$358 \stackrel{?}{=} 39(9.62) - 17$
$358 \approx 358.18$

**27.** $37 - 58b = 204$
$-58b = 167$
$b \approx -2.88$

Check:
$37 - 58(-2.88) \stackrel{?}{=} 2.04$
$204.04 \approx 204$

**28.** $3(31 - 12t) = 82$
$93 - 36t = 82$
$-36t = -11$
$t \approx 0.31$

Check:
$3[31 - 12(0.31)] \stackrel{?}{=} 82$
$81.84 \approx 82$

**29.** $4(-7y + 13) = 49$
$-28y + 52 = 49$
$-28y = -3$
$y \approx 0.11$

Check:
$4[-7(0.11) + 13] \stackrel{?}{=} 49$
$48.92 \approx 49$

**30.** $2(-5a + 7) = -a$
$-10a + 14 = -a$
$14 = 9a$
$1.56 \approx a$

Check:
$2[-5(1.56) + 7] \stackrel{?}{=} -1.56$
$-1.6 \approx -1.56$

**31.** $-(d - 3) = 2(3d + 1)$
$-d + 3 = 6d + 2$
$3 = 7d + 2$
$1 = 7d$
$0.14 \approx d$

Check:
$-(0.14 - 3) \stackrel{?}{=} 2[3(0.14) + 1]$
$2.86 \approx 2.84$

**32.** $12.67 + 42.35x = 5.34x + 26.58$
$12.67 + 37.01x = 26.58$
$37.01x = 13.91$
$x \approx 0.38$

**33.** $4.65x - 4.79 = 13.57 - 6.84x$
$11.49x - 4.79 = 13.57$
$11.49x = 18.36$
$x \approx 1.60$

**34.** $7.45x - 8.81 = 5.29 + 9.47x$
$-8.81 = 5.29 + 2.02x$
$-14.1 = 2.02x$
$-6.98 \approx x$

**35.** $39.21x + 2.65 = -31.68 + 42.03x$
$2.65 = -31.68 + 2.82x$
$34.33 = 2.82x$
$12.17 \approx x$

**36.** $5.86x - 31.94 = 27.51x - 3.21$
$-31.94 = 21.65x - 3.21$
$-28.73 = 21.65x$
$-1.33 \approx x$

**37.** $-2(4.36 - 6.92x) = 9.27x + 3.87$
$-8.72 + 13.84x = 9.27x + 3.87$
$-8.72 + 4.57x = 3.87$
$4.57x = 12.59$
$x \approx 2.75$

**38.** $6.1(3.1 + 2.5x) = 15.3x - 3.9$
$18.91 + 15.25x = 15.3x - 3.9$
$18.91 = 0.05x - 3.9$
$22.81 = 0.05x$
$456.20 = x$

**39.** $4.21x + 5.39 = 12.07(2.01 - 4.72x)$
$4.21x + 5.39 = 24.2607 - 56.9704x$
$61.1804x + 5.39 = 24.2607$
$61.1804x = 18.8707$
$x \approx 0.31$

**40.** $2.5x + 0.7 = 4.6 - 1.3x$
$25x + 7 = 46 - 13x$

**41.** $-0.625y - 0.184 = 2.506y$
$-625y - 184 = 2506y$

**42.** $1.67 + 2.43x = 3.29(x - 5)$
$167 + 243x = 329(x - 5)$

**43.** $4.5n - 0.375 = 0.75n + 2.0$
$4500n - 375 = 750n + 2000$

**44.** Because masses are not usually described to the nearest hundred-millionth of a kilogram, B is a more reasonable answer.

**45.** a situation involving physical objects that cannot be divided

**46.** a problem involving metric measures

**47.** a problem involving money

# Chapter 3 *continued*

**48.** Method 1:

**a.** $\dfrac{1000 \text{ m}}{76.51 \text{ sec}} \approx 13.1 \text{ m/sec}$

**b.** $\dfrac{13.1 \text{ m}}{1 \text{ sec}} \cdot \dfrac{60 \text{ sec}}{1 \text{ min}} = 786 \text{ m/min}$

**c.** $\dfrac{786 \text{ m}}{1 \text{ min}} \cdot \dfrac{60 \text{ min}}{1 \text{ h}} = 47{,}160 \text{ m/h}$

**d.** $\dfrac{47{,}160 \text{ m}}{1 \text{ h}} \cdot \dfrac{1 \text{ km}}{1000 \text{ m}} \approx 47.2 \text{ km/h}$

Method 2:

**a–d.** $\dfrac{1000 \text{ m}}{76.51 \text{ sec}} \approx \dfrac{13.0701869 \text{ m}}{1 \text{ sec}} \cdot \dfrac{60 \text{ sec}}{1 \text{ min}} \approx$

$\dfrac{784.2112142 \text{ m}}{1 \text{ min}} \cdot \dfrac{60 \text{ min}}{1 \text{ h}} \approx$

$\dfrac{47052.67285 \text{ m}}{1 \text{ h}} \cdot \dfrac{1 \text{ km}}{1000 \text{ m}} \approx 47.1 \text{ km/h}$

The two methods do not give the same final result.

**49.** $1.15x = 8.39$

$x = 7.296$

Your limit is $7.29.

**50.** C $0.93x = 186$

$x = 200$

About 200 people were surveyed.

**51.** 0.24 or 24%

**52.** $0.02x = 42$

$x = 2100$

The group contains about 2100 people.

**53.** $440x + 500 = 2975$

$440x = 2475$

$x \approx 5.6$

It will take about 6 min.

**54.** B

**55.** $4.95 + 2.5(x - 3) = 21.83$

$4.95 + 2.5x - 7.5 = 21.83$

$2.5x - 2.55 = 21.83$

$2.5x = 24.38$

$x \approx 9.75 \text{ h}$

You used the Internet for about 9.75 hours.

**56.** The temperature rose 8°C.

The gap decreased by $8(0.37) \approx 3.0$ mm. The new width of the gap is $16.8 - 3.0 \approx 13.8$ mm.

**57.** Temperature rise: $(t - 10)°C$

Decrease in width of gap: $0.37(t - 10)$ mm

New width of gap: $16.8 - 0.37(t - 10)$ mm

**58.** $16.8 - 0.37(t - 10) = 9.4$

$16.8 - 0.37t + 3.7 = 9.4$

$20.5 - 0.37t = 9.4$

$-0.37t = -11.1$

$t = 30$

The temperature would be 30°C.

**59. a.** $t - 2$ min

**b.** $7.8t = 12.3(t - 2)$

$7.8t = 12.3t - 24.6$

$-4.5t = -24.6$

$t \approx 5.4\overline{6}$ min

**c.** The hot water runs for about 5.5 min and the cold water for 3.5 min. Rounding to tenths is most appropriate. It is easy to time 5.5 min.

**60.** Answers may vary.

### 3.6 Mixed Review (p. 172)

**61.**

| Input | Output |
|-------|--------|
| 2 | 13 |
| 3 | 15.5 |
| 4 | 18 |
| 5 | 20.5 |
| 6 | 23 |

$A = 8 + 2.5t$

domain: 2, 3, 4, 5, 6

range: 13, 15.5, 18, 20.5, 23

**62.** $-8$  **63.** 3  **64.** 4.9  **65.** $-7.9$  **66.** $\frac{1}{26} \approx 0.038$

**67.** $\frac{16}{30} = \frac{8}{15} \approx 0.533$  **68.** $\frac{3}{6} = \frac{1}{2} = 50\%$  **69.** It increased by $85.25.  **70.** $697.45  **71.** $-$234.87

### Quiz 2 (p. 172)

**1.** $3x + 1 = 5x$

$1 = 2x$

$\frac{1}{2} = x$

**2.** $8 - 2y = 21 - 6y$

$8 + 4y = 21$

$4y = 13$

$y = 3\frac{1}{4}$

**3.** $3n = (6 - n)(-3)$

$3n = -18 + 3n$

$0 \neq -18$

no solution

**4.** $\frac{1}{2}(14 + 8a) = 9a$

$7 + 4a = 9a$

$7 = 5a$

$1\frac{2}{5} = a$

**5.** $7 - 6d = 3(5 - 2d)$

$7 - 6d = 15 - 6d$

$7 \neq 15$

no solution

**6.** $-7(b + 1) = 5(b - 2)$

$-7b - 7 = 5b - 10$

$-7 = 12b - 10$

$3 = 12b$

$\frac{1}{4} = b$

**7.** $15y - 8 = 4y - 3$

$11y - 8 = -3$

$11y = 5$

$y \approx 0.45$

**8.** $2(2n + 11) = 31 - 3n$

$4n + 22 = 31 - 3n$

$7n + 22 = 31$

$7n = 9$

$n \approx 1.29$

# Chapter 3 *continued*

**9.** $7.6x + 3.7 = 2.8 - 1.6x$
$9.2x + 3.7 = 2.8$
$9.2x = -0.9$
$x \approx -.10$

**10.** $4.9(3.7x + 1.4) = 34.7x$
$18.13x + 6.86 = 34.7x$
$6.86 = 16.57x$
$0.41 \approx x$

**11.**

| cats | + | birds | + | hamsters | = | dogs | + | turtles |
|------|---|-------|---|----------|---|------|---|---------|
| $n$ | + | 3 | + | 5 | = | $2n$ | + | 3 |

C models the situation.

**12.** $n + 8 = 2n + 3$
$8 = n + 3$
$5 = n$
There are 10 dogs.

## Technology Activity 3.6 (p. 173)

**1.** $19.65x + 2.2(x - 6.05) = 255.65$
$x \approx 12.3$

**2.** $16.2(3.1 - x) - 31.55x = -19.5$
$x \approx 1.5$

**3.** $3.56x + 2.43 = 6.17x - 11.40$
$x \approx 5.3$

**4.** $3.5(x - 5.6) + 0.03x = 4.2x - 25.5$
$x \approx 8.8$

## Lesson 3.7

### 3.7 Guided Practice (p. 177)

**1.** *Sample answer:* $A = \frac{1}{2}bh$; $A$ represents the area of a triangle, $b$ represents the base, and $h$ represents the height.

**2.** yes   **3.** no   **4.** yes   **5.** yes   **6.** no   **7.** no   **8.** $s = \frac{1}{4}P$

**9.**
$2x + 2y = 10$
$2x + 2y - 2x = 10 - 2x$
$2y = 10 - 2x$
$\frac{2y}{2} = \frac{10}{2} - \frac{2x}{2}$
$y = 5 - x$

**10.** $x = -2$
$y = 5 - (-2)$
$y = 7$

$x = 0$
$y = 5 - (0)$
$y = 5$

$x = 2$
$y = 5 - (2)$
$y = 3$

$x = -1$
$y = 5 - (-1)$
$y = 6$

$x = 1$
$y = 5 - (1)$
$y = 4$

### 3.7 Practice and Applications (pp. 177–179)

**11.** $A = \frac{1}{2}bh$
$\frac{2A}{h} = b$

**12.** $V = lwh$
$\frac{V}{lw} = h$

**13.** $V = \pi r^2 h$
$\frac{V}{\pi r^2} = h$

**14.** $A = \frac{1}{2}h(b_1 + b_2)$
$2A = h(b_1 + b_2)$
$\frac{2A}{h} = b_1 + b_2$
$\frac{2A}{h} - b_1 = b_2$

**15.** $2x + y = 5$
$y = 5 - 2x$

**16.** $3x + 5y = 7$
$5y = 7 - 3x$
$y = \frac{7}{5} - \frac{3}{5}x$

**17.** $13 = 12x - 2y$
$13 - 12x = -2y$
$6x - \frac{13}{2} = y$

**18.** $2x = -3y + 10$
$2x - 10 = -3y$
$-\frac{2}{3}x + \frac{10}{3} = y$

**19.** $9 - y = 1.5x$
$-y = 1.5x - 9$
$y = -1.5x + 9$

**20.** $1 + 7y = 5x - 2$
$7y = 5x - 3$
$y = \frac{5}{7}x - \frac{3}{7}$

**21.** $\frac{y}{5} - 7 = -2x$
$y - 35 = -10x$
$y = -10x + 35$

**22.** $\frac{1}{4}y + 3 = -5x$
$y + 12 = -20x$
$y = -20x - 12$

**23.** $-3x + 4y - 5 = -14$
$-3x + 4y = -9$
$4y = 3x - 9$
$y = \frac{3}{4}x - \frac{9}{4}$

**24.** $7x - 4x + 12 = 36 - 5y$
$3x + 12 = 36 - 5y$
$3x - 24 = -5y$
$-\frac{3}{5}x + \frac{24}{5} = y$

**25.** $\frac{1}{3}(y + 2) + 3x = 7x$
$y + 2 + 9x = 21x$
$y + 2 = 12x$
$y = 12x - 2$

**26.** $5(y - 3x) = 8 - 2x$
$5y - 15x = 8 - 2x$
$5y = 13x + 8$
$y = \frac{13}{5}x + \frac{8}{5}$

**27.** $4x - 3(y - 2) = 15 + y$
$4x - 3y + 6 = 15 + y$
$4x - 4y + 6 = 15$
$-4y = -4x + 15 - 6$
$-4y = -4x + 9$
$y = x - \frac{9}{4}$

**28.** $3(x - 2y) = -12(x + 2y)$
$3x - 6y = -12x - 24y$
$-6y = -15x - 24y$
$18y = -15x$
$y = -\frac{5}{6}x$

**29.** $\frac{1}{5}(25 - 5y) = 4x - 9y + 13$
$5 - y = 4x - 9y + 13$
$5 + 8y = 4x + 13$
$8y = 4x + 8$
$y = \frac{1}{2}x + 1$

**30.** $2x + y = 5$
$2x = -y + 5$
$x = -\frac{1}{2}y + \frac{5}{2}$
$x = -\frac{1}{2}(-2) + \frac{5}{2}$
$x = 3\frac{1}{2}$
$x = -\frac{1}{2}(0) + \frac{5}{2}$
$x = 2\frac{1}{2}$

$x = -\frac{1}{2}(-1) + \frac{5}{2}$
$x = 3$
$x = -\frac{1}{2}(1) + \frac{5}{2}$
$x = 2$

**31.** $3y - x = 12$
$-x = -3y + 12$
$x = 3y - 12$
$x = 3(-2) - 12$
$x = -18$
$x = 3(0) - 12$
$x = -12$

$x = 3(-1) - 12$
$x = -15$
$x = 3(1) - 12$
$x = -9$

**32.** $4(5 - y) = 14x + 3$
$20 - 4y = 14x + 3$
$-4y + 17 = 14x$
$-\frac{2}{7}y + \frac{17}{14} = x$
$x = -\frac{2}{7}(-2) + \frac{17}{14}$
$x = \frac{4}{7} + \frac{17}{14}$
$x = 1\frac{11}{14}$
$x = -\frac{2}{7}(0) + \frac{17}{14}$
$x = 1\frac{3}{14}$

$x = -\frac{2}{7}(-1) + \frac{17}{14}$
$x = 1\frac{1}{2}$
$x = -\frac{2}{7}(1) + \frac{17}{14}$
$x = \frac{13}{14}$

**33.** $5y - 2(x - 7) = 20$
$-2(x - 7) = 20 - 5y$
$-2x + 14 = 20 - 5y$
$-2x = -5y + 6$
$x = \frac{5}{2}y - 3$
$x = \frac{5}{2}(-2) - 3$
$x = -8$
$x = \frac{5}{2}(0) - 3$
$x = -3$

$x = \frac{5}{2}(-1) - 3$
$x = -5\frac{1}{2}$
$x = \frac{5}{2}(1) - 3$
$x = -\frac{1}{2}$

**34.** $S = L - rL$
$S - L = -rL$
$\frac{S - L}{L} = -r$
$\frac{S}{L} - 1 = -r$
$1 - \frac{S}{L} = r$

**35. a.** $r = 1 - \frac{80}{100}$
$r = 0.20$
20%
**b.** $r = 1 - \frac{72}{120}$
$r = 0.40$
40%
**c.** $r = 1 - \frac{12.50}{25}$
$r = 0.50$
50%

**36.** Given $r$ and $t$, use $d = rt$; given $d$ and $t$, use $r = \frac{d}{t}$; given $r$ and $d$, use $t = \frac{d}{r}$.

**37.** $P = 64d + 2112$
$P - 2112 = 64d$
$\frac{P - 2112}{64} = d$

**38.** $d = \frac{(4032) - 2112}{64}$
$d = 30$
The diver is at 30 ft.

**39.** $P$ is a function of $d$; $P$ is the isolated variable.

**40.** Circumference is a function of the radius.

**41.** $2\pi X$ **42.** $\frac{Y_1}{2\pi}$ **43.** Tables may vary. The first and third columns are the same.

**44.** $C = 2\pi r = \pi d$
$C = \pi d$
$\frac{C}{\pi} = d$
$d = \frac{102.6}{\pi}$
$d = 32.7$ ft

**45.** D **46.** B

**47.** $C + d = 2\pi(r + h)$
$C + d = 2\pi r + 2\pi h$
$d = 2\pi h$

**48.** $d = 2\pi h$

**49.** $d = 2\pi(500)$
$d \approx 3141.6$ ft

### 3.7 Mixed Review (p. 179)

**50.** $y - 5 > 4$
$9 - 5 \overset{?}{>} 4$
$4 \not> 4$
9 is not a solution.

**51.** $4 + 7y > 12$
$4 + 7(1) \overset{?}{>} 12$
$11 \not> 12$
1 is not a solution.

**52.** $3x - 3 \geq x + 3$
$3(5) - 3 \overset{?}{\geq} (5) + 3$
$15 - 3 \overset{?}{\geq} 5 + 3$
$12 \geq 8$
5 is a solution.

**53.** $-(2)^3(b)$
$-8b$

**54.** $(-2)^3(b)$
$-8b$

**55.** $3(-x)(-x)(-x)$
$-3x^3$

**56.** $(-4)(-c)(-c)(-c)$
$4c^3$

**57.** $(-4)^2(t)(t)$
$16t^2$

**58.** $(-5)^2(-y)(-y)$
$25y^2$

**59.** pro sports

**60.** 28%

**61.** 31

## Lesson 3.8

### 3.8 Guided Practice (p. 183)

**1.** A ratio compares quantities measured in the same units; a rate compares quantities measured in different units.

**2.** B  $\text{yards} \cdot \dfrac{\text{feet}}{\text{yard}} = \text{feet}$

3. Let $x$ = the number of students in the school who have a pet at home. Use the equation

$$\frac{18}{31} = \frac{x}{1746}.$$

4. Clothing and shoes: $\frac{\$688}{12} \approx \$57.33$ per person

   Restaurants: $\frac{\$714}{12} \approx \$59.50$ per person

   Food at home: $\frac{\$1128}{12} \approx \$94$ per person

5. euros $\cdot \dfrac{\text{dollars}}{\text{euro}} =$ dollars

   $e \cdot \dfrac{1.066}{1} = D$

   $200 \cdot 1.066 = D$

   $\$213 \approx D$

6. rand $\cdot \dfrac{\text{dollars}}{\text{rand}} =$ dollars

   $R \cdot \dfrac{0.607}{1} = D$

   $0.607R = 340$

   $R \approx 560$ rand

7. $0.45(280) = x$

   $126 = x$

8. $0.075(340) = x$

   $25.5 = x$

9. $0.20x = 15$

   $x = 75$

10. $x(12.50) = 1.75$

    $x = .14$

    The tip was 14% of the bill.

### 3.8 Practice and Applications (p. 183–185)

11. $\frac{2}{5} = \$0.40$ per can

12. $\frac{121.50}{18} = \$6.75$ per hour

13. $\frac{1.39}{1.5} \approx \$0.93$ per quart

14. $\frac{6}{2.5} = 2.4$ oz per serving

15. $\frac{1200}{4} = 300$ miles per hour

16. $\frac{52}{3} = 17\frac{1}{3}$ miles per day

17. $\frac{2}{40} = 0.05$ mile per minute

18. $\frac{69}{\frac{3}{4}} = 92$ km/h

19. $\frac{1.20}{5} = 0.24$

    $\frac{1.59}{7} \approx 0.23$

    7 for $1.59 is a better buy; the price per bar for the smaller box is $0.24, the price per bar for the larger box is about $0.23.

20. $\frac{38,000,000}{365} \approx 104,000$ changes per day

21. $\frac{1,100,000,000}{52} \approx 21,000,000$ mi/week

22. $4500 \times 32 \div 50 =$

23. ounces $\cdot \dfrac{\text{cups}}{\text{ounce}} =$ cups

    $12 \cdot \dfrac{1}{8} = c$

    $1\frac{1}{2}$ cups $= c$

24. inches $\cdot \dfrac{\text{cm}}{\text{inch}} =$ cm

    $21 \cdot \dfrac{2.54}{1} = 53.3$ cm

25. miles $\cdot \dfrac{\text{km}}{\text{mile}} =$ km

    $56 \cdot \dfrac{1.609}{1} = 90.104$ km

26. yards $\cdot \dfrac{\text{meters}}{\text{yard}} =$ meters

    $100 \cdot \dfrac{0.9144}{1} = 91.4$ meters

27. gallons $\cdot \dfrac{\text{miles}}{\text{gallon}} =$ miles

    $20 \cdot 15 = 300$ miles

28. $\frac{14,588}{313} \approx 46.6$

    about 47 books per shelf

29. $50 \cdot 47 = x$

    $2350 = x$

    The library can hold about 2350 more books.

30. dollars $\cdot \dfrac{\text{yen}}{\text{dollar}} =$ yen

    $325 \cdot \dfrac{114.52}{1} = 37,219$ yen

31. yen $\cdot \dfrac{\text{dollars}}{\text{yen}} =$ dollars

    $605 \cdot \dfrac{1}{114.52} \approx 5.28$

    You will get $5 back.

32. $320 - 288 = 32$ made calls

    $\frac{32}{320} = \frac{x}{3500}$

    $350 = x$

    Out of 3500, 350 would have made calls.

33. $\frac{2.25}{14} \approx 0.16$

    16%

34. $\frac{0.68}{11.29} \approx 0.06$

    6%

35. $\frac{292}{450} \approx 0.65$

    65%

36. $\frac{11.7}{382} \approx 0.031$

    3.1%

37. $\frac{110.5}{382} \approx 0.289$

    28.9%

38. $0.075(382) = 28.65$

    28.7 kilograms

39. $0.71x = 719$

    $x \approx 1012.68$

    about 1013 people

40. $0.839(1400) = 1174.6$

    about 1175 students

41. $0.772x = 270$

    $x \approx 349.74$

    about 350 students

42. Answers may vary.

43. Both involve setting a ratio equal to another number. In example 4, the other number is also a ratio.

44. C    45. B    46. A

47. $(0.18)(0.31)(0.43)(0.19)(0.44)(0.27)x \approx 0.0005x$

# Chapter 3 *continued*

**48.** $0.0005x = 1000$

$x \approx 2,000,000$ credits

### 3.8 Mixed Review (p. 186)

**49.**
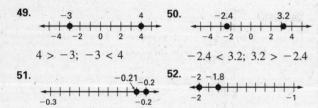
$4 > -3; \; -3 < 4$

**50.**
$-2.4 < 3.2; \; 3.2 > -2.4$

**51.**
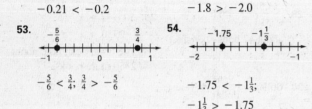
$-0.2 > -0.21;$
$-0.21 < -0.2$

**52.**
$-2.0 < -1.8;$
$-1.8 > -2.0$

**53.**
$-\frac{5}{6} < \frac{3}{4}; \; \frac{3}{4} > -\frac{5}{6}$

**54.**
$-1.75 < -1\frac{1}{3};$
$-1\frac{1}{3} > -1.75$

**55.** $-11a - 16 = 5$

$-11a = 21$

$a \approx -1.91$

Check:

$-11(-1.91) - 16 \overset{?}{=} 5$

$5.01 \approx 5$

**56.** $2(4 - 3x) = -x$

$8 - 6x = -x$

$8 = 5x$

$1.6 = x$

Check:

$2[4 - 3(1.6)] \overset{?}{=} -1.6$

$-1.6 = -1.6$

**57.** $2.62x - 4.03 = 7.65 - 5.34x$

$7.96x - 4.03 = 7.65$

$7.96x = 11.68$

$x \approx 1.47$

Check:

$2.62(1.47) - 4.03 \overset{?}{=} 7.65 - 5.34(1.47)$

$-0.1786 \approx -0.1998$

**58.** $2.6(4.2y - 6.5) = -7.1y - 2.8$

$10.92y - 16.9 = -7.1y - 2.8$

$18.02y - 16.9 = -2.8$

$18.02y = 14.1$

$y \approx 0.78$

Check:

$2.6[4.2(0.78) - 6.5] \overset{?}{=} -7.1(0.78) - 2.8$

$-8.38 \approx -8.34$

**59.**
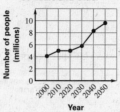

**Projected Number of People 85 Years and Older**

### Quiz 3 (p. 186)

**1.** $3x + 2y = 12$

$2y = -3x + 12$

$y = -\frac{3}{2}x + 6$

$y = -\frac{3}{2}(-2) + 6$

$y = 9$

$y = -\frac{3}{2}(0) + 6$

$y = 6$

$y = -\frac{3}{2}(2) + 6$

$y = 3$

**2.** $\frac{y}{3} - 4 = x$

$\frac{y}{3} = x + 4$

$y = 3x + 12$

$y = 3(-2) + 12$

$y = 6$

$y = 3(0) + 12$

$y = 12$

$y = 3(2) + 12$

$y = 18$

**3.** $\frac{1}{2}(2x + 10y) = 8$

$2x + 10y = 16$

$10y = -2x + 16$

$y = -\frac{1}{5}x + \frac{8}{5}$

$y = -\frac{1}{5}(-2) + \frac{8}{5}$

$y = 2$

$y = -\frac{1}{5}(0) + \frac{8}{5}$

$y = 1\frac{3}{5}$

$y = -\frac{1}{5}(2) + \frac{8}{5}$

$y = 1\frac{1}{5}$

**4.** $V = lwh$

$\frac{V}{lh} = w$

**5.** $A = \frac{1}{2}bh$

$\frac{2A}{b} = h$

**6.** $P = C + rC$

$\frac{P - C}{C} = r$

$\frac{P}{C} - 1 = r$

**7.** $\text{francs} \cdot \dfrac{\text{dollars}}{\text{franc}} = \text{dollars}$

$125 \cdot \dfrac{0.746}{1} = 93.25$

about $93

**8.** $\text{cm} \cdot \dfrac{\text{in.}}{\text{cm}} = \text{in.}$

$50.8 \cdot \dfrac{1}{2.54} = 20$ inches

**9.** $33 \cdot 12 = x$

$396$ mi $= x$

**10.** $0.03x = 42$

$x = 1400$ students

**11.** $\dfrac{210}{525} = \dfrac{x}{900}$

$360$ people $= x$

**12.** $525x = 210$

$x = 0.4$

$40\%$

# Chapter 3 *continued*

## Chapter 3 Extension *(p. 188)*

**1.** inductive reasoning; the conclusion is reached based on observation of a pattern.

**2.** deductive reasoning; the conclusion is reached because the hypothesis is true.

**3.** deductive reasoning; the conclusion is reached because the hypothesis is true.

**4.** inductive reasoning; the conclusion is reached based on observation of a pattern.

**5.** replace $x$ with 1: $(1)^2 = 1$

**6.** Answers may vary.

**7.** 21, 24, 27

**8.** hypothesis: "you add two odd numbers"
conclusion: "the answer will be an even number"

**9.** hypothesis: "you are in Minnesota in January"
conclusion: "you will be cold"

**10.** $(z + 2) + (-2) = z + (2 + (-2))$ Associative Property
$= z + 0$        Inverse Property
$= z$            Identity Property

## Chapter 3 Review *(pp. 190–192)*

**1.** $y - 15 = -4$
$y = 11$

**2.** $-7 + x = -3$
$x = 4$

**3.** $25 = -35 - c$
$60 = -c$
$-60 = c$

**4.** $-11 = z - (-15)$
$-11 = z + 15$
$-26 = z$

**5.** $36 = \dfrac{h}{-12}$
$-432 = h$

**6.** $-\dfrac{2}{3}w = -70$
$w = 105$

**7.** $6m = -72$
$m = -12$

**8.** $\dfrac{y}{4} = \dfrac{15}{6}$
$y = 10$

**9.** $26 - 9p = -1$
$-9p = -27$
$p = 3$

**10.** $\dfrac{4}{5}c - 12 = -32$
$\dfrac{4}{5}c = -20$
$c = -25$

**11.** $\dfrac{y}{4} + 2 = 0$
$\dfrac{y}{4} = -2$
$y = -8$

**12.** $-2(4 - x) - 7 = 5$
$-2(4 - x) = 12$
$-8 + 2x = 12$
$2x = 20$
$x = 10$

**13.** $6r - 2 - 9r = 1$
$-3r - 2 = 1$
$-3r = 3$
$r = -1$

**14.** $16 = 5(1 - x)$
$16 = 5 - 5x$
$11 = -5x$
$-\dfrac{11}{5} = x$

**15.** $-\dfrac{2}{3}(6 - 2a) = 6$
$6 - 2a = -9$
$-2a = -15$
$a = \dfrac{15}{2}$

**16.** $n - 4(1 + 5n) = -2$
$n - 4 - 20n = -2$
$-19n - 4 = -2$
$-19n = 2$
$n = -\dfrac{2}{19}$

**17.** $9z + 24 = -3z$
$24 = -12z$
$-2 = z$

**18.** $12 - 4h = -18 + 11h$
$12 = -18 + 15h$
$30 = 15h$
$2 = h$

**19.** $24a - 8 - 10a = -2(4 - 7a)$
$14a - 8 = -8 + 14a$
$-8 = -8$
true for all values of $a$

**20.** $9(-5 - r) = -10 - 2r$
$-45 - 9r = -10 - 2r$
$-45 = -10 + 7r$
$-35 = 7r$
$-5 = r$

**21.** $\dfrac{2}{3}(3x - 9) = 4(x + 6)$
$2x - 6 = 4x + 24$
$-6 = 2x + 24$
$-30 = 2x$
$-15 = x$

**22.** $6m - 3 = 10 - 6(2 - m)$
$6m - 3 = 10 - 12 + 6m$
$-3 \neq -2$
no solution

**23.** $12\left(\dfrac{5}{12}\right) = x\left(\dfrac{1}{2}\right)$
$5 = \dfrac{1}{2}x$
$10 \text{ km/h} = x$
The speed of the second runner is 10 km/h.

**24.** $12 + 1\dfrac{1}{2}(x) = 6 + 2x$
$12 = 6 + \dfrac{1}{2}x$
$6 = \dfrac{1}{2}x$
$12 = x$
The plants will be the same height in 12 weeks.
Check:

| Week | 0 (now) | 4 | 8 | 12 |
|---|---|---|---|---|
| *Height of first* | 12 | 18 | 24 | 30 |
| *Height of second* | 6 | 14 | 22 | 30 |

equal

**25.** $13.7t - 4.7 = 9.9 + 8.1t$
$5.6t - 4.7 = 9.9$
$5.6t = 14.6$
$t \approx 2.6$

**26.** $4.6(2a + 3) = 3.7a - 0.4$
$9.2a + 13.8 = 3.7a - 0.4$
$5.5a + 13.8 = -0.4$
$5.5a = -14.2$
$a \approx -2.6$

**27.** $-6(5.61x - 3.21) = 4.75$
$-33.66x + 19.26 = 4.75$
$-33.66x = -14.51$
$x \approx 0.4$

# Chapter 3 continued

**28.** $S = 2\pi rh$

$\dfrac{S}{2\pi r} = h$

**29.** $S = \pi s(R + r)$

$\dfrac{S}{\pi s} = R + r$

$\dfrac{S}{\pi s} - R = r$

**30.** $R = \dfrac{pV}{nT}$

$RnT = pV$

$\dfrac{RnT}{V} = p$

**31.** $3x + 2y - 4 = 2(5 - y)$

$3x + 2y - 4 = 10 - 2y$

$3x + 4y - 4 = 10$

$3x + 4y = 14$

$4y = -3x + 14$

$y = -\frac{3}{4}x + \frac{7}{2}$

$y = -\frac{3}{4}(-2) + \frac{7}{2}$
$y = 5$

$y = -\frac{3}{4}(0) + \frac{7}{2}$
$y = 3\frac{1}{2}$

$y = -\frac{3}{4}(1) + \frac{7}{2}$
$y = 2\frac{3}{4}$

$y = -\frac{3}{4}(5) + \frac{7}{2}$
$y = -\frac{1}{4}$

**32.** $\dfrac{210}{40} = \dfrac{x}{55}$

$\$288.75 = x$

**33.** $0.15x = 30$

$x = \$200$

**34.** $470 \cdot \dfrac{1}{7.821} \approx \$60$

## Chapter 3 Test (p. 193)

**1.** $2 + x = 8$
$x = 6$

**2.** $19 = a - 4$
$23 = a$

**3.** $-3y = -18$
$y = 6$

**4.** $\dfrac{x}{4} = 5$
$x = 20$

**5.** $17 = 5 - 3p$
$12 = -3p$
$-4 = p$

**6.** $-\dfrac{3}{4}x - 2 = -8$

$-\dfrac{3}{4}x = -6$

$x = 8$

**7.** $\frac{5}{3}(9 - w) = -10$
$9 - w = -6$
$-w = -15$
$w = 15$

**8.** $-3(x - 2) = x$
$-3x + 6 = x$
$6 = 4x$
$\frac{3}{2} = x$

**9.** $-5r - 6 + 4r = -r + 2$
$-r - 6 = -r + 2$
$-6 \neq 2$
no solution

**10.** $-4y - (5y + 6) = -7y + 3$
$-4y - 5y - 6 = -7y + 3$
$-9y - 6 = -7y + 3$
$-6 = 2y + 3$
$-9 = 2y$
$-4\frac{1}{2} = y$

**11.** $13.2x + 4.3 = 2(2.7x - 3.6)$
$13.2x + 4.3 = 5.4x - 7.2$
$7.8x + 4.3 = -7.2$
$7.8x = -11.5$
$x \approx -1.47$

**12.** $-4(2.5x + 8.7) = (1.4 - 9.2x)(6)$
$-10x - 34.8 = 8.4 - 55.2x$
$-34.8 = 8.4 - 45.2x$
$-43.2 = -45.2x$
$0.96 \approx x$

**13.** $C = 2\pi r$

$\dfrac{C}{2\pi} = r$

**14.** $S = B + \dfrac{1}{2}Pl$

$S - B = \dfrac{1}{2}Pl$

$2(S - B) = Pl$

$\dfrac{2(S - B)}{P} = l$

**15.** $3x + 4y = 15 + 6y$
$3x - 2y = 15$
$-2y = -3x + 15$
$y = \dfrac{3}{2}x - \dfrac{15}{2}$

**16.** $y = \frac{3}{2}(-1) - \frac{15}{2}$
$y = -\frac{3}{2} - \frac{15}{2}$
$y = -\frac{18}{2}$
$y = -9$

$y = \frac{3}{2}(0) - \frac{15}{2}$
$y = -\frac{15}{2}$
$y = -7\frac{1}{2}$

$y = \frac{3}{2}(2) - \frac{15}{2}$
$y = \frac{6}{2} - \frac{15}{2}$
$y = -\frac{9}{2}$
$y = -4\frac{1}{2}$

**17.** $3.5 \cdot \dfrac{3281}{1} = 11,483.5 \text{ ft}$

**18.** C

**19.** $(25x + 5) + 11 = 15x + 12x$
$25x + 16 = 27x$
$16 = 2x$
$8 = x$

8 weeks

**20.**

| Week | 1 | 4 | 8 |
|---|---|---|---|
| *Cousin's earnings* | 30 | 105 | 205 |
| *Your earnings* | 27 | 108 | 216 |

# Chapter 3 *continued*

**21.**
$$A = P + Prt$$
$$414.40 = 400 + 400(r)(1)$$
$$414.40 = 400 + 400r$$
$$14.40 = 400r$$
$$0.036 = r$$
$$3.6\%$$

**22.** $\dfrac{0.75}{108} = \dfrac{2}{x}$

$0.75x = 216$

$x = 288$ envelopes

**23.** $0.08x = 0.94$

$x = \$11.75$

## Chapter 3 Standardized Test *(pp. 194–195)*

**1.** $4 - x = -5$

$-x = -9$

$x = 9$

D

**2.** D

**3.** $2(3x) + 2(x - 1) = 30$

$6x + 2x - 2 = 30$

$8x - 2 = 30$

$8x = 32$

$x = 4$

D

**4.** $9x - 4(3x - 2) = 4$

$9x - 12x + 8 = 4$

$-3x + 8 = 4$

$-3x = -4$

$x = \dfrac{4}{3}$

A

**5.** $-2y + 3(4 - y) = 12 - 5y$

$-2y + 12 - 3y = 12 - 5y$

$-5y + 12 = -5y + 12$

$12 = 12$

E

**6.** $\frac{1}{3}(27x + 18) = 12 + 6(x - 4)$

$9x + 6 = 12 + 6x - 24$

$9x + 6 = 6x - 12$

$3x + 6 = -12$

$3x = -18$

$x = -6$

A

**7.** $0.75t = 12$

$t = 16$

E

**8.** C

**9.** $13.56y - 14.76 = 3(4.12y - 6.72)$

$13.56y - 14.76 = 12.36y - 20.16$

$1.2y - 14.76 = -20.16$

$1.2y = -5.4$

$y = -4.5$

A

**10.** $2(2 - y) = 3x$

$2 - y = \frac{3}{2}x$

$-y = \frac{3}{2}x - 2$

$y = -\frac{3}{2}x + 2$

| | |
|---|---|
| $y = -\frac{3}{2}(-4) + 2$ | $y = -\frac{3}{2}(0) + 2$ |
| $y = 8$ | $y = 2$ |
| $y = -\frac{3}{2}\left(\frac{2}{3}\right) + 2$ | $y = -\frac{3}{2}(3) + 2$ |
| $y = 1$ | $y = -\frac{5}{2}$ |

C

**11.** $0.06x = 0.87$

$x \doteq 14.5$

E

**12.** $2(-5) + t(-5) - 5 = 30$

$-10 - 5t - 5 = 30$

$-15 - 5t = 30$

$-5t = 45$

$t = -9$

A

**13.** C  **14.** D  **15.** B

**16. a.** cost of one child's visits:  $6.5r$

cost of Karim's visits:  $10r$

total cost of family visits:  $29.5r$

**b.** $29.5r = 650$

$r \approx 22$

**c.** $29.5(27) = 796.5$

$796.5 - 650 = 146.5$

Ans: Save $146.50

$29.5(12) = 354$

$650 - 354 = 296$

Ans: Lose $296.00

Karim will save $146.50 if they visit 27 times and he will lose $296 if they visit once per month.

**d.** According to the given information, Karim is able to visit at most twice per month. If he is sure he will be able to visit that often, he will save money by purchasing the annual membership. His decision should be based on a realistic estimate of how often he can get to the rink.

## Cumulative Practice, Chs. 1–3 *(pp. 196–197)*

**1.** $5 + 3(18 \div 6) = 5 + 3(3) = 5 + 9 = 14$

**2.**

$3.6 \div (2 + 4) + 1.2 = 3.6 \div 6 + 1.2 = 0.6 + 1.2 = 1.8$

**3.** $[2(9 - 5) + 1] \cdot 3^2 = [2(4) + 1] \cdot 9 = [8 + 1] \cdot 9$

$= 9 \cdot 9 = 81$

**4.** $|6 - 14| - 2.5(7) = 8 - 17.5 = -9.5$

**5.** $2(3.5) - |13.2 - 7.21| = 7 - 5.99 = 1.01$

**6.** $4(12.7 - 31.2) + 3.6 = 4(-18.5) + 3.6$

$= -74 + 3.6 = -70.4$

# Chapter 3 *continued*

**7.** $(20 - 3 + 7) \div (-8) = (24) \div (-8) = -3$

**8.** $\dfrac{4(9-2)}{7(8)} = \dfrac{4(7)}{7(8)} = \dfrac{1}{2}$

**9.** $\dfrac{2^3 - 2(9)}{-5} = \dfrac{8 - 18}{-5} = \dfrac{-10}{-5} = 2$

**10.** $-20 - 4(-3) = -20 + 12 = -8$

**11.** $10 + 6(5 - 10) = 10 + (-30) = -20$

**12.** $\dfrac{22 - 7}{3} = \dfrac{15}{3} = 5$

**13.** $\dfrac{15(3) - 21}{6} = \dfrac{45 - 21}{6} = \dfrac{24}{6} = 4$

**14.** $\dfrac{10(-2)^2 - 8}{11} = \dfrac{40 - 8}{11} = \dfrac{32}{11}$

**15.** $\dfrac{(10)^2\left(-\frac{1}{2}\right)^2}{4(10) - 10} = \dfrac{100\left(\frac{1}{4}\right)}{30} = \dfrac{25}{30} = \dfrac{5}{6}$

**16.** $4 + 2(2) \overset{?}{=} 12$
$8 \neq 12$
no

**17.** $6(3) - 5 \overset{?}{=} 13$
$13 = 13$
yes

**18.** $3(8) + 7 \overset{?}{=} 4(8) - 2$
$31 \neq 30$
no

**19.** $9 - 4 \overset{?}{<} 6$
$5 < 6$
yes

**20.** $5(1) + 3 \overset{?}{>} 8$
$8 \not> 8$
no

**21.** $9 - 3 \overset{?}{\leq} 3 + 3$
$6 \leq 6$
yes

**22.** $n^3 - 8$

**23.** $4n + 17$

**24.** $2n - 4 = 10$

**25.** $(-3)n > 12$

**26.** $\dfrac{20}{n} < 1$

**27.** $\begin{bmatrix} 8 & -2 \\ 1 & -5 \end{bmatrix} + \begin{bmatrix} -3 & 6 \\ -7 & 0 \end{bmatrix} = \begin{bmatrix} 5 & 4 \\ -6 & -5 \end{bmatrix}$

**28.** $\begin{bmatrix} -5 & 5 \\ -2 & -1 \\ 8 & 4 \end{bmatrix} - \begin{bmatrix} -2 & 0 \\ 4 & -5 \\ 1 & -10 \end{bmatrix} = \begin{bmatrix} -3 & 5 \\ -6 & 4 \\ 7 & 14 \end{bmatrix}$

**29.** $\begin{bmatrix} 23 & -6 & 1 \\ -47 & 15 & 4 \end{bmatrix} + \begin{bmatrix} 3 & 20 & -7 \\ -7 & -18 & 31 \end{bmatrix} =$
$\begin{bmatrix} 26 & 14 & -6 \\ -54 & -3 & 35 \end{bmatrix}$

**30.** $\begin{bmatrix} 4 & -2 & -7 & 1 \end{bmatrix} - \begin{bmatrix} 6 & -4 & -8 & -1 \end{bmatrix} =$
$\begin{bmatrix} -2 & 2 & 1 & 2 \end{bmatrix}$

**31.** $\dfrac{2}{11} \approx 0.18$

**32.** $\dfrac{1}{8} = 0.125$

**33.** $\dfrac{2}{12} = \dfrac{1}{6} \approx 0.17$

**34.** $\dfrac{3}{10} = 0.3$

**35.** $x + 11 = 19$
$x = 8$

**36.** $-7 - x = -2$
$-x = 5$
$x = -5$

**37.** $9b = 135$
$b = 15$

**38.** $35 = 3c - 19$
$54 = 3c$
$18 = c$

**39.** $\dfrac{p}{2} - 9 = -1$
$\dfrac{p}{2} = 8$
$p = 16$

**40.** $4(2x - 9) = 6(10x - 6)$
$8x - 36 = 60x - 36$
$8x = 60x$
$0 = 52x$
$0 = x$

**41.** $3(q - 12) = 5q + 2$
$3q - 36 = 5q + 2$
$-36 = 2q + 2$
$-38 = 2q$
$-19 = q$

**42.** $-\dfrac{3}{4}(2x + 5) = 6$
$2x + 5 = -8$
$2x = -13$
$x = -\dfrac{13}{2}$

**43.** $9(2p + 1) - 3p = 4p - 6$
$18p + 9 - 3p = 4p - 6$
$15p + 9 = 4p - 6$
$11p + 9 = -6$
$11p = -15$
$p = -\dfrac{15}{11}$

**44.** $-3.46y = -5.78$
$y \approx 1.67$

**45.** $4.17n + 3.29 = 2.74n$
$3.29 = -1.43n$
$-2.30 \approx n$

**46.** $4.2(0.3 + x) = 8.7$
$1.26 + 4.2x = 8.7$
$4.2x = 7.44$
$x \approx 1.77$

**47.** $23.5a + 12.5 = 5.2(9.3a - 4.8)$
$23.5a + 12.5 = 48.36a - 24.96$
$12.5 = 24.86a - 24.96$
$37.46 = 24.86a$
$1.51 \approx a$

**48.** $x + \frac{1}{2}y = -3$
$x = -\frac{1}{2}y - 3$

$x = -\frac{1}{2}(-2) - 3$
$x = 1 - 3$
$x = -2$

$x = -\frac{1}{2}(0) - 3$
$x = -3$

$x = -\frac{1}{2}(1.5) - 3$
$x = -0.75 - 3$
$x = -3.75$

$x = -\frac{1}{2}(3) - 3$
$x = -4\frac{1}{2}$

**49.** $2(3y - 1) = 4x$
$4x = 6y - 2$
$x = \frac{3}{2}y - \frac{1}{2}$

$x = \frac{3}{2}(-2) - \frac{1}{2}$
$x = -3\frac{1}{2}$

$x = \frac{3}{2}(0) - \frac{1}{2}$
$x = -\frac{1}{2}$

$x = \frac{3}{2}(1.5) - \frac{1}{2}$
$x = 1\frac{3}{4}$

$x = \frac{3}{2}(3) - \frac{1}{2}$
$x = 4$

**Algebra 1**
Chapter 3 Worked-out Solution Key

# Chapter 3 *continued*

**50.** 
$$-3(x + y) + 4 = 7y$$
$$-3x - 3y + 4 = 7y$$
$$-3x + 4 = 10y$$
$$-3x = 10y - 4$$
$$x = \frac{-10}{3}y + \frac{4}{3}$$

$x = \frac{-10}{3}(-2) + \frac{4}{3}$     $x = \frac{-10}{3}(0) + \frac{4}{3}$

$x = 8$                $x = \frac{4}{3}$

$x = \frac{-10}{3}(1.5) + \frac{4}{3}$     $x = \frac{-10}{3}(3) + \frac{4}{3}$

$x = \frac{-11}{3}$           $x = \frac{-26}{3}$

**51.** $C = \frac{5}{9}(82 - 32)$

$C = 27\frac{7}{9}°C$

**52.**

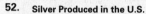

**Silver Produced in the U.S.**

*(graph: Amount of silver (metric tons) vs. Year 1991–1996)*

**53.** 
$$2x + 150 = 750$$
$$2x = 600$$
$$x = 300$$

300 rolls

**54.** 
$$V = \pi r^2 h$$
$$42.4 = \pi(1.5)^2 h$$
$$42.4 = 7.069h$$
$$6.0 \approx h$$

about 6.0 in.

**55.** $0.75(620)(7.827) \approx 3639$ kronor

**56.** $1255\left(\frac{1}{7.827}\right) \approx 160$

about $160

**57.** $(620)(7.827) \approx 4852$ kronor

## Project, Chs. 1–3 *(p. 198)*

**1.** $\frac{0.90}{60}(2) + \frac{1.98}{33}(0.5) = \$0.06$    **2.** $C = .06x$

**3.**

| Number of windows | 5 | 9 | 15 | 25 |
|---|---|---|---|---|
| Cost for one side | $0.30 | $0.54 | $0.90 | $1.50 |
| Cost for both sides | $0.60 | $1.08 | $1.80 | $3.00 |

**4.** $I = 1.1x$

**5.** $I = 1(15) = \$15$

$C = 0.06(30) = \$1.80$

$P = 15 - 1.8 = 13.2$

Your profit would be $13.20.

**6.** $P = 1.1(12) - (0.06)(24) = 11.76;$   $11.76

$P = 1.1(13) - (0.06)(26) = 12.74;$   $12.74

$P = 1.1(14) - (0.06)(28) = 13.72;$   $13.72

$P = 1(17) - (0.06)(34) = 14.96;$   $14.96

$P = 1.1(9) - (0.06)(18) = 8.82;$   $8.82

$P = 1.1(10) - (0.06)(20) = 9.8;$   $9.80

**7.** Answers may vary.

**8.** The rates you decided to charge are not high enough to cover your expenses.

# CHAPTER 4

## Think & Discuss (p. 201)

1. *Sample answer:* Steepness is an important factor in ski slopes and mountain roads. It is helpful for downhill skiers. It is a problem for motorists on icy roads.

2. *Sample answer:* Section 1 is not very steep and slants down from left to right; section 2 is very steep and slants up; section 3 is even steeper and slants up.

## Skill Review (p. 202)

1. $50\% = 0.50 = \frac{1}{2}$

2. $75\% = 0.75 = \frac{75}{100} = \frac{3}{4}$

3. $1\% = 0.01 = \frac{1}{100}$

4. $20\% = 0.20 = \frac{20}{100} = \frac{1}{5}$

5.

| x | y |
|---|---|
| 0 | 70 |
| 1 | 75 |
| 2 | 80 |
| 3 | 85 |
| 5 | 95 |
| 10 | 120 |

6.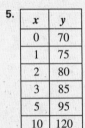

7. domain: $x \geq 0$; range: $y \geq 70$

8. $\dfrac{x - y}{2} = \dfrac{-3 - (-1)}{2}$

$= \dfrac{-3 + 1}{2}$

$= \dfrac{-2}{2}$

$= -1$

9. $\dfrac{x + 2y}{x} = \dfrac{6 + 2(3)}{6}$

$= \dfrac{6 + 6}{6}$

$= \dfrac{12}{6}$

$= 2$

## Lesson 4.1

### 4.1 Guided Practice (p. 206)

1. The x-coordinate is 4.   2. true

3. $A(1, 1), B(-2, 1), C(0, -2)$

4.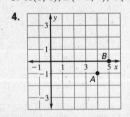

5.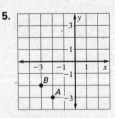

6. The point $(-2, 5)$ lies in Quadrant II.

7.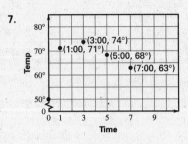

8.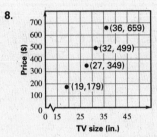

9. *Sample answer:* yes, the amount spent on snowmobiles increased but not rapidly.

### 4.1 Practice and Applications (pp. 206–208)

10. $A(-3, 2), B(-1, -2), C(2, 0), D(2, 3)$

11. $A(2, 4), B(0, -1), C(-1, 0), D(-2, -1)$

12. $A(-3, 0), B(-4, -3), C(0, 2), D(1, -2)$

13.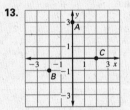

14.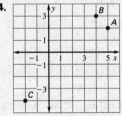

15.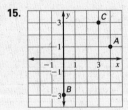

16.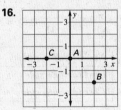

17.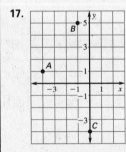

18.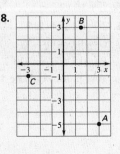

# Chapter 4 continued

19. Quadrant IV    20. Quadrant II    21. Quadrant I

22. Quadrant IV    23. Quadrant III    24. Quadrant I

25. Quadrant III    26. Quadrant II

27. On the Weight vs. Length graph, the horizontal axis is pounds and the vertical axis is inches.

28. The coordinates are approximately (4000, 210).

29. C    30. $w$ is 2010 and $G$ is 29.

31. Gas mileage decreases as weight increases.

32. As the length of a car increases the gas mileage is expected to decrease because more weight is added to the car.

33. Answers may vary.    34. Answers may vary.

35.
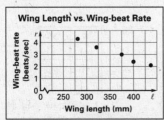

36. The slowest wing-beat rate is the Great Egret with 2.1 beats per second, located at the bottom of the scatter plot. The fastest wing-beat rate is the Velvet Scoter with 4.3 beats per second, located at the top of the scatter plot.

37. The smaller the wing length, the faster the wing-beat rate.

38. **a.** The number of rolls of film developed increased for every year of the Winter Olympics.

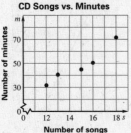

**b.** There will be approximately 83,000 rolls of film developed for the Winter Olympics in 2002. This prediction was made because if you sketch a line through the points in the scatter plot and extend it to $x = 22$, the $y$-coordinate is about 83, 000.

39. The data suggests that the more the number of songs recorded on a CD, the more minutes of music.

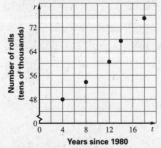

40. The data suggests that the more songs on a CD the less the minutes of music recorded.

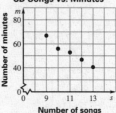

41. The relationship between the number of songs and the number of minutes on a CD depends on data that is collected. Testing whether there is a relationship requires more data to be collected. (Explanations may vary.)

## 4.1 Mixed Review (p. 208)

42. $3x + 9 = 3(2) + 9 = 6 + 9 = 15$

43. $-(y + 2) = 13 - (4 + 2) = 13 - 6 = 7$

44. $4.2t + 17.9 = 4.2(3) + 17.9 = 12.6 + 17.9 = 30.5$

45. $-3x - 9y = (-3)(-2) - (9)(-1)$
    $= 6 - (-9) = 6 + 9 = 15$

46. $x^2 - 3 = (4)^2 - 3 = 16 - 3 = 13$

47. $12 + y^3 = 12 + (3)^3 = 12 + 27 = 39$

48. $|-2.6| = 2.6$    49. $|1.07| = 1.07$

50. $\left|\frac{9}{10}\right| = \frac{9}{10}$    51. $\left|\frac{-2}{3}\right| = \frac{2}{3}$    52. $\frac{345}{500} = \frac{69}{100}$ or 69%

## Technology Activity 4.1 (p. 209)

1. As age increases, time decreases.

2.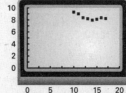

3. From ages 10–14, as age increases, time decreases. From ages 14–16, time increases, then decreases at age 17.

4. The scatter plot for the boys' times is more linear than the scatter plot for the girls' times.

## Lesson 4.2

### 4.2 Guided Practice (p. 214)

1. An ordered pair that makes an equation in two variables true is called a solution of an equation.

2. Choosing different values of $x$ will give solutions that lie on the same line because for every $x$-value that is chosen, there is a corresponding $y$-value. So, the graph of the line has infinitely many solutions.

3. False. The graph of $x = 3$ consists of all points with $x$-coordinate 3. This is a vertical line.

# Chapter 4 *continued*

**4.**

| x | y |
|---|---|
| 1 | −2 |
| 2 | 0 |
| 3 | 2 |

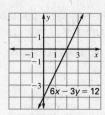

$6x − 3y = 12$

**5.**

| x | y |
|---|---|
| 1.5 | −1 |
| 1.5 | 0 |
| 1.5 | 1 |

$x = 1.5$

**6.**

| x | y |
|---|---|
| −2 | −2 |
| 0 | −2 |
| 2 | −2 |

$y = −2$

**7.** no  **8.** yes  **9.** no  **10.** no

**11.** 35.18 million households

### 4.2 Practice and Applications (pp. 214–217)

**12. a.** $(2, −1)$

$$3x − 4y = 10$$
$$3(2) − 4(−1) \overset{?}{=} 10$$
$$6 − (−4) \overset{?}{=} 10$$
$$10 = 10$$

**b.** $(−1, 2)$

$$3x − 4y = 10$$
$$3(−1) − 4(2) \overset{?}{=} 10$$
$$−3 − 8 \overset{?}{=} 10$$
$$−11 ≠ 10$$

**13. a.** $(5, 0)$

$$y \overset{?}{=} 5$$
$$0 ≠ 5$$

**b.** $(0, 5)$

$$y \overset{?}{=} 5$$
$$5 = 5$$

**14. a.** $(0, 14)$

$$x \overset{?}{=} 0$$
$$0 = 0$$

**b.** $(14, 0)$

$$x \overset{?}{=} 0$$
$$14 ≠ 0$$

**15.** no  **16.** yes  **17.** yes  **18.** no

**19.** no  **20.** no

**21.**

| x | y |
|---|---|
| −1 | −8 |
| 0 | −5 |
| 1 | −2 |

$(−1, −8)$, $(0, −5)$, $(1, −2)$

**22.**

| x | y |
|---|---|
| −1 | 11 |
| 0 | 7 |
| 1 | 3 |

$(−1, 11)$, $(0, 7)$, $(1, 3)$

**23.**

| x | y |
|---|---|
| −2 | −2 |
| 0 | −6 |
| 2 | −10 |

$(−2, −2)$, $(0, −6)$, $(2, −10)$

**24.**

| x | y |
|---|---|
| 2 | −1 |
| 2 | 0 |
| 2 | 1 |

$(2, −1)$, $(2, 0)$, $(2, 1)$

**25.**

| x | y |
|---|---|
| $\frac{1}{2}$ | −1 |
| $\frac{1}{2}$ | 0 |
| $\frac{1}{2}$ | 1 |

$\left(\frac{1}{2}, −1\right)$, $\left(\frac{1}{2}, 0\right)$, $\left(\frac{1}{2}, 1\right)$

**26.**

| x | y |
|---|---|
| −1 | −6 |
| 0 | −6 |
| 1 | −6 |

$(−1, −6)$, $(0, −6)$, $(1, −6)$

**27.**

| x | y |
|---|---|
| −1 | 3 |
| 0 | 2 |
| 1 | 1 |

$(−1, 3)$, $(0, 2)$, $(1, 1)$

**28.**

| x | y |
|---|---|
| −1 | −21 |
| 0 | −3 |
| 1 | 15 |

$(−1, −21)$, $(0, −3)$, $(1, 15)$

**29.**

| x | y |
|---|---|
| −2 | −8 |
| 0 | −4 |
| 2 | 0 |

$(−2, −8)$, $(0, −4)$, $(2, 0)$

**30.** $−3x + y = 12$

$$y = 3x + 12$$

**31.** $2x + 3y = 6$

$$3y = −2x + 6$$
$$y = −\frac{2}{3}x + 2$$

**32.** $x + 4y = 48$

$$4y = −x + 48$$
$$y = −\frac{1}{4}x + 12$$

**33.** $5x + 5y = 19$

$$5y = −5x + 19$$
$$y = −x + \frac{19}{5}$$

**34.** $\frac{1}{2}x + \frac{5}{2}y = 1$

$$\frac{5}{2}y = −\frac{1}{2}x + 1$$
$$y = −\frac{1}{5}x + \frac{2}{5}$$

**35.** $−x − y = 5$

$$−y = x + 5$$
$$y = −x − 5$$

**36.**

| x | y |
|---|---|
| −1 | 5 |
| 0 | 4 |
| 1 | 3 |

$y = −x + 4$

**37.**

| x | y |
|---|---|
| 0 | 5 |
| 1 | 3 |
| 2 | 1 |

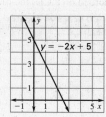

$y = −2x + 5$

**Algebra 1**
Chapter 4  Worked-out Solution Key

# Chapter 4 *continued*

**38.**

| x | y |
|---|---|
| −1 | −4 |
| 0 | −3 |
| 1 | −2 |

$y = -(3 - x)$

**39.**

| x | y |
|---|---|
| 3 | 6 |
| 4 | 4 |
| 5 | 2 |

$y = -2(x - 6)$

**40.**

| x | y |
|---|---|
| −2 | −4 |
| −1 | −1 |
| 0 | 2 |

$y = 3x + 2$

**41.**

| x | y |
|---|---|
| −1 | −5 |
| 0 | −1 |
| 1 | 3 |

$y = 4x - 1$

**42.**

| x | y |
|---|---|
| −1 | $\frac{2}{3}$ |
| 0 | 2 |
| 1 | $\frac{10}{3}$ |

$y = \frac{4}{3}x + 2$

**43.**

| x | y |
|---|---|
| −1 | $\frac{7}{4}$ |
| 0 | 1 |
| 1 | $\frac{1}{4}$ |

$y = -\frac{3}{4}x + 1$

**44.**

| x | y |
|---|---|
| 9 | −2 |
| 9 | 0 |
| 9 | 2 |

$x = 9$

**45.**

| x | y |
|---|---|
| −2 | −1 |
| 0 | −1 |
| 2 | −1 |

$y = -1$

**46.**

| x | y |
|---|---|
| −2 | 0 |
| 0 | 0 |
| 2 | 0 |

$y = 0$

**47.**

| x | y |
|---|---|
| −1 | 4 |
| 0 | 1 |
| 1 | −2 |

$y = -3x + 1$

**48.**

| x | y |
|---|---|
| 0 | −1 |
| 0 | 0 |
| 0 | 1 |

$x = 0$

**49.**

| x | y |
|---|---|
| −1 | $-\frac{7}{2}$ |
| 0 | −3 |
| 1 | $-\frac{5}{2}$ |

$y = \frac{1}{2}x - 3$

**50.**

| x | y |
|---|---|
| $-\frac{1}{2}$ | 1 |
| 0 | $\frac{1}{2}$ |
| $\frac{1}{2}$ | 0 |

$y = -x + \frac{1}{2}$

**51.**

| x | y |
|---|---|
| −2 | −3 |
| 0 | −4 |
| 2 | −5 |

$y = -\frac{1}{2}x - 4$

# Chapter 4 *continued*

**52.** B   **53.** D   **54.** A   **55.** C

**56.** With at least three different ordered pairs, a pattern for a graph is usually recognized.

**57.** 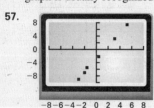   **58.** $(-1.6, -5.5)$

**59.**
$$y = 2x - 3$$
$$-5.5 \overset{?}{=} 2(-1.6) - 3$$
$$-5.5 \overset{?}{=} -3.2 - 3$$
$$-5.5 \neq -6.2$$

**60.** 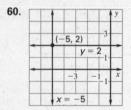   **61.**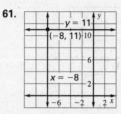

**62.** 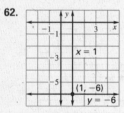   **63.**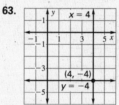

**64.** 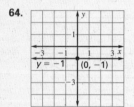   **65.**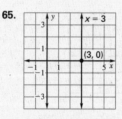

**66.**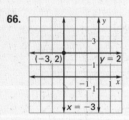

**67.**
$$20x + 10y = 300$$
$$10y = -20x + 300$$
$$y = -2x + 30$$

**68.**

| x | y |
|---|---|
| 5 | 20 |
| 10 | 10 |
| 15 | 0 |

**69.** Yes, the line passes through the points.

**70.**
$$20x + 10y = 300$$
$$20x + 10(0) = 300$$
$$20x = 300$$
$$x = 15 \text{ lawns}$$

**71. Verbal Model:**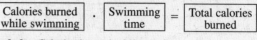

**Labels:** Calories burned while running = 7.1 (calories/minute)

Running time = $x$ (minutes)

Calories burned while swimming = 10.1 (calories/minute)

Swimming time = $y$ (minutes)

Total calories burned = 800 (calories)

**Algebraic Model:** $7.1x + 10.1y = 800$

**72.**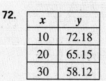

| x | y |
|---|---|
| 10 | 72.18 |
| 20 | 65.15 |
| 30 | 58.12 |

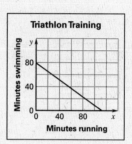

**73.** Approximately 47.57 minutes

**74.** Neither. The number of books read by Avery may not be the same amount every year.

**75.** Horizontal. The number of senators in the United States Congress does not change from year to year.

**76.** D   **77.** C   **78.** A

**79.**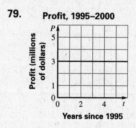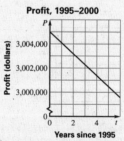

**80.** Sample answer: The first graph might be used to imply that profits remained steady. (The actual decrease was about 0.030% per year.) The second graph might be used to imply that profits decreased steeply. The second graph is more likely to be misinterpreted.

### 4.2 Mixed Review (p. 217)

**81.** $5 + 2 + (-3) = 7 + (-3) = 4$

**82.** $-6 + (-14) + 8 = -20 + 8 = -12$

**83.** $-18 + (-10) + (-1) = -28 + (-1) = -29$

# Chapter 4 *continued*

**84.** $\frac{-1}{3} + 6 + \frac{1}{3} = 0 + 6 = 6$

**85.** $\frac{4}{7} + \frac{3}{7} - 1 = \frac{7}{7} - 1 = 0$

**86.** $\frac{-7}{9} + \frac{1}{3} + 2 = \frac{-4}{9} + 2 = \frac{14}{9}$

**87.** $\begin{bmatrix} 1 + 15 & 6 + -3 \\ -4 + 0 & 2 + 16 \end{bmatrix} = \begin{bmatrix} 16 & 3 \\ -4 & 18 \end{bmatrix}$

**88.** $\begin{bmatrix} 2 + -14 & 5 + -5 \\ 3 + 12 & 10 + 7 \end{bmatrix} = \begin{bmatrix} -12 & 0 \\ 15 & 17 \end{bmatrix}$

**89.** $\begin{bmatrix} 4 + -20 & 10 + 40 \\ -1 + -8 & 9 + 10 \end{bmatrix} = \begin{bmatrix} -16 & 50 \\ -9 & 19 \end{bmatrix}$

**90.** $\begin{bmatrix} 5 + -1 & -2 + -16 \\ 5 + 11 & -1 + -3 \end{bmatrix} = \begin{bmatrix} 4 & -18 \\ 16 & -4 \end{bmatrix}$

**91.** $-2z = -26$

$\frac{-2z}{-2} = \frac{-26}{-2}$

$z = 13$

**92.** $9x = 3$

$\frac{9x}{9} = \frac{3}{9}$

$x = \frac{1}{3}$

**93.** $6p = -96$

$\frac{6p}{6} = \frac{-96}{6}$

$p = -16$

**94.** $24 = 8c$

$\frac{24}{8} = \frac{8c}{8}$

$3 = c$

**95.** $\frac{2}{3}t = -10$

$\left(\frac{3}{2}\right)\left(\frac{2}{3}t\right) = (-10)\left(\frac{3}{2}\right)$

$t = -15$

**96.** $\frac{-p}{7} = -9$

$\left(\frac{-7}{1}\right)\left(\frac{-p}{7}\right) = (-9)\left(\frac{-7}{1}\right)$

$p = 63$

**97.** $\frac{n}{15} = \frac{3}{5}$

$\left(\frac{15}{1}\right)\left(\frac{n}{15}\right) = \left(\frac{3}{5}\right)\left(\frac{15}{1}\right)$

$n = 9$

**98.** $\frac{c}{6} = \frac{2}{3}$

$\left(\frac{6}{1}\right)\left(\frac{c}{6}\right) = \left(\frac{2}{3}\right)\left(\frac{6}{1}\right)$

$c = 4$

## Lesson 4.3

### 4.3 Guided Practice (p. 221)

**1.** 2 is the $y$-intercept.

$y = 2x + 2$

$y = 2(0) + 2$

$y = 2$

**2.** Two points are needed to determine a line.

**3.** A line that has no $x$-intercept is a horizontal line where $y = a$ for any number $a$ except 0.

**4.** $y = 2x + 20$

$0 = 2x + 20$

$-20 = 2x$

$-10 = x$

**5.** $y = 0.1x + 0.3$

$0 = 0.1x + 0.3$

$-0.3 = 0.1x$

$-3 = x$

**6.** $y = x - \frac{1}{4}$

$0 = x - \frac{1}{4}$

$\frac{1}{4} = x$

**7.** $x$-intercept

$y = x + 2$

$0 = x + 2$

$-2 = x$

$y$-intercept

$y = x + 2$

$y = 0 + 2$

$y = 2$

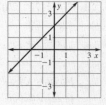

**8.** $x$-intercept

$y - 2x = 3$

$0 - 2x = 3$

$-2x = 3$

$x = \frac{-3}{2}$

$y$-intercept

$y - 2x = 3$

$y - 2(0) = 3$

$y = 3$

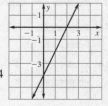

**9.** $x$-intercept

$2x - y = 4$

$2x - 0 = 4$

$2x = 4$

$x = 2$

$y$-intercept

$2x - y = 4$

$2(0) - y = 4$

$-y = 4$

$y = -4$

**10.** $x$-intercept

$3y = -6x + 3$

$3(0) = -6x + 3$

$6x = 3$

$x = \frac{1}{2}$

$y$-intercept

$3y = -6x + 3$

$3y = -6(0) + 3$

$3y = 3$

$y = 1$

**11.** $x$-intercept

$5y = 5x + 15$

$5(0) = 5x + 15$

$-15 = 5x$

$-3 = x$

$y$-intercept

$5y = 5x + 15$

$5y = 5(0) + 15$

$5y = 15$

$y = 3$

**12.** $x$-intercept

$x - y = 1$

$x - 0 = 1$

$x = 1$

$y$-intercept

$x - y = 1$

$0 - y = 1$

$-y = 1$

$y = -1$

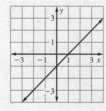

**13.** $200x + 100y = 2000$
$\qquad 2x + y = 20$

*x*-intercept $\qquad$ *y*-intercept
$2x + y = 20 \qquad 2x + y = 20$
$2x + 0 = 20 \qquad 2(0) + y = 20$
$\qquad 2x = 20 \qquad\qquad y = 20$
$\qquad\quad x = 10$

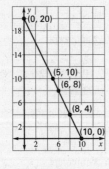

| Possible prices to raise $2000 | |
|---|---|
| **Adult** | **Student** |
| $0.00 | $20.00 |
| $5.00 | $10.00 |
| $6.00 | $8.00 |
| $8.00 | $4.00 |
| $10.00 | $0.00 |

One reasonable price to charge is $8 for adults and $4 for students.

### 4.3 Practice and Applications (pp. 221–223)

**14.** *x*-intercept: 2
*y*-intercept: 3

**15.** *x*-intercept: $-2$
*y*-intercept: 4

**16.** *x*-intercept: $-4$
*y*-intercept: $-1$

**17.** $x + 3y = 5$
$\quad x + 3(0) = 5$
$\qquad\qquad x = 5$

**18.** $x - 2y = 6$
$\quad x - 2(0) = 6$
$\qquad\quad x = 6$

**19.** $2x + 2y = -10$
$\quad 2x + 2(0) = -10$
$\qquad\quad 2x = -10$
$\qquad\qquad x = -5$

**20.** $3x + 4y = 12$
$\quad 2x + 4(0) = 12$
$\qquad\quad 3x = 12$
$\qquad\qquad x = 4$

**21.** $5x - y = 45$
$\quad 5x - 0 = 45$
$\qquad 5x = 45$
$\qquad\quad x = 9$

**22.** $-x + 3y = 27$
$\quad -x + 3(0) = 27$
$\qquad\qquad -x = 27$
$\qquad\qquad\quad x = -27$

**23.** $-7x - 3y = 42$
$\quad -7x - 3(0) = 42$
$\qquad\qquad -7x = 42$
$\qquad\qquad\quad x = -6$

**24.** $2x + 6y = -24$
$\quad 2x + 6(0) = -24$
$\qquad\quad 2x = -24$
$\qquad\qquad x = -12$

**25.** $12x - 20y = 60$
$\quad -12x - 20(0) = 60$
$\qquad\quad -12x = 60$
$\qquad\qquad x = -5$

**26.** $y = -2x + 5$
$\quad y = -2(0) + 5$
$\quad y = 5$

**27.** $y = 3x - 4$
$\quad y = 3(0) - 4$
$\quad y = -4$

**28.** $y = 8x + 27$
$\quad y = 8(0) + 27$
$\quad y = 27$

**29.** $y = 7x - 15$
$\quad y = 7(0) - 15$
$\quad y = -15$

**30.** $4x - 5y = -35$
$\quad 4(0) - 5y = -35$
$\qquad\quad -5y = -35$
$\qquad\qquad y = 7$

**31.** $6x - 9y = 72$
$\quad 6(0) - 9y = 72$
$\qquad\quad -9y = 72$
$\qquad\qquad y = -8$

**32.** $3x + 12y = -84$
$\quad 3(0) + 12y = -84$
$\qquad\quad 12y = -84$
$\qquad\qquad y = -7$

**33.** $-x + 1.7y = 5.1$
$\quad 0 + 1.7y = 5.1$
$\qquad\qquad y = 3$

**34.** $2x - 6y = -18$
$\quad 2(0) - 6y = -18$
$\qquad\quad -6y = -18$
$\qquad\qquad y = 3$

**35.**

**36.**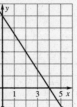

**37.**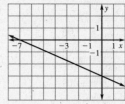

**38.**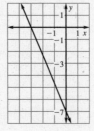

**39.**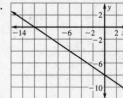

**40.**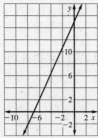

**41.** A $\qquad$ **42.** C $\qquad$ **43.** B

**44.** *x*-intercept $\qquad$ *y*-intercept
$\quad y = x + 2 \qquad y = x + 2$
$\quad 0 = x + 2 \qquad y = 0 + 2$
$\quad -2 = x \qquad\qquad y = 2$

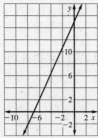

# Chapter 4 *continued*

**45.** *x*-intercept     *y*-intercept

$y = x - 3$        $y = x - 3$

$0 = x - 3$        $y = 0 - 3$

$3 = x$            $y = -3$

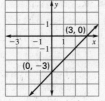

**46.** *x*-intercept     *y*-intercept

$y = 4x + 8$      $y = 4x + 8$

$0 = 4x + 8$      $y = 4(0) + 8$

$-8 = 4x$        $y = 8$

$-2 = x$

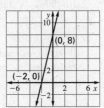

**47.** *x*-intercept     *y*-intercept

$y = -6 + 3x$    $y = -6 + 3x$

$0 = -6 + 3x$    $y = -6 + 3(0)$

$6 = 3x$          $y = -6$

$2 = x$

**48.** *x*-intercept     *y*-intercept

$y = 5x + 15$    $y = 5x + 15$

$0 = 5x + 15$    $y = 5(0) + 15$

$-15 = 5x$      $y = 15$

$-3 = x$

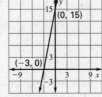

**49.** *x*-intercept     *y*-intercept

$2x + 4y = 16$    $2x + 4y = 16$

$2x + 4(0) = 16$   $2(0) + 4y = 16$

$2x = 16$         $4y = 16$

$x = 8$           $y = 4$

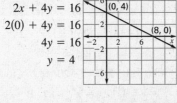

**50.** *x*-intercept

$-4x + 3y = 24$

$-4x + 3(0) = 24$

$-4x = 24$

$x = -6$

*y*-intercept

$-4x + 3y = 24$

$-4(0) + 3y = 24$

$3y = 24$

$y = 8$

**51.** *x*-intercept     *y*-intercept

$x - 7y = 21$    $x - 7y = 21$

$x - 7(0) = 21$   $0 - 7y = 21$

$x = 21$       $-7y = 21$

                $y = -3$

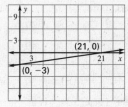

**52.** *x*-intercept     *y*-intercept

$6x - y = 36$    $6x - y = 36$

$6x - 0 = 36$    $6(0) - y = 36$

$6x = 36$       $-y = 36$

$x = 6$          $y = -36$

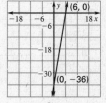

**53.** *x*-intercept

$2x + 9y = -36$

$2x + 9(0) = -36$

$2x = -36$

$x = -18$

*y*-intercept

$2x + 9y = -36$

$2(0) + 9y = -36$

$9y = -36$

$y = -4$

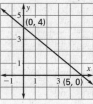

**54.** *x*-intercept     *y*-intercept

$4x + 5y = 20$    $4x + 5y = 20$

$4x + 5(0) = 20$   $4(0) + 5y = 20$

$4x = 20$        $5y = 20$

$x = 5$           $y = 4$

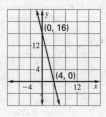

**55.** *x*-intercept

$0.5y = -2x + 8$

$(0.5)(0) = -2x + 8$

$2x = 8$

$x = 4$

*y*-intercept

$0.5y = -2x + 8$

$0.5y = -2(0) + 8$

$0.5y = 8$

$y = 16$

**56.** false

$3x + 5y = 30$

$3(0) + 5y = 30$

$5y = 30$

$y = 6$

The *y*-intercept is 6.

**57.** true

$3x + 5y = 30$

$3x + 5(0) = 30$

$3x = 30$

$x = 10$

**58.** false

$$3x + 5y = 30$$
$$3(3) + 5(5) \stackrel{?}{=} 30$$
$$9 + 25 \stackrel{?}{=} 30$$
$$34 \neq 30$$

**59.** False, $x = 4$ is a vertical line.

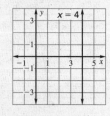

**60.**

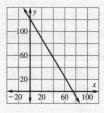

**61.** $x$-intercept

$$8x + 5y = 6$$
$$8x + 5(0) = 600$$
$$8x = 600$$
$$x = 75$$

The $x$-intercept is the number of adult tickets sold to reach the goal when no student tickets were sold.

**62.** $y$-intercept

$$8x + 5y = 600$$
$$8(0) + 5y = 600$$
$$5y = 600$$
$$y = 120$$

The $y$-intercept is the number of student tickets sold to reach the goal when no adult tickets were sold.

**63.**

| Possible number of tickets | |
|---|---|
| **Adult** | **Student** |
| 40 | 56 |
| 50 | 40 |
| 60 | 24 |

**64.** Let $x$ = hours running and $y$ = hours walking.
$$8x + 4y = 26.2$$

**65.**

Running Time vs. Walking Time

| Possible running and walking times | |
|---|---|
| **Running** | **Walking** |
| 1 | 4.55 |
| 2 | 2.55 |
| 3 | 0.55 |

**66.** 2.775 hours

**67.**

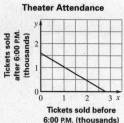

Theater Attendance

**68.** 1604 people attended after 6:00 P.M., which is the $y$-intercept.

**69. a.** $y$-intercept

$$y = -6.61x + 229$$
$$y = -6.61(0) + 229$$
$$y = 229$$

The $y$-intercept represents the number of railroad employees (in thousands) in 1989.

**b.** $x$-intercept

$$y = -6.61x + 229$$
$$0 = -6.61x + 229$$
$$6.61x = 229$$
$$x \approx 34.64$$

The $x$-intercept represents the year in which there will be no railroad employees.

**c.** Approximately 189,340 railroad employees

**d.** The line in the graph will only be a good model for the years shown because the more data represented the more accurate the model, and after 2024 the number of employees would be negative.

**70.** Numbers that are divisible by both 8 and 6 will produce $x$- and $y$-intercepts that are integers.

### 4.3 Mixed Review (p. 223)

**71.** $5 - 9 = -4$      **72.** $17 - 6 = 11$

**73.** $-8 - 9 = -17$

**74.** $|8| - 12.6 = 8 - 12.6 = -4.6$

**75.** $\frac{-2}{3} - \left(\frac{-7}{3}\right) = \frac{-2}{3} + \frac{7}{3} = \frac{5}{3}$    **76.** $13.8 - 6.9 = 6.9$

**77.** $7 - |-1| = 7 - 1 = 6$

**78.** $-4.1 - (-5.1) = -4.1 + 5.1 = 1$

**79.** $54 \div 9 = (54)\left(\frac{1}{9}\right) = 6$

**80.** $-72 \div 8 = (-72)\left(\frac{1}{8}\right) = -9$

**81.** $12 \div \left(\frac{-1}{5}\right) = 12 \cdot (-5) = -60$

**82.** $3 \div \frac{1}{4} = 3 \cdot 4 = 12$

**83.** $26 \div (-13) = (26)\left(\frac{-1}{13}\right) = -2$

**84.** $-1 \div 8 = (-1)\left(\frac{1}{8}\right) = \frac{-1}{8}$

**85.** $-20 \div \left(-2\frac{1}{2}\right) = (-20)\left(\frac{-2}{5}\right) = 8$

**86.** $\frac{1}{8} \div \frac{1}{2} = \left(\frac{1}{8}\right)(2) = \frac{1}{4}$    **87.** B

### Quiz 1 (p. 224)

**1.**

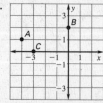

**2.**

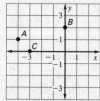

**3.**

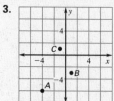

**4.**

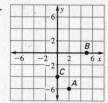

# Chapter 4 *continued*

**5.**

| x | y |
|---|---|
| −2 | −10 |
| 0 | −6 |
| 2 | −2 |

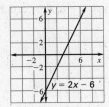

$y = 2x - 6$

**6.**

| x | y |
|---|---|
| −1 | −5 |
| 0 | −1 |
| 1 | 3 |

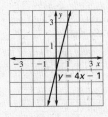

$y = 4x - 1$

**7.**

| x | y |
|---|---|
| −1 | 8 |
| 0 | 2 |
| 1 | −4 |

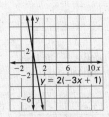

$y = 2(-3x + 1)$

**8.**

| x | y |
|---|---|
| 3 | −1 |
| 3 | 0 |
| 3 | 1 |

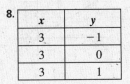

$x = 3$

**9.**

| x | y |
|---|---|
| −1 | 15 |
| 0 | 12 |
| 1 | 9 |

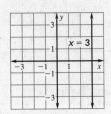

$y = -3(x - 4)$

**10.**

| x | y |
|---|---|
| −1 | −5 |
| 0 | −5 |
| 1 | −5 |

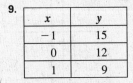

$y = -5$

**11.** x-intercept    y-intercept

$y = 4 - x$    $y = 4 - x$
$0 = 4 - x$    $y = 4 - 0$
$x = 4$    $y = 4$

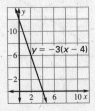

(0, 4)

(4, 0)

**12.** x-intercept    y-intercept

$y = -5 + 2x$    $y = -5 + 2x$
$0 = -5 + 2x$    $y = -5 + 2(0)$
$5 = 2x$    $y = -5$
$\frac{5}{2} = x$

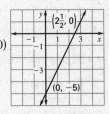

$\left(2\frac{1}{2}, 0\right)$

(0, −5)

**13.** x-intercept    y-intercept

$y = 3x + 12$    $y = 3x + 12$
$0 = 3x + 12$    $y = 3(0) + 12$
$-12 = 3x$    $y = 12$
$-4 = x$

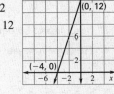

(0, 12)

(−4, 0)

**14.** x-intercept    y-intercept

$3x + 3y = 27$    $3x + 3y = 27$
$3x + 3(0) = 27$    $3(0) + 3y = 27$
$3x = 27$    $3y = 27$
$x = 9$    $y = 9$

(0, 9)

(9, 0)

**15.** x-intercept

$-6x + y = -3$
$-6x + 0 = -3$
$x = \frac{1}{2}$

y-intercept

$-6x + y = -3$
$-6(0) + y = -3$
$y = -3$

$\left(\frac{1}{2}, 0\right)$

(0, −3)

**16.** x-intercept    y-intercept

$y = 10x + 50$    $y = 10x + 50$
$0 = 10x + 50$    $y = 10(0) + 50$
$-50 = 10x$    $y = 50$
$-5 = x$

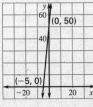

(0, 50)

(−5, 0)

## Math and History (p. 224)

**1.** yes    **2.** yes    **3.** no    **4.** no

## Lesson 4.4

**Developing Concepts Activity 4.4 (p. 225)**
**Drawing Conclusions**

1. The slope decreases. If the slope is $\frac{7}{12}$, and the run is increased to 14, then the slope is $\frac{1}{2}$, which is less than $\frac{7}{12}$.

2. The slope increases. If the slope is $\frac{7}{12}$, and the rise is increased to 8, then the slope is $\frac{2}{3}$, which is greater than $\frac{7}{12}$.

3. The rise and run are equal.

# Chapter 4 *continued*

**4.** Yes. The rise is larger than the run.

**5.** No. The rise is smaller than the run.

## *4.4 Guided Practice (p. 230)*

**1.** Ramps will vary. The slope of a ramp describes its steepness.

**2.** The subtraction order is different. The correct slope is
$\frac{4-2}{1-3} = -1$.

**3.** Denominator becomes zero, so slope is undefined.

**4.** The slope is negative because the line falls from left to right.

**5.** $m = \frac{2-0}{1-0} = 2$    **6.** $m = \frac{-1-0}{-1-0} = \frac{-1}{-1} = 1$

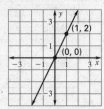    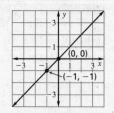

**7.** $m = \frac{1-2}{2-1} = \frac{-1}{1} = -1$    **8.** $m = \frac{4-2}{1-3} = \frac{2}{-2} = -1$

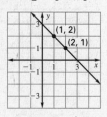    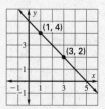

**9.** $m = \frac{5-2}{3-2} = \frac{3}{1} = 3$    **10.** $m = \frac{6-(-2)}{1-(-3)} = \frac{8}{4} = 2$

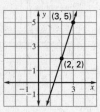    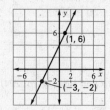

**11.**    $m = \frac{y_2 - y_1}{x_2 - x_1}$

$-3 = \frac{y-3}{4-0}$

$-3 = \frac{y-3}{4}$

$-12 = y - 3$

$-9 = y$

## *4.4 Practice and Applications (pp. 230–233)*

**12.** 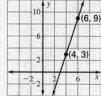    **13.**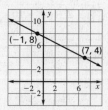

Positive. The line rises to the right.

Negative. The line falls to the right.

**14.** 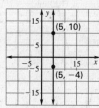    **15.**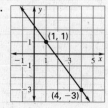

Undefined. The line is vertical.

Negative. The line falls to the right.

**16.** 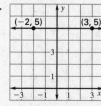    **17.**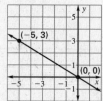

Zero. The line is horizontal.

Negative. The line falls to the right.

**18.** 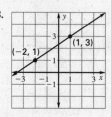    **19.**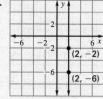

Positive. The line rises to the right.

Undefined. The line is vertical.

**20.** $m = \frac{3-5}{2-4} = \frac{-2}{-2} = 1$    **21.** $m = \frac{2-5}{5-1} = -\frac{3}{4}$

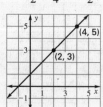    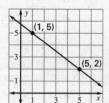

# Chapter 4 continued

**22.** $m = \dfrac{0 - 3}{-3 - 2} = \dfrac{-3}{-5} = \dfrac{3}{5}$ **23.** $m = \dfrac{0 - (-6)}{8 - 0} = \dfrac{6}{8} = \dfrac{3}{4}$

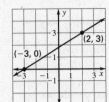
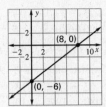

**24.** $m = \dfrac{0 - 6}{8 - 0} = \dfrac{-6}{8} = -\dfrac{3}{4}$ **25.** $m = \dfrac{-4 - 4}{4 - 2} = \dfrac{-8}{2} = -4$

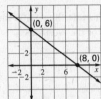
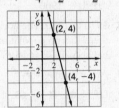

**26.** $m = \dfrac{4 - (-1)}{-6 - (-6)} = \dfrac{5}{0}$ **27.** $m = \dfrac{0 - (-10)}{-4 - 0}$

(undefined)
$= \dfrac{10}{-4} = \quad -\dfrac{5}{2}$

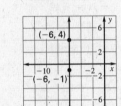
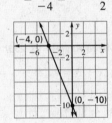

**28.** $m = \dfrac{2 - (-2)}{-2 - 1} = -\dfrac{4}{3}$ **29.** $m = \dfrac{0 - 6}{3 - 3} = \dfrac{-6}{0}$

undefined

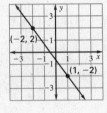
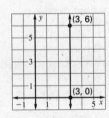

**30.** $m = \dfrac{-2 - 2}{4 - (-6)}$ **31.** $m = \dfrac{-6 - (-1)}{-3 - (-1)}$

$= \dfrac{-4}{10} = -\dfrac{2}{5}$ $= \dfrac{-5}{-2} = \dfrac{5}{2}$

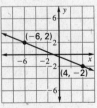
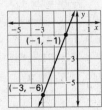

**32.** $m = \dfrac{0 - \frac{1}{2}}{0 - 0} = \dfrac{\frac{1}{2}}{0}$ **33.** $m = \dfrac{5 - 2}{-3 - 2} = -\dfrac{3}{5}$

(undefined)

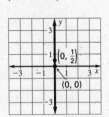
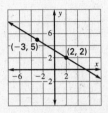

**34.** $m = \dfrac{1 - 1}{6 - 4} = \dfrac{0}{2} = 0$ **35. a.** $-1$ **b.** $1$

**36. a.** $-\dfrac{1}{3}$ **b.** $0$

**37. a.** $-2$ **b.** undefined

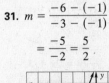

**38.** $2 = \dfrac{y - 5}{2 - 4}$

$2 = \dfrac{y - 5}{-2}$

$-2 \cdot 2 = -2 \cdot \dfrac{y - 5}{-2}$

$-4 = y - 5$

$1 = y$

**39.** $3 = \dfrac{y - (-2)}{2 - 0}$

$3 = \dfrac{y + 2}{2}$

$2 \cdot 3 = 2 \cdot \dfrac{y + 2}{2}$

$6 = y + 2$

$4 = y$

**40.** $-\dfrac{1}{2} = \dfrac{y - 3}{3 - (-5)}$

$-\dfrac{1}{2} = \dfrac{y - 3}{8}$

$8 \cdot -\dfrac{1}{2} = 8 \cdot \dfrac{y - 3}{8}$

$-4 = y - 3$

$-1 = y$

**41.** $2 = \dfrac{y - 5}{0 - 2}$

$2 = \dfrac{y - 5}{-2}$

$-2 \cdot 2 = -2 \cdot \dfrac{y - 5}{-2}$

$-4 = y - 5$

$1 = y$

**42.** $5 = \dfrac{y - 5}{3 - (-1)}$

$5 = \dfrac{y - 5}{4}$

$4 \cdot 5 = 4 \cdot \dfrac{y - 5}{4}$

$20 = y - 5$

$25 = y$

**43.** $-1 = \dfrac{y - 3}{5 - (-1)}$

$-1 = \dfrac{y - 3}{6}$

$6 \cdot -1 = 6 \cdot \dfrac{y - 3}{6}$

$-6 = y - 3$

$-3 = y$

**44.** $\dfrac{4}{3} = \dfrac{y - 7}{8 - 5}$

$\dfrac{4}{3} = \dfrac{y - 7}{3}$

$3 \cdot \dfrac{4}{3} = 3 \cdot \dfrac{y - 7}{3}$

$4 = y - 7$

$11 = y$

**45.** $-\dfrac{1}{2} = \dfrac{y - 4}{3 - 1}$

$-\dfrac{1}{2} = \dfrac{y - 4}{2}$

$2 \cdot -\dfrac{1}{2} = 2 \cdot \dfrac{y - 4}{2}$

$-1 = y - 4$

$3 = y$

**Algebra 1**
Chapter 4 Worked-out Solution Key

# Chapter 4 continued

**46.** $\dfrac{4}{5} = \dfrac{y-(-15)}{5-2}$

$\dfrac{4}{5} = \dfrac{y+15}{3}$

$3 \cdot \dfrac{4}{5} = 3 \cdot \dfrac{y+15}{3}$

$\dfrac{12}{5} = y + 15$

$-\dfrac{63}{5} = y$

**47.** $A(-4,-1)$, $B(-2,0)$

$m = \dfrac{0-(-1)}{-2-(-4)} = \dfrac{1}{2}$

$C(0,1)$, $D(2,2)$

$m = \dfrac{2-1}{2-0} = \dfrac{1}{2}$

$D(2,2)$, $E(4,3)$

$m = \dfrac{3-2}{4-2} = \dfrac{1}{2}$

The slopes are equal.

**48.** $A(-6,3)$, $B(-3,2)$

$m = \dfrac{2-3}{-3-(-6)} = -\dfrac{1}{3}$

$C(0,1)$, $D(3,0)$

$m = \dfrac{0-1}{3-0} = -\dfrac{1}{3}$

$D(3,0)$, $E(6,-1)$

$m = \dfrac{-1-0}{6-3} = -\dfrac{1}{3}$

The slopes are equal.

**49.** $m = \dfrac{-2-4}{2-(-2)} = \dfrac{-6}{4} = -\dfrac{3}{2}$

$m = \dfrac{0-(-2)}{6-2} = \dfrac{2}{4} = \dfrac{1}{2}$

$m = \dfrac{0-4}{6-(-2)} = \dfrac{-4}{8} = -\dfrac{1}{2}$

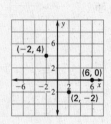

The points do not lie on the same line because the slope of the lines between them are not equal.

**50.** Any two points on a line can be chosen to find slope. The x- and y-intercepts of the line are convenient points to calculate slope.

**51.** $m = \dfrac{3-0}{50-0} = \dfrac{3}{50}$; Road grade and slope are equal.

**52.** $m = \dfrac{23-2}{9-2} = \dfrac{21}{7} = 3$ inches per minute

**53.** $m = \dfrac{69-5}{11-3} = \dfrac{64}{8} = 8$ dollars per year

**54.** $m = \dfrac{14-44}{32-53} = \dfrac{-30}{-21} = \dfrac{10}{7}$ liters per second

**55.** **1.** The slope is $-\dfrac{15}{300}$ or $-\dfrac{1}{20}$.

**2.** The slope is $\dfrac{45}{300}$ or $\dfrac{3}{20}$.

**3.** The slope is $\dfrac{50}{300}$ or $\dfrac{1}{6}$.

**56.** $\dfrac{2309.6-0}{9:31-9:23} = \dfrac{2309.6}{8}$ or 288.7 miles per minute

**57.** $\dfrac{206{,}000{,}000 - 173{,}000{,}000}{1996-1990} = \dfrac{33{,}000{,}000}{6} =$
5,500,000 dollars per year

**58.** $\dfrac{4.40-0.90}{1995-1960} = \dfrac{3.50}{35} = 0.10$ cents per year

**59.** $\dfrac{4.25-3.50}{1990-1985} = \dfrac{0.75}{5} = 0.15$ cents per year

**60.** 1990 to 1995    **61.** Answers may vary.

**62.** Estimates may vary; 1970 to 1980

$\dfrac{62-75}{1980-1970} = \dfrac{13}{10} = 1.3$ million people per year

**63.** Estimates may vary; 1960 to 1970

$\dfrac{58-51}{1970-1960} = 0.7$ million people per year

**64.** 1970 to 1980. The horizontal part of the line graph means there was no population change for the Northeast between 1970 and 1980.

**65.** Answers may vary.    **66.** No. Examples may vary.

**67.** D    **68.** B    **69.** C

**70.** $m = \dfrac{-9-3}{x-0} = \dfrac{-12}{x}$

$\dfrac{-12}{x} = -2$

$-12 = -2x$

$6 = x$

**71.** $m = \dfrac{-12}{x}$

$-3 = \dfrac{-12}{x}$

$-3x = -12$

$x = 4$

## 4.4 Mixed Review (p. 233)

**72.** $12x + 3 = 12(5) + 3 = 60 + 3 = 63$

**73.** $\dfrac{3}{4}p = \dfrac{3}{4}(16) = 12$

**74.** $8n = 8\left(\dfrac{3}{16}\right) = \dfrac{3}{2}$ or $1\dfrac{1}{2}$

**75.** $\dfrac{7}{2}y - 3 = \dfrac{7}{2}(4) - 3 = 14 - 3 = 11$

**76.** $5x + x = 5(7) + 7 = 35 + 7 = 42$

**77.** $\dfrac{6}{5}z + 2 = \dfrac{6}{5}(5) + 2 = 6 + 2 = 8$

**78.** $(-6)(y) = -6y$    **79.** $(-1)(x)(-x) = x^2$

**80.** $-(-3)^2(-y) = 9y$    **81.** $\left(\dfrac{2}{5}\right)(5)(-x)(x) = -2x^2$

**82.** $(y)(-23)(-y^2) = 23y^3$

**83.** $\left(\dfrac{1}{8}\right)(-4)(-x)(-x) = -\dfrac{x^2}{2}$

**84.** $9x - 2y = 14$
$-2y = 14 - 9x$
$y = \dfrac{9}{2}x - 7$

**85.** $5x + 9y = 18$
$9y = 18 - 5x$
$y = -\dfrac{5}{9}x + 2$

**86.** $-2x - 2y = 7$
$-2y = 2x + 7$
$y = -x - \dfrac{7}{2}$

**87.** $-x + 4y = 36$
$4y = x + 36$
$y = \dfrac{1}{4}x + 9$

**88.** $6x - 3y = 21$
$-3y = 21 - 6x$
$y = 2x - 7$

**89.** $3x + 5y = 17$
$5y = 17 - 3x$
$y = -\dfrac{3}{5}x + \dfrac{17}{5}$

# Chapter 4 *continued*

**90.** 
| x-intercept | y-intercept |
|---|---|
| $y = x + 8$ | $y = x + 8$ |
| $0 = x + 8$ | $y = 0 + 8$ |
| $-8 = x$ | $y = 8$ |

**91.** 
| x-intercept | y-intercept |
|---|---|
| $y = 2x + 12$ | $y = 2x + 12$ |
| $0 = 2x + 12$ | $y = 2(0) + 12$ |
| $-12 = 2x$ | $y = 12$ |
| $-6 = x$ | |

**92.** 
| x-intercept | y-intercept |
|---|---|
| $y = 3x + 9$ | $y = 3x + 9$ |
| $0 = 3x + 9$ | $y = 3(0) + 9$ |
| $-9 = 3x$ | $y = 9$ |
| $-3 = x$ | |

**93.** 
| x-intercept | y-intercept |
|---|---|
| $y = -6 + 2x$ | $y = -6 + 2x$ |
| $0 = -6 + 2x$ | $y = -6 + 2(0)$ |
| $6 = 2x$ | $y = -6$ |
| $3 = x$ | |

**94.** 
| x-intercept | y-intercept |
|---|---|
| $y = -2x + 16$ | $y = -2x + 16$ |
| $0 = -2x + 16$ | $y = -2(0) + 16$ |
| $2x = 16$ | $y = 16$ |
| $x = 8$ | |

**95.** 
| x-intercept | y-intercept |
|---|---|
| $3y = 6x + 3$ | $3y = 6x + 3$ |
| $3(0) = 6x + 3$ | $3y = 6(0) + 3$ |
| $-3 = 6x$ | $3y = 3$ |
| $-\frac{1}{2} = x$ | $y = 1$ |

## Lesson 4.5

### 4.5 Guided Practice (p. 237)

**1.** Two variables $x$ and $y$ vary directly if there is a nonzero number $k$ such that $y = kx$.

**2.** The constant of variation and slope are equal.

**3.**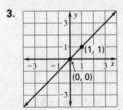

Direct variation; $k = 1$

$m = \dfrac{1-0}{1-0} = 1$

**4.**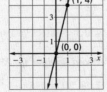

Direct variation; $k = 4$

$m = \dfrac{4-0}{1-0} = 4$

**5.**

Direct variation; $k = \frac{1}{2}$

$m = \dfrac{1-0}{2-0} = \dfrac{1}{2}$

**6.**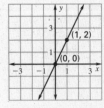

Direct variation; $k = 2$

$m = \dfrac{2-0}{1-0} = 2$

**7.**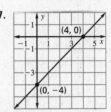

No direct variation

**8.**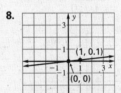

Direct variation; $k = 0.1$

$m = \dfrac{0.1-0}{1-0} = 0.1$

**9.**

| Total pay, $p$ | $18 | $42 | $48 | $30 |
|---|---|---|---|---|
| Hours worked, $h$ | 3 | 7 | 8 | 5 |
| Ratio | 6 | 6 | 6 | 6 |

**10.** $k = \dfrac{p}{h}$

$6 = \dfrac{p}{h}$

$p = 6h$

**11.** $k = \dfrac{p}{h}$

$6 = \dfrac{p}{6}$

$p = \$36$

### 4.5 Practice and Applications (pp. 237–239)

**12.** $k = 3$

$m = \dfrac{3-0}{1-0} = 3$

**13.** $k = -\dfrac{2}{5}$

$m = \dfrac{-2-0}{5-0} = -\dfrac{2}{5}$

**14.** $k = 0.75$

$m = \dfrac{3-0}{4-0} = \dfrac{3}{4} = 0.75$

**15.**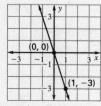

$k = -3$

$m = \dfrac{-3 - 0}{1 - 0} = -3$

**16.**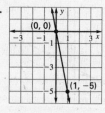

$k = -5$

$m = \dfrac{-5 - 0}{1 - 0} = -5$

**17.**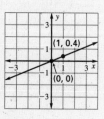

$k = 0.4$

$m = \dfrac{0.4 - 0}{1 - 0} = 0.4$

**18.**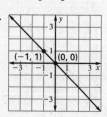

$k = -1$

$m = \dfrac{1 - 0}{-1 - 0} = -1$

**19.**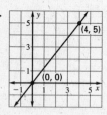

$k = \dfrac{5}{4}$

$m = \dfrac{5 - 0}{4 - 0} = \dfrac{5}{4}$

**20.**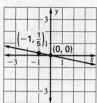

$k = -\dfrac{1}{5}$

$m = \dfrac{\frac{1}{5} - 0}{-1 - 0} = -\dfrac{1}{5}$

**21.** direct variation

**22.** direct variation

**23.** $y = kx$
$12 = k(4)$
$3 = k$
$y = 3x$

**24.** $y = kx$
$35 = k(7)$
$5 = k$
$y = 5x$

**25.** $y = kx$
$4 = k(18)$
$\frac{2}{9} = k$
$y = \frac{2}{9}x$

**26.** $y = kx$
$11 = k(22)$
$\frac{1}{2} = k$
$y = \frac{1}{2}x$

**27.** $y = kx$
$1.1 = k(5.5)$
$0.2 = k$
$y = 0.2x$

**28.** $y = kx$
$3.3 = k(16.5)$
$0.2 = k$
$y = 0.2x$

**29.** $y = kx$
$-1 = k(-1)$
$1 = k$
$y = x$

**30.** $y = kx$
$-9 = k\left(7\frac{1}{5}\right)$
$-1\frac{1}{4} = k$
$y = -\frac{5}{4}x$

**31.** $y = kx$
$3 = k(-9)$
$-\frac{1}{3} = k$
$y = -\frac{1}{3}x$

**32. a.** $V = kp$
$V = 0.06p$

**b.** Answers may vary.

**33.** $\dfrac{x}{54} = \dfrac{360}{60}$

$x = 324$ pounds

**34.**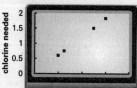

Chlorine Needed vs. Amount of Water in Pool

**35.** The scatter plot does show direct variation.

**36.** Because the ratio is about 0.6 for each instrument, you can use direct variation to create a model.

$k = \dfrac{b}{t}$

$0.62 = \dfrac{b}{t}$

$0.62t = b$

**37.** $b = 0.62t$
$28 = 0.62t$
$t \approx 45$ inches

Its tone is lower than a viola's tone.

**38.** D   **39.** A   **40.** A   **41.** B

**42.** $P = 6x$

Yes, the equation models direct variation.

**43.** Yes. If $a = 6$ and $b = 2$, then
$6 = k(2)$
$3 = k$
$a = 3b$.

If you solve for $b$, then

$b = \dfrac{a}{3}$,

which is direct variation.

### 4.5 Mixed Review (p. 239)

**44.** $7z + 30 = -5$
$7z = -35$
$z = -5$

**45.** $4b = 26 - 9b$
$13b = 26$
$b = 2$

**46.** $2(w - 2) = 2$
$w - 2 = 1$
$w = 3$

**47.** $9x + 65 = -4x$
$13x + 65 = 0$
$13x = -65$
$x = -5$

**Algebra 1**
Chapter 4 Worked-out Solution Key

# Chapter 4 *continued*

**48.** $55 - 5y = 9y + 27$
$55 = 14y + 27$
$28 = 14y$
$2 = y$

**49.** $7c - 3 = 4(c - 3)$
$7c - 3 = 4c - 12$
$3c - 3 = -12$
$3c = -9$
$c = -3$

**50.** $\quad 15 = 7(x - y) + 3x$
$15 = 7x - 7y + 3x$
$15 = 10x - 7y$
$7y + 15 = 10x$
$\frac{1}{10}(7y + 15) = x$ or $x = 0.7y + 1.5$

**51.** $\quad 3x + 12 = 5(x + y)$
$3x + 12 = 5x + 5y$
$12 = 2x + 5y$
$12 - 5y = 2x$
$\frac{1}{2}(12 - 5y) = x$ or $x = 6 - \frac{5}{2}y$

**52.** $\dfrac{152.25 \text{ dollars}}{21 \text{ hours}} = \$7.25$ per hour

**53.**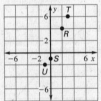

**54.** $\quad x - y = 10$
$5 - (-5) \overset{?}{=} 10$
$5 + 5 \overset{?}{=} 10$
$10 = 10$
The point lies on the line.

**55.** $\quad 3x - 6y = -2$
$3(-4) - 6(-2) \overset{?}{=} -2$
$-12 + 12 \overset{?}{=} -2$
$0 \neq -2$
The point does not lie on the line.

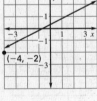

**56.** $\quad 5x + 6y = -1$
$5(1) + 6(-1) \overset{?}{=} -1$
$5 - 6 \overset{?}{=} -1$
$-1 = -1$
The point lies on the line.

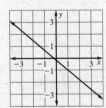

**57.** $\quad -4x - 3y = -8$
$-4(-4) - 3(2) \overset{?}{=} -8$
$16 - 6 \overset{?}{=} -8$
$10 \neq -8$
The point does not lie on the line.

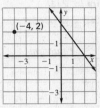

**58.** x-intercept
$2x + y = 6$
$2x + 0 = 6$
$2x = 6$
$x = 3$

y-intercept
$2x + y = 6$
$2(0) + y = 6$
$y = 6$

**59.** x-intercept
$x - 1.1y = 10$
$x - 1.1(0) = 10$
$x = 10$

y-intercept
$x - 1.1y = 10$
$0 - 1.1y = 10$
$-1.1y = 10$
$y = -\frac{100}{11}$ or $-9\frac{1}{11}$

**60.** x-intercept
$x - 6y = 4$
$x - 6(0) = 4$
$x = 4$

y-intercept
$x - 6y = 4$
$0 - 6y = 4$
$-6y = 4$
$y = -\frac{2}{3}$

**61.** x-intercept
$x + 5y = 10$
$x + 5(0) = 10$
$x = 10$

y-intercept
$x + 5y = 10$
$0 + 5y = 10$
$5y = 10$
$y = 2$

**62.** x-intercept
$y = x - 5$
$0 = x - 5$
$5 = x$

y-intercept
$y = x - 5$
$y = 0 - 5$
$y = -5$

**63.** x-intercept
$y = \frac{1}{3}x - \frac{7}{3}$
$0 = \frac{1}{3}x - \frac{7}{3}$
$\frac{7}{3} = \frac{1}{3}x$
$7 = x$

y-intercept
$y = \frac{1}{3}x - \frac{7}{3}$
$y = \frac{1}{3}(0) - \frac{7}{3}$
$y = -\frac{7}{3}$

## Lesson 4.6

### Developing Concepts Activity 4.6 (p. 240)
### Drawing Conclusions

**1.** Both are in the form $y = mx + b$, where $b$ is the height in centimeters of the desk; their graphs are lines that pass through the point $(0, b)$.

**2.** Let $h$ be the height of the book; $y = hx + 74$ and $y = hx + 68$; in both equations, the coefficient of $x$ is the same; the graphs have the same slope.

**3.** Their slopes are equal and their graphs are parallel.

**4.** The only common characteristic is the graphs share the same y-intercept.

# Chapter 4 *continued*

## Activity (p. 241)

| | Equation | Two Solutions |
|---|---|---|
| 1. | $y = 2x + 1$ | $(0, 1) \left(-\frac{1}{2}, 0\right)$ |
| 2. | $y = -2x - 3$ | $(0, -3) \left(-\frac{3}{2}, 0\right)$ |
| 3. | $y = x + 4$ | $(0, 4) (-4, 0)$ |
| 4. | $y = 0.5x - 2.5$ | $(0, -2.5) (5, 0)$ |

| | Equation | Slope | y-intercept |
|---|---|---|---|
| 1. | $y = 2x + 1$ | 2 | 1 |
| 2. | $y = -2x - 3$ | $-2$ | $-3$ |
| 3. | $y = x + 4$ | 1 | 4 |
| 4. | $y = 0.5x - 2.5$ | 0.5 | $-2.5$ |

## 4.6 Guided Practice (p. 244)

1. The equation $y = mx + b$ is called the slope-intercept form because $m$ is the slope of the line and $b$ is the y-intercept of the line and they can be taken directly from the equation.

2. First, the equation needs to written in slope-intercept form:

$$2x - 3y = 6$$
$$-3y = -2x + 6$$
$$y = \frac{2}{3}x - 2$$

Next, find the slope and y-intercept.

$$m = \frac{2}{3} \quad b = -2$$

Plot $(0, -2)$.

Draw a slope triangle to locate a second point on the line. Finally, draw a line through the two points.

3. The graphs are parallel.      4. B

5. Slope: 2            6. Slope: 11
   y-intercept: 1         y-intercept: 0

7. Slope: 1            8. Slope: $-1.5$
   y-intercept: 3         y-intercept: 0

9. Slope: 5            10. Slope: $-1$
   y-intercept: $-3$      y-intercept: $-2$

11. The slope is 30 and the y-intercept is 50.

12.
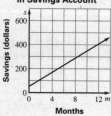

**Amount of Money in Savings Account**

## 4.6 Practice and Applications (pp. 244–246)

13. Slope: 6          14. Slope: 3
    y-intercept: 4        y-intercept: 1

15. Slope: 2          16. Slope: 9
    y-intercept: $-3$      y-intercept: 0

17. Slope: 0          18. Slope: $-\frac{3}{4}$
    y-intercept: $-2$      y-intercept: 4

19. Slope: $\frac{1}{4}$          20. Slope: $-\frac{1}{3}$
    y-intercept: $\frac{1}{2}$        y-intercept: 2

21. Slope: $-3$
    y-intercept: $\frac{1}{2}$

22.
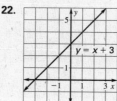

23.
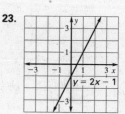

24.
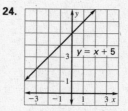

25.
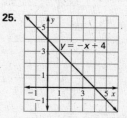

26.
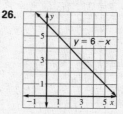

27.
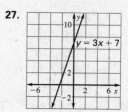

28.
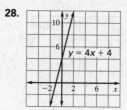

29.
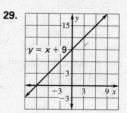

30.
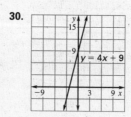

31.
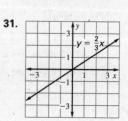

# Chapter 4 *continued*

**32.**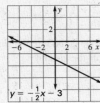

$y = -\frac{1}{2}x - 3$

**33.**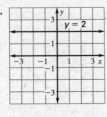

$y = 2$

**34.** $y = -2$

$y = 0x - 2$

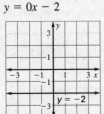

$y = -2$

**35.** $3x - 6y = 9$

$-6y = -3x + 9$

$y = \frac{1}{2}x - \frac{3}{2}$

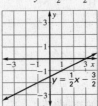

$y = \frac{1}{2}x - \frac{3}{2}$

**36.** $4x + 5y = 15$

$5y = -4x + 15$

$y = -\frac{4}{5}x + 3$

$y = -\frac{4}{5}x + 3$

**37.** $4x - y - 3 = 0$

$-y = -4x + 3$

$y = 4x - 3$

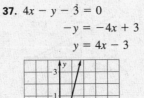

$y = 4x - 3$

**38.** $x - y + 4 = 0$

$-y = -x - 4$

$y = x + 4$

$y = (1)x + 4$

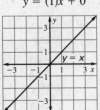

$y = x + 4$

**39.** $x + y = 0$

$y = -x$

$y = (-1)x + 0$

$y = -x$

**40.** $x - y = 0$

$-y = -x$

$y = x$

$y = (1)x + 0$

$y = x$

**41.** $x + 3y - 3 = 0$

$3y = -x + 3$

$y = -\frac{1}{3}x + 1$

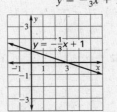

$y = -\frac{1}{3}x + 1$

**42.** $2x - 3y - 6 = 0$

$-3y = -2x + 6$

$y = \frac{2}{3}x - 2$

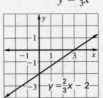

$y = \frac{2}{3}x - 2$

**43.** $y - 0.5 = 0$

$y = 0.5$

$y = 0x + 0.5$

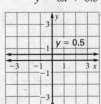

$y = 0.5$

**44.** $5(x + 3 + y) = 10x$

$x + 3 + y = 2x$

$y = x - 3$

$y = (1)x - 3$

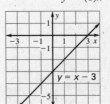

$y = x - 3$

**45.** $2x + 3y - 4 = x + 5$

$3y = -x + 9$

$y = -\frac{1}{3}x + 3$

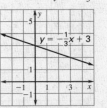

$y = -\frac{1}{3}x + 3$

**46.** The lines are parallel because they have the same slope.

$y = -3x + 2$

$y = -3x - 4$

**47.** The lines are parallel because they have the same slope.

$y = 2x - 12$

$y = 2x + 10$

**48.** The lines are not parallel because they have different slopes.

$y = -6x + 8$

$y = 6x - 2$

**49.** The lines are not parallel because they have different slopes.

$y = \frac{4}{3}x + 1$

$y = -\frac{4}{3}x + 3$

**50.** The lines are parallel because they have the same slope.

$y = 3x - 1$

$y = 3x - \frac{7}{3}$

**51.** The lines are parallel because they have the same slope.

$y = 3x$

$y = 3x - 4$

**52.** D    **53.** C    **54.** A    **55.** B

**56.** Slope: $\frac{1}{2}$; $y$-intercept: 6

**57.** The slope is the rate of the snow falling in inches per hour. The $y$-intercept means there were already 6 inches of snow on the ground before the snowstorm.

**58.**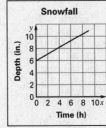

**59.** School would not be cancelled because when 9 is substituted for $x$, the depth $y$ of snow is 10.5 inches.

# Chapter 4 *continued*

**60.**

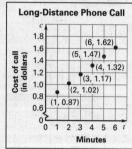

**61.**

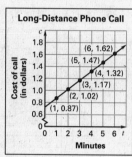

The slope is 0.15. It represents the long-distance rate for each additional minute.

**62.** None of the lines are parallel because their slopes are not equal.

**63.**

The lines form a triangle.
Area $= \frac{1}{2}bh$
$b = 6$ units; $h = 3$ units
Area $= \frac{1}{2}(6)(3)$
Area $= \frac{1}{2}(18)$
Area $= 9$ square units

**64.**

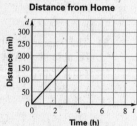

**65.**

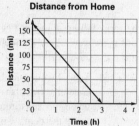

**66.**

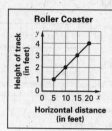

$m = \dfrac{2-1}{10-5} = \dfrac{1}{5}$

The slope is the rate the track rises off the ground.

**67.** 4 feet    **68.** The roller coaster will pass inspection.

**69.** B    $1.6x - 3.2y = 16$
$- 3.2y = -1.6x + 16$
$y = 0.5x - 0.5$

**70.** C

**71.** $2 \cdot \left(-\frac{1}{2}\right) = -1;$
perpendicular

**72.** $\left(-\frac{1}{3}\right) \cdot 3 = -1;$
perpendicular

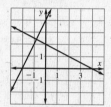

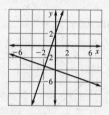

**73.** $\frac{2}{3} \cdot \frac{3}{2} \neq -1;$
not perpendicular

**74.** $y = -\frac{1}{3}x - 4$
Answers may vary.

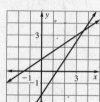

**75.** $y = \frac{1}{2}x + 1$
Answers may vary.

**76.** $y = -\frac{3}{4}x - 3$
Answers may vary.

**77.** No, the given line passes through $\left(0, \frac{1}{9}\right)$.

**4.6 Mixed Review (p. 247)**

**78.** $x + 6 = 14$
$x = 8$

**79.** $9 - y = 4$
$5 = y$

**80.** $7b = 21$
$\dfrac{7b}{7} = \dfrac{21}{7}$
$b = 3$

**81.** $\dfrac{a}{4} = 3$
$(4)\dfrac{a}{4} = 3(4)$
$a = 12$

**82.** $\frac{1}{3}g - 2 = 1$
$\frac{1}{3}g = 3$
$(3)\left(\frac{1}{3}g\right) = (3)(3)$
$g = 9$

**83.** $3p - 12 = 6$
$3p = 18$
$\dfrac{3p}{3} = \dfrac{18}{3}$
$p = 6$

# Chapter 4 *continued*

**84.** $2(v + 1) = 4$
$2v + 2 = 4$
$2v = 2$
$v = 1$

**85.** $3(r - 1) = 2(r - 2)$
$3r - 3 = 2r - 4$
$r = -1$

**86.** $5(w - 5) = 25$
$5w - 25 = 25$
$5w = 50$
$w = 10$

**87.** A

**88.**

**89.** 2001

## Quiz 2 (p. 247)

**1.** $m = \dfrac{2 - 0}{5 - 0} = \dfrac{2}{5}$

**2.** $m = \dfrac{-5 - (-3)}{-4 - 1} = \dfrac{-2}{-5} = \dfrac{2}{5}$

**3.** $m = \dfrac{-4 - 3}{-6 - 3} = \dfrac{-7}{-9} = \dfrac{7}{9}$

**4.** $m = \dfrac{-2 - 2}{-5 - (-3)} = \dfrac{-4}{-2} = 2$

**5.** $m = \dfrac{-4 - 4}{5 - 0} = \dfrac{0}{5} = 0$

**6.** $m = \dfrac{6 - (-2)}{-7 - 1} = \dfrac{8}{-8} = -1$

**7.** $y = kx$
$10 = k(2)$
$5 = k$
$y = 5x$

**8.** $y = kx$
$64 = k(8)$
$8 = k$
$y = 8x$

**9.** $y = kx$
$72 = k(12)$
$6 = k$
$y = 6x$

**10.** $y = kx$
$12 = k(16)$
$\frac{3}{4} = k$
$y = \frac{3}{4}x$

**11.** $y = kx$
$7 = k(10)$
$\frac{7}{10} = k$
$y = \frac{7}{10}x$
$y = 0.7x$

**12.** $y = kx$
$8 = k(18)$
$\frac{4}{9} = k$
$y = \frac{4}{9}x$

**13.** $y = 2x - 4$

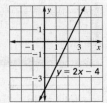

**14.** $6x + 12y + 4 = 0$
$12y = -6x - 4$
$y = -\frac{1}{2}x - \frac{1}{3}$

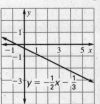

**15.** $x - 2y = 12$
$-2y = -x + 12$
$y = \frac{1}{2}x - 6$

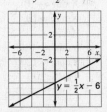

**16.** $x - y + 2 = 0$
$-y = -x - 2$
$y = x + 2$
$y = (1)x + 2$

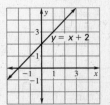

**17.** $x + 2y - 2 = 0$
$2y = -x + 2$
$y = -\frac{1}{2}x + 1$

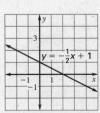

**18.** $\frac{3}{2}x + \frac{3}{2}y = \frac{3}{4}$
$\frac{3}{2}y = -\frac{3}{2}x + \frac{3}{4}$
$\frac{2}{3}\left(\frac{3}{2}y\right) = \frac{2}{3}\left(-\frac{3}{2}x + \frac{3}{4}\right)$
$y = -x + \frac{1}{2}$
$y = (-1)x + \frac{1}{2}$

**19.** $\dfrac{215{,}400 - 365{,}800}{6 - 1} = \dfrac{-150{,}400}{5}$
$= -30{,}080$ dollars per month

## Technology Activity 4.6 (pp. 248–249)

**1.**

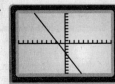

**2.**

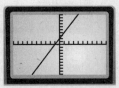

**3.**

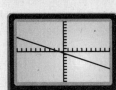

**4.**

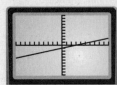

# Chapter 4 *continued*

**5.**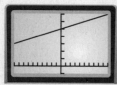

**6.**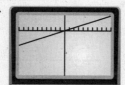

**7.**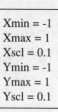

**8.**
Xmin = -10
Xmax = 10
Xscl = 1
Ymin = -340
Ymax = -320
Yscl = 1

**9.**
Xmin = -1
Xmax = 1
Xscl = 0.1
Ymin = -1
Ymax = 1
Yscl = 0.1

**10.**
Xmin = -10
Xmax = 10
Xscl = 1
Ymin = -1200
Ymax = 1200
Yscl = 100

**11.**
Xmin = -10
Xmax = 10
Xscl = 1
Ymin = 0
Ymax = 55000
Yscl = 10000

**12.** Window settings may vary. The following window setting can be used with the *trace* feature to approximate the value of $y$

 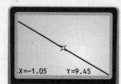

$y \approx 9.45$

**13.** Window settings may vary. The following window setting can be used with the *trace* feature to approximate the value of $y$

 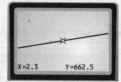

$y \approx 662.5$

**14.** Window settings may vary. The following window setting can be used with the *trace* feature to approximate the value of $y$

 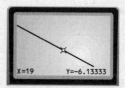

$y \approx -6.1\overline{3}$

**15.** Window settings may vary. The following window setting can be used with the *trace* feature to approximate the value of $y$

 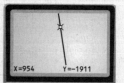

$y \approx -1911$

## Lesson 4.7

### 4.7 Guided Practice (p. 253)

**1.** The solution of a linear equation can be found graphically using the following steps:
   **a.** Write the equation in the form $ax + b = 0$.
   **b.** Write the related function $y = ax + b$.
   **c.** Graph the equation $y = ax + b$.
   The solution of $ax + b = 0$ is the $x$-intercept of $y = ax + b$.

**2.** The $x$-intercept of $y = -\frac{2}{3}x + 3$ is the value of $x$ when $y = 0$. This value of $x$ is the solution of $0 = -\frac{2}{3}x + 3$.

**3.** $x - 3 = 7$
$x - 10 = 0$
$y = x - 10$

Check:
$x - 3 = 7$
$10 - 3 \stackrel{?}{=} 7$
$7 = 7$

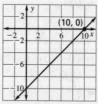

**4.** $2 - x = -5$
$-x + 7 = 0$
$y = -x + 7$

Check:
$2 - x = -5$
$2 - 7 \stackrel{?}{=} -5$
$-5 = -5$

**5.** $5x + 6 = -9$
$5x + 15 = 0$
$y = 5x + 15$

Check:
$5x + 6 = -9$
$5(-3) + 6 \overset{?}{=} -9$
$-15 + 6 \overset{?}{=} -9$
$-9 = -9$

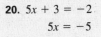

$(-3, 0)$

**6.** C    **7.** A    **8.** D    **9.** B    **10.** F

## 4.7 Practice and Applications (pp. 253–255)

**11.** C    **12.** A    **13.** B

**14.** $7x - 3 = 3$
$7x - 6 = 0$
$y = 7x - 6$

**15.** $6 - 4x = 13$
$-4x - 7 = 0$
$y = -4x - 7$

**16.** $9 + 5x = 19$
$5x - 10 = 0$
$y = 5x - 10$

**17.** $12x + 5 = 8x$
$4x + 5 = 0$
$y = 4x + 5$

**18.** $-4x - 5 = 3x + 8$
$-7x - 13 = 0$
$y = -7x - 13$

**19.** $6 - \frac{4}{7}x = 13 + \frac{3}{7}x$
$-\frac{7}{7}x - 7 = 0$
$-x - 7 = 0$
$x + 7 = 0$
$y = x + 7$

**20.** $5x + 3 = -2$
$5x = -5$
$x = -1$
Check:
$5x + 3 = -2$
$5x + 5 = 0$
$y = 5x + 5$

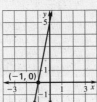

$(-1, 0)$

**21.** $-3x + 11 = 2$
$-3x = -9$
$x = 3$
Check:
$-3x + 11 = 2$
$-3x + 9 = 0$
$y = -3x + 9$

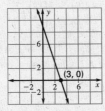

$(3, 0)$

**22.** $-x = -2$
$x = 2$
Check:
$-x = -2$
$-x + 2 = 0$
$y = -x + 2$

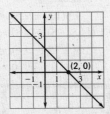

$(2, 0)$

**23.** $2x + 7 = 10$
$2x = 3$
$x = \frac{3}{2}$
Check:
$2x + 7 = 10$
$2x - 3 = 0$
$y = 2x - 3$

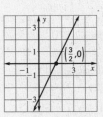

$\left(\frac{3}{2}, 0\right)$

**24.** $\frac{1}{4}x + 1 = -\frac{1}{2}$
$\frac{1}{4}x = -\frac{3}{2}$
$x = -6$
Check:
$\frac{1}{4}x + 1 = -\frac{1}{2}$
$\frac{1}{4}x + \frac{3}{2} = 0$
$y = \frac{1}{4}x + \frac{3}{2}$

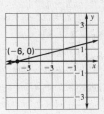

$(-6, 0)$

**25.** $\frac{1}{2}x + 5 = 3$
$\frac{1}{2}x = -2$
$x = -4$
Check:
$\frac{1}{2}x + 5 = 3$
$\frac{1}{2}x + 2 = 0$
$y = \frac{1}{2}x + 2$

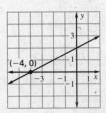

$(-4, 0)$

**26.** $-\frac{2}{3}x - 6 = -4$
$-\frac{2}{3}x = 2$
$x = -3$
Check:
$-\frac{2}{3}x - 6 = -4$
$-\frac{2}{3}x - 2 = 0$
$y = -\frac{2}{3}x - 2$

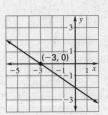

$(-3, 0)$

**27.** $-\frac{3}{4}x + 3 = \frac{9}{4}$
$-\frac{3}{4}x = -\frac{3}{4}$
$x = 1$
Check:
$-\frac{3}{4}x + 3 = \frac{9}{4}$
$-\frac{3}{4}x + \frac{3}{4} = 0$
$y = -\frac{3}{4}x + \frac{3}{4}$

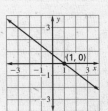

$(1, 0)$

**28.** $\frac{2}{3}x - \frac{2}{3} = 2$
$\frac{2}{3}x = \frac{8}{3}$
$x = 4$
Check:
$\frac{2}{3}x - \frac{2}{3} = 2$
$\frac{2}{3}x - \frac{8}{3} = 0$
$y = \frac{2}{3}x - \frac{8}{3}$

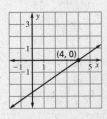

$(4, 0)$

# Chapter 4 continued

**29.** $2x - 7 = -5$
$2x - 2 = 0$
$y = 2x - 2$
Check:
$2x - 7 = -5$
$2(1) - 7 \stackrel{?}{=} -5$
$2 - 7 \stackrel{?}{=} -5$
$-5 = -5$

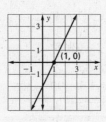

**30.** $5x - 2 = 8$
$5x - 10 = 0$
$y = 5x - 10$
Check:
$5x - 2 = 8$
$5(2) - 2 \stackrel{?}{=} 8$
$10 - 2 \stackrel{?}{=} 8$
$8 = 8$

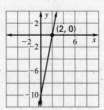

**31.** $\quad -4x = -12$
$-4x + 12 = 0$
$\quad\quad y = -4x + 12$
Check:
$\quad -4x = -12$
$-4(3) \stackrel{?}{=} -12$
$-12 = -12$

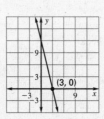

**32.** $6x + 9 = 3$
$6x + 6 = 0$
$y = 6x + 6$
Check:
$6x + 9 = 3$
$6(-1) + 9 \stackrel{?}{=} 3$
$-6 + 9 \stackrel{?}{=} 3$
$3 = 3$

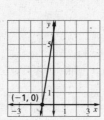

**33.** $\quad 7 - 9x = -11$
$-9x + 18 = 0$
$\quad\quad y = -9x + 18$
Check:
$7 - 9x = -11$
$7 - 9(2) \stackrel{?}{=} -11$
$7 - 18 \stackrel{?}{=} -11$
$-11 = -11$

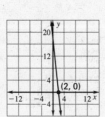

**34.** $3x + 10 = -2x$
$5x + 10 = 0$
$y = 5x + 10$
Check:
$3x + 10 = 2x$
$3(-2) + 10 \stackrel{?}{=} -2(-2)$
$-6 + 10 \stackrel{?}{=} 4$
$4 = 4$

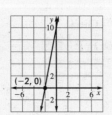

**35.** $-5x + 4 = 12 - 3x$
$-2x - 8 = 0$
$\quad\quad y = -2x - 8$
Check:
$\quad -5x + 4 = 12 - 3x$
$-5(-4) + 4 \stackrel{?}{=} 12 - (3)(-4)$
$\quad 20 + 4 \stackrel{?}{=} 12 + 12$
$\quad\quad 24 = 24$

**36.** $\frac{1}{3}x + 1 = 4$
$\frac{1}{3}x - 3 = 0$
$y = \frac{1}{3}x - 3$
Check:
$\frac{1}{3}x + 1 = 4$
$\frac{1}{3}(9) + 1 \stackrel{?}{=} 4$
$3 + 1 \stackrel{?}{=} 4$
$4 = 4$

**37.** $\frac{1}{2}x + 5 = 3$
$\frac{1}{2}x + 2 = 0$
$y = \frac{1}{2}x + 2$
Check:
$\frac{1}{2}x + 5 = 3$
$\frac{1}{2}(-4) + 5 \stackrel{?}{=} 3$
$-2 + 5 \stackrel{?}{=} 3$
$3 = 3$

**38.** $\quad\quad 4(x + 2) = 3(x + 5)$
$4(x + 2) - 3(x + 5) = 0$
$\quad\quad\quad y = 4(x + 2) - 3(x + 5)$
Check:
$4(x + 2) = 3(x + 5)$
$4(7 + 2) \stackrel{?}{=} 3(7 + 5)$
$4(9) \stackrel{?}{=} 3(12)$
$36 = 36$

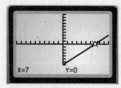

**39.** $\quad\quad\quad -1.6(1.5x + 7.5) = 0.6(6x + 30)$
$-1.6(1.5x + 7.5) - 0.6(6x + 30) = 0$
$\quad\quad\quad\quad y = 1.6(1.5x + 7.5) -$
$\quad\quad\quad\quad\quad\quad 0.6(6x + 30)$
Check:
$-1.6(1.5 + 7.5) = 0.6(6x + 30)$
$-1.6[1.5(-5) + 7.5] = 0.6[6(-5) + 30]$
$-1.6(-7.5 + 7.5) = 0.6(-30 + 30)$
$0 = 0$

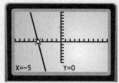

**Algebra 1**
Chapter 4 Worked-out Solution Key

# Chapter 4 *continued*

**40.**
$$\tfrac{1}{2}(x + 7) = \tfrac{1}{3}(10x + 2)$$
$$\tfrac{1}{2}(x + 7) - \tfrac{1}{3}(10x + 2) = 0$$
$$y = \tfrac{1}{2}(x + 7) - \tfrac{1}{3}(10x + 2)$$

Check:
$$\tfrac{1}{2}(x + 7) = \tfrac{1}{3}(10x + 2)$$
$$\tfrac{1}{2}(1 + 7) \overset{?}{=} \tfrac{1}{3}[10(1) + 2]$$
$$\tfrac{1}{2}(8) \overset{?}{=} \tfrac{1}{3}(12)$$
$$4 = 4$$

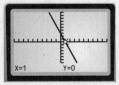
X=1    Y=0

**41.**
$$0.7(3x - 20) = 22 - 3.9x$$
$$0.7(3x - 20) - 22 + 3.9x = 0$$
$$y = 0.7(3x - 20) - 22 + 3.9x$$

Check:
$$0.7(3x - 20) = 22 - 3.9x$$
$$0.7[3(6) - 20] \overset{?}{=} 22 - (3.9)(6)$$
$$0.7(18 - 20) \overset{?}{=} 22 - 23.4$$
$$-1.4 = -1.4$$

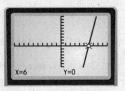
X=6    Y=0

**42.**
$$\tfrac{3}{4}(4x - 15) = -\tfrac{3}{2}(4x - 18)$$
$$\tfrac{3}{4}(4x - 15) + \tfrac{3}{2}(4x - 18) = 0$$
$$y = \tfrac{3}{4}(4x - 15) + \tfrac{3}{2}(4x - 18)$$

Check:
$$\tfrac{3}{4}(4x - 15) = -\tfrac{3}{2}(4x - 18)$$
$$\tfrac{3}{4}[4(\tfrac{17}{4}) - 15] \overset{?}{=} -\tfrac{3}{2}[4(\tfrac{17}{4}) - 18]$$
$$\tfrac{3}{4}(17 - 15) \overset{?}{=} -\tfrac{3}{2}(17 - 18)$$
$$\tfrac{3}{4}(2) \overset{?}{=} -\tfrac{3}{2}(-1)$$
$$\tfrac{3}{2} = \tfrac{3}{2}$$

X=4.25    Y=0

**43.**
$$\tfrac{1}{3}(10x + 27) = -\tfrac{1}{5}(\tfrac{55}{3}x + 25)$$
$$\tfrac{1}{3}(10x + 27) + \tfrac{1}{5}(\tfrac{55}{3}x + 25) = 0$$
$$y = \tfrac{1}{3}(10x + 27) + \tfrac{1}{5}(\tfrac{55}{3}x + 25)$$

Check:
$$\tfrac{1}{3}(10x + 27) = -\tfrac{1}{5}(\tfrac{55}{3}x + 25)$$
$$\tfrac{1}{3}[10(-2) + 27] \overset{?}{=} -\tfrac{1}{5}[\tfrac{55}{3}(-2) + 25]$$
$$\tfrac{1}{3}(-20 + 27) \overset{?}{=} -\tfrac{1}{5}(-\tfrac{110}{3} + 25)$$
$$\tfrac{1}{3}(7) \overset{?}{=} -\tfrac{1}{5}(-\tfrac{35}{3})$$
$$\tfrac{7}{3} = \tfrac{7}{3}$$

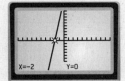
X=−2    Y=0

**44.**
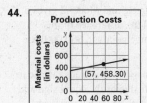
Production Costs

$$y = 1.9x + 350$$
$$458.30 = 1.9x + 350$$
$$108.3 = 1.9x$$
$$x = 57 \text{ hats}$$

**45.**
$$v = 559{,}100t + 6{,}423{,}000$$
$$11{,}000{,}000 = 559{,}100t + 6{,}423{,}000$$
$$4{,}577{,}000 = 559{,}100t$$
$$8.2 \sim t$$
Sometime in 1999

**46.**
$$h = 0.57t + 54.85$$
$$70 = 0.57t + 54.85$$
$$15.15 = 0.57t$$
$$26.6 \approx t$$
Sometime in 2011

**47.**
$$h = 0.35t + 4.06$$
$$10 = 0.35t + 4.06$$
$$5.94 = 0.35t$$
$$16.971 \approx t$$
In 2001

**48.**
$$h = -0.2t + 3.4$$
$$0 = -0.2t + 3.4$$
$$0.2t = 3.4$$
$$t = 17$$
In 2002

**49.** Answers may vary.

**50. a.** Model 1
$$m = 13{,}272t + 736{,}000$$
$$800{,}000 = 13{,}272t + 736{,}000$$
$$64{,}000 = 13{,}272t$$
$$4.8 \approx t$$
Sometime in 1998

Model 2
$$m = 9455t + 736{,}000$$
$$800{,}000 = 9455t + 736{,}000$$
$$64{,}000 = 9455t$$
$$6.8 \approx t$$
Sometime in 2000

Model 3
$$m = 11{,}455t + 736{,}000$$
$$800{,}000 = 11{,}455t + 736{,}000$$
$$64{,}000 = 11{,}455t$$
$$5.6 \approx t$$
Sometime in 1999

**b.**
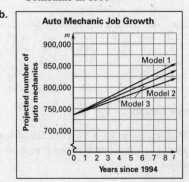
Auto Mechanic Job Growth

# Chapter 4 continued

c. Model 1 gives a high estimate and Model 2 gives a low estimate. You can tell the difference of the estimates by where the three points $(t, 800{,}000)$ are located.

51. The only difference between methods is Tanya chose to write the equation on the left side of the equal sign and Dwight chose to write the equation on the right side. Either choice gives the correct solution.

## 4.7 Mixed Review (p. 255)

52. $x + 3 > 5$
$7 + 3 \overset{?}{>} 5$
$10 > 5$
7 is a solution.

53. $-2 - x \le -9$
$-2 - (-7) \overset{?}{\le} -9$
$-2 + 7 \overset{?}{\le} -9$
$5 \not\le -9$
$-7$ is not a solution.

54. $3x - 11 \ge 2$
$3(4) - 11 \overset{?}{\ge} 2$
$12 - 11 \overset{?}{\ge} 2$
$1 \not\ge 2$
4 is not a solution.

55. $8 + 5x < -2$
$8 + 5(-2) \overset{?}{<} -2$
$8 - 10 \overset{?}{<} -2$
$-2 \not< -2$
$-2$ is not a solution.

56. $4x - 1 > 10$
$4(3) - 1 \overset{?}{>} 10$
$12 - 1 \overset{?}{>} 10$
$11 > 10$
3 is a solution.

57. $2x + 4 \le 3x - 3$
$2(5) + 4 \overset{?}{\le} 3(5) - 3$
$10 + 4 \overset{?}{\le} 15 - 3$
$14 \not\le 12$
5 is not a solution.

58. $4(x + 7) = 4(x) + 4(7) = 4x + 28$

59. $-8(8 - y) = -8(8) - (-8)(y) = -64 + 8y$

60. $(3 - b)9 = 3(9) - b(9) = 27 - 9b$

61. $(q + 4)(-3q) = (q)(-3q) + (4)(-3q) = -3q^2 - 12q$

62. $10(6x + 2) = 10(6x) + 10(2) = 60x + 20$

63. $-2(d - 5) = -2(d) - (-2)(5) = -2d + 10$

64. $(7 - 2a)(4a) = 7(4a) - (2a)(4a) = 28a - 8a^2$

65. $-5w(-3 + 2w) = (-5w)(-3) + (-5w)(2w)$
$= 15w - 10w^2$

66.

| Number of club members, $n$ | Income, $I$ |
|---|---|
| 2 | $48 |
| 4 | $96 |
| 6 | $144 |
| 8 | $192 |
| 10 | $240 |

## Lesson 4.8

### 4.8 Guided Practice (p. 259)

1. Every function is a relation but not every relation is a function. A relation is a function if for each input there is exactly one output.

2. A vertical line cannot be the graph of a linear function.

3. $f(x) = 3x - 10$
$f(0) = 3(0) - 10 = -10$

4. $f(x) = 3x - 10$
$f(20) = 3(20) - 10 = 60 - 10 = 50$

5. $f(x) = 3x - 10$
$f(-2) = 3(-2) - 10 = -6 - 10 = -16$

6. $f(x) = 3x - 10$
$f\left(\frac{2}{3}\right) = 3\left(\frac{2}{3}\right) - 10 = 2 - 10 = -8$

7. Function. The domain of the function is the set of input values 10, 20, 30, 40, and 50. The range is the set of output values 100, 200, 300, 400, and 500.

8. Not a function   9. Not a function   10. $d(t) = 40t$

## 4.8 Practice and Applications (pp. 259–261)

11. The graph represents a function because no vertical line passes through two points on the graph.

12. The graph does not represent a function because a vertical line passes through two points on the graph.

13. The graph does not represent a function because a vertical line passes through two points on the graph.

14. Function. The domain of the function is the set of input values 1, 2, 3, and 4. The range is the set of output values 5, 4, 3, and 2.

15. Function. The domain of the function is the set of input values 1, 2, 3, and 4. The range is the output value 0.

16. Not a function

17. Function. The domain of the function is the set of input values 0, 1, 2, and 3. The range is the set of output values 2, 4, 6, and 8.

18. Not a function

19. Function. The domain of the function is the set of input values 1, 3, 5, and 7. The range is the set of output values 1, 2, and 3.

20. $f(x) = 10x + 1$
$f(2) = 10(2) + 1 = 20 + 1 = 21$
$f(0) = 10(0) + 1 = 1$
$f(-3) = 10(-3) + 1 = -30 + 1 = -29$

21. $g(x) = 8x - 2$
$g(2) = 8(2) - 2 = 16 - 2 = 14$
$g(0) = 8(0) - 2 = -2$
$g(-3) = 8(-3) - 2 = -24 - 2 = -26$

22. $h(x) = 3x + 6$
$h(2) = 3(2) + 6 = 6 + 6 = 12$
$h(0) = 3(0) + 6 = 6$
$h(-3) = 3(-3) + 6 = -9 + 6 = -3$

# Chapter 4 *continued*

**23.** $g(x) = 1.25x$

$g(2) = 1.25(2) = 2.5$

$g(0) = 1.25(0) = 0$

$g(-3) = 1.25(-3) = -3.75$

**24.** $h(x) = 0.75x + 8$

$h(2) = 0.75(2) + 8 = 1.5 + 8 = 9.5$

$h(0) = 0.75(0) + 8 = 8$

$h(-3) = 0.75(-3) + 8 = -2.25 + 8 = 5.75$

**25.** $f(x) = 0.33x - 2$

$f(2) = 0.33(2) - 2 = 0.66 - 2 = -1.34$

$f(0) = 0.33(0) - 2 = -2$

$f(-3) = 0.33(-3) - 2 = -0.99 - 2 = -2.99$

**26.** $h(x) = \frac{3}{4}x - 4$

$h(2) = \frac{3}{4}(2) - 4 = \frac{3}{2} - 4 = -\frac{5}{2} = -2.5$

$h(0) = \frac{3}{4}(0) - 4 = -4$

$h(-3) = \frac{3}{4}(-3) - 4 = -\frac{9}{4} - 4 = -\frac{25}{4} = -6.25$

**27.** $g(x) = \frac{2}{5}x + 7$

$g(2) = \frac{2}{5}(2) + 7 = \frac{4}{5} + 7 = \frac{39}{5} = 7.8$

$g(0) = \frac{2}{5}(0) + 7 = 7$

$g(-3) = \frac{2}{5}(-3) + 7 = -\frac{6}{5} + 7 = \frac{29}{5} = 5.8$

**28.** $f(x) = \frac{2}{7}x + 4$

$f(2) = \frac{2}{7}(2) + 4 = \frac{4}{7} + 4 = \frac{32}{7} = 4\frac{4}{7}$

$f(0) = \frac{2}{7}(0) + 4 = 4$

$f(-3) = \frac{2}{7}(-3) + 4 = -\frac{6}{7} + 4 = \frac{22}{7} = 3\frac{1}{7}$

**29.** C    **30.** A    **31.** B

**32.**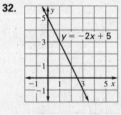

**33.**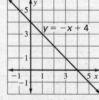

**34.**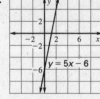

**35.**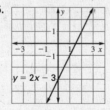

**36.**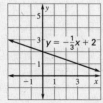

**37.**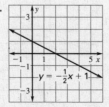

**38.**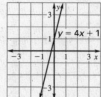

**39.**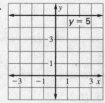

**40.**

**41.** Points on the linear function are $(1, 2)$ and $(3, -1)$. Use the slope formula to determine the slope, which is $-\frac{3}{2}$.

**42.** $m = \dfrac{5 - (-3)}{-2 - 2} = -\dfrac{8}{4} = -2$

**43.** $m = \dfrac{0 - 4}{4 - 0} = -\dfrac{4}{4} = -1$   **44.** $m = \dfrac{9 - (-9)}{3 - (-3)} = \dfrac{18}{6} = 3$

**45.** $m = \dfrac{8 - (-1)}{3 - 6} = -\dfrac{9}{3} = -3$

**46.** $m = \dfrac{2 - (-1)}{-1 - 9} = -\dfrac{3}{10} = -0.3$

**47.** $m = \dfrac{6 - (-1)}{2 - (-2)} = \dfrac{7}{4} = 1.75$

**48.** $m = \dfrac{3 - 2}{3 - 2} = \dfrac{1}{1} = 1$   **49.** $m = \dfrac{2 - 2}{3 - (-1)} = \dfrac{0}{4} = 0$

**50.** Both the projected high school enrollment and projected college enrollment are functions of the year because no two ordered pairs have the same year (first coordinate).

**51.** $f(2005) \approx 16$ million    **52.** $f(2005) \approx 16.9$ million

**53.** The number of people who attended the festival is a function of the year because no two ordered pairs have the same year (first coordinate).

**54.**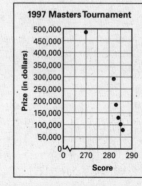

The money earned is a function of the score because two ordered pairs do not have the same score (first coordinate).

**55.** $f(t) = \dfrac{25}{4.25}t$, so $f(t) = \dfrac{100}{17}t$ or $f(t) \approx 5.88t$

**56.** B; $-7(-4) + 4 = 32$, $-7(-5) + 4 = 39$, $32 < 39$

**57.** C; $\frac{1}{2}(7) - 3 = \frac{1}{2}$, $\frac{1}{2}(11) - 5 = \frac{1}{2}$, $\frac{1}{2} = \frac{1}{2}$

**58.** C; $6\left(\frac{2}{7}\right) - \frac{1}{7} = \frac{11}{7}$, $6\left(\frac{2}{7}\right) - \frac{1}{7} = \frac{11}{7}$, $\frac{11}{7} = \frac{11}{7}$

**59.** $y = \dfrac{1}{x}$          $y = \dfrac{1}{x + 3}$

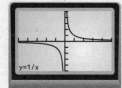
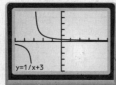

**60.** $y = \dfrac{1}{x}$ is not a function or all real values of $x$ because when $x = 0$, the equation is undefined. $y = \dfrac{1}{x + 3}$ is not a function for all real values of $x$ because when $x = -3$, the equation is undefined.

**61.** The domain can be restricted by adding $x \neq 0$ and $x \neq -3$ to the respective equations.

**62.** For $x < 0$ but close to 0, the graph falls sharply. For $x > 0$ but close to 0, the graph rises sharply.

### 4.8 Mixed Review (p. 262)

**63.** $\begin{bmatrix} 4 & -8 \\ 7 & 0 \end{bmatrix} - \begin{bmatrix} -5 & 3 \\ -5 & -7 \end{bmatrix} = \begin{bmatrix} 4 - (-5) & -8 - 3 \\ 7 - (-5) & 0 - (-7) \end{bmatrix}$

$= \begin{bmatrix} 9 & -11 \\ 12 & 7 \end{bmatrix}$

**64.** $\begin{bmatrix} -6.5 & -4.2 \\ 0 & 3.7 \end{bmatrix} + \begin{bmatrix} 2.4 & -5.1 \\ 4.3 & -3 \end{bmatrix}$

$= \begin{bmatrix} -6.5 + 2.4 & -4.2 - 5.1 \\ 0 + 4.3 & 3.7 - 3 \end{bmatrix} = \begin{bmatrix} -4.1 & -9.3 \\ 4.3 & 0.7 \end{bmatrix}$

**65.** $\begin{bmatrix} 6.2 & -12 \\ -2.5 & -4.4 \\ 3.4 & -5.8 \end{bmatrix} - \begin{bmatrix} -3.6 & 5.9 \\ 9.8 & -4.3 \\ -9 & 7.4 \end{bmatrix}$

$= \begin{bmatrix} 6.2 - (-3.6) & -12 - 5.9 \\ -2.5 - 9.8 & -4.4 - (-4.3) \\ 3.4 - (-9) & -5.8 - 7.4 \end{bmatrix}$

$= \begin{bmatrix} 9.8 & -17.9 \\ -12.3 & -0.1 \\ 12.4 & -13.2 \end{bmatrix}$

**66.** $\begin{bmatrix} 9 & 1 & 6 \\ -4 & -7 & 1 \\ -5 & 0 & -1 \end{bmatrix} + \begin{bmatrix} -6 & 3 & -5 \\ -2 & 4 & -4 \\ 0 & 5 & 1 \end{bmatrix}$

$= \begin{bmatrix} 9 - 6 & 1 + 3 & 6 - 5 \\ -4 - 2 & -7 + 4 & 1 - 4 \\ -5 + 0 & 0 + 5 & -1 + 1 \end{bmatrix} = \begin{bmatrix} 3 & 4 & 1 \\ -6 & -3 & -3 \\ -5 & 5 & 0 \end{bmatrix}$

**67.** $4x + 8 = 24$

$4x = 16$

$x = 4$

Check:

$4x + 8 = 24$

$4(4) + 8 = 24$

$16 + 8 = 24$

$24 = 24$

**68.** $3n = 5n - 12$

$12 = 2n$

$6 = n$

Check:

$3n = 5n - 12$

$3(6) = 5(6) - 12$

$18 = 30 - 12$

$18 = 18$

**69.** $9 - 5z = -8z$

$3z = -9$

$z = -3$

Check:

$9 - 5z = -8z$

$9 - 5(-3) = 24$

$9 + 15 = 24$

$24 = 24$

**70.** $-5y + 6 = 4y + 3$

$3 = 9y$

$\dfrac{3}{9} = y$

$\dfrac{1}{3} = y$

Check:

$-5y + 6 = 4y + 3$

$-5\left(\dfrac{1}{3}\right) + 6 = 4\left(\dfrac{1}{3}\right) + 3$

$-\dfrac{5}{3} + 6 = \dfrac{4}{3} + 3$

$\dfrac{13}{3} = \dfrac{13}{3}$

**71.** $3b + 8 = 9b - 7$

$15 = 6b$

$\dfrac{15}{6} = b$

$\dfrac{5}{2} = b$

Check:

$3b + 8 = 9b - 7$

$3\left(\dfrac{15}{6}\right) + 8 = 9\left(\dfrac{15}{6}\right) - 7$

$\dfrac{15}{2} + 8 = \dfrac{45}{2} - 7$

$\dfrac{31}{2} = \dfrac{31}{2}$

**72.** $-7q - 13 = 4 - 7q$

$0 - 13 = 4$

$-13 \neq 4$

There is no solution.

**73.** $2x - y + 3 = 0$

$-y = -2x - 3$

$y = 2x + 3$

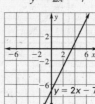

**74.** $x + 2y - 6 = 0$

$2y = -x + 6$

$y = -\dfrac{1}{2}x + 3$

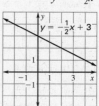

**75.** $y - 2x = -7$

$y = 2x - 7$

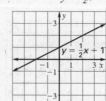

**76.** $5x - y = 4$

$-y = -5x + 4$

$y = 5x - 4$

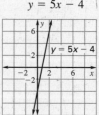

**77.** $x - 2y + 4 = 2$

$-2y = -x - 2$

$y = \dfrac{1}{2}x + 1$

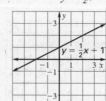

**78.** $4y + 12 = 0$

$4y = -12$

$y = -3$

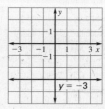

**Algebra 1**
Chapter 4  Worked-out Solution Key

# Chapter 4 *continued*

**79.** $d = 5 - rt$
$d = 5 - 1.5t$

| $t$ | $d$ |
|-----|-----|
| 1 | 3.5 |
| 2 | 2 |

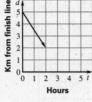

**Quiz 3 (p. 262)**

**1.** $4x + 3 = -5$
$4x = -8$
$x = -2$
Check:
$4x + 3 = -5$
$4x + 8 = 0$
$y = 4x + 8$

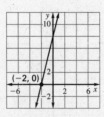

**2.** $6x - 12 = -9$
$6x = 3$
$x = \frac{1}{2}$
Check:
$6x - 12 = -9$
$6x - 3 = 0$
$y = 6x - 3$

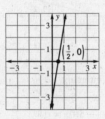

**3.** $8x - 7 = x$
$7x = 7$
$x = 1$
Check:
$8x - 7 = x$
$7x - 7 = 0$
$y = 7x - 7$

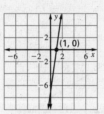

**4.** $-5x - 4 = 3x$
$-4 = 8x$
$-\frac{1}{2} = x$
Check:
$-5x - 4 = 3x$
$-8x - 4 = 0$
$y = -8x - 4$

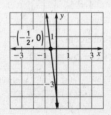

**5.** $\frac{1}{3}x + 5 = -\frac{2}{3}x - 8$
$x = -13$
Check:
$\frac{1}{3}x + 5 = -\frac{2}{3}x - 8$
$x + 13 = 0$
$y = x + 13$

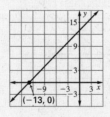

**6.** $\frac{3}{4}x + 2 = -\frac{3}{4}x - 6$
$\frac{3}{2}x = -8$
$x = -\frac{16}{3} = -5\frac{1}{3}$
Check:
$\frac{3}{4}x + 2 = \frac{3}{4}x - 6$
$\frac{3}{2}x + 8 = 0$
$y = \frac{3}{2}x + 8$

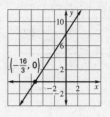

**7.** $h(x) = 5x - 9$
$h(3) = 5(3) - 9 = 15 - 9 = 6$
$h(0) = 5(0) - 9 = -9$
$h(-4) = 5(-4) - 9 = -20 - 9 = -29$

**8.** $g(x) = -4x + 3$
$g(3) = -4(3) + 3 = -12 + 3 = -9$
$g(0) = -4(0) + 3 = 3$
$g(-4) = (-4)(-4) + 3 = 16 + 3 = 19$

**9.** $f(x) = 1.75x - 2$
$f(3) = 1.75(3) - 2 = 5.25 - 2 = 3.25$
$f(0) = 1.75(0) - 2 = -2$
$f(-4) = 1.75(-4) - 2 = -7 - 2 = -9$

**10.** $h(x) = -1.4x$
$h(3) = -1.4(3) = -4.2$
$h(0) = -1.4(0) = 0$
$h(-4) = (-1.4)(-4) = 5.6$

**11.** $f(x) = \frac{1}{4}x + 9$
$f(3) = \frac{1}{4}(3) + 9 = \frac{3}{4} + 9 = \frac{39}{4} = 9\frac{3}{4}$
$f(0) = \frac{1}{4}(0) + 9 = 9$
$f(-4) = \frac{1}{4}(-4) + 9 = -1 + 9 = 8$

**12.** $g(x) = \frac{4}{7}x + \frac{2}{7}$
$g(3) = \frac{4}{7}(3) + \frac{2}{7} = \frac{12}{7} + \frac{2}{7} = \frac{14}{7} = 2$
$g(0) = \frac{4}{7}(0) + \frac{2}{7} = \frac{2}{7}$
$g(-4) = \frac{4}{7}(-4) + \frac{2}{7} = -\frac{16}{7} + \frac{2}{7} = -\frac{14}{7} = -2$

**13.**

**14.**
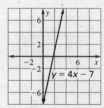

**15.**
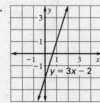

**16.**
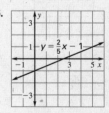

# Chapter 4 *continued*

**17.**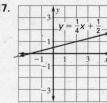

**18.**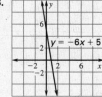

**9.** $x - 10 = 2y$

$$y = \frac{x - 10}{2}$$

| $x$ | $y$ |
|---|---|
| $-2$ | $-6$ |
| $0$ | $-5$ |
| $2$ | $-4$ |

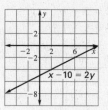

**19.**  $y = 2.1x + 75$
$600 = 2.1x + 75$
$525 = 2.1x$
$x = 250$ lunches

**10.**  *x*-intercept                    *y*-intercept
$-x + 4y = 8$                 $-x + 4y = 8$
$-x + 4(0) = 8$            $0 + 4y = 8$
$-x = 8$                      $4y = 8$
$x = -8$                     $y = 2$

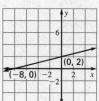

## Chapter 4 Review *(pp. 264–266)*

**1.**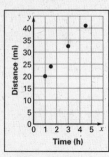

**2–5.**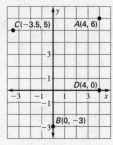

**6.** $y = 2x + 2$

| $x$ | $y$ |
|---|---|
| $-1$ | $0$ |
| $0$ | $2$ |
| $1$ | $4$ |

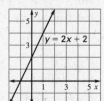

**11.**  *x*-intercept                    *y*-intercept
$3x + 5y = 15$              $3x + 5y = 15$
$3x + 5(0) = 15$         $3(0) + 5y = 15$
$3x = 15$                     $5y = 15$
$x = 5$                        $y = 3$

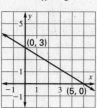

**7.** $y = 7 - \frac{1}{2}x$

| $x$ | $y$ |
|---|---|
| $-2$ | $8$ |
| $0$ | $7$ |
| $2$ | $6$ |

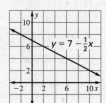

**12.**  *x*-intercept                    *y*-intercept
$4x - 5y = -20$           $4x - 5y = -20$
$4x - 5(0) = -20$       $4(0) - 5y = -20$
$4x = -20$                   $-5y = -20$
$x = -5$                      $y = 4$

**8.** $y = -4(x + 1) = -4x - 4$

| $x$ | $y$ |
|---|---|
| $-1$ | $0$ |
| $0$ | $-4$ |
| $1$ | $-8$ |

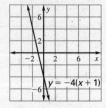

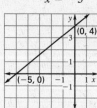

**Algebra 1**
Chapter 4 Worked-out Solution Key

# Chapter 4 *continued*

**13.** *x*-intercept

$$2x + 3y = 12$$
$$2x + 3(0) = 12$$
$$2x = 12$$
$$x = 6$$

*y*-intercept

$$2x + 3y = 12$$
$$2(0) + 3y = 12$$
$$3y = 12$$
$$y = 4$$

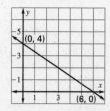

**14.**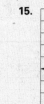

$$m = \frac{4 - 1}{3 - 2} = \frac{3}{1} = 3$$

**15.**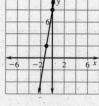

$$m = \frac{2 - 8}{-1 - 0} = \frac{-6}{-1} = 6$$

**16.**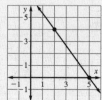

$$m = \frac{0 - 4}{5 - 2} = -\frac{4}{3}$$

**17.**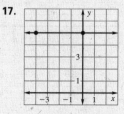

$$m = \frac{5 - 5}{-4 - 0} = \frac{0}{-4} = 0$$

**18.** $y = kx$

$$35 = k(7)$$
$$5 = k$$
$$y = 5x$$

**19.** $y = kx$

$$-4 = k(12)$$
$$-\frac{4}{12} = k$$
$$-\frac{1}{3} = k$$
$$y = -\frac{1}{3}x$$

**20.** $y = kx$

$$-16 = k(4)$$
$$-4 = k$$
$$y = -4x$$

**21.** $y = kx$

$$10.5 = k(3)$$
$$3.5 = k$$
$$y = 3.5x$$

**22.** $y = -x - 2$

**23.** $x - 4y = 12$

$$-4y = -x + 12$$
$$y = \frac{1}{4}x - 3$$

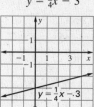

**24.** $-x + 6y = -24$

$$6y = x - 24$$
$$y = \frac{1}{6}x - 4$$

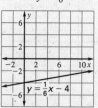

**25.** $3x - 6 = 0$

$$y = 3x - 6$$

Check:

$$3x - 6 = 0$$
$$3(2) - 6 \overset{?}{=} 0$$
$$6 - 6 \overset{?}{=} 0$$
$$0 = 0$$

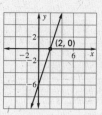

**26.** $-5x - 3 = 0$

$$y = -5x - 3$$

Check:

$$-5x - 3 = 0$$
$$-5\left(-\frac{3}{5}\right) - 3 \overset{?}{=} 0$$
$$3 - 3 \overset{?}{=} 0$$
$$0 = 0$$

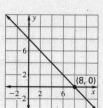

**27.** $3x + 8 = 4x$

$$-x + 8 = 0$$
$$y = 8 - x$$

Check:

$$3x + 8 = 4x$$
$$3(8) + 8 \overset{?}{=} 4(8)$$
$$24 + 8 \overset{?}{=} 32$$
$$32 = 32$$

# Chapter 4 *continued*

**28.** $-4x - 1 = 7$
$-4x - 8 = 0$
$\qquad y = -4x - 8$
Check:
$\qquad -4x - 1 = 7$
$\qquad -4(-2) - 1 \overset{?}{=} 7$
$\qquad\qquad 8 - 1 \overset{?}{=} 7$
$\qquad\qquad\qquad 7 = 7$

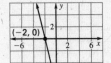

**29.** $f(x) = x - 7$
$f(-2) = -2 - 7 = -9$

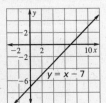

**30.** $f(x) = -x + 4$
$f(4) = -4 + 4 = 0$

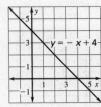

**31.** $f(x) = 2x + 6$
$f(-3) = 2(-3) + 6 = -6 + 6 = 0$

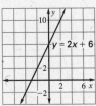

**32.** $f(x) = 1.5x - 4.2$
$f(-9) = 1.5(-9) - 4.2 = -13.5 - 4.2 = -17.7$

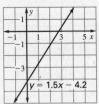

## Chapter 4 Test *(p. 267)*

**1.**

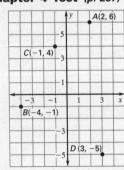

**2.**

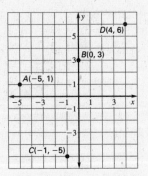

**3.**

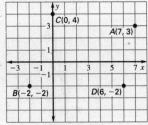

**4.**

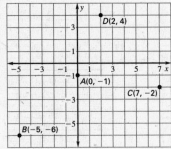

**5.**

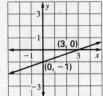

**6.**

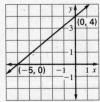

**7.**

**8.**

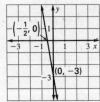

**9.**

| $x$ | $y$ |
|-----|-----|
| $-1$ | $4$ |
| $0$ | $3$ |
| $1$ | $2$ |

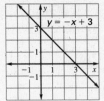

# Chapter 4 *continued*

**10.**

| x | y |
|----|----|
| −2 | 4 |
| 0 | 4 |
| 2 | 4 |

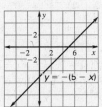

**11.**

| x | y |
|----|----|
| −1 | −6 |
| 0 | −5 |
| 1 | −4 |

**12.**

| x | y |
|----|----|
| 6 | −2 |
| 6 | 0 |
| 6 | 2 |

**13.**

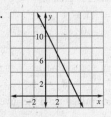

Methods may vary.

**14.**

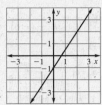

Methods may vary.

**15.**

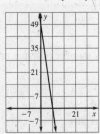

Methods may vary.

**16.**

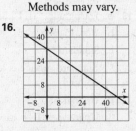

Methods may vary.

**17.** $m = \dfrac{-6 - 1}{-2 - 0} = \dfrac{-7}{-2} = \dfrac{7}{2}$

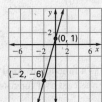

**18.** $m = \dfrac{-1 - (-7)}{-4 - 5} = \dfrac{6}{-9} = -\dfrac{2}{3}$

**19.** $m = \dfrac{-2 - 5}{2 - (-3)} = -\dfrac{7}{5}$

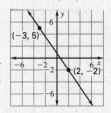

**20.** $m = \dfrac{-1 - (-1)}{2 - (-3)} = \dfrac{0}{5} = 0$

**21.** $y = kx$
$-2 = k(-2)$
$1 = k$
$y = x$

**22.** $y = kx$
$10 = k(2)$
$5 = k$
$y = 5x$

**23.** $y = kx$
$7 = k(-3)$
$-\dfrac{7}{3} = k$
$y = -\dfrac{7}{3}x$

**24.** $y = kx$
$6 = k\left(\dfrac{1}{2}\right)$
$12 = k$
$y = 12x$

**25.** $y = kx$
$3.9 = k(1.3)$
$3 = k$
$y = 3x$

**26.** $y = kx$
$3.2 = k(1.6)$
$0.2 = k$
$y = 0.2x$

**27.** The lines are not parallel because they have different slopes.
$y = 4x + 3$
$y = -4x - 5$

**28.** The lines are parallel because their slopes are equal.
$y = \dfrac{3}{5}x - 2$
$y = \dfrac{3}{5}x + 7$

# Chapter 4 *continued*

**29.** $x - 2 = -3x$

$4x - 2 = 0$

$y = 4x - 2$

Check:

$x - 2 = -3x$

$\frac{1}{2} - 2 \overset{?}{=} -3\left(\frac{1}{2}\right)$

$-\frac{3}{2} = -\frac{3}{2}$

**30.** $f(x) = 6x$

$f(3) = 6(3) = 18$

$f(0) = 6(0) = 0$

$f(-4) = 6(-4) = -24$

**31.** $f(x) = -(x - 2)$

$f(3) = -3(3 - 2) = -1$

$f(0) = -(0 - 2) = -(-2) = 2$

$f(-4) = -(-6) = -(-6) = 6$

**32.** $g(x) = 3.2x + 2.8$

$g(3) = 3.2(3) + 2.8 = 9.6 + 2.8 = 12.4$

$g(0) = 3.2(0) + 2.8 = 2.8$

$g(-4) = 3.2(-4) + 2.8 = -12.8 + 2.8 = -10$

**33.**

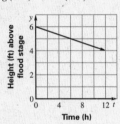

**34.** Shoe size is a function of foot length because no shoe sizes have the same foot length.

## Chapter 4 Standardized Test *(pp. 268–269)*

**1.** E

**2.** A

$-4x - \frac{1}{2}y = 10$

$-4(0) - \frac{1}{2}y = 10$

$-\frac{1}{2}y = 10$

$y = -20$

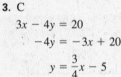

**3.** C

$3x - 4y = 20$

$-4y = -3x + 20$

$y = \frac{3}{4}x - 5$

**4.** D

$m = \frac{1 - 2}{2 - 1} = \frac{-1}{1} = -1$

**5.** C

$m = \frac{0 - 3}{-5 - 0} = \frac{-3}{-5} = \frac{3}{5}$

**6.** B

$5x - y = -2$

$-y = -5x - 2$

$y = 5x + 2$

**7.** D

$(-1, -12)$

**8.** D

$-13x - y = -65$

$-13x - 0 = -65$

$-13x = -65$

$x = 5$

**9.** C

$f(x) = -x^2 - 6x - 7$

$f(-2) = -(-2^2) - 6(-2) - 7 = -4 + 12 - 7 = 1$

**10.** C

**a.** $m = \frac{-3 - (-3)}{-12 - 4} = \frac{0}{-16} = 0$

The two numbers are equal.

**b.** 0

**11.** A

**a.** $m = \frac{4 - 6}{-7 - 4.5} = \frac{-2}{-11.5} = \frac{2}{11.5}$

The slope of the line in column A is greater.

**b.** $m = \frac{-7 - 4.5}{4 - (-6)} = \frac{-11.5}{10} = -\frac{11.5}{10}$

**12.** D

**a.** $m = \frac{4 - y}{6.8 - 3.5}$  **b.** $m = \frac{4 - q}{6.8 - 3.5}$

The relationship cannot be determined from the given information.

**13. a.** I. Upgrade  II. Unlimited  III. Standard

**b.** $T = 15 + b$

**c.** Standard

Standard service is less for any number of hours up to 15 hours.

**d.** Upgrade

Upgrade $= T = 15 + (b - 20)$

$= 15 + (24 - 20)$

$= 15 + 4$

$= \$19$

Standard $= T = 10 + (b - 10)$

$= 10 + (24 - 10)$

$= 10 + 14$

$= \$24$

Unlimited $= T = \$20$

**e.** The model in part a gives cost as a function of total time. The model $c = 10 + a$ gives cost as a function of time beyond 10 hours.

# CHAPTER 5

## Think & Discuss (p. 271)
1. about 5800 B.C.
2. about 5200 years
3. from about 1000 B.C. to A.D. 2000

## Skill Review (p. 272)
1. $4(x + 8) = 20x$
   $4x + 32 = 20x$
   $32 = 16x$
   $x = 2$

2. $2x + 2 = 3x - 8$
   $10 = x$

3. $9a = 4a - 24 + a$
   $9a = 5a - 24$
   $4a = -24$
   $a = -6$

4.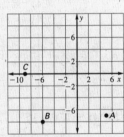

5.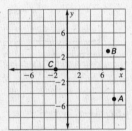

6. $4x + 9y = 18$
   x-intercept: $4x + 9(0) = 18$
   $\quad\quad\quad 4x = 18$
   $\quad\quad\quad\quad x = 4.5$
   y-intercept: $4(0) + 9y = 18$
   $\quad\quad\quad\quad 9y = 18$
   $\quad\quad\quad\quad\quad y = 2$

7. $-x - 5y = 15 + 2x$
   $-3x - 5y = 15$
   x-intercept: $-3x - 5(0) = 15$
   $\quad\quad\quad\quad\quad x = -5$
   y-intercept: $-3(0) - 5y = 15$
   $\quad\quad\quad\quad\quad y = -3$

8. $-6x - 2y = 8$
   x-intercept: $-6x - 2(0) = 8$
   $\quad\quad\quad\quad -6x = 8$
   $\quad\quad\quad\quad\quad x = -\frac{4}{3}$
   y-intercept: $-6(0) - 2y = 8$
   $\quad\quad\quad\quad\quad -2y = 8$
   $\quad\quad\quad\quad\quad\quad y = -4$

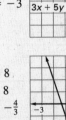

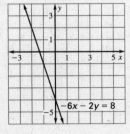

## Lesson 5.1

### 5.1 Guided Practice (p. 276)
1. 5 is the y-intercept; the equation is in slope-intercept form and $b = 5$.

2. the increase in the number of millions of people per year

3. $y = x$
4. $y = -7x - \frac{2}{3}$
5. $y = -x + 3$
6. $y = -2x$
7. $y = -3x + \frac{1}{2}$
8. $y = 4x - 6$
9. $y = 1.5x + 4$

10.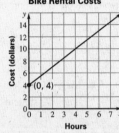

11. $y = 1.5(12) + 4$
    $\quad = 18 + 4$
    $\quad = \$22$

### 5.1 Practice and Applications (pp. 276–278)
12. $y = 3x - 2$
13. $y = x + 2$
14. $y = 4$
15. $y = 2x - 1$
16. $y = \frac{3}{2}x + 3$
17. $-\frac{1}{4}x + 1$
18. $y = -6x + \frac{3}{4}$
19. $y = -3x - \frac{1}{2}$
20. $y = 2x + 2$
21. $y = x - 3$
22. $y = -\frac{2}{3}x + 2$
23. a. $y = 2x + 3$  b. $y = 2x$  c. $y = 2x - 4$
24. a. $y = -x + 4$  b. $y = -x + 2$  c. $y = -x$
25. a. $y = 4$  b. $y = 1$  c. $y = -2$
26. $P = 37{,}400t + 3{,}486{,}000$
27. $P = 37{,}400(6) + 3{,}486{,}000$
    $\quad = 224{,}400 + 3{,}486{,}000$
    $\quad = 3{,}710{,}400$
28. $y = 0.25x + 30$
29. $y = 0.25(25) + 30 = \$36.25$
    $y = 0.25(50) + 30 = \$42.50$
    $y = 0.25(75) + 30 = \$48.75$
    $y = 0.25(100) + 30 = \$55.00$

| Miles (x) | 25 | 50 | 75 | 100 |
|-----------|-----|-----|-----|-----|
| Cost (y) | \$36.25 | \$42.50 | \$48.75 | \$55.00 |

30. $y = 40 - 5x$; dollars $- \dfrac{\text{dollars}}{\text{week}} \cdot$ weeks $=$ dollars

31. $y = 180 - 2x$; lb $- \dfrac{\text{lb}}{\text{month}} \cdot$ months $=$ lb

32. $y = 2x + 5$; mi $+ \dfrac{\text{mi}}{\text{hour}} \cdot$ hours $=$ mi

33. $y = 200 - 50t$; The distance from home decreases as time increases.

# Chapter 5 *continued*

**34.** $C = 2.5x + 500$     **35.** $T = 5x$

**36.**

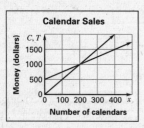

**Calendar Sales**

The point of intersection is at (200, 1000) which means the cost of producing 200 calendars is $1000. Also, there would be a $1000 income from the sales of 200 calendars.

**37.** Answers may vary. Because the amount raised would equal the amount paid, no money would be either lost or made.

**38.** C;   $P = 0.22t + 17.4$

**39.** C;   $P = 0.22(1) + 17.4 = 17.62$

**40.** **a.** $y = 2x + 12$

**b.**

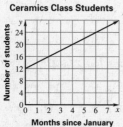

**Ceramics Class Students**

**c.** May:  20 students
    June:  22 students

### 5.1 Mixed Review (p. 278)

**41.** $\dfrac{3x}{x + y} = \dfrac{3(-3)}{-3 + 6} = \dfrac{-9}{3} = -3$

**42.** $\dfrac{x}{x + 2} = \dfrac{-3}{-3 + 2} = \dfrac{-3}{-1} = 3$

**43.** $\dfrac{y^2 + x}{x} = \dfrac{6^2 + (-3)}{-3} = \dfrac{36 + (-3)}{-3} = \dfrac{33}{-3} = -11$

**44.**

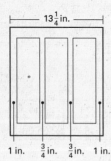

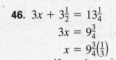

1 in.   $\frac{3}{4}$ in.   $\frac{3}{4}$ in.   1 in.

**45.** $3x + 1 + 1 + \frac{3}{4} + \frac{3}{4} = 13\frac{1}{4}$
$3x + 3\frac{1}{2} = 13\frac{1}{4}$

**46.** $3x + 3\frac{1}{2} = 13\frac{1}{4}$
$3x = 9\frac{3}{4}$
$x = 9\frac{3}{4}\left(\frac{1}{3}\right)$
$x = \frac{13}{4} = 3\frac{1}{4}$ in.

**47.** $y = 10x$

**Sailboard Rental Cost**

**48.** $y = 10$

**Life Jacket Rental Cost**

**49.** $y + 2x = 2$
$y = -2x + 2$
$-2, 2$

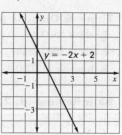

**50.** $3x - y = -5$
$y = 3x + 5$
$3, 5$

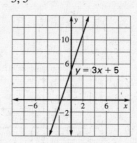

**51.** $2y - 3x = 6$
$2y = 3x + 6$
$y = \frac{3}{2}x + 3$
$\frac{3}{2}, 3$

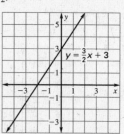

**52.** $4x + 2y = 6$
$2y = -4x + 6$
$y = -2x + 3$
$-2, 3$

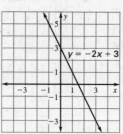

**53.** $4y + 12x = 16$
$4y = -12x + 16$
$y = -3x + 4$
$-3, 4$

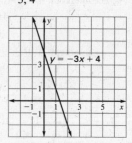

**54.** $25x - 5y = 30$
$-5y = -25x + 30$
$y = 5x - 6$
$5, -6$

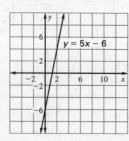

**55.** $x + 3y = 15$
$3y = -x + 15$
$y = -\frac{1}{3}x + 5$
$-\frac{1}{3}, 5$

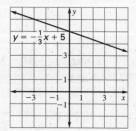

**Algebra 1**
Chapter 5 Worked-out Solution Key

# Chapter 5 continued

**56.** $x + 6y = 12$
$6y = -x + 12$
$y = -\frac{1}{6}x + 2$
$-\frac{1}{6}, 2$

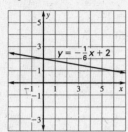

**57.** $x - y = 10$
$-y = -x + 10$
$y = x - 10$
$1, -10$

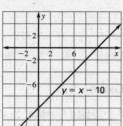

## Lesson 5.2

### 5.2 Guided Practice (p. 282)

**1.** parallel

**2.** Substitute the slope and the coordinates into $y = mx + b$ and solve for $b$, the $y$-intercept. Substitute $b$ and $m$, the slope, back into $y = mx + b$ to get the equation of a line.

**3.** $y = mx + b$
$4 = \frac{1}{2}(3) + b$
$4 = \frac{3}{2} + b$
$\frac{5}{2} = b$
$y = \frac{1}{2}x + \frac{5}{2}$

**4.** $y = mx + b$
$-4 = -5(2) + b$
$-4 = -10 + b$
$6 = b$
$y = -5x + 6$

**5.** $y = mx + b$
$-10 = \frac{2}{3}(10) + b$
$-10 = \frac{20}{3} + b$
$-\frac{50}{3} = b$
$y = \frac{2}{3}x - \frac{50}{3}$

**6.** $y = mx + b$
$2 = -2(-12) + b$
$2 = 24 + b$
$-22 = b$
$y = -2x - 22$

**7.** $y = mx + b$
$8 = 5(4) + b$
$8 = 20 + b$
$-12 = b$
$y = 5x - 12$

**8.** $y = mx + b$
$-5 = 0(0) + b$
$-5 = b$
$y = -5$

**9.** $y = mx + b$
$4 = \frac{1}{2}(-6) + b$
$4 = -3 + b$
$7 = b$
$y = \frac{1}{2}x + 7$

**10.** $y = mx + b$
$-3 = -3(-4) + b$
$-3 = 12 + b$
$-15 = b$
$y = -3x - 15$

**11.** $y = -8.25x + 616.50$

### 5.2 Practice and Applications (pp. 282–284)

**12.** $y = mx + b$
$4 = 3(1) + b$
$4 = 3 + b$
$1 = b$
$y = 3x + 1$

**13.** $y = mx + b$
$-3 = 2(2) + b$
$-3 = 4 + b$
$-7 = b$
$y = 2x - 7$

**14.** $y = mx + b$
$2 = -1(-4) + b$
$2 = 4 + b$
$-2 = b$
$y = -x - 2$

**15.** $y = mx + b$
$-3 = -4(1) + b$
$-3 = -4 + b$
$1 = b$
$y = -4x + 1$

**16.** $y = mx + b$
$-5 = -2(-3) + b$
$-5 = 6 + b$
$-11 = b$
$y = -2x - 11$

**17.** $y = mx + b$
$3 = 4(1) + b$
$3 = 4 + b$
$-1 = b$
$y = 4x - 1$

**18.** $y = mx + b$
$5 = \frac{1}{2}(2) + b$
$5 = 1 + b$
$4 = b$
$y = \frac{1}{2}x + 4$

**19.** $y = mx + b$
$2 = \frac{1}{3}(-3) + b$
$2 = -1 + b$
$3 = b$
$y = \frac{1}{3}x + 3$

**20.** $y = mx + b$
$2 = 3(0) + b$
$2 = b$
$y = 3x + 2$

**21.** $y = mx + b$
$-2 = 4(0) + b$
$-2 = b$
$y = 4x - 2$

**22.** $y = mx + b$
$4 = 0(3) + b$
$4 = b$
$y = 4$

**23.** $y = mx + b$
$4 = 0(-2) + b$
$4 = b$
$y = 4$

**24.** $y = -\frac{2}{3}x + b$
$2 = -\frac{2}{3}(-2) + b$
$2 = \frac{4}{3} + b$
$\frac{2}{3} = b$
$y = -\frac{2}{3}x + \frac{2}{3}$

**25.** $y = \frac{1}{2}x + b$
$3 = \frac{1}{2}(3) + b$
$3 = \frac{3}{2} + b$
$\frac{3}{2} = b$
$y = \frac{1}{2}x + \frac{3}{2}$

**26.** $y = \frac{2}{3}x + b$
$1 = \frac{2}{3}(1) + b$
$1 = \frac{2}{3} + b$
$\frac{1}{3} = b$
$y = \frac{2}{3}x + \frac{1}{3}$

**27.** $y = \frac{3}{5}x + b$
$0 = \frac{3}{5}(0) + b$
$0 = b$
$y = \frac{3}{5}x$

**28.** $y = b$
$-3 = b$
$y = -3$

**29.** $x = 2$

**30.** $y = mx + b$
$0 = -\frac{2}{3}(2) + b$
$0 = -\frac{4}{3} + b$
$\frac{4}{3} = b$
$y = -\frac{2}{3}x + \frac{4}{3}$

**31.** $y = mx + b$
$0 = 3(4) + b$
$-12 = b$
$y = 3x - 12$

**32.** $y = mx + b$
$2 = 2(3) + b$
$2 = 6 + b$
$-4 = b$
$y = 2x - 4$

**33.** $y = mx + b$
$0 = 1(-2) + b$
$0 = -2 + b$
$2 = b$
$y = x + 2$

# Chapter 5 *continued*

**34.** $y = mx + b$
$4 = -2(4) + b$
$4 = -8 + b$
$12 = b$
$y = -2x + 12$

**35.** $y = mx + b$
$2 = -3(4) + b$
$2 = -12 + b$
$14 = b$
$y = -3x + 14$

**36.** $y = mx + b$
$1 = \frac{2}{3}(2) + b$
$1 = \frac{4}{3} + b$
$-\frac{1}{3} = b$
$y = \frac{2}{3}x - \frac{1}{3}$

**37.** $y = mx + b$
$1 = -\frac{1}{3}(4) + b$
$1 = -\frac{4}{3} + b$
$\frac{7}{3} = b$
$y = -\frac{1}{3}x + \frac{7}{3}$

**38.** $y = mx + b$
$3 = -4(5) + b$
$3 = -20 + b$
$23 = b$
$y = -4x + 23$

**39.** $y = mx + b$
$-3 = 6(5) + b$
$-3 = 30 + b$
$-33 = b$
$y = 6x - 33$

**40.** $y = mx + b$
$-2 = -9(7) + b$
$-2 = -63 + b$
$61 = b$
$y = -9x + 61$

**41.** $y = mx + b$
$1 = \frac{1}{2}(6) + b$
$1 = 3 + b$
$-2 = b$
$y = \frac{1}{2}x - 2$

**42.** $y = 2000x + 132,000$

**43.** $y = 2000(15) + 132,000 = 162,000$

**44.** $s = 750n + 17,250$

**45.** $s = 750(6) + 17,250 = \$21,750$

**46.** $C = 0.23m + 26$

**47.** $C = 0.23(60) + 26 = 13.8 + 26 = \$39.80$

**48.** $y = 1.5x + 2$     **49.** $\$2.00$

**50. a.** $m = 2n + 23$
**b.** $45 = 2n + 23$
$22 = 2n$
$n = 11$ weeks
**c.** $m = 2(52) + 23 = 127$ miles
No, he would have to, on average, run about 18 miles/day. The increase would not go on indefinitely.

**51. a.** $y = 0.08x + 30$; $y = 40$
**b.**

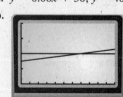

**c.** $40 = 0.08x + 30$
$10 = 0.08x$
$125 = x$
If you drive exactly 125 miles.

**d.** If you drive less than 125 miles, Car Rental Agency A is cheaper. If you drive more than 125 miles, Car Rental Agency B is cheaper.

### 5.2 Mixed Review (p. 284)

**52.** 78,125     **53.** 32,768     **54.** 531,441     **55.** 531,441

**56.** 256     **57.** 64     **58.** 4096     **59.** 10,000

**60.** 5,764,801     **61.** 10,077,696     **62.** 48,828,125

**63.** about 2,541,865,828,000     **64.** $-8$     **65.** $-\frac{1}{2}$

**66.** 2     **67.** $-11$     **68.** $-4$     **69.** $-20$

**70.** $6 + (-8) - 4 = 6 + (-8) + (-4) = -2 + (-4) = -6$

**71.** $4 - (-7) + 3 = 4 + 7 + 3 = 14$

**72.** $5 + (-3) + (-5) = 2 + (-5) = -3$

**73.** $\frac{3}{4}$     **74.** $-2$     **75.** $-\frac{2}{3}$

## Lesson 5.3

### 5.3 Guided Practice (p. 288)

**1.** *Sample answer*: Given two points, you first need to find the slope. Then you can choose one of the points and proceed as if you were given the slope and one point.

**2.** You can use one of the points with the slope in the equation $y = mx + b$ to find the y-intercept.

**3.** $\frac{1}{4}$     **4.** $-2$     **5.** $-1$

**6.** $m = \frac{5 - (-3)}{-1 - 3} = \frac{8}{-4} = -2$
$y = -2x + b$
$5 = -2(-1) + b$
$5 = 2 + b$
$3 = b$
$y = -2x + 3$

**7.** $m = \frac{3 - (-1)}{4 - (-1)} = \frac{4}{5}$
$y = \frac{4}{5}x + b$
$-1 = \frac{4}{5}(-1) + b$
$-1 = -\frac{4}{5} + b$
$-\frac{1}{5} = b$
$y = \frac{4}{5}x - \frac{1}{5}$

**8.** $m = \frac{16 - 1}{0 - (-4)} = \frac{15}{4}$
$y = \frac{15}{4}x + b$
$1 = \frac{15}{4}(-4) + b$
$1 = -15 + b$
$16 = b$
$y = \frac{15}{4}x + 16$

**9.** $m = \frac{5 - 1}{4 - (-1)} = \frac{4}{5}$
$y = \frac{4}{5}x + b$
$1 = \frac{4}{5}(-1) + b$
$1 = -\frac{4}{5} + b$
$\frac{9}{5} = b$
$y = \frac{4}{5}x + \frac{9}{5}$

**10.** $m = \frac{4 - (-2)}{-6 - 3} = \frac{6}{-9} = \frac{-2}{3}$
$y = -\frac{2}{3}x + b$
$-2 = -\frac{2}{3}(3) + b$
$-2 = -2 + b$
$0 = b$
$y = -\frac{2}{3}x$

# Chapter 5 *continued*

**11.** $m = \dfrac{-7 - 3}{-1 - (-4)} = \dfrac{-10}{3}$

$y = -\dfrac{10}{3}x + b$

$3 = -\dfrac{10}{3}(-4) + b$

$3 = \dfrac{40}{3} + b$

$-\dfrac{31}{3} = b$

$y = -\dfrac{10}{3}x - \dfrac{31}{3}$

**12.** $m = \dfrac{-8 - 5}{-6 - (-2)} = \dfrac{-13}{-4} = \dfrac{13}{4}$

$y = \dfrac{13}{4}x + b$

$5 = \dfrac{13}{4}(-2) + b$

$5 = -\dfrac{13}{2} + b$

$\dfrac{23}{2} = b$

$y = \dfrac{13}{4}x + \dfrac{23}{2}$

**13.** $m = \dfrac{2 - (-4)}{4 - (-8)} = \dfrac{6}{12} = \dfrac{1}{2}$

$y = \dfrac{1}{2}x + b$

$2 = \dfrac{1}{2}(4) + b$

$2 = 2 + b$

$0 = b$

$y = \dfrac{1}{2}x$

**14.** $m = \dfrac{-9 - (-3)}{-8 - (-1)} = \dfrac{-6}{-7} = \dfrac{6}{7}$

$y = \dfrac{6}{7}x + b$

$-3 = \dfrac{6}{7}(-1) + b$

$-3 = -\dfrac{6}{7} + b$

$-\dfrac{15}{7} = b$

$y = \dfrac{6}{7}x - \dfrac{15}{7}$

**15.** $m = \dfrac{-3 - 3}{4 - 5} = \dfrac{-6}{-1} = 6$

$y = 6x + b$

$3 = 6(5) + b$

$3 = 30 + b$

$-27 = b$

$y = 6x - 27$

**16.** $m = \dfrac{-8 - (-10)}{-3 - 6} = \dfrac{2}{-9}$

$y = -\dfrac{2}{9}x + b$

$-10 = -\dfrac{2}{9}(6) + b$

$-10 = -\dfrac{4}{3} + b$

$-\dfrac{26}{3} = b$

$y = -\dfrac{2}{9}x - \dfrac{26}{3}$

**17.** $m = \dfrac{2 - 2}{7 - 12} = \dfrac{0}{-5} = 0$

$y = 0x + b$

$2 = 0(7) + b$

$2 = b$

$y = 2$

## 5.3 Practice and Applications (pp. 288–290)

**18.** $m = \dfrac{-1 - \frac{1}{2}}{0 - (-3)} = \dfrac{-1\frac{1}{2}}{3} = -\dfrac{1}{2}$

$y = -\dfrac{1}{2}x + b$

$-1 = -\dfrac{1}{2}(0) + b$

$-1 = b$

$y = -\dfrac{1}{2}x - 1$

**19.** $m = \dfrac{-3 - 0}{0 - (-1)} = \dfrac{-3}{1} = -3$

$y = -3x + b$

$0 = -3(-1) + b$

$0 = 3 + b$

$-3 = b$

$y = -3x - 3$

**20.** $m = \dfrac{-4 - 2}{-2 - 4} = \dfrac{-6}{-6} = 1$

$y = 1x + b$

$2 = 1(4) + b$

$2 = 4 + b$

$-2 = b$

$y = x - 2$

# Chapter 5 *continued*

**21.** $m = \dfrac{6-(-2)}{2-(-1)} = \dfrac{8}{3}$

$y = \dfrac{8}{3}x + b$

$-2 = \dfrac{8}{3}(-1) + b$

$-2 = -\dfrac{8}{3} + b$

$\dfrac{2}{3} = b$

$y = \dfrac{8}{3}x + \dfrac{2}{3}$

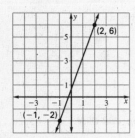

**22.** $m = \dfrac{-1-4}{5-1} = -\dfrac{5}{4}$

$y = -\dfrac{5}{4}x + b$

$4 = -\dfrac{5}{4}(1) + b$

$4 = -\dfrac{5}{4} + b$

$\dfrac{21}{4} = b$

$y = -\dfrac{5}{4}x + \dfrac{21}{4}$

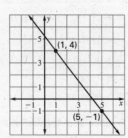

**23.** $m = \dfrac{-2-(-2)}{3-(-1)} = \dfrac{0}{4} = 0$

$y = 0x + b$

$-2 = 0(-1) + b$

$-2 = b$

$y = -2$

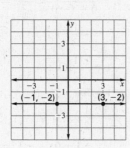

**24.** $m = \dfrac{6-0}{-2-2} = \dfrac{6}{-4} = -\dfrac{3}{2}$

$y = -\dfrac{3}{2}x + b$

$0 = -\dfrac{3}{2}(2) + b$

$0 = -3 + b$

$3 = b$

$y = -\dfrac{3}{2}x + 3$

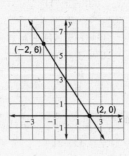

**25.** $m = \dfrac{-3-7}{2-(-3)} = \dfrac{-10}{5} = -2$

$y = -2x + b$

$-3 = -2(2) + b$

$-3 = -4 + b$

$1 = b$

$y = -2x + 1$

**26.** $m = \dfrac{-5-4}{0-3} = \dfrac{-9}{-3} = 3$

$y = 3x + b$

$-5 = 3(0) + b$

$-5 = b$

$y = 3x - 5$

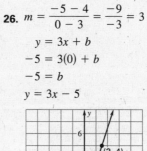
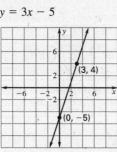

**27.** $m = \dfrac{-4-2}{6-(-1)} = -\dfrac{6}{7}$

$y = -\dfrac{6}{7}x + b$

$2 = -\dfrac{6}{7}(-1) + b$

$2 = \dfrac{6}{7} + b$

$\dfrac{8}{7} = b$

$y = -\dfrac{6}{7}x + \dfrac{8}{7}$

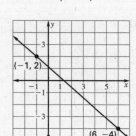

**28.** $m = \dfrac{8-(-1)}{8-(-2)} = \dfrac{9}{10}$

$y = \dfrac{9}{10}x + b$

$-1 = \dfrac{9}{10}(-2) + b$

$-1 = -\dfrac{18}{10} + b$

$\dfrac{8}{10} = b$

$\dfrac{4}{5} = b$

$y = \dfrac{9}{10}x + \dfrac{4}{5}$

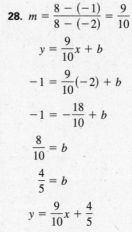

**29.** $m = \dfrac{4-1}{7-1} = \dfrac{3}{6} = \dfrac{1}{2}$

$y = \dfrac{1}{2}x + b$

$1 = \dfrac{1}{2}(1) + b$

$1 = \dfrac{1}{2} + b$

$\dfrac{1}{2} = b$

$y = \dfrac{1}{2}x + \dfrac{1}{2}$

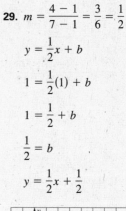

**30.** $m = \dfrac{-2-4}{1-2} = \dfrac{-6}{-1} = 6$

$y = 6x + b$

$4 = 6(2) + b$

$4 = 12 + b$

$-8 = b$

$y = 6x - 8$

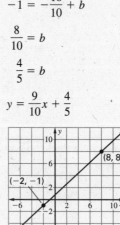
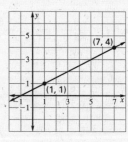
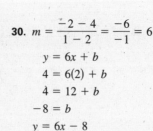
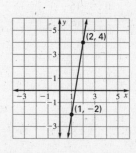

# Chapter 5 *continued*

**31.** $m = \dfrac{-6 - (-3)}{5 - 5}$

$\quad = \dfrac{-3}{0}$

$\quad = $ undefined

$\quad x = 5$

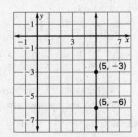

**32.** $m = \dfrac{-5 - 9}{-3 - 1} = \dfrac{-14}{-4} = \dfrac{7}{2}$

$\quad y = \dfrac{7}{2}x + b$

$\quad 9 = \dfrac{7}{2}(1) + b$

$\quad 9 = \dfrac{7}{2} + b$

$\quad \dfrac{11}{2} = b$

$\quad y = \dfrac{7}{2}x + \dfrac{11}{2}$

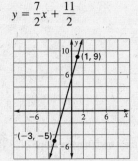

**33.** $m = \dfrac{1 - 2}{6 - (-5)} = -\dfrac{1}{11}$

$\quad y = -\dfrac{1}{11}x + b$

$\quad 1 = -\dfrac{1}{11}(6) + b$

$\quad 1 = -\dfrac{6}{11} + b$

$\quad \dfrac{17}{11} = b$

$\quad y = -\dfrac{1}{11}x + \dfrac{17}{11}$

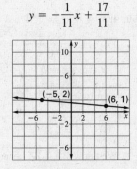

**34.** $m = \dfrac{2 - 11}{-6 - (-4)} = \dfrac{-9}{-2} = \dfrac{9}{2}$

$\quad y = \dfrac{9}{2}x + b$

$\quad 2 = \dfrac{9}{2}(-6) + b$

$\quad 2 = -27 + b$

$\quad 29 = b$

$\quad y = \dfrac{9}{2}x + 29$

**35.** $m = \dfrac{-4 - 10}{12 - (-1)} = -\dfrac{14}{13}$

$\quad y = -\dfrac{14}{13}x + b$

$\quad 10 = -\dfrac{14}{13}(-1) + b$

$\quad 10 = \dfrac{14}{13} + b$

$\quad \dfrac{116}{13} = b$

$\quad y = -\dfrac{14}{13}x + \dfrac{116}{13}$

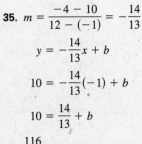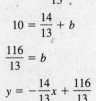

**36.** $m = \dfrac{4 - (-5)}{1 - (-6)} = \dfrac{9}{7}$

$\quad y = \dfrac{9}{7}x + b$

$\quad 4 = \dfrac{9}{7}(1) + b$

$\quad 4 = \dfrac{9}{7} + b$

$\quad \dfrac{19}{7} = b$

$\quad y = \dfrac{9}{7}x + \dfrac{19}{7}$

**37.** $m = \dfrac{3 - 3}{4 - 2} = \dfrac{0}{2} = 0$

$\quad y = 0x + b$

$\quad 3 = 0(2) + b$

$\quad 3 = b$

$\quad y = 3$

**38.** $m = \dfrac{-7 - (-10)}{12 - 5} = \dfrac{3}{7}$

$\quad y = \dfrac{3}{7}x + b$

$\quad -10 = \dfrac{3}{7}(5) + b$

$\quad -10 = \dfrac{15}{7} + b$

$\quad -\dfrac{85}{7} = b$

$\quad y = \dfrac{3}{7}x - \dfrac{85}{7}$

**39.** $m = \dfrac{9 - (-3)}{-6 - 14} = \dfrac{12}{-20} = -\dfrac{3}{5}$

$\quad y = -\dfrac{3}{5}x + b$

$\quad 9 = -\dfrac{3}{5}(-6) + b$

$\quad 9 = \dfrac{18}{5} + b$

$\quad \dfrac{27}{5} = b$

$\quad y = -\dfrac{3}{5}x + \dfrac{27}{5}$

**40.** $m = \dfrac{8 - 9}{-3 - (-7)} = -\dfrac{1}{4}$

$\quad y = -\dfrac{1}{4}x + b$

$\quad 8 = -\dfrac{1}{4}(-3) + b$

$\quad 8 = \dfrac{3}{4} + b$

$\quad \dfrac{29}{4} = b$

$\quad y = -\dfrac{1}{4}x + \dfrac{29}{4}$

**41.** $m = \dfrac{-3 - 9}{10 - (-8)} = \dfrac{-12}{18} = -\dfrac{2}{3}$

$\quad y = -\dfrac{2}{3}x + b$

$\quad -3 = -\dfrac{2}{3}(10) + b$

$\quad -3 = -\dfrac{20}{3} + b$

$\quad \dfrac{11}{3} = b$

$\quad y = -\dfrac{2}{3}x + \dfrac{11}{3}$

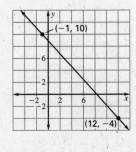

# Chapter 5 *continued*

**42.** $m = \dfrac{\frac{2}{3} - 2}{-5 - \frac{1}{4}} = \dfrac{-\frac{4}{3}}{-\frac{21}{4}} = -\dfrac{4}{3} \cdot -\dfrac{4}{21} = \dfrac{-16}{-63} = \dfrac{16}{63}$

$y = \dfrac{16}{63}x + b$

$2 = \dfrac{16}{63}\left(\dfrac{1}{4}\right) + b$

$2 = \dfrac{4}{63} + b$

$\dfrac{122}{63} = b$

$y = \dfrac{16}{63}x + \dfrac{122}{63}$

**43.** $m = \dfrac{\frac{3}{9} - \left(-\frac{1}{2}\right)}{\frac{1}{9} - \frac{1}{2}} = \dfrac{\frac{15}{18}}{-\frac{7}{18}} = \dfrac{15}{18} \cdot -\dfrac{18}{7} = -\dfrac{15}{7}$

$y = -\dfrac{15}{7}x + b$

$-\dfrac{1}{2} = -\dfrac{15}{7}\left(\dfrac{1}{2}\right) + b$

$-\dfrac{1}{2} = -\dfrac{15}{14} + b$

$\dfrac{8}{14} = b$

$\dfrac{4}{7} = b$

$y = -\dfrac{15}{7}x + \dfrac{4}{7}$

**44.** $m = \dfrac{-9.75 - 6.75}{3.33 - (-8.5)} = -\dfrac{16.5}{11.83} = -\dfrac{1650}{1183}$

$y = -\dfrac{1650}{1183}x + b$

$-9.75 = -\dfrac{1650}{1183}(3.33) + b$

$-9.75 = -\dfrac{5494.50}{1183} + b$

$-\dfrac{24{,}159}{4732} = b$

$y = -\dfrac{1650}{1183}x - \dfrac{24{,}159}{4732}$

**45.** Line $p$ and line $r$; their slopes are negative reciprocals.

**46.** $y = \frac{1}{4}x + b$

$2 = \frac{1}{4}(0) + b$

$2 = b$

$y = \frac{1}{4}x + 2$

**47.** $y = -2x + b$

$5 = -2(4) + b$

$5 = -8 + b$

$13 = b$

$y = -2x + 13$

**48.** $\overline{WX} \perp \overline{XY}$

$\overline{WX}$: $m = \dfrac{0 - 4}{3 - 0} = -\dfrac{4}{3}$

$\overline{XY}$: $m = \dfrac{-3 - 0}{-1 - 3} = \dfrac{-3}{-4} = \dfrac{3}{4}$

Mathematically, we know the 2 lines are perpendicular because their slopes are negative reciprocals.

**49.** $\overline{WX}$: $m = -\frac{4}{3}$ $\quad (0, 4)$

$y = -\frac{4}{3}x + b$

$4 = -\frac{4}{3}(0) + b$

$4 = b$

$y = -\frac{4}{3}x + 4$

$\overline{XY}$: $m = \frac{3}{4}$ $\quad (3, 0)$

$y = \frac{3}{4}x + b$

$0 = \frac{3}{4}(3) + b$

$0 = \frac{9}{4} + b$

$-\frac{9}{4} = b$

$y = \frac{3}{4}x - \frac{9}{4}$

**50.** $\overline{WX}$: $y = -\frac{4}{3}x + 4$

$\overline{YZ}$: $m = \dfrac{-3 - 3}{-1 + 5\frac{1}{2}} = \dfrac{-6}{4\frac{1}{2}} = -\dfrac{4}{3}$

$y = -\frac{4}{3}x + b$

$-3 = -\frac{4}{3}(-1) + b$

$-3 = \frac{4}{3} + b$

$-\frac{13}{3} = b$

$y = -\frac{4}{3}x - \frac{13}{3}$

We know $\overline{WX} \parallel \overline{YZ}$ because they both have a slope of $-\frac{4}{3}$.

**51.** $m = \dfrac{-70 - 60}{15 - 0} = \dfrac{-130}{15} = -\dfrac{26}{3}$

$y = -\dfrac{26}{3}x + b$

$60 = -\dfrac{26}{3}(0) + b$

$60 = b$

$y = -\dfrac{26}{3}x + 60$

**52.** $m = \dfrac{50 - (-90)}{50 - 38} = \dfrac{140}{12} = \dfrac{35}{3}$

$y = \dfrac{35}{3}x + b$

$50 = \dfrac{35}{3}(50) + b$

$50 = \dfrac{1750}{3} + b$

$-\dfrac{1600}{3} = b$

$y = \dfrac{35}{3}x - \dfrac{1600}{3}$

It is steeper on the French side.

# Chapter 5 *continued*

**53.** $m = \dfrac{340 - 352}{15 - 35} = \dfrac{-12}{-20} = \dfrac{3}{5}$

$y = \dfrac{3}{5}x + b$

$340 = \dfrac{3}{5}(15) + b$

$340 = 9 + b$

$331 = b$

$s = \dfrac{3}{5}T + 331$

**54.** $s = \dfrac{3}{5}T + 331$

$s = \dfrac{3}{5}(25) + 331 = 15 + 331 = 346 \text{ m/sec}$

**55.** $s = \dfrac{3}{5}T + 331$

$346 = \dfrac{3}{5}T + 331$

$15 = \dfrac{3}{5}T$

$25 = T$

Temperature is 25°C.

**56.** $s = \dfrac{3}{5}T + 331$

$343 = \dfrac{3}{5}T + 331$

$12 = \dfrac{3}{5}T$

$20°C = T$

**57.** $343 \text{ m/sec} \cdot 2 \text{ sec} = 686 \text{ m}$

**58.** D

$m = \dfrac{-2 - 4}{-5 - 7} = \dfrac{-6}{-12} = \dfrac{1}{2}$

$y = \dfrac{1}{2}x + b$

$4 = \dfrac{1}{2}(7) + b$

$4 = \dfrac{7}{2} + b$

$\dfrac{1}{2} = b$

$y = \dfrac{1}{2}x + \dfrac{1}{2}$

**59.** D

$m = \dfrac{3 - 2}{4 - (-1)} = \dfrac{1}{5}$

$y = \dfrac{1}{5}x + b$

$2 = \dfrac{1}{5}(-1) + b$

$2 = -\dfrac{1}{5} + b$

$\dfrac{11}{5} = b$

$y = \dfrac{1}{5}x + \dfrac{11}{5}$

**60.** C

Line $p$: $m = \dfrac{4 - 0}{6 - 4} = \dfrac{4}{2} = 2$

Line $q$: $m = \dfrac{4 - 4}{6 - 0} = \dfrac{0}{6} = 0$

Line $r$: $m = \dfrac{0 - 4}{0 - 0} = \dfrac{-4}{0} = \text{undefined}$

Line $q$ and line $r$ are perpendicular.

**61.**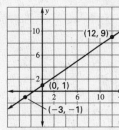

Yes; $m = \dfrac{9 - 1}{12 - 0} = \dfrac{8}{12} = \dfrac{2}{3}$

$m = \dfrac{-1 - 1}{-3 - 0} = \dfrac{-2}{-3} = \dfrac{2}{3}$

$y = \dfrac{2}{3}x + b$

$1 = \dfrac{2}{3}(0) + b$

$1 = b$

$y = \dfrac{2}{3}x + 1$

**62.**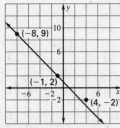

No; $m = \dfrac{2 - (-2)}{-1 - 4} = -\dfrac{4}{5}$

$m = \dfrac{9 - 2}{-8 - (-1)} = \dfrac{7}{-7} = -1$

The slope of the line through $(4, -2)$ and $(-1, 2)$ is $-\dfrac{4}{5}$ while the slope of the line through $(-1, 2)$ and $(-8, 9)$ is $-1$.

**63.**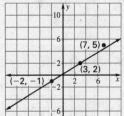

No; $m = \dfrac{2 - (-1)}{3 - (-2)} = \dfrac{3}{5}$

$m = \dfrac{5 - 2}{7 - 3} = \dfrac{3}{4}$

The slope of the line through $(-2, -1)$ and $(3, 2)$ is $\dfrac{3}{5}$ while the slope of the line through $(3, 2)$ and $(7, 5)$ is $\dfrac{3}{4}$.

**64.**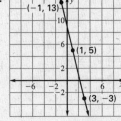

Yes; $m = \dfrac{5 - 13}{1 - (-1)} = \dfrac{-8}{2} = -4$

$m = \dfrac{-3 - 5}{3 - 1} = \dfrac{-8}{2} = -4$

$y = -4x + b$

$13 = -4(-1) + b$

$13 = 4 + b$

$9 = b$

$y = -4x + 9$

**65.** $m = \dfrac{-\frac{9}{2} - 4}{-5 - 3\frac{1}{2}} = \dfrac{-\frac{17}{2}}{-8\frac{1}{2}} = 1$

$y = 1x + b$

$4 = 1\left(3\frac{1}{2}\right) + b$

$4 = 3\frac{1}{2} + b$

$\frac{1}{2} = b$

$y = x + \frac{1}{2}$

### 5.3 Mixed Review (p. 291)

**66.** $\dfrac{6x + 12y}{24} = \dfrac{6(x + 2y)}{24} = \dfrac{x + 2y}{4}$

**67.** $\dfrac{48a - 56b}{16} = \dfrac{8(6a - 7b)}{16} = \dfrac{6a - 7b}{2}$

**68.** $\dfrac{14x}{-28y - 1}$ in simplest form

**69.** $4x - 11 = -31$

$4x = -20$

$x = -5$

**70.** $5x - 7 + x = 19$

$6x - 7 = 19$

$6x = 26$

$x = 4\frac{1}{3}$

**71.** $18 = 4 - \dfrac{2x}{5}$

$\dfrac{2x}{5} = -14$

$2x = -70$

$x = -35$

**72.** $2x - 6 = 20$

$2x = 26$

$x = 13$

**73.** $\frac{1}{2}a + 8\frac{1}{2}a = 3$

$9a = 3$

$a = \frac{1}{3}$

**74.** $7y = 9y - 8$

$-2y = -8$

$y = 4$

**75.**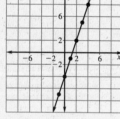

positive

**76.**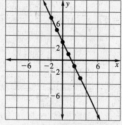

negative

### Quiz 1 (p. 291)

**1.** $m = \dfrac{4 - 0}{0 - (-3)} = \dfrac{4}{3}$

$y = \dfrac{4}{3}x + b$

$0 = \dfrac{4}{3}(-3) + b$

$4 = b$

$y = \dfrac{4}{3}x + 4$

**2.** $m = \dfrac{2 - 0}{0 - 1} = \dfrac{2}{1} = -2$

$y = -2x + b$

$2 = -2(0) + b$

$2 = b$

$y = -2x + 2$

**3.** $m = \dfrac{-4 - 0}{0 - 3} = \dfrac{-4}{-3} = \dfrac{4}{3}$

$y = \dfrac{4}{3}x + b$

$-4 = \dfrac{4}{3}(0) + b$

$-4 = b$

$y = \dfrac{4}{3}x - 4$

**4.** $y = 2x + b$

$4 = 2(-5) + b$

$4 = -10 + b$

$14 = b$

$y = 2x + 14$

**5.** $y = -3x + b$

$1 = -3(2) + b$

$1 = -6 + b$

$7 = b$

$y = -3x + 7$

**6.** $y = 1x + b$

$-6 = 1(-3) + b$

$-6 = -3 + b$

$-3 = b$

$y = x - 3$

**7.** $m = 1$

$y = 1x + b$

$-4 = 1(-2) + b$

$-4 = -2 + b$

$-2 = b$

$y = x - 2$

**8.** $m = \dfrac{-2 - (-3)}{5 - 8} = \dfrac{1}{-3} = -\dfrac{1}{3}$

$y = -\dfrac{1}{3}x + b$

$-3 = -\dfrac{1}{3}(8) + b$

$-3 = -\dfrac{8}{3} + b$

$-\dfrac{1}{3} = b$

$y = -\dfrac{1}{3}x - \dfrac{1}{3}$

**9.** $m = \dfrac{-7 - 6}{-5 - (-2)} = \dfrac{-13}{-3} = \dfrac{13}{3}$

$y = \dfrac{13}{3}x + b$

$6 = \dfrac{13}{3}(-2) + b$

$6 = -\dfrac{26}{3} + b$

$\dfrac{44}{3} = b$

$y = \dfrac{13}{3}x + \dfrac{44}{3}$

**10.** $m = \dfrac{4 - 4}{-7 - 4} = \dfrac{0}{-11} = 0$

$y = 0x + b$

$4 = 0(4) + b$

$4 = b$

$y = 4$

**11.** $y = 28x + 10$

**Algebra 1**
Chapter 5 Worked-out Solution Key

# Chapter 5 *continued*

**12.**

**Canoe Rental Cost**

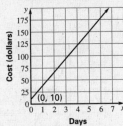

**13.** $y = 28x + 10 = 28(3) + 10 = 84 + 10 = \$94.00$

## Lesson 5.4

### Activity (p. 292)

**1.**

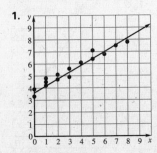

**2.** The line I drew comes close to as many points as possible.

**3.** *Sample answer:* (0, 3.6) and (6, 7)

**4.** *Sample answer:* $m = \dfrac{7 - 3.6}{6 - 0} = \dfrac{3.4}{6} \approx 0.57$

$$y = mx + b$$
$$7 = 6\left(\dfrac{3.4}{6}\right) + b$$
$$7 = 3.4 + b$$
$$3.6 = b$$
$$y = 0.57x + 3.6$$

### 5.4 Guided Practice (p. 296)

**1.** best-fitting line

**2.**

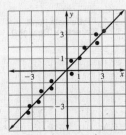

**3.**

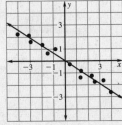

**4.**

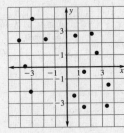

---

**5.** $y = mx + b$

$y = 1.05(105) + 129.6$

$y = 239.85$ ft

For the year 2005,
$x = 2005 - 1900 = 105$, so
$y = 1.05(105) + 129.6 = 239.85$ ft

**6.**

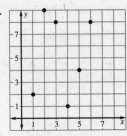

No correlation

**7.** *Sample answer:*

$$m = \dfrac{0 - 8}{2 - (-3)} = -\dfrac{8}{5}$$
$$y = -\dfrac{8}{5}x + b$$
$$0 = -\dfrac{8}{5}(2) + b$$
$$0 = -\dfrac{16}{5} + b$$
$$\dfrac{16}{5} = b$$
$$y = -1.6x + 3.2$$

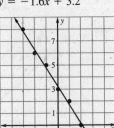

Negative correlation

**8.** *Sample answer:*

$$m = \dfrac{7.5 - 5.5}{3.3 - 1.7} - \dfrac{2}{1.6} = 1.25$$
$$y = 1.25x + b$$
$$5.5 = 1.25(1.7) + b$$
$$5.5 = 2.125 + b$$
$$3.375 = b$$
$$y = 1.25x + 3.375$$

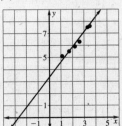

Positive correlation

**9.** *Sample answer:*

$$m = \dfrac{4.3 - 1}{9.2 - 6.2} = \dfrac{3.3}{3} = 1.1$$
$$y = 1.1x + b$$
$$1 = 1.1(6.2) + b$$
$$1 = 6.82 + b$$
$$-5.82 = b$$
$$y = 1.1x - 5.82$$

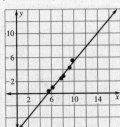

Positive correlation

---

# Chapter 5 *continued*

**5.4 Practice and Applications (p. 296–298)**

**10.** *Sample answer:*

$$m = \frac{\frac{1}{2} - (-3)}{3 - (-2)} = \frac{3\frac{1}{2}}{5} = 0.7$$

$$y = 0.7x + b$$
$$-3 = 0.7(-2) + b$$
$$-3 = -1.4 + b$$
$$-1.6 = b$$
$$y = 0.7x - 1.6$$

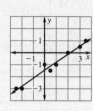

**11.** *Sample answer:*

$$m = \frac{0.2 - \left(-3\frac{1}{2}\right)}{-0.8 - (-4)} = \frac{3.7}{3.2} \approx 1.16$$

$$y = 1.16x + b$$
$$-3\frac{1}{2} = 1.16(-4) + b$$
$$-3.5 = -4.64 + b$$
$$1.14 = b$$
$$y = 1.16x + 1.14$$

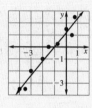

**12.** *Sample answer:*

$$m = \frac{-2 - 1\frac{1}{2}}{3 - (-1)} = \frac{-3.5}{4} \approx -0.88$$

$$y = -0.88x + b$$
$$-2 = -0.88(3) + b$$
$$-2 = -2.64 + b$$
$$0.64 = b$$
$$y = -0.88x + 0.64$$

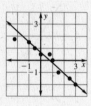

**13.** *Sample answer:*

$$m = \frac{5.8 - 3.8}{2.5 - 1} = \frac{2}{1.5} \approx 1.33$$

$$y = 1.33x + b$$
$$3.8 = 1.33(1) + b$$
$$3.8 = 1.33 + b$$
$$2.47 = b$$
$$y = 1.33x + 2.47$$

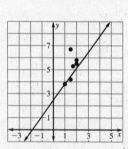

**14.** *Sample answer:*

$$m = \frac{8.9 - 8.1}{4.1 - 3.4} = \frac{0.8}{0.7} \approx 1.14$$

$$y = 1.14x + b$$
$$8.1 = 1.14(3.4) + b$$
$$8.1 = 3.88 + b$$
$$4.22 = b$$
$$y = 1.14x + 4.22$$

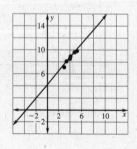

**15.** *Sample answer:*

$$m = \frac{9.9 - 8.6}{3 - 3.7} = \frac{1.3}{-0.7} \approx -1.86$$

$$y = -1.86x + b$$
$$9.9 = -1.86(3) + b$$
$$9.9 = -5.58 + b$$
$$15.48 = b$$
$$y = -1.86x + 15.48$$

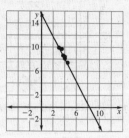

**16.** *Sample answer:*

$$m = \frac{4.3 - 6.8}{6.8 - 5.0} = \frac{-2.5}{1.8} \approx -1.39$$

$$y = -1.39x + b$$
$$6.8 = -1.39(5) + b$$
$$6.8 = -6.95 + b$$
$$13.75 = b$$
$$y = -1.39x + 13.75$$

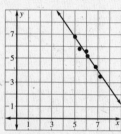

**17.** positive correlation    **18.** no correlation

**19.** negative correlation    **20.** no correlation

**21.** positive correlation    **22.** negative correlation

**23.** *Sample answer:*
(4, 140), (14, 440)

$$m = \frac{440 - 140}{14 - 4} = \frac{300}{10} = 30$$

$$y = 30x + b$$
$$140 = 30(4) + b$$
$$140 = 120 + b$$
$$20 = b$$
$$y = 30x + 20$$

**24.** $y = 30x + 20$
$y = 30(30) + 20$
$y = 900 + 20$
$y = 920$
about $920,000

**25.**

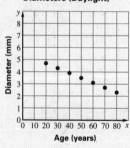

**Sample Pupil Diameters (Daylight)**

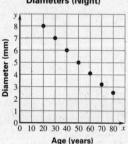

**Sample Pupil Diameters (Night)**

# Chapter 5 *continued*

**26.** daylight:

$$m = \frac{2.7 - 4.3}{70 - 30} = -\frac{1.6}{40} = -0.04$$

$y = -0.04x + b$

$4.3 = -0.04(30) + b$

$4.3 = -1.2 + b$

$5.5 = b$

$y = -0.04x + 5.5$

$$m = \frac{2.5 - 7}{80 - 30} = -\frac{4.5}{50} = -0.09$$

$y = -0.09x + b$

$7 = -0.09(30) + b$

$7 = -2.7 + b$

$9.7 = b$

$y = -0.09x + 9.7$

**27.** No. The graph of the night diameters is much steeper than the graph of the day diameters. This means that the night diameter of the pupil decreases with age much more dramatically than the day diameter. The eye's ability to adjust to lower light decreases with age.

**28.** (3, 16.9), (12, 18)

$$m = \frac{18 - 16.9}{12 - 3} = \frac{1.1}{9} \approx 0.12$$

$y = 0.12x + b$

$18 = 0.12(12) + b$

$18 = 1.44 + b$

$16.56 = b$

$B = 0.12t + 16.56$

**29.** (3, 22.3), (12, 41)

$$m = \frac{41 - 22.3}{12 - 3} = \frac{18.7}{9} \approx 2.08$$

$y = 2.08x + b$

$41 = 2.08(12) + b$

$41 = 24.96 + b$

$16.04 = b$

$P = 2.08t + 16.04$

**30.** The average book price is increasing much more rapidly than the library's book budget.

**31.** Answers may vary.

**32.** The line through those points does not approximate the data points very well.

**33.** No. The two endpoints could be significantly far away from the majority of other points.

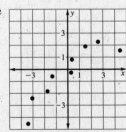

**34.** Once a line has been chosen, choose two points on the line whose coordinates can be closely approximated.

Copyright © McDougal Littell Inc.
All rights reserved.

---

**5.4 Mixed Review (p. 298)**

**35.** horizontal

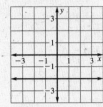

**36.** horizontal

**37.** vertical

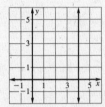

**38.** vertical

**39.** positive   **40.** negative   **41.** zero   **42.** negative

**43.** $m = \dfrac{4 - 5}{6 - 2} = -\dfrac{1}{4}$

$y = -\dfrac{1}{4}x + b$

$5 = -\dfrac{1}{4}(2) + b$

$5 = -\dfrac{1}{2} + b$

$\dfrac{11}{2} = b$

$y = -\dfrac{1}{4}x + \dfrac{11}{2}$

**44.** $m = \dfrac{7 - 4}{3 - 1} = \dfrac{3}{2}$

$y = \dfrac{3}{2}x + b$

$4 = \dfrac{3}{2}(1) + b$

$4 = \dfrac{3}{2} + b$

$\dfrac{5}{2} = b$

$y = \dfrac{3}{2}x + \dfrac{5}{2}$

**45.** $m = \dfrac{3 - 7}{7 - 3} = \dfrac{-4}{4} = -1$

$y = -1x + b$

$7 = -1(3) + b$

$7 = -3 + b$

$10 = b$

$y = -x + 10$

**46.** $m = \dfrac{3 - 2}{4 - 5} = \dfrac{1}{-1} = -1$

$y = -1x + b$

$2 = -1(5) + b$

$2 = -5 + b$

$7 = b$

$y = -x + 7$

**47.** $m = \dfrac{-2 - 1}{4 - (-3)} = -\dfrac{3}{7}$

$y = -\dfrac{3}{7}x + b$

$1 = -\dfrac{3}{7}(-3) + b$

$1 = \dfrac{9}{7} + b$

$-\dfrac{2}{7} = b$

$y = -\dfrac{3}{7}x - \dfrac{2}{7}$

**48.** $m = \dfrac{4 - (-3)}{0 - (-2)} = \dfrac{7}{2}$

$y = \dfrac{7}{2}x + b$

$4 = \dfrac{7}{2}(0) + b$

$4 = b$

$y = \dfrac{7}{2}x + 4$

---

# Chapter 5 continued

**49.** $\dfrac{-6-1}{3-5} = \dfrac{-7}{-2} = \dfrac{7}{2}$

$y = \dfrac{7}{2}x + b$

$1 = \dfrac{7}{2}(5) + b$

$1 = \dfrac{35}{2} + b$

$\dfrac{-33}{2} = b$

$y = \dfrac{7}{2}x - \dfrac{33}{2}$

**50.** $m = \dfrac{7-(-6)}{-1-0} = \dfrac{13}{-1} = -13$

$y = -13x + b$

$-6 = -13(0) + b$

$-6 = b$

$y = -13x - 6$

## Technology Activity 5.4 (p. 299)

**1.** $y = 0.47x + 2.01$

**2.** $y = 1.86x + 61.88$

**3.** $y = 0.95x + 1.41$

**4.** $y = -1.07x + 8.13$

## Lesson 5.5

### Activity (p. 301)

**1.** *Sample answers:* $(-5, 4), (-2, 2)$

$m = \dfrac{2-4}{-2-(-5)} = -\dfrac{2}{3}$

$y = mx + b$

$2 = -\dfrac{2}{3}(-2) + b$

$2 = \dfrac{4}{3} + b$

$\dfrac{2}{3} = b$

$y = -\dfrac{2}{3}x + \dfrac{2}{3}$

**2.** Always true; no matter which two points you use, you can rewrite the equations in slope-intercept form and you'll get the same equation.

## Lesson 5.5

### 5.5 Guided Practice (p. 303)

**1.** point-slope form   **2.** $3, 2, -1$   **3.** $y - 4 = \frac{1}{2}(x - 3)$

**4.** $y + 5 = \frac{2}{3}(x - 1)$   **5.** $y + 2 = 0$

**6.** $y + 1 = 3(x - 2)$   **7.** $y - 4 = \frac{1}{2}(x - 3)$

**8.** $y + 7 = -2(x + 5)$   **9.** $y - 10 = 4(x + 3)$

**10.** $y + 9 = 4(x + 2)$   **11.** $y + 7 = -8(x - 7)$

**12.** $m = \dfrac{-2-12}{6-5} = \dfrac{-14}{1} = -14$

$y - 12 = -14(x - 5)$   or   $y + 2 = -14(x - 6)$

**13.** $m = \dfrac{2-(-1)}{-9-(-4)} = -\dfrac{3}{5}$

$y + 1 = -\dfrac{3}{5}(x + 4)$   or   $y - 2 = -\dfrac{3}{5}(x + 9)$

**14.** $m = \dfrac{-4-(-8)}{-7-(-17)} = \dfrac{4}{10} = \dfrac{2}{5}$

$y + 8 = \dfrac{2}{5}(x + 17)$   or   $y + 4 = \dfrac{2}{5}(x + 7)$

**15.** $m = \dfrac{2-12}{7-2} = \dfrac{-10}{5} = -2$   **16.** $m = \dfrac{4-(-12)}{8-3} = \dfrac{16}{5}$

$y - 12 = -2(x - 2)$   or
$y - 2 = -2(x - 7)$

$y + 12 = \dfrac{16}{5}(x - 3)$ or

$y - 4 = \dfrac{16}{5}(x - 8)$

**17.** $m = \dfrac{2-(-7)}{2-(-4)} = \dfrac{9}{6} = \dfrac{3}{2}$

$y + 7 = \dfrac{3}{2}(x + 4)$   or   $y - 2 = \dfrac{3}{2}(x - 2)$

### 5.5 Practice and Applications (pp. 303–305)

**18.** $m = \dfrac{2}{1} = 2$   **19.** $m = \dfrac{-1}{2} = -\dfrac{1}{2}$

$y - 2 = 2(x + 4)$

$y + 3 = -\dfrac{1}{2}(x + 1)$

**20.** $m = \dfrac{-3}{-1} = 3$   **21.** $m = \dfrac{-5-5}{-1-1} = \dfrac{-10}{-2} = 5$

$y - 5 = 3(x - 3)$

$y - 5 = 5(x - 1)$ or
$y + 5 = 5(x + 1)$

**22.** $m = \dfrac{-6-(-5)}{7-(-2)} = -\dfrac{1}{9}$   **23.** $m = \dfrac{-3-10}{-4-(-9)} = -\dfrac{13}{5}$

$y + 5 = -\dfrac{1}{9}(x + 2)$ or

$y - 10 = -\dfrac{13}{5}(x + 9)$ or

$y + 6 = -\dfrac{1}{9}(x - 7)$

$y + 3 = -\dfrac{13}{5}(x + 4)$

**24.** $m = \dfrac{-7-(-5)}{-2-4} = \dfrac{-2}{-6} = \dfrac{1}{3}$

$y + 5 = \dfrac{1}{3}(x - 4)$ or $y + 7 = \dfrac{1}{3}(x + 2)$

**25.** $m = \dfrac{-2-10}{-4-(-5)} = \dfrac{-12}{1} = -12$

$y - 10 = -12(x + 5)$ or $y + 2 = -12(x + 4)$

**26.** $m = \dfrac{-8-(-2)}{-3-(-3)} = \dfrac{-6}{0} = $ undefined

$x = -3$

**27.** $m = \dfrac{-8-(-9)}{-6-(-3)} = -\dfrac{1}{3}$

$y + 9 = -\dfrac{1}{3}(x + 3)$ or $y + 8 = -\dfrac{1}{3}(x + 6)$

## Chapter 5 *continued*

**28.** $m = \dfrac{-8 - (-7)}{-4 - (-3)} = \dfrac{-1}{-1} = 1$

$y + 7 = x + 3$ or $y + 8 = x + 4$

**29.** $m = \dfrac{-5 - (-7)}{-1 - 1} = \dfrac{2}{-2} = -1$

$y + 7 = -(x - 1)$ or $y + 5 = -(x + 1)$

**30.** $y + 3 = 4(x + 1)$    **31.** $y - 2 = -5(x + 6)$

**32.** $y = 2(x + 10)$    **33.** $y + 2 = 2(x + 8)$

**34.** $y - 3 = -6(x + 4)$    **35.** $y - 4 = 6(x + 3)$

**36.** $y - 2 = -7(x - 12)$    **37.** $y + 1 = 0$

**38.** $y + 12 = -11(x - 5)$

**39.** $y - 4 = 2(x - 1)$
$y - 4 = 2x - 2$
$y = 2x + 2$

**40.** $y - 4 = 3(x + 2)$
$y - 4 = 3x + 6$
$y = 3x + 10$

**41.** $y - 2 = \frac{1}{2}(x - 6)$
$y - 2 = \frac{1}{2}x - 3$
$y = \frac{1}{2}x - 1$

**42.** $y + 1 = 4(x + 3)$
$y + 1 = 4x + 12$
$y = 4x + 11$

**43.** $y + 1 = -1(x - 5)$
$y + 1 = -x + 5$
$y = -x + 4$

**44.** $y + 5 = -2(x + 5)$
$y + 5 = -2x - 10$
$y = -2x - 15$

**45.** $y - 1 = -\frac{1}{8}(x + 1)$
$y - 1 = -\frac{1}{8}x - \frac{1}{8}$
$y = -\frac{1}{8}x + \frac{7}{8}$

**46.** $y + 2 = \frac{1}{4}(x - 4)$
$y + 2 = \frac{1}{4}x - 1$
$y = \frac{1}{4}x - 3$

**47.** $y + 6 = -\frac{2}{3}(x + 9)$

$y + 6 = -\frac{2}{3}x - 6$

$y = -\frac{2}{3}x - 12$

**48.** $m = \dfrac{2 - (-2)}{0 - 3} = -\dfrac{4}{3}$

$y - 2 = -\frac{4}{3}(x - 0)$

$y = -\frac{4}{3}x + 2$

**49.** $m = \dfrac{0 - (-2)}{-4 - 2} = \dfrac{2}{-6} = -\dfrac{1}{3}$

$y - 0 = -\frac{1}{3}(x + 4)$

$y = -\frac{1}{3}x - \frac{4}{3}$

**50.** $m = \dfrac{0 - (-3)}{5 - 0} = \dfrac{3}{5}$

$y - 0 = \frac{3}{5}(x - 5)$

$y = \frac{3}{5}x - 3$

**51.** $m = \dfrac{5 - 3}{2 - 1} = \dfrac{2}{1} = 2$

$y - 3 = 2(x - 1)$ or $y - 5 = 2(x - 2)$

**52.** $m = \dfrac{-5 - 3}{2 - (-2)} = \dfrac{-8}{4} = -2$

$y - 3 = -2(x + 2)$ or $y + 5 = -2(x - 2)$

**53.** $m = \dfrac{-22 - (-10)}{15 - 7} = \dfrac{-12}{8} = -\dfrac{3}{2}$

$y + 10 = -\frac{3}{2}(x - 7)$ or $y + 22 = -\frac{3}{2}(x - 15)$

**54.** $m = \dfrac{1 - 6}{5 - 2} = -\dfrac{5}{3}$

$y - 6 = -\frac{5}{3}(x - 2)$ or $y - 1 = -\frac{5}{3}(x - 5)$

**55.** $m = \dfrac{-1 - (-4)}{1 - (-3)} = \dfrac{3}{4}$

$y + 4 = \frac{3}{4}(x + 3)$ or $y + 1 = \frac{3}{4}(x - 1)$

**56.** $m = \dfrac{5 - (-2)}{-9 - 4} = -\dfrac{7}{13}$

$y + 2 - \frac{7}{13}(x - 4)$ or $y - 5 = -\frac{7}{13}(x + 9)$

**57.** $d = -0.76t + 120$

**58.** $d = -0.76(120) + 120 = -91.2 + 120 = 28.8$ miles

**59.**    $0 = -0.76t + 120$    **60.** $y = -10.2x + 2340$
$-120 = -0.76t$
$157.9 \approx t$
at about 11:38 A.M.

**61.**    $0 = -10.2x + 2340$    **62.** $d = -124t + 300$
$-2340 = -10.2x$
$x \approx 229$ minutes
at about 11:49 A.M.

**63.**    $0 = -124t + 300$
$-300 = -124t$
$2.4 \approx t$
at about 2:25 P.M.

**64.** B    **65.** A    **66.** B

**67.** $2 = \dfrac{k - 3}{1 - 2k}$
$2(1 - 2k) = k - 3$
$2 - 4k = k - 3$
$-5k = -5$
$k = 1$

**68.** $k + 1 = \dfrac{-k - (k - 1)}{k - (k + 1)}$
$k + 1 = \dfrac{-k - k + 1}{k - k - 1}$
$k + 1 = \dfrac{-2k + 1}{-1}$
$-k - 1 = -2k + 1$
$k = 2$

**69.** $3 = \dfrac{2 - (k + 1)}{3 - k} = \dfrac{-k + 1}{3 - k}$
$3(3 - k) = -k + 1$
$9 - 3k = -k + 1$
$-2k = -8$
$k = 4$

**70.** $k = \dfrac{1 - k - (3 + 2k)}{k - 1 - (k + 1)}$
$k = \dfrac{1 - k - 3 - 2k}{k - 1 - k - 1}$
$k = \dfrac{-2 - 3k}{-2}$
$-2k = -2 - 3k$
$k = -2$

# Chapter 5 *continued*

**5.5 Mixed Review (p. 305)**

**71.** $2(8) \overset{?}{<} 24$
$16 < 24$
8 is a solution.

**72.** $7(3) + 6 \overset{?}{\geq} 10$
$21 + 6 \overset{?}{\geq} 10$
$27 \geq 10$
3 is a solution.

**73.** $16(5) - 9 \overset{?}{\leq} 71$
$80 - 9 \overset{?}{\leq} 71$
$71 \leq 71$
5 is a solution.

**74.** $12(-2) \overset{?}{\leq} -2 - 9$
$-24 \leq -11$
$-2$ is a solution.

**75.** $4(7) \overset{?}{\leq} 28$
$28 \leq 28$
7 is a solution.

**76.** $6(3) - 4 \overset{?}{>} 14$
$18 - 4 \overset{?}{>} 14$
$14 \not> 14$
3 is not a solution.

**77.** $f(x) = -3x + 4$

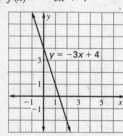

**78.** $g(x) = -x - 7$

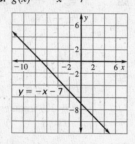

**79.** $h(x) = 2x - 8$

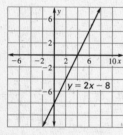

**80.** $f(x) = -5x - 7$

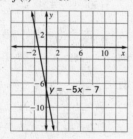

**81.** $f(x) = 6x + 7$

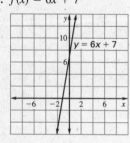

**82.** $f(x) = 3 - 7x$

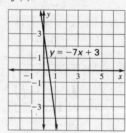

**83.** $g(x) = -\frac{1}{2}x - 3$

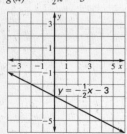

**84.** $f(x) = 8x + \frac{2}{3}$

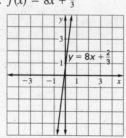

**85.** $h(x) = \frac{6}{5}x + 5$

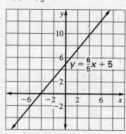

**86.** $\frac{460}{40} = \frac{46}{4} = \frac{23}{2}$   23 to 2

**87.** $\frac{7}{21} = .\overline{33}$ or $33\frac{1}{3}\%$

**Quiz 2 (p. 306)**

**1.**

Sample answer:

$m = \dfrac{-1 - (-4)}{2 - (-1)} = \dfrac{3}{3} = 1$

$y = 1x + b$

$-1 = 1(2) + b$

$-1 = 2 + b$

$-3 = b$

$y = x - 3$

positive correlation

**2.**

Sample answer:

$m = \dfrac{-2 - (-1)}{3 - 1} = -\dfrac{1}{2}$

$y = -\dfrac{1}{2}x + b$

$-1 = -\dfrac{1}{2}(1) + b$

$-1 = -\dfrac{1}{2} + b$

$-\dfrac{1}{2} = b$

$y = -\dfrac{1}{2}x - \dfrac{1}{2}$

negative correlation

**3.**

Sample answer:

$m = \dfrac{3 - (-2)}{1.8 - (-2)} = \dfrac{5}{3.8} \approx 1.32$

$y = 1.32x + b$

$-2 = 1.32(-2) + b$

$-2 = -2.64 + b$

$0.64 = b$

$y = 1.32x + 0.64$

positive correlation

**4.** $y + 4 = -(x - 1)$

**5.** $y - 5 = 2(x - 2)$

**6.** $y + 2 = -2(x + 3)$

**7.** $y + 3 = 0$

**8.** $y - 6 = \frac{1}{2}(x + 6)$

**9.** $y - 8 = 7(x - 3)$

**Algebra 1**
Chapter 5 Worked-out Solution Key

# Chapter 5 continued

**10.** $m = \dfrac{-1 - 0}{-3 - 0} = \dfrac{-1}{-3} = \dfrac{1}{3}$

$y - 0 = \dfrac{1}{3}(x - 0)$

$y = \dfrac{1}{3}x$ or $y + 1 = \dfrac{1}{3}(x + 3)$

**11.** $m = \dfrac{-2 - (-2)}{-6 - (-5)} = \dfrac{0}{-1} = 0$

$y + 2 = 0(x + 5)$ or $y + 2 = 0(x + 6)$

$y + 2 = 0$

**12.** $m = \dfrac{6 - (-3)}{1 - (-5)} = \dfrac{9}{6} = \dfrac{3}{2}$

$y + 3 = \dfrac{3}{2}(x + 5)$ or $y - 6 = \dfrac{3}{2}(x - 1)$

## Math & History (p. 306)

**1.** about 1000 B.C.

**2.** later

## Lesson 5.6

### Developing Concepts Activity 5.6 (p. 307)
### Exploring the Concept

**1.**

| Number of GFE Shares | 0 | 1 | 2 | 3 | 4 | 5 |
|---|---|---|---|---|---|---|
| Number of JIH Shares | 1 | 1 | 1 | 1 | 1 | 1 |
| Total cost (dollars) | 20 | 28 | 36 | 44 | 52 | 60 |

**2.** They are located on the line $y = 1$.

**3.** Cost is \$80; the points that satisfy $8x + 20y = 80$

**4.** $8x + 20y = 80$

$20y = -8x + 80$

$y = -\dfrac{2}{5}x + 4$

### Drawing Conclusions

**1.** Answers may vary.

**2.** For each amount $n$, the equation is $y = -\dfrac{2}{5}x + \dfrac{n}{20}$. Lines may vary.

**3.** For each amount $n$, the equation is $8x + 20y = n$.

**4.** The graphs are parallel; in Question 2, $m$ is the same for each equation; in Question 3, the coefficients of $x$ and $y$ are the same for each equation.

## Activity (p. 308)

**1. a.–c.**

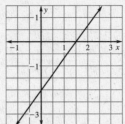

**2.** The three equations are equivalent because they all have the same graph.

## 5.6 Guided Practice (p. 311)

**1.** slope-intercept form      **2.** standard form

**3.** To graph $y = -\dfrac{2}{3}x + 3$, I would use the $y$-intercept to plot $(0, 3)$, then use a slope triangle to plot a second point. To graph $2x + 3y = 9$, I would make a table of values and then plot the points.

**4.** $2x - y = 9$      **5.** $5x + y = 6$      **6.** $-x + y = 9$

$\phantom{4.}-2x + y = -9$      $\phantom{5.}$      $\phantom{6.}x - y = -9$

**7.** $-\dfrac{1}{2}x + y = 8$      **8.** $-7x + y = 3$

$\phantom{7.}x - 2y = -16$      $\phantom{8.}7x - y = -3$

**9.** $-\dfrac{3}{2}x + y = -2$      **10.** $y - 4 = -4(x + 3)$

$\phantom{9.}3x - 2y = 4$      $\phantom{10.}y - 4 = -4x - 12$

$\phantom{10.....}y = -4x - 8$

$\phantom{10.....}4x + y = -8$

**11.** $y + 2 = 5(x - 1)$      **12.** $y - 5 = 3(x - 2)$

$\phantom{11.}y + 2 = 5x - 5$      $\phantom{12.}y - 5 = 3x - 6$

$\phantom{11....}y = 5x - 7$      $\phantom{12....}y = 3x - 1$

$\phantom{11.}-5x + y = -7$      $\phantom{12.}-3x + y = -1$

$\phantom{11....}5x - y = 7$      $\phantom{12....}3x - y = 1$

**13.** $y + 8 = -1(x - 5)$      **14.** $y + 6 = 3(x + 5)$

$\phantom{13.}y + 8 = -x + 5$      $\phantom{14.}y + 6 = 3x + 15$

$\phantom{13....}y = -x - 3$      $\phantom{14....}y = 3x + 9$

$\phantom{13....}x + y = -3$      $\phantom{14.}-3x + y = 9$

$\phantom{14....}3x - y = -9$

**15.** $y + 3 = -2(x - 4)$      **16.** $2x + 1.25y = 10$

$\phantom{15.}y + 3 = -2x + 8$

$\phantom{15....}y = -2x + 5$

$\phantom{15.}2x + y = 5$

**17.**

| Tomatoes (lb), x | 0 | 1.25 | 2.5 | 3.75 | 5 |
|---|---|---|---|---|---|
| Carrots (lb), y | 8 | 6 | 4 | 2 | 0 |

# Chapter 5 *continued*

## 5.6 *Practice and Applications (pp. 311–314)*

**18.** $4x - y = 7$    **19.** $x + 3y = 4$

**20.** $-4x + 5y = -16$    **21.** $2x - 3y = 14$
$4x - 5y = 16$

**22.** $x = 5$    **23.** $y = -3$    **24.** $5x + y = 2$

**25.** $-3x + y = -8$
$3x - y = 8$

**26.** $0.4x + y = 1.2$
$2x + 5y = 6$

**27.** $3x - \frac{7}{2}y = -9$
$6x - 7y = -18$

**28.** $-9x + y = \frac{1}{2}$
$18x - 2y = -1$

**29.** $-\frac{5}{2}x + y = 9$
$5x - 2y = -18$

**30.** $\frac{3}{4}x + y = \frac{5}{4}$
$3x + 4y = 5$

**31.** $\frac{1}{7}x + y = \frac{6}{7}$
$x + 7y = 6$

**32.** $\frac{1}{3}x + y = -4$
$x + 3y = -12$

**33.** $y - 3 = 2(x + 8)$
$y - 3 = 2x + 16$
$y = 2x + 19$
$-2x + y = 19$
$2x - y = -19$

**34.** $y - 7 = -4(x + 2)$
$y - 7 = -4x - 8$
$y = -4x - 1$
$4x + y = -1$

**35.** $y - 4 = -3(x + 1)$
$y - 4 = -3x - 3$
$y = -3x + 1$
$3x + y = 1$

**36.** $y + 7 = -1(x + 6)$
$y + 7 = -x - 6$
$y = -x - 13$
$x + y = -13$

**37.** $y + 2 = 5(x - 3)$
$y + 2 = 5x - 15$
$y = 5x - 17$
$-5x + y = -17$
$5x - y = 17$

**38.** $y - 6 = 7(x - 10)$
$y - 6 = 7x - 70$
$y = 7x - 64$
$-7x + y = -64$
$7x - y = 64$

**39.** $y - 9 = -7(x - 2)$
$y - 9 = -7x + 14$
$y = -7x + 23$
$7x + y = 23$

**40.** $y + 8 = 10(x - 5)$
$y + 8 = 10x - 50$
$y = 10x - 58$
$-10x + y = -58$
$10x - y = 58$

**41.** $y - 3 = -2(x - 7)$
$y - 3 = -2x + 14$
$y = -2x + 17$
$2x + y = 17$

**42.** $y - 2 = -2(x + 4)$
$y - 2 = -2x - 8$
$y = -2x - 6$
$2x + y = -6$

**43.** $y - 3 = 1(x - 0)$
$y - 3 = x$
$-x + y = 3$
$x - y = -3$

**44.** $y - 3 = 4(x + 3)$
$y - 3 = 4x + 12$
$y = 4x + 15$
$-4x + y = 15$
$4x - y = -15$

**45.** $m = \frac{7 - 4}{5 - 1} = \frac{3}{4}$
$y - 4 = \frac{3}{4}(x - 1)$
$y - 4 = \frac{3}{4}x - \frac{3}{4}$
$y = \frac{3}{4}x + \frac{13}{4}$
$4y = 3x + 13$
$-3x + 4y = 13$
$3x - 4y = -13$

**46.** $m = \frac{2 - (-3)}{7 - (-3)} = \frac{5}{10} = \frac{1}{2}$
$y + 3 = \frac{1}{2}(x + 3)$
$y + 3 = \frac{1}{2}x + \frac{3}{2}$
$2y + 6 = x + 3$
$-x + 2y = -3$
$x - 2y = 3$

**47.** $m = \frac{-5 - 1}{2 - (-4)} = \frac{-6}{6} = -1$
$y - 1 = -1(x + 4)$
$y - 1 = -x - 4$
$x + y = -3$

**48.** $m = \frac{2 - (-2)}{-3 - 9} = \frac{4}{-12} = -\frac{1}{3}$
$y + 2 = -\frac{1}{3}(x - 9)$
$y + 2 = -\frac{1}{3}x + 3$
$3y + 6 = -x + 9$
$x + 3y = 3$

**49.** $m = \frac{0 - 0}{2 - 0} = \frac{0}{2} = 0$
$y - 0 = 0(x - 0)$
$y = 0$

**50.** $m = \frac{-1 - (-7)}{5 - 4} = \frac{6}{1} = 6$
$y + 7 = 6(x - 4)$
$y + 7 = 6x - 24$
$y = 6x - 31$
$-6x + y = -31$
$6x - y = 31$

**51.** $y = -4, x = 2$    **52.** $y = 4, x = -1$

**53.** $y = -3, x = 0$    **54.** $y = 3, x = 1$

**55.** $y = 4, x = -4$    **56.** $y = -5, x = -2$

**57.** $y = -1, x = 6$    **58.** $y = 7, x = 0$

**59.** $y = 0, x = -9$    **60.** $y = 7, x = -3$

**61.** $y = -3, x = 10$

**62.** They did not multiply the left side by 3.

**63.** Inside the brackets, $x - (-6)$ was simplified incorrectly;
$x - (-6) = x + 6$.

**64.** $2x + 1.50y = 20$
$4x + 3y = 40$

**65.** $4x + 3y = 40$

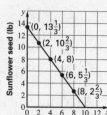

*Thistle seeds (lb)*

**Algebra 1**
Chapter 5 Worked-out Solution Key

# Chapter 5 *continued*

**66.**

| Thistleseed (lb), x | 0 | 2 | 4 | 6 | 8 |
|---|---|---|---|---|---|
| Dark oil sunflower seed (lb), y | $13.\overline{3}$ | $10.\overline{6}$ | 8 | $5.\overline{3}$ | $2.\overline{6}$ |

$$4x + 3y = 40 \qquad 4(0) + 3y = 40$$
$$y = 13.3$$

$$4(2) + 3y = 40 \qquad 4(4) + 3y = 40$$
$$3y = 32 \qquad\qquad 3y = 24$$
$$y = 10.\overline{6} \qquad\qquad y = 8$$

$$4(6) + 3y = 40 \qquad 4(8) + 3y = 40$$
$$3y = 16 \qquad\qquad 3y = 8$$
$$y = 5.\overline{3} \qquad\qquad y = 2.\overline{6}$$

**67.** $2x + 3y = 10$  **68.** 5 students  **69.** 3 students

**70.** $x + y = 60$

**71.** 0% blue, 60% red; 10% blue, 50% red; 20% blue, 40% red; 30% blue, 30% red; 40% blue, 20% red; 50% blue, 10% red; 60% blue, 0% red

**72.** $x + y = 80$

**73.** 0% blue, 80% red; 10% blue, 70% red; 20% blue, 60% red; 30% blue, 50% red; 40% blue, 40% red; 50% blue, 30% red; 60% blue, 20% red; 70% blue, 10% red; 80% blue, 0% red

**74.** Each square on the palette would represent a point on the graph of the equation on a coordinate grid.

**75.** B
$$y + 4 = 2(x + 1)$$
$$y + 4 = 2x + 2$$
$$y = 2x - 2$$
$$-2x + y = -2$$

**76.** D
$$y - 4 = -\tfrac{1}{3}(x + 3)$$
$$y - 4 = -\tfrac{1}{3}x - 1$$
$$y = -\tfrac{1}{3}x + 3$$
$$3y = -x + 9$$
$$x + 3y = 9$$
$$-x - 3y = -9$$

**77.** $2x + 7y = 14$
$$2x + 7(0) = 14$$
$$2x = 14$$
$$x = 7$$

**78.** $2(0) + 7y = 14$
$$0 + 7y = 14$$
$$y = 2$$

**79.** $2x\left(\dfrac{1}{14}\right) + 7y\left(\dfrac{1}{14}\right) = 14\left(\dfrac{1}{14}\right)$
$$\tfrac{1}{7}x + \tfrac{1}{2}y = 1$$

**80.** The denominator of the *x*-term is the *x*-intercept; the denominator of the *y*-term is the *y*-intercept.

**81.** $\dfrac{x}{2} + \dfrac{y}{3} = 1$  **82.** $3x + 2y = 6$  **83.** $y = -\dfrac{3}{2}x + 3$

---

**5.6 Mixed Review (p. 314)**

**84.** $8 + y = 3$
$$y = -5$$

**85.** $y - 9 = 2$
$$y = 11$$

**86.** $-6(q + 22) = 5q$
$$-6q - 132 = 5q$$
$$-11q = 132$$
$$q = -12$$

**87.** $2(x + 5) = 18$
$$2x + 10 = 18$$
$$2x = 8$$
$$x = 4$$

**88.** $7 - 2a = -14$
$$-2a = -21$$
$$a = \tfrac{21}{2}$$
$$a = 10\tfrac{1}{2}$$

**89.** $-2 + 4c = 19$
$$4c = 21$$
$$c = 5\tfrac{1}{4}$$

**90.** $y = -x + 8$

| x | y |
|---|---|
| −1 | 9 |
| 0 | 8 |
| 1 | 7 |
| 8 | 0 |

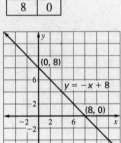

**91.** $y = 4x - 4$

| x | y |
|---|---|
| 0 | −4 |
| 1 | 0 |
| −1 | −8 |
| 2 | 4 |

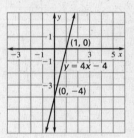

**92.** $y = x + 5$

| x | y |
|---|---|
| 0 | 5 |
| −5 | 0 |

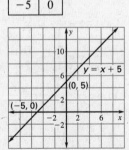

**93.** $y = -9 + 3x$
$$y = 3x - 9$$

| x | y |
|---|---|
| 0 | −9 |
| 3 | 0 |

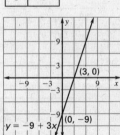

# Chapter 5 *continued*

**94.** $y = -x - 1$

| x | y |
|---|---|
| 0 | -1 |
| -1 | 0 |

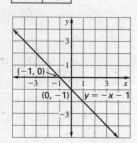

**95.** $y = 10 - x$

| x | y |
|---|---|
| 0 | 10 |
| 10 | 0 |

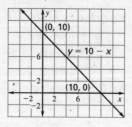

**96.**

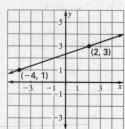

$m = \dfrac{1 - 3}{-4 - 2} = \dfrac{-2}{-6} = \dfrac{1}{3}$

$y - 3 = \dfrac{1}{3}(x - 2)$

$y - 3 = \dfrac{1}{3}x - \dfrac{2}{3}$

$y = \dfrac{1}{3}x + \dfrac{7}{3}$

**97.**

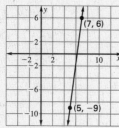

$m = \dfrac{-9 - 6}{5 - 7} = \dfrac{-15}{-2} = 7\frac{1}{2}$

$y - 6 = 7\frac{1}{2}(x - 7)$

$y - 6 = 7\frac{1}{2}x - \dfrac{105}{2}$

$y = \dfrac{15}{2}x - \dfrac{93}{2}$

**98.**

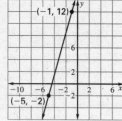

$m = \dfrac{-2 - 12}{-5 - (-1)} = \dfrac{-14}{-4} = \dfrac{7}{2}$

$y - 12 = \dfrac{7}{2}(x + 1)$

$y - 12 = \dfrac{7}{2}x + \dfrac{7}{2}$

$y = \dfrac{7}{2}x + \dfrac{31}{2}$

**99.**

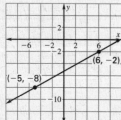

$m = \dfrac{-2 - (-8)}{6 - (-5)} = \dfrac{6}{11}$

$y + 2 = \dfrac{6}{11}(x - 6)$

$y + 2 = \dfrac{6}{11}x - \dfrac{36}{11}$

$y = \dfrac{6}{11}x - \dfrac{58}{11}$

**100.**

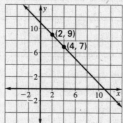

$m = \dfrac{7 - 9}{4 - 2} = \dfrac{-2}{2} = -1$

$y - 9 = -1(x - 2)$

$y - 9 = -x + 2$

$y = -x + 11$

**101.**

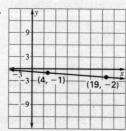

$m = \dfrac{-2 - (-1)}{19 - 4} = -\dfrac{1}{15}$

$y + 1 = -\dfrac{1}{15}(x - 4)$

$y + 1 = -\dfrac{1}{15}x + \dfrac{4}{15}$

$y = -\dfrac{1}{15}x - \dfrac{11}{15}$

**102.**

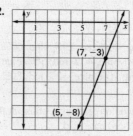

$m = \dfrac{-3 - (-8)}{7 - 5} = \dfrac{5}{2}$

$y + 3 = \dfrac{5}{2}(x - 7)$

$y + 3 = \dfrac{5}{2}x - \dfrac{35}{2}$

$y = \dfrac{5}{2}x - \dfrac{41}{2}$

**103.**

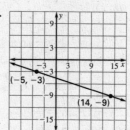

$m = \dfrac{-9 - (-3)}{14 - (-5)} = -\dfrac{6}{19}$

$y + 3 = -\dfrac{6}{19}(x + 5)$

$y + 3 = -\dfrac{6}{19}x - \dfrac{30}{19}$

$y = -\dfrac{6}{19}x - \dfrac{87}{19}$

**104.**

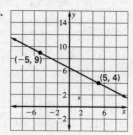

$m = \dfrac{4 - 9}{5 - (-5)} = \dfrac{-5}{10} = -\dfrac{1}{2}$

$y - 4 = -\dfrac{1}{2}(x - 5)$

$y - 4 = -\dfrac{1}{2}x + \dfrac{5}{2}$

$y = -\dfrac{1}{2}x + \dfrac{13}{2}$

**105.** $1.06a = 125.79$
$a = \$118.67$

**106.** $1.085a = 15$
$a = \$13.82$

**107.** $1.50 = 6n$
$0.25 = n$
$25\% = n$

**108.** $10 = 84n$
$0.119 = n$
$11.9\% = n$

# Chapter 5 *continued*

## Lesson 5.7

### Developing Concepts Activity 5.7 (p. 315)
### Exploring the Concept

**1.**

| No. of squares in figure | 1 | 2 | 3 | 4 | 5 |
|---|---|---|---|---|---|
| No. of toothpicks | 4 | 7 | 10 | 13 | 16 |

**2.**

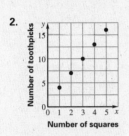

All of the points fall on the same line.

**3.** $(1, 4)$ and $(2, 7)$

$$m = \frac{7 - 4}{2 - 1} = \frac{3}{1} = 3$$

**4.** $y - 4 = 3(x - 1)$
$y - 4 = 3x - 3$
$y = 3x + 1$

**5.** $y = 3(10) + 1 = 30 + 1 = 31$

### Drawing Conclusions

**1. A.**

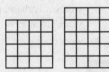

**B.**

**C.**

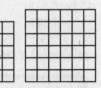

**2. A.**

| Number of squares | 1 | 4 | 9 | 16 | 25 | 36 |
|---|---|---|---|---|---|---|
| Number of toothpicks | 4 | 12 | 24 | 40 | 60 | 84 |

**B.**

| Number of triangles | 1 | 3 | 5 | 7 | 9 | 11 |
|---|---|---|---|---|---|---|
| Number of toothpicks | 3 | 7 | 11 | 15 | 19 | 23 |

**C.**

| Number of triangles | 1 | 4 | 9 | 16 | 25 | 36 |
|---|---|---|---|---|---|---|
| Number of toothpicks | 3 | 9 | 18 | 30 | 45 | 63 |

**3. A.**

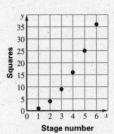

**4. A.**

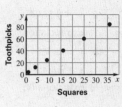

**B.**

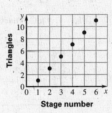

**B.**

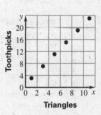

**C.**

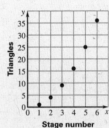

**C.**

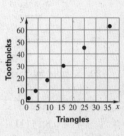

**5.** Answers may vary. The scatter plots for A and C of Question 3 cannot be modeled with linear models. All the others can.

**6.** For sequences 4A and 4C, the sample answer given was obtained using a graphing calculator.
3B: $y = 2x - 1$
4A: $y = 2.27x + 2.85$
4B: $y = 2x + 1$
4C: $y = 1.71x + 2.14$

### 5.7 Guided Practice (p. 319)

Throughout, answers may vary depending upon the points chosen to find the linear model.

**1.** Linear interpolation is a method of estimating the coordinates of a point that lies between two given data points. Linear extrapolation is a method of estimating the coordinates of a point that lies to the right or left of all the given points.

**2.** You could make a scatter plot and decide whether the points fall on a straight line.

**3.** yes  **4.** yes  **5.** no

**6.**

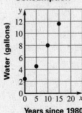

**Bottled Water Consumption**

**7.** *Sample answer:*

$$m = \frac{8 - 2.4}{10 - 0} = \frac{5.6}{10} = 0.56$$
$$y - 8 = 0.56(x - 10)$$
$$y - 8 = 0.56x - 56$$
$$y = 0.56x + 2.4$$

**8.** $y = 0.56(14) + 2.4 = 10.24$ gal; interpolation

# Chapter 5 *continued*

**9.** $y = 0.56(30) + 2.4 = 19.2$ gal

extrapolation

**10.** $y = 0.56(-1) + 2.4$

$y = 1.84$ gal

extrapolation

## 5.7 Practice and Applications (pp. 319–321)

**11.** no **12.** yes **13.** yes **14.** no **15.** yes **16.** yes

**17.**

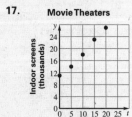
Movie Theaters

**18.**

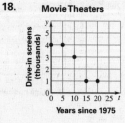
Movie Theaters

**19.** indoor screens

**20.** *Sample answer:*

$m = \dfrac{23 - 11}{15 - 0} = \dfrac{12}{15} = \dfrac{4}{5} = 0.8$

$y = 0.8t + 11$

**21.** $y = 0.8(30) + 11 = 35$

about 35,000 theaters

**22.** $y = 0.8(14) + 11 = 22.2$

about 22,200 theaters;
interpolation

**23.** *Sample answer:*

$m = \dfrac{20.1 - 16.5}{4 - 0} = \dfrac{3.6}{4} = 0.9$ $(0, 16.5), (4, 20.1)$

$y = 0.9x + 16.5$

**24.** $y = 0.9(15) + 16.5 = 13.5 + 16.5 = \$30$ billion

extrapolation

**25.** *Sample answer:*

$(1, 32.8), (4, 40.1)$

$m = \dfrac{40.1 - 32.8}{4 - 1} = \dfrac{7.3}{3} = 2.4\overline{3}$

$y = 2.4\overline{3}x + 31.6$

**26.** $y = 2.4\overline{3}(2) + 31.6 = 4.86 + 31.6 = \$36.46$ billion

interpolation

**27.** $y = 2.4\overline{3}(15) + 31.6 = \$68.05$ billion

extrapolation

**28.**

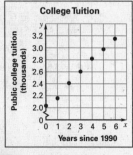
College Tuition

*Sample answer:*

$(0, 2035), (5, 2977)$

$m = \dfrac{2977 - 2035}{5 - 0} = \dfrac{942}{5} = 188.4$

$y = 188.4x + 2035$

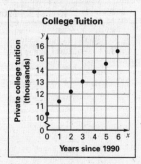
College Tuition

*Sample answer:*

$(0, 10{,}348), (5, 14{,}537)$

$m = \dfrac{14{,}537 - 10{,}348}{5 - 0} = \dfrac{4189}{5} = 837.8$

$y = 837.8x + 10{,}348$

**29.** Answers may vary. **30.** Answers may vary.

**31.** $27.4 \div 52 \approx 0.53$ lb. **32.** $y = 0.01x + 0.48$,

$27.4 \div 12 \approx 2.28$ lb. where $x$ is the number of
years since 1990.

**33.** Answers may vary. **34.** Answers may vary.

**35.**

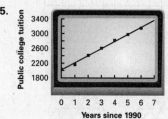

**36.** The $x$-intercept represents the year tuition would be zero.

**37.** The absolute lower limit is $-10$, since tuition values for
$x < -10$ are negative. Reasonable limits might be $-5$
and 10, since most economic trends do not continue
indefinitely.

## 5.7 Mixed Review (p. 322)

**38.**

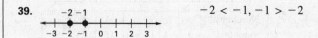

$-7 < 3, 3 > -7$

**39.**

$-2 < -1, -1 > -2$

**40.**

$-12 < 12, 12 > -12$

**41.**

$-\dfrac{1}{2} < \dfrac{3}{2}, \dfrac{3}{2} > -\dfrac{1}{2}$

# Chapter 5 *continued*

**42.**

$$-7 < 2, 2 > -7$$

(number line showing points at $-7$ and $2$, marks at $-8, -4, 0, 4$)

**43.**

$$-\tfrac{7}{2} < -3, -3 > -\tfrac{7}{2}$$

(number line showing points at $-\tfrac{7}{2}$ and $-3$, marks at $-4, -2, 0, 2$)

**44.** $y = 2x - 5$     **45.** $y = \frac{1}{2}x - 8$     **46.** $y = -4x + 7$

**47.** $y = -2x - 6$

**48.**
$$5 = 2(6) + b$$
$$5 = 12 + b$$
$$-7 = b$$
$$y = 2x - 7$$

**49.**
$$7 = 7(-3) + b$$
$$7 = -21 + b$$
$$28 = b$$
$$y = 7x + 28$$

**50.**
$$2 = 7(8) + b$$
$$2 = 56 + b$$
$$-54 = b$$
$$y = 7x - 54$$

**51.**
$$4 = -5(8) + b$$
$$4 = -40 + b$$
$$44 = b$$
$$y = -5x + 44$$

**52.**
$$1 = 7(3) + b$$
$$1 = 21 + b$$
$$-20 = b$$
$$y = 7x - 20$$

**53.**
$$7 = 2(-8) + b$$
$$7 = -16 + b$$
$$23 = b$$
$$y = 2x + 23$$

**54.** $(9, 17), (30, 150)$
$$m = \frac{150 - 17}{30 - 9} = \frac{133}{21} = \frac{19}{3}$$
$$17 = \frac{19}{3}(9) + b$$
$$17 = 57 + b$$
$$-40 = b$$
$$y = \frac{19}{3}x - 40$$

## Quiz 3 (p. 322)

**1.**
$$2y = x + 6$$
$$x - 2y = -6$$

**2.** $9x + 3y = 18$

**3.**
$$-2x + 6y = 8$$
$$2x - 6y = -8$$

**4.**
$$-9x + y = 12$$
$$9x - y = -12$$

**5.**
$$3y = 2x + 18$$
$$-2x + 3y = 18$$
$$2x - 3y = -18$$

**6.**
$$-16y = 1 - 14x$$
$$14x - 16y = 1$$

**7.**
$$y + 8 = -2(x - 6)$$
$$y + 8 = -2x + 12$$
$$y = -2x + 4$$
$$2x + y = 4$$

**8.**
$$y - 1 = -\tfrac{1}{2}(x - 4)$$
$$y - 1 = -\tfrac{1}{2}x + 2$$
$$y = -\tfrac{1}{2}x + 3$$
$$\tfrac{1}{2}x + y = 3$$
$$x + 2y = 6$$

**9.**
$$y - 5 = -3(x - 1)$$
$$y - 5 = -3x + 3$$
$$y = -3x + 8$$
$$3x + y = 8$$

**10.**
$$y + 1 = 2(x + 3)$$
$$y + 1 = 2x + 6$$
$$y = 2x + 5$$
$$-2x + y = 5$$
$$2x - y = -5$$

**11.**
$$y - 4 = 3(x + 12)$$
$$y - 4 = 3x + 36$$
$$y = 3x + 40$$
$$-3x + y = 40$$
$$3x - y = -40$$

**12.**
$$y + 2 = 6(x + 16)$$
$$y + 2 = 6x + 96$$
$$y = 6x + 94$$
$$-6x + y = 94$$
$$6x - y = -94$$

**13.** $m = \dfrac{77.3 - 75.4}{20 - 5} = \dfrac{1.9}{15} = 0.13$
$$75.4 = 0.13(5) + b$$
$$75.4 = 0.65 + b$$
$$74.8 = b$$
$y = 0.13x + 74.8$, where $x =$ the number of years since 1985

**14.** $y = 0.13(4) + 74.8 = 75.32$ years

**15.** $y = 0.13(30) + 74.8 = 78.7$ years

## Chapter 5 Review (pp. 324–326)

**1.** $y = 2x - 2$     **2.** $y = -\tfrac{1}{2}x + 5$     **3.** $y = -8x - 3$

**4.**
$$-3 = 6(4) + b$$
$$-3 = 24 + b$$
$$-27 = b$$
$$y = 6x - 27$$

**5.**
$$4 = 2(-9) + b$$
$$4 = -18 + b$$
$$22 = b$$
$$y = 2x + 22$$

**6.**
$$2 = -1(-3) + b$$
$$2 = 3 + b$$
$$-1 = b$$
$$y = -x - 1$$

**7.** $m = \dfrac{2 - (-9)}{-3 - 4} = \dfrac{11}{-7}$
$$-9 = \frac{11}{-7}(4) + b$$
$$-9 = -\frac{44}{7} + b$$
$$-\frac{19}{7} = b$$
$$y = -\frac{11}{7}x - \frac{19}{7}$$

**8.** $m = \dfrac{-1 - 8}{-2 - 1} = \dfrac{-9}{-3} = 3$
$$8 = 3(1) + b$$
$$8 = 3 + b$$
$$5 = b$$
$$y = 3x + 5$$

**9.** $m = \dfrac{2 - 5}{-8 - 2} = \dfrac{-3}{-10} = \dfrac{3}{10}$
$$5 = \frac{3}{10}(2) + b$$
$$5 = \frac{6}{10} + b$$
$$\frac{22}{5} = \frac{44}{10} = b$$
$$y = \frac{3}{10}x + \frac{22}{5}$$

**10.** $y = \dfrac{7}{11}x - 3$

# Chapter 5 *continued*

**11.**

Sample answer:

$$m = \frac{25 - (-1)}{12 - (-1)} = \frac{26}{13} = 2$$
$$-1 = 2(-1) + b$$
$$-1 = -2 + b$$
$$1 = b$$
$$y = 2x + 1$$

**12.**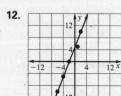

Sample answer:

$$m = \frac{10 - (-10)}{2 - (-6)} = \frac{20}{8} = \frac{10}{4} = \frac{5}{2}$$
$$10 = \frac{5}{2}(2) + b$$
$$10 = 5 + b$$
$$5 = b$$
$$y = \frac{5}{2}x + 5$$

**13.** $m = \dfrac{5 - 4}{2 - (-4)} = \dfrac{1}{6}$    **14.** $m = \dfrac{0 - 3}{-5 - (-2)} = \dfrac{-3}{7}$

$y - 4 = \frac{1}{6}(x + 4)$   or     $y - 3 = \frac{-3}{7}(x + 2)$   or

$y - 5 = \frac{1}{6}(x - 2)$          $y = -\frac{3}{7}(x - 5)$

$y - 4 = \frac{1}{6}x + \frac{2}{3}$          $y - 3 = \frac{-3}{7}x - \frac{6}{7}$

$y = \frac{1}{6}x + \frac{14}{3}$            $y = -\frac{3}{7}x + \frac{15}{7}$

**15.** $m = \dfrac{8 - (-2)}{-1 - 1} = \dfrac{10}{-2} = -5$   **16.** $-2x + y = 9$

$\qquad\qquad\qquad\qquad\qquad\qquad\qquad\quad 2x - y = -9$

$y + 2 = -5(x - 1)$   or

$y - 8 = -5(x + 1)$

$y + 2 = -5x + 5$

$y = -5x + 3$

**17.** $8x + 3y = 2$     **18.** $2x + 2y = 6$     **19.** $\frac{1}{3}x + y = \frac{2}{3}$

$\qquad\qquad\qquad\qquad\qquad\qquad\qquad\qquad\qquad x + 3y = 2$

**20.** $-\frac{3}{4}x + y = \frac{1}{2}$       **21.** $-\frac{2}{3}x + \frac{1}{2}y = -2$

$\quad -3x + 4y = 2$            $-4x + 3y = -12$

$\quad 3x - 4y = -2$             $4x - 3y = 12$

**22.** $y = 16x + 18$      **23.** $y = 16(14) + 18 = 242;$

$\quad y = 16(-1) + 18$        about 242 million; linear

$\quad y = -16 + 18 = 2$      extrapolation

about 2 million; linear
extrapolation

**24.** $y = 16(6) + 18 = 114;$   about 114 million; linear
extrapolation

**25.** $y = 16(10) + 18 = 178;$   about 178 million; linear
extrapolation

## Chapter 5 Test (p. 327)

**1.** $y = 2x - 1$     **2.** $y = -4x + 3$     **3.** $y = 6x + 9$

**4.** $y = \frac{1}{4}x - 3$     **5.** $y = -3x + 3$     **6.** $y = 4$

**7.** $6 = 2(2) + b$          **8.** $-9 = -5(3) + b$

$\quad 6 = 4 + b$                $-9 = -15 + b$

$\quad 2 = b$                     $6 = b$

$\quad y = 2x + 2$              $y = -5x + 6$

**9.** $-6 = -3(-5) + b$   **10.** $8 = -4(1) + b$

$\quad -6 = 15 + b$              $8 = -4 + b$

$\quad -21 = b$                  $12 = b$

$\quad y = -3x - 21$          $y = -4x + 12$

**11.** $-2 = \frac{1}{2}(4) + b$     **12.** $-5 = 8\left(\frac{1}{3}\right) + b$

$\quad -2 = 2 + b$             $-5 = \frac{8}{3} + b$

$\quad -4 = b$                 $-\frac{23}{3} = b$

$\quad y = \frac{1}{2}x - 4$           $y = 8x - \frac{23}{3}$

**13.** $m = \dfrac{-1 - 2}{4 - (-3)} = \dfrac{-3}{7}$   **14.** $m = \dfrac{-4 - 2}{8 - 6} = \dfrac{-6}{2} = -3$

$\quad 2 = \frac{-3}{7}(-3) + b$        $2 = -3(6) + b$

$\quad 2 = \frac{9}{7} + b$               $2 = -18 + b$

$\quad \frac{5}{7} = b$                  $20 = b$

$\qquad\qquad\qquad\qquad\qquad\quad y = -3x + 20$

$\quad y = -\frac{3}{7}x + \frac{5}{7}$

**15.** $m = \dfrac{4 - 5}{2 - (-2)} = \dfrac{-1}{4}$   **16.** $m = \dfrac{0 - (-8)}{-1 - (-2)} = \dfrac{8}{1} = 8$

$\quad 4 = -\frac{1}{4}(2) + b$         $-8 = 8(-2) + b$

$\quad 4 = -\frac{1}{2} + b$           $-8 = -16 + b$

$\quad \frac{9}{2} = b$                  $8 = b$

$\qquad\qquad\qquad\qquad\qquad\quad y = 8x + 8$

$\quad y = -\frac{1}{4}x + \frac{9}{2}$

**17.** $m = \dfrac{4 - 2}{2 - (-5)} = \dfrac{2}{7}$    **18.** $m = \dfrac{-9 - (-1)}{1 - 9} = \dfrac{-8}{-8} = 1$

$\quad 2 = \frac{2}{7}(-5) + b$        $-1 = 1(9) + b$

$\quad 2 = \frac{-10}{7} + b$          $-1 = 9 + b$

$\qquad\qquad\qquad\qquad\qquad\quad -10 = b$

$\quad \frac{24}{7} = b$               $y = x - 10$

$\quad y = \frac{2}{7}x + \frac{24}{7}$

**19.** $m = \frac{1}{2}; (-4, 7)$      **20.** $-2x + 4y = 24$

$\quad 7 = \frac{1}{2}(-4) + b$         $2x - 4y = -24$

$\quad 7 = -2 + b$

$\quad 9 = b$

$\quad y = \frac{1}{2}x + 9$

**21.** $-7x + y = 8$      **22.** $18x + 6y = 3$

$\quad 7x - y = -8$

**Algebra 1**
Chapter 5 Worked-out Solution Key

# Chapter 5 *continued*

**23.**
$$1 - 2x = 18y$$
$$-2x - 18y = -1$$
$$2x + 18y = 1$$

**24.**
$$-25x + 5y = 0$$
$$25x + -5y = 0$$

**25.**
$$-4y + x = 8$$
$$x - 4y = 8$$

**26.**
$$13y = 5x + 52$$
$$-5x + 13y = 52$$
$$5x - 13y = -52$$

**27.** $5x + 10y = 220$

**28.** $y = 0.04x + 1250$

**29.**

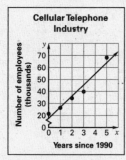

**Cellular Telephone Industry**

Number of employees (thousands)

Years since 1990

**30.** *Sample answer:*
$$m = \frac{64{,}000 - 26{,}300}{5 - 1} = \frac{37{,}700}{4} = 9425$$
$$64{,}000 = 9425(5) + b$$
$$64{,}000 = 47{,}125 + b$$
$$16{,}875 = b$$
$$y = 9425x + 16{,}875$$

**31.** $y = 9425(4) + 16{,}875 = 54{,}575$
employees; interpolation

**32.** $y = 9425(14) + 16{,}875 = 148{,}825$
employees; extrapolation

**33.**
$$0 = 9425(x) + 16{,}875$$
$$-16{,}875 = 9425x$$
$$-2 \approx x$$

1988; extrapolation; no; cellular phones were available before 1988, so clearly there were employes at that time.

### Chapter 5 Standard Test (pp. 328–329)

**1.** A
$$m = \frac{2 - 6}{-4 - 6} = \frac{-4}{-10} = \frac{2}{5}$$
$$2 = \frac{2}{5}(-4) + b$$
$$2 = \frac{-8}{5} + b$$
$$\frac{18}{5} = b$$
$$y = \frac{2}{5}x + \frac{18}{5}$$

**2.** D
$$m = \frac{1}{2}$$
$$y = \frac{1}{2}x - \frac{3}{4}$$

**3.** E
$$-1 = -1(2) + b$$
$$-1 = -2 + b$$
$$1 = b$$
$$y = -x + 1$$
$$y = -(-4) + 1 = 4 + 1$$
$$y = 5$$

**4.** A
$$(3, 2), (1, 1)$$
$$m = \frac{1 - 2}{1 - 3} = \frac{-1}{-2} = \frac{1}{2}$$
$$1 = \frac{1}{2}(1) + b$$
$$1 = \frac{1}{2} + b$$
$$\frac{1}{2} = b$$
$$y = \frac{1}{2}x + \frac{1}{2}$$

**5.** B
$$12.5 = 8 + 1.50h$$
$$4.5 = 1.50h$$
$$3 = \text{half hours} = 1.5 \text{ hours}$$

**6.** E
$$-5 = \frac{1}{2}(4) + b$$
$$-5 = 2 + b$$
$$-7 = b$$
$$y = \frac{1}{2}x - 7$$

**7.** C
$$y = -\frac{5}{3}(0) + 5$$
$$y = 5$$
$$0 = -\frac{5}{3}x + 5$$
$$-5 = -\frac{5}{3}x$$
$$3 = x$$

**8.** D
$$(-1, 9), (3, 1)$$
$$y = -2x + 7$$
$$9 = -2(-1) + 7$$
$$9 = 2 + 7$$
$$9 = 9$$
$$1 = -2(3) + 7$$
$$1 = -6 + 7$$
$$1 = 1$$

**9.** E
$$-2x + y = -5$$

**10.** C
$$1 = -2(-6) + b$$
$$1 = 12 + b$$
$$-11 = b$$
$$y = -2x - 11$$
$$2x + y = -11$$

**11.** D
$$m = \frac{1 - (-4)}{6 - 3} = \frac{5}{3} \qquad m = -\frac{3}{5}$$
$$y - 1 = -\frac{3}{5}(x - 6)$$
$$y - 1 = -\frac{3}{5}x + \frac{18}{5}$$
$$y = -\frac{3}{5}x + \frac{23}{5}$$
$$3x + 5y = 23$$

**12.** A
$$2x + 3y = 12 \qquad\qquad -5y = 6 + 10x$$
$$3y = -2x + 12 \qquad y = \frac{6}{-5} - 2x$$
$$y = -\frac{2}{3}x + 4 \qquad y = -2x - \frac{6}{5}$$

**13.** A
$$3y = -2x + 12$$
$$y = -\frac{2}{3}x + 4$$
$$y = -2x - \frac{6}{5}$$

**14.** A
$$\frac{2}{3}x + 6(0) = 8 \qquad 3(0) = -7 + 2x$$
$$\frac{2}{3}x = 8 \qquad\qquad 0 = -7 + 2x$$
$$x = 12 \qquad\qquad 7 = 2x$$
$$\frac{7}{2} = x$$

**15.**

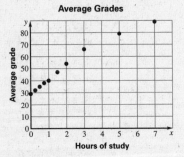

**Average Grades**

**16.** *Sample answer:*

$$m = \frac{79 - 40}{5 - 1} = \frac{39}{4} = 9.75$$

$$40 = 9.75(1) + b$$

$$40 = 9.75 + b$$

$$30.25 = b$$

$$y = 9.75x + 30.25$$

**17.** $y = 9.75(4) + 30.25 = 39 + 30.25 = 69.25$

**18.** $93 = 9.75x + 30.25$

$$62.75 = 9.75x$$

$$6.4 = x$$

about 6.4 hours

**19.** $y = 9.75(4.5) + 30.25 = 43.875 + 30.25 = 74.125$

**20.** The material on this test must be very difficult. In order to receive a 93%, a student had to study nearly 7 hours.

# CHAPTER 6

## Think & Discuss *(p. 331)*

**1.** *Sample answer:* flute: $260 \leq x \leq 2300$;

soprano clarinet: $135 \leq x \leq 1800$;

soprano saxophone: $200 \leq x \leq 1350$

**2.** flute   **3.** *Sample answer:* $1800 \leq x \leq 2300$

## Skill Review *(p. 332)*

**1.** $3(5) + 2 \overset{?}{<} 18$

$17 < 18$

solution

**2.** $5 - 3 \overset{?}{>} 2$

$2 \overset{?}{>} 2$

not a solution

**3.** $4 + 4(5) \overset{?}{\geq} 24$

$24 \geq 24$

solution

**4.** $1 + 2(5) \overset{?}{\leq} 5$

$11 \overset{?}{\leq} 5$

not a solution

**5.** $8 + 5 \overset{?}{<} 12$

$13 \overset{?}{<} 12$

not a solution

**6.** $2(5)^2 \overset{?}{>} 45$

$50 > 45$

solution

**7.** $\frac{1}{2}x + 10 = 15$

$\frac{1}{2}x = 5$

$x = 10$

**8.** $\frac{2}{3}x - 4 = 12$

$\frac{2}{3}x = 16$

$x = 24$

**9.** $20 = 2(x + 1)$

$20 = 2x + 2$

$18 = 2x$

$9 = x$

**10.** $4x + 5x - 1 = 53$

$9x = 54$

$x = 6$

**11.** $6x - 2x + 8 = 36$

$4x = 28$

$x = 7$

**12.** $4(5 - x) = 12$

$20 - 4x = 12$

$-4x = -8$

$x = 2$

**13.** $y = -4x$

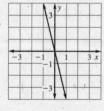

**14.** $y = -2x + 5$

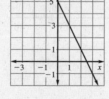

**15.** $2y - x = 4$

$y = \frac{1}{2}x + 2$

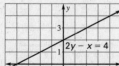

**16.** $-2x + y = 3$

$y = 2x + 3$

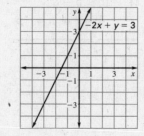

## Lesson 6.1

**Developing Concepts Activity 6.1 *(p. 333)***
*Exploring the Concept*

**1.** *Sample answer:* $4 > -8$

**2.** *Sample answers:*

  **a.** $4 + 4 > -8 + 4; 8 > -4$

  **b.** $4 - 4 > -8 - 4; 0 > -12$

  **c.** $4 \cdot 4 > -8 \cdot 4; 16 > -32$

  **d.** $\frac{4}{4} > \frac{-8}{4}; 1 > -2$

  **e.** $4 \cdot (-4) < -8 \cdot (-4); -16 < 32$

  **f.** $\frac{4}{-4} < \frac{-8}{-4}; -1 < 2$

**3.** in parts (e) and (f)

**4.** when multiplying or dividing by a negative number

*Drawing Conclusions*

**1.** no

$4 + 7 \overset{?}{<} 9 + 7$

$11 < 16$

**2.** no

$15 - (-4) \overset{?}{>} 12 - (-4)$

$19 > 16$

**3.** no

$4 \cdot 5 \overset{?}{>} -3 \cdot 5$

$20 > -15$

**4.** no

$2 + (-7) \overset{?}{>} -11 + (-7)$

$-5 > -18$

**5.** yes

$\frac{-6}{-3} \overset{?}{>} \frac{2}{-3}$

$2 > -\frac{2}{3}$

**6.** yes

$1(-10) > 8(-10)$

$-10 > -80$

**7.**

|  | a positive number | a negative number |
|---|---|---|
| Add | no | no |
| Subtract | no | no |
| Multiply by | no | yes |
| Divide by | no | yes |

**8.** $x + 3 > 9$

$x > 6$

**9.** $x + 7 \leq 12$

$x \leq 5$

**10.** $4x \geq 15$

$x \geq \frac{15}{4}$

**11.** $-3x > 11$

$x < -\frac{11}{3}$

**12.** $2x < 11$

$x < \frac{11}{2}$

**13.** $-\frac{1}{3}x \leq 12$

$x \geq -36$

**14.** $x + 6 < 15$

$x < 9$

**15.** $x - 2 \geq 90$

$x \geq 92$

**16.** $5x \leq 25$

$x \leq 5$

**17.** $-6x > 30$

$x < -5$

# Chapter 6 *continued*

## 6.1 Guided Practice (p. 337)

1. equivalent

2. all real numbers greater than $-7$

3. all real numbers less than 1

4. all real numbers greater than or equal to 9

5. all real numbers less than or equal to 10

6. When dividing by a negative number, reverse the inequality symbol.
$$-3x \geq 15$$
$$x \leq -5$$

7. Do not reverse the inequality symbol when subtracting a number from each side.
$$x + 4 < 1$$
$$x < -3$$

8. open dot    9. open dot    10. closed dot

11. open dot    12. closed dot    13. closed dot

14. $-2 + x < 5$
$$x < 7$$

15. $-3 \leq y + 2$
$$-5 \leq y \text{ or } y \geq -5$$

16. $-2b \leq -8$
$$b \geq 4$$

17. $x - 4 > 10$
$$x > 14$$

18. $5x > -45$
$$x > -9$$

19. $-6y \leq 36$
$$y \geq -6$$

20. $b \leq 19{,}550$

21.

## 6.1 Practice and Applications (pp. 337–339)

22.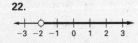
$$x > -2$$

23.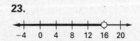
$$x < 16$$

24.
$$x \leq 7$$

25.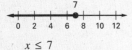
$$x \geq -6$$

26.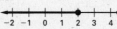
$$2 \geq x \text{ or } x \leq 2$$

27.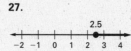
$$2.5 \leq x \text{ or } x \geq 2.5$$

28.
$$-1 \leq x \text{ or } x \geq -1$$

29.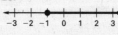
$$10 \geq x \text{ or } x \leq 10$$

30.
$$x < -0.5$$

31. $x + 6 < 8$
$$x < 2$$

32. $-5 < 4 + x$
$$-9 < x \text{ or } x > -9$$

33. $-4 + x < 20$
$$x < 24$$

34. $8 + x \leq -9$
$$x \leq -17$$

35. $p - 12 \geq -1$
$$p \geq 11$$

36. $-2 > b - 5$
$$3 > b \text{ or } b < 3$$

37. $x - 3 > 2$
$$x > 5$$

38. $x - 5 \geq 1$
$$x \geq 6$$

39. $6 \leq c + 2$
$$4 \leq c \text{ or } c \geq 4$$

40. $-8 \leq x - 14$
$$6 \leq x \text{ or } x \geq 6$$

41. $m + 7 \geq -10$
$$m \geq -17$$

42. $-6 > x - 4$
$$-2 > x \text{ or } x < -2$$

43. $-2 + x < 0$
$$x < 2$$

44. $-10 > a - 6$
$$-4 > a \text{ or } a < -4$$

45. $5 + x \geq -5$
$$x \geq -10$$

# Chapter 6 *continued*

**46.** $15p < 60$
$p < 4$

**47.** $-10a > 100$
$a < -10$

**48.** $-\dfrac{n}{5} < 17$
$n > -85$

**49.** $11 \geq -2.2m$
$-5 \leq m$

**50.** $-18.2x \geq -91$
$x \leq 5$

**51.** $2.1x \leq -10.5$
$x \leq -5$

**52.** $13 \leq -\dfrac{x}{3}$
$-39 \geq x$

**53.** $-\dfrac{a}{10} \leq -2$
$a \geq 20$

**54.** $\dfrac{x}{4} \leq -9$
$x \leq -36$

**55.** $x - 3.2 < 1.8$
$x < 5$
E

**56.** $x + 4 \leq 6$
$x \leq 2$
D

**57.** $10x > 50$
$x > 5$
C

**58.** $x - 2 \geq -10$
$x \geq -8$
B

**59.** $-\dfrac{x}{2} \leq 0$
$x \geq 0$
A

**60.** $-3.5x \geq 28$
$x \leq -8$
F

**61.** $\dfrac{3}{27.5} = \dfrac{6}{55}$ mi/min
$x < \dfrac{6}{55}$

**62.** $p > -38.87$

**63.** $M < 1376$

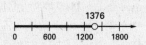

**64.** $x + 475 \geq 684$
$x \geq 209$

The lowest score she can get is 209.

**65.** $l < 1700$

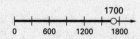

**66.** $73 \leq f \leq 698$  **67.** $131 \leq f \leq 587$

**68. a.** Wind Chime Sales

**b.** $4x - 48 \geq 0$
$4x \geq 48$
$x \geq 12$
12 wind chimes

**c.** $48 + 3.5x \leq 7.5x$

**69.** An inequality such as $x < y$ means that $x$ is to the left of $y$ on the number line. Since a number and its opposite are on opposite sides of zero, if $x < y$ then $-x > -y$. For example, $2 < 3$ and $-2 > -3$ and $-4 < -1$ and $4 > 1$.

**70.** You don't know if $x$ is positive or negative so you don't know whether or not to reverse the inequality symbol. (Note: $x \neq 0$)

***6.1 Mixed Review (p. 339)***

**71.** $45b - 3 = 2$
$45b = 5$
$b = \dfrac{1}{9}$

**72.** $-5x + 50 = 300$
$-5x = 250$
$x = -50$

**73.** $-3s - 2 = -44$
$-3s = -42$
$s = 14$

**74.** $\dfrac{1}{3}x + 5 = -4$
$\dfrac{1}{3}x = -9$
$x = -27$

**75.** $\dfrac{x}{4} + 4 = 18$
$\dfrac{x}{4} = 14$
$x = 56$

**76.** $-8 = \dfrac{3}{5}a - 5$
$-3 = \dfrac{3}{5}a$
$-5 = a$

**77.** $x + 2x + 5 = 14$
$3x + 5 = 14$
$3x = 9$
$x = 3$

**78.** $3(x - 6) = 12$
$x - 6 = 4$
$x = 10$

**79.** $9 = -\dfrac{3}{2}(x - 2)$
$-6 = x - 2$
$-4 = x$

**80.** $(1, 2), (4, -1)$
$m = \dfrac{-1 - 2}{4 - 1} = \dfrac{-3}{3} = -1$
equation: $y = -x + 3$
crosses $y$-axis at $(0, 3)$
crosses $x$-axis at $(3, 0)$

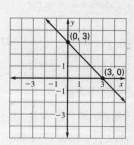

**81.** $(2, 0), (-4, -3)$
$m = \dfrac{-3 - 0}{-4 - 2} = \dfrac{-3}{-6} = \dfrac{1}{2}$
equation: $y = \dfrac{1}{2}x - 1$
crosses $y$-axis at $(0, -1)$
crosses $x$-axis at $(2, 0)$

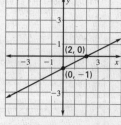

**82.** $(1, 1), (-3, 5)$
$m = \dfrac{5 - 1}{-3 - 1} = \dfrac{4}{-4} = -1$
equation: $y = -x + 2$
crosses $y$-axis at $(0, 2)$
crosses $x$-axis at $(2, 0)$

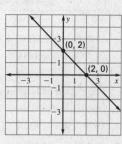

# Chapter 6 *continued*

**83.** $(-1, 4), (2, 4)$

$m = \dfrac{4 - 4}{2 - (-1)} = \dfrac{0}{3} = 0$

equation: $y = 4$

crosses $y$-axis at $(0, 4)$

does not cross the $x$-axis

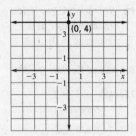

**84.** $(-1, -3), (2, 3)$

$m = \dfrac{3 - (-3)}{2 - (-1)} = \dfrac{6}{3} = 2$

equation: $y = 2x - 1$

crosses $y$-axis at $(0, -1)$

crosses $x$-axis at $\left(\frac{1}{2}, 0\right)$

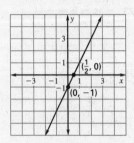

**85.** $(5, 0), (5, -2)$

$m = \dfrac{-2 - 0}{5 - 5} = \dfrac{-2}{0}$

no slope

equation: $x = 5$

crosses $x$-axis at $(5, 0)$

does not cross the $y$-axis

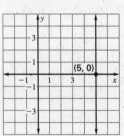

**86.** $p = 5n$
$p = 5(17) = 85$

**87.** $d = 0.12t$

## Lesson 6.2

### Activity (p. 341)

**1.** $L = 853,000 + 58,000t$

**2.**

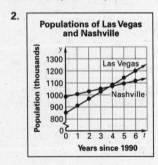

Populations of Las Vegas and Nashville

Populations will be the same in 1993.

**3.** $58,000t + 853,000 > 985,000 + 22,000t$

$36,000t > 132,000$

$t > 3.67$

Population of Las Vegas exceeded the population of Nashville in 1993.

**4.** 1990, 1991, 1992

**5.** $58,000y + 853,000 < 22,000y + 985,000$

---

### 6.2 Guided Practice (p. 343)

**1.** There are multiple steps involved in the solution.
*Sample answer:* $3x - 4 < 8$

**2.** Subtract 2 from each side, then divide each side by $-3$ and reverse the inequality symbol.

**3.** The inequality symbol be reversed when dividing by $-4$. $y > -\frac{5}{4}$

**4.** Four was added to the left side, but subtracted from the right side.

$6x - 4 \geq 2x + 1$

$6x \geq 2x + 5$

$4x \geq 5$

$x \geq \frac{5}{4}$

**5.** $y + 2 > -1$
$y > -3$

**6.** $-2x < -14$
$x > 7$

**7.** $-4x \geq -12$
$x \leq 3$

**8.** $4y - 3 < 13$
$4y < 16$
$y < 4$

**9.** $5x + 12 \leq 62$
$5x \leq 50$
$x \leq 10$

**10.** $10 - c \geq 6$
$-c \geq -4$
$c \leq 4$

**11.** $2 - x > 6$
$-x > 4$
$x < -4$

**12.** $3x + 2 \leq 7x$
$2 \leq 4x$
$\frac{1}{2} \leq x$
$x \geq \frac{1}{2}$

**13.** $2x - 1 > 6x + 2$
$-1 > 4x + 2$
$-3 > 4x$
$-\frac{3}{4} > x$ or $x < -\frac{3}{4}$

**14.** $0.58x - (0.15x + 15) \geq 150$
$0.43x - 15 \geq 150$
$0.43x \geq 165$
$x \geq 383.72$

You must sell 384 flies.

---

### 6.2 Practice and Applications (pp. 343–345)

**15.** $x + 5 > -13$
$x > -18$

**16.** $15 - x < 7$
$-x < -8$
$x > 8$

**17.** $-5 \leq 6x - 12$
$7 \leq 6x$
$\frac{7}{6} \leq x$ or $x \geq \frac{7}{6}$

**18.** $-6 + 5x < 19$
$5x < 25$
$x < 5$

**19.** $7 - 3x \leq 16$
$-3x \leq 9$
$x \geq -3$

**20.** $-3x - 0.4 > 0.8$
$-3x > 12$
$x < -0.4$

**21.** $6x + 5 < 23$
$6x < 18$
$x < 3$

**22.** $-17 > 5x - 2$
$-15 > 5x$
$-3 > x$ or $x < -3$

**23.** $-x + 9 \geq 14$
$-x \geq 5$
$x \leq -5$

**24.** $4x - 1 \leq -17$
$4x \leq -16$
$x \leq -4$

---

# Chapter 6 *continued*

**25.** $12 > -2x - 6$
$18 > -2x$
$-9 < x$ or $x > -9$

**26.** $-x - 4 > 3x - 2$
$-4 > 4x - 2$
$-2 > 4x$
$-\frac{1}{2} > x$ or $x < -\frac{1}{2}$

**27.** $\frac{2}{3}x + 3 \geq 11$
$\frac{2}{3}x \geq 8$
$x \geq 12$

**28.** $6 \geq \frac{7}{3}x - 1$
$7 \geq \frac{7}{3}x$
$3 \geq x$ or $x \leq 3$

**29.** $-\frac{1}{2}x + 3 < 7$
$-\frac{1}{2}x < 4$
$x > -8$

**30.** $3x + 1.2 < -7x - 1.3$
$10x + 1.2 < -1.3$
$10x < -2.5$
$x < -0.25$

**31.** $x + 3 \leq 2(x - 4)$
$x + 3 \leq 2x - 8$
$3 \leq x - 8$
$11 \leq x$ or $x \geq 11$

**32.** $2x + 10 \geq 7(x + 1)$
$2x + 10 \geq 7x + 7$
$10 \geq 5x + 7$
$3 \geq 5x$
$\frac{3}{5} \geq x$ or $x \leq \frac{3}{5}$

**33.** $-x + 4 < 2(x - 8)$
$-x + 4 < 2x - 16$
$4 < 3x - 16$
$20 < 3x$
$\frac{20}{3} < x$ or $x > \frac{20}{3}$

**34.** $-x + 6 > -(2x + 4)$
$-x + 6 > -2x - 4$
$x + 6 > -4$
$x > -10$

**35.** $-2(x + 3) < 4x - 7$
$-2x - 6 < 4x - 7$
$-6 < 6x - 7$
$1 < 6x$
$\frac{1}{6} < x$ or $x > \frac{1}{6}$

**36.** Let $p$ = population of Tampa
Let $n$ = number of years since 1990
$p = 850n + 280,015$

**37.** Let $x$ = number of rides
$25 \geq 1.25x + 5$
$20 \geq 1.25x$
$16 \geq x$
You can go on 16 rides.

**38.** 1991 to 1995

**39.** Let $t$ = number of toppings
$14 + 0.75t \leq 18.50$
$0.75t \leq 4.50$
$t \leq 6$
The pizza can have 6 toppings.

**40.** $n \geq \left(\dfrac{24}{1 \text{ sec}}\right)\left(\dfrac{3 \text{ h}}{2}\right)\left(\dfrac{60 \text{ min}}{1 \text{ h}}\right)\left(\dfrac{60 \text{ sec}}{1 \text{ min}}\right)$
$n \geq 129,600$
It would take at least 129,600 hours.

**41.** $t \geq \dfrac{129,600}{36(40)(45)}$
$t \geq 2$
It would take 2 years.

**42.** If all artists worked the same number of hours per year as the original team, the time would be cut in half.

**43.** $9x > 36$
$x > 4$ m

**44.** $\frac{1}{2}x(10) \geq 25$
$5x \geq 25$
$x \geq 5$ m

**45.** $\frac{1}{2}(2 + 4)x < 30$
$3x < 30$
$x < 10$ ft

**46.** $\frac{1}{2}(24)(x) < 144$
$12x < 144$
$x < 12$ in.

**47.** $5x \leq 25$
$x \leq 5$ in.

**48.** $\frac{1}{2}(12)x > 60$
$6x > 60$
$x > 10$ cm

**49.** $2.39 + 0.69x \leq 5$
$0.69x \leq 2.61$
$x \leq 3.78$
A

**50.** $250 \geq 1.75x$
$142.9 \geq x$
C

**51.** $15x \leq 170$
$x \leq 11.3$
B

**52.** To find the solution of $\frac{2}{3}x - 2 = 0$, graph the equations $y = \frac{2}{3}x - 2$ and $y = 0$ and find their intersection, $(3, 0)$. The $x$-coordinate, 3, is the solution of the original equation. On the graph of $y = \frac{2}{3}x - 2$, for points to the left of $(3, 0)$, $y < 0$ and for points to the right of $(3, 0)$, $y > 0$. Therefore, the solution of $\frac{2}{3}x - 2 < 0$ is $x < 3$.

**53.** To find the solution of $-2x + 4 = 0$, graph the equations $y = -2x + 4$ and $y = 0$ and find their intersection, $(2, 0)$. The $x$-coordinate, 2, is the solution of the original equation. On the graph of $y = -2x + 4$, for points to the right of $(2, 0)$, $y < 0$ and for points to the left of $(2, 0)$, $y > 0$. Therefore, the solution of $-2x + 4 > 0$ is $x < 2$.

## 6.2 Mixed Review (p. 345)

**54.** $(9 + 4) - 8 = (7 + 4) - 8 = 3$

**55.** $3x + 2 = 3(-4) + 2 = -10$

**56.** $b^3 - 5 = (2)^3 - 5 = 3$

**57.** $2(r + s) = 2(2 + 4) = 12$

**58.** $x^2 - 3x = 5^2 - 3(5) = 10$

**59.** $\dfrac{a^2}{8 - b} = \dfrac{4^2}{8 - (-9)} = \dfrac{16}{17}$

**60.** $S = J + 4$

**61.** $t > 2\ell$   **62.** $-12$   **63.** $3$

**64.** $(42.99 + 14.50 + 29.99) - 10 = \$77.48$

**65.** $\frac{15}{20} = 0.75$

## Lesson 6.3

### 6.3 Guided Practice (p. 349)

**1.** *Sample answer:*
$-1 < x < 4$
This is a compound inequality because it consists of two inequalities connected by *and*.

**2.** $-4 \leq x < 0$

**3.** $2 < x < 5$

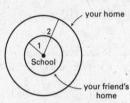

**4.** $x > 3$ or $x < -1$

**5.** $7 < 4 + x < 8$
$3 < x < 4$

**6.** $-1 < 2x + 3 \leq 13$
$-4 < 2x \leq 10$
$-2 < x \leq 5$

**7.** $6 < 4x - 2 \leq 14$
$8 < 4x \leq 16$
$2 < x \leq 4$

**8.** $4x - 1 > 7$ or $5x - 1 < -6$
$4x > 8$ or $5x < -5$
$x > 2$ or $x < -1$

**9.** $2x + 3 < -1$ or $3x - 5 > -2$
$2x < -4$ or $3x > 3$
$x < -2$ or $x > 1$

**10.** $4 \leq -8 - x < 7$
$12 \leq -x < 15$
$-12 \geq x > -15$

**11.**

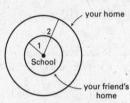

your home
2
1
School
your friend's home
$1 \leq d \leq 3$

## 6.3 Practice and Applications *(pp. 349–351)*

**12.** $3 < x < 8$

**13.** $x > 7$ or $x < 5$

**14.** $0 \leq x < 5$

**15.** $-9 \leq x < 4$

**16.** $-4 \leq x < -2$

**17.** $-6 < x < -1$

**18.** $6 < x - 6 \leq 8$
$12 < x \leq 14$
$x$ is greater than 12 and less than or equal to 14.

**19.** $-5 < x - 3 < 6$
$-2 < x < 9$
$x$ is greater than $-2$ and less than 9.

**20.** $-4 < 2 + x < 1$
$-6 < x < -1$
$x$ is greater than $-6$ and less than $-1$.

**21.** $8 \leq 2x + 6 \leq 18$
$2 < 2x \leq 12$
$1 \leq x \leq 6$
$x$ is greater than or equal to 1 and less than or equal to 6.

**22.** $6 + 2x > 20$ or $8 + x \leq 0$
$2x > 14$ or $x \leq -8$
$x > 7$
$x$ is greater than 7 or less than or equal to $-8$

**23.** $-3x - 7 \geq 8$ or $-2x - 11 \leq -31$
$-3x \geq 15$ or $-2x \leq -20$
$x \leq -5$ or $x \geq 10$
$x$ is less than or equal to $-5$ or greater than or equal to 10.

**24.** $-4 \leq -3x - 13 \leq 26$
$9 \leq -3x \leq 39$
$-3 \geq x \geq -13$
$x$ is greater than or equal to $-13$ and less than or equal to $-3$.

**25.** $-13 \leq 5 - 2x < 9$
$-18 \leq -2x < 4$
$9 \geq x > -2$
$x$ is less than or equal to 9 and greater than $-2$.

**26.** $3 \leq 2x < 7$
$\frac{3}{2} \leq x < \frac{7}{2}$

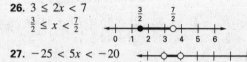

**27.** $-25 < 5x < -20$
$-5 < x < -4$

**28.** $-5 < 3x + 4 < 19$
$-9 < 3x < 15$
$-3 < x < 5$

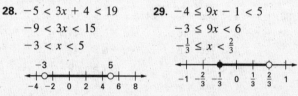

**29.** $-4 \leq 9x - 1 < 5$
$-3 \leq 9x < 6$
$-\frac{1}{3} \leq x < \frac{2}{3}$

**30.** $2x + 7 < 3$ or $5x + 5 \geq 10$
$2x < -4$ or $5x \geq 5$
$x < -2$ or $x \geq 1$

**31.** $3x + 8 > 17$ or $2x + 5 \leq 7$
$3x > 9$ or $2x \leq 2$
$x > 3$ or $x \leq 1$

**32.** $-4 < 4x - 8 < 12$
$4 < 4x < 20$
$1 < x < 5$

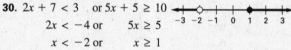

1 is not a solution.

**33.** $-1 < 7x - 15 \leq 20$
$14 < 7x \leq 35$
$2 < x \leq 5$

5 is a solution.

**34.** $-2x \geq 6$ or $2x + 1 > 5$
$x \leq -3$ or $2x > 4$
$x > 2$

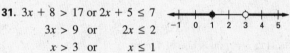

0 is not a solution

**35.** $7 \leq -2x + 21 < 31$
$-14 \leq -2x < 10$
$7 \geq x > -5$

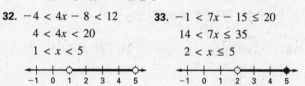

0 is a solution.

**36.** $252{,}000 \leq x \leq 28{,}600{,}000$  **37.** $85{,}000 \leq x \leq 2{,}600{,}000$

**38.** $7 \leq x \leq 2000$  **39.** A solution would have to be less than 1 and greater than 3. No such number exists.

**40.** Every point that is not on the graph of $x < 2$ would be on the graph of $x < 1$.

**41.** A

**42.** $10^7 < d < 10^8$

**43.** $10^8 < d < 10^9$  **44.** $220{,}000 < d < 250{,}000$

# Chapter 6 *continued*

**45. a.** $\frac{1}{2} \le d \le 1\frac{1}{2}$     **46.** $4 + 6 > x$ and $4 + x > 6$

**b.** $2 \le d \le 3$           $10 > x$ and     $x > 2$

                    $10 > x > 2$

**47.** $x + 3\frac{1}{4} > 7\frac{1}{2}$ and $3\frac{1}{4} + 7\frac{1}{2} > x$

$x > 4\frac{1}{4}$ and $10\frac{3}{4} > x$

$10\frac{3}{4} > x > 4\frac{1}{4}$

**48.** $x + 5.5 > 11.75$   and $5.5 + 11.75 > x$

$x > 6.25$   and     $17.25 > x$

$17.25 > x > 6.25$

### 6.3 Mixed Review (p. 351)

**49.** $x + 5 = 2 + 5 = 7$     **50.** $6.5a = 6.5(4) = 26$

**51.** $m - 20 = 30 - 20 = 10$     **52.** $\frac{x}{15} = \frac{30}{15} = 2$

**53.** $5x = 5(3.3) = 16.5$     **54.** $4.2p = 4.2(4.1) = 17.22$

**55.** $x + 17 = 9$     **56.** $8 = x + 2\frac{1}{2}$     **57.** $x - 4 = 12$

$x = -8$     $5\frac{1}{2} = x$     $x = 16$

**58.** $x - (-9) = 15$     **59.** $\frac{1}{2}x = -6$     **60.** $-3x = -27$

$x + 9 = 15$     $x = -12$     $x = 9$

$= 6$

**61.** $4x = -28$     **62.** $-\frac{3}{4}x = 21$     **63.** $x + 3.2 = 11$

$x = -7$     $x = -28$     $x = 7.8$

**64.** $x - 4 = 16.7$     **65.** $\frac{x}{-6} = -\frac{1}{2}$     **66.** $\frac{5}{6}x = -25$

$x = 20.7$     $x = 3$     $x = -30$

**67.** You will save $4.75 - 3.25 = 1.50$ per visit.

$1.50x \ge 45$

$x \ge 30$

You must go at least 30 times.

**68.** $d = 6 - 2t$     **69.** $d = 6 - 2(1)$

              $d = 4$ mi

### Quiz 1 (p. 352)

**1.** $x + 2 < 7$

$x < 5$

**2.** $-3 + x \le -11$

$x \le -8$

**3.** $3.4x \le 13.6$

$x \le 4$

**4.** $5 \le -\frac{x}{2}$

$-10 \ge x$

**5.** $-4x - 2 \ge 14$

$-4x \ge 16$

$x \le -4$

**6.** $-5 < x - 8 < 4$

$3 < x < 12$

**7.** $-x - 4 > 3x - 12$

$-4 > 4x - 12$

$8 > 4x$

$2 > x$ or $x < 2$

**8.** $x + 3 \le 2(x - 7)$

$x + 3 \le 2x - 14$

$3 \le x - 14$

$17 \le x$ or $x \ge 17$

**9.** $-10 \le -4x - 18 \le 30$

$8 \le -4x \le 48$

$-2 \ge x \ge -12$

**10.** $-3 < x + 6$ or $-\frac{x}{3} > 4$

$-9 < x$     or   $x < -12$

**11.** $2 - x < -3$ or $2x + 14 < 12$

$-x < -5$ or     $2x < -2$

$x > 5$   or       $x < -1$

**12.** $2x - 6 < -8$ or $10 - 5x < -19$

$2x < -2$ or   $-5x < -29$

$x < -1$ or       $x > 5\frac{4}{5}$

**13.** $6x - 2 > -7$ or $-3x - 1 > 11$

$6x > -5$ or     $-3x > 12$

$x > -\frac{5}{6}$ or       $x < -4$

**14.** $x \ge 52$     **15.** $-128.6 < T < 136$

### Math & History (p. 352)

**1.** $10 \le t \le 11$     **2.** $t \ge 108$     **3.** $t < 1$

## Lesson 6.4

### Activity (p. 354)

**1.**     $-2 < x < 2$

**2.**     $x \le -3$ or $x \ge -1$

**3.**     $1 \le x \le 5$

# Chapter 6 *continued*

## 6.4 Guided Practice (p. 356)

1. absolute-value equation; absolute-value inequality

2. Write 2 equations

   $x + 3 = 8$ and $x + 3 = -8$

   Solve each equation by subtracting 3 from each side.

   $x = 5$ and $x = -11$

3. $|x - 5| < 2$ can be written as 2 inequalities $x - 5 < 2$ and $x - 5 > -2$ or $-2 < x - 5 < 2$ which means that $x - 5$ is between $-2$ and $2$.

4. and
5. or
6. $|n| = 5$

   $n = 5, n = -5$

7. $|a| = 0$

   $a = 0$

8. $|x + 3| = 6$

   $\begin{aligned} x + 3 &= 6 & x + 3 &= -6 \\ x &= 3 & x &= -9 \end{aligned}$

9. $|x - 4| = 10$

   $\begin{aligned} x - 4 &= 10 & x - 4 &= -10 \\ x &= 14 & x &= -6 \end{aligned}$

10. $|2n - 3| + 4 = 8$

    $\begin{aligned} |2n - 3| &= 4 & 2n - 3 &= -4 \\ 2n &= 7 & 2n &= -1 \\ n &= \tfrac{7}{2} & n &= -\tfrac{1}{2} \end{aligned}$

11. $|3x + 2| = 3$

    $\begin{aligned} 3x + 2 &= 3 & 3x + 2 &= -3 \\ 3x &= 1 & 3x &= -5 \\ x &= \tfrac{1}{3} & x &= -\tfrac{5}{3} \end{aligned}$

12. $|x + 2| \geq 1$

    $x + 2 \geq 1$ or $x + 2 \leq -1$

    $x \geq -1$ or $x \leq -3$

    B

13. $|x + 2| \leq 1$

    $x + 2 \leq 1$ and $x + 2 \geq -1$

    $x \leq -1$ and $x \geq -3$

    C

14. $|x + 2| > 1$

    $x + 2 > 1$ or $x + 2 < -1$

    $x > -1$ or $x < -3$

    A

15. $|x + 6| < 4$

    $x + 6 < 4$ and $x + 6 > -4$

    $x < -2$ and $x > -10$

    $-10 < x < -2$

16. $|x - 2| > 9$

    $x - 2 > 9$ or $x - 2 < -9$

    $x > 11$ or $x < -7$

17. $|3x + 1| \leq 5$

    $3x + 1 \leq 5$ and $3x + 1 \geq -5$

    $3x \leq 4$ and $3x \geq -6$

    $x \leq \tfrac{4}{3}$ and $x \geq -2$

    $-2 \leq x \leq \tfrac{4}{3}$

18. $|d - 0.5| \leq 0.005$

## 6.4 Practice and Applications (p. 356)

19. $|x| = 7$

    $x = 7 \quad x = -7$

20. $|x| = 10$

    $x = 10 \quad x = -10$

21. $|x| = 25$

    $x = 25 \quad x = -25$

22. $|x - 4| = 6$

    $\begin{aligned} x - 4 &= 6 & x - 4 &= -6 \\ x &= 10 & x &= -2 \end{aligned}$

23. $|x + 5| = 11$

    $\begin{aligned} x + 5 &= 11 & x + 5 &= -11 \\ x &= 6 & x &= -16 \end{aligned}$

24. $|x + 8| = 9$

    $\begin{aligned} x + 8 &= 9 & x + 8 &= -9 \\ x &= 1 & x &= -17 \end{aligned}$

25. $|x - 5| = 2$

    $\begin{aligned} x - 5 &= 2 & x - 5 &= -2 \\ x &= 7 & x &= 3 \end{aligned}$

26. $|x - 1| = 4$

    $\begin{aligned} x - 1 &= 4 & x - 1 &= -4 \\ x &= 5 & x &= -3 \end{aligned}$

27. $|x + 3| = 9$

    $\begin{aligned} x + 3 &= 9 & x + 3 &= -9 \\ x &= 6 & x &= -12 \end{aligned}$

28. $|x + 1| = 6$

    $\begin{aligned} x + 1 &= 6 & x + 1 &= -6 \\ x &= 5 & x &= -7 \end{aligned}$

29. $|x - 3.2| = 7$

    $\begin{aligned} x - 3.2 &= 7 & x - 3.2 &= -7 \\ x &= 10.2 & x &= -3.8 \end{aligned}$

30. $|x + 5| = 6.5$

    $\begin{aligned} x + 5 &= 6.5 & x + 5 &= -6.5 \\ x &= 1.5 & x &= -11.5 \end{aligned}$

31. $|4x - 2| = 22$

    $\begin{aligned} 4x - 2 &= 22 & 4x - 2 &= -22 \\ 4x &= 24 & 4x &= -20 \\ x &= 6 & x &= -5 \end{aligned}$

32. $|6x - 4| = 2$

    $\begin{aligned} 6x - 4 &= 2 & 6x - 4 &= -2 \\ 6x &= 6 & 6x &= 2 \\ x &= 1 & x &= \tfrac{1}{3} \end{aligned}$

**Algebra 1**
Chapter 6 Worked-out Solution Key

# Chapter 6 *continued*

**33.** $|3x + 5| = 23$

$3x + 5 = 23$      $3x + 5 = -23$

$3x = 18$      $3x = -28$

$x = 6$      $x = -\frac{28}{3}$

**34.** $|5 - 4x| = 7$

$5 - 4x = 7$      $5 - 4x = -7$

$-4x = 2$      $-4x = -12$

$x = -\frac{1}{2}$      $x = 3$

**35.** $|2x - 4| = 18$

$2x - 4 = 18$      $2x - 4 = -18$

$2x = 22$      $2x = -14$

$x = 11$      $x = -7$

**36.** $|7x + 3| = 31$

$7x + 3 = 31$      $7x + 3 = -31$

$7x = 28$      $7x = -34$

$x = 4$      $x = -\frac{34}{7}$

**37.** $|x + 3.6| = 4.6$

$x + 3.6 = 4.6$      $x + 3.6 = -4.6$

$x = 1$      $x = -8.2$

**38.** $|x - 1.2| = 7$

$x - 1.2 = 7$      $x - 1.2 = -7$

$x = 8.2$      $x = -5.8$

**39.** $\left|x - \frac{1}{2}\right| = \frac{5}{2}$

$x - \frac{1}{2} = \frac{5}{2}$      $x - \frac{1}{2} = -\frac{5}{2}$

$x = 3$      $x = -2$

**40.** $|x + 3| < 8$

$x + 3 < 8$ and $x + 3 > -8$

$x < 5$ and    $x > -11$

$-11 < x < 5$

**41.** $|2x - 9| \leq 11$

$2x - 9 \leq 11$ and $2x - 9 \geq -11$

$2x \leq 20$ and    $2x \geq -2$

$x \leq 10$ and    $x \geq -1$

$-1 \leq x \leq 10$

**42.** $|x + 10| \geq 20$

$x + 10 \geq 20$ or $x + 10 \leq -20$

$x \geq 10$ or    $x \leq -30$

**43.** $|x - 2.2| > 3$

$x - 2.2 > 3$   or $x - 2.2 < -3$

$x > 5.2$ or    $x < -0.8$

**44.** $|4x + 2| < 6$

$4x + 2 < 6$ and $4x + 2 > -6$

$4x < 4$ and    $4x > -8$

$x < 1$ and    $x < -2$

$-2 < x < 1$

**45.** $|5x - 15| \geq 25$

$5x - 15 \geq 25$ or $5x - 15 \leq -25$

$5x \geq 40$ or    $5x \leq -10$

$x \geq 8$ or    $x \leq -2$

**46.** $|9 + x| \leq 7$

$9 + x \leq 7$   and $9 + x \geq -7$

$x \leq -2$ and    $x \geq -16$

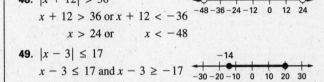

**47.** $|4 - x| < 5$

$4 - x < 5$   and $4 - x > -5$

$-x < 1$   and   $-x > -9$

$x > -1$ and    $x < 9$

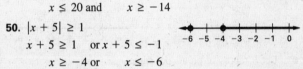

**48.** $|x + 12| > 36$

$x + 12 > 36$ or $x + 12 < -36$

$x > 24$ or    $x < -48$

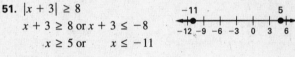

**49.** $|x - 3| \leq 17$

$x - 3 \leq 17$ and $x - 3 \geq -17$

$x \leq 20$ and    $x \geq -14$

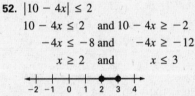

**50.** $|x + 5| \geq 1$

$x + 5 \geq 1$   or $x + 5 \leq -1$

$x \geq -4$ or    $x \leq -6$

**51.** $|x + 3| \geq 8$

$x + 3 \geq 8$ or $x + 3 \leq -8$

$x \geq 5$ or    $x \leq -11$

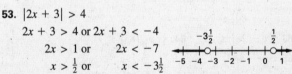

**52.** $|10 - 4x| \leq 2$

$10 - 4x \leq 2$   and $10 - 4x \geq -2$

$-4x \leq -8$ and    $-4x \geq -12$

$x \geq 2$   and    $x \leq 3$

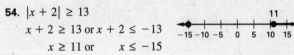

**53.** $|2x + 3| > 4$

$2x + 3 > 4$ or $2x + 3 < -4$

$2x > 1$ or    $2x < -7$

$x > \frac{1}{2}$ or    $x < -3\frac{1}{2}$

**54.** $|x + 2| \geq 13$

$x + 2 \geq 13$ or $x + 2 \leq -13$

$x \geq 11$ or    $x \leq -15$

**55.** $|3 + x| < 3$

$3 + x < 3$ and $3 + x > -3$

$x < 0$ and    $x > -6$

**56.** $|3x + 2| \geq 11$

$3x + 2 \geq 11$ or $3x + 2 \leq -11$

$3x \geq 9$   or    $3x \leq -13$

$x \geq 3$   or    $x \leq -4\frac{1}{3}$

**57.** $|5x + 1| \leq 24$

$5x + 1 \leq 24$ and $5x + 1 \geq -24$

$5x \leq 23$ and $5x \geq -25$

$x \leq 4\frac{3}{5}$ and $x \geq -5$

**58.** $|5x + 3| \geq 13$

$5x + 3 \geq 13$ or $5x + 3 \leq -13$

$5x \geq 10$ or $5x \leq -16$

$x \geq 2$ or $x \leq -3\frac{1}{5}$

**59.** $|2x + 5| < 7$

$2x + 5 < 7$ and $2x + 5 > -7$

$2x < 2$ and $2x > -12$

$x < 1$ and $x > -6$

**60.** $|3x - 9| > 5$

$3x - 9 > 5$ or $3x - 9 < -5$

$3x > 14$ or $3x < 4$

$x > 4\frac{2}{3}$ or $x < 1\frac{1}{3}$

**61.** Let $P$ represent points per game

$8 \leq P \leq 22$

Halfway point: 15

$8 - 15 \leq P - 15 \leq 22 - 15$

$-7 \leq P - 15 \leq 7$

$|P - 15| \leq 7$

**62.** Let $x =$ test scores

$60 \leq x \leq 100$

Halfway point: 80

$60 - 80 \leq x - 80 \leq 100 - 80$

$-20 \leq x - 80 \leq 20$

$|x - 80| \leq 20$

**63.** $|x - 28| \leq 4$    **64.** $|x - 183| \leq 7$

**65.** $|w - 643| < 38$

$w - 643 < 38$ and $w - 643 > -38$

$w < 681$ and $w > 605$

$605 < w < 681$

orange or red

**66.** $|w - 455| < 23$

$w - 455 < 23$ and $w - 455 > -23$

$w < 478$ and $w > 432$

$432 < w < 478$

blue

**67.** $|w - 519.5| < 12.5$

$w - 519.5 < 12.5$ and $w - 519.5 > -12.5$

$w < 532$ and $w > 507$

$507 < w < 532$

green

**68.** $|w - 600| < 5$

$w - 600 < 5$ and $w - 600 > -5$

$w < 605$ and $w > 595$

$595 < w < 605$

orange

**69.** $|x - 7| < 6$

$x - 7 < 6$ and $x - 7 > -6$

$x < 13$ and $x > 1$

$1 < x < 13$

C

**70.** $|3x + 3| > 12$

$3x + 3 > 12$ or $3x + 3 < -12$

$3x > 9$ or $3x < -15$

$x > 3$ or $x < -5$

C

**71.** $|2x - 4| \leq 3$

$2x - 4 \leq 3$ and $2x - 4 \geq -3$

$2x \leq 7$ and $2x \geq 1$

$x \leq \frac{7}{2}$ and $x \geq \frac{1}{2}$

$\frac{1}{2} \leq x \leq \frac{7}{2}$

B

**72.**     $440 \leq w \leq 570$

$440 - 505 \leq w - 505 \leq 570 - 505$

$-65 \leq w - 505 \leq 65$

$|w - 505| \leq 65$

**73.**     $470 \leq w \leq 510$

$470 - 490 \leq w - 490 \leq 510 - 490$

$-20 \leq w - 490 \leq 20$

$|w - 490| \leq 20$

### 6.4 Mixed Review (p. 358)

**74.** $\begin{bmatrix} -2 & 7 \\ 0 & 4 \end{bmatrix} + \begin{bmatrix} 3 & -6 \\ -1 & -5 \end{bmatrix} = \begin{bmatrix} 1 & 1 \\ -1 & -1 \end{bmatrix}$

**75.** $\begin{bmatrix} -4 & -9 \\ -1 & 0 \end{bmatrix} - \begin{bmatrix} -12 & 8 \\ -10 & -5 \end{bmatrix} = \begin{bmatrix} 8 & -17 \\ 9 & 5 \end{bmatrix}$

**76.** $x + 5y = 20$          **77.** $6x + 9y = 36$

$5y = -x + 20$          $9y = -6x + 36$

$y = -\frac{1}{5}x + 4$          $y = -\frac{2}{3}x + 4$

**78.** $3x - 7y = 42$

$-7y = -3x + 42$

$y = \frac{3}{7}x - 6$

**79.** $x = -1$

# Chapter 6 continued

**80.** $3y = 15$
$y = 5$

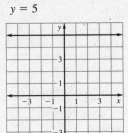

**81.** $x + y = 7$
$y = -x + 7$

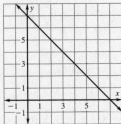

**82.** Point-slope: $y - 4 = 3(x - 0)$
Slope-intercept: $y - 4 = 3x$
$y = 3x + 4$

**83.** Point-slope: $y + 5 = -2(x - 2)$
Slope-intercept: $y + 5 = -2x + 4$
$y = -2x - 1$

**84.** Point-slope: $y - 1 = \frac{2}{3}(x + 3)$
Slope-intercept: $y - 1 = \frac{2}{3}x + 2$
$y = \frac{2}{3}x + 3$

**85.** $368 \div 6\frac{1}{2} \approx 57$ mi/h

### Technology Activity 6.4 (p. 359)

**1.** $|x + 3| - 1 = 3$
$|x + 3| = 4$

| | |
|---|---|
| $x + 3 = 4$ | $x + 3 = -4$ |
| $x = 1$ | $x = -7$ |

**2.** $|4x - 4| + 1 = 6$
$|4x - 4| = 5$

| | |
|---|---|
| $4x - 4 = 5$ | $4x - 4 = -5$ |
| $4x = 9$ | $4x = -1$ |
| $x = \frac{9}{4}$ | $x = -\frac{1}{4}$ |
| $x = 2.25$ | $x = -0.25$ |

**3.** $\left|\frac{1}{2}x - 2\right| + 1 = 8$
$\left|\frac{1}{2}x - 2\right| = 7$

| | |
|---|---|
| $\frac{1}{2}x - 2 = 7$ | $\frac{1}{2}x - 2 = -7$ |
| $\frac{1}{2}x = 9$ | $\frac{1}{2}x = -5$ |
| $x = 18$ | $x = -10$ |

**4.** $|x - 1.67| - 3.24 = -1.1$
$|x - 1.67| = 2.14$

| | |
|---|---|
| $x - 1.67 = 2.14$ | $x - 1.67 = -2.14$ |
| $x = 3.81$ | $x = -0.47$ |

**5.** $|2x - 60| - 55 = 13$
$|2x - 60| = 68$

| | |
|---|---|
| $2x - 60 = 68$ | $2x - 60 = -68$ |
| $2x = 128$ | $2x = -8$ |
| $x = 64$ | $x = -4$ |

**6.** $\left|\frac{1}{3}x + 1\right| + 2 = 4$
$\left|\frac{1}{3}x + 1\right| = 2$

| | |
|---|---|
| $\frac{1}{3}x + 1 = 2$ | $\frac{1}{3}x + 1 = -2$ |
| $\frac{1}{3}x = 1$ | $\frac{1}{3}x = -3$ |
| $x = 3$ | $x = -9$ |

**7.** $|x - 4.3| + 2.8 = 5.3$
$|x - 4.3| = 2.5$

| | |
|---|---|
| $x - 4.3 = 2.5$ | $x - 4.3 = -2.5$ |
| $x = 6.8$ | $x = 1.8$ |

**8.** $|x - 7.2| - 7.6 = 10.9$
$|x - 7.2| = 18.5$

| | |
|---|---|
| $x - 7.2 = 18.5$ | $x - 7.2 = -18.5$ |
| $x = 25.7$ | $x = -11.3$ |

**9.** $|x + 36| + 35 = 49$
$|x + 36| = 14$

| | |
|---|---|
| $x + 36 = 14$ | $x + 36 = -14$ |
| $x = -22$ | $x = -50$ |

## Lesson 6.5

### 6.5 Guided Practice (p. 363)

**1.** an ordered pair $(x, y)$ for which substituting the values of $x$ and $y$ in the inequality gives a true statement

**2.** The problem deals with coins. It is impossible to have a negative number of coins. Shading the graph in Quadrants II, III and IV would produce negative solutions.

**3.** D; the solid line is the graph of $x - y = 4$. Since $(0, 0)$ is in the shaded region and satisfies the inequality $x - y \leq 4$, the shaded region with the solid boundary line is the graph of $x - y \leq 4$.

**4.** $0 \overset{?}{<} -2$
not a solution

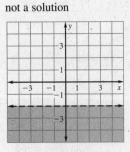

**5.** $0 \overset{?}{>} -2$
solution

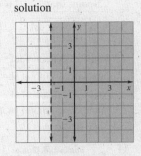

**6.** $0 + 0 \overset{?}{\geq} -1$
solution

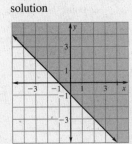

**7.** $0 - 0 \overset{?}{\leq} -2$
not a solution

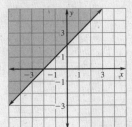

# Chapter 6 *continued*

**8.** $0 + 0 \overset{?}{<} 4$

solution

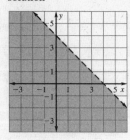

**9.** $0 - 0 \overset{?}{\leq} 5$

solution

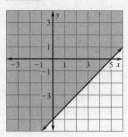

**10.** $0 + 0 \overset{?}{>} 3$

not a solution

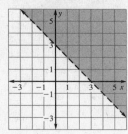

**11.** $3(0) - 0 \overset{?}{<} 3$

solution

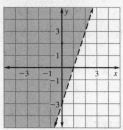

**12.** $0 - 3(0) \overset{?}{\geq} 12$

not a solution

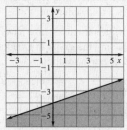

**13.**

| Value of 2-point shot | · | Number of 2-point shots | + | Value of 3-point shot | · |

| Number of 3-point shots | ≥ | Total points |

Value of 2-point shot = 2     Inequality: $2x + 3y \geq 12$

Number of 2-point shots = $x$

Value of 3-point shot = 3

Number of 3-point shots = $y$

Total points = 12

**14.**

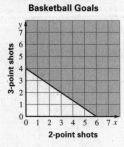

**Basketball Goals**

Your team could score four 3-point shots or two 2-point shots and three 3-point shots. There are many other possible solutions.

---

**15.** $0 + 0 > -3$

solution

$-6 + 3 \overset{?}{>} -3$

$-3 \not> -3$

not a solution

**16.** $2(-1) + 2(-1) \overset{?}{\leq} 0$

$-4 \leq 0$

solution

$2(1) + 2(1) \overset{?}{\leq} 0$

$-4 \not\leq 0$

not a solution

**17.** $2(1) + 5(2) \overset{?}{\geq} 10$

$12 \geq 10$

solution

$2(6) + 5(1) \overset{?}{\geq} 10$

$17 \geq 10$

solution

**18.** $4(3) + 7(2) \overset{?}{\leq} 26$

$12 + 14 \overset{?}{\leq} 26$

$26 \leq 26$

solution

$4(2) + 7(3) \overset{?}{\leq} 26$

$8 + 21 \overset{?}{\leq} 26$

$29 \not\leq 26$

not a solution

**19.** $0.6(2) + 0.6(2) \overset{?}{>} 2.4$

$1.2 + 1.2 \overset{?}{>} 2.4$

$2.4 \not> 2.4$

not a solution

$0.6(3) + 0.6(-3) \overset{?}{>} 2.4$

$1.8 + (1.8) \overset{?}{>} 2.4$

$0 \not> 2.4$

not a solution

**20.** $1.8(0) - 3.8(0) \overset{?}{\geq} 5$

$0 \not\geq 5$

not a solution

$1.8(1) - 3.8(-1) \overset{?}{\geq} 5$

$1.8 + 3.8 \overset{?}{\geq} 5$

$5.6 \not\geq 5$

solution

**21.** $\frac{3}{4}(8) - \frac{3}{4}(8) \overset{?}{<} 2$

$6 - 6 \overset{?}{<} 2$

$0 < 2$

solution

$\frac{3}{4}(8) - \frac{3}{4}(-8) \overset{?}{<} 2$

$6 + 6 \overset{?}{<} 2$

$12 \not< 2$

not a solution

**22.** $\frac{5}{6}(6) + \frac{5}{3}(-12) \overset{?}{>} 4$

$5 - 20 \overset{?}{>} 4$

$-15 \not> 4$

not a solution

$\frac{5}{6}(8) + \frac{5}{3}(-8) \overset{?}{>} 4$

$\frac{20}{3} - \frac{40}{3} \overset{?}{>} 4$

$-\frac{20}{3} \not> 4$

not a solution

**23.** $x \geq 4$

**24.** $x \leq 5$

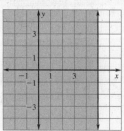

---

**Algebra 1**
Chapter 6 Worked-out Solution Key

# Chapter 6 *continued*

**25.** $y > -3$

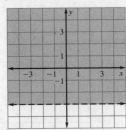

**26.** $y < 9$

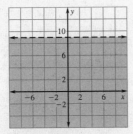

**27.** $x + 3 > -2$
  $x > -5$

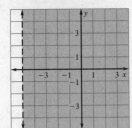

**28.** $7 - x \le 16$
  $-x \le 9$
  $x \ge -9$

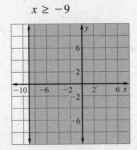

**29.** $y + 6 > 5$
  $y > -1$

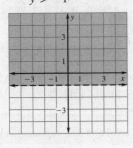

**30.** $8 - y \le 0$
  $-y \le -8$
  $y \ge 8$

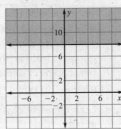

**31.** $4x < -12$
  $x < -3$

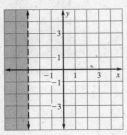

**32.** $-2x \ge 10$
  $x \le -5$

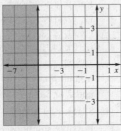

**33.** $5y \le -25$
  $y \le -5$

**34.** $8y > 24$
  $y > 3$

**35.** $-3x \ge 15$
  $x \le -5$

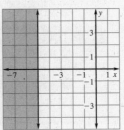

**36.** $6y \le 24$
  $y \le 4$

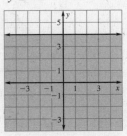

**37.** $-4y < 8$
  $y > -2$

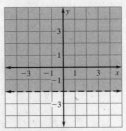

**38.** $-5x > -10$
  $x < 2$

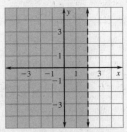

**39.** $x < \frac{1}{2}$

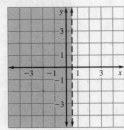

**40.** $y \ge 1.5$

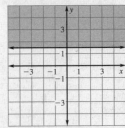

**41.** $2y > 1$

**42.** $x \le 3.5$

**43.** D   **44.** A   **45.** E   **46.** B   **47.** C   **48.** F

**49.** $x + y > -8$
  $y > -x - 8$

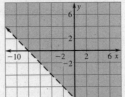

**50.** $x - y \ge 4$
  $-y \ge -x + 4$
  $y \le x - 4$

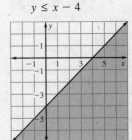

# Chapter 6 continued

**51.** $y - x \le 11$
$y \le x + 11$

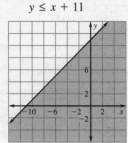

**52.** $-x - y < 3$
$-y < x + 3$
$y > -x - 3$

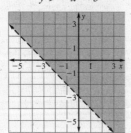

**53.** $x + 6y \le 12$
$6y \le -x + 12$
$y \le -\frac{1}{6}x + 2$

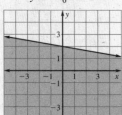

**54.** $3x - y \ge 5$
$-y \ge -3x + 5$
$y \le 3x - 5$

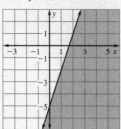

**55.** $2x + y > 6$
$y > -2x + 6$

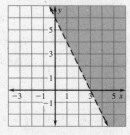

**56.** $y - 5x < 0$
$y < 5x$

**57.** $4x + 3y < 24$
$3y < -4x + 24$
$y < -\frac{4}{3}x + 8$

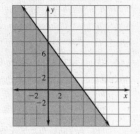

**58.** $9x - 3y \ge 18$
$-3y \ge -9x + 18$
$y \le 3x - 6$

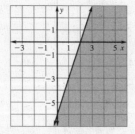

**59.** $\frac{1}{4}x + \frac{1}{2}y < 1$
$\frac{1}{2}y < -\frac{1}{4}x + 1$
$y < -\frac{1}{2}x + 2$

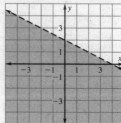

**60.** $\frac{1}{3}x - \frac{2}{3}y \ge 2$
$x - 2y \ge 6$
$-2y \ge -x + 6$
$y \le \frac{1}{2}x - 3$

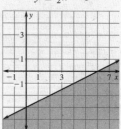

**61.** $y \ge -2x + 2$   **62.** $y \ge \frac{2}{5}x - 2$   **63.** $y < 2x$

**64.** $x + 1.5y \le 12$
$\frac{3}{2}y \le -x + 12$
$y \le -\frac{2}{3}x + 8$

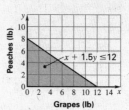

**65. a.** $3x + 7y < 21$
$7y < -3x + 21$
$y < -\frac{3}{7}x + 3$

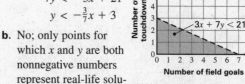

**b.** No; only points for which $x$ and $y$ are both nonnegative numbers represent real-life solutions. Some examples are $(0, 0)$, $(1, 1)$, $(2, 2)$, . . .

**66.** $11{,}000x + 22{,}000y \le 1{,}408{,}000$ or $11x + 22y \le 1{,}408$

**67. a.** Let $x$ represent cups of cereal and let $y$ represent cups of milk.
$88x + 85y + 43 \le 500$
$88x + 85y \le 457$
Possible solutions are:
$(1, 1), (1, 2), (1, 3), (1, 4),$
$(2, 1), (2, 2), (2, 3), (3, 1),$
$(3, 2), (4, 1).$

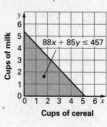

**b.** *Sample answer:* 1 glass of tomato juice, 2 cups of cereal, and 2 or 3 cups of milk

**c.** $88x + 85y \le 557$
Possible solutions are all those from part (a) as well as $(1, 5), (2, 4), (3, 3), (4, 2),$ and $(5, 1)$.

*Sample answer:* one glass of tomato juice, 2 cups of cereal and 3 cups of milk. Other answers are possible.

**68. a.** Let $x$ represent teaspoons of margarine and let $y$ represent glasses of apple juice.
$33x + 120y + 200 \le 500$
$33x + 120y \le 300$
Possible solutions are:
$(1, 1), (1, 2), (2, 1), (3, 1),$
$(4, 1), (5, 1).$

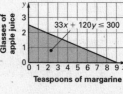

# Chapter 6 *continued*

**b.** *Sample answer:* 1 bagel, 2 tsp of margarine, and 1 glass of apple juice

**69.** Graph the inequality $33x + 120y + 200 \leq 500$ and find specific solutions from the graph or make a table and try different combinations.

**70.** C **71.** B

**72.** $|y| < 2$

$y < 2$ and $y > -2$

Domain: all real numbers

Range: all real numbers between $-2$ and 2

**73.** $|x + 3| \leq y$

$x + 3 \leq y$ and $x + 3 \geq -y$

$y \geq x + 3$ and $-y \leq x + 3$

$y \geq -x - 3$

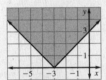

### 6.5 Mixed Review (p. 366)

**74.** $\dfrac{16 + 11 + 18}{3} = \dfrac{45}{3} = 15$

**75.** $\dfrac{20 + 15 + 22 + 19}{4} = \dfrac{76}{4} = 19$

**76.** $\dfrac{37 + 65 + 89 + 72 + 82}{5} = \dfrac{345}{5} = 69$

**77.** $d = rt$

$r = \dfrac{d}{t}$

**78.** $V = \dfrac{1}{3}Bh$

$\dfrac{3V}{B} = h$

**79.** $y = -6x + 7$

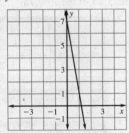

| x | y |
|---|---|
| 0 | 7 |
| 1 | 1 |
| 2 | -5 |

**80.** $y = 3x - 7$

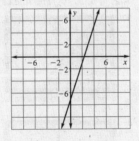

| x | y |
|---|---|
| 0 | -7 |
| 1 | -4 |
| 2 | -1 |

**81.** $-2x + 2y = 5$

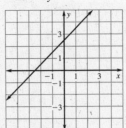

| x | y |
|---|---|
| 0 | $2\frac{1}{2}$ |
| 1 | $3\frac{1}{2}$ |
| 2 | $4\frac{1}{2}$ |

**82.** $3x + 4y = 20$

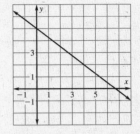

| x | y |
|---|---|
| 0 | 5 |
| 1 | $4\frac{1}{4}$ |
| 2 | $3\frac{1}{2}$ |

**83.** $y = -5x + 2$

$m = -5$

$y$-intercept $= 2$

**84.** $y = \dfrac{x}{2} - 2$

$m = \dfrac{1}{2}$

$y$-intercept $= -2$

**85.** $5x - 5y = 1$

$-5y = -5x + 1$

$y = x - \dfrac{1}{5}$

$m = 1$

$y$-intercept $= -\dfrac{1}{5}$

**86.** $6x + 2y = 14$

$2y = -6x + 14$

$y = -3x + 7$

$m = -3$

$y$-intercept $= 7$

**87.** $y = -2$

$m = 0$

$y$-intercept $= -2$

**88.** $y = 5$

$m = 0$

$y$-intercept $= 5$

**89.** $y = 3x - 6.5$

$m = 3$

$y$-intercept $= -6.5$

**90.** $y = -4x + \dfrac{1}{2}$

$m = -4$

$y$-intercept $= \dfrac{1}{2}$

**91.** $3x + 2y = 10$

$2y = -3x + 10$

$y = -\dfrac{3}{2}x + 5$

$m = -\dfrac{3}{2}$

$y$-intercept $= 5$

### Quiz 2 (p. 366)

**1.** $|x| = 15$

$x = 15 \quad x = -15$

**2.** $|x| = 22$

$x = 22 \quad x = -22$

**3.** $|x + 3| = 6$

$x + 3 = 6 \qquad x + 3 = -6$

$x = 3 \qquad x = -9$

**4.** $|x - 2| = 10$

$x - 2 = 10 \qquad x - 2 = -10$

$x = 12 \qquad x = -8$

# Chapter 6 *continued*

**5.** $|2x + 7| = 7$

$2x + 7 = 7 \quad 2x + 7 = -7$

$\quad 2x = 0 \qquad 2x = -14$

$\qquad x = 0 \qquad\quad x = -7$

**6.** $|3x - 2| - 2 = 5$

$|3x - 2| = 7$

$3x - 2 = 7 \quad 3x - 2 = -7$

$\quad 3x = 9 \qquad 3x = -5$

$\qquad x = 3 \qquad\quad x = -\frac{5}{3}$

**7.** $|x - 4| > 1$

$x - 4 > \quad \text{or } x - 4 < -1$

$\quad x > 5 \text{ or} \qquad x < 3$

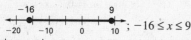

**8.** $|x + 7| < 2$

$x + 7 < 2 \quad \text{and } x + 7 > -2$

$\quad x < -5 \text{ and} \qquad x > -9; \ -9 < x < -5$

**9.** $|x - 12| \leq 9$

$x - 12 \leq 9 \quad \text{and } x - 12 \geq -9$

$\quad x \leq 21 \text{ and} \qquad x \geq 3 \ ; \ 3 \leq x \leq 21$

**10.** $\left|x - \frac{1}{4}\right| > \frac{9}{4}$

$x - \frac{1}{4} > \frac{9}{4} \text{ or } x - \frac{1}{4} < -\frac{9}{4}$

$\quad x > \frac{5}{2} \text{ or} \qquad x < -2$

**11.** $|2x + 7| \leq 25$

$2x + 7 \leq 25 \text{ and } 2x + 7 \geq -25$

$\quad 2x \leq 18 \text{ and} \qquad 2x \geq -32$

$\qquad x \leq 9 \text{ and} \qquad\quad x \geq -16$

$-16 \leq x \leq 9$

**12.** $|4x + 2| - 5 > 17$

$|4x + 2| > 22$

$4x + 2 > 22 \text{ or } 4x + 2 < -22$

$\quad 4x > 20 \text{ or} \qquad 4x < -24$

$\qquad x > 5 \text{ or} \qquad\quad x < -6$

**13.** $0 + (-1) \overset{?}{\leq} 4$

$\qquad -1 \leq 4$

solution

$2 + 2 \overset{?}{\leq} 4$

$\qquad 4 \leq 4$

solution

**14.** $0 - 3(0) \overset{?}{>} 0$

$\qquad 0 \not> 0$

not a solution

$1 - 3(-4) \overset{?}{>} 0$

$1 + 12 \overset{?}{>} 0$

$\qquad 13 > 0$

solution

**15.** $-2(2) + 5(1) \overset{?}{\geq} 5$

$\qquad -4 + 5 \overset{?}{\geq} 5$

$\qquad\quad 1 \not\geq 5$

not a solution

$-2(-1) + 5(2) \overset{?}{\geq} 5$

$\qquad 2 + 10 \overset{?}{\geq} 5$

$\qquad\quad 12 \geq 5$

solution

**16.** $-1 - 2(-1) \overset{?}{<} 4$

$\qquad 1 < 4$

solution

$-2 - 2(-3) \overset{?}{<} 4$

$\qquad 4 \not< 4$

not a solution

**17.** $y - 5x > 0$

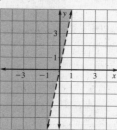

**18.** $y \geq 3$

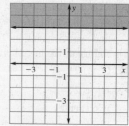

**19.** $x \leq -4$

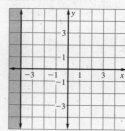

**20.** $y < -2x$

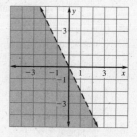

**21.** $2x - y \geq 10$

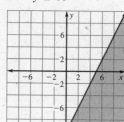

**22.** $3x + y > 15$

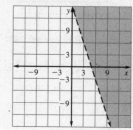

**23.** $\qquad 65 < x < 143$

$65 - 104 < x - 104 < 143 - 104$

$\qquad -39 < x - 104 < 39$

$|x - 104| \leq 39$

*Technology Activity (p. 367)*

**1, 2, 5, 6, 9, 10.** The lines are not part of the solution.

**1.**
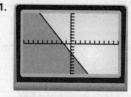

**2.**
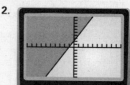

**3.** **4.**
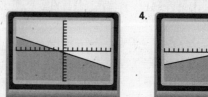

# Chapter 6 continued

**5.**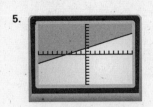

**6.**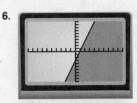

**7.**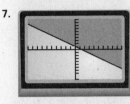

**8.**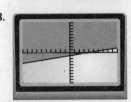

**9.**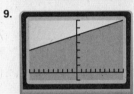

**10.**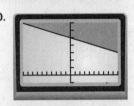

**11.**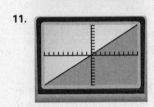

**12.**

**13.** $y > x$   **14.** $y < x + 2$

## Lesson 6.6

### Guided Practice (p. 371)

**1.** The mean of $n$ numbers is the sum of the numbers divided by $n$. The median of $n$ numbers is the middle number when the numbers are written in order. If $n$ is even, the median is the average of two middle numbers. The *mode* of $n$ numbers is the number that occurs most frequently.

**2.** A stem-and-leaf plot makes it simple to: order data; group data into subdivisions; determine the median of the data, determine the range of the data; compare all data values to a particular value.

**3.** *Sample answer:* 1, 2, 2, 4, 4, 4, 144   The mean is 23.

**4.**

```
0 |
1 | 1  3
2 | 1  2  2
3 | 0  4  8  9
4 | 6  8
5 | 5        Key: 1|1 = 11
```

**5.** Mean: 3
   Median: 2
   Mode: 2

**6.** Mean: 6.375
   Median: 6
   Mode: no mode

**7.** Mean: 1
   Median: $-\frac{1}{2}$
   Mode: no mode

**8.** Mean: 4.94
   Median: 3.2
   Mode: 3.2

**9.** 11, 13, 13, 15, 16, 17, 18, 20, 20, 22, 25, 27, 27, 28, 31, 33, 34, 37, 38

**10.** If the class is large and the teacher is young, the teacher's age may have been included. If the class is small and the teacher is much older than the students, the teacher's age was probably not included.

### 6.6 Practice and Applications (p. 371)

**11.**
```
5 | 0  2  4
6 | 0  3  5
7 | 0  4  4  4  5  8  8     Key: 6|0 = 60
```
50, 52, 54, 60, 63, 65, 70, 74, 74, 74, 75, 78, 78

**12.**
```
1 | 7  8  9
2 | 0  2  3  4  9
3 | 0  9
4 | 0  5
5 | 0  1              Key: 1|7 = 17
```
17, 18, 19, 20, 22, 23, 24, 29, 30, 39, 40, 45, 50, 51

**13.**
```
0 | 1  2  3  4  5
1 |
2 | 0  0  2  4  6  8  8
3 | 0  1  7  9         Key: 0|1 = 1
```
1, 2, 3, 4, 5, 20, 20, 22, 24, 26, 28, 28, 28, 30, 31, 37, 39

**14.**
```
0 | 1  9
1 | 0  2  3  5  7
2 | 3
3 | 0  1  2  2  2  8  9
4 | 1  5  6              Key: 1|0 = 10
```
1, 9, 10, 12, 13, 15, 17, 23, 30, 31, 32, 32, 32, 38, 39, 41, 45, 46

**15.**
```
3 | 0  3  7  9
4 | 4  8
5 | 2  4
6 | 1  1  2  8  8
7 | 6  6  6  7  9
8 | 0  1  2  4  6  7  7     Key: 4|4 = 44
```
30, 33, 37, 39, 44, 48, 52, 54, 61, 61, 62, 68, 68, 76, 76, 76, 77, 79, 80, 81, 82, 84, 86, 87, 87

**16.**
```
1 | 0  2  3  3  7  8  9
2 | 1  1  3  5  5
3 | 3  5
4 | 0  2  4  6  8  8
5 | 0  0  7              Key: 3|5 = 35
```
10, 12, 13, 13, 17, 18, 19, 21, 21, 23, 25, 25, 33, 35, 40, 42, 44, 46, 48, 48, 50, 50, 57

# Chapter 6 *continued*

**17.** Mean: 7  **18.** Mean: $2\frac{2}{9}$  **19.** Mean: 6
Median: 7   Median: 2   Median: 6
Mode: 10   Mode: 1 and 3   Mode: 6

**20.** Mean: 4  **21.** Mean: 8  **22.** Mean: 9
Median: 4   Median: 7.5   Median: 9
Mode: 4   Mode: 5 and 10   Mode: 12

**23.** Mean: 4.19  **24.** Mean: 153.5
Median: 3.98   Median: 153
Mode: 1.2   Mode: 150

**25.** Mean: 72.8  **26.** 110, 90, 80, 76, 69, 63, 61, 61, 60, 58
Median: 66

**27.** yes; 61  **28.** You could add any number (except 61) in the list. Then there would be 2 modes, 61, and the added number.

**29.** No; the mean salary of the other 99 adults is $25,000. The median and mode (both $25,000) are much more representative of the "average" salary.

**30.** The median (619) and mode (619) are more representative of the data than the mean (1902), due to California's large population.

**31.** The median (619) and mode (619) are slightly more representative of the data than the mean (753) because they are closer to the majority of the data.

**32.** *Sample answers:*
(1) 10, 12, 12, 12, 18, 18, 18, 20, 20, 20
(2) 12, 12, 15, 15, 16, 16, 18, 18, 29, 29

**33.**
| | |
|---|---|
| 1 | 3 24 25 |
| 2 | 3 20 22 |
| 3 | 17 |
| 4 | 1 8 14 17 30 30 |
| 5 | 10 |
| 6 | 3 5 13 24 |
| 7 | 31 |
| 8 | 21 26 |
| 9 | 12 |
| 10 | 11 17 |
| 11 | 4 11 28 |
| 12 | 9 15 28  Key: 3|17 = March 17 |

Birthdays: Jan. 3, Jan. 24, Jan. 25, Feb. 3, Feb. 20, Feb. 22, March 17, April 1, April 8, April 14, April 17, April 30, April 30, May 10, June, 3, June 5, June 13, June 24, July 31, Aug. 21, Aug. 26, Sept. 12, Oct. 11, Oct. 17, Nov. 4, Nov. 11, Nov. 28, Dec. 9, Dec. 15, Dec. 28

**34.**
| | |
|---|---|
| 10 | 6 8 |
| 11 | 8 |
| 12 | 6 |
| 13 | |
| 14 | 3 6 8 |
| 15 | 8 |
| 16 | 4 |
| 17 | 7 |
| 18 | |
| 19 | 5 |
| 20 | 6  Key: 19|5 = 195 |

Weights: 106, 108, 118, 126, 143, 146, 148, 158, 164, 177, 195, 206

**35.**
| | |
|---|---|
| 20 | 44 |
| 21 | 93 |
| 22 | 14 17 26 52 |
| 23 | 93 |
| 24 | 78 |
| 25 | 70 74  Key: 22|14 = 221.4 |

Distances: 204.4, 219.3, 221.4, 221.7, 222.6, 225.2, 239.3, 247.8, 257.0, 257.4

**36.** 300  **37.** 300

**38.** It would not affect either. There would be 51 numbers in the data set and the median would be the twenty-sixth, which would be 300. The number 400 would still appear fewer times than the number 300, which would still be the mode.

**39.** About half of the top 17 golfers had fewer than 40 wins. For males and females, the most common number of wins was between 30 and 39. There were 4 males and 2 females with more than 60 wins.

**40.** The positions of the stems and leaves on the left-hand plot must be reversed. The leaves must be rewritten in increasing order. The order of the stems must also be reversed.

| Female | | Male | |
|---|---|---|---|
| 6 9 9 | 2 | 2 | 9 9 |
| 1 1 1 2 5 8 | 3 | 3 | 0 1 1 2 3 6 8 |
| 2 2 4 | 4 | 4 | 0 0 |
| 0 5 7 | 5 | 5 | 1 2 |
| | 6 | 6 | 0 3 |
| | 7 | 7 | 0 |
| 2 | 8 | 8 | 1 |

Key: 8|2 = 82  Key: 8|1 = 81

**41.** B  **42.** A

**136**  **Algebra 1**
Chapter 6 Worked-out Solution Key

Copyright © McDougal Littell Inc.
All rights reserved.

# Chapter 6 *continued*

**43.** a. Mean = $\dfrac{84 + 92 + 76 + x}{4}$

$\qquad\qquad = \dfrac{252 + x}{4}$

b. $\dfrac{252 + x}{4} \geq 85$

$\qquad 252 + x \geq 340$

$\qquad\qquad x \geq 88$

You must score at least 88.

c. *Sample answer:* The average of the three tests is 84. Subtract $3 \times 84$ from $4 \times 85$; the result is 88.

### 6.6 Mixed Review (p. 374)

**44.** 55    **45.** 358    **46.** 190

**47.** $6 + 5 \not> 11$    **48.** $2(4) - 3 \overset{?}{<} 4$    **49.** A

not a solution

$\qquad\qquad 5 \not< 4$

$\qquad\qquad$ not a solution

**50.** $(2, -1), (5, 5)$

$m = \dfrac{5 - (-1)}{5 - 2} = \dfrac{6}{3} = 2$

**51.** $(0, 5), (5, 0)$

$m = \dfrac{0 - 5}{5 - 0} = \dfrac{-5}{5} = -1$

**52.** $(-3, -12), (3, 12)$

$m = \dfrac{12 - (-12)}{3 - (-3)} = \dfrac{24}{6} = 4$

**53.** $(0, 1), (4, -1)$

$m = \dfrac{-1 - 1}{4 - 0} = \dfrac{-2}{4} = -\dfrac{1}{2}$

**54.** $-x + 7 > 13$

$\qquad -x > 6$

$\qquad\quad x < -6$

**55.** $16 - x \leq 7$

$\qquad -x \leq -9$

$\qquad\quad x \geq 9$

**56.** $-4 < x + 3 < 7$

$\qquad -7 < x < 4$

**57.** $7 < 3 - 2x \leq 19$

$\quad 4 < -2x \leq 16$

$\quad -2 > x \geq -8$

**58.** $|x - 3| < 12$

$x - 3 < 12$ and $x - 3 > -12$

$\qquad x < 15$ and $\qquad x > -9$ ; $-9 < x < 15$

**59.** $|x + 16| \geq 9$

$x + 16 \geq 9$ or $x + 16 \leq -9$

$\qquad x \geq -7$ or $\qquad x \leq -25$

## Lesson 6.7

### Guided Practice (p. 378)

**1.** second quartile

**2.** The box extends from the first quartile to the third quartile. It represents the middle half of the data.

**3.** The lower whisker connects the least number to the first quartile and the upper whisker connects the greatest number to the third quartile. Each whisker represents a quarter of the data.

**4.** 1, 1, 1, 1, 2, 2, 2, 2, 3

First Quartile: 1

Second Quartile: 2

Third Quartile: 2

**5.** 1, 2, 3, 4, 5, 6, 7

First Quartile: 2

Second Quartile: 4

Third Quartile: 6

**6.** 12, 15, 18, 19, 22, 30

First Quartile: 15

Second Quartile: 18.5

Third Quartile: 22

**7.** 2, 4, 6, 8, 10, 12, 14, 16, 18, 20

First Quartile: 6

Second Quartile: 11

Third Quartile: 16

**8.** 2, 3, 3, 4, 5, 5, 7, 8, 10, 11, 12 ,12, 18, 23, 24

First Quartile: 4

Second Quartile: 8

Third Quartile: 12

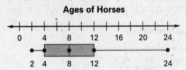

**Ages of Horses**

**9.** C    **10.** No; the precipitation was less than 8.1 inches in 25% of years shown.

### Practice and Applications (p. 378)

**11.** 3, 5, 5, 6, 7, 8, 10, 12

First Quartile: 5

Second Quartile: 6.5

Third Quartile: 9

**12.** 20, 22, 31, 47, 53, 64, 73

First Quartile: 22

Second Quartile: 47

Third Quartile: 64

**13.** 1, 3, 4, 4, 5, 5, 6, 7, 10, 12

First Quartile: 4

Second Quartile: 5

Third Quartile: 7

**14.** 1.2, 2.3, 3.4, 3.8, 4.5, 5.6, 9.7

First Quartile: 2.3

Second Quartile: 3.8

Third Quartile: 5.6

**15.** 2, 6, 6, 7, 7, 7, 8, 8, 10

**16.** 3, 4, 5, 6, 9, 10, 10, 15, 20, 50

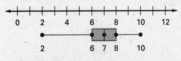

# Chapter 6 *continued*

**17.** 5, 6, 7, 10, 10, 12, 13, 14, 15, 16, 25, 25, 29

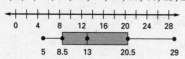

**18.** 1, 1, 5, 6, 8, 8, 8, 9, 10, 10, 10, 12

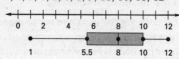

**19.** 12, 15, 19, 22, 22, 26, 27, 29, 34, 35, 36, 37, 37, 38, 38, 39, 40, 44, 46, 46, 47, 49, 51, 55, 57, 58, 69, 69, 72, 75

### Days of Rainfall

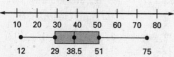

**20.** 20, 22, 23, 35, 39, 46, 46, 57, 59, 69, 69, 72, 75, 75, 75, 77, 77, 78, 79, 79, 82, 83, 83, 84, 88, 89, 91, 91, 97, 98

### Weights of Dogs

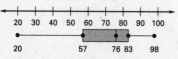

**21.** 20, 24, 34, 35, 35, 36, 40, 40, 41, 45, 47, 48, 48, 48, 49, 51, 52, 53, 53, 57, 60, 61, 63, 64, 65, 65, 65, 65, 68

### Movie Attendance

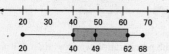

**22.** 154, 155, 157, 158, 160, 161, 163, 166, 167, 167, 169, 172, 174, 176, 180, 182, 184, 188, 190, 191, 192

### Weekly Tips

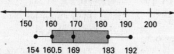

**23.** *Sample answer:* 10, 11, 11, 12, 14, 15, 16, 18, 20, 21, 25, 27, 27, 29, 30, 38
Answers may vary.

**24.** *Sample answer:* 106, 110, 118, 124, 124, 130, 140, 149, 151, 155, 159, 160, 164, 180, 182, 193.
Answers may vary.

**25.**
```
3 | 6  8
4 | 8
5 |
6 | 2  7
7 | 0  1  7
8 | 4  8        Key: 3|6 = 3.6
```
3.6, 3.8, 4.8, 6.2, 6.7, 7.0, 7.1, 7.7, 8.4, 8.8

**26.**

### Phone Weights (oz)

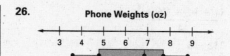

**27.**
```
0 | 60  70  75  80  90  90  90
1 | 35  40
2 | 60                    Key: 1|35 = 135
```
60, 70, 75, 80, 90, 90, 90, 135, 140, 260

**28.**

### Operating Times (min)

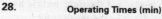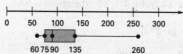

**29.** The lowest 25% of the data are clumped closely together while the upper 25% of the data extend far to the right due to an outlier. The range of the lower three fourths of the data is less than half the range of all the data. The median operating time is less than the mean. At least one phone has a high operating time, but most do not.

**30.**

### Milk Production Before (pt)

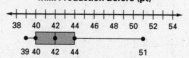

### Milk Production After (pt)

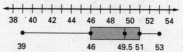

The feed change increased milk production significantly. It increased all three quartiles and the greatest number. The quartiles increased by amounts from 6 to 7.5. The first part of the graph from the first quartile to the right represents 75% of the data as does the part of the graph from the third quartile to the left. The second graph indicates that 75% of the data from the second group are higher than 75% of the data from the first.

**31.** 0.75 h   **32.** The numbers are about the same. Each whisker represents about 25% of the data.

**33.** No; more people travel 0.5–1 hour; the box represents about 50% of the data and each whisker represents about 25%.

**34.**

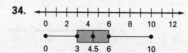

**35.**

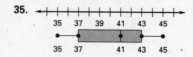

# Chapter 6 *continued*

**36. a.**

| 27 | 4 5 9 |
|----|-------|
| 28 | 0 7 8 9 |
| 29 | |
| 30 | 0 5 5 |
| 31 | 0 |
| 32 | |
| 33 | |
| 34 | 3 6 |
| 35 | |
| 36 | |
| 37 | |
| 38 | 1 |

Key: $27|4 = 274$

381, 346, 343, 310, 305, 305, 300, 289, 288, 287, 280, 279, 275, 274

**b.**

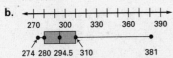

**c.** It is easier to determine the median from the box-and-whisker plot since the median and the second quartile are the same and the second quartile is indicated on a box-and-whisker plot.

**37.** The comparison should include the fact that conclusions about specific data can be drawn from a stem-and-leaf plot, but not a box-and-whisker plot (except for the least and greatest numbers). Conclusions about how the data are distributed are more easily obtained from a box-and-whisker plot. The median can be read directly from the box-and-whisker plot, while it can be found by counting on the stem-and-leaf plot to locate the middle item. The mean can be calculated from the stem-and-leaf plot and the mode can be observed. Neither the mean nor the mode can be derived from a box-and-whisker plot.

## 6.7 Mixed Review (p. 381)

**38.** $3x + y = 7$
$3x = -y + 7$
$x = -\frac{1}{3}y + \frac{7}{3}$
$x = -\frac{1}{3}(-2) + \frac{7}{3}$
$x = 3$
$x = -\frac{1}{3}(-1) + \frac{7}{3}$
$x = \frac{8}{3}$
$x = -\frac{1}{3}(0) + \frac{7}{3}$
$x = \frac{7}{3}$
$x = -\frac{1}{3}(1) + \frac{7}{3}$
$x = 2$

**39.** $4x - y = 4$
$4x = y + 4$
$x = \frac{1}{4}y + 1$
$x = \frac{1}{4}(-2) + 1$
$x = \frac{1}{2}$
$x = \frac{1}{4}(-1) + 1$
$x = \frac{3}{4}$
$x = \frac{1}{4}(0) + 1$
$x = 1$
$x = \frac{1}{4}(1) + 1$
$x = 1\frac{1}{4}$

**40.** $3(2 - y) = -9x + 8$
$6 - 3y = -9x + 8$
$-2 - 3y = -9x$
$\frac{2}{9} + \frac{1}{3}y = x$
$x = \frac{1}{3}(-2) + \frac{2}{9}$
$x = -\frac{4}{9}$
$x = \frac{1}{3}(-1) + \frac{2}{9}$
$x = -\frac{1}{9}$
$x = \frac{1}{3}(0) + \frac{2}{9}$
$x = \frac{2}{9}$
$x = \frac{1}{3}(1) + \frac{2}{9}$
$x = \frac{5}{9}$

**41.** $6y - 4(x + 3) = -2$
$6y - 4x - 12 = -2$
$4x = 6y - 10$
$x = \frac{3}{2}y - \frac{5}{2}$
$x = \frac{3}{2}(-2)\frac{5}{2}$
$x = -\frac{11}{2}$
$x = \frac{3}{2}(-1)\frac{5}{2}$
$x = -4$
$x = \frac{3}{2}(0) - \frac{5}{2}$
$x = -\frac{5}{2}$
$x = \frac{3}{2}(1) - \frac{5}{2}$
$x = -1$

**42.** $\dfrac{12 \text{ oz}}{4 \text{ servings}} = 3 \text{ oz/serving}$

**43.** $\dfrac{\$420}{40 \text{ h}} = \$10.50/\text{h}$

**44.** $3(1) - 2(3) \overset{?}{=} 2$
$-3 \neq 2$
not a solution

**45.** $5(-2) + 4(4) \overset{?}{=} 6$
$6 = 6$
solution

**46.** $-2(4) - 2(-1) \overset{?}{=} -6$
$-6 = -6$
solution

**47.** $-3 \neq 3$
not a solution

**48.** 1980

**49.** $y = 1430x + 64{,}500$ (in thousands)
Answers may vary.

**50.** $y = 1430(35) + 64{,}500$
$y = 114{,}550$ households (approx.)
Number of households $\approx 114{,}550{,}000$

## Quiz 3 (p. 381)

**1.**

| 0 | 5 7 |
|---|-----|
| 1 | 0 6 |
| 2 | 0 3 4 5 9 |
| 3 | 1 2 7 8 |

Key: $3|1 = 31$

5, 7, 10, 16, 20, 23, 24, 25, 29, 31, 32, 37, 38

**2.**

| 1 | 2 8 |
|---|-----|
| 2 | 6 7 |
| 3 | 3 3 |
| 4 | 2 4 6 7 |
| 5 | 9 |
| 6 | 1 |

Key: $1|2 = 12$

12, 18, 26, 27, 33, 33, 42, 44, 46, 47, 59, 61

**3.** Mean: 8
Median: 9
Mode: 9 and 10

**4.** Mean: 37.5
Median: 41
Mode: none

**5.**

First Quartile: 8.5
Second Quartile: 13
Third Quartile: 18

# Chapter 6 *continued*

**6.**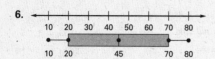

First Quartile: 20

Second Quartile: 45

Third Quartile: 70

**7.**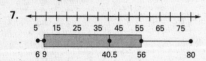

First Quartile: 9

Second Quartile: 40.5

Third Quartile: 56

**8.**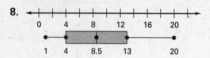

First Quartile: 4

Second Quartile: 8.5

Third Quartile: 13

**9.** Mean: 39

Median: 40

Mode: 57

### Technology Activity 6.7 (p. 382)

**1.** 4, 6, 9, 11, 12, 13, 14, 14, 17, 19, 19, 19

First Quartile: 10

Second Quartile: 13.5

Third Quartile: 18

**2.** 6, 8, 10, 11, 12, 14, 14, 14, 16, 17, 20

First Quartile: 10

Second Quartile: 14

Third Quartile: 16

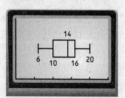

**3.** 20.1, 22.5, 23.9, 24.1, 25.7, 33.5, 34.9, 42.4, 43.4

First Quartile: 23.2

Second Quartile: 25.7

Third Quartile: 38.65

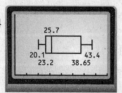

**4.** 3.2, 37.1, 58.8, 65.6, 75.5, 76.3, 77.8, 78.4, 78.5, 90.3, 93.2, 97.1, 107.5

First Quartile: 62.2

Second Quartile: 77.8

Third Quartile: 91.75

### Chapter 6 Review (pp. 384–386)

**1.** $x - 5 \le -3$

$x \le 2$

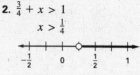

**2.** $\frac{3}{4} + x > 1$

$x > \frac{1}{4}$

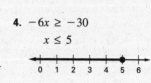

**3.** $\frac{x}{3} < -5$

$x < -15$

**4.** $-6x \ge -30$

$x \le 5$

**5.** $6x + 8 > 4$

$6x > -4$

$x > -\frac{2}{3}$

**6.** $10 - 3x < 5$

$-3x < -5$

$x > \frac{5}{3}$

**7.** $9 - 4x \ge -11$

$-4x \ge -20$

$x \le 5$

**8.** $5 - \frac{1}{2}x \le -3$

$-\frac{1}{2} \le -8$

$x \ge 16$

**9.** $3 < x + 1 < 6$

$2 < x < 5$

$x$ is greater than 2 and less than 5.

**10.** $6 \le 2x + 3 \le 10$

$3 \le 2x \le 7$

$\frac{3}{2} \le x \le \frac{7}{2}$

$x$ is greater than or equal to $\frac{3}{2}$ and less than or equal to $\frac{7}{2}$.

**11.** $x + \frac{1}{2} \le -\frac{3}{4}$ or $\frac{1}{4}x - 1 > \frac{3}{4}$

$x \le -\frac{5}{4}$ or $\frac{1}{4}x > \frac{7}{4}$

$x > 7$

$x$ is less than or equal to $-\frac{5}{4}$ or $x$ is greater than 7.

**12.** $|x| = 12$

$x = 12 \quad x = -12$

**13.** $|x - 5| = 11$

$x - 5 = 11 \quad x - 5 = -11$

$x = 16 \qquad x = -6$

**14.** $|x + 0.5| = 0.25$

$x + 0.5 = 0.25 \quad x + 0.5 = -0.25$

$x = -0.25 \qquad x = -0.75$

**15.** $|5 + x| \le 6$

$5 + x \le 6$ and $5 + x \ge -6$

$x \le 1$ and $x \ge -11$

**16.** $|7 - x| > 2$

$7 - x > 2 \qquad 7 - x < -2$

$-x > -5$ or $-x < -9$

$x < 5$ or $x > 9$

**17.** $|3x + 2| - 4 \le 8$

$|3x + 2| \le 12$

$3x + 2 \le 12$ and $3x + 2 \ge -12$

$3x \le 10$ and $3x \ge -14$

$x \le \frac{10}{3}$ and $x \ge -\frac{14}{3}$

# Chapter 6 continued

**18.** $x < 2$

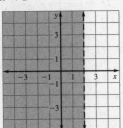

**19.** $-2x + y > 4$

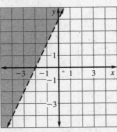

**20.** $\frac{1}{3}x - 3y \le 3$

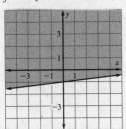

**21.** $2y - 6x \ge -2$

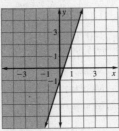

**22.**

| 5 | 6 8 8 |
|---|---|
| 6 | 3 5 |
| 7 | 1 8 8 |
| 8 | 2 4 5 6 |

Key: $6|3 = 63$

**23.** Mean: 72
Median: 74.5
Mode: 58 and 78

**24.** 3.5, 4.2, 6.6, 7.3, 8.9, 9.5, 12.2, 13.2, 14.7, 15.9, 16.0, 16.3

First Quartile: 6.6

Second Quartile: 9.5

Third Quartile: 15.9

**Paris Temperatures**

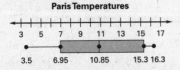

**25.** *Sample answer:* The temperatures in Tokyo are much higher. The highest Paris temperature is only slightly higher than the median Tokyo temperature. Only 25% of the Paris temperatures are at least 15.3°, while half of the Tokyo temperatures are at least 15.85°.

## Chapter 6 Test *(p. 387)*

**1.** $x - 3 < 10$
$\phantom{xxx}x < 13$

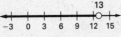

**2.** $-6 > x + 5$
$-11 > x$ or $x < -11$

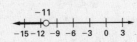

**3.** $-32x > 64$
$\phantom{xxx}x < -2$

**4.** $\frac{x}{4} \le 8$

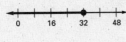

$\phantom{xx}x \le 32$

**5.** $\frac{2}{3}x + 2 \le 4$
$\phantom{xx}\frac{2}{3}x \le 2$
$\phantom{xxxx}x \le 3$

**6.** $6 - x > 15$
$\phantom{xx}-x > 9$
$\phantom{xxx}x < -9$

**7.** $3x + 5 \le 2x - 1$
$3x + 6 \le 2x$
$\phantom{xx}6 \le -x$
$\phantom{xxx}x \le -6$

**8.** $(x + 6) \ge 2(1 - x)$
$\phantom{x}x + 6 \ge 2 - 2x$
$3x + 6 \ge 2$
$\phantom{xx}3x \ge -4$
$\phantom{xxx}x \ge -\frac{4}{3}$

**9.** $-2x + 8 > 3x + 10$
$\phantom{xxxx}8 > 5x + 10$
$\phantom{xxx}-2 > 5x$
$\phantom{xx}-\frac{2}{5} > x$ or $x < -\frac{2}{5}$

**10.** $-15 \le 5x < 20$
$\phantom{xx}-3 \le x < 4$
$x$ is greater than or equal to $-3$ and less than 4.

**11.** $-3 \le 4x + 5 \le 7$
$-8 \le 4x \le 2$
$-2 \le x \le \frac{1}{2}$
$x$ is greater than or equal to $-2$ and less than or equal to $\frac{1}{2}$.

**12.** $8x - 11 < 5$ or $4x - 7 > 13$
$\phantom{xx}8x < 16$ or $\phantom{xx}4x > 20$
$\phantom{xxx}x < 2$ or $\phantom{xxxx}x > 5$
$x$ is less than 2 or greater than 5.

**13.** $-2x > 8$ or $3x + 1 \ge 7$
$\phantom{xx}x < 4$ or $\phantom{xx}3x \ge 6$
$\phantom{xxxxxxxxxxx}x \ge 2$
$x$ is less than $-4$ or $x$ is greater than or equal to 2.

**14.** $12 > 4 - x > -5$
$\phantom{xx}8 > -x > -9$
$-8 < x < 9$
$x$ is greater than $-8$ and less than 9.

**15.** $6x + 9 \ge 21$ or $9x - 5 \le 4$
$\phantom{xx}6x \ge 12$ or $\phantom{xx}9x \le 9$
$\phantom{xxx}x \ge 2$ or $\phantom{xxx}x \le 1$
$x$ is at least 2 or $x$ is less than or equal to 1.

**16.** $x \le -2$ or $x \ge 2$   **17.** $-4 < x < 0$

**18.** $|x + 7| = 11$
$x + 7 = 11 \phantom{xx} x + 7 = -11$
$\phantom{xxx}x = 4 \phantom{xxxxx} x = -18$

**19.** $|x - 8| - 3 \le 10$
$|x - 8| \le 13$
$x - 8 \le 13$ and $x - 8 \ge -13$
$\phantom{x}x \le 21$ and $\phantom{xxx}x \ge -5$; $-5 \le x \le 21$

**Algebra 1**
Chapter 6 Worked-out Solution Key

# Chapter 6 *continued*

**20.** $|x + 4.2| + 3.6 = 16.2$

$|x + 4.2| = 12.6$

$x + 4.2 = 12.6 \quad x + 4.2 = -12.6$

$x = 8.4 \qquad\quad x = -16.8$

**21.** $|2x - 6| > 14$

$2x - 6 > 14$ or $2x - 6 < -14$

$2x > 20$ or $\quad 2x < -8$

$x > 10$ or $\quad\; x < -4$

**22.** $|4x + 5| - 6 \le 1$

$|4x + 5| \le 7$

$4x + 5 \le 7$ and $4x + 5 \ge -7$

$4x \le 2$ and $\quad 4x \ge -12$

$x \le \frac{1}{2}$ and $\quad x \ge -3$

**23.** $|3x - 9| + 6 = 18$

$|3x - 9| = 12$

$3x - 9 = 12 \quad 3x - 9 = -12$

$3x = 21 \qquad 3x = -3$

$x = 7 \qquad\; x = -1$

**24.** $x > -1$

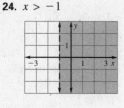

**25.** $x - 1 \le -3$

$x \le -2$

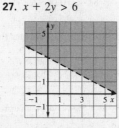

**26.** $-4x \le 8$

$x \ge -2$

**27.** $x + 2y > 6$

**28.** $3x + 4y \ge 12$

**29.** $7y - 2x + 3 < 17$

$7y - 2x < 14$

**30.**

| | |
|---|---|
| 1 | 40 45 70 80 85 |
| 2 | 20 |
| 3 | 50 |
| 4 | |
| 5 | 00 |
| 6 | 30 95 |
| 7 | |
| 8 | |
| 9 | 00 |
| 10 | |
| 11 | 30 |

Key: $6|30 = 630$

140, 145, 170, 180, 185, 220, 350, 500, 630, 695, 900, 1130

**31.** Mean: 437.08

Median: 285

Mode: none

**32.** First quartile: 175

Second quartile: 285

Third quartile: 662.5

The second quartile is the median.

**33.**

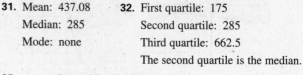

Costs of Exercise Bikes

**34.** $12 \le x \le 33$ **35.** $d < 2520$

**36.** $36 \le t \le 48$

$36 - 42 \le t - 42 \le 48 - 42$

$-6 \le t - 42 \le 6$

$|t - 42| \le 6$

## Chapter 6 Standardized Test (pp. 388-389)

**1.** $x + 10 < 17$

$x < 7$

D

**2.** $x + 3 \le 7$

$x \le 4$

C

**3.** $2 - 3x \ge -4$

$-3x \ge -6$

$x \le 2$

B

**4.** $-3x + 2 > 11$ or $5x + 1 > 6$

$-3x > 9$ or $\quad 5x > 5$

$x < 3$ or $\quad x > 1$

A

**5.** $5(3x + 4) \le 5x - 10$

$15x + 20 \le 5x - 10$

$10x \le -30$

$x \le -3$

B

**6.** $|2x - 10| \ge 6$

$2x - 10 \ge 6$ or $2x - 10 \le -6$

$2x \ge 16$ or $\quad 2x \le 4$

$x \ge 8$ or $\quad x \le 2$

D

# Chapter 6 *continued*

**7.** $|x - 7| + 5 = 17$

$|x - 7| = 12$

$x - 7 = 12 \qquad x - 7 = -12$

$x = 19 \qquad\qquad x = -5$

E

**8.** C    **9.** C    **10.** E    **11.** D    **12.** D

**13.** C    **14.** D

**15. a.** Company A: $4.5x + 25$

Company B: $4x + 75$

**b.** For 60 reams, Company A charges $295. Company B charges $315.

**c.**

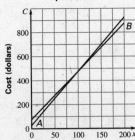

Paper Costs

For any number of reams, the lower graph represents the company that charges the lower price.

**d.** $x < 100; x > 100$

## Cumulative Practice, Chs. 1-6 *(pp. 390-391)*

**1.** $x + (-1) = 7$

**2.** $3(7) - 2 = 21 - 2$

$= 19$

**3.** $5(4 + 5) = 5(9)$

$= 45$

**4.** $(1 - 5)^2 = (-4)^2$

$= 16$

**5.** $2\left(\frac{4 + 8}{4}\right) = 2\left(\frac{12}{4}\right)$

$= 2(3)$

$= 6$

**6.** $(2)^3 - 3(2) + 1 = 8 - 6 + 1$

$= 3$

**7.** $2 - (-6) + (-14) = 8 - 14$

$= -6$

**8.** $3.1 + (-3.3) - 1.8 = -2$

**9.** $20 - |-5.5| = 20 - 5.5$

$= 14.5$

**10.** $\begin{bmatrix} -1 & 0 \\ 7 & -2 \end{bmatrix}$

**11.** $\begin{bmatrix} 6 & 7 & -6 \\ -5 & -1 & 5 \end{bmatrix}$

**12.** $4y - 16$

**13.** $18 + 3x$    **14.** $10y + 2y^2$    **15.** $-15t + 5t^2$

**16.** $3x$    **17.** $11b + 7$    **18.** $2x^2$    **19.** $5.3y$

**20.** $x + 4 = -1$

$x = -5$

**21.** $-3 = n - 15$

$12 = n$

**22.** $5b = -25$

$b = -5$

**23.** $\frac{x}{4} = 6$

$x = 24$

**24.** $3x + 4 = 13$

$3x = 9$

$x = 3$

**25.** $6 + \frac{2}{3}x = 14$

$\frac{2}{3}x = 8$

$x = 12$

**26.** $5x - 10 = 15$

$5x = 25$

$x = 5$

**27.** $14x + 50 = 75$

$14x = 25$

$x = \frac{25}{14}$

**28.** $x + 8 = 3x - 12$

$20 = 2x$

$10 = x$

**29.** $-(x - 7) = \frac{1}{2}x + 1$

$-x + 7 = \frac{1}{2}x + 1$

$-2x + 14 = x + 2$

$12 = 3x$

$4 = x$

**30.** $3x - 15.6 = 75.3$

$3x = 90.9$

$x = 30.3$

**31.** $5.5x + 2.1 = 7.6$

$5.5x = 5.5$

$x = 1$

**32.** $x = y + 3$

$y = x - 3$

**33.** $4x - 5y = 13$

$-5y = -4x + 13$

$y = \frac{4}{5}x - \frac{13}{5}$

**34.** $3(y - x) = 10 - 4x$

$3y - 3x = 10 - 4x$

$3y = -x + 10$

$y = -\frac{1}{3}x + \frac{10}{3}$

**35.** $\dfrac{1.00}{2 \text{ cans}} = \dfrac{\$.50}{\text{can}}$

**36.** $\dfrac{440}{40} = \dfrac{\$11}{h}$

**37.**

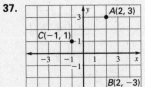

**38.**

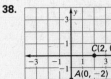

**39.**

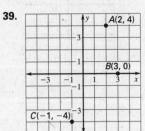

**40.**

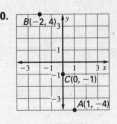

**41.**

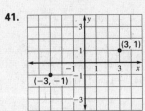

$m = \dfrac{1 - (-1)}{3 - (-3)} = \dfrac{2}{6} = \dfrac{1}{3}$

**42.**

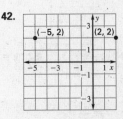

$m = \dfrac{2 - 2}{2 - (-5)} = 0$

**43.**

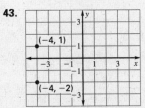

$m = \dfrac{-2 - 1}{-4 - (-4)} = \dfrac{-3}{0}$

undefined

**44.**

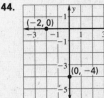

$m = \dfrac{-4 - 0}{0 - (-2)} = \dfrac{-4}{2}$

$= -2$

**Algebra 1**

Chapter 6 Worked-out Solution Key

# Chapter 6 continued

**45.**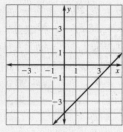

| x | y |
|---|---|
| 0 | 4 |
| -4 | 0 |

**46.**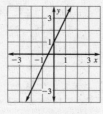

| x | y |
|---|---|
| 0 | 1 |
| $-\frac{1}{2}$ | 0 |

**47.**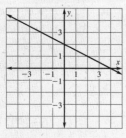

| x | y |
|---|---|
| 0 | 2 |
| 4 | 0 |

**48.**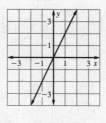

| x | y |
|---|---|
| 0 | 0 |
| 1 | 2 |

**49.**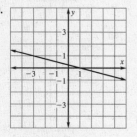

| x | y |
|---|---|
| 0 | $\frac{1}{4}$ |
| 1 | 0 |

**50.**

**51.** $y = x - 3$    **52.** $y = -2x + 5$

**53.** $1 = 2(-1) + b$
$3 = b$
$y = 2x + 3$

**54.** $-1 = \frac{1}{4}(3) + b$
$-\frac{7}{4} = b$
$y = \frac{1}{4}x - \frac{7}{4}$

**55.** $6 = (-3)(-5) + b$
$-9 = b$
$y = -5x - 9$

**56.** $m = \dfrac{1 - (-7)}{-2 - (-1)}$
$= \dfrac{8}{-1}$
$= -8$
$1 = -2(-8) + b$
$-15 = b$
$y = -8x - 15$

**57.** $m = \dfrac{4 - 3}{2 - 0}$
$= \dfrac{1}{2}$
$4 = \dfrac{1}{2}(2) + b$
$3 = b$
$y = \dfrac{1}{2}x + 3$

**58.** $m = \dfrac{3.4 - (-3.6)}{7 - 4.2}$
$= \dfrac{7}{2.8}$
$= 2.5$
$3.4 = 7(2.5) + b$
$-14.1 = b$
$y = 2.5x - 14.1$

**59.** $6 > 3x$
$2 > x \text{ or } x < 2$

**60.** $-6 \le x + 12$
$-18 \le x$
$x \ge -18$

**61.** $-\dfrac{x}{6} \ge 8$
$x \le -48$

**62.** $-4 - 5x \le 31$
$-5x \le 35$
$x \ge -7$

**63.** $-4x + 3 > -21$
$-4x > -24$
$x < 6$

**64.** $-x + 2 < 2(x - 5)$
$-x + 2 < 2x - 10$
$-3x < -12$
$x > 4$

**65.** $-3.2 + x \ge 6.9$
$x \ge 10.1$

**66.** $-4 \le -2x \le 10$
$2 \ge x \ge -5$

**67.** $5 < 4x - 11 < 13$
$16 < 4x < 24$
$4 < x < 6$

**68.** $-2 < x - 7 \le 15$
$5 < x \le 22$

**69.** $|x - 8| > 10$
$x - 8 > 10 \text{ or } x - 8 < -10$
$x > 18 \text{ or } \qquad x < -2$

**70.** $|2x + 5| \le 7$
$2x + 5 \le 7 \text{ and } 2x + 5 \ge -7$
$2x \le 2 \text{ and } \qquad 2x \ge -12$
$x \le 1 \text{ and } \qquad x \ge -6; -6 \le x \le 1$

**71.** 5, 5, 5, 10, 10, 15, 15, 20, 25, 50
Mean: 16
Median: 12.5
Mode: 5

**72.** 1, 2, 2, 2, 2, 3, 3, 5, 5, 7, 7, 8
Mean: 3.92
Median: 3
Mode: 2

**73.** $25 - 2(9.99) = 5.02$
No

**74.** $65 + 3n = C$

| Input | Output |
|-------|--------|
| 0 | 65 |
| 1 | 68 |
| 2 | 71 |
| 3 | 74 |
| 4 | 77 |
| 5 | 80 |
| 6 | 83 |

**75.** Velocity $= -100$ ft/min
Speed $= 100$ ft/min

# Chapter 6 *continued*

**76.** $33 + 50 = 83°F$

**77.** $1.06x \leq 10.50$

$x \leq 9.90$

Your limit is \$9.90.

**78.**

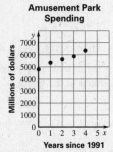

Amusement Park Spending

**79.** $y = 365x + 4896$

Answers may vary.

**80.** $y = 365(14) + 4896$

$y = 10,006$

about 10 billion dollars

## Project, Chs. 4–6 *(pp. 392–393)*

**1–8.** Answers may vary.

**1.** Plots and patterns may vary. As the number of pennies increases, the distance from the top of the rubber band to the bottom of the cup increases.

**2.** Answers may vary.

**3.** The $y$-intercept is the distance from the top of the rubber band to the bottom of the cup before any pennies are added. The slope is the change in distance per added penny.

**4.** The domain is all integers greater than or equal to zero and less than or equal to 50 (the number of pennies): $0 \leq x \leq 50$. The range is all real numbers greater than or equal to the initial distance from the top of the rubber band to the bottom of the cup and less than or equal to the maximum distance.

**5.** Predictions can be made using the scatter plot in Ex. 1, the equation in Ex. 2, or the relationship described in the answer to Ex. 6.

**6.** The initial distance is the same, but the additional distance due to stretching is multiplied by the same factor that the number of pennies is multiplied by. For example, the length for 100 pennies is twice the length for 50 pennies.

**7.** The initial distance will be longer for a longer rubber band so the equation will be different.

**8.** The graph will have a negative slope because the distance from the floor to the cup will decrease as pennies are added.

# CHAPTER 7

## Think and Discuss (p. 395)

1. 6,100,000 units

2. $10,500,000 - 6,100,000 = 4,400,000$ renters

## Skill Review (p. 396)

1. $13x + 8x = 21x$  2. $9r + (-45r) = -36r$

3. $0.1d - 1.1d = -d$  4. $\frac{1}{2}w + \frac{3}{4}w = \frac{2}{4}w + \frac{3}{4}w = \frac{5}{4}w$

5. $-\frac{3}{7}g + \left(-\frac{1}{3}g\right) = -\frac{9}{21}g + \left(-\frac{7}{21}g\right) = -\frac{16}{21}g$

6. $-\frac{3}{10}y + \frac{7}{8}y = -\frac{12}{40} + \frac{35}{40} = \frac{23}{40}y$

7. $2x + 6(x + 1) = -2$   8. $5y + 8 = 5y$

$\qquad 8x + 6 = -2 \qquad\qquad 8 \neq 0$

$\qquad\quad 8x = -8 \qquad\qquad$ no solution

$\qquad\quad\ x = -1$

9. $1 + 4x = 4x + 1$   10. $2(4y - 1) + 2y = 3$

$\quad 4x = 4x \qquad\qquad\ 8y - 2 + 2y = 3$

all real numbers $\qquad\qquad\quad 10y = 5$

$\qquad\qquad\qquad\qquad\qquad\qquad\ y = \frac{1}{2}$

11. $\qquad 2x + y = 1,\ \ (1, -1)$

$2(1) + (-1) \overset{?}{=} 1$

$\qquad\qquad 1 = 1$   solution

12. $\qquad 5x + 5y = 10,\ \ (2, 1)$

$5(2) + 5(1) \overset{?}{=} 10$

$\qquad\qquad 15 \neq 10$   not a solution

13. $\qquad 3y - x \geq 9,\ \ (3, -2)$

$3(-2) - 3 \geq 9$

$\qquad -9 \not\geq 9$   not a solution

14. $\qquad 7y - 8x > 56,\ \ (112, 7)$

$7(7) - 8(112) \overset{?}{>} 56$

$\qquad -847 \not> 56$   not a solution

## Lesson 7.1

### Developing Concepts Activity 7.1 (p. 397)

**Drawing Conclusions (p. 397)**

1. All the students whose ordered pairs are solutions of Equation 1 are in a line, as are all students whose ordered pairs are solutions of Equation 2. The ordered pair belonging to the student in both lines is the only ordered pair that is a solution of both equations. If two lines intersect, they intersect in only one point, so there is no other solution of both equations.

2. (3, 4)  $3 - 4 = -1$   3. (2, 2)  $2 + 2 = 4$

(2, 1)  $2 + 1 = 3$   (4, 2)  $-4 + 2 = -2$

They intersect at (1, 2).   They intersect at (3, 1).

4. (4, 4)  $-4 + 4 = 0$   5. $3 = 3$

(1, 5)  $1 + 5 = 6$   $2 = 2$

They intersect at (3, 3).   They intersect at (3, 2).

6. The solution of such a system is the ordered pair whose graph is the intersection of the graphs of the linear equations.

## 7.1 Guided Practice (p. 401)

1. to find an ordered pair that satisfies each of the equations in the system

2. Determine the coordinates of the point of intersection and substitute them into each equation. The intersection is (0, 2). Since $2 = -0 + 2$ and $2 = 0 + 2$, (0, 2) is the solution of the system.

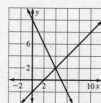

3. not a solution

4. not a solution

5. not a solution

6. solution

7. $-(-4) + (-2) = 2; 2(-4) + (-2) = -10; (-4, -2)$ is not a solution of either equation; $-4 + (-2) = -6$; $2(4) + (-2) = 6; (4, -2)$ is not a solution of either equation; $-(-4) + 2 = 6; 2(-4) + 2 = -6; (-4, 2)$ is not a solution of either equation; $-4 + 2 = -2$; $2(4) + 2 = 10; (4, 2)$ is a solution of both equations so it is a solution of the system.

8.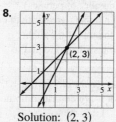

Solution: (2, 3)

Check:  $y = 2x - 1$

$3 \overset{?}{=} 2(2) - 1$

$3 = 3$

$y = x + 1$

$3 \overset{?}{=} 2 + 1$

$3 = 3$

9. Solution: $(2, -1)$

Check:  $y = -2x + 3$

$-1 \overset{?}{=} -2(2) + 3$

$-1 = -1$

$y = x - 3$

$-1 \overset{?}{=} 2 - 3$

$-1 = -1$

10. Solution: (2, 3)

Check:  $y = \frac{1}{2}x + 2$

$3 \overset{?}{=} \frac{1}{2}(2) + 2$

$3 = 3$

$y = -x + 5$

$3 \overset{?}{=} -2 + 5$

$3 = 3$

# Chapter 7 *continued*

**7.1 Practice and Applications (pp. 401–403)**

**11.**  $3x - 2y = 11$
$3(5) - 2(2) \overset{?}{=} 11$
$11 = 11$

$-x + 6y = 7$
$-(5) + 6(2) \overset{?}{=} 7$
$7 = 7$

$(5, 2)$ is a solution.

**12.**  $6x - 3y = -15$
$6(-2) - 3(1) \overset{?}{=} -15$
$-15 = -15$

$2x + y = -3$
$2(-2) + 1 \overset{?}{=} -3$
$-3 = -3$

$(-2, 1)$ is a solution.

**13.**  $x + 3y = 15$
$3 + 3(-6) \overset{?}{=} 15$
$-15 \neq 15$

$(3, -6)$ is not a solution.

**14.**  $5x + y = 19$
$-5(-4) + (-1) \overset{?}{=} 19$
$19 = 19$

$x - 7y = 3$
$-4 - 7(-1) \overset{?}{=} 3$
$3 = 3$

$(-4, -1)$ is a solution.

**15.**  $-15x + 7y = 1$
$-15(3) + 7(5) \overset{?}{=} 1$
$-10 \neq 1$

$(3, 5)$ is not a solution.

**16.**  $-2x + y = -11$
$-2(6) + 1 \overset{?}{=} -11$
$-11 = -11$

$-x - 9y = -15$
$-6 - 9(1) \overset{?}{=} -15$
$-15 = -15$

$(6, 1)$ is a solution.

**17.** Solution: $(4, 5)$
Check:  $-x + 2y = 6$
$-4 + 2(5) \overset{?}{=} 6$
$6 = 6$

$x + 4y = 24$
$4 + 4(5) \overset{?}{=} 24$
$24 = 24$

**18.** Solution: $(-2, -2)$
Check:  $2x - y = -2$
$2(-2) - (-2) \overset{?}{=} -2$
$-2 = -2$

$4x - y = -6$
$4(-2) - (-2) \overset{?}{=} -6$
$-6 = -6$

**19.** Solution: $(3, 0)$
Check:  $x + y = 3$
$3 + 0 \overset{?}{=} 3$
$3 = 3$

$-2x + y = -6$
$-2(3) + 0 \overset{?}{=} -6$
$-6 = -6$

**20.**

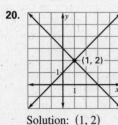

Check:  $y = -x + 3$
$2 \overset{?}{=} -(1) + 3$
$2 = 2$

$y = x + 1$
$2 \overset{?}{=} 1 + 1$
$2 = 2$

Solution: $(1, 2)$

**21.**

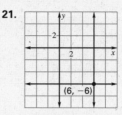

Check:  $6 = 6$
$-6 = -6$
Solution: $(6, -6)$

**22.**

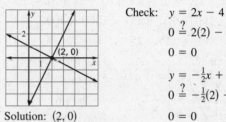

Check:  $y = 2x - 4$
$0 \overset{?}{=} 2(2) - 4$
$0 = 0$

$y = -\frac{1}{2}x + 1$
$0 \overset{?}{=} -\frac{1}{2}(2) + 1$
$0 = 0$

Solution: $(2, 0)$

**23.**

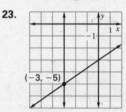

Check:  $2x - 3y = 9$ $\qquad$ $x = -3$
$2(-3) - 3(-5) \overset{?}{=} 9$ $\qquad$ $-3 = -3$
$9 = 9$

Solution: $(-3, -5)$

**24.**

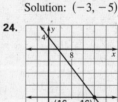

Check:  $5x + 4y = 16$ $\qquad$ $-16 = -16$
$5(16) + 4(-16) \overset{?}{=} 16$
$16 = 16$

Solution: $(16, -16)$

**25.**

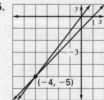

Check:  $x - y = 1$ $\qquad$ $5x - 4y = 0$
$-4 - (-5) \overset{?}{=} 1$ $\qquad$ $5(-4) - 4(-5) \overset{?}{=} 0$
$1 = 1$ $\qquad\qquad$ $0 = 0$

Solution: $(-4, -5)$

# Chapter 7 *continued*

**26.**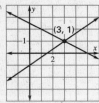

Check: $\quad 3x + 6y = 15 \qquad -2x + 3y = -3$

$\qquad 3(3) + 6(1) \stackrel{?}{=} 15 \quad -2(3) + 3(1) \stackrel{?}{=} -3$

$\qquad\qquad\qquad 15 = 15 \qquad\qquad\qquad -3 = -3$

Solution: $(3, 1)$

**27.**

$7y = -14x + 42$

$y = -2x + 6$

$7y = 14x + 14$

$y = 2x + 2$

Check: $\quad y = -2x + 6 \qquad y = 2x + 2$

$\qquad 4 \stackrel{?}{=} -2(1) + 6 \qquad 4 \stackrel{?}{=} 2(1) + 2$

$\qquad 4 = 4 \qquad\qquad\qquad 4 = 4$

Solution: $(1, 4)$

**28.**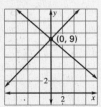

Check: $\quad 0.5x + 0.6y = 5.4 \qquad -x + y = 9$

$\qquad 0.5(0) + 0.6(9) \stackrel{?}{=} 5.4 \qquad -0 + 9 \stackrel{?}{=} 9$

$\qquad\qquad\qquad 5.4 = 5.4 \qquad\qquad\qquad 9 = 9$

Solution: $(0, 9)$

**29.**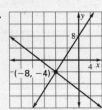

$15x - 10y = -80$

$3x - 2y = -16$

$6x + 8y = -80$

$3x + 4y = -40$

Check: $\quad 3x - 2y = -16$

$\qquad 3(-8) - 2(-4) \stackrel{?}{=} -16$

$\qquad\qquad\qquad -16 = -16$

$\qquad\qquad 3x + 4y = -40$

$\qquad 3(-8) + 4(-4) \stackrel{?}{=} -40$

$\qquad\qquad\qquad -40 = -40$

Solution: $(-8, -4)$

**30.**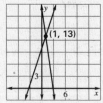

Check: $\quad -3x + y = 10 \qquad 7x + y = 20$

$\qquad -3(1) + 13 \stackrel{?}{=} 10 \quad 7(1) + 13 \stackrel{?}{=} 20$

$\qquad\qquad 10 = 10 \qquad\qquad 20 = 20$

Solution: $(1, 13)$

**31.**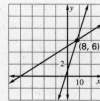

Check: $\quad x - 8y = -40 \qquad -5x + 8y = 8$

$\qquad 8 - 8(6) \stackrel{?}{=} -40 \quad -5(8) + 8(6) \stackrel{?}{=} 8$

$\qquad\qquad -40 = -40 \qquad\qquad 8 = 8$

Solution: $(8, 6)$

**32.**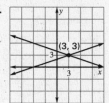

$\frac{1}{5}x + \frac{3}{5}y = \frac{12}{5} \qquad -\frac{1}{5}x + \frac{3}{5}y = \frac{6}{5}$

$x + 3y = 12 \qquad -x + 3y = 6$

Check: $\quad x + 3y = 12 \qquad -x + 3y = 6$

$\qquad 3 + 3(3) \stackrel{?}{=} 12 \qquad -3 + 3(3) \stackrel{?}{=} 6$

$\qquad\qquad 12 = 12 \qquad\qquad 6 = 6$

Solution: $(3, 3)$

**33.**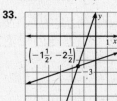

$\frac{3}{4}x - \frac{1}{4}y = -\frac{1}{2} \qquad \frac{1}{4}x - \frac{3}{4}y = \frac{3}{2}$

$3x - y = -2 \qquad x - 3y = 6$

Check: $\quad 3x - y = -2 \qquad\qquad x - 3y = 6$

$\quad 3\left(-\frac{3}{2}\right) - \left(-\frac{5}{2}\right) \stackrel{?}{=} -2 \qquad -\frac{3}{2} - 3\left(-\frac{5}{2}\right) \stackrel{?}{=} 6$

$\qquad\qquad\qquad -2 = -2 \qquad\qquad\qquad 6 = 6$

Solution: $\left(-1\frac{1}{2}, -2\frac{1}{2}\right)$

# Chapter 7 *continued*

**34.**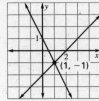

$2.8x + 1.4y = 1.4$

$2x + y = 1$

$0.7x - 0.7y = 1.4$

$x - y = 2$

Check:

$2x + y = 1$      $x - y = 2$

$2(1) + (-1) \overset{?}{=} 1$     $1 - (-1) \overset{?}{=} 2$

$1 = 1$           $2 = 2$

Solution: $(1, -1)$

**35.** 125,000 mi

**36.** $y = 2x + 40$: morning class

$y = 8x + 22$: evening class

The number of people will be the same in 3 months.

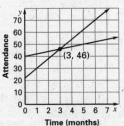

**Aerobics Classes**

**37.** coastal: $y = 0.12t + 46$

inland: $y = -0.12t + 54$

**38.**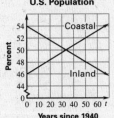

**U.S. Population**

**39.** 1973

**40.** $y = 5x$

$y = x + 400$

$5x = x + 400$

$4x = 400$

$x = 100$ weeks

400 loads

**41.** C    $x + y = 0.5$    $(0, 0.5)$

$0 + 0.5 = 0.5$

$x + 2y = 1$    $(0, 0.5)$

$0 + 2(0.5) \overset{?}{=} 1$

$1 = 1$

**42.** A    $y - x = -3$    $4x + y = 2$

$y = x - 3$     $y = -4x + 2$

$-4x + 2 = x - 3$

$-5x = -5$

$x = 1$

**43. a.** The linear system represents the equations relating to both sides of the original equation. The *x*-coordinate of the intersection of the graphs is the solution of the original equation.

**b.**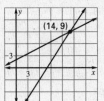

**c.** $\frac{1}{2}x + 2 = \frac{3}{2}x - 12$

$\frac{1}{2}(14) + 2 \overset{?}{=} \frac{3}{2}(14) - 12$

$9 = 9$

Solution: $(14, 9)$

**44.** $y = \frac{4}{5}x$

$y = 3x - 11$

Check:   $\frac{4}{5}x = 3x - 11$

$\frac{4}{5}(5) \overset{?}{=} 3(5) - 11$

$4 = 4$

Solution: $x = 5$

**45.** $y = x + 1$

$y = \frac{3}{2}x + 2$

Check:   $x + 1 = \frac{3}{2}x + 2$

$-2 + 1 \overset{?}{=} \frac{3}{2}(-2) + 2$

$-1 = -1$

Solution: $x = -2$

**46.** $y = 1.2x - 2$

$y = 3.4x - 13$

Check:

$1.2x - 2 = 3.4x - 13$

$1.2(5) - 2 \overset{?}{=} 3.4(5) - 13$

$4 = 4$

Solution: $x = 5$

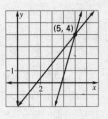

## 7.1 Mixed Review (p. 403)

**47.** $3x + 7 = -2$

$3x = -9$

$x = -3$

**48.** $15 - 2a = 7$

$-2a = -8$

$a = 4$

**49.** $21 = 7(w - 2)$

$3 = w - 2$

$5 = w$

**50.** $2y + 5 = 3y$

$5 = y$

**51.** $2(z - 3) = 12$

$z - 3 = 6$

$z = 9$

**52.** $-(3 - x) = -7$

$3 - x = 7$

$-x = 4$

$x = -4$

**53.** $y = -4x + b$

$0 = -4(3) + b$

$12 = b$

$y = -4x + 12$

**54.** $y = 1x + b$

$3 = 1(-4) + b$

$7 = b$

$y = x + 7$

# Chapter 7 *continued*

**55.** $y = 4x + b$
$-5 = 4(1) + b$
$-9 = b$
$y = 4x - 9$

**56.** $y = -2x + b$
$-1 = -2(-4) + b$
$-9 = b$
$y = -2x - 9$

**57.** $y = 2x + b$
$3 = 2(2) + b$
$-1 = b$
$y = 2x - 1$

**58.** $y = \frac{2}{3}x + b$
$5 = \frac{2}{3}(-1) + b$
$\frac{17}{3} = b$
$y = \frac{2}{3}x + \frac{17}{3}$

**59.** $x < 4260$

```
          4260
 ←─┼────┼────┼──○─┼──→
   0   2000  4000 6000
```

## Technology Activity 7.1 (p. 404)

**1.** $(-3.5, 2.5)$

Check:  $y = x + 6$                $y = -x - 1$
$2.5 \overset{?}{=} -3.5 + 6$      $2.5 \overset{?}{=} -(-3.5) - 1$
$2.5 = 2.5$                        $2.5 = 2.5$

Solution: $(-3.5, 2.5)$

**2.** $(-2.5, 5.5)$

Check:  $y = -3x - 2$             $y = x + 8$
$5.5 \overset{?}{=} -3(-2.5) - 2$  $5.5 \overset{?}{=} -2.5 + 8$
$5.5 = 5.5$                        $5.5 = 5.5$

Solution: $(-2.5, 5.5)$

**3.** $(-1, -2)$

Check:  $y = -0.25x - 2.25$       $y = x - 1$
$-2 \overset{?}{=} -0.25(-1) - 2.25$  $-2 \overset{?}{=} -1 - 1$
$-2 = -2$                          $-2 = -2$

Solution: $(-1, -2)$

**4.** $(23, 10.7)$

Check:  $y = \frac{4}{3}x - 20$       $y = \frac{5}{6}x - 8\frac{1}{2}$
$10.7 \overset{?}{=} \frac{4}{3}(23) - 20$   $10.7 \overset{?}{=} \frac{5}{6}(23) - 8.5$
$10.7 \approx 10.67$              $10.7 \approx 10.67$

Approximate solution: $(23, 10.7)$

**5.** The lines are parallel. Since they do not intersect, the system has no solution.

## Lesson 7.2

### 7.2 Guided Practice (p. 408)

**1.** You may choose either equation. It is equally simple to isolate $y$ in either equation.

**2.** Equation 1: $y = x + 5$; Equation 2: $y = -\frac{1}{2}x + 8$

**3.** Equation 2:
$\frac{1}{2}x + (x + 5) = 8$
$\frac{3}{2}x = 3$
$x = 2$

Equation 1:
$-x + (-\frac{1}{2}x + 8) = 5$
$-\frac{3}{2}x = -3$
$x = 2$

**4.** Equation 1:
$-2 + y = 5$
$y = 7$

Equation 2:
$\frac{1}{2}(2) + y = 8$
$y = 7$

Solution: $(2, 7)$

**5.** To check the solution algebraically, substitute 2 for $x$ and 7 for $y$ into each of the original equations. To check graphically, graph both equations. The lines should intersect at $(2, 7)$.

**6.** $\begin{cases} 3x + y = 3 \\ 7x + 2y = 1 \end{cases}$

$y = -3x + 3$
$7x + 2y = 1$
$7x + 2(-3x + 3) = 1$
$7x - 6x + 6 = 1$
$x = -5$
$y = -3(-5) + 3$
$y = 18$

Solution: $(-5, 18)$

**7.** $\begin{cases} 2x - y = -1 \\ 2x + y = -7 \end{cases}$

$y = -2x - 7$
$2x - (-2x - 7) = -1$
$2x + 2x + 7 = -1$
$4x = -8$
$x = -2$
$y = -2(-2) - 7$
$y = -3$

Solution: $(-2, -3)$

**8.** $\begin{cases} 3x - y = 0 \\ 5y = 15 \end{cases}$

$y = 3$
$3x - 3 = 0$
$3x = 3$
$x = 1$

Solution: $(1, 3)$

**9.** $\begin{cases} 2x + y = 4 \\ -x + y = 1 \end{cases}$

$y = -2x + 4$
$-x - 2x + 4 = 1$
$-3x = -3$
$x = 1$
$y = -2(1) + 4$
$y = 2$

Solution: $(1, 2)$

**10.** $\begin{cases} x - y = 0 \\ x + y = 2 \end{cases}$

$x = y$
$y + y = 2$
$2y = 2$
$y = 1$
$x = 1$

Solution: $(1, 1)$

**11.** $\begin{cases} x + y = 1 \\ 2x - y = 2 \end{cases}$

$x = -y + 1$
$2(-y + 1) - y = 2$
$-2y + 2 - y = 2$
$-3y = 0$
$y = 0$
$x = 0 + 1$
$x = 1$

Solution: $(1, 0)$

# Chapter 7 *continued*

**12.** $\begin{cases} -x + 4y = 10 \\ x - 3y = 11 \end{cases}$

$$x = 3y + 11$$
$$-(3y + 11) + 4y = 10$$
$$-3y - 11 + 4y = 10$$
$$y = 21$$
$$x = 3(21) + 11$$
$$x = 74 \qquad \text{Solution: } (74, 21)$$

**13.** $\begin{cases} x + y = 1 \\ x - y = 2 \end{cases}$

$$x = -y + 1$$
$$-y + 1 - y = 2$$
$$-2y = 1$$
$$y = -\tfrac{1}{2}$$
$$x = \tfrac{1}{2} + 1$$
$$x = \tfrac{3}{2} \qquad \text{Solution: } \left(\tfrac{3}{2}, -\tfrac{1}{2}\right)$$

## 7.2 Practice and Applications (pp. 408–410)

**14.** *Sample answer:* I would use equation 2 to isolate $y$ because it is easy to do so and the value is easy to substitute into the other equation.

**15.** *Sample answer:* I would isolate $m$ in the second equation because it is easy to do so and the value is easy to substitute in the other equation.

**16.** *Sample answer:* I would use equation 2 to isolate $d$ because $d$ has a coefficient of $-1$. It would be simple to divide by $-1$ to get a coefficient of 1.

**17.** $\begin{cases} y = x - 4 \\ 4x + y = 26 \end{cases}$

$$4x + x - 4 = 26$$
$$5x = 30$$
$$x = 6$$
$$y = 6 - 4$$
$$y = 2$$

Solution: (6, 2)

**18.** $\begin{cases} s = t + 4 \\ 2t + s = 19 \end{cases}$

$$2t + t + 4 = 19$$
$$3t = 15$$
$$t = 5$$
$$s = 5 + 4$$
$$s = 9$$

Solution: (9, 5)

**19.** $\begin{cases} 2c - d = -2 \\ 4c + d = 20 \end{cases}$

$$d = -4c + 20$$
$$2c - (-4c + 20) = -2$$
$$6c - 20 = -2$$
$$6c = 18$$
$$c = 3$$
$$d = -4(3) + 20$$
$$d = 8$$

Solution: (3, 8)

**20.** $\begin{cases} 2a = 8 \\ a + b = 2 \end{cases}$

$$a = 4$$
$$4 + b = 2$$
$$b = -2$$

Solution: (4, −2)

**21.** $\begin{cases} 2x + 3y = 31 \\ y = x + 7 \end{cases}$

$$2x + 3(x + 7) = 31$$
$$2x + 3x + 21 = 31$$
$$5x = 10$$
$$x = 2$$
$$y = 2 + 7$$
$$y = 9$$

Solution: (2, 9)

**22.** $\begin{cases} p + q = 4 \\ 4p + q = 1 \end{cases}$

$$p = -q + 4$$
$$4(-q + 4) + q = 1$$
$$-4q + 16 + q = 1$$
$$-3q = -15$$
$$q = 5$$
$$p = -5 + 4$$
$$p = -1$$

Solution: (−1, 5)

**23.** $\begin{cases} x - 2y = -25 \\ 3x - y = 0 \end{cases}$

$$x = 2y - 25$$
$$3(2y - 25) - y = 0$$
$$6y - 75 - y = 0$$
$$5y = 75$$
$$y = 15$$
$$x = 2(15) - 25$$
$$x = 5$$

Solution: (5, 15)

**24.** $\begin{cases} u - v = 0 \\ 7u + v = 0 \end{cases}$

$$u = v$$
$$7v + v = 0$$
$$8v = 0$$
$$v = 0$$
$$0 = 0$$

Solution: (0, 0)

**25.** $\begin{cases} x - y = 0 \\ 12x - 5y = -21 \end{cases}$

$$x = y$$
$$12y - 5y = -21$$
$$7y = -21$$
$$y = -3$$
$$x = -3$$

Solution: (−3, −3)

**26.** $\begin{cases} m + 2n = 1 \\ 5m + 3n = -23 \end{cases}$

$$m = -2n + 1$$
$$5(-2n + 1) + 3n = -23$$
$$-10n + 5 + 3n = -23$$
$$-7n = -28$$
$$n = 4$$
$$m = -2(4) + 1$$
$$m = -7$$

Solution: (−7, 4)

**27.** $\begin{cases} x - y = -5 \\ x + 4 = 16 \end{cases}$

$$x = 12$$
$$12 - y = -5$$
$$-y = -17$$
$$y = 17$$

Solution: (12, 17)

**28.** $\begin{cases} -3a + b = 4 \\ -9a + 5b = -1 \end{cases}$

$$b = 3a + 4$$
$$-9a + 5(3a + 4) = -1$$
$$-9a + 15a + 20 = -1$$
$$6a = -21$$
$$a = -\tfrac{7}{2}$$
$$b = 3\left(-\tfrac{7}{2}\right) + 4$$
$$b = -\tfrac{13}{2}$$

Solution: $\left(-\tfrac{7}{2}, -\tfrac{13}{2}\right)$

**29.** $\begin{cases} 3w - 2u = 12 \\ w - u = 60 \end{cases}$

$$w = u + 60$$
$$3(u + 60) - 2u = 12$$
$$3u + 180 - 2u = 12$$
$$u = -168$$
$$w = -168 + 60$$
$$w = -108$$

Solution: (−108, −168)

**30.** $\begin{cases} y = 3x \\ x = 3y \end{cases}$

$$x = 3(3x)$$
$$x = 9x$$
$$0 = 8x$$
$$0 = x$$
$$y = 3(0)$$
$$y = 0$$

Solution: (0, 0)

# Chapter 7 continued

**31.** $\begin{cases} x + y = 5 \\ 0.5x + 6y = 8 \end{cases}$

$$x = -y + 5$$
$$0.5(-y + 5) + 6y = 8$$
$$-0.5y + 2.5 + 6y = 8$$
$$5.5y = 5.5$$
$$y = 1$$

$x = -1 + 5$
$x = 4$      Solution: $(4, 1)$

**32.** $\begin{cases} x + y = 12 \\ x + \frac{3}{2}y = \frac{3}{2} \end{cases}$

$$x = -y + 12$$
$$-y + 12 + \frac{3}{2}y = \frac{3}{2}$$
$$\frac{1}{2}y = -\frac{21}{2}$$
$$y = -21$$

$x = 21 + 12$

$x = 33$      Solution: $(33, -21)$

**33.** $\begin{cases} 7g + h = -2 \\ g - 2h = 9 \end{cases}$

$$g = 2h + 9$$
$$7(2h + 9) + h = -2$$
$$14h + 63 + h = -2$$
$$15h = -65$$
$$h = -\frac{13}{3}$$

$g = 2\left(-\frac{13}{3}\right) + 9$
$g = \frac{1}{3}$      Solution: $\left(\frac{1}{3}, -\frac{13}{3}\right)$

**34.** $\begin{cases} \frac{1}{8}p + \frac{3}{4}q = 7 \\ \frac{3}{2}p - q = 4 \end{cases}$

$$-q = -\frac{3}{2}p + 4$$
$$q = \frac{3}{2}p - 4$$
$$\frac{1}{8}p + \frac{3}{4}\left(\frac{3}{2}p - 4\right) = 7$$
$$\frac{1}{8}p + \frac{9}{8}p - 3 = 7$$
$$\frac{10}{8}p = 10$$
$$p = 8$$

$q = \frac{3}{2}(8) - 4$

$q = 8$

Solution: $(8, 8)$

**35.** The variables are eliminated, leaving a statement that is always true. You solved for one equation in one variable, and substituted for that variable in the *same* equation; substitute $y = -3x + 9$ into Equation 2 to get $(1, 6)$ for the solution.

**36.** $\begin{cases} x - y = 2 \\ 2x + y = 1 \end{cases}$

$$x = y + 2$$
$$2(y + 2) + y = 1$$
$$2y + 4 + y = 1$$
$$3y = -3$$
$$y = -1$$

$x = -1 + 2$
$x = 1$

Solution: $(1, -1)$

**37.** $\begin{cases} 2y = x \\ 4y = 300 - x \end{cases}$

$$4y = 300 - 2y$$
$$6y = 300$$
$$y = 50$$
$$2(50) = x$$
$$100 = x$$

Solution: $(100, 50)$

**38.** $\begin{cases} x - 2y = 9 \\ 1.5x + 0.5y = 6.5 \end{cases}$

$$x = 2y + 9$$
$$1.5(2y + 9) + 0.5y = 6.5$$
$$3y + 13.5 + 0.5y = 6.5$$
$$3.5y = -7$$
$$y = -2$$

$x = 2(-2) + 9$
$x = 5$      Solution: $(5, -2)$

**39.** $\begin{cases} 0.5x + 0.25y = 2 \\ x + y = 1 \end{cases}$

$$x = -y + 1$$
$$0.5(-y + 1) + 0.25y = 2$$
$$-0.5y + 0.5 + 0.25y = 2$$
$$-0.25y = 1.5$$
$$y = -6$$

$x = 6 + 1$
$x = 7$      Solution: $(7, -6)$

**40.** $\begin{cases} x + y = 20 \\ \frac{1}{5}x + \frac{1}{2}y = 8 \end{cases}$

$$x = -y + 20$$
$$\frac{1}{5}(-y + 20) + \frac{1}{2}y = 8$$
$$-\frac{1}{5}y + 4 + \frac{1}{2}y = 8$$
$$\frac{3}{10}y = 4$$
$$y = \frac{40}{3}$$

$x = -\frac{40}{3} + 20$
$x = \frac{20}{3}$      Solution: $\left(\frac{20}{3}, \frac{40}{3}\right)$

**41.** $\begin{cases} 1.5x - y = 40 \\ 0.5x + 0.5y = 10 \end{cases}$

$$-y = -1.5x + 40$$
$$y = 1.5x - 40$$
$$0.5x + 0.5(1.5x - 40) = 10$$
$$0.5x + 0.75x - 20 = 10$$
$$1.25x = 30$$
$$x = 24$$

$y = 1.5(24) - 40$
$y = -4$      Solution: $(24, -4)$

**42.** $\begin{cases} x + y = 525 \\ 4x + 6y = 2876 \end{cases}$

$$x = -y + 525$$
$$4(-y + 525) + 6y = 2876$$
$$-4y + 2100 + 6y = 2876$$
$$2y = 776$$
$$y = 388$$

$x = -388 + 525$
$x = 137$

I sold 137 student tickets and 388 general admission tickets.

# Chapter 7 *continued*

**43.** $\begin{cases} x + y = 80 \\ 2.75x + 3.25y = 245 \end{cases}$

$$x = -y + 80$$
$$2.75(-y + 80) + 3.25y = 245$$
$$-2.75y + 220 + 3.25y = 245$$
$$0.5y = 25$$
$$y = 50$$

$$x = -50 + 80$$
$$x = 30$$

I ordered 30 11-inch softballs and 50 12-inch softballs.

**44.** $\begin{cases} x + y = 38 \\ 5x + 2y = 100 \end{cases}$

$$x = -y + 38$$
$$5(-y + 38) + 2y = 100$$
$$-5y + 190 + 2y = 100$$
$$-3y = -90$$
$$y = 30$$

$$x = -30 + 38$$
$$x = 8$$

There are 8 5-point questions and 30 2-point questions.

**45.** $\begin{cases} x = 3y \\ x + y = 4500 \end{cases}$

$$3y + y = 4500$$
$$4y = 4500$$
$$y = 1125$$
$$x = 3(1125)$$
$$x = 3375$$

There is $3375 invested in EFG and $1125 invested in PQR.

**46.** Let $u$ = meters uphill
Let $d$ = meters downhill
$$u + d = 1557$$
$$\frac{u}{180} + \frac{d}{250} = 7.6$$

**47.** $\begin{cases} u + d = 1557 \\ \frac{u}{180} + \frac{d}{250} = 7.6 \end{cases}$

$$u = -d + 1557$$
$$\frac{-d + 1557}{180} + \frac{d}{250} = 7.6$$
$$25(-d + 1557) + 18d = 34{,}200$$
$$-25d + 38{,}925 + 18d = 34{,}200$$
$$-7d = -4725$$
$$d = 675$$

$$u + 675 = 1557$$
$$u = 882$$

I ran 882 m uphill and 675 m downhill.

**48.** 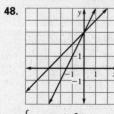 The graphs of the equations intersect at $(0, 3)$ which means that $(0, 3)$ is the solution of the system.

**49.** $\begin{cases} y = x + 3 \\ y = 2x + 3 \end{cases}$

$$x + 3 = 2x + 3$$
$$0 = x$$
$$y = 0 + 3$$
$$y = 3$$

Solution: $(0, 3)$; The point $(0, 3)$ is a solution of each equation in the system.

**50.** substitution

**51.** The graphing method provides a visual interpretation of the solution, but if the coordinates of the solution are not integers, they may be difficult to estimate without a graphing calculator. The substitution method gives an exact solution and may be quicker and simpler in some cases, but may involve complicated arithmetic in others.

**52.** $\begin{cases} y = -\frac{3}{2}x - 215 \\ y = 7x + 1026 \end{cases}$

$$-\frac{3}{2}x - 215 = 7x + 1026$$
$$-3x - 430 = 14x + 2052$$
$$-17x = 2482$$
$$x = -146$$
$$y = 7(-146) + 1026$$
$$y = 4$$

Solution: $(-146°, 4°)$

**53.**

$$36 + 7y = -5483$$
$$y = -\frac{36}{7}x - \frac{5483}{7}$$

$$-\frac{3}{2}x - 215 = -\frac{36}{7}x - \frac{5483}{7} \qquad 7x + 1026 = -\frac{36}{7}x - \frac{5483}{7}$$
$$-21x - 3010 = -72x - 10{,}966 \qquad 49x + 7182 = -36x - 5483$$
$$51x = -7956 \qquad 85x = -12{,}665$$
$$x = 156 \qquad x = -149$$
$$y = -\frac{3}{2}(156) - 215 \qquad y = 7(-149) + 1026$$
$$y = 19 \qquad y = -17$$

Hawaii: $(-156°, 19°)$      Tahiti: $(-149°, -17°)$

## 7.2 Mixed Review (p. 410)

**54.** $4g + 3h + 2g - 3h$
$= 6g$

**55.** $3x + 2y - (5x + 2y)$
$= 3x + 2y - 5x - 2y$
$= -2x$

**56.** $6(2p - m) - 3m - 12p$
$= 12p - 6m - 3m - 12p$
$= -9m$

**57.** $4(3x + 5y) + 3(-4x + 2y)$
$= 12x + 20y - 12x + 6y$
$= 26y$

**58.** $6x + y = 0$
$y = -6x$

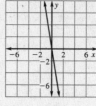

**59.** $8x - 4y + 16 = 0$
$-4y = -8x - 16$
$y = 2x + 4$

**60.** $3x + y + 5 = 0$
$y = -3x - 5$

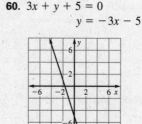

**61.** $5x + 3y = 3$
$3y = -5x + 3$
$y = -\frac{5}{3}x + 1$

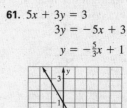

# Chapter 7 *continued*

**62.** $x + y = 0$
$y = -x$

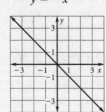

**63.** $y = -2$

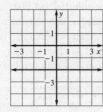

**64.** $-5 < -x \le -1$
$5 > x \ge 1$

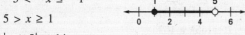

**65.** $|x + 5| \le 14$
$x + 5 \le 14$ and $x + 5 \ge -14$
$x \le 9$ and $\quad x \ge -19$

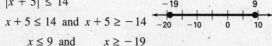

**66.** $3 > -x > -1$
$-3 < x < 1$

**67.** $2x - 6 < -7$ or $2x - 6 > 5$
$2x < -1$ or $\quad 2x > 11$
$x < -\frac{1}{2}$ or $\quad x > 5\frac{1}{2}$

**68.** $3x - 2 > 4$ or $3x - 2 < -5$
$3x > 6$ or $\quad 3x < -3$
$x > 2$ or $\quad x < -1$

## Lesson 7.3

### 7.3 Guided Practice (p. 414)

**1.** to obtain coefficients that are opposites for one of the variables, so that the variable can be eliminated by addition

**2.** When multiplying the second equation by $-2$, $4y$ becomes $-8y$, not $8y$.

$$10x + 10y = \quad 10$$
$$\underline{-10x - 8y = -28}$$
$$2y = -18$$
$$y = \quad -9$$
$$x + (-9) = \quad 1$$
$$x = \quad 10 \qquad \text{Solution: } (10, -9)$$

**3.** The right side of the first equation was not multiplied by 3. Also, $9x + 7x = 16x$, not $2x$.

$$9x + 3y = 72$$
$$\underline{7x - 3y = \quad 8}$$
$$16x = 80$$
$$x = \quad 5$$
$$3(5) + y = 24$$
$$y = \quad 9 \qquad \text{Solution: } (5, 9)$$

**4.** Add the equations, solve for $x$, then substitute in either original equation to solve for $y$.

$$x + 3y = \quad 6 \qquad 9 + 3y = 6$$
$$\underline{+ \; x - 3y = 12} \qquad 3y = -3$$
$$2x = 18 \qquad \quad y = -1$$
$$x = \quad 9 \qquad \text{Solution: } (9, -1)$$

**5.** Multiply either equation by $-1$, solve for $y$, then substitute in either original equation to solve for $x$.

$$x - 3y = 0 \quad \text{Multiply by } -1 \quad -x + 3y = 0$$
$$x + 10y = 13 \qquad\qquad\qquad \underline{x + 10y = 13}$$
$$13y = 13$$
$$y = 1$$
$$x - 3(1) = 0$$
$$x = 3$$
$$\text{Solution: } (3, 1)$$

**6.** Multiply the second equation by $-4$, solve for $x$, then substitute in either original equation to solve for $y$.

$$3x - 4y = 7 \qquad\qquad\qquad\qquad 3x - 4y = \quad 7$$
$$2x - y = 3 \quad \text{Multiply by } -4 \quad \underline{-8x + 4y = -12}$$
$$-5x = \quad -5$$
$$x = \quad 1$$
$$2(1) - y = \quad 3$$
$$-y = \quad 1$$
$$y = \quad -1$$
$$\text{Solution: } (1, -1)$$

**7.** Subtract $2x$ from each side of the first equation, solve for $y$, then substitute into either original equation to solve for $x$.

$$2y = -2 + 2x \quad \text{Subtract} \quad -2x + 2y = -2$$
$$2x + 3y = 12 \qquad \; 2x \text{ from} \quad \underline{2x + 3y = \quad 12}$$
$$\text{each side} \qquad 5y = \quad 10$$
$$y = \quad 2$$
$$2(2) = -2 + 2x$$
$$6 = 2x$$
$$3 = x$$
$$\text{Solution: } (3, 2)$$

### 7.3 Practice and Applications (pp. 414–416)

**8.** $2x + y = 4$
$$\underline{x - y = 2}$$
$$3x = 6$$
$$x = 2$$
$$2 - y = 2$$
$$-y = 0$$
$$y = 0$$
Solution: $(2, 0)$

**9.** $a - b = 8$
$$\underline{a + b = 20}$$
$$2a = 28$$
$$a = 14$$
$$14 + b = 20$$
$$b = 6$$
Solution: $(14, 6)$

**10.** $y - 2x = 0$
$$\underline{6y + 2x = 0}$$
$$7y = 0$$
$$y = 0$$
$$0 - 2x = 0$$
$$x = 0$$
Solution: $(0, 0)$

**11.** $m + 3n = 2$
$$\underline{-m + 2n = 3}$$
$$5n = 5$$
$$n = 1$$
$$m + 3(1) = 2$$
$$m = -1$$
Solution: $(-1, 1)$

# Chapter 7 *continued*

**12.** $p + 4q = 23$
$$\underline{-p + q = 2}$$
$$5q = 25$$
$$q = 5$$
$$p + 4(5) = 23$$
$$p = 3$$
Solution: $(3, 5)$

**13.** $3v - 2w = 1$
$$\underline{2v + 2w = 4}$$
$$5v = 5$$
$$v = 1$$
$$3(1) - 2w = 1$$
$$-2w = -2$$
$$w = 1$$
Solution: $(1, 1)$

**14.** $\frac{1}{2}g + h = 2$
$$\underline{-g - h = 2}$$
$$-\frac{1}{2}g = 4$$
$$g = -8$$
$$8 - h = 2$$
$$-h = -6$$
$$h = 6$$
Solution: $(-8, 6)$

**15.** $6.5x - 2.5y = 4$
$$\underline{1.5x + 2.5y = 4}$$
$$8x = 8$$
$$x = 1$$
$$1.5(1) + 2.5y = 4$$
$$2.5y = 2.5$$
$$y = 1$$
Solution: $(1, 1)$

**16.** $x - y = 0$   Multiply by $-1 \rightarrow$   $-x + y = 0$
$$-3x - y = 2 \qquad\qquad\qquad \underline{-3x - y = 2}$$
$$-4x = 2$$
$$x = -\frac{1}{2}$$
$$-\frac{1}{2} - y = 0$$
$$-y = \frac{1}{2}$$
$$y = -\frac{1}{2}$$
Solution: $\left(-\frac{1}{2}, -\frac{1}{2}\right)$

**17.** $v - w = -5$   Multiply by $-1 \rightarrow$   $-v + w = 5$
$$v + 2w = 4 \qquad\qquad\qquad \underline{v + 2w = 4}$$
$$3w = 9$$
$$w = 3$$
$$v - 3 = -5$$
$$v = -2$$
Solution: $(v, w) = (-2, 3)$

**18.** $x + 3y = 3$   Multiply by $-1 \rightarrow$   $-x - 3y = -3$
$$x + 6y = 3 \qquad\qquad\qquad \underline{x + 6y = 3}$$
$$3y = 0$$
$$y = 0$$
$$x + 3(0) = 3$$
$$x = 3$$
Solution: $(3, 0)$

**19.** $2g - 3h = 0$   Multiply by $3 \rightarrow$   $6g - 9h = 0$
$$3g - 2h = 5 \quad \text{Multiply by } -2 \rightarrow \underline{-6g + 4h = -10}$$
$$-5h = -10$$
$$h = 2$$
$$2g - 3(2) = 0$$
$$2g = 6$$
$$g = 3$$
Solution: $(3, 2)$

**20.** $2p - q = 2$   Multiply by $-1 \rightarrow$   $-2p + q = -2$
$$2p + 3q = 22 \qquad\qquad\qquad \underline{2p + 3q = 22}$$
$$4q = 20$$
$$q = 5$$
$$2p - 5 = 2$$
$$2p = 7$$
$$p = \frac{7}{2}$$
Solution: $\left(\frac{7}{2}, 5\right)$

**21.** $2a + 6z = 4$   Multiply by $3 \rightarrow$   $6a + 18z = 12$
$$3a - 7z = 6 \quad \text{Multiply by } -2 \rightarrow \underline{-6a + 14z = -12}$$
$$32z = 0$$
$$z = 0$$
$$2a + 6(0) = 4$$
$$2a = 4$$
$$a = 2$$
Solution: $(2, 0)$

**22.** $5e + 4f = 9$   Multiply by $4 \rightarrow$   $20e + 16f = 36$
$$4e + 5f = 9 \quad \text{Multiply by } -5 \rightarrow \underline{-20e - 25f = -45}$$
$$-9f = -9$$
$$f = 1$$
$$5e + 4(1) = 9$$
$$5e = 5$$
$$e = 1$$
Solution: $(1, 1)$

**23.** $10m + 16n = 140 \qquad\qquad 10m + 16n = 140$
$$5m - 8n = 60 \quad \text{Multiply by } 2 \rightarrow \underline{10m - 16n = 120}$$
$$20m = 260$$
$$m = 13$$
$$10(13) + 16n = 140$$
$$16n = 10$$
$$n = \frac{5}{8}$$
Solution: $\left(13, \frac{5}{8}\right)$

**24.** $9x - 3z = 20$   Multiply by $2 \rightarrow$   $18x - 6z = 40$
$$3x + 6z = 2 \qquad\qquad\qquad \underline{3x + 6z = 2}$$
$$21x = 42$$
$$x = 2$$
$$9(2) - 3z = 20$$
$$-3z = 2$$
$$z = -\frac{2}{3}$$
Solution: $\left(2, -\frac{2}{3}\right)$

**25.** $x + 3y = 12 \qquad\qquad\qquad x + 3y = 12$
$$-3y + x = 30 \quad \text{Rearrange} \rightarrow \underline{x - 3y = 30}$$
$$2x = 42$$
$$x = 21$$
$$21 + 3y = 12$$
$$3y = -9$$
$$y = -3$$
Solution: $(21, -3)$

# Chapter 7 *continued*

**26.** $3b + 2c = 46$

$5c + b = 11$    Rearrange

$b + 5c = 11$    Multiply by $-3 \rightarrow$

$$-3b - 15c = -33$$
$$\underline{3b + 2c = 46}$$
$$-13c = 13$$
$$c = -1$$

$$b + 5(-1) = 11$$
$$b = 16$$

Solution: $(16, -1)$

**27.** $y = x - 9$    Subtract $x \rightarrow$

$x + 8y = 0$

$$-x + y = -9$$
$$\underline{x + 8y = 0}$$
$$9y = -9$$
$$y = -1$$

$$-1 = x - 9$$
$$8 = x$$

Solution: $(8, -1)$

**28.** $2q = 7 - 5p$

$4p - 16 = q$    Multiply by $-2 \rightarrow$

$$2q = 7 - 5p$$
$$\underline{-2q = 32 - 8p}$$
$$0 = 39 - 13p$$
$$13p = 39$$
$$p = 3$$

$$2q = 7 - 5(3)$$
$$q = -4$$

Solution: $(3, -4)$

**29.** $2v = 150 - u$    Add $u \rightarrow$   $2v + u = 150$

$2u = 150 - v$    Add $v \rightarrow$   $v + 2u = 150$

$2v + u = 150$ Multiply by $-2 \rightarrow$   $-4v - 2u = -300$

$v + 2u = 150$

$$\underline{v + 2u = 150}$$
$$-3v = -150$$
$$v = 50$$

$$2(50) = 150 - u$$
$$100 - 150 = -u$$
$$50 = u$$

Solution: $(50, 50)$

**30.** $0.1g - h + 4.3 = 0$ Subtract $4.3 \rightarrow$   $0.1g - h = -4.3$

$3.6 = -0.2g + h$    Rearrange $\rightarrow$   $\underline{-0.2g + h = 3.6}$

$$-0.1g = -0.7$$
$$g = 7$$

$$0.1(7) - h = -4.3$$
$$-h = -5$$
$$h = 5$$

Solution: $(7, 5)$

**31.** $x + 2y = 5$        $x + 2y = 5$

$5x - y = 3$    Multiply by $2 \rightarrow$   $\underline{10x - 2y = 6}$

$$11x = 11$$
$$x = 1$$

$$1 + 2y = 5$$
$$2y = 4$$
$$y = 2$$

Solution; $(1, 2)$

**32.** $3p - 2 = -q$    Rearrange $\rightarrow$   $q + 3p = 2$

$-q + 2p = 3$

$$\underline{-q + 2p = 3}$$
$$5p = 5$$
$$p = 1$$

$$3(1) - 2 = -q$$
$$1 = -q$$
$$-1 = q$$

Solution: $(1, -1)$

**33.** $3g - 24 = -4h$ Rearrange $\rightarrow$   $3g + 4h = 24$

$-2 + 2h = g$    Rearrange $\rightarrow$   $-g + 2h = 2$

$$3g + 4h = 24$$

Multiply by $3 \rightarrow$   $\underline{-3g + 6h = 6}$

$$10h = 30$$
$$h = 3$$

$$-2 + 2(3) = g$$
$$4 = g$$    Solution: $(4, 3)$

**34.** $t + r = 1$        $t + r = 1$

$2r - t = 2$    Rearrange $\rightarrow$   $\underline{-t + 2r = 2}$

$$3r = 3$$
$$r = 1$$

$$t + 1 = 1$$
$$t = 0$$    Solution: $(1, 0)$

**35.** $x + 1 - 3y = 0$   Subtract $1 \rightarrow$   $x - 3y = -1$

$2x = 7 - 3y$    Add $3y \rightarrow$   $\underline{2x + 3y = 7}$

$$3x = 6$$
$$x = 2$$

$$2 + 1 - 3y = 0$$
$$-3y = -3$$
$$y = 1$$   Solution: $(2, 1)$

**36.** $3a + 9b = 8b - a$     Combine
like terms $\rightarrow$   $4a + b = 0$

$5a - 10b = 4a - 9b + 5$    Combine
like terms $\rightarrow$   $\underline{a - b = 5}$

$$5a = 5$$
$$a = 1$$

$$4(1) + b = 0$$
$$b = -4$$

Solution: $(1, -4)$

**37.** $2m - 4 = 4n$        $2m - 4 = 4n$

$m - 2 = n$    Multiply by $-2 \rightarrow$   $\underline{-2m + 4 = -2n}$

$$0 = 2n$$
$$0 = n$$

$$m - 2 = 0$$
$$m = 2$$    Solution: $(2, 0)$

**38.** $3y = -5x + 15$        $3y = -5x + 15$

$-y = -3x + 9$ Multiply by $3 \rightarrow$   $\underline{-3y = -9x + 27}$

$$0 = -14x + 42$$
$$14x = 42$$
$$x = 3$$

$$3y = -5(3) + 15$$
$$3y = 0$$
$$y = 0$$    Solution: $(3, 0)$

**39.** $3j + 5k = 19$    Multiply by $4 \rightarrow$   $12j + 20k = 76$

$4j - 8k = -4$    Multiply by $-3 \rightarrow$   $\underline{-12j + 24k = 12}$

$$44k = 88$$
$$k = 2$$

$$3j + 5(2) = 19$$
$$3j = 9$$
$$j = 3$$    Solution: $(3, 2)$

**Algebra 1**
Chapter 7 Worked-out Solution Key

# Chapter 7 continued

**40.** $1.5v - 6.5w = 3.5$          $1.5v - 6.5w = 3.5$

$0.5v + 2w = -3$   Multiply

             by $-3 \rightarrow$       $\underline{-1.5v - 6w = 9}$

                                $-12.5w = 12.5$

                                      $w = -1$

$1.5v - 6.5(-1) = 3.5$

             $1.5v = -3$

                $v = -2$

Solution: $(-2, -1)$

**41.** $5y - 20 = -4x$   Rearrange $\rightarrow$    $4x + 5y = 20$

    $y = -\frac{5}{4}x + 4$   Rearrange $\rightarrow$    $\frac{5}{4}x + y = 4$

                                    $4x + 5y = 20$

     Multiply by $-5 \rightarrow$    $\underline{-\frac{25}{4}x - 5y = -20}$

                                   $-\frac{9}{4}x = 0$

                                     $x = 0$

$y = -\frac{5}{4}(0) + 4$

    $y = 4$            Solution: $(0, 4)$

**42.** $9g - 7h = \frac{2}{3}$             $9g - 7h = \frac{2}{3}$

$3g + h = \frac{1}{3}$   Multiply by $7 \rightarrow$   $\underline{21g - 7h = \frac{7}{3}}$

                                   $30g = 3$

                                     $g = \frac{1}{10}$

$3\left(\frac{1}{10}\right) + h = \frac{1}{3}$

            $h = \frac{1}{30}$    Solution: $\left(\frac{1}{10}, \frac{1}{30}\right)$

**43.** $x + y = 15$   Multiply by $-9 \rightarrow$   $-9x - 9y = -135$

$19.3x + 9y = 238$                 $\underline{19.3x + 9y = 238}$

                                   $10.3x = 103$

                                   $x = 10$

        $10 + y = 15$

             $y = 5$ grams of copper

**44.**   $y = \frac{9}{7}x$            **45.** $w$ = wind speed

                         $s$ = speed in still air

   $y = -3x + 12$           $s - w = 300$

   $\frac{9}{7}x = -3x + 12$        $s + w = 450$

   $\frac{30}{7}x = 12$

     $x = \frac{14}{5}$

     $y = \frac{9}{7}\left(\frac{14}{5}\right)$

     $y = \frac{18}{5}$

coordinates of the hive: $\left(\frac{14}{5}, \frac{18}{5}\right)$

**46.**    $s - w = 300$         **47.** 375 mi/h; 75 mi/h

     $\underline{s + w = 450}$

       $2s = 750$

        $s = 375$

$375 - w = 300$

      $w = 75$

Solution: $(375, 75)$

**48.** Current speed = $c$ (miles per hour)

Boat speed in still water = $b$ (miles per hour)

$\begin{cases} b - c = 8 \\ (b + c)0.5 = 8 \rightarrow 0.5b + 0.5c = 8 \end{cases}$

$b - c = 8$    Multiply by $0.5 \rightarrow$   $0.5b - 0.5c = 4$

                                  $\underline{0.5b + 0.5c = 8}$

                                      $b = 12$

     $12 - c = 8$

        $-c = -4$

          $c = 4$

boat speed: 12 mi/h

current speed: 4 mi/h

**49.** $x$; only one equation must be multiplied

**50.** $x$; only one equation must be multiplied

**51.** $x$ or $y$; only one equation must be multiplied

**52.** First determine whether one variable may be eliminated without multiplication. If not, determine whether either variable may be eliminated by multiplying only one equation. If not, you may eliminate either variable.

**53.** $x + y = 4$    Multiply by $2 \rightarrow$    $2x + 2y = 8$

$x - 2y = 1$                      $\underline{x - 2y = 1}$

                                  $3x = 9$

                                   $x = 3$

         $3 + y = 4$

            $y = 1$

       Solution: $(3, 1)$   A

**54.**    $x - 5y + 1 = 6$      **55.**      $3x + 5y = -8$

          $x = 12y$                      $x - 2y = 1$

$12y - 5y + 1 = 6$                 $x = 2y + 1$

        $7y = 5$         $3(2y + 1) + 5y = -8$

         $y = \frac{5}{7}$          $6y + 3 + 5y = -8$

$x = 12\left(\frac{5}{7}\right)$                $11y = -11$

$x = \frac{60}{7}$                         $y = -1$

Solution: $\left(\frac{60}{7}, \frac{5}{7}\right)$ A     $x - 2(-1) = 1$

                           $x = -1$

              Solution: $(-1, -1)$ C

**56.** $3x + 2y + z = 42$       $3x + 2y + z = 42$

$2y + z + 12 = 3x$    $\rightarrow$   $-3x + 2y + z = -12$

   $x - 3y = 0$

   $9 - 3y = 0$

     $-3y = -9$

        $y = 3$

                           $3x + 2y + z = 42$

     Multiply by $-1 \rightarrow$   $\underline{3x - 2y - z = 12}$

                               $6x = 54$

                                 $x = 9$

$3(9) + 2(3) + z = 42$

$27 + 6 + z = 42$

$z = 9$

*continued*

# Chapter 7 continued

Solution: $(9, 3, 9)$; rearrange the second equation so like terms in the first and second equations are aligned in columns. Multiply either the first or second equation by $-1$ to eliminate the $y$ and $z$ terms. Solve for $x$, then substitute that value into the third equation to solve for $y$. Substitute the $x$ and $y$ values into either the first or second equation to solve for $z$. Check that $(9, 3, 9)$ is a solution to all three equations.

## 7.3 Mixed Review (p. 416)

**57.** $m = \dfrac{2 - (-1)}{4 - (-2)} = \dfrac{3}{6} = \dfrac{1}{2}$

$y = \frac{1}{2}x + b$

$2 = \frac{1}{2}(4) + b$

$0 = b$

Equation: $y = \frac{1}{2}x$

**58.** $y = 3x + b$

$4 = 3(-2) + b$

$10 = b$

Equation: $y = 3x + 10$

**59.** $m = \dfrac{1 - 5}{2 - 6} = \dfrac{-4}{-4} = 1$

$y = 1x + b$

$1 = 1(2) + b$

$-1 = b$

Equation: $y = x - 1$

**60.** $y = 5x + b$

$1 = 5(5) + b$

$-24 = b$

Equation: $y = 5x - 24$

**61.** $y = -\dfrac{1}{3}x + b$

$3 = -\dfrac{1}{3}(9) + b$

$6 = b$

Equation: $y = -\dfrac{1}{3}x + 6$

**62.** $m = \dfrac{-3 - (-5)}{-1 - 4} = \dfrac{2}{-5}$

$y = -\dfrac{2}{5}x + b$

$-3 = -\dfrac{2}{5}(-1) + b$

$-\dfrac{15}{2} - \dfrac{2}{5} = b$

$-\dfrac{17}{5} = b$

Equation: $y = -\dfrac{2}{5}x - \dfrac{17}{5}$

**63.** $3x - 2y < 2$

$3(1) - 2(3) \overset{?}{<} 2$

$-3 < 2$

$(1, 3)$ is a solution.

$3(2) - 2(0) \overset{?}{<} 2$

$4 \not< 2$

$(2, 0)$ is not a solution.

**64.** $5x + 4y \geq 6$

$5(-2) + 4(4) \overset{?}{\geq} 6$

$6 \geq 6$

$(-2, 4)$ is a solution.

$5(5) + 4(5) \overset{?}{\geq} 6$

$45 \geq 6$

$(5, 5)$ is a solution.

**65.** $5x + y > 5$

$5(5) + 5 \overset{?}{>} 5$

$30 > 5$

$(5, 5)$ is a solution.

$5(-5) + (-5) \overset{?}{>} 5$

$-30 \not> 5$

$(-5, -5)$ is not a solution.

**66.** $12y - 3x \leq 3$

$12(4) - 3(-2) \overset{?}{\leq} 3$

$54 \not\leq 3$

$(-2, 4)$ is not a solution.

$12(-1) - 3(1) \overset{?}{\leq} 3$

$-15 \leq 3$

$(1, -1)$ is a solution.

**67.** $-6x - 5y = 28$

$x - 2y = 1$

$x = 2y + 1$

$-6(2y + 1) - 5y = 28$

$-12y - 6 - 5y = 28$

$-17y = 34$

$y = -2$

$x - 2(-2) = 1$

$x = -3$

Solution: $(-3, -2)$

**68.** $m + 2n = 1$

$5m - 4n = -23$

$m = -2n + 1$

$5(-2n + 1) - 4n = -23$

$-10n + 5 - 4n = -23$

$-14n = -28$

$n = 2$

$m + 2(2) = 1$

$m = -3$

Solution: $(-3, 2)$

**69.** $g - 5h = 20$

$4g + 3h = 34$

$g = 5h + 20$

$4(5h + 20) + 3h = 34$

$20h + 80 + 3h = 34$

$23h = -46$

$h = -2$

$g - 5(-2) = 20$

$g = 10$

Solution: $(10, -2)$

**70.** $p + 4q = -9$

$2p - 3q = 4$

$p = -4q - 9$

$2(-4q - 9) - 3q = 4$

$-8q - 18 - 3q = 4$

$-11q = 22$

$q = -2$

$p + 4(-2) = -9$

$p = -1$

Solution: $(-1, -2)$

**71.** $\frac{3}{5}b - a = 0$

$1 + b = 2a$

$b = 2a - 1$

$\frac{3}{5}(2a - 1) - a = 0$

$\frac{6}{5}a - \frac{3}{5} - a = 0$

$\frac{1}{5}a = \frac{3}{5}$

$a = 3$

$1 + b = 2(3)$

$b = 5$

Solution: $(3, 5)$

**72.** $d - e = 8$

$\frac{1}{5}d = e + 4$

$d = 5e + 20$

$5e + 20 - e = 8$

$4e = -12$

$e = -3$

$d - (-3) = 8$

$d = 5$

Solution: $(5, -3)$

## Quiz 7.3 (p. 417)

**1.** $3x + y = 5$

$y = -3x + 5$

$-x + y = -7$

$y = x - 7$

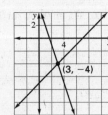

$(3, -4)$

Check: $3x + y = 5$

$3(3) + (-4) \overset{?}{=} 5$

$5 = 5$

$-x + y = -7$

$-3 + (-4) \overset{?}{=} -7$

$-7 = -7$

Solution: $(3, -4)$

# Chapter 7 *continued*

**2.** $\frac{1}{2}x + \frac{3}{4}y = 9$      $-2x + y = -4$

$2x + 3y = 36$          $y = 2x - 4$

$3y = -2x + 36$

$y = -\frac{2}{3}x + 12$

Check:   $\frac{1}{2}x + \frac{3}{4}y = 9$

$\frac{1}{2}(6) + \frac{3}{4}(8) \overset{?}{=} 9$

$3 + 6 = 9$

$-2x + y = -4$

$-2(6) + 8 \overset{?}{=} -4$

$-4 = -4$

Solution: $(6, 8)$

**3.** $x - 2y = 0$        $3x - y = 0$

$-2y = -x$        $-y = -3x$

$y = \frac{1}{2}x$         $y = 3x$

Check:   $x - 2y = 0$

$0 - 2(0) \overset{?}{=} 0$

$0 = 0$

$3x - y = 0$

$3(0) - 0 \overset{?}{=} 0$

$0 = 0$

Solution: $(0, 0)$

**4.** $4x + 3y = 31$      **5.** $-12x + y = 15$

$y = 2x + 7$              $3x + 2y = 3$

$4x + 3(2x + 7) = 31$      $y = 12x + 15$

$4x + 6x + 21 = 31$

$10x = 10$      $3x + 2(12x + 15) = 3$

$x = 1$        $3x + 24x + 30 = 3$

$y = 2(1) + 7$           $27x = -27$

$y = 9$                  $x = -1$

Solution: $(1, 9)$      $-12(-1) + y = 15$

                          $y = 3$

**6.** $x + \frac{1}{2}y = 7$      Solution: $(-1, 3)$

$3x + 2y = 18$

$x = -\frac{1}{2}y + 7$

$3\left(-\frac{1}{2}y + 7\right) + 2y = 18$

$-\frac{3}{2}y + 21 + 2y = 18$

$\frac{1}{2}y = -3$

$y = -6$

$x + \frac{1}{2}(-6) = 7$

$x = 10$

Solution: $(10, -6)$

**7.** $x + 7y = 12$   Multiply by $-3$ $\rightarrow$   $-3x - 21y = -36$

$3x - 5y = 10$                        $\underline{3x - 5y = 10}$

                               $-26y = -26$

                                 $y = 1$

$x + 7(1) = 12$

$x = 5$

Solution: $(5, 1)$

**8.** $3x - 5y = -4$   Multiply by 3 $\rightarrow$   $9x - 15y = -12$

$-9x + 7y = 8$                        $\underline{-9x + 7y = 8}$

                               $-8y = -4$

                                  $y = \frac{1}{2}$

$3x - 5\left(\frac{1}{2}\right) = -4$

$3x = -\frac{3}{2}$

$x = -\frac{1}{2}$

Solution: $\left(-\frac{1}{2}, \frac{1}{2}\right)$

**9.** $\frac{2}{3}x + \frac{1}{6}y = \frac{2}{3}$      Multiply by 6 $\rightarrow$   $4x + y = 4$

$-y = 12 - 2x$    Rearrange $\rightarrow$   $\underline{2x - y = 12}$

                                     $6x = 16$

                                      $x = \frac{8}{3}$

$-y = 12 - 2\left(\frac{8}{3}\right)$

$-y = \frac{36}{3} - \frac{16}{3}$

$y = -\frac{20}{3}$

Solution: $\left(2\frac{2}{3}, -6\frac{2}{3}\right)$

**10.** Let $x =$ number of compact discs at \$10.50

Let $y =$ number of compact discs at \$8.50

$x + y = 10$

$10.5x + 8.5y = 93$

$x = -y + 10$

$10.5(-y + 10) + 8.5y = 93$

$-10.5y + 105 + 8.5y = 93$

$-2y = -12$

$y = 6$

$x + 6 = 10$

$x = 4$

4 compact discs at \$10.50;    6 compact discs at \$8.50

## Math and History (p. 417)

**1.** $x =$ the number of rolls of cotton; $y =$ the number of men to be clothed

**2.**        $-x = -\frac{8y}{6} + 160$

          $x = \frac{9y}{7} + 560$

            $0 = \frac{9y}{7} - \frac{8y}{6} + 720$

     $-720 = \frac{54y - 56y}{42}$

     $-720 = \frac{-2y}{42}$

$-30,240 = -2y$

$15,120 = y$

$x = \frac{9}{7}(15,120) + 560$

$x = 20,000$

Solution: $(20,000, 15,120)$

20,000 rolls of cotton will be used and there are 15,120 men.

# Chapter 7 *continued*

## 7.4 Guided Practice (p. 421)

**1.** $y = 30,000 + 0.01x$
$y = 24,000 + 0.02x$

$30,000 + 0.01x = 24,000 + 0.02x$
$6000 = 0.01x$
$600,000 = x$

$y = 30,000 + 0.01(600,000)$
$y = 36,000$

Solution: $(600,000, 36,000)$

**2.** Personal preferences may vary.

**3.** Substitution or linear combinations; it would be simple to write either variable in terms of the other or to eliminate $x$ by multiplying either equation by $-1$.

**4.** Linear combinations; neither variable has a coefficient of 1 or $-1$ in either equation.

**5.** Any of the three methods would be reasonable; it would be simple to write either variable in terms of the other, both would be simple to graph, and $y$ could be eliminated by multiplying either equation by $-1$.

**6.**

| Number of gallons of regular | · | Price per gallon of regular | + |
|---|---|---|---|

| Number of gallons of premium | · | Price per gallon of premium | = | Total cost |
|---|---|---|---|---|

| Cost of 1 gal of premium | = | Cost of 1 gal of regular | + 0.20 |
|---|---|---|---|

**7.** Let $x$ = price per gallon of regular.
Let $y$ = price per gallon of premium.

**8.** $10x + 15y = 32.75$
$y = x + 0.20$

**9.** $10x + 15y = 32.75$
$y = x + 0.20$

$10x + 15(x + 0.20) = 32.75$
$10x + 15x + 3 = 32.75$
$25x = 29.75$
$x = 1.19$

$y = 1.19 + 0.20$
$y = 1.39$

regular: \$1.19
premium: \$1.39

## 7.4 Practice and Applications (pp. 421–424)

**10.** $x + y = 2$
$6x + y = 2$

Substitution:

$x = -y + 2$
$6(-y + 2) + y = 2$
$-6y + 12 + y = 2$
$-5y = -10$
$y = 2$

$x + 2 = 2$
$x = 0$

Solution: $(0, 2)$

Linear Combinations:

$x + y = 2$    Multiply by $-1 \rightarrow$    $-x - y = -2$
$6x + y = 2$                     $\underline{6x + y = 2}$
                                  $5x = 0$
                                    $x = 0$

$0 + y = 2$
$y = 2$

Solution: $(0, 2)$

Graphing Method:

$x + y = 2$                    $6x + y = 2$
$y = -x + 2$               $y = -6x + 2$

Check: $x + y = 2$
$0 + 2 \overset{?}{=} 2$
$2 = 2$

$6x + y = 2$
$6(0) + 2 \overset{?}{=} 2$
$2 = 2$

**11.** $x - y = 1$
$x + y = 5$

Substitution:

$x = y + 1$
$y + 1 + y = 5$
$2y = 4$
$y = 2$

$x - 2 = 1$
$x = 3$

Solution: $(3, 2)$

Linear Combinations:

$x - y = 1$
$\underline{x + y = 5}$
$2x = 6$
$x = 3$

$3 - y = 1$
$-y = -2$
$y = 2$

Solution: $(3, 2)$

Graphing Method:

$x - y = 1$                 $x + y = 5$
$-y = -x + 1$          $y = -x + 5$
$y = x - 1$

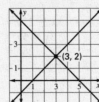

Check: $x - y = 1$
$3 - 2 \overset{?}{=} 1$
$1 = 1$

$x + y = 5$
$3 + 2 \overset{?}{=} 5$
$5 = 5$

# Chapter 7 *continued*

**12.** $3x - y = 3$
$-x + y = 3$

**Substitution:**
$$y = x + 3$$
$$3x - (x + 3) = 3$$
$$3x - x - 3 = 3$$
$$2x = 6$$
$$x = 3$$
$$-3 + y = 3$$
$$y = 6$$

Solution: $(3, 6)$

**Graphing Method:**

$3x - y = 3$  $\qquad -x + y = 3$
$-y = -3x + 3$ $\qquad y = x + 3$
$y = 3x - 3$

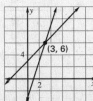

Check:  $3x - y = 3$
$3(3) - 6 \overset{?}{=} 3$
$3 = 3$

$-x + y = 3$
$-3 + 6 \overset{?}{=} 3$
$3 = 3$

**Linear Combinations:**

$3x - y = 3$
$\underline{-x + y = 3}$
$2x = 6$
$x = 3$

$-3 + y = 3$
$y = 6$

Solution: $(3, 6)$

**13–18 Sample answers are given.**

**13.** Linear combinations; the $y$ terms in both equations have coefficients of 1 and $-1$ which will eliminate $y$.

**14.** Linear combinations; neither variable has a coefficient of 1 or $-1$ in either equation.

**15.** Substitution; the first equation can be easily solved for $x$.

**16.** Substitution; the variables in the second equation have coefficients of 1 and $-1$.

**17.** Linear combinations; neither variable has a coefficient of 1 or $-1$ in either equation.

**18.** Linear combinations; neither variable has a coefficient of 1 or $-1$ in either equation.

**19.** Linear combinations; the $y$-terms have coefficients that will add to zero.

$2x + y = 5$  $\qquad 2 - y = 1$
$\underline{x - y = 1}$ $\qquad -y = -1$
$3x = 6$ $\qquad y = 1$
$x = 2$ $\qquad$ Solution: $(2, 1)$

**20.** Linear combinations; the $y$-terms can be eliminated by multiplying the first equation by 3.

$2x - y = 3$  Multiply by 3 $\rightarrow$  $6x - 3y = 9$
$4x + 3y = 21$ $\qquad\qquad\qquad \underline{4x + 3y = 21}$
$\qquad\qquad\qquad\qquad\qquad\qquad 10x = 30$
$\qquad\qquad\qquad\qquad\qquad\qquad x = 3$

$2(3) - y = 3$
$y = 3$

Solution: $(3, 3)$

**21.** Linear combinations; the $y$-terms have coefficients that will add to zero.

$x - 2y = 4$  $\qquad 2 - 2y = 4$
$\underline{6x + 2y = 10}$ $\qquad -2y = 2$
$7x = 14$ $\qquad y = -1$
$x = 2$ $\qquad$ Solution: $(2, -1)$

**22.** Linear combinations; the $x$-terms can be eliminated by multiplying the first equation by 2.

$3x + 6y = 8$  Multiply by 2 $\rightarrow$  $6x + 12y = 16$
$-6x + 3y = 2$ $\qquad\qquad\qquad \underline{-6x + 3y = 2}$
$\qquad\qquad\qquad\qquad\qquad\qquad 15y = 18$
$\qquad\qquad\qquad\qquad\qquad\qquad y = \frac{6}{5}$

$-6x + 3\left(\frac{6}{5}\right) = 2$
$-6x + \frac{18}{5} = \frac{10}{5}$
$-6x = -\frac{8}{5}$
$x = \frac{4}{15}$

Solution: $\left(\frac{4}{15}, \frac{6}{5}\right)$

**23.** Substitution; $x$ and $y$ have coefficients of 1 in the first equation.

$x + y = 0$  $\quad 3(-y) + 2y = 1$  $\quad x + (-1) = 0$
$3x + 2y = 1$ $\qquad -y = 1$ $\qquad\quad x = 1$
$x = -y$ $\qquad\quad y = -1$ $\quad$ Solution: $(1, -1)$

**24.** Linear combinations; the $y$-terms can be eliminated by multiplying the second equation by 3.

$2x - 3y = -7$ $\qquad\qquad\qquad 2x - 3y = -7$
$3x + y = -5$  Multiply by 3 $\rightarrow$ $\underline{9x + 3y = -15}$
$\qquad\qquad\qquad\qquad\qquad\qquad 11x = -22$
$\qquad\qquad\qquad\qquad\qquad\qquad x = -2$

$3(-2) + y = -5$
$y = 1$

Solution: $(-2, 1)$

**25.** Substitution; $x$ and $y$ both have coefficients of 1 in the second equation.

$3y + 4x = 5$ $\qquad 3y + 4(-y + 1) = 5$
$x + y = 1$ $\qquad 3y - 4y + 4 = 5$
$x = -y + 1$ $\qquad\qquad -y = 1$
$\qquad\qquad\qquad\qquad y = -1$

$x - 1 = 1$
$x = 2$

Solution: $(2, -1)$

**26.** Substitution; in the first equation, $y$ has a coefficient of 1.

$8x + y = 15$ $\qquad 9 = 2(-8x + 15) + 2x$
$9 = 2y + 2x$ $\qquad 9 = -16x + 30 + 2x$
$y = -8x + 15$ $\qquad -21 = -14x$
$\qquad\qquad\qquad\qquad \frac{3}{2} = x$

$8\left(\frac{3}{2}\right) + y = 15$
$12 + y = 15$
$y = 3$

Solution: $\left(\frac{3}{2}, 3\right)$

# Chapter 7 *continued*

**27.** Linear combinations; after both equations are rearranged, the $y$-terms will add to zero.

$$100 - 9x = 5y \quad \text{Rearrange} \rightarrow \quad -9x - 5y = -100$$
$$0 = 5y - 9x \qquad\qquad\qquad\quad \underline{-9x + 5y = 0}$$
$$-18x = -100$$
$$x = \tfrac{50}{9}$$

$$0 = 5y - 9\left(\tfrac{50}{9}\right)$$
$$50 = 5y$$
$$10 = y \qquad\qquad \text{Solution: } \left(\tfrac{50}{9}, 10\right)$$

**28.** Linear combinations; the $x$-terms can be eliminated by multiplying either equation by $-1$.

$$x + 2y = 2 \quad \text{Multiply by } -1 \rightarrow \quad -x - 2y = -2$$
$$x + 4y = -2 \qquad\qquad\qquad\qquad \underline{x + 4y = -2}$$
$$2y = -4$$
$$y = -2$$

$$x + 2(-2) = 2$$
$$x = 6 \qquad\qquad \text{Solution: } (6, -2)$$

**29.** Substitution; the first equation is already solved for $-y$.

$$-y = -4 \qquad\qquad x + 2(4) = 4$$
$$x + 2y = 4 \qquad\qquad\quad x = -4$$
$$y = 4 \qquad\qquad \text{Solution: } (-4, 4)$$

**30.** Linear combinations; neither variable has a coefficient of 1 or $-1$ in either equation.

$$0.2x - 0.5y = -3.8 \quad \begin{matrix}\text{Multiply}\\ \text{by } 40 \rightarrow\end{matrix} \quad 8x - 20y = -152$$
$$0.3x + 0.4y = 10.4 \quad \begin{matrix}\text{Multiply}\\ \text{by } 50 \rightarrow\end{matrix} \quad \underline{15x + 20y = 520}$$
$$23x = 368$$
$$x = 16$$

$$0.2(16) - 0.5y = -3.8$$
$$-0.5y = -7$$
$$y = 14 \qquad\qquad \text{Solution: } (16, 14)$$

**31.**
$$6x - y = 18$$
$$\underline{8x + y = 24}$$
$$14x = 42$$
$$x = 3$$

$$6(3) - y = 18$$
$$-y = 0$$
$$y = 0$$

Solution: $(3, 0)$

**32.**
$$x - y = -4$$
$$2y + x = 5$$
$$x = y - 4$$

$$2y + y - 4 = 5$$
$$3y = 9$$
$$y = 3$$

$$x - 3 = -4$$
$$x = -1$$

Solution: $(-1, 3)$

**33.**
$$x + 2y = 8$$
$$\underline{3x - 2y = 8}$$
$$4x = 16$$
$$x = 4$$

$$4 + 2y = 8$$
$$2y = 4$$
$$y = 2 \qquad\qquad \text{Solution: } (4, 2)$$

**34.**
$$x + 2y = 1 \quad \text{Multiply by } 2 \rightarrow \quad 2x + 4y = 2$$
$$5x - 4y = -23 \qquad\qquad\qquad\quad \underline{5x - 4y = -23}$$
$$7x = -21$$
$$x = -3$$

$$-3 + 2y = 1$$
$$2y = 4$$
$$y = 2 \qquad\qquad \text{Solution: } (-3, 2)$$

**35.**
$$8x + 4y = 8 \qquad\qquad\qquad\qquad\qquad 8x + 4y = 8$$
$$-2x + 3y = 12 \quad \text{Multiply by } 4 \rightarrow \quad \underline{-8x + 12y = 48}$$
$$16y = 56$$
$$y = \tfrac{7}{2}$$

$$8x + 4\left(\tfrac{7}{2}\right) = 8$$
$$8x = -6$$
$$x = -\tfrac{3}{4} \qquad\qquad \text{Solution: } \left(-\tfrac{3}{4}, \tfrac{7}{2}\right)$$

**36.**
$$3x - 5y = 8 \quad \text{Multiply by } 2 \rightarrow \quad 6x - 10y = 16$$
$$-2x + 3y = 3 \quad \text{Multiply by } 3 \rightarrow \quad \underline{-6x + 9y = 9}$$
$$-y = 25$$
$$y = -25$$

$$3x - 5(-25) = 8$$
$$3x = -117$$
$$x = -39 \quad \text{Solution: } (-39, -25)$$

**37.**
$$2x - y = 6 \qquad\qquad\qquad\qquad 2x - y = 6$$
$$y - x = 0 \quad \text{Rearrange} \rightarrow \quad \underline{-x + y = 0}$$
$$x = 6$$

$$y - 6 = 0$$
$$y = 6 \qquad\qquad \text{Solution: } (6, 6)$$

**38.**
$$8x + 9y = 42 \qquad\qquad\qquad\qquad\qquad 8x + 9y = 42$$
$$6x - y = 16 \quad \text{Multiply by } 9 \rightarrow \quad \underline{54x - 9y = 144}$$
$$62x = 186$$
$$x = 3$$

$$6(3) - y = 16$$
$$-y = -2$$
$$y = 2 \qquad\qquad \text{Solution: } (3, 2)$$

**39.**
$$7x + 4y = 22 \quad \text{Multiply by } 5 \rightarrow \quad 35x + 20y = 110$$
$$-5x - 9y = 15 \quad \text{Multiply by } 7 \rightarrow \quad \underline{-35x - 63y = 105}$$
$$-43y = 215$$
$$y = -5$$

$$7x + 4(-5) = 22$$
$$7x = 42$$
$$x = 6 \qquad\qquad \text{Solution: } (6, -5)$$

**Algebra 1**
Chapter 7 Worked-out Solution Key

# Chapter 7 *continued*

**40.** $3x - 5y = 3$  Multiply by $-3 \rightarrow$  $-9x + 15y = -9$

$9x - 20y = 6$  $\underline{\hspace{1em} 9x - 20y = 6 \hspace{1em}}$

$-5y = -3$

$y = \frac{3}{5}$

$3x - 5\left(\frac{3}{5}\right) = 3$

$3x = 6$

$x = 2$    Solution: $\left(2, \frac{3}{5}\right)$

**41.** $0.5x + 2.2y = 9$  Multiply

by $-12 \rightarrow$  $-6x - 26.4y = -108$

$6x + 0.4y = -22$  $\underline{\hspace{1em} 6x + 0.4y = -22 \hspace{1em}}$

$-26y = -130$

$y = 5$

$6x + 0.4(5) = -22$

$6x = -24$

$x = -4$    Solution: $(-4, 5)$

**42.** $1.5x - 2.5y = 8.5$  Multiply

by $-4 \rightarrow$  $-6x + 10y = -34$

$6x + 30y = 24$  $\underline{\hspace{1em} 6x + 30y = 24 \hspace{1em}}$

$40y = -10$

$y = -\frac{1}{4}$

$6x + 30\left(-\frac{1}{4}\right) = 24$

$6x = \frac{63}{2}$

$x = \frac{21}{4}$    Solution: $\left(\frac{21}{4}, -\frac{1}{4}\right)$

**43.** $3x + 9y = 1$  $3x + 9y = 1$

$2x + 3y = \frac{2}{3}$  Multiply by $-3 \rightarrow$  $\underline{\hspace{1em} -6x - 9y = -2 \hspace{1em}}$

$-3x = -1$

$x = \frac{1}{3}$

$3\left(\frac{1}{3}\right) + 9y = 1$

$9y = 0$

$y = 0$    Solution: $\left(\frac{1}{3}, 0\right)$

**44.** $3x - 2y = 8$  $3x - 2y = 8$

$x + \frac{3}{2}y = 20$  Multiply by $-3 \rightarrow$  $\underline{\hspace{1em} -3x - \frac{9}{2}y = -60 \hspace{1em}}$

$-\frac{13}{2}y = -52$

$y = 8$

$3x - 2(8) = 8$

$3x = 24$

$x = 8$    Solution: $(8, 8)$

**45.** $\frac{1}{2}x - y = -5$  Multiply by $-2 \rightarrow$  $-x + 2y = 10$

$x - \frac{1}{3}y = 0$  $\underline{\hspace{1em} x - \frac{1}{3}y = 0 \hspace{1em}}$

$\frac{5}{3}y = 10$

$y = 6$

$x - \frac{1}{3}(6) = 0$

$x = 2$    Solution: $(2, 6)$

**46.** $x + y = 6$

$5x + 8y = 36$

$x = -y + 6$

$5(-y + 6) + 8y = 36$    $x + 2 = 6$

$-5y + 30 + 8y = 36$    $x = 4$

$3y = 6$    4 8-in. pots; 2 10-in. pots

$y = 2$

**47.** $x + y = 60$

$0.05x + 0.02y = 0.03(60)$

$x = -y + 60$

$0.05(-y + 60) + 0.02y = 1.8$    $x + 40 = 60$

$-0.05y + 3 + 0.02y = 1.8$    $x = 20$

$-0.03y = -1.2$

$y = 40$

20 mL of the 5% solution
40 mL of the 2% solution

**48.** $y = 14 + 4x$   **49.** $y = 30 + 8x$

$y = 8 + 6x$    $y = 14 + 12x$

$8 + 6x = 14 + 4x$    $30 + 8x = 14 + 12x$

$2x = 6$    $16 = 4x$

$x = 3$    $4 = x$

$y = 14 + 4(3)$    4 children

$y = 26$

3 years; 26 in.

**50.** $30 + 8x < 14 + 12x$

$16 < 4x$

$4 < x$

more than 4 children

**51.** $m = \dfrac{10{,}500 - 6200}{25 - 0} = 172$    $m = \dfrac{6100 - 6500}{25 - 0} = -16$

$y = 172x + 6200$    $y = -16x + 6500$

System:  $y = 172x + 6200$

$y = -16x + 6500$

$172x + 6200 = 16x + 6500$

$188x = 300$

$x = 1\frac{28}{47}$

$y = 172\left(\frac{300}{188}\right) + 6200$

$y = 6474\frac{22}{47}$

Solution: $\left(1\frac{28}{47}, 6474\frac{22}{47}\right)$; the point of intersection represents the number of years after 1970 (about 1.6) when the demand for low-income housing and the availability were equal (about 6,474,500).

**52.** $y = 12x$   **53.** $y = 7x + 50$   **54.** $12x = 7x + 50$

$5x = 50$

$x = 10$

10 sketches

**55.** $7 = 4x + 6y$   $7 = 4(1.5 - y) + 6y$   $x + 0.5 = 1.5$

$x + y = 1.5$   $7 = 6 - 4y + 6y$   $x = 1$

$x = 1.5 - y$   $1 = 2y$

$0.5 = y$

1 h at 4 mi/h; $\frac{1}{2}$ h at 6 mi/h

**56.** $\dfrac{\text{Total distance}}{45} = \text{Total time}$

Let $x$ = total time
Let $y$ = total distance

$40(2) + 55(x - 2) = y$   $80 + 55x - 110 = 45x$

$45x = y$   $-30 = -10x$

$3 = x$

You traveled 1 h at 55 mi/h.

# Chapter 7 *continued*

**57.** substitution, graphing, linear combinations

**58.** about $-275°C$

**59.** No; solving a system involving any two of the equations will give the solution to the system of three equations.

**60.**
$$-2T + 302V = 546.4 \qquad -2T + 302V = 546.4$$
$$-T + 228V = 273.2 \quad \text{Mult. by}$$
$$-2 \rightarrow \quad \underline{2T - 456V = -546.4}$$
$$-154V = 0$$
$$V = 0$$

$$-T + 228(0) = 273.2$$
$$T = -273.2°C$$

**61.** $r + b = 12$
$$3b = r$$

**62.** $3b + b = 12 \qquad 3(3) = r$
$$4b = 12 \qquad 9 = r$$
$$b = 3 \qquad \text{9 red marbles}$$

**63.** $\frac{9}{12} = 0.75$

**64.** D $\;3x - 2y = 0 \quad 3(0) - 2y = 0$
$$\underline{5x + 2y = 0} \qquad -2y = 0$$
$$8x = 0 \qquad y = 0$$
$$x = 0$$

**65.** B $\;y = 2x - 2$
$$y = 3x + 1$$
$$2x - 2 = 3x + 1$$
$$-3 = x$$

**66.** $x + y = 19$
$$285x + 335y = 5765$$
$$x = -y + 19$$
$$285(-y + 19) + 335y = 5765$$
$$-285y + 5415 + 335y = 5765$$
$$50y = 350$$
$$y = 7$$
$$x + 7 = 19$$
$$x = 12 \text{ min}$$

### 7.4 Mixed Review (p. 424)

**67.** $y = 4x + 3$
$$2y - 8x = -3$$
$$2y = 8x - 3$$
$$y = 4x - \tfrac{3}{2}$$
parallel

**68.** $4y + 5x = 1 \qquad\qquad 10x + 2y = 2$
$$4y = -5x + 1 \qquad\quad 2y = -10x + 2$$
$$y = -\tfrac{5}{4}x + \tfrac{1}{4} \qquad\quad y = -5x + 1$$
not parallel

**69.** $3x + 9y + 2 = 0 \qquad 2y = -6x + 3$
$$9y = -3x - 2 \qquad y = -3x + \tfrac{3}{2}$$
$$y = -\tfrac{1}{3}x - \tfrac{2}{9}$$
not parallel

**70.** $4y - 1 = 5 \qquad 6y + 2 = 8$
$$4y = 6 \qquad\quad 6y = 6$$
$$y = \tfrac{3}{2} \qquad\qquad y = 1$$
parallel

**71.** $f(x) = 2x + 3$

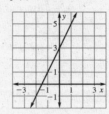

**72.** $h(x) = x + 5$

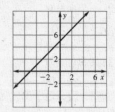

**73.** $g(x) = -3x - 1$

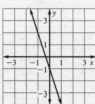

**74.** $(1, 10) \; (4, 30)$

$$m = \frac{30 - 10}{4 - 1} = \frac{20}{3}$$
$$y = \frac{20}{3}x + b$$
$$10 = \frac{20}{3}(1) + b$$
$$\frac{10}{3} = b$$
$$y = \frac{20}{3}x + \frac{10}{3}$$

**75.** $y = \frac{20}{3}(10) + \frac{10}{3}$
$$y = 70$$
70 out of 100 households

### Developing Concepts Activity 7.5 (p. 425)

### Exploring the Concept (p.425)

**1. a.**

$$x + y = 0$$
$$y = -x$$
$$3x - 2y = 10$$
$$-2y = -3x + 10$$
$$y = \tfrac{3}{2}x - 5$$

**b.**

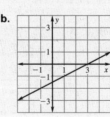

$$2x - 4y = 6$$
$$-4y = -2x + 6$$
$$y = \tfrac{1}{2}x - \tfrac{3}{2}$$
$$x - 2y = 3$$
$$-2y = -x + 3$$
$$y = \tfrac{1}{2}x - \tfrac{3}{2}$$

**c.**

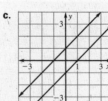

$$x - y = 1$$
$$-y = -x + 1$$
$$y = x - 1$$
$$-3x + 3y = 3$$
$$3y = 3x + 3$$
$$y = x + 1$$

# Chapter 7 *continued*

**2.** The graph for the first system has two lines that intersect in one point; for the second, there is only one line; for the third, there are two parallel lines.

**3. a.** $y = -x$; $y = \frac{3}{2}x - 5$

**b.** $y = \frac{1}{2}x - \frac{3}{2}$; $y = \frac{1}{2}x - \frac{3}{2}$

**c.** $y = x - 1$; $y = x + 1$

**4.** In the first pair of equations, both the slopes and the *y*-intercepts are different; in the second, both the slopes and the *y*-intercepts are the same; in the third, the slopes are the same and the *y*-intercepts are different.

## Drawing Conclusions 7.5

**1. a.** $x - 3y = 9$

$-3y = -x + 9$

$y = \frac{1}{3}x - 3$

$-2x + 6y = -18$

$6y = 2x - 18$

$y = \frac{1}{3}x - 3$

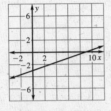

**b.** $x - \frac{1}{4}y = 5$

$-\frac{1}{4}y = -x + 5$

$y = 4x - 20$

$5x + \frac{1}{4}y = 7$

$\frac{1}{4}y = -5x + 7$

$y = -20x + 28$

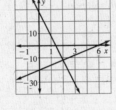

**c.** $x + 2y = 3$

$2y = -x + 3$

$y = -\frac{1}{2}x + \frac{3}{2}$

$x + 2y = 6$

$2y = -x + 6$

$y = -\frac{1}{2}x + 3$

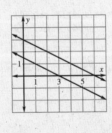

The graph of the first system has only one line; for the second, there are two lines that intersect in one point; for the third, there are two parallel lines.

Slope-intercept form of all systems.:
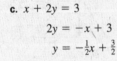
**a.** $y = \frac{1}{3}x - 3$    **b.** $y = 4x - 20$    **c.** $y = -\frac{1}{2}x + \frac{3}{2}$

$y = \frac{1}{3}x - 3$     $y = -20x + 28$     $y = -\frac{1}{2}x + 3$

In the first pair of equations, the slopes and the *y*-intercepts are the same; in the second pair, the slopes and the *y*-intercepts are different; in the third pair, the slopes are the same but the *y*-intercepts are different.

**2.** $y = -\frac{1}{2}x + 4$    **3.** $y = \frac{1}{3}x + 1$    **4.** $y = 2$

$y = -\frac{1}{2}x - 4$     $2y = \frac{2}{3}x + 2$     $y = 4x - 2$

**5.** Many solutions; the graphs are the same line, so every point on the line is a solution of both equations.

**6.** Exactly one solution; the graphs intersect in one point.

**7.** No solution; the graphs are parallel and do not intersect.

**8.** *Sample answer:* $y = 2x - 3$ and $y = 2x + 3$

**9.** *Sample answer:* $y = 3x - 4$ and $y = 2x + 5$

**10.** *Sample answer:* $y = 2x + 1$ and $2y = 4x + 2$

## Lesson 7.5

### 7.5 Guided Practice (p. 429)

**1.** The graphs of the equations are parallel lines.

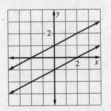

**2.** The graphs of the equations are the same line.

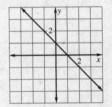

**3.** The graphs intersect in exactly one point.

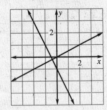

**4.** If the graphs were extended, the lines would intersect.

**5.** Write both equations in slope-intercept form. If the slopes and *y*-intercepts are both different, the system has exactly one solution. If the slopes and the *y*-intercepts are both the same, the system has many solutions. If the slopes are the same but the *y*-intercepts are different, there is no solution.

$x - y = 2$      $4x - 4y = 8$

$-y = -x + 2$     $-4y = -4x + 8$

$y = x - 2$       $y = x - 2$

The graphs are the same line. There are many solutions.

**6.** $2x + y = 5$

$y = -2x + 5$

$-6x - 3y = -15$

$-3y = 6x - 15$

$y = -2x + 5$

infinitely many solutions

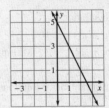

# Chapter 7 *continued*

**7.** $-6x + 2y = 4$
$\qquad 2y = 6x + 4$
$\qquad y = 3x + 2$

$-9x + 3y = 12$
$\qquad 3y = 9x + 12$
$\qquad y = 3x + 4$

no solution

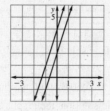

**8.** $2x + y = 7$
$\qquad y = -2x + 7$

$3x - y = -2$
$\qquad -y = -3x - 2$
$\qquad y = 3x + 2$

exactly one solution

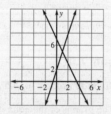

**9.** $-x + y = 7$ Multiply by 2 → $\quad -2x + 2y = 14$
$2x - 2y = -18$ $\qquad\qquad\qquad \underline{2x - 2y = -18}$
$\qquad\qquad\qquad\qquad\qquad\qquad\qquad 0 \neq -4$

no solution

**10.** $-4x + y = -8$ Multiply
$\qquad\qquad\qquad$ by $-3$ → $\quad 12x - 3y = 24$
$-12x + 3y = -24$ $\qquad\qquad \underline{-12x + 3y = -24}$
$\qquad\qquad\qquad\qquad\qquad\qquad 0 = 0$

infinitely many solutions

**11.** $-4x + y = -8$ $\qquad\qquad -4x + y = -8$
$2x - 2y = -14$ Multiply by 2 → $\underline{4x - 4y = -28}$
$\qquad\qquad\qquad\qquad\qquad\qquad -3y = -36$
$\qquad\qquad\qquad\qquad\qquad\qquad\quad y = 12$

$-4x + 12 = -8$
$\qquad -4x = -20$
$\qquad\quad x = 5$

exactly one solution, $(5, 12)$

## 7.5 Practice and Applications (pp. 429–431)

**12.** E; exactly one solution   **13.** D; no solution

**14.** F; exactly one solution   **15.** B; infinitely many solutions

**16.** A; no solution   **17.** C; infinitely many solutions

**18.** $-7x + 7y = 7$ Multiply by 2 → $-14x + 14y = 14$
$2x - 2y = -18$ Multiply by 7 → $\underline{14x - 14y = -126}$
$\qquad\qquad\qquad\qquad\qquad\qquad\qquad 0 \neq -112$

no solution

**19.** $4x + 4y = -8$ $\qquad\qquad 4x + 4y = -8$
$2x + 2y = -4$ Multiply by $-2$ → $\underline{-4x - 4y = 8}$
$\qquad\qquad\qquad\qquad\qquad\qquad\qquad 0 = 0$

infinitely many solutions

**20.** $2x + y = -4$ Multiply by 2 → $\quad 4x + 2y = -8$
$4x - 2y = 8$ $\qquad\qquad\qquad\qquad \underline{4x - 2y = 8}$
$2(0) + y = -4$ $\qquad\qquad\qquad\qquad\quad 8x = 0$
$\qquad\quad y = -4$ $\qquad\qquad\qquad\qquad\quad x = 0$

exactly one solution, $(0, -4)$

**21.** $15x - 5y = -20$ $\qquad\qquad\qquad 15x - 5y = -20$
$-3x + y = 4$ Multiply by 5 → $\underline{-15x + 5y = 20}$
$\qquad\qquad\qquad\qquad\qquad\qquad\qquad\qquad 0 = 0$

infinitely many solutions

**22.** $-6x + 2y = -2$ $\qquad\qquad -6x + 2y = -2$
$-4x - y = 8$ Multiply by 2 → $\underline{-8x - 2y = 16}$
$\qquad\qquad\qquad\qquad\qquad\qquad -14x = 14$
$\qquad\qquad\qquad\qquad\qquad\qquad\quad x = -1$

$-6(-1) + 2y = -2$
$\qquad\qquad 2y = -8$
$\qquad\qquad\; y = -4$

exactly one solution, $(-1, -4)$

**23.** $2x + y = -1$ Multiply by 3 → $\quad 6x + 3y = -3$
$-6x - 3y = -15$ $\qquad\qquad\qquad\quad \underline{-6x - 3y = -15}$
$\qquad\qquad\qquad\qquad\qquad\qquad\qquad 0 \neq -18$

no solution

**24.** $x + y = 8$
$\qquad y = -x + 8$

$x + y = -1$
$\qquad y = -x - 1$

no solution

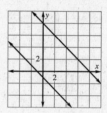

**25.** $3x - 2y = 3$
$\qquad -2y = -3x + 3$
$\qquad\quad y = \frac{3}{2}x - \frac{3}{2}$

$-6x + 4y = -6$
$\qquad 4y = 6x - 6$
$\qquad\; y = \frac{3}{2}x - \frac{3}{2}$

infinitely many solutions

**26.** $x - y = 2$
$\qquad -y = -x + 2$
$\qquad\; y = x - 2$

$-2x + 2y = 2$
$\qquad 2y = 2x + 2$
$\qquad\; y = x + 1$

no solution

**27.** $-x + 4y = -20$
$\qquad 4y = x - 20$
$\qquad\; y = \frac{1}{4}x - 5$

$3x - 12y = 48$
$\qquad -12y = -3x + 48$
$\qquad\quad y = \frac{1}{4}x - 4$

no solution

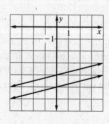

**Algebra 1**
Chapter 7 Worked-out Solution Key

# Chapter 7 *continued*

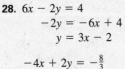

**28.** $6x - 2y = 4$

$-2y = -6x + 4$

$y = 3x - 2$

$-4x + 2y = -\frac{8}{3}$

$2y = 4x - \frac{8}{3}$

$y = 2x - \frac{4}{3}$

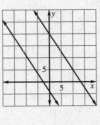

exactly one solution, $\left(\frac{2}{3}, 0\right)$

**29.** $\frac{3}{4}x + \frac{1}{2}y = 10$

$\frac{1}{2}y = -\frac{3}{4}x + 10$

$y = -\frac{3}{2}x + 20$

$-\frac{3}{2}x - y = 4$

$-y = \frac{3}{2}x + 4$

$y = -\frac{3}{2}x - 4$

no solution

**30.** $30x + 6y = 3.6$      $30x + 6y = 3.6$

$10x + 2y = 1.2$ Multiply by $-3 \rightarrow \underline{-30x - 6y = -3.6}$

                                                 $0 = 0$

No; the system of equations that describes the situation has infinitely many solutions.

**31.** $4x + 2y = 99.62$     Multiply

                 by $-3 \rightarrow$   $-12x - 6y = -298.86$

$12x + 6y = 298.86$       $\underline{12x + 6y = 298.86}$

                                          $0 = 0$

No; the system of equations that describes the situation has infinitely many solutions.

**32.** $8x + y = 139.69$

$4x + 2y = 99.62$

$y = -8x + 139.69$

$4x + 2(-8x + 139.69) = 99.62$

$4x - 16x + 279.38 = 99.62$

$-12x = -179.76$

$x = 14.98$

$8(14.98) + y = 139.69$

$y = 19.85$

Yes; you can combine the new equation $8x + y = 139.69$ with either of the original equations to produce a system that has a solution $(14.98, 19.85)$. A sheet of oak paneling costs $14.98.

**33.** $2x + 3y = 25$     $2x + 3x = 25$     $y = 5$

        $x = y$            $5x = 25$

                          $x = 5$       Solution: $(5, 5)$

**34.** The system has no solution; the graphs of the equations are parallel; at every $x$-value, the difference between the $y$-values is the same.

**35.** Ex. 6

|   | A | B | C | D |
|---|---|---|---|---|
| 1 | $x$ | $y = -2x + 5$ | $y = -2x + 5$ | Col. B − Col C. |
| 2 | −2 | 9 | 9 | 0 |
| 3 | −1 | 7 | 7 | 0 |
| 4 | 0 | 5 | 5 | 0 |
| 5 | 1 | 3 | 3 | 0 |
| 6 | 2 | 1 | 1 | 0 |

infinitely many solutions

Ex. 7

|   | A | B | C | D |
|---|---|---|---|---|
| 1 | $x$ | $y = 3x + 2$ | $y = 3x + 4$ | Col. B − Col C. |
| 2 | −2 | −4 | −2 | −2 |
| 3 | −1 | −1 | 1 | −2 |
| 4 | 0 | 2 | 4 | −2 |
| 5 | 1 | 5 | 7 | −2 |
| 6 | 2 | 8 | 10 | −2 |

no solution

Ex. 8

|   | A | B | C | D |
|---|---|---|---|---|
| 1 | $x$ | $y = -2x + 7$ | $y = 3x + 2$ | Col. B − Col C. |
| 2 | −2 | 11 | −4 | 15 |
| 3 | −1 | 9 | −1 | 10 |
| 4 | 0 | 7 | 2 | 5 |
| 5 | 1 | 5 | 5 | 0 |
| 6 | 2 | 3 | 8 | −5 |

exactly one solution

**36.** $y = \frac{1}{2}[2(x + 10) - 18] - x$

**37.** $y = 1$

$y = \frac{1}{2}[2(x + 10) - 18] - x$

$1 = \frac{1}{2}[2(x + 10) - 18] - x$

$1 = \frac{1}{2}[2x + 2] - x$

$1 = x + 1 - x$

$1 = 1$

The system has infinitely many solutions, all ordered pairs whose graphs are on the line $y = 1$. This means that every real number $x$ is a solution.

**38.** Answers may vary.

**39. a.** $x - y = 3$   Multiply by $-4 \rightarrow -4x + 4y = -12$

     $4x - 4y = n$                         $\underline{4x - 4y = n}$

                                       $0 = n - 12$

                                       $n = 12$

**b.** any real number $n$, $n \neq 12$

—CONTINUED—

# Chapter 7 *continued*

**39.** **—CONTINUED—**

**c.**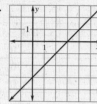

$$x - y = 3$$
$$-y = -x + 3$$
$$y = x - 3$$

$$4x - 4y = 12$$
$$-4y = -4x + 12$$
$$y = x - 3$$

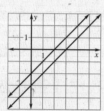

$$x - y = 3$$
$$-y = -x + 3$$
$$y = x - 3$$

$$4x - 4y = 8$$
$$-4y = -4x + 8$$
$$y = x - 2$$

**40.** **a.** $x - y = 4$   Multiply by 2 → $2x - 2y = 8$
$-2x + 2y = n$   $\underline{-2x + 2y = n}$
$$0 = 8 + n$$
$$-8 = n$$

**b.** any real number $n, n \neq -8$

**c.**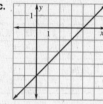

$$x - y = 4$$
$$-y = -x + 4$$
$$y = x - 4$$

$$-2x + 2y = -8$$
$$2y = 2x - 8$$
$$y = x - 4$$

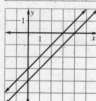

$$x - y = 4$$
$$-y = -x + 4$$
$$y = x - 4$$

$$-2x + 2y = -6$$
$$2y = 2x - 6$$
$$y = x - 3$$

**41.** **a.** $6x - 9y = n$   $6x - 9y = n$
$-2x + 3y = 3$ Multiply by 3 → $\underline{-6x + 9y = 9}$
$$0 = n + 9$$
$$-9 = n$$

**b.** any real number $n, n \neq -9$

**c.**

$$6x - 9y = -9$$
$$-9y = -6x - 9$$
$$y = \tfrac{2}{3}x + 1$$

$$-2x + 3y = 3$$
$$3y = 2x + 3$$
$$y = \tfrac{2}{3}x + 1$$

**—CONTINUED—**

---

**41.** **—CONTINUED—**

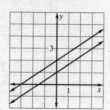

$$6x - 9y = -18$$
$$-9y = -6x - 18$$
$$y = \tfrac{2}{3}x + 2$$

$$-2x + 3y = 3$$
$$3y = 2x + 3$$
$$y = \tfrac{2}{3}x + 1$$

**42.** **a.** $9x + 6y = n$   $9x + 6y = n$
$1.8x + 1.2y = 3$   Multiply
by $-5$ →   $\underline{-9x - 6y = -15}$
$$0 = n - 15$$
$$15 = n$$

**b.** any real number $n, n \neq 15$

**c.**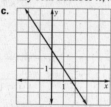

$$9x + 6y = 15$$
$$6y = -9x + 15$$
$$y = -\tfrac{3}{2}x + \tfrac{5}{2}$$

$$1.8x + 1.2y = 3$$
$$1.2y = -1.8x + 3$$
$$y = -\tfrac{3}{2}x + \tfrac{5}{2}$$

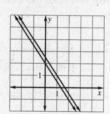

$$9x + 6y = 12$$
$$6y = -9x + 12$$
$$y = -\tfrac{3}{2}x + 2$$

$$1.8x + 1.2y = 3$$
$$1.2y = -1.8x + 3$$
$$y = -\tfrac{3}{2}x + \tfrac{5}{2}$$

### 7.5 Mixed Review (p. 431)

**43.** A   **44.** C   **45.** D   **46.** B   **47.** $\dfrac{290 - 125}{4 - 1} = 55$ ft/hr

**48.** $55t = 210$
$t = 3.82$ hrs

You will reach the top at approximately 7:49 P.M.

## Lesson 7.6

### Guided Practice (p.435)

**1.** False; a solution of a system of linear inequalities is an ordered pair that is a solution of each of the inequalities in the system.

**2.** You must graph the boundary lines, which are graphs of linear equations; you must determine whether the boundary lines are dashed or solid and which half-plane determined by the boundary line must be shaded.

**3.** Select a point in the shaded region and substitute it into each inequality. A solution will yield a true statement in each of the inequalities.

**Algebra 1**
Chapter 7 Worked-out Solution Key

# Chapter 7 *continued*

**4.** The graphs of $y > -1$ and $y > x - 4$ should have dotted lines, not solid; $y > 1$ is graphed instead of $y > -1$; the right side of $x = 2$ should be shaded; the left side of $y = x - 4$ should be shaded.

**5.**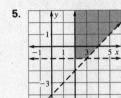

**6.**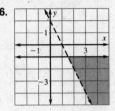

**7.**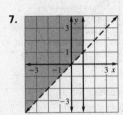

**8.**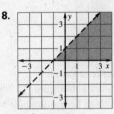

### 7.6 Practice and Applications (pp. 435–437)

**9.** C **10.** A **11.** B **12.** $y \le -\frac{5}{2}x + 4$
$y > -\frac{1}{2}x - 2$

**13.** $y \le \frac{1}{2}x + 2$ **14.** $y \le -\frac{1}{2}x + 2$
$y \ge \frac{1}{2}x - 2$ $\qquad y < \frac{1}{2}x + 2$

**15.**

$2x + y > 2$
$\qquad y > -2x + 2$

$6x + 3y < 12$
$\qquad 3y < -6x + 12$
$\qquad y < -2x + 4$

**16.**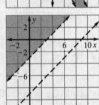

$2x - 2y < 6$
$\qquad -2y < -2x + 6$
$\qquad y > x - 3$

$x - y < 9$
$\qquad -y < -x + 9$
$\qquad y > x - 9$

**17.**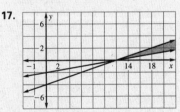

$x - 3y \ge 12$
$\qquad -3y \ge -x + 12$
$\qquad y \le \frac{1}{3}x - 4$

$x - 6 \le 12$
$\qquad -6y \le -x + 12$
$\qquad y \ge \frac{1}{6}x - 2$

**18.**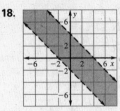

$x + y < 4$
$\qquad y < -x + 4$

$x + y > -2$
$\qquad y > -x - 2$

**19.**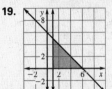

$x + y \le 6$
$\qquad y \le -x + 6$

**20.**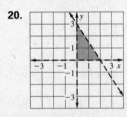

$\frac{3}{2}x + y < 3$
$\qquad y < -\frac{3}{2}x + 3$

**21.** 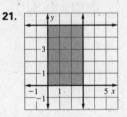 **22.**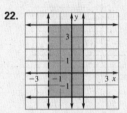

**23.**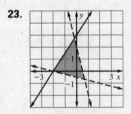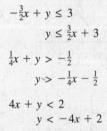

$-\frac{3}{2}x + y \le 3$
$\qquad y \le \frac{3}{2}x + 3$

$\frac{1}{4}x + y > -\frac{1}{2}$
$\qquad y > -\frac{1}{4}x - \frac{1}{2}$

$4x + y < 2$
$\qquad y < -4x + 2$

**24.**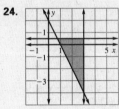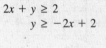

$2x + y \ge 2$
$\qquad y \ge -2x + 2$

**25.**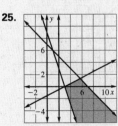

$x - 2y \ge 3$
$\qquad -2y \ge -x + 3$
$\qquad y \le \frac{1}{2}x - \frac{3}{2}$

$3x + y \ge 9$
$\qquad y \ge -3x + 9$

$x + y \le 7$
$\qquad y \le -x + 7$

# Chapter 7 *continued*

**26.**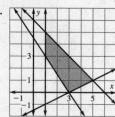

$x - 2y \le 3$
$-2y \le -x + 3$
$y \ge \frac{1}{2}x - \frac{3}{2}$

$3x + 2y \ge 9$
$2y \ge -3x + 9$
$y \ge -\frac{3}{2}x + \frac{9}{2}$

$x + y \le 6$
$y \le -x + 6$

**27.** $(0, 11), (0, -5), (6, 5)$   **28.** $(1, 4), (-2, -1), (3, -2)$

**29.** $(0, 0), (-2, 1), (2, 5)$

**30.**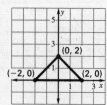

$y \ge 0$
$y \le -x + 2$
$y \le x + 2$

**31.**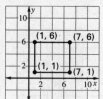

$y \le 6$
$y \ge 1$
$x \le 7$
$x \ge 1$

**32.**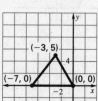

$y \ge 0$
$y \ge -\frac{5}{3}x$
$y \le \frac{5}{4}x + \frac{35}{4}$

**33.**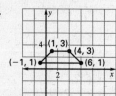

$y \le 3$
$y \ge 1$
$y \le -x + 7$
$y \le x + 2$

**34.** $x + y \ge 240$   $5x + 3y \le 1200$   $x \ge 0$
$y > x$   $y \ge 0$

**35.** 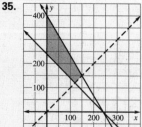   **36.** $x \ge 4$   $4x + y \le 9$
$y \ge 1$

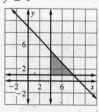

**37.**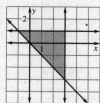

$A = \frac{1}{2}bh$
$A = \frac{1}{2}(4)(4)$
$A = 8$ square units

**38.** $y \ge 0$   $5x + 6y \ge 90$
$x \ge 0$   $x + y \le 20$

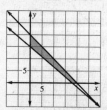

**39.** *Sample answers:*
5 hours babysitting, 15 hours at cashier job
10 hours babysitting, 10 hours at cashier job

**40.**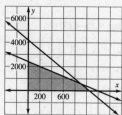

**41. a.** $(0, 2400), (0, 0), (1050, 0)$
$(900, 600)$

$-2x + 2400 = -4x + 4200$
$2x = 1800$
$x = 900$

$y = -2(900) + 2400$
$y = 600$

**b.** $P = 10x + 30y$   $P = 10(1050) + 30(0)$
$P = 10(0) + 30(2400)$   $P = 10,500$
$P = 72,000$

$P = 10(0) + 30(0)$   $P = 10(900) + 30(600)$
$P = 0$   $P = 27,000$

**c.** $\$72,000$

**42. a.** $x \ge 0$   $y \ge x + 4$
$y \le 8$

**b.** $y \ge 0$   $y \le -x + 4$
$x \ge 0$

**c.** $y \ge -x + 4$   $y \le -x + 12$
$y \le x + 4$   $y \ge x - 4$

**d.** 32 square units; the area of the large square is 64 square units and the area of each triangular region is 8 square units; $64 - 4(8) = 32$.

# Chapter 7 *continued*

**43.**

$$2x - y > 4$$
$$-y > -2x + 4$$
$$y < 2x - 4$$
$$y > 2x - 2$$

The half-planes determined by the inequalities do not intersect.

**44.**
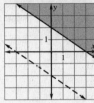

the line with equation $2x + 3y = 6$ and the half-plane above it

### 7.6 Mixed Review (p. 438)

**45.** $3^5 = 243$   **46.** $64 - 17 = 47$   **47.** $125 + 12 = 137$

**48.** $(27 - 20)^2 = 49$   **49.** $64 - 31 = 33$

**50.** $25 + 27 = 52$   **51.** $(5 + 2)^2 = 49$   **52.** $(2 - 1)^2 = 1$

**53.** $4 - 64 = -60$   **54.** $64 + 12 = 76$   **55.** $8 + 1 = 9$

**56.** $27 + 5 = 32$   **57.** 30, 30, 40, 40, 56, 56, 57, 57, 57
mean $= 47$
median $= 56$
mode $= 57$

**58.** $x + y = 68$
$$5x + 2y = 250$$
$$y = -x + 68$$

$$5x + 2(-x + 68) = 250 \qquad 38 + y = 68$$
$$5x - 2x + 136 = 250 \qquad\quad y = 30$$
$$3x = 114$$
$$x = 38$$

38 5-point questions

30 2-point questions

### Quiz 2 (p. 438)

**1.** $2l + 2w = 22$
$$5 + \tfrac{1}{2}l + w = 12$$
$$\tfrac{1}{2}l + w = 7$$
$$w = -\tfrac{1}{2}l + 7$$
$$2l + 2(-\tfrac{1}{2}l + 7) = 22$$
$$2l - l + 14 = 22$$
$$l = 8$$
$$2(8) + 2w = 22$$
$$2w = 6$$
$$w = 3$$
length $= 8$ ft
width $= 3$ ft

**2.** $5x - 2y = 0$
$$-2y = -5x$$
$$y = \tfrac{5}{2}x$$
$$5x - 2y = -4$$
$$-2y = -5x - 4$$
$$y = \tfrac{5}{2}x + 2$$

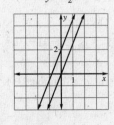

no solution

**3.** $3x + 2y = 12$
$$2y = -3x + 12$$
$$y = -\tfrac{3}{2}x + 6$$
$$9x + 6y = 18$$
$$6y = -9x + 18$$
$$y = -\tfrac{3}{2}x + 3$$

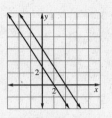

no solution

**4.** $-4x + 11y = 44$
$$11y = 4x + 44$$
$$y = \tfrac{4}{11}x + 4$$
$$4x - 11y = -44$$
$$-11y = -4x - 44$$
$$y = \tfrac{4}{11}x + 4$$

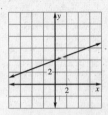

infinitely many solutions

**5.** $3x + y = 9$
$$y = -3x + 9$$
$$2x + y = 4$$
$$y = -2x + 4$$

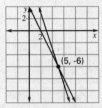

exactly one solution $(5, -6)$

**6.** $4x + 8y = 8$
$$8y = -4x + 8$$
$$y = -\tfrac{1}{2}x + 1$$
$$x + y = 1$$
$$y = -x + 1$$

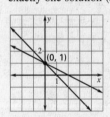

exactly one solution $(0, 1)$

**7.** $3x + 6y = 30$
$$6y = -3x + 30$$
$$y = -\tfrac{1}{2}x + 5$$
$$4x + 8y = 40$$
$$8y = -4x + 40$$
$$y = -\tfrac{1}{2}x + 5$$

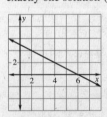

infinitely many solutions

**8.** $x - 2y < -6$
$$-2y < -x + 6$$
$$y > \tfrac{1}{2}x + 3$$
$$5x - 3y < -9$$
$$-3y < -5x - 9$$
$$y > \tfrac{5}{3}x + 3$$

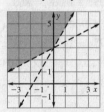

# Chapter 7 continued

**9.** $-3x + 2y < 6$

$\qquad 2y < 3x + 6$

$\qquad y < \frac{3}{2}x + 3$

$x - 4y > -2$

$\qquad -4y > -x - 2$

$\qquad y < \frac{1}{4}x + \frac{1}{2}$

$2x + y < 3$

$\qquad y < -2x + 3$

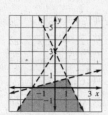

**10.** $x + y < 3$

$\qquad y < -x + 3$

$x \geq 0$

$y \geq 1$

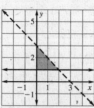

**11.** $x \geq 0$

$y \geq 1$

$x \leq 3$

$y \leq 5$

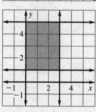

**12.** $x + y \leq \frac{1}{2}$

$\qquad y \leq -x + \frac{1}{2}$

$-x + y \leq \frac{1}{2}$

$\qquad y \leq x + \frac{1}{2}$

$y \geq 0$

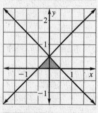

**13.** $\qquad x + y < 26$

$\qquad\qquad y < -x + 26$

$-x + 4y \geq 40$

$\qquad 4y \geq x + 40$

$\qquad y \geq \frac{1}{4}x + 10$

$2x > 4$

$\qquad x > 2$

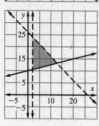

**14.** $x + y = 16,000$

$0.05x + 0.06y = 860$

$y = 16,000 - x$

$0.05x + 0.06(16,000 - x) = 860$

$\qquad 0.05x + 960 - 0.06x = 860$

$\qquad\qquad\qquad\qquad -0.01x = -100$

$\qquad\qquad\qquad\qquad\qquad x = 10,000$

$10,000 + y = 16,000$

$\qquad\qquad y = 6,000$

$\$10,000$ at 5%

$\$6,000$ at 6%

## Chapter 7 Review (pp. 440–442)

**1.**

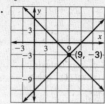

$x + y = 6$

$\qquad y = -x + 6$

$x - y = 12$

$\qquad -y = -x + 12$

$\qquad y = x - 12$

Check: $x + y = 6 \qquad x - y = 12$

$\qquad\quad 9 - 3 = 6 \qquad 9 - (-3) = 12$

Solution: $(9, -3)$

**2.**

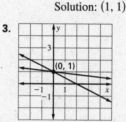

$4x - y = 3$

$\qquad -y = -4x + 3$

$\qquad y = 4x - 3$

$3x + y = 4$

$\qquad y = -3x + 4$

Check: $4x - y = 3 \qquad 3x + y = 4$

$\qquad 4(1) - 1 \overset{?}{=} 3 \qquad 3(1) + 1 \overset{?}{=} 4$

$\qquad\qquad\quad 3 = 3 \qquad\qquad\quad 4 = 4$

Solution: $(1, 1)$

**3.**

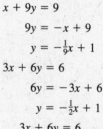

$x + 9y = 9$

$\qquad 9y = -x + 9$

$\qquad y = -\frac{1}{9}x + 1$

$3x + 6y = 6$

$\qquad 6y = -3x + 6$

$\qquad y = -\frac{1}{2}x + 1$

Check: $x + 9y = 9 \qquad 3x + 6y = 6$

$\qquad 0 + 9(1) \overset{?}{=} 9 \qquad 3(0) + 6(1) \overset{?}{=} 6$

$\qquad\qquad\quad 9 = 9 \qquad\qquad\quad 6 = 6$

Solution: $(0, 1)$

**4.**

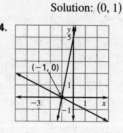

$5x - y = -5$

$\qquad -y = -5x - 5$

$\qquad y = 5x + 5$

$3x + 6y = -3$

$\qquad 6y = -3x - 3$

$\qquad y = -\frac{1}{2}x - \frac{1}{2}$

Check: $5x - y = -5 \qquad 3x + 6y = -3$

$\quad 5(-1) - 0 \overset{?}{=} -5 \quad 3(-1) + 6(0) \overset{?}{=} -3$

$\qquad\qquad -5 = -5 \qquad\qquad\quad -3 = -3$

Solution: $(-1, 0)$

# Chapter 7 *continued*

**5.**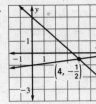

$$7x + 8y = 24$$
$$8y = -7x + 24$$
$$y = -\frac{7}{8}x + 3$$
$$x - 8y = 8$$
$$-8y = -x + 8$$
$$y = \frac{1}{8}x - 1$$

Check:
$$7x + 8y = 24 \qquad x - 8y = 8$$
$$7(4) + 8\left(-\tfrac{1}{2}\right) \overset{?}{=} 24 \qquad 4 - 8\left(-\tfrac{1}{2}\right) \overset{?}{=} 8$$
$$28 - 4 = 24 \qquad\quad 4 + 4 = 8$$

Solution: $\left(4, -\tfrac{1}{2}\right)$

**6.**

$$2x - 3y = -3$$
$$-3y = -2x - 3$$
$$y = \frac{2}{3}x + 1$$
$$x + 6y = -9$$
$$6y = -x - 9$$
$$y = -\frac{1}{6}x - \frac{3}{2}$$

Check:
$$2x - 3y = -3 \qquad x + 6y = -9$$
$$2(-3) - 3(-1) \overset{?}{=} -3 \quad -3 + 6(-1) \overset{?}{=} -9$$
$$-6 + 3 = -3 \qquad\quad -3 - 6 = -9$$

Solution: $(-3, -1)$

**7.**
$$x + 3y = 9$$
$$4x - 2y = -6$$
$$x = -3y + 9$$
$$4(-3y + 9) - 2y = -6$$
$$-12y + 36 - 2y = -6$$
$$-14y = -42$$
$$y = 3$$
$$x + 3(3) = 9$$
$$x = 0$$
Solution: $(0, 3)$

**8.**
$$-2x - 5y = 7$$
$$7x + y = -8$$
$$y = -7x - 8$$
$$-2x - 5(-7x - 8) = 7$$
$$-2x + 35x + 40 = 7$$
$$33x = -33$$
$$x = -1$$
$$7(-1) + y = -8$$
$$y = -1$$
Solution: $(-1, -1)$

**9.**
$$4x - 3y = -2$$
$$4x + y = 4$$
$$y = -4x + 4$$
$$4x - 3(-4x + 4) = -2$$
$$4x + 12x - 12 = -2$$
$$16x = 10$$
$$x = \frac{5}{8}$$
$$4\left(\tfrac{5}{8}\right) + y = 4$$
$$y = \frac{3}{2}$$
Solution: $\left(\tfrac{5}{8}, \tfrac{3}{2}\right)$

**10.**
$$-x + 3y = 24$$
$$5x + 8y = -5$$
$$-x = -3y + 24$$
$$x = 3y - 24$$
$$5(3y - 24) + 8y = -5$$
$$15y - 120 + 8y = -5$$
$$23y = 115$$
$$y = 5$$
$$-x + 3(5) = 24$$
$$-x = 9$$
$$x = -9$$
Solution: $(-9, 5)$

**11.**
$$4x + 9y = 2$$
$$2x + 6y = 1$$
$$2x = -6y + 1$$
$$x = -3y + \frac{1}{2}$$
$$4\left(-3y + \tfrac{1}{2}\right) + 9y = 2$$
$$-12y + 2 + 9y = 2$$
$$-3y = 0$$
$$y = 0$$
$$2x + 6(0) = 1$$
$$2x = 1$$
$$x = \frac{1}{2}$$
Solution: $\left(\tfrac{1}{2}, 0\right)$

**12.**
$$9x + 6y = 3$$
$$3x - 7y = -26$$
$$3x = 7y - 26$$
$$x = \frac{7}{3}y - \frac{26}{3}$$
$$9\left(\tfrac{7}{3}y - \tfrac{26}{3}\right) + 6y = 3$$
$$21y - 78 + 6y = 3$$
$$27y = 81$$
$$y = 3$$
$$9x + 6(3) = 3$$
$$9x = -15$$
$$x = -\frac{5}{3}$$
Solution: $\left(-\tfrac{5}{3}, 3\right)$

**13.**
$$-4x - 5y = 7$$
$$\underline{x + 5y = 8}$$
$$-3x \qquad = 15$$
$$x = -5$$
$$-5 + 5y = 8$$
$$5y = 13$$
$$y = \frac{13}{5}$$
Solution: $\left(-5, \tfrac{13}{5}\right)$

**14.** $2x + y = 0$ Multiply by $4 \rightarrow$
$$5x - 4y = 26$$

$$8x + 4y = 0$$
$$\underline{5x - 4y = 26}$$
$$13x \qquad = 26$$
$$x = 2$$
$$2(2) + y = 0$$
$$y = -4$$
Solution: $(2, -4)$

**15.**
$$3x + 5y = -16 \text{ Multiply by } 2 \rightarrow 6x + 10y = -32$$
$$-2x + 6y = -36 \text{ Multiply by } 3 \rightarrow \underline{-6x + 18y = -108}$$
$$28y = -140$$
$$y = -5$$
$$-3x + 5(-5) = -16$$
$$3x = 9$$

Solution: $(3, -5)$ $\quad x = 3$

**16.**
$$9x + 6y = 3$$
$$9x + 6y = 3$$
$$3y + 6x = 18 \text{ Multiply by } -2 \rightarrow \underline{-12x - 6y = -36}$$
$$-3x = -33$$
$$x = 11$$
$$9(11) + 6y = 3$$
$$6y = -96$$

Solution: $(11, -16)$ $\quad y = -16$

**17.**
$$2 - 7x = 9y \quad \text{Rearrange} \rightarrow -7x - 9y = -2$$
$$2y - 4x = 6 \quad \text{Rearrange} \rightarrow -4x + 2y = 6$$
$$\text{Multiply by } -4 \rightarrow 28x + 36y = 8$$
$$\text{Multiply by } 7 \rightarrow \underline{-28x + 14y = 42}$$
$$50y = 50$$
$$y = 1$$
$$-4x + 2(1) = 6$$
$$-4x = 4$$

Solution: $(-1, 1)$ $\quad x = -1$

**18.**
$$4x - 9y = 1 \text{ Multiply by } 5 \rightarrow 20x - 45y = 5$$
$$-5x + 6y = 4 \text{ Multiply by } 4 \rightarrow \underline{-20x + 24y = 16}$$
$$-21y = 21$$
$$y = -1$$
$$4x - 9(-1) = 1$$
$$4x = -8$$

Solution: $(-2, -1)$ $\quad x = -2$

**19.**
$$x + y = 12$$
$$3x + 5y = 50$$
$$x = -y + 12$$
$$3(-y + 12) + 5y = 50$$
$$-3y + 36 + 5y = 50$$
$$2y = 14$$
$$y = 7$$
$$x + 7 = 12$$
$$x = 5$$

Ferris wheel: 5 times

roller coaster: 7 times

**20.**
$$x + y = 5$$
$$3x + 2y = 13$$
$$x = -y + 5$$
$$3(-y + 5) + 2y = 13$$
$$-3y + 15 + 2y = 13$$
$$-y = -2$$
$$y = 2$$
$$x + 2 = 5$$
$$x = 3$$

regular movies: 2

new releases: 3

**21.**
$$\tfrac{1}{3}x + y = 2 \qquad\qquad 2x + 6\left(-\tfrac{1}{3}x + 2\right) = 12$$
$$2x + 6y = 12 \qquad\qquad 2x - 2x + 12 = 12$$
$$y = -\tfrac{1}{3}x + 2 \qquad\qquad\quad 0 = 0$$

infinitely many solutions

**22.**
$$2x - 3y = 1$$
$$\underline{-2x + 3y = 1}$$
$$0 \neq 2$$

no solution

**23.**
$$-6x + 5y = 18 \text{ Multiply by } 7 \rightarrow -42x + 35y = 126$$
$$7x + 2y = 26 \text{ Multiply by } 6 \rightarrow \underline{42x + 12y = 156}$$
$$47y = 182$$
$$y = 6$$
$$-6x + 5(6) = 18$$
$$-6x = -12$$

Exactly one solution: $(2, 6)$ $\quad x = 2$

**24.**
$$10x + 4y = 25 \text{ Multiply by } -2 \rightarrow -20x - 8y = -50$$
$$5x + 8y = 11 \qquad\qquad\qquad\qquad \underline{5x + 8y = \quad 11}$$
$$-15x = -39$$
$$x = \tfrac{39}{15}$$
$$10\left(\tfrac{39}{15}\right) + 4y = 25$$
$$4y = -1$$

Solution: $\left(2\tfrac{3}{5}, -\tfrac{1}{4}\right)$ $\quad y = -\tfrac{1}{4}$

**25.**
$$14x + 7y = 0 \qquad\qquad\quad 14x + 7y = 0$$
$$-2x + y = 13 \text{ Multiply by } -7 \rightarrow \underline{-14x + 7y = 91}$$
$$14y = 91$$
$$y = 6\tfrac{1}{2}$$
$$-2x + 6\tfrac{1}{2} = 13$$
$$-2x = 6\tfrac{1}{2}$$

Exactly one solution: $\left(-3\tfrac{1}{4}, 6\tfrac{1}{2}\right)$ $\quad x = -3\tfrac{1}{4}$

**26.**
$$21x + 28y = 14 \text{ Multiply by } -\tfrac{3}{7} \rightarrow -9x - 12y = -6$$
$$9x + 12y = 6 \qquad\qquad\qquad\qquad\quad 9x + 12y = 6$$

infinitely many solutions $\qquad\qquad 0 = 0$

**27.**
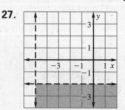

# Chapter 7 *continued*

**28.**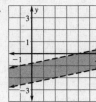

$$2x - 10y > 8$$
$$-10y > -2x + 8$$
$$y < \tfrac{1}{5}x - \tfrac{4}{5}$$
$$x - 5y < 12$$
$$-5y < -x + 12$$
$$y > \tfrac{1}{5}x - \tfrac{12}{5}$$

**29.**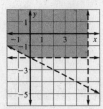

$$-x + 3y \le 15$$
$$3y \le x + 15$$
$$y \le \tfrac{1}{3}x + 5$$
$$9x \ge 27$$
$$x \ge 3$$

**30.**

$$x + 2y > -4$$
$$2y > -x - 4$$
$$y > -\tfrac{1}{2}x - 2$$

**31.**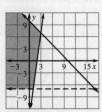

$$x + y < 8$$
$$y < -x + 8$$
$$x - y < 0$$
$$-y < -x$$
$$y > x$$

**32.** 

$$7y > -49$$
$$y > -7$$
$$-7x + y \ge -14$$
$$y \ge 7x - 14$$
$$x + y \le 10$$
$$y \le -x + 10$$

## Chapter 7 Test *(p. 443)*

**1.** $y = 2x - 3$ $\qquad$ $-y = 2x - 1$
$$y = -2x + 1$$

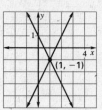

Check: $\quad y = 2x - 3 \qquad\qquad -y = 2x - 1$
$$-1 \overset{?}{=} 2(1) - 3 \qquad -(-1) \overset{?}{=} 2(1) - 1$$
$$-1 = -1 \qquad\qquad\quad 1 = 1$$
Solution: $(1, -1)$

**2.** $6x + 2y = 16 \qquad\qquad -2x + y = -2$
$$2y = -6x + 16 \qquad\qquad y = 2x - 2$$
$$y = -3x + 8$$

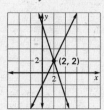

Check: $\quad 6x + 2y = 16 \qquad -2x + y = -2$
$$6(2) + 2(2) \overset{?}{=} 16 \qquad -2(2) + 2 \overset{?}{=} -2$$
$$12 + 4 = 16 \qquad\qquad -2 = -2$$
Solution: $(2, 2)$

**3.** $4x - y = 10 \qquad\qquad -2x + 4y = 16$
$$-y = -4x + 10 \qquad\qquad 4y = 2x + 16$$
$$y = 4x - 10 \qquad\qquad y = \tfrac{1}{2}x + 4$$

Check: $\quad 4x - y = 10 \qquad -2x + 4y = 16$
$$4(4) - 6 \overset{?}{=} 10 \qquad -2(4) + 4(6) = 16$$
$$16 - 6 = 10 \qquad\qquad -8 + 24 = 16$$
Solution: $(4, 6)$

# Chapter 7 *continued*

**4.** $-4x + y = -10$   $\qquad$ $6x + 2y = 22$

$\qquad$ $y = 4x - 10$ $\qquad\qquad$ $2y = -6x + 22$

$\qquad\qquad\qquad\qquad\qquad\qquad\qquad$ $y = -3x + 11$

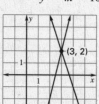

Check: $\qquad$ $6x + 2y = 22$ $\qquad\qquad$ $-4x + y = -10$

$\qquad\qquad$ $6(3) + 2(2) \overset{?}{=} 22$ $\qquad$ $-4(3) + 2 \overset{?}{=} -10$

$\qquad\qquad\qquad$ $18 + 4 = 22$ $\qquad\qquad$ $-12 + 2 = -10$

Solution: $(3, 2)$

**5.** $3x - 5y = -10$ $\qquad\qquad$ $-x + 2y = 18$

$\qquad\qquad$ $5y = -3x - 10$ $\qquad\qquad$ $2y = x + 18$

$\qquad\qquad$ $y = -\frac{3}{5}x - 2$ $\qquad\qquad$ $y = \frac{1}{2}x + 9$

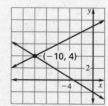

Check: $\qquad$ $-x + 2y = 18$ $\qquad\qquad$ $3x + 5y = -10$

$\qquad$ $-(-10) + 2(4) \overset{?}{=} 18$ $\qquad$ $3(-10) - 5(4) \overset{?}{=} -10$

$\qquad\qquad\qquad$ $10 + 8 = 18$ $\qquad\qquad$ $-30 + 20 = -10$

Solution: $(-10, 4)$

**6.** $2x - 3y = 12$ $\qquad\qquad$ $-x - 3y = -6$

$\qquad\qquad$ $-3y = -2x + 12$ $\qquad\qquad$ $-3y = x - 6$

$\qquad\qquad$ $y = \frac{2}{3}x - 4$ $\qquad\qquad$ $y = -\frac{1}{3}x + 2$

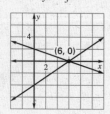

Check: $\qquad$ $2x - 3y = 12$ $\qquad\qquad$ $-x - 3y = -6$

$\qquad\qquad$ $2(6) - 3(0) \overset{?}{=} 12$ $\qquad\qquad$ $-6 - 3(0) \overset{?}{=} -6$

$\qquad\qquad\qquad$ $12 = 12$ $\qquad\qquad\qquad$ $-6 = -6$

Solution: $(6, 0)$

**7.** $-4x + 7y = -2$ $\qquad\qquad$ $-(-3) - y = 5$

$\qquad$ $-x - y = 5$ $\qquad\qquad\qquad\qquad$ $-y = 2$

$\qquad$ $-y = x + 5$ $\qquad\qquad\qquad\qquad$ $y = -2$

$\qquad\qquad$ $y = -x - 5$ $\qquad\qquad$ Solution: $(-3, -2)$

$\qquad$ $-4x + 7(-x - 5) = -2$

$\qquad\qquad$ $-4x - 7x - 35 = -2$

$\qquad\qquad\qquad\qquad$ $-11x = 33$

$\qquad\qquad\qquad\qquad\qquad$ $x = -3$

**8.** $7x + 4y = 5$ $\qquad\qquad$ $x - 6(3) = -19$

$\qquad$ $x - 6y = -19$ $\qquad\qquad\qquad$ $x = -1$

$\qquad\qquad$ $x = 6y - 19$

$\qquad$ $7(6y - 19) + 4y = 5$

$\qquad\qquad$ $42y - 133 + 4y = 5$

$\qquad\qquad\qquad\qquad$ $46y = 138$

$\qquad\qquad\qquad\qquad\qquad$ $y = 3$

Solution: $(-1, 3)$

**9.** $-3x + 6y = 24$ $\qquad\qquad$ $-2(-2) - y = 1$

$\qquad$ $-2x - y = 1$ $\qquad\qquad\qquad\qquad$ $-y = -3$

$\qquad\qquad$ $-y = 2x + 1$ $\qquad\qquad\qquad\qquad$ $y = 3$

$\qquad\qquad$ $y = -2x - 1$

$\qquad$ $-3x + 6(-2x - 1) = 24$

$\qquad\qquad$ $-3x - 12x - 6 = 24$

$\qquad\qquad\qquad\qquad$ $-15x = 30$

$\qquad\qquad\qquad\qquad\qquad$ $x = -2$

Solution: $(-2, 3)$

**10.** $5x - y = 7$ $\qquad\qquad$ $5(1) - y = 7$

$\qquad$ $4x + 8y = -12$ $\qquad\qquad\qquad$ $-y = 2$

$\qquad\qquad$ $-y = -5x + 7$ $\qquad\qquad\qquad$ $y = -2$

$\qquad$ $4x + 8(5x - 7) = -12$

$\qquad\qquad$ $4x + 40x - 56 = -12$

$\qquad\qquad\qquad\qquad$ $44x = 44$

$\qquad\qquad\qquad\qquad\qquad$ $x = 1$

Solution: $(1, -2)$

**11.** $x + 6y = 9$ $\qquad\qquad$ $x + 6(2) = 9$

$\qquad$ $-x + 4y = 11$ $\qquad\qquad\qquad$ $x = -3$

$\qquad\qquad$ $x = -6y + 9$

$\qquad$ $-(-6y + 9) + 4y = 11$

$\qquad\qquad$ $6y - a + 4y = 11$

$\qquad\qquad\qquad\qquad$ $10y = 20$

$\qquad\qquad\qquad\qquad\qquad$ $y = 2$

Solution: $(-3, 2)$

**Algebra 1**
Chapter 7 Worked-out Solution Key

# Chapter 7 *continued*

**12.** $8x + 3y = 0$

$-x - 9y = 92$

$-x = 9y + 92$

$x = -9y - 92$

$8(-9y - 92) + 3y = 0$

$-72y - 736 + 3y = 0$

$-69y = 736$

$y = -10\frac{2}{3}$

$8x + 3\left(-\frac{32}{3}\right) = 0$

$8x = 32$

$x = 4$

Solution: $\left(4, -10\frac{2}{3}\right)$

**13.** $6x + 7y = 5$    Multiply by $2 \rightarrow 12x + 14y = 10$

$4x - 2y = -10$ Multiply by $7 \rightarrow \underline{28x - 14y = -70}$

$40x = -60$

$x = -\frac{3}{2}$

$6\left(-\frac{3}{2}\right) + 7y = 5$

$7y = 14$

Solution: $\left(-\frac{3}{2}, 2\right)$        $y = 2$

**14.** $-7x + 2y = -5$

$\underline{10x - 2y = 6}$

$3x = 1$

$x = \frac{1}{3}$

$-7\left(\frac{1}{3}\right) + 2y = -5$

$2y = -\frac{8}{3}$

$y = -\frac{4}{3}$

Solution: $\left(\frac{1}{3}, -\frac{4}{3}\right)$

**15.** $-3x + 3y = 12$ Multiply by $4 \rightarrow -12x + 12y = 48$

$4x + 2y = 20$    Multiply by $3 \rightarrow \underline{12x + 6y = 60}$

$18y = 108$

$y = 6$

$-3x + 3(6) = 12$

$-3x = -6$

Solution: $(2, 6)$        $x = 2$

**16.** $3x + 4y = 9$          $3x + 4y = 9$

$4y - 3x = -1$ Rearrange $\rightarrow$ $\underline{-3x + 4y = -1}$

$8y = 8$

$y = 1$

$3x + 4(1) = 9$

$3x = 5$

Solution: $\left(\frac{5}{3}, 1\right)$      $x = \frac{5}{3}$

**17.** $8x - 2 + y = 0$ Rearrange $\rightarrow$ $8x + y = 2$

$9x - y = 219$          $\underline{9x - y = 219}$

$17x = 221$

$x = 13$

$8(13) + y = 2$

Solution: $(13, -102)$     $y = 102$

**18.** $5y - 3x = 1$   Multiply by $2 \rightarrow 10y - 6x = 2$

$4y + 2x = 80$ Multiply by $3 \rightarrow \underline{12y + 6x = 240}$

$22y = 242$

$y = 11$

$5(11) - 3x = 11$

$-3x = -54$

Solution: $(18, 11)$      $x = 18$

**19.** $8x + 4y = -4$          $8x + 4y = -4$

$2x - y = -3$ Multiply by $4 \rightarrow$ $\underline{8x - 4y = -12}$

$16x = -16$

$x = -1$

$8(-1) + 4y = -4$

$4y = 4$

Exactly one solution: $(-1, 1)$     $y = 1$

**20.** $-6x + 3y = -6$         $-6x + 3y = -6$

$2x + 6y = 30$ Multiply by $3 \rightarrow$ $\underline{6x + 18y = 90}$

$21y = 84$

$y = 4$

$-6x + 3(4) = -6$

$-6x = -18$

Exactly one solution: $(3, 4)$     $x = 3$

**21.** $-x + \frac{1}{3}y = -6$      $3\left(\frac{1}{3}y + 6\right) - y = -16$

$3x - y = -16$       $y + 18 - y = -16$

$-x = -\frac{1}{3}y - 6$         $18 \neq -16$

$x = \frac{1}{3}y + 6$       no solution

**22.** $3x + y = 8$ Multiply by $-6 \rightarrow -18x - 6y = -48$

$4x + 6y = 6$            $\underline{4x + 6y = 6}$

$-14x = -42$

$x = 3$

$3(3) + y = 8$

Exactly one solution: $(3, -1)$     $y = -1$

# Chapter 7 *continued*

**23.** $3x - 4y = 8$ Multiply by $-3 \to -9x + 12y = -24$

$\frac{9}{2}x - 6y = 12$ Multiply by $2 \to$ $9x - 12y = 24$

infinitely many solutions $\qquad\qquad 0 = 0$

**24.** $6x + y = 12$ Multiply by $2 \to$ $\quad 12x + 2y = 24$

$\quad -4x - 2y = 0 \qquad\qquad \underline{-4x - 2y = \;\;0}$

$\qquad\qquad\qquad\qquad\qquad\qquad 8x \qquad\quad = 24$

$\qquad\qquad\qquad\qquad\qquad\qquad\quad x = \;\;3$

$\qquad\qquad\qquad\qquad\qquad\quad 6(3) + y = 12$

Exactly one solution: $(3, -6)$ $\qquad y = -6$

**25.**   **26.**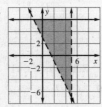

**27.**

**28.** $x \geq 0$

$y \geq 0$

$y \leq \frac{2}{3}x + 4$

$x \leq 6$

**29.** $y \leq 9$ $\qquad\qquad$ **30.** $x \geq 0$ $\qquad y \leq 2x + 5$

$x \leq -4$ $\qquad\qquad\qquad\quad y \geq 0$ $\qquad y \geq 2x - 1$

$y > -2x - 7$

**31.** $x + y = 6$

$4x + 4.45y = 25.80$

$x = -y + 6$

$4(-y + 6) + 4.45y = 25.80$

$-4y + 24 + 4.45y = 25.80$

$\qquad\qquad\quad 0.45y = 1.80$

$\qquad\qquad\qquad\quad y = 4$

$x + 4 = 6$

$\quad x = 2$

2 bags of oyster shell grit

4 bags of sunflower seeds

## Chapter 7 Standardized Test *(pp. 444–445)*

**1.** D

**2.** C $\frac{4}{5}x + \frac{1}{2}y = 16$ $\qquad$ **3.** C $\quad x + y = 15$

$\qquad x + y = 24$ $\qquad\qquad\qquad\quad x - y = -19$

$\qquad x = -y + 24$ $\qquad\qquad\qquad x = -y + 15$

$\frac{4}{5}(-y + 24) + \frac{1}{2}y = 16$ $\qquad -y + 15 - y = -19$

$-\frac{4}{5} + \frac{96}{5} + \frac{1}{2}y = 16$ $\qquad\qquad -2y = -34$

$\qquad\qquad -\frac{3}{10}y = -\frac{16}{5}$ $\qquad\qquad\qquad y = 17$

$\qquad\qquad\qquad y = \frac{32}{3}$ $\qquad\quad x + 17 = 15$

$\qquad\qquad\qquad\qquad\qquad\qquad\qquad x = -2$

**4.** A $-2x + 7y = -8$ $\qquad -2x + 7\left(-\frac{12}{5}\right) = -8$

$\qquad x - 6y = 10$ $\qquad\qquad -2x - \frac{84}{5} = -8$

$\qquad x = 6y + 10$ $\qquad\qquad\qquad -2x = \frac{44}{5}$

$-2(6y + 10) + 7y = -8$ $\qquad\qquad\qquad x = -\frac{22}{5}$

$-12y - 20 + 7y = -8$

$\qquad\qquad -5y = 12$ $\qquad -\frac{22}{5} + \frac{12}{5} = -2$

$\qquad\qquad\quad y = -\frac{12}{5}$

**5.** E $-\frac{3}{4}x + \frac{7}{10}y = 5$ $\qquad\qquad -\frac{3}{4}x + \frac{7}{10}y = 5$

$\frac{1}{4}x - \frac{3}{10}y = -5$ Multiply by $3 \to \frac{3}{4}x - \frac{9}{10}y = -15$

$\qquad\qquad\qquad\qquad\qquad\qquad\qquad -\frac{1}{5}y = -10$

$\qquad\qquad\qquad\qquad\qquad\qquad\qquad\quad y = 50$

$\qquad\qquad\qquad\qquad -\frac{3}{4}x + \frac{7}{10}(50) = 5$

$\qquad\qquad\qquad\qquad\quad -\frac{3}{4}x + 35 = 5$

$\qquad\qquad\qquad\qquad\qquad\qquad\quad x = 40$

$\qquad\qquad\qquad x + y = 40 + 50 = 90$

**6.** E $-3(5x - 2) = -15x + 6$

**7.** B $\quad x + y = 42$

$\quad 5x + 2y = 150$

$\qquad\quad x = -y + 42$

$5(-y + 42) + 2y = 150$

$-5y + 210 + 2y = 150$

$\qquad\qquad -3y = -60$

$\qquad\qquad\quad y = 20$

**8.** B $4x - 2y = 6$ $\qquad\qquad\qquad 4x - 2y = \;\;6$

$\quad 7x + y = 15$ Multiply by $2 \to \underline{14x + 2y = 30}$

$\qquad\qquad\qquad\qquad\qquad\qquad 18x \qquad\quad = 36$

$\qquad\qquad\qquad\qquad\qquad\qquad\quad x = \;\;2$

$\qquad\qquad\qquad\qquad\quad 4(2) - 2y = 6$

$\qquad\qquad\qquad\qquad\qquad -2y = -2$

$\qquad\qquad\qquad\qquad\qquad\quad y = 1$

**9.** B $\qquad$ **10.** A $\qquad$ **11.** D $\qquad$ **12.** C

# Chapter 7 *continued*

**13.** E $\quad 4x - y = 10$

$\qquad x + y = 5$

$\qquad x = -y + 5$

$\qquad 4(-y + 5) - y = 10$

$\qquad -4y + 20 - y = 10$

$\qquad\qquad\qquad -5y = -10$

$\qquad\qquad\qquad\quad y = 2$

$\qquad x + 2 = 5$

$\qquad\quad x = 3$

**14.** B

**15.** A $\quad 2x + 6y = 13 \qquad\qquad 2x + 6\left(\frac{1}{2}\right) = 13$

$\qquad\quad x - 4y = 3 \qquad\qquad\qquad\quad 2x = 10$

$\qquad\qquad x = 4y + 3 \qquad\qquad\qquad\quad x = 5$

$\qquad 2(4y + 3) + 6y = 13$

$\qquad\quad 8y + 6 + 6y = 13$

$\qquad\qquad\qquad 14y = 7$

$\qquad\qquad\qquad\quad y = \frac{1}{2}$

**16. a.** $7.5x + 4.5y = 3150$

**b.** $x + y = 500$

**c.** $7.5x + 4.5y = 3150$

$\qquad x + y = 500$

$\qquad x = -y + 500$

$\qquad 7.5(-y + 500) + 4.5y = 3150$

$\qquad -7.5y + 3750 + 4.5y = 3150$

$\qquad\qquad\qquad\qquad -3y = -600$

$\qquad\qquad\qquad\qquad\quad y = 200$

$\quad x + 200 = 500$

$\qquad\quad x = 300$

300 adult tickets

200 student tickets

**d.** $x + y = 400 \qquad 3x = y$

$\qquad 4x = 400$

$\qquad\quad x = 100$

total raised $= 2700$

$3150 - 2700 = \$450$ below

**e.** $7.5x \geq 3150$

$\qquad x \geq 420$

At least 420 tickets; possible numbers are solutions of the inequality $7.5x \geq 3150$

# CHAPTER 8

## Think & Discuss (p. 447)

**1.**
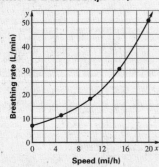

**2.** For each increase of 5 miles per hour, the racer's breathing rate increases by an increasing amount: 4.3 L/min, 7.4 L/min, and so on. The amounts of increase are not the same, but the percents are approximately equal (about 70%).

## Skill Review (p. 448)

**1.** $6^2$     **2.** $4^3$     **3.** $(2y)^5$     **4.** $\frac{20}{80} = \frac{1}{4}$

**5.** $\frac{5}{64} = \frac{1}{9}$     **6.** $\frac{36}{48} = \frac{3}{4}$     **7.** $\frac{25}{50} = \frac{1}{2}$

**8.** $8.25/hr     **9.** $.79/cantaloupe

**10.** $.50/can     **11.** 45 miles/gallon

## Lesson 8.1

### Developing Concepts Activity (p. 449)
### Exploring the concept

**1.**

| Product of Powers | Expanded Product |
|---|---|
| $7^3 \cdot 7^2$ | $(7 \cdot 7 \cdot 7) \cdot (7 \cdot 7)$ |
| $2^4 \cdot 2^4$ | $(2 \cdot 2 \cdot 2 \cdot 2) \cdot (2 \cdot 2 \cdot 2 \cdot 2)$ |
| $x^4 \cdot x^5$ | $(x \cdot x \cdot x \cdot x) \cdot (x \cdot x \cdot x \cdot x \cdot x)$ |

| Product of Powers | No. of factors | Product as a power |
|---|---|---|
| $7^3 \cdot 7^2$ | 5 | $7^5$ |
| $2^4 \cdot 2^4$ | 8 | $2^8$ |
| $x^4 \cdot x^5$ | 9 | $x^9$ |

**2.**

| Sum of Exponents |
|---|
| 5 |
| 8 |
| 9 |

The sum of the exponents is the same as the number of factors.

**3.**

| Power of a Power | Expanded Product |
|---|---|
| $(5^2)^3$ | $(5^2) \cdot (5^2) \cdot (5^2)$ |
| $[(-3)^2]^2$ | $[(-3)^2] \cdot [(-3)^2]$ |
| $(b^2)^4$ | $(b^2) \cdot (b^2) \cdot (b^2) \cdot (b^2)$ |

| Power of a Power | Expanded Product |
|---|---|
| $(5^2)^3$ | $(5 \cdot 5) \cdot (5 \cdot 5) \cdot (5 \cdot 5)$ |
| $[(-3)^2]^2$ | $[(-3) \cdot (-3)] \cdot [(-3) \cdot (-3)]$ |
| $(b^2)^4$ | $(b \cdot b) \cdot (b \cdot b) \cdot (b \cdot b) \cdot (b \cdot b)$ |

| Power of a Power | No. of factors | Product as a power |
|---|---|---|
| $(5^2)^3$ | 6 | $5^6$ |
| $[(-3)^2]^2$ | 4 | $(-3)^4$ |
| $(b^2)^4$ | 8 | $b^8$ |

**4.**

| Product of Exponents |
|---|
| 6 |
| 4 |
| 8 |

The product of the exponents is the same as the number of factors.

### Drawing Conclusions

**1.** $6^3 \cdot 6^2 = (6 \cdot 6 \cdot 6) \cdot (6 \cdot 6)$
$= 6^5$

**2.** $(-2) \cdot (-2)^4 = (-2) \cdot (-2 \cdot -2 \cdot -2 \cdot -2)$
$= (-2)^5$

**3.** $p^4 \cdot p^6 = (p \cdot p \cdot p \cdot p) \cdot (p \cdot p \cdot p \cdot p \cdot p \cdot p)$
$= p^{10}$

**4.** $x^{12} \cdot x^7 = (x \cdot x \cdot x \cdot x \cdot x \cdot x \cdot x \cdot x \cdot x \cdot x \cdot x \cdot x) \cdot$
$(x \cdot x \cdot x \cdot x \cdot x \cdot x \cdot x)$
$= x^{19}$

**5.** $(4^2)^6 = (4^2) \cdot (4^2) \cdot (4^2) \cdot (4^2) \cdot (4^2) \cdot (4^2)$
$= 4 \cdot 4 \cdot 4 \cdot 4 \cdot 4 \cdot 4 \cdot 4 \cdot 4 \cdot 4 \cdot 4 \cdot 4 \cdot 4$
$= 4^{12}$

**6.** $[(-5)^2]^4 = (-5)^2 \cdot (-5)^2 \cdot (-5)^2 \cdot (-5)^2$
$= (-5 \cdot -5) \cdot (-5 \cdot -5) \cdot (-5 \cdot -5) \cdot (-5 \cdot -5)$
$= (-5)^8$

**7.** $(d^5)^5 = d^5 \cdot d^5 \cdot d^5 \cdot d^5 \cdot d^5$
$= d \cdot d \cdot d \cdot d \cdot d \cdot d \cdot d \cdot d \cdot d \cdot d \cdot d \cdot d \cdot d \cdot d \cdot$
$d \cdot d \cdot d \cdot d \cdot d \cdot d \cdot d \cdot d \cdot d \cdot d \cdot d$
$= d^{25}$

**8.** $[(-n)^3]^8 = (-n)^3 \cdot (-n)^3 \cdot (-n)^3 \cdot (-n)^3 \cdot$
$(-n)^3 \cdot (-n)^3 \cdot (-n)^3 \cdot (-n)^3$
$= -n \cdot -n \cdot -n \cdot -n \cdot -n \cdot -n \cdot -n \cdot -n \cdot$
$-n \cdot -n \cdot -n \cdot -n \cdot -n \cdot -n \cdot -n \cdot -n \cdot$
$-n \cdot -n \cdot -n \cdot -n \cdot -n \cdot -n \cdot -n \cdot -n$
$= (-n)^{24}$

**9.** Addition; *Sample answer:* $6^2 \cdot 6^3 = 6^5$, $2^3 \cdot 2^3 = 2^6$

# Chapter 8 continued

**10.** Multiplication; *Sample Answer:*
$(3^2)^3 = 3^6, [(-2)^2]^4 = (-2)^8$

**11.** No; $x^3 \cdot y^5 = x \cdot x \cdot x \cdot y \cdot y \cdot y \cdot y \cdot y$,
while $xy^8 = x \cdot y \cdot y \cdot y \cdot y \cdot y \cdot y \cdot y \cdot y$

## 8.1 Guided Practice (p. 453)

**1.** base

**2.** $x^7 \cdot x^3 = x \cdot x \cdot x \cdot x \cdot x \cdot x \cdot x \cdot x \cdot x \cdot x = x^{10}$
$(x^7)^3 = x \cdot x \cdot x \cdot x \cdot x \cdot x \cdot x \cdot x \cdot x \cdot x \cdot x \cdot x \cdot x \cdot x \cdot x \cdot x \cdot x \cdot x \cdot x \cdot x \cdot x = x^{21}$

The first expression uses the product of powers property. The second uses the power of a power property.

**3.** No; the product of powers property can only be used if the bases of the powers are the same.

**4.** $c^3$   **5.** $m^3$   **6.** $2^5$, or 32   **7.** $3^7$, or 2187   **8.** $a^{10}$

**9.** $x^9$   **10.** $3^2$, or 9   **11.** $(-2)^2$, or 4

**12.** $2^{12}$, or 4096   **13.** $4^9$, or 262,144   **14.** $y^{20}$

**15.** $m^{32}$   **16.** $8m^6$   **17.** $a^2b^4$   **18.** $25x^2$   **19.** $x^{12}y^{20}$

**20.** $x^{15}y^{40}$   **21.** $(-2)^3x^9$, or $-8x^9$

## 8.1 Practice and Applications (pp. 453–455)

**22.** $3^{10}$, or 59,049   **23.** $5^{11}$, or 48,828,125

**24.** $2^6$, or 64   **25.** $7^8$, or 5,764,801   **26.** $x^7$

**27.** $3^2 \cdot 7^2$, or 441   **28.** $2^2x^2$, or $4x^2$

**29.** $(-5)^3a^3$, or $-125a^3$   **30.** $(-2)^2m^8n^{12}$, or $4m^8n^{12}$

**31.** $(-4)^6$, or 4096   **32.** $(-5)^{10}x^{10}y^{10}$, or $9,765,625x^{10}y^{10}$

**33.** $(5 + x)^{18}$   **34.** $(2x + 3)^6$

**35.** $3^3b^3 \cdot b = 3^3b^4$, or $27b^4$

**36.** $5^3 \cdot 5^2 \cdot a^8 = 5^5a^8$, or $3125a^8$   **37.** $4x \cdot x^2 \cdot x^6 = 4x^9$

**38.** $(-3)^5a^5 \cdot 4^2a^2 = (-3)^5 \cdot 4^2 \cdot a^7$, or $-3888a^7$

**39.** $-3^2 \cdot x^2 \cdot 7^2x^8 = -3^2 \cdot 7^2 \cdot x^{10}$, or $-441x^{10}$

**40.** $2x^3 \cdot 3^2 \cdot x^2 = 2 \cdot 3^2 \cdot x^5$, or $18x^5$

**41.** $3y^2 \cdot 2^3y^3 = 3 \cdot 2^3 \cdot y^5$, or $24y^2$

**42.** $-ab(a^4)(b^2) = -a^5b^3$

**43.** $-rsr^2s^6 = -r^3s^7$

**44.** $(-2)^3x^3y^3(-x^2) = -(-2)^3x^5y^3$, or $8x^5y^3$

**45.** $(-3)^3c^3d^3(-d^2) = -(-3)^3c^3d^5$, or $27c^3d^5$

**46.** $5^3b^6 \cdot \left(\frac{1}{2}\right)^2b^6 = 5^3\left(\frac{1}{2}\right)^2b^{12}$, or $\frac{125}{4}b^{12}$

**47.** $6^2a^8\left(\frac{1}{4}\right)^2a^6 = 6^2\left(\frac{1}{4}\right)^2a^{14}$, or $\frac{9}{4}a^{14}$

**48.** $2^3t^3(-t^2) = -2^3t^5$, or $-8t^5$

**49.** $-(w^3)3^2w^4 = -3^2w^7$, or $-9w^7$

**50.** $(-y)^3(-y)^4(-y)^5 = (-1 \cdot y)^{12} = (-1)^{12} \cdot y^{12} = y^{12}$

**51.** $(-x)^4(-x)^3(-x)^2 = (-1 \cdot x)^9 = (-1)^9 \cdot x^9 = -x^9$

**52.** $a^3b^3c^6a^4b^2 = a^7b^5c^6$

**53.** $-(r^4s^2t^6)s^{12}t^3 = -r^4s^{14}t^9$

**54.** $(-3)^3x^3y^6(-2)^2x^4y^2 = (-3)^3 \cdot (-2)^2x^7y^8$, or $-108x^7y^8$

**55.** $a^5$; $1^5 = 1$   **56.** $b^5$; $2^5 = 32$

**57.** $a^6$; $1^6 = 1$   **58.** $-b^5$; $-(2)^5 = -32$

**59.** $a^2b^4$; $(1)^2(2)^4 = 16$   **60.** $a^8b^4$; $(1)^8(2)^4 = 16$

**61.** $-a^2b^6$; $-(1)^2(2)^6 = -64$   **62.** $b^{13}$; $2^{13} = 8192$

**63.** $(5 \cdot 6)^4$ ___ $5 \cdot 6^4$
    $30^4$ ___ $5 \cdot 1296$
   $810,000$ __>__ $6480$

**64.** $5^2 \cdot 3^6$ ___ $(5 \cdot 3)^6$
    $25 \cdot 729$ ___ $15^6$
   $18,225$ __<__ $11,390,625$

**65.** $(3^6 \cdot 3^{12})$ ___ $3^{72}$
    $3^{18}$ __<__ $3^{72}$

**66.** $4^2 \cdot 4^8$ ___ $4^{16}$
    $4^{10}$ __<__ $4^{16}$

**67.** $(7^2)^3$ ___ $7^5$
    $7^6$ __>__ $7^5$

**68.** $(6^2 \cdot 3)^3$ ___ $6^5 \cdot 3^3$
    $6^6 \cdot 3^3$ __>__ $6^5 \cdot 3^3$

**69.** $(2.1 \cdot 4.4)^3 = 9.24^3 = 788.9$

**70.** $6.5^3 \cdot 6.5^4 = 6.5^7 = 490,222.8$

**71.** $2.6^4 \cdot 2.6^2 = 2.6^6 = 308.9$

**72.** $(5.0 \cdot 4.9)^2 = 24.5^2 = 600.3$

**73.** $(3.7^3)^5 = 3.7^{15} = 333,446,267.9$

**74.** $(8.4^2)^4 = 8.4^8 = 24,787,589.1$

**75.** $V = \frac{4}{3}\pi r^3 \approx \frac{4}{3}(3.14)(3a)^3 \approx \frac{4}{3}(3.14)27a^3 \approx 113.04a^3$

**76.** $V = \frac{1}{3}\pi r^2h \approx \frac{1}{3}(3.14)(2b^2)^2(24) \approx \frac{1}{3}(3.14)4b^4(24)$
       $\approx 100.48b^4$

**77.** $W = 0.015s^3 = 0.015(20)^3 = 120$
   $W = 0.015s^3$
   $W = 0.015(10)^3 = 15$
   120 to 15 $\Rightarrow$ 8 to 1

**78.** If the wind speed is doubled, the power generated is multiplied by 8.

**79. a.** $2^{10} \cdot 4^{10} = 1024 \cdot 1,048,576 = 1,073,741,824$ ways
    **b.** $\left(\frac{1}{2}\right)^{30} \approx 0.0000000009$

**80.** $\frac{1}{3^8} \approx 0.00015$   **81.** $3^{14}$; $\frac{1}{3^{14}} \approx 0.0000002$

**82. a.**

| x | 0 | 1 | 2 | 3 | 4 |
|---|---|---|---|---|---|
| 2x | 0 | 2 | 4 | 6 | 8 |
| $2^x$ | 1 | 2 | 4 | 8 | 16 |

**b.**

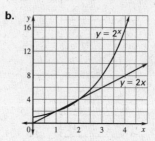

$y = 2x$      $y = 2^x$

| 0 | 0 |
|---|---|
| 1 | 2 |
| 2 | 4 |
| 3 | 6 |
| 4 | 8 |

| 0 | 1 |
|---|---|
| 1 | 2 |
| 2 | 4 |
| 3 | 8 |
| 4 | 16 |

**c.** Both graphs increase from left to right and both pass through (2, 4). The graph of $y = 2^x$ is a curve, while the graph of $y = 2x$ is a line.

# Chapter 8 *continued*

**83.** If you work more than 2 h, the pay is much better at the rate of $2^x$ dollars per hour. For example, at $2^x$ dollars per hour, you would earn \$256 for 8 hours, while at $2x$ dollars per hour, you would earn \$16.

**84.** $(a^2)^3 = a^2 \cdot a^2 \cdot a^2$      definition of $x^3$

$= a \cdot a \cdot a \cdot a \cdot a \cdot a$      definition of $x^2$

$= a^6$      definition of $x^6$

Explanations may vary.

**85.** $(a \cdot b)^m = (a \cdot b) \cdot (a \cdot b) \cdot \ldots \cdot (a \cdot b)$, with $a \cdot b$ as a factor $m$ times. The associative and commutative properties of multiplication allow you to rewrite this expression as $a \cdot a \cdot \ldots \cdot a \cdot b \cdot b \cdot \ldots \cdot b$ with each factor appearing $m$ times. This expression is equal to $a^m \cdot b^m$.

## 8.1 Mixed Review *(p. 455)*

**86.** $8^2 = 8 \cdot 8 = 64$     **87.** $(5(2))^4 = 10^4 = 10,000$

**88.** $\frac{1}{2}(-2)^3 = \frac{1}{2}(-8) = \frac{-8}{2} = -4$    **89.** $\frac{1}{5^2} = \frac{1}{25}$

**90.** $\frac{24}{2^3} = \frac{24}{8} = 3$    **91.** $\frac{45}{2^2} = \frac{45}{4}$

**92.** $y = x + 2$

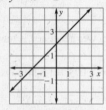

| x | y |
|---|---|
| 0 | 2 |
| 1 | 3 |
| −1 | 1 |
| 2 | 4 |

**93.** $y = -(x - 4)$

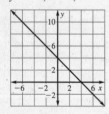

| x | y |
|---|---|
| 0 | 4 |
| 1 | 3 |
| 2 | 2 |
| −1 | 5 |

**94.** $y = \frac{1}{2}x - 5$

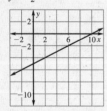

| x | y |
|---|---|
| 0 | −5 |
| 1 | $-4\frac{1}{2}$ |
| 2 | −4 |
| −1 | $-5\frac{1}{2}$ |

**95.** $y = \frac{3}{4}x + 2$

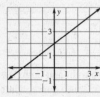

| x | y |
|---|---|
| 0 | 2 |
| 1 | $2\frac{3}{4}$ |
| 2 | $3\frac{1}{2}$ |
| −1 | $1\frac{1}{4}$ |

**96.** $y = 2$

**97.** $x = -3$

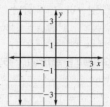

**98.** $x < 4$

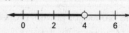

**99.** $x > 15$

**100.** $x \geq -9$

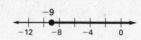

**101.** $x \leq 3$

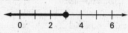

**102.**   $-x - 2 < -5$

         $-x < -3$

            $x > 3$

**103.** $8 + 3x \geq -2$

         $3x \geq -10$

          $x \geq -\frac{10}{3}$

**104.**   $2 < 2x + 7$

     $-5 < 2x$

   $-\frac{5}{2} < x$

**105.** $\$0.07 \leq p \leq \$0.71$ where $p$ is the price of a quart of milk.

## Lesson 8.2

### Activity *(p. 456)*

**1.**

| Exponent, $n$ | 3 | 2 | 1 | 0 | −1 | −2 | −3 |
|---|---|---|---|---|---|---|---|
| Power, $2^n$ | 8 | 4 | 2 | 1 | $\frac{1}{2}$ | $\frac{1}{4}$ | $\frac{1}{8}$ |
| Power, $3^n$ | 27 | 9 | 3 | 1 | $\frac{1}{3}$ | $\frac{1}{9}$ | $\frac{1}{27}$ |
| Power, $4^n$ | 64 | 16 | 4 | 1 | $\frac{1}{4}$ | $\frac{1}{16}$ | $\frac{1}{64}$ |

**2.** 1      **3.** when $a \neq 0$, $a^{-n} = \frac{1}{a^n}$

### 8.2 Guided Practice *(p. 459)*

**1.** exponential

**2.** The exponent applies only to the variable.

$5x^{-3} = \frac{5}{x^3}$

**3.** $\frac{1}{3}$    **4.** not defined    **5.** not defined

# Chapter 8 *continued*

**6.** $6 \cdot 3^0 = 6 \cdot 1 = 6$   **7.** $\dfrac{1}{m^2}$   **8.** $\dfrac{3}{c^5}$   **9.** $\dfrac{a^5}{b^8}$

**10.** $8x^3$   **11.** yes

**12.** Yes; if $a$ is positive, then any power $a^n$ is a product of positive factors and so, is positive. The reciprocal of a positive number is positive, so $a^{-n}$ is positive for every number $n$.

**13.** $P = 18.5 \cdot 1.038^{10} \approx 26.9$ points per game

### 8.2 Practice and Applications (pp. 459–461)

**14.** $\dfrac{1}{4^2} = \dfrac{1}{16}$   **15.** $\dfrac{1}{3^4} = \dfrac{1}{81}$   **16.** $\left(\dfrac{1}{5}\right)^{-1} = 5$

**17.** $8\left(\dfrac{1}{4}\right)^{-1} = 8\left(\dfrac{4}{1}\right) = 32$   **18.** $4\left(\dfrac{1}{4^2}\right) = \dfrac{4}{16} = \dfrac{1}{4}$

**19.** $\left(\dfrac{1}{10}\right)^{-2} = \left(\dfrac{10}{1}\right)^2 = 100$

**20.** $-6^0 \cdot \dfrac{1}{3^{-2}} = -1 \cdot \dfrac{3^2}{1} = -1 \cdot 9 = -9$

**21.** $\dfrac{1}{2^3} \cdot 2^2 = \dfrac{1}{8} \cdot 4 = \dfrac{4}{8} = \dfrac{1}{2}$   **22.** $8^3 \cdot 0^{-1}$   not defined

**23.** $7^4 \cdot 7^{-4} = 7^4 \cdot \dfrac{1}{7^4} = 1$   **24.** $\dfrac{1}{8^7} \cdot \dfrac{8^7}{1} = \dfrac{8^7}{8^7} = 1$

**25.** $-4 \cdot (-4)^{-1} = -4 \cdot \dfrac{1}{-4} = 1$

**26.** $(5^{-3})^2 = \left(\dfrac{1}{5^3}\right)^2 = \left(\dfrac{1}{125}\right)^2 = \dfrac{1}{15,625}$

**27.** $(-3^{-2})^{-1} = \left(\dfrac{1}{-3^2}\right)^{-1} = \dfrac{-3^2}{1} = -9$

**28.** $11 \cdot \dfrac{1}{11} = \dfrac{11}{11} = 1$   **29.** $4^0 \cdot \dfrac{1}{5^3} = 1 \cdot \dfrac{1}{5^3} = \dfrac{1}{125}$

**30.** $\dfrac{1}{x^5}$   **31.** $\dfrac{3}{x^4}$   **32.** $\dfrac{x^5}{2}$   **33.** $\dfrac{y^4}{x^2}$   **34.** $\dfrac{x^4}{y^7}$

**35.** $\dfrac{8}{x^2 y^6}$   **36.** $\dfrac{x^3 y^1}{9}$   **37.** $\dfrac{x^{10}}{4y^{14}}$

**38.** $(-9)^0 x = 1(x) = x$   **39.** $(-4x)^{-3} = \dfrac{1}{(-4x)^3} = -\dfrac{1}{64x^3}$

**40.** $(-10a)^0 = 1$   **41.** $(3xy)^{-2} = \dfrac{1}{(3xy)^2} = \dfrac{1}{9x^2 y^2}$

**42.** $(6a^{-3})^3 = \left(\dfrac{6}{a^3}\right)^3 = \dfrac{216}{a^9}$   **43.** $8m^2$

**44.** $\dfrac{1}{(4x)^{-5}} = (4x)^5 = 1024x^5$

**45.** $\left(\dfrac{-4x^2}{2x^{-1}}\right)^{-1} = \left(\dfrac{-4x^2 x}{2}\right)^{-1} = -\dfrac{2}{4x^3} = -\dfrac{1}{2x^3}$

**46.** B   **47.** C   **48.** A   **49.** 0.03125   **50.** 0.82645

**51.** 0.00352   **52.** 0.00457   **53.** no   **54.** yes

**55.** no   **56.** yes

**57.** $y = \left(\dfrac{1}{3}\right)^x$
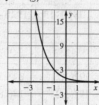

**58.** $y = \left(\dfrac{1}{5}\right)^x$
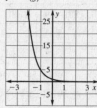

**59.** $y = 4^{-x}$
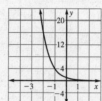

**60.** $y = 5^x$
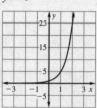

**61.** Both graphs are curves that pass through $(0, 1)$ and both increase indefinitely in one direction and get very close to the $x$-axis in the other; $y = \left(\dfrac{1}{2}\right)^x$ decreases from left to right, while $y = 2^x$ increases from left to right.

$y = \left(\dfrac{1}{2}\right)^x$     $y = 2^x$

**62.** $y = \left(\dfrac{1}{3}\right)^x$     $y = 3^x$
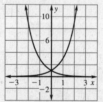

Both graphs are curves that pass through $(0, 1)$ and are mirror images, or reflections of each other over the $y$-axis.

**63.** $(0, 1)$; yes; $(0, 2)$

**64.** 1990: $A = 450(1.06)^{-10} = \$251.28$
2000: $A = 450(1.06)^0 = \$450$
2010: $A = 450(1.06)^{10} = \$805.88$

# Chapter 8 *continued*

**65. a.**

|        | 1870–79 | 1880–89 | 1900–09 | 1910–19 | 1920–29 |
|--------|---------|---------|---------|---------|---------|
| $t$    | $-5$    | $-4$    | $-2$    | $-1$    | $0$     |
| $S$    | $117$   | $141$   | $203$   | $243$   | $292$   |

$S = 292(1.2)^0 = 292$  $\qquad$ $S = 292(1.2)^{-1} = 243.3$

$S = 292(1.2)^{-2} = 202.8$ $\qquad$ $S = 292(1.2)^{-4} = 140.8$

$S = 292(1.2)^{-5} = 117.3$

**b.**

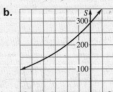

**66.** B

$y = 2(3)$ $\quad$ $y = 2^3$

$y = 6$ $\qquad$ $y = 8$

**67.** A

$y = 2^1$ $\quad$ $y = 2^{-1}$

$y = 2$ $\quad$ $y = 0.5$

**68.** C

$y = \left(\frac{1}{2}\right)^0$ $\quad$ $y = \left(\frac{1}{2}\right)^{-0}$

$y = 1$ $\qquad$ $y = 1$

**69.** C

**70.** By the product of powers property, $b^2 \cdot b^0 = b^{2+0} = b^2$.
Since $0^0$ is not defined, $b \neq 0$. Then

$\dfrac{b^2 b^0}{b^2} = \dfrac{b^2}{b^2} = 1$, so $b^0 = 1$.

### 8.2 Mixed Review (p. 461)

**71.** $\left(\frac{2}{5}\right)^2 = \frac{4}{25}$ $\quad$ **72.** $\left(\frac{1}{2}\right)^3 = \frac{1}{8}$ $\quad$ **73.** $\left(-\frac{9}{10}\right)^3 = -\frac{729}{1000}$

**74.** $\left(\frac{1}{5}\right)^4 = \frac{1}{625}$

**75.** $|5 + x| + 4 \leq 11$

$\qquad |5 + x| \leq 7$

$\qquad 5 + x \leq 7$ and $5 + x \geq -7$

$\qquad\qquad x \leq 2$ and $\qquad x \geq -12$

$-12 \leq x \leq 2$

**76.** $|3x + 7| - 4 > 9$

$\qquad |3x + 7| > 13$

$\qquad 3x + 7 > 13$ or $3x + 7 < -13$

$\qquad\quad 3x > 6$ or $\qquad 3x < -20$

$\qquad\qquad x > 2$ or $\qquad x < -6\frac{2}{3}$

**77.** $|x + 2| - 1 \leq 8$

$\qquad |x + 2| \leq 9$

$\qquad x + 2 \leq 9$ and $x + 2 \geq -9$

$\qquad\quad x \leq 7$ and $\qquad x \geq -11$

$-11 \leq x \leq 7$

**78.** $|3 - x| - 6 > -4$

$\qquad |3 - x| > 2$

$\qquad 3 - x > 2$ or $3 - x < -2$

$\qquad -x > -1$ or $\quad -x < -5$

$\qquad\quad x < 1$ or $\qquad x > 5$

**79.** $|9 - 2x| + 3 < 4$

$\qquad |9 - 2x| < 1$

$\qquad 9 - 2x < 1$ and $9 - 2x > -1$

$\qquad -2x < -8$ and $\quad -2x > -10$

$\qquad\quad x > 4$ and $\qquad x < 5$

$\qquad\qquad\qquad\qquad 4 < x < 5$

**80.** $|3x + 2| + 9 \geq -1$

$\qquad |3x + 2| \geq -10$

$\qquad 3x + 2 \geq -10$ or $3x + 2 \leq 10$

$\qquad\quad 3x \geq -12$ or $\qquad 3x \leq 8$

$\qquad\qquad x \geq -4$ or $\qquad x \leq 2\frac{2}{3}$

$\qquad\qquad\qquad\qquad$ all real numbers

**81.** 10, 13, 17, 21, 23, 25, 40, 42, 44, 48, 48, 50

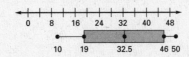

**82.** 30, 55, 61, 68, 76, 76, 78, 79, 82, 85, 86, 87

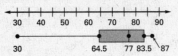

**83.** $2x - y = -2$

$4x + y = 5$

$2x = -2 + y$ $\qquad\qquad 4\left(-1 + \frac{1}{2}y\right) + y = 5$

$x = -1 + \frac{1}{2}y$ $\qquad\qquad\quad -4 + 2y + y = 5$

$\qquad\qquad\qquad\qquad\qquad\qquad -4 + 3y = 5$

$2x - 3 = -2$ $\qquad\qquad\qquad\qquad 3y = 9$

$\quad 2x = 1$ $\qquad\qquad\qquad\qquad\qquad y = 3$

$\qquad x = \frac{1}{2}$

Solution: $\left(\frac{1}{2}, 3\right)$

# Chapter 8 *continued*

**84.** $-3x + y = 4$

$-9x + 5y = 10$

$y = 4 + 3x$

$\qquad -9x + 5(4 + 3x) = 10$

$\qquad -9x + 20 + 15x = 10$

$-3\left(-\frac{10}{6}\right) + y = 4 \qquad 6x + 20 = 10$

$\qquad 5 + y = 4 \qquad\qquad 6x = -10$

$\qquad\qquad y = -1 \qquad\qquad x = \frac{-10}{6}$

$\qquad\qquad\qquad\qquad\qquad x = -1\frac{2}{3}$

Solution: $\left(-1\frac{2}{3}, -1\right)$

**85.** $x + 4y = 300$

$x - 2y = 0$

$x = 0 + 2y \qquad 0 + 2y + 4y = 300$

$\qquad\qquad\qquad 6y = 300$

$x - 2(50) = 0 \qquad y = 50$

$x - 100 = 0$

$\qquad x = 100$

Solution: $(100, 50)$

**86.** $2x - 3y = 10$

$3x + 3y = 15$

$2x = 10 + 3y \qquad 3\left(5 + \frac{3}{2}y\right) + 3y = 15$

$x = 5 + \frac{3}{2}y \qquad 15 + \frac{9}{2}y + 3y = 15$

$\qquad\qquad\qquad\qquad \frac{15}{2}y = 0$

$2x - 3(0) = 10 \qquad\qquad y = 0$

$\qquad 2x = 10$

$\qquad x = 5$

Solution: $(5, 0)$

**87.** $x + 15y = 6$

$-x - 5y = 84$

$x = 6 - 15y \qquad -(6 - 15y) - 5y = 84$

$\qquad\qquad\qquad -6 + 15y - 5y = 84$

$-x - 5(9) = 84 \qquad -6 + 10y = 84$

$-x - 45 = 84 \qquad\qquad 10y = 90$

$\qquad -x = 129 \qquad\qquad y = 9$

$\qquad x = -129$

Solution: $(-129, 9)$

**88.** $4x - y = 5$

$2x + 4y = 15$

$2x = 15 - 4y \qquad 4\left(\frac{15}{2} - 2y\right) - y = 5$

$x = \frac{15}{2} - 2y \qquad 30 - 8y - y = 5$

$\qquad\qquad\qquad 30 - 9y = 5$

$\qquad\qquad\qquad -9y = -25$

$\qquad\qquad\qquad y = \frac{25}{9}$

$\qquad\qquad\qquad y = 2\frac{7}{9}$

$x = \frac{15}{2} - 2\left(\frac{25}{9}\right) = \frac{15}{2} - \frac{50}{9} = \frac{135}{18} - \frac{100}{18} = \frac{35}{18} = 1\frac{17}{18}$

Solution: $\left(1\frac{17}{18}, 2\frac{7}{9}\right)$

## Technology Activity 8.2 *(p. 462)*

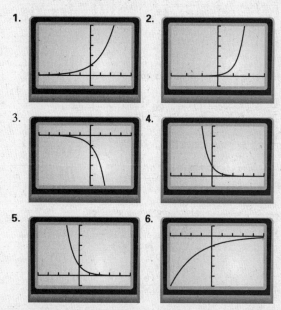

**7.** The graph is close to the negative $x$-axis and curves upward to the right.

**8.** The graph curves upward to the left and is close to the $x$-axis on the right.

**9.** The graph is close to the negative $x$-axis and curves downward to the right.

**10.** The graph is close to the positive $x$-axis and curves downward to the left.

## Lesson 8.3

### 8.3 Guided Practice *(p. 466)*

**1.** quotient of powers

**2.** No; the quotient of powers property only applies when the powers have the same base.

**3.** $\dfrac{5^4}{5^1} = 5^{4-1} = 5^3 = 125$

**4.** $\dfrac{7^6}{7^9} = 7^{6-9} = 7^{-3} = \dfrac{1}{7^3} = \dfrac{1}{343}$

**5.** $\dfrac{a^{12}}{a^9} = a^{12-9} = a^3$    **6.** $\dfrac{m^5}{m^{11}} = m^{5-11} = m^{-6} = \dfrac{1}{m^6}$

**7.** $\dfrac{a^5}{a^2} = a^{5-2} = a^3$    **8.** $\dfrac{(-2)^8}{(-2)^3} = (-2)^5 = -32$

**9.** $\dfrac{5^3 \cdot 5^5}{5^9} = \dfrac{5^8}{5^9} = 5^{8-9} = 5^{-1} = \dfrac{1}{5}$

**10.** $\dfrac{x^7 \cdot x}{x^{-2}} = \dfrac{x^8}{x^{-2}} = x^{8-(-2)} = x^{10}$    **11.** $\left(\dfrac{1}{2}\right)^5 = \dfrac{1^5}{2^5} = \dfrac{1}{32}$

**12.** $\left(\dfrac{3}{5}\right)^3 = \dfrac{3^3}{5^3} = \dfrac{27}{125}$    **13.** $\left(\dfrac{5}{m}\right)^2 = \dfrac{5^2}{m^2} = \dfrac{25}{m^2}$

**14.** $\left(\dfrac{2}{b}\right)^4 = \dfrac{2^4}{b^4} = \dfrac{16}{b^4}$    **15.** $\left(\dfrac{5}{4}\right)^{-3} = \dfrac{5^{-3}}{4^{-3}} = \dfrac{4^3}{5^3} = \dfrac{64}{125}$

**16.** $\left(\dfrac{x^4}{2^3}\right)^2 = \dfrac{(x^4)^2}{(2^3)^2} = \dfrac{x^8}{2^6} = \dfrac{x^8}{64}$   **17.** $\left(\dfrac{x^3}{y^5}\right)^6 = \dfrac{(x^3)^6}{(y^5)^6} = \dfrac{x^{18}}{y^{30}}$

**18.** $\left(\dfrac{a^6}{b^9}\right)^5 = \dfrac{(a^6)^5}{(b^9)^5} = \dfrac{a^{30}}{b^{45}}$

### 8.3 Practice and Applications (pp. 466–468)

**19.** $\dfrac{5^6}{5^3} = \dfrac{15625}{125} = 125$   **20.** $\dfrac{8^3}{8^1} = \dfrac{512}{8} = 64$

**21.** $\dfrac{(-3)^6}{-3^6} = \dfrac{729}{-729} = -1$   **22.** $\dfrac{(-3)^9}{(-3)^9} = \dfrac{-19683}{-19683} = 1$

**23.** $\dfrac{3^3}{3^{-4}} = \dfrac{3^3 \cdot 3^4}{1} = \dfrac{3^{3+4}}{1} = \dfrac{3^7}{1} = 2187$

**24.** $\dfrac{8^3 \cdot 8^2}{8^5} = \dfrac{8^{3+2}}{8^5} = \dfrac{8^5}{8^5} = 8^{5-5} = 8^0 = 1$

**25.** $\dfrac{5 \cdot 5^4}{5^8} = \dfrac{5^{1+4}}{5^8} = \dfrac{5^5}{5^8} = 5^{5-8} = 5^{-3} = \dfrac{1}{5^3} = \dfrac{1}{125}$

**26.** $\left(\dfrac{3}{4}\right)^2 = \dfrac{3^2}{4^2} = \dfrac{9}{16}$   **27.** $\left(\dfrac{6}{2}\right)^3 = \dfrac{6^3}{2^3} = \dfrac{216}{8} = 27$

**28.** $\left(-\dfrac{2}{3}\right)^3 = \dfrac{(-2)^3}{3^3} = -\dfrac{8}{27}$   **29.** $\left(-\dfrac{3}{5}\right)^2 = \dfrac{(-3)^2}{5^2} = \dfrac{9}{25}$

**30.** $\left(\dfrac{9}{6}\right)^{-1} = \dfrac{6}{9} = \dfrac{2}{3}$   **31.** $\left(\dfrac{3}{x}\right)^4 = \dfrac{3^4}{x^4} = \dfrac{81}{x^4}$

**32.** $\dfrac{x^4}{x^5} = x^{4-5} = x^{-1} = \dfrac{1}{x}$   **33.** $\left(\dfrac{1}{x}\right)^5 = \dfrac{1^5}{x^5} = \dfrac{1}{x^5}$

**34.** $x^3 \cdot \dfrac{1}{x^2} = \dfrac{x^3}{x^2} = x^{3-2} = x^1 = x$

**35.** $x^5 \cdot \dfrac{1}{x^8} = \dfrac{x^5}{x^8} = x^{5-8} = x^{-3} = \dfrac{1}{x^3}$

**36.** $\left(\dfrac{a^9}{a^5}\right)^{-1} = \dfrac{a^5}{a^9} = a^{5-9} = a^{-4} = \dfrac{1}{a^4}$

**37.** $\left(\dfrac{y^2}{y^3}\right)^{-2} = \left(\dfrac{y^3}{y^2}\right)^2 = \dfrac{y^6}{y^4} = y^{6-4} = y^2$

**38.** $\dfrac{m^3 \cdot m^5}{m^2} = \dfrac{m^{3+5}}{m^2} = \dfrac{m^8}{m^2} = m^{8-2} = m^6$

**39.** $\dfrac{(r^3)^4}{(r^3)^8} = \dfrac{r^{12}}{r^{24}} = r^{12-24} = r^{-12} = \dfrac{1}{r^{12}}$

**40.** $\left(\dfrac{-6x^2y}{2xy^3}\right)^3 = \dfrac{-216x^6y^3}{8x^3y^9} = \dfrac{-216}{8}x^{6-3}y^{3-9}$
$= -27x^3y^{-6} = -\dfrac{27x^3}{y^6}$

**41.** $\left(\dfrac{2x^3y^4}{3xy}\right)^3 = \dfrac{8x^9y^{12}}{27x^3y^3} = \dfrac{8}{27}x^{9-3}y^{12-3} = \dfrac{8x^6y^9}{27}$

**42.** $\dfrac{16x^3y}{-4xy^3} \cdot \dfrac{-2xy}{-x^{-1}} = -4x^{3-1}y^{1-3} \cdot -2xy(-x)$
$= -4x^2y^{-2} \cdot 2x^2y = -8x^{2+2}y^{-2+1}$
$= -8x^4y^{-1} = \dfrac{-8x^4}{y}$

**43.** $\dfrac{4x^3y^3}{2xy} \cdot \dfrac{5xy^2}{2y} = 2x^{3-1}y^{3-1} \cdot \dfrac{5}{2}xy^{2-1} = 2x^2y^2 \cdot \dfrac{5}{2}xy = 5x^3y^3$

**44.** $\dfrac{36a^8b^2}{ab} \cdot \left(\dfrac{6}{ab^2}\right)^{-1} = 36a^{8-1}b \cdot \left(\dfrac{ab^2}{6}\right) = 36a^7b \cdot \dfrac{ab^2}{6}$
$= \dfrac{36a^{7+1}b^{1+2}}{6} = 6a^8b^3$

**45.** $\dfrac{16x^5y^{-8}}{x^7y^4} \cdot \left(\dfrac{x^3y^2}{8xy}\right)^4 = 16x^{5-7}y^{-8-4} \cdot \dfrac{x^{12}y^8}{4096x^4y^4}$
$= 16x^{-2}y^{-12} \cdot \dfrac{x^{12-4}y^{8-4}}{4096}$
$= 16x^{-2}y^{-12} \cdot \dfrac{x^8y^4}{4096} = \dfrac{x^6y^{-8}}{256} = \dfrac{x^6}{256y^8}$

**46.** $\dfrac{6x^{-2}y^2}{xy^{-3}} \cdot \dfrac{(4x^2y)^{-2}}{xy^2} = 6x^{-2-1}y^{2-(-3)} \cdot \dfrac{1}{(4x^2y)^2xy^2}$
$= 6x^{-3}y^5 \cdot \dfrac{1}{16x^{4+1}y^{2+2}} = 6x^{-3}y^5 \cdot \dfrac{1}{16x^5y^4}$
$= \dfrac{6y^5}{x^3} \cdot \dfrac{1}{16x^5y^4} = \dfrac{6y^5}{16x^8y^4} = \dfrac{6y^{5-4}}{16x^8}$
$= \dfrac{6y}{16x^8} = \dfrac{3y}{8x^8}$

**47.** $\dfrac{5x^{-3}y^2}{x^5y^{-1}} \cdot \dfrac{(2xy^3)^{-2}}{xy} = \dfrac{5y^2y^1}{x^3x^5} \cdot \dfrac{1}{xy(2xy^3)^2} = \dfrac{5y^3}{x^8} \cdot \dfrac{1}{xy(4x^2y^6)}$
$= \dfrac{5y^3}{x^8} \cdot \dfrac{1}{4x^3y^7} = \dfrac{5y^3}{4x^{11}y^7} = \dfrac{5}{4x^{11}y^4}$

**48.** $\left(\dfrac{2xy^{-2}y^4}{3x^{-1}y}\right)^{-2} \cdot \left(\dfrac{4xy}{2x^{-1}y^{-3}}\right) = \left(\dfrac{2xxy^4}{3y^2y}\right)^{-2} \cdot \left(\dfrac{4xyxy^3}{2}\right)^2$
$= \left(\dfrac{3y^3}{2x^2y^4}\right)^2 \cdot \dfrac{16x^4y^8}{4}$
$= \dfrac{9y^6}{4x^4y^8} \cdot \dfrac{16x^4y^8}{4}$
$= \dfrac{9}{4x^4y^2} \cdot 4x^4y^8$
$= \dfrac{36x^4y^8}{4x^4y^2} = 9y^6$

**49.** $\dfrac{6^3}{6} = 6^{3-1} = 6^2 = 36$

**50.** $\dfrac{x^{-9}}{x^{-3}} = x^{-9-(-3)} = x^{-6} = \dfrac{1}{x^6}$

**51.** Moon: $V = 1.33(\pi)\left(\dfrac{r}{4}\right)^3 = 1.33\pi\left(\dfrac{r^3}{64}\right) = \dfrac{1.33\pi r^3}{64}$

Earth: $V = 1.33\pi(r)^3 = 1.33\pi r^3$

The volume of Earth is about 64 times the volume of the moon.

**52.** 1998: $S = 3723\left(\dfrac{6}{5}\right)^4$

1995: $S = 3723\left(\dfrac{6}{5}\right)^1$

Ratio $= \dfrac{3723\left(\frac{6}{5}\right)^4}{3723\left(\frac{6}{5}\right)^1} = \left(\dfrac{6}{5}\right)^3 = \dfrac{216}{125}$

# Chapter 8 *continued*

**53.** 1988: $y = 283(1.2)^4$

1994: $y = 283(1.2)^{10}$

Ratio: $\dfrac{283(1.2)^4}{283(1.2)^{10}} = \dfrac{1}{1.2^6} \approx \dfrac{1}{3}$

**54.** 5 year old: $w = 1.16(1.44)^5$

2 year old: $w = 1.16(1.44)^2$

Ratio: $\dfrac{1.16(1.44)^5}{1.16(1.44)^2} = 1.44^3$

**55.** about 3 times  **56.** $w = 1.16(1.44)^0 = 1.16$  1.16 lbs.

**57.** 1997: $S = 476(1.13)^2$

1999: $S = 476(1.13)^4$

Ratio: $\dfrac{476(1.13)^2}{476(1.13)^4} = \dfrac{1}{1.13^2} \approx 0.8$

**58.** 8 rower shell: $s = 16.3(1.0285)^8$

2 rower shell: $s = 16.3(1.0285)^2$

Ratio: $\dfrac{16.3(1.0285)^8}{16.3(1.0285)^2} = 1.0285^6 \approx 1.2$

**59. a.**

| Weeks, n | 0 | 1 | 2 | 3 | 4 | 5 | 6 |
|---|---|---|---|---|---|---|---|
| Words, s | 200 | 160 | 128 | 102 | 82 | 66 | 52 |

$s = 200\left(\frac{4}{5}\right)^0 = 200 \qquad s = 200\left(\frac{4}{5}\right)^1 = 200\left(\frac{4}{5}\right) = 160$

$s = 200\left(\frac{4}{5}\right)^2 = 200\left(\frac{16}{25}\right) = 128$

$s = 200\left(\frac{4}{5}\right)^3 = 200\left(\frac{64}{125}\right) = 102.4$

$s = 200\left(\frac{4}{5}\right)^4 = 200\left(\frac{256}{625}\right) = 81.9$

$s = 200\left(\frac{4}{5}\right)^5 = 200\left(\frac{1024}{3125}\right) = 65.5$

$s = 200\left(\frac{4}{5}\right)^6 = 200\left(\frac{4096}{15625}\right) = 52.4$

**b.** 19 weeks. *Sample answer*: Since the number of words remembered each week is 0.8 times the number remembered the previous week, I did repeated multiplications on my calculator, beginning with 0.8(200) and counted the number of times it took to get to 3.

**60.** $\left(\frac{1}{6}\right)^8 = \dfrac{1}{1,679,616} \approx .0000006$  **61.** $\left(\frac{1}{2}\right)^6 = \dfrac{1}{64} \approx 0.02$

**62.** 2025: $P = 4264(1.0208)^{30}$

2000: $P = 4264(1.0208)^5$

Ratio: $\dfrac{4264(1.0208)^{30}}{4264(1.0208)^5} = 1.0208^{25} \approx 1.67$

**63.** 2025: $P = 9540(1.0026)^{30}$

2000: $P = 9540(1.0026)^5$

Ratio: $\dfrac{9540(1.0026)^{30}}{9540(1.0026)^5} = (1.0026)^{25} \approx 1.07$

**64.** Arizona

**65.** While Arizona's population is growing more rapidly, Michigan's population will be greater than that of Arizona for nearly 50 years. Answers may vary.

**66.** $s = 0.0032(2)^{25} \approx 107{,}374$ in. $\approx 8947.8$ ft

## 8.3 Mixed Review (p. 469)

**67.** $10^5 = 100{,}000$  **68.** $10^3 = 1000$

**69.** $10^{-4} = \dfrac{1}{10{,}000}$  **70.** $10^{-8} = \dfrac{1}{100{,}000{,}000}$

**71.** $x \geq 5$

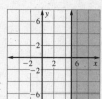

**72.** $x + 3 < 4$

$x < 1$

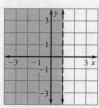

**73.** $y > -2$

**74.** $y \leq -1.5$

**75.** $x \geq 2.5$

**76.** $3x - y < 0$

$y > 3x$

**77.** $y \leq \dfrac{x}{2}$

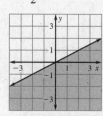

**78.** $\dfrac{3}{4}x + \dfrac{1}{4}y \geq 1$

$3x + y \geq 4$

$y \geq 4 - 3x$

**79.** $2(-3) + 4(2) = 2$

$-6 + 8 = 2$

$2 = 2$

$-(-3) + 5(2) = 13$

$3 + 10 = 13$

$13 = 13$

solution

**80.** $1 - 5(-4) = 9$

$1 + 20 = 9$

$21 \neq 9$

not a solution

# Chapter 8 *continued*

**81.** $8\left(\frac{3}{4}\right) + 4(0) = 6$

$6 + 0 = 6$

$6 = 6$

$4\left(\frac{3}{4}\right) + 0 = 3$

$3 + 0 = 3$

$3 = 3$

solution

**82.** $3(5) - 8\left(-\frac{1}{2}\right) = 11$

$15 + 4 = 11$

$19 \neq 11$

not a solution

**83.** $x - y = 4 \qquad 8 + y = 12$

$\dfrac{x + y = 12}{2x = 16} \qquad y = 4$

$x = 8 \qquad$ Solution: $(8, 4)$

**84.** $-x + 2y = 12 \qquad x + 6(4) = 20$

$\dfrac{x + 6y = 20}{8y = 32} \qquad x + 24 = 20$

$y = 4 \qquad$ Solution: $(-4, 4)$

**85.** $2a + 3b = 17 \qquad -6a - 9b = -51$

$3a + 4b = 24 \qquad \dfrac{6a + 8b = 48}{-b = -3}$

$2a + 3(3) = 17 \qquad b = 3$

$2a + 9 = 17$

$2a = 8$

$a = 4 \qquad$ Solution: $(4, 3)$

## Quiz 1 *(p. 469)*

**1.** $3^3 \cdot 3^4 = 27 \cdot 81 = 2187$ **2.** $(2^2)^4 = 2^8 = 256$

**3.** $[(8 + 2)^2]^2 = [10^2]^2 = 10^4 = 10{,}000$

**4.** $7^{-4} = \dfrac{1}{7^4} = \dfrac{1}{2401}$

**5.** $4^{-3} \cdot 4^{-4} = \dfrac{1}{4^3} \cdot \dfrac{1}{4^4} = \dfrac{1}{64} \cdot \dfrac{1}{256} = \dfrac{1}{16{,}384}$

**6.** $\left(\dfrac{6}{7}\right)^{-1} = \dfrac{7}{6}$ **7.** $\dfrac{5^{-3}}{5^2} = \dfrac{1}{5^5} = \dfrac{1}{3125}$

**8.** $\dfrac{3^4 \cdot 3^6}{3^3} = \dfrac{3^{4+6}}{3^3} = \dfrac{3^{10}}{3^3} = 3^{10-3} = 3^7 = 2187$

**9.** $\left(\dfrac{5}{4}\right)^{-3} = \left(\dfrac{4}{5}\right)^3 = \dfrac{4^3}{5^3} = \dfrac{64}{125}$

**10.** $\dfrac{(-2)^9}{(-2)^2} = (-2)^7 = -128$

**11.** $6^0 \cdot \dfrac{1}{4^{-3}} = 1 \cdot 4^3 = 1 \cdot 64 = 64$

**12.** $\dfrac{2^3 \cdot 2^{-4}}{2^{-3}} = \dfrac{2^3 \cdot 2^3}{2^4} = \dfrac{2^6}{2^4} = 2^2 = 4$

**13.** $x^4 \cdot x^5 = x^{4+5} = x^9$ **14.** $(-2x)^5 = -32x^5$

**15.** $-\dfrac{3}{a^{-5}} = -3a^5$ **16.** $200^0 c^5 = 1c^5 = c^5$

**17.** $\dfrac{x^6}{x^4} = x^{6-4} = x^2$ **18.** $\dfrac{x^{-5}}{x^{-6}} = x^{-5-(-6)} = x$

**19.** $\left(\dfrac{-2m^2n}{3mn^2}\right)^4 = \dfrac{16m^8n^4}{81m^4n^8} = \dfrac{16}{81}m^{8-4}n^{4-8} = \dfrac{16m^4}{81n^4}$

**20.** $x^4 \cdot \dfrac{1}{x^3} = \dfrac{x^4}{x^3} = x^{4-3} = x$

**21.** $(3a)^3 \cdot (-4a)^3 = 27a^3 \cdot -64a^3 = -1728a^6$

**22.** $(8m^3)^2\left(\frac{1}{2}m^2\right)^2 = 64m^6\left(\frac{1}{4}m^4\right) = 16m^{10}$

**23.** $\dfrac{20x^3y}{4xy^2} \cdot \dfrac{-6xy}{-x} = 5x^{3-1}y^{1-2} \cdot 6x^{1-1}y = 5x^2y^{-1} \cdot 6x^0y$

$= \dfrac{5x^2}{y} \cdot 6y = 30x^2y^{1-1} = 30x^2$

**24.** 1994: $A = 250(1.08)^{-7} = \$145.87$

1999: $A = 250(1.08)^{-2} = \$214.33$

2001: $A = 250(1.08)^0 = \$250.00$

## Lesson 8.4

### Activity *(p. 470)*

**1. a.** 64,300 **b.** 3,072,000 **c.** 0.042 **d.** 0.00152

**2.** Move the decimal point $n$ places to the right if $n > 0$ and $|n|$ places left if $n < 0$.

### 8.4 Guided Practice *(p. 473)*

**1.** No; 12.38 is not between 1 and 10. **2.** right

**3.** 430 **4.** 8110 **5.** 0.245 **6.** 938,000

**7.** $3.96 \times 10$ **8.** $7.2 \times 10^{-1}$ **9.** $1.2 \times 10^3$

**10.** $3 \times 10^{-4}$ **11.** $6.9 \times 10^6$ **12.** $2.05 \times 10^{-5}$

**13.** $7.2 \times 10^7$ **14.** $6 \times 10^{-9}$

**15.** about $1.97 \times 10^4$ sec or about 5.5 h

### 8.4 Practice and Applications *(p. 473–475)*

**16.** 21,400 **17.** 0.98 **18.** 7.75 **19.** 8652.1

**20.** 0.000465 **21.** 0.000006002 **22.** 433,200,000

**23.** 100,012,000 **24.** 11,098,000,000 **25.** $5 \times 10^{-2}$

**26.** $9.52 \times 10$ **27.** $4.22 \times 10^{-2}$ **28.** $3.70207 \times 10^2$

**29.** $7 \times 10^8$ **30.** $1.9314 \times 10$ **31.** $8.551 \times 10^{-3}$

**32.** $2.73 \times 10^9$ **33.** $4.59 \times 10^{-4}$ **34.** $3.2954 \times 10^{-4}$

**35.** $8.8 \times 10^7$ **36.** $2.88 \times 10^{-5}$ **37.** $1.2 \times 10^5$; 120,000

**38.** $5.6 \times 10^{-6}$; 0.0000056 **39.** $1.5 \times 10^5$; 150,000

**40.** $2.76 \times 10^{-2}$; 0.0276 **41.** $\frac{8}{5} \times 10^2 = 1.6 \times 10^2$; 160

**42.** $\frac{1.4}{3.5} \times 10^3 = 0.4 \times 10^3 = 4.0 \times 10^2$; 400

**43.** $\frac{6.6}{1.1} \times 10^0 = 6.0 \times 10^0$; 6

**44.** $9.0 \times 10^{-6}$; 0.000009

**45.** $81 \times 10^6 = 8.1 \times 10^7$; 81,000,000

**46.** $81 \times 10^{-8} = 8.1 \times 10^{-7}$; 0.00000081

**47.** $2.4 \times 10^{10}$; 24,000,000,000

**48.** $1.944 \times 10^{12}$; 1,944,000,000,000

**49.** $1.09926 \times 10^6$; 1,099,260 **50.** $2.7468 \times 10$; 27.468

**51.** $5.76 \times 10^{-8}$; 0.0000000576

**52.** $8.30584 \times 10^{-13}$; 0.000000000000830584

**Algebra 1**
Chapter 8 Worked-out Solution Key

# Chapter 8 continued

**53.** $1.2 \times 10^8$   **54.** $5.852 \times 10^9$   **55.** $5.0819 \times 10^{13}$

**56.** $2 \times 10^{-23}$

**57.** $V = \frac{4}{3}\pi r^3 = \frac{4}{3}\pi (4.4 \times 10^4)^3$

$\approx \frac{4}{3}(3.14)(4.4)^3 \times 10^{12}$

$\approx 357 \times 10^{12}$

$\approx 3.57 \times 10^{14} \, \text{mi}^3$

**58.** Louisiana Purchase:

$\dfrac{1.5 \times 10^7}{8.28 \times 10^5} = \dfrac{1.5}{8.28} \times 10^2 \approx \$18.12/\text{mi}^2$

Gadsden Purchase:

$\dfrac{1 \times 10^7}{2.94 \times 10^4} = \dfrac{1}{2.94} \times 10^3 \approx \$340.14/\text{mi}^2$

**59.** $\dfrac{1.5 \times 10^7}{5.2992 \times 10^8} = \dfrac{1.5}{5.2992} \times 10^{-1}$

Louisiana Purchase $\approx$ \$0.03/acre

Gadsden Purchase $= \dfrac{1 \times 10^7}{1.8816 \times 10^7} = \dfrac{1}{1.8816} \times 10^0$

$\approx \$0.53/\text{acre}$

**60.** Answers may vary.

*Sample answer:* As time passed, land values increased.

**61.** $1.7 \times 10^4(2.592 \times 10^6)$

$4.4064 \times 10^{10} \, \text{m}^3$

**62.** $85(365)(24)(60) = 3,127,320,000$

$3.12732 \times 10^9$ times

**63.** Texas: $\dfrac{3.9 \times 10^{10}}{1.8 \times 10^7} = \dfrac{3.9}{1.8} \times 10^3 \approx 2167$

Minnesota: $\dfrac{7.0 \times 10^9}{4.6 \times 10^6} = \dfrac{7}{4.6} \times 10^3 \approx 1522$

Pennsylvania: $\dfrac{1.9 \times 10^{10}}{1.2 \times 10^7} = \dfrac{1.9}{1.2} \times 10^3 \approx 1583$

Vermont: $\dfrac{4.7 \times 10^8}{5.8 \times 10^5} = \dfrac{4.7}{5.8} \times 10^3 \approx 810$

California: $\dfrac{5.6 \times 10^{10}}{3.1 \times 10^7} = \dfrac{5.6}{3.1} \times 10^3 \approx 1806$

**64.** D   **65.** C   $\dfrac{1.1 \times 10^{-1}}{5.5 \times 10^{-5}} = \dfrac{1.1}{5.5} \times 10^4 = 2 \times 10^3$

**66. a.** $\dfrac{60.5 \, \text{ft}}{0.5 \, \text{sec}} \cdot \dfrac{1 \, \text{m}}{3.3 \, \text{ft}} \cdot \dfrac{1000 \, \text{mm}}{\text{m}} \approx 3.67 \times 10^4 \, \text{mm/sec}$

**b.** About 0.005 sec; The player has the 0.005 sec. it takes the ball to travel those 200 mm.

**c.** No; In 0.006 sec, the ball would have traveled farther than 200 mm and I would not be able to hit it.

## 8.4 Mixed Review (p. 475)

**67.** 0.22   **68.** 0.875   **69.** 0.0007   **70.** 0.0842

**71.** 0.005   **72.** 0.0075   **73.** 2.55   **74.** 0.0125

**75.** $4x + 2y = 12 \qquad\qquad -6x + 3y = 6$

$\quad\ 2y = 12 - 4x \qquad\quad\ 3y = 6 + 6x$

$\quad\ \ y = 6 - 2x \qquad\qquad y = 2 + 2x$

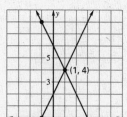

The solution is $(1, 4)$.

**76.** $3x - 2y = 0 \qquad\qquad 3x - 2y = -4$

$\quad -2y = 0 - 3x \qquad\quad -2y = -4 - 3x$

$\qquad y - 0 + \frac{3}{2}x \qquad\qquad\ y = 2 + \frac{3}{2}x$

no solution

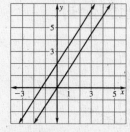

**77.** $2x + y \le 1 \qquad\qquad -2x + y \le 1$

$\qquad\ y \le 1 - 2x \qquad\qquad\ y \le 1 + 2x$

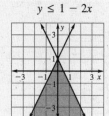

**78.** $x + 2y < 3 \qquad\qquad x - 3y > 1$

$\quad\ 2y < 3 - x \qquad\qquad -3y > 1 - x$

$\qquad y < \frac{3}{2} - \frac{1}{2}x \qquad\qquad\ y < -\frac{1}{3} + \frac{1}{3}x$

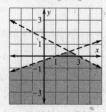

**79.** $2x + y \ge 2 \qquad x \le 2$

$\qquad\ y \ge 2 - 2x$

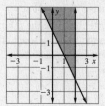

**80.** $2^{-4} = \dfrac{1}{2^4} = \dfrac{1}{16}$   **81.** $\left(\dfrac{1}{10}\right)^{-3} = \left(\dfrac{10}{1}\right)^3 = 1000$

# Chapter 8 *continued*

**82.** $\dfrac{1}{(2x)^{-2}} = \dfrac{(2x)^2}{1} = 4x^2$

**83.** $\dfrac{7^4 \cdot 7}{7^7} = \dfrac{7^{4+1}}{7^7} = \dfrac{7^5}{7^7} = 7^{5-7} = 7^{-2} = \dfrac{1}{7^2} = \dfrac{1}{49}$

## Lesson 8.5

### Developing Concepts Activity 8.5 (p. 476)
**Exploring the Concept**

**1.**

| $x$ | 0 | 1 | 2 | 3 | 4 | 5 |
|---|---|---|---|---|---|---|
| $y$ | 20 | 25 | 30 | 35 | 40 | 45 |

**2.** $y = 5x + 20$

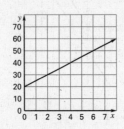

| $x$ | $y$ |
|---|---|
| 0 | 20 |
| 1 | 25 |
| 2 | 30 |
| 5 | 45 |

**3.**

| $x$ | 0 | 1 | 2 | 3 | 4 | 5 |
|---|---|---|---|---|---|---|
| $y$ | 1 | 5 | 25 | 125 | 625 | 3125 |

**4.** $y = 5^x$

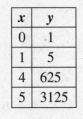 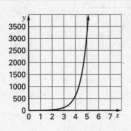

| $x$ | $y$ |
|---|---|
| 0 | 1 |
| 1 | 5 |
| 4 | 625 |
| 5 | 3125 |

**5.** b; a; The points of graph (b) lie on a straight line while the points of graph (a) lie on an exponential curve.

### Drawing Conclusions

**1.** linear growth model   **2.** exponential growth model

**3.** linear growth model   **4.** exponential growth model

**5.** linear growth model   **6.** exponential growth model

**7.** same

**8.** *Sample answer:* The rate of increase is increasing.

**9.** *Sample answer:* For every year after the first, the salary is higher with the 3% annual raise. In the first few years after that, the differences are only a few hundred dollars. After 10 years, the difference is nearly $2000.

## Lesson 8.5

### 8.5 Guided Practice (p. 480)

**1.** initial amount; growth factor

**2.** $A = C(1 + r)^t$
  $A = 500(1 + 0.12)^t$
  $= 500(1.12)^t$

**3.** $1 + r = 2$
  Percent of increase = 100%
  Growth factor = 2

**4.** $A = 500(1 + 0.04)^5 = 500(1.04)^5 \approx \$608.33$
  $A = 500(1 + 0.04)^{10} = 500(1.04)^{10} \approx \$740.12$

**5.** B

### 8.5 Practice and Applications (pp. 480–482)

**6.** $A = 1400(1 + 0.06)^5 = 1400(1.06)^5 \approx \$1{,}873.52$

**7.** $A = 1400(1 + 0.06)^8 = 1400(1.06)^8 \approx \$2{,}231.39$

**8.** $A = 1400(1 + 0.06)^{12} = 1400(1.06)^{12} \approx \$2{,}817.08$

**9.** $A = 1400(1 + 0.06)^{20} = 1400(1.06)^{20} \approx \$4{,}489.99$

**10.** $A = 250(1 + 0.048)^5 \approx \$316.04$

**11.** $A = 300(1 + 0.048)^5 \approx \$379.25$

**12.** $A = 350(1 + 0.048)^5 \approx \$442.46$

**13.** $A = 400(1 + 0.048)^5 \approx \$505.67$

**14.** $A = 200(1 + 0.042)^5 = 200(1.042)^5 \approx \$245.68$

**15.**   $400 = C(1 + 0.035)^6$  **16.**   $1000 = C(1 + 0.06)^8$
  $400 = C(1.035)^6$     $1000 = C(1.06)^8$
  $400 \approx C(1.229255)$   $1000 \approx C(1.59385)$
  $\$325.40 \approx C$     $\$627.41 \approx C$

**17.** $y = 6.37(1.11)^x$     $y = 6.37(1.11)^x$
  $y = 6.37(1.11)^{19}$     $y = 6.37(1.11)^{25}$
  $y \approx 46.3$ L/min     $y \approx 86.5$ L/min

**18.** C   **19.** A   **20.** B

**21.** $P = 10{,}000(1 + 0.25)^t$   **22.** $P = 20{,}000(1.2)^t$
  $P = 10{,}000(1.25)^t$

**23.** $P = 10{,}000(1.25)^t$

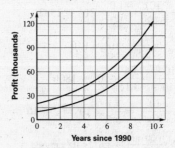

Years since 1990

Answers may vary; items to consider–The profits for the business in Example 22 started higher than those for the other business and would remain higher for 16 years. After that, the profits of the other business would increase more rapidly and be much higher.

**24.** $P = C(1 + r)^t = 30(1 + 1)^4 = 30(2)^4 = 480$
  $1 + r = 2$
  $r = 1$

**25.** B   **26.** Estimates may vary; about 17 cm

# Chapter 8 *continued*

**27.** $L = 12 + (0.09)(1.23)^A = 12 + 0.09(1.23)^{19} \approx$
$12 + 4.60 \approx 16.60$ cm; comparisons may vary

**28.** $L = 12 + (0.09)(1.23)^{13} \approx 12 + 1.33 = 13.33$
$L = 12 + (0.09)(1.23)^{25} \approx 12 + 15.92 = 27.92$
Ratio: about $\frac{1}{2}$

**29.** C $\quad A = 5(1 + 0.09)^5 = 5(1.09)^5 \approx \$7.69$ $\quad$ **30.** C

**31. a.** 2%

**b.** 20 periods

**c.** $A = 8000(1 + 0.02)^{20} = 8000(1.02)^{20} \approx \$11{,}887.58$

## 8.5 Mixed Review *(p. 482)*

**32.** $0.12 \cdot 56 = 6.72$ $\quad$ **33.** $0.75 \cdot 235 = 176.25$

**34.** $0.0125 \cdot 90 = 1.125$ $\quad$ **35.** $2 \cdot 130 = 260$

**36.** $0.02 \cdot 105 = 2.1$ $\quad$ **37.** $0.008 \cdot 120 = 0.96$

**38.** $24 + m^3 = 24 + 5^3 = 24 + 125 = 149$

**39.** $\dfrac{a^2 - b^2}{ab} = \dfrac{3^2 - 5^2}{3(5)} = \dfrac{9 - 25}{15} = -\dfrac{16}{15}$

**40.** $x^6 - 1 = 1.2^6 - 1 \approx 2.985984 - 1 = 1.985984$

**41.** $3y^4 + 15y = 3(-0.02)^4 + 15(-0.02) \approx -0.29999952$

**42.** $(1 - x)^t = (1 - 0.5)^3 = 0.5^3 = 0.125$

**43.** $\dfrac{(1 - x)^t}{2} = \dfrac{(1 - .09)^2}{2} = \dfrac{0.8281}{2} = 0.41405$

**44.** $6B + 8D = 4.10$ $\qquad$ $6B + 8(0.25) = 4.10$
$\phantom{6B + }3B + 3D = 1.80$ $\qquad$ $6B = 4.10 - 2$
$\phantom{xx}6B + 8D = 4.10$ $\qquad$ $6B = 2.10$
$\underline{-6B - 6D = -3.6}$ $\qquad$ $B = 0.35$
$\phantom{xxxxx}2D = 0.5$
$\phantom{xxxxxx}D = 0.25$
Donut $= \$.25$
Bagel $= \$.35$

**45.** $-2(7 - 5x) = 10$
$-14 + 10x = 10$
$10x = 24$
$x = 2.4$

**46.** $25 - (6x + 5) = 4(3x - 5) + 4$
$25 - 6x - 5 = 12x - 20 + 4$
$20 - 6x = 12x - 16$
$36 = 18x$
$2 = x$

**47.** $\frac{3}{2}(8m - 30) = -3m$
$3(8m - 30) = -6m$
$24m - 90 = -6m$
$-90 = -30m$
$3 = m$

**48.** $1.4(6.4y - 3.5) = -9.54y + 22.85$
$8.96y - 4.9 = -9.54y + 22.85$
$18.5y = 27.75$
$y = 1.5$

## Lesson 8.6

### Developing Concepts Activity 8.6 *(p. 483)*
### Drawing Conclusions

**1.** *Sample answer:* The number of pennies remaining appears to decrease exponentially.

**2.** half

**3.** $y = 100\left(\frac{1}{2}\right)^x$, where $x$ is the number of the toss.

**4.** *Sample answer:* All the graphs go through (0, 100). All the points lie along more or less smooth curves that at first descend rapidly and then level off rapidly. The graphs show that by the time there were 7 tosses, most of the cups had no more pennies in them.

### 8.6 Guided Practice *(p. 488)*

**1.** $1 - r$

**2.** No; it is an exponential growth model; 1.02 is a growth factor, not a decay factor.

**3.** $y = 18{,}000(0.8)^t$

**4.** $y = 7000(1 - 0.06)^2 = 7000(0.94)^2 = \$6185.20$

**5.** $y = 7000(1 - 0.06)^5 = 7000(0.94)^5 \approx \$5137.33$

**6.** $y = 7000(1 - 0.06)^8 = 7000(0.94)^8 \approx \$4266.98$

**7.** $y = 7000(1 - 0.06)^{10} = 7000(0.94)^{10} \approx \$3770.31$

**8.** $E = 85{,}000(1 - 0.02)^t$ $\quad$ **9.** C
$E = 85{,}000(0.98)^t$

### 8.6 Practice and Applications *(p. 488–490)*

**10.** $y = 20{,}000(1 - 0.15)^3 = 20{,}000(0.85)^3 \approx \$12{,}283$

**11.** $y = 20{,}000(1 - 0.15)^8 = 20{,}000(0.85)^8 \approx \$5{,}450$

**12.** $y = 20{,}000(1 - 0.15)^{10} = 20{,}000(0.85)^{10} \approx \$3{,}937$

**13.** $y = 20{,}000(1 - 0.15)^{12} = 20{,}000(0.85)^{12} \approx \$2{,}845$

**14.** C $\quad$ **15.** B $\quad$ **16.** A

**17.** exponential growth; 1.18; 18%

**18.** exponential decay; 0.98; 2%

**19.** exponential growth; $\frac{5}{4}$; 25%

**20.** exponential decay; 0.4; 60%

**21.** exponential decay; $\frac{2}{5}$; 60%

**22.** exponential growth; 1.01; 1%

**23.** $y = A(0.8)^t$
$y = 250(0.8)^2 = 160$ mg

**24.** $y = 500(0.8)^{3.5} \approx 229$ mg

**25.** $y = 750(0.8)^5 \approx 246$ mg

**26.** $y = 10{,}500(0.9)^t$ $\qquad$ **27.** $N = 64\left(\frac{1}{2}\right)^t$
$y = 10{,}500(0.9)^{10} \approx \$3{,}661$

**28.** $N = 64\left(\frac{1}{2}\right)^3 = 8$ teams
$N = 64\left(\frac{1}{2}\right)^4 = 4$ teams

**29.** Yes; there are only 6 rounds so the team won each one, including the final one.

# Chapter 8 *continued*

**30.** $E = 320(0.98)^1 \approx 314$  $\quad E = 320(0.98)^2 \approx 307$
$E = 320(0.98)^3 \approx 301$  $\quad E = 320(0.98)^4 \approx 295$
$E = 320(0.98)^5 \approx 289$

| Year | 1995 | 1996 | 1997 | 1998 | 1999 | 2000 |
|---|---|---|---|---|---|---|
| Enrollment | 320 | 314 | 307 | 301 | 295 | 289 |

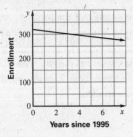

**31.** $M = 302(0.9)^t$

**32.**

| Year | 1894 | 1896 | 1898 | 1899 | 1900 | 1901 | 1903 |
|---|---|---|---|---|---|---|---|
| Miles of Track | 302 | 245 | 198 | 178 | 160 | 144 | 117 |

$M = 302((0.9)^2 \approx 245$  $\quad M = 302((0.9)^4 \approx 198$
$M = 302(0.9)^5 \approx 178$  $\quad M = 302(0.9)^6 \approx 160$
$M = 302(0.9)^7 \approx 144$  $\quad M = 302(0.9)^9 \approx 117$

**33.**

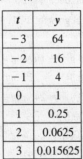

**34.** No; the sweater will never be free because there is no number $n$ other than 0 for which 80% of $n$ is 0. If the price reduction did continue at 20% each day, and the sweater did not sell, the price would eventually be under $0.01. Example: After 38 days, a $40 sweater would be less than $0.01. This is not a realistic possibility however.

**35.** C  $\quad V = 8000(1 - 0.04)^7 = 8000(0.96)^7 \approx \$6,012$

**36.** B

**37.** $y = 4^t$ $\qquad\qquad\qquad y = \left(\frac{1}{4}\right)^t$

| $t$ | $y$ |
|---|---|
| $-3$ | 0.015625 |
| $-2$ | 0.0625 |
| $-1$ | 0.25 |
| 0 | 1 |
| 1 | 4 |
| 2 | 16 |
| 3 | 64 |

| $t$ | $y$ |
|---|---|
| $-3$ | 64 |
| $-2$ | 16 |
| $-1$ | 4 |
| 0 | 1 |
| 1 | 0.25 |
| 2 | 0.0625 |
| 3 | 0.015625 |

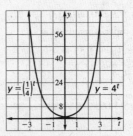

**38.** Each graph is the mirror image of the other reflected in the $y$-axis.

**39.** The graph will be the mirror image of the given graph reflected in the $y$-axis. It will curve up to the left and come very close to the $t$-axis on the right.

### 8.6 Mixed Review (p. 490)

**40.** $4a^2 + 11 = 4(5^2) + 11 = 4(25) + 11 = 111$

**41.** $c^3 + 6cd = 2^3 + 6(2)(1) = 8 + 12 = 20$

**42.** $b^2 - 4ac = 3^2 - 4(1)(5) = 9 - 20 = -11$

**43.** $\dfrac{a^2 - b^2}{2c^2} + 9 = \dfrac{(-3)^2 - 5^2}{2(-2)^2} + 9 = \dfrac{9 - 25}{2(4)} + 9$
$\qquad = \dfrac{-16}{8} + \dfrac{72}{8} = \dfrac{56}{8} = 7$

**44.** $12m - 9 = 5m - 2$ **45.** $5(2x + 2.3) - 11.2 = 6x - 5$
$\qquad 7m = 7$ $\qquad\qquad\quad 10x + 11.5 - 11.2 = 6x - 5$
$\qquad\quad m = 1$ $\qquad\qquad\qquad\quad 4x + 0.3 = -5$
$\qquad\qquad\qquad\qquad\qquad\qquad\qquad\quad 4x = -5.3$
$\qquad\qquad\qquad\qquad\qquad\qquad\qquad\quad\; x \approx -1.3$

**46.** $-1.3y + 3.7 = 4.2 - 5.4y$
$\qquad 4.1y = 0.5$
$\qquad\quad y \approx 0.1$

**47.** $2.5(3.5p + 6.4) = 18.2p - 6.5$
$\qquad 8.75p + 16 = 18.2p - 6.5$
$\qquad\qquad 22.5 = 9.45p$
$\qquad\qquad 2.4 \approx p$

**48.** $y = 50,000(1.02)^t$

### Quiz 2 (p. 491)

**1.** $1.1205 \times 10^{-2}$ **2.** $1.4 \times 10^8$ **3.** $6.7 \times 10^{-8}$

**4.** $3.072 \times 10^{10}$ **5.** $4820$ **6.** $5,000,000,000$

**7.** $0.00000704$ **8.** $0.01112$

**9.** $y = 50(1.05)^t$; $y = 50(1.05)^7 \approx \$70.36$

**10.** $y = 20,000(0.85)^t$; $y = 20,000(0.85)^5 \approx \$8,874.11$

### Math and History (p. 491)

**1.** $5 \times 10^{-7}$ **2.** $1.0032 \times 10^6$ ft

**3.** $400(2500) = 1,000,000$
$\qquad 2.5 \times 10^3$ times

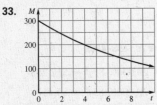

# Chapter 8 *continued*

**Technology Activity 8.6** *(p. 492)*

**1.** $y = 1.0339(1.6293)^x$   **2.** $y = 0.4439(1.3476)^x$

## Chapter 8 Review *(p. 494–496)*

**1.** $2^2 \cdot 2^7 = 2^{2+7} = 2^9$, or 512

**2.** $(4^3)^2 = 4^{3 \cdot 2} = 4^6$, or 4096

**3.** $(3a)^3 \cdot (2a)^2 = 3^3 a^3 \cdot 2^2 \cdot a^2 = 3^3 \cdot 2^2 \cdot a^5$, or $108a^5$

**4.** $(w^3 x^4 y)^2 \cdot (wx^2 y^3)^4 = w^6 x^8 y^2 \cdot w^4 x^8 y^{12} = w^{10} \cdot x^{16} \cdot y^{14}$

**5.** $s^3 \cdot s^4 = s^{3+4} = s^7; \; 2^7 = 128$

**6.** $s^4 \cdot (-t)^3 = -s^4 t^3; \; -(2)^4(3)^3 = -16 \cdot 27 = -432$

**7.** $(s^3 \cdot t)^2 = s^6 t^2; \; 2^6 \cdot 3^2 = 64 \cdot 9 = 576$

**8.** $-(st^2)^2 = -s^2 t^4; \; -(2^2)(3^4) = -4(81) = -324$

**9.** $5^{-3} = \dfrac{1}{5^3} = \dfrac{1}{125}$   **10.** $7^{-4} \cdot 7^6 = 7^{-4+6} = 7^2 = 49$

**11.** $16\left(\dfrac{1}{2}\right)^{-1} = \dfrac{16}{1} \cdot \dfrac{2}{1} = 32$

**12.** $2^0 \cdot \left(\dfrac{1}{4^{-2}}\right) = 1 \cdot 4^2 = 1 \cdot 16 = 16$

**13.** $x^6 y^{-6} = x^6 \cdot \dfrac{1}{y^6} = \dfrac{x^6}{y^6}$   **14.** $\dfrac{1}{5p^8 q^{-3}} = \dfrac{q^3}{5p^8}$

**15.** $(a^2 b)^0 = 1$   **16.** $(-2y)^{-4} = \dfrac{1}{(-2y)^4} = \dfrac{1}{16y^4}$

**17.** $y = 4^x$

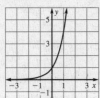

**18.** $y = \left(\dfrac{1}{2}\right)^x$

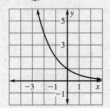

**19.** $y = 3^{-x}$

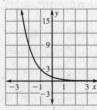

**20.** $y = \left(\dfrac{2}{3}\right)^{-x}$

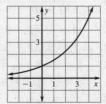

**21.** $\dfrac{3^2}{3^5} = 3^{2-5} = 3^{-3} = \dfrac{1}{3^3} = \dfrac{1}{27}$

**22.** $\dfrac{5^2}{5^{-2}} = 5^2 \cdot 5^2 = 5^{2+2} = 5^4 = 625$

**23.** $\left(-\dfrac{4}{9}\right)^2 = \dfrac{(-4)^2}{9^2} = \dfrac{16}{81}$   **24.** $\left(\dfrac{10}{7}\right)^{-1} = \dfrac{7}{10}$

**25.** $\left(\dfrac{9}{b}\right)^6 = \dfrac{9^6}{b^6} = \dfrac{531,441}{b^6}$   **26.** $\dfrac{x^{12}}{x^6} = x^{12-6} = x^6$

**27.** $\left(\dfrac{m^7}{m^4}\right)^2 = \dfrac{m^{14}}{m^8} = m^{14-8} = m^6$

**28.** $\dfrac{(p^2)^3}{(p^2)^5} = \dfrac{p^6}{p^{10}} = p^{6-10} = p^{-4} = \dfrac{1}{p^4}$

**29.** $\left(-\dfrac{9a^2 b^2}{3ab}\right)^3 = -\dfrac{9^3 a^6 b^6}{3^3 a^3 b^3} = -\dfrac{729 a^6 b^6}{27 a^3 b^3} = -27 a^{6-3} b^{6-3}$

$$= -27 a^3 b^3$$

**30.** $\left(\dfrac{25 a^4 b^5}{-5a^2 b}\right)^3 = \dfrac{25^3 a^{12} b^{15}}{(-5)^3 a^6 b^3} = \dfrac{15625 a^{12} b^{15}}{-125 a^6 b^3}$

$$= -125 a^{12-6} b^{15-3} = -125 a^6 b^{12}$$

**31.** $\dfrac{32 a^4 b^{-2}}{2a^3 b^3} \cdot \dfrac{3a^2 b^7}{-2a} = 16 a^{4-3} b^{-2-3} \cdot -\dfrac{3}{2} a^{2-1} b^7$

$$= 16 ab^{-5} \cdot -\dfrac{3}{2} ab^7 = -\dfrac{24 a^2 b^7}{b^5}$$

$$= -24 a^2 b^{7-5} = -24 a^2 b^2$$

**32.** $\dfrac{9x^{-3} y^6}{x^4 y^{-5}} \cdot \dfrac{(3x^2 y)^{-2}}{xy^3} = 9 x^{-3-4} y^{6-(-5)} \cdot \dfrac{1}{3^2 x^4 y^2 \cdot xy^3}$

$$= \dfrac{9x^{-7} y^{11}}{9x^5 y^5} = x^{-7-5} y^{11-5} = x^{-12} y^6$$

$$= x^{-12} y^6 = \dfrac{y^6}{x^{12}}$$

**33.** $S = 1686(1.17)^0 = 1686$   $S = 1686(1.17)^5 = 3696.47$

Ratio: $\dfrac{1686(1.17)^0}{1686(1.17)^5} = (1.17)^{-5} = \dfrac{1}{(1.17)^5} \approx 0.46$

**34.** 0.006667   **35.** 768,000   **36.** 0.375   **37.** 0.0002

**38.** $5.23 \times 10^8$   **39.** $6.79 \times 10^{-4}$   **40.** $2.33 \times 10^{-8}$

**41.** $7.52 \times 10^7$

**42.** $\dfrac{4.07 \times 10^{13}}{3.0 \times 10^5} \approx 1.36 \times 10^8$

about $1.36 \times 10^8$ sec or about 4.3 years

**43.** $w = 2(1.05)^x$   **44.** $w = 2(1.05)^9 \approx 3.3$ mi

**45.** $e = 125(0.97)^x$   **46.** $e = 125(0.97)^7 \approx 101$ people

## Chapter 8 Test *(p. 497)*

**1.** $x^3 \cdot x^4 = x^{3+4} = x^7$   **2.** $a^0 \cdot a^4 = a^{0+4} = a^4$

**3.** $b^2 \cdot b^{-5} = b^{2+(-5)} = b^{-3} = \dfrac{1}{b^3}$   **4.** $5y^{-4} = \dfrac{5}{y^4}$

**5.** $(x^3)^7 = x^{21}$   **6.** $(a^{-2})^3 = a^{-6} = \dfrac{1}{a^6}$

**7.** $\dfrac{n^3}{n^5} = n^{3-5} = n^{-2} = \dfrac{1}{n^2}$

**8.** $(2b)^3(b^{-4}) = 2^3 b^3 b^{-4} = 2^3 b^{3+(-4)} = 2^3 b^{-1} = \dfrac{2^3}{b} = \dfrac{8}{b}$

**9.** $(mn)^2 \cdot n^4 = m^2 n^2 \cdot n^4 = m^2 n^6$

**10.** $3a^5 \cdot 5a^{-2} \cdot a^3 = 15 a^{5+-2+3} = 15a^6$

**11.** $\left(\dfrac{x^3}{xy^4}\right) \cdot \left(\dfrac{y}{x}\right)^5 = \dfrac{x^3}{xy^4} \cdot \dfrac{y^5}{x^5} = \dfrac{x^3 y^5}{x^6 y^4} = x^{3-6} y^{5-4} = x^{-3} y = \dfrac{y}{x^3}$

**12.** $\dfrac{a^{-1} b^2}{ab} \cdot \dfrac{a^2 b^3}{b^{-2}} = \dfrac{a^{-1+2} b^{2+3}}{ab^{1+-2}}$

$$= \dfrac{ab^5}{ab^{-1}} = a^{1-1} b^{5-(-1)} = a^0 b^6 = b^6$$

# Chapter 8 continued

**13.** $5^4 \cdot 5^{-1} = 5^{4+-1} = 5^3 = 125$     **14.** $4^{-3} = \dfrac{1}{4^3} = \dfrac{1}{64}$

**15.** $(425^2)^0 = 1$     **16.** $\left(\dfrac{5}{2}\right)^{-2} = \left(\dfrac{2}{5}\right)^2 = \dfrac{2^2}{5^2} = \dfrac{4}{25}$

**17.** $\dfrac{3 \cdot 3^5}{3^4} = \dfrac{3^6}{3^4} = 3^{6-4} = 3^2 = 9$

**18.** $\left(\dfrac{3}{4}\right)^3 \cdot 4^2 \cdot 3^0 = \dfrac{3^3}{4^3} \cdot 16 \cdot 1 = \dfrac{27}{64} \cdot 16 = \dfrac{27}{4}$

**19.** $(5 \cdot 4)^3 \cdot 5^{-2} = \dfrac{20^3}{5^2} = \dfrac{8000}{25} = 320$

**20.** $[(-2)^5]^2 = [(-32)]^2 = 1024$     **21.** $427,000$

**22.** $0.000000006283$     **23.** $45,600,000,000$

**24.** $0.000000000005$     **25.** $9.875 \times 10^6$     **26.** $1.25 \times 10^{-3}$

**27.** $6.557 \times 10^9$     **28.** $3.17 \times 10^{-8}$

**29.** $y = 2^x$     **30.** $y = \left(\frac{1}{3}\right)^x$

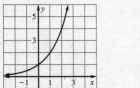
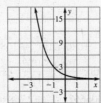

**31.** $y = 10(1.4)^x$

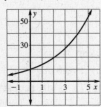

**32.** $V = s^3$
$V = (3a)^3 = 27a^3 = 27(2^3) = 27(8) = 216$

**33.** 2000
$A = 400(1.1)^4 = \$585.64$
2003
$A = 400(1.1)^7 = \$779.49$

**34.** $y = 88,500(1.2)^x$; $y = 88,500(1.2)^5 \approx \$220,216$

**35.** $W = 16(0.5)^h$

| Half-life Periods, h | 0 | 1 | 2 | 3 | 4 |
|---|---|---|---|---|---|
| Grams remaining, W | 16 | 8 | 4 | 2 | 1 |

$w = 16(0.5)^0 = 16$     $w = 16(0.5)^1 = 8$
$w = 16(0.5)^2 = 4$     $w = 16(0.5)^3 = 2$
$w = 16(0.5)^4 = 1$

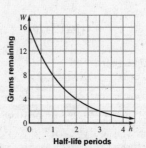

## Chapter 8 Standardized Test (pp. 498–499)

**1.** B     $7^4 \cdot 7^7 = 7^{4+7} = 7^{11}$

**2.** E     $[(a + 1)^2]^2 \cdot a^3 = [(2 + 1)^2]^2 \cdot 2^3$
$\qquad = [(3)^2]^2 \cdot 8$
$\qquad = [9]^2 \cdot 8$
$\qquad = 81 \cdot 8$
$\qquad = 648$

**3.** D     $(5^{-3})^2 = 5^{-6} = \dfrac{1}{5^6} = \dfrac{1}{15625}$

**4.** A     $-8^0 \cdot 2^x \cdot 10^y = -8^0 \cdot 2^{-2} \cdot 10^{-3} = \dfrac{-1}{2^2 \cdot 10^3}$
$\qquad = \dfrac{-1}{4 \cdot 1000} = -\dfrac{1}{4000}$

**5.** A     $\dfrac{4x^2y^2}{4xy} \cdot \dfrac{8xy^3}{4y} = \dfrac{32x^3y^5}{16xy^2} = 2x^{3-1}y^{5-2} = 2x^2y^3$

**6.** C     $\dfrac{x^5}{x^2} = x^{5-2} = x^3$     **7.** C   $21.2 \times 10^{-5}$

**8.** E     $0.0000036$

**9.** B     $(6.2 \times 10^4) \cdot (2.4 \times 10^5) = 14.88 \times 10^9$
$\qquad\qquad = 1.488 \times 10^{10}$

**10.** A     $\dfrac{2.3622 \times 10^4}{3.81 \times 10^{-3}} = 0.62 \times 10^7 = 6.2 \times 10^6$     **11.** A

**12.** C     $A = 450(1 + 0.06)^6 = 450(1.06)^6 \approx 638.33$

**13.** D     $V = 500(1 + 0.12)^8 = 500(1.12)^8 \approx \$1237.98$

**14.** B     $y = 3x = 3(3) = 9$
$\qquad y = 3^x = 3^3 = 27$

**15.** B     $y = 3^x = 3^{-2} = \dfrac{1}{9}$
$\qquad y = \left(\frac{1}{3}\right)^x = \left(\frac{1}{3}\right)^{-2} = 3^2 = 9$

**16.** C     $y = 3^{-x} = 3^{-1} = \dfrac{1}{3}$
$\qquad y = \left(\frac{1}{3}\right)^x = \left(\frac{1}{3}\right)^1 = \dfrac{1}{3}$

**17.** D     $y = 263(0.92)^4 \approx 188.41$ ng/mL

**18.** C     $E = 142,000(1 - 0.08)^t = 142,000(0.92)^t$

**19.** E     $y = 0.97^t$ and $y = \left(\frac{2}{3}\right)^t$

**20. a.** $P = 2(1.175)^0 = 2$; 2000
   **b.** $P = 2(1.175)^5 \approx 4.48$ ; 4480
   **c.** $P = 2(1.175)^{-2} \approx 1.45$ ; 1450
   **d.** Answers may vary.

# CHAPTER 9

## Think & Discuss *(p. 501)*

**1.** about 100 ft   **2.** about 400 ft

## Skill Review *(p. 502)*

**1.** $3x^2 - 108$

$3(-4)^2 - 108 = -60$

**2.** $8x^2 \div \frac{2}{3}$

$8(-1)^2 \times \frac{3}{2} = 12$

**3.** $x^2 - 4xy$

$(-2)^2 - 4(-2)(5) = 44$

**4.** $-\dfrac{x}{2y}$

$-\dfrac{12}{2(-3)} = 2$

**5.**

| $x$ | 0 | 2 | $-6$ | $-4$ | $-2$ |
|-----|---|---|------|------|------|
| $y$ | 3 | 4 | 0 | 1 | 2 |

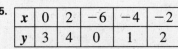

**6.**

| $x$ | 0 | 1 | 2 | $-2$ | $-1$ |
|-----|---|---|---|------|------|
| $y$ | $-1$ | $-3$ | $-5$ | 3 | 1 |

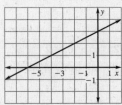

**7.**

| $x$ | 0 | 1 | 2 | 3 | 4 |
|-----|---|---|---|---|---|
| $y$ | 2 | $\frac{13}{7}$ | $\frac{12}{7}$ | $\frac{11}{7}$ | $\frac{10}{7}$ |

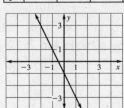

**8.** $3(-1) + 4(2) \overset{?}{<} 5$

$5 \not< 5$

not a solution

**9.** $\frac{1}{2}(0) - \frac{2}{3}(0) \overset{?}{\geq} -6$

$0 \geq -6$

solution

**10.** $6(2) - 2(-3) \overset{?}{>} -8$

$18 > -8$

solution

## Lesson 9.1

### 9.1 Guided Practice *(p. 507)*

**1.** the principal (or positive) square root; the negative square root; both the positive and the negative square root

**2.** *Sample answer:* 16; $\sqrt{5}$

**3.** Subtract $c$ from both sides, then divide both sides by $a$. If $-\dfrac{c}{a}$ is positive, find both square roots of $-\dfrac{c}{a}$. If $-\dfrac{c}{a}$ is zero, the only solution is 0. If $-\dfrac{c}{a}$ is negative, there are no real solutions.

**4.** True; $\sqrt{-3}$   **5.** False; $\sqrt{0} = 0$

**6.** True; $-4$ and 4 are both square roots of 16.

**7.** False; *Sample answer:* $\sqrt{6+3} = 3$; $\sqrt{6} + \sqrt{3} \approx 4.8$

**8.** 6   **9.** 0.9

**10.** $-0.2$   **11.** $\pm 3$

**12.** $\sqrt{4^2 + 10(2)} = \sqrt{36} = 6$

**13.** $\dfrac{10 \pm 2\sqrt{4}}{2}$

$\dfrac{10 + 4}{2} = 7$ and $\dfrac{10 - 4}{2} = 3$

solutions: 7, 3

**14.** $\sqrt{4^2 - 8(2)} = 0$

**15.** $2x^2 - 8 = 0$

$2x^2 = 8$

$x^2 = 4$

$x = 2, -2$

**16.** $x^2 + 25 = 0$

$x^2 = -25$

There is no solution; negative numbers have no real square roots.

**17.** $x^2 - 1.44 = 0$

$x^2 = 1.44$

$x = 1.2, -1.2$

**18.** $5x^2 = -15$

$x^2 = -3$

There is no solution; negative numbers have no real square roots.

**19.** $0 = -16t^2 + 48$

$t^2 = 3$

$t \approx 1.7$ sec.

**20.** $0 = -16t^2 + 96$

$t^2 = 6$

$t \approx 2.4$ sec

**21.** $0 = -16t^2 + 192$

$t^2 = 12$

$t \approx 3.5$ sec

**22.** No; the height is a function of the square of the falling time rather than of the falling time.

# Chapter 9 *continued*

**23.** $(-7)^2 = 49$
$(7)^2 = 49$
$\quad 7, -7$

**24.** $(-12)^2 = 144$
$12^2 = 144$
$\quad 12, -12$

**25.** no square roots

**26.** no square roots

**27.** $0^2 = 0$
$\quad 0$

**28.** $\quad 10^2 = 100$
$(-10)^2 = 100$
$\quad 10, -10$

**29.** $\quad (0.3)^2 = 0.09$
$(-0.3)^2 = 0.09$
$\quad 0.3, -0.3$

**30.** $\quad 0.4^2 = 0.16$
$(-0.4)^2 = 0.16$
$\quad 0.4, -0.4$

**31.** $-13$

**32.** $8$

**33.** $3.61$

**34.** $-11.18$

**35.** $0.2$

**36.** $0.87, -0.87$

**37.** $-0.32$

**38.** $2.5, -2.5$

**39.** $\sqrt{5^2 - 4(4)(1)} = \sqrt{9} = 3$

**40.** $\sqrt{8^2 - 4(-2)(-8)} = \sqrt{0} = 0$

**41.** $\sqrt{(-7)^2 - 4(3)(6)} = \sqrt{-23}$; undefined

**42.** $\sqrt{4^2 - 4(2)(0.5)} = \sqrt{12} = 2\sqrt{3}$

**43.** $\sqrt{7^2 - 4(-3)(5)} = \sqrt{109}$

**44.** $\sqrt{(-8)^2 - 4(6)(4)} = \sqrt{-32}$; undefined

**45.** $4.27, -2.27$

**46.** $-11.66, -0.34$

**47.** $2.13, -1.33$

**48.** $3.00, -2.66$

**49.** $-11.24, -2.76$

**50.** $7.12, -5.12$

**51.** $5.13, -1.80$

**52.** $2.99, -1.49$

**53.** $-1.34, -0.99$

**54.** $x^2 = 36$
$x = 6, -6$

**55.** $b^2 = 64$
$b = 8, -8$

**56.** $5x^2 = 500$
$x^2 = 100$
$x = 10, -10$

**57.** $x^2 = 16$
$x = 4, -4$

**58.** $x^2 = 0$
$x = 0$

**59.** $x^2 = -9$
no solution

**60.** $3x^2 = 6$
$x^2 = 2$
$x = \sqrt{2}, -\sqrt{2}$

**61.** $a^2 + 3 = 12$
$a^2 = 9$
$a = 3, -3$

**62.** $x^2 - 7 = 57$
$x^2 = 64$
$x = 8, -8$

**63.** $2x^2 - 5 = 27$
$2x^2 = 32$
$x^2 = 16$
$x = 4, -4$

**64.** $5x^2 + 5 = 20$
$5x^2 = 15$
$x^2 = 3$
$x = \sqrt{3}, -\sqrt{3}$

**65.** $7x^2 + 30 = 9$
$7x^2 = -21$
$x^2 = -3$
no solution

**66.** $x^2 + 4 = 0$
$x^2 = -4$
no solution

**67.** $6x^2 - 54 = 0$
$6x^2 = 54$
$x^2 = 9$
$x = 3, -3$

**68.** $7x^2 - 63 = 0$
$7x^2 = 63$
$x^2 = 9$
$x = 3, -3$

**69.** $4x^2 - 3 = 57$
$4x^2 = 60$
$x^2 = 15$
$x = 3.87, -3.87$

**70.** $6y^2 + 22 = 34$
$6y^2 = 12$
$y^2 = 2$
$y = 1.41, -1.41$

**71.** $2x^2 - 4 = 10$
$2x^2 = 14$
$x^2 = 7$
$x = 2.65, -2.65$

**72.** $\frac{2}{3}n^2 - 6 = 2$
$\frac{2}{3}n^2 = 8$
$n^2 = 12$
$n = 3.46, -3.46$

**73.** $\frac{4}{5}x^2 + 12 = 5$
$\frac{4}{5}x^2 = -7$
$x^2 = -\frac{35}{4}$
no solution

**74.** $\frac{1}{2}x^2 + 3 = 8$
$\frac{1}{2}x^2 = 5$
$x^2 = 10$
$x = 3.16, -3.16$

**75.** $3x^2 + 7 = 31$
$3x^2 = 24$
$x^2 = 8$
$x = 2.83, -2.83$

**76.** $6s^2 - 12 = 0$
$6s^2 = 12$
$s^2 = 2$
$s = 1.41, -1.41$

**77.** $5a^2 + 10 = 20$
$5a^2 = 10$
$a^2 = 2$
$a = 1.41, -1.41$

**78.** **a.** $x^2 = d$, for any number $d < 0$

**b.** $x^2 = 0$

**c.** $x^2 = d$, for any number $d > 0$

**79.** $h = -16t^2 + 60$

**80.**

| $t$ | 0 | 1 | 1.5 | 1.75 | 1.85 | 1.9 | 1.95 |
|---|---|---|---|---|---|---|---|
| $h$ | 60 | 44 | 24 | 11 | 5.24 | 2.24 | $-0.84$ |

about 1.95 sec

**Algebra 1**
Chapter 9 Worked-out Solution Key

# Chapter 9 *continued*

**81.** $0 = -16t^2 + 60$

$t^2 = 3.75$

$t \approx 1.94$ sec

**82.** *Sample answer:* I prefer solving the equation. It requires fewer calculations and provides a more accurate answer.

**83.** $0 = -16t^2 + 144$

$t^2 = 9$

$t = 3$ sec

**84.** $0 = -16t^2 + 256$

$t^2 = 16$

$t = 4$ sec

**85.** $0 = -16t^2 + 400$

$t^2 = 25$

$t = 5$ sec

**86.** $0 = -16t^2 + 600$

$t^2 = 37.5$

$t \approx 6.12$ sec

**87.** $36,000 = 145.63t^2 + 3327.56$

$t^2 \approx 224.35$

$t \approx 14.98$

in the year 2003

**88.** $7200 = 61.98t^2 + 1001.15$

$t^2 \approx 100$

$t \approx 10$

in the year 2000

**89.** $12d^2 = 1.89$

$d \approx 0.40$ mm

**90.** $50d^2 = 1.89$

$d = 0.19$ mm

**91.** $80d^2 = 1.89$

$d \approx 0.15$ mm

**92.** $125d^2 = 1.89$

$d \approx 0.12$ mm

**93.** $140d^2 = 1.89$

$d \approx 0.12$ mm

**94.** $755d^2 = 1.89$

$d \approx 0.05$ mm

**95.** A

**96.** C

**97.** $132 = \frac{1}{2}(9.8)t^2$

$t \approx 5.19$ sec

$24 = \frac{1}{2}(9.8)t^2$

$t \approx 2.21$ sec

**98.** $490 = \frac{1}{2}(9.8)t^2$

$t = 10$ sec

**99.** 4 times; the distance that an object is dropped is a function of the time squared. So, $(2t)^2 = 2^2 \cdot t^2 = 4t^2$.

**100.** The first equation represents the distance a falling object falls in a given time $t$, and the second represents the height of the same object at time $t$. Then $h = s - d$ and so $-16t^2 = -\frac{1}{2}g(t^2)$; that is, the same exponential term in both equations is the same.

### 9.1 Mixed Review (p. 510)

**101.** 11

**102.** $2^3 \cdot 3$

**103.** $2^3 \cdot 3^2$

**104.** $2^2 \cdot 3^3$

**105.**

$-3x + 4y = -5 \qquad\qquad 4x + 2y = -8$

$\qquad 4y = 3x - 5 \qquad\qquad 2y = -4x - 8$

$\qquad y = \frac{3}{4}x - \frac{5}{4} \qquad\qquad y = -2x - 4$

Check: $-3(-1) + 4(-2) \overset{?}{=} -5 \quad 4(-1) + 2(-2) \overset{?}{=} -8$

$\qquad\qquad\qquad 3 - 8 = -5 \qquad\qquad -4 - 4 = -8$

Solution: $(-1, -2)$

**106.**

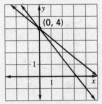

$4x + 5y = 20 \qquad\qquad \frac{5}{4}x + y = 4$

$\qquad 5y = -4x + 20 \qquad\qquad y = -\frac{5}{4}x + 4$

$\qquad y = -\frac{4}{5}x + 4$

Check: $4(0) + 5(4) \overset{?}{=} 20$

$\qquad\qquad\qquad 20 = 20$

$\qquad\qquad \frac{5}{4}(0) + 4 \overset{?}{=} 4$

$\qquad\qquad\qquad 4 = 4$

Solution: $(0, 4)$

**107.**

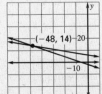

$\frac{1}{2}x + 3y = 18 \qquad\qquad 2x + 6y = -12$

$\qquad 3y = -\frac{1}{2}x + 18 \qquad\qquad 6y = -2x - 12$

$\qquad y = -\frac{1}{6}x + 6 \qquad\qquad y = -\frac{1}{3}x - 2$

Check: $\frac{1}{2}(-48) + 3(14) \overset{?}{=} 18$

$\qquad\qquad -24 + 42 = 18$

$\qquad 2(-48) + 6(14) \overset{?}{=} -12$

$\qquad\qquad -96 + 84 = -12$

Solution: $(-48, 14)$

# Chapter 9 *continued*

**108.** $12x - 4y = -32$

$x + 3y = 4 \qquad 12x - 4y = -32$

Multiply by $-12 \rightarrow \underline{-12x - 36y = -48}$

$\qquad\qquad\qquad\qquad -40y = -80$

$\qquad\qquad\qquad\qquad\quad y = \quad 2$

$\qquad\qquad x + 3(2) = 4$

$\qquad\qquad\qquad\qquad x = -2$

Solution: $(-2, 2)$

**109.** $10x - 3y = 17$

$-7x + y = 9 \qquad 10x - 3y = 17$

Multiply by $3 \rightarrow \underline{-21x + 3y = 27}$

$\qquad\qquad\qquad -11x \quad\quad = 44$

$\qquad\qquad\qquad\quad x \quad\quad = -4$

$\qquad\qquad 10(-4) - 3y = 17$

$\qquad\qquad\qquad\qquad -3y = 57$

$\qquad\qquad\qquad\qquad\quad y = -19$

Solution: $(-4, -19)$

**110.** $8x - 5y = 100$

$2x + \frac{1}{2}y = 4 \qquad 8x - 5y = \quad 100$

Multiply by $-4 \rightarrow \underline{-8x - 2y = -16}$

$\qquad\qquad\qquad\quad -7y = \quad 84$

$\qquad\qquad\qquad\qquad\quad y = -12$

$\qquad\qquad 2x + \frac{1}{2}(-12) = 4$

$\qquad\qquad\qquad 2x - 6 = 4$

$\qquad\qquad\qquad\quad 2x = 10$

$\qquad\qquad\qquad\quad x = 5$

Solution: $(5, -12)$

**111.** $x + y = 2342$

$2x + 3y = 5801$

Multiply by $-2 \rightarrow -2x - 2y = -4684$

$\qquad\qquad\qquad\quad \underline{2x + 3y = \quad 5801}$

$\qquad\qquad\qquad\qquad\quad y = \quad 1117$

$\qquad\qquad x + 1117 = 2342$

$\qquad\qquad\qquad\qquad x = 1225$

1225 student tickets and 1117 general admission tickets

**112.** $x + y = 25 \quad x = -y + 25$

$x + 0.5y = 20 \quad (-y + 25) + 0.5y = 20$

$\qquad\qquad\qquad -y + 25 + 0.5y = 20$

$\qquad\qquad\qquad\qquad\qquad -0.5y = -5$

$\qquad\qquad\qquad\qquad\qquad\quad y = 10$

$\qquad\qquad x + 10 = 25$

$\qquad\qquad\qquad\quad x = 15$

15 irises and 10 white tulips

## Lesson 9.2

### Activity (p. 511)

**1.** a. $\sqrt{(4)(9)} = \sqrt{36} = 6$

$\sqrt{4} \cdot \sqrt{9} = 2 \cdot 3 = 6$

b. $\sqrt{(64)(100)} = \sqrt{6400} = 80$

$\sqrt{64} \cdot \sqrt{100} = 8 \cdot 10 = 80$

c. $\sqrt{(25)(4)} = \sqrt{100} = 10$

$\sqrt{25} \cdot \sqrt{4} = 5 \cdot 2 = 10$

d. $\sqrt{(36)(16)} = \sqrt{576} = 24$

$\sqrt{36} \cdot \sqrt{16} = 6 \cdot 4 = 24$

e. $\sqrt{(100)(625)} = \sqrt{62,500} = 250$

$\sqrt{100} \cdot \sqrt{625} = 10 \cdot 25 = 250$

f. $\sqrt{(121)(49)} = \sqrt{5929} = 77$

$\sqrt{121} \cdot \sqrt{49} = = 11 \cdot 7 = 77$

**2.** If both $a$ and $b$ are positive, $\sqrt{ab} = \sqrt{a} \cdot \sqrt{b}$.

**3.** a. $\sqrt{\dfrac{4}{49}} = \dfrac{2}{7}$  b. $\sqrt{\dfrac{16}{64}} = \sqrt{\dfrac{1}{4}} = \dfrac{1}{2}$

$\dfrac{\sqrt{4}}{\sqrt{49}} = \dfrac{2}{7}$  $\dfrac{\sqrt{16}}{\sqrt{64}} = \dfrac{4}{8} = \dfrac{1}{2}$

c. $\sqrt{\dfrac{25}{36}} = \dfrac{5}{6}$  d. $\sqrt{\dfrac{225}{4}} = \dfrac{15}{2}$

$\dfrac{\sqrt{25}}{\sqrt{36}} = \dfrac{5}{6}$  $\dfrac{\sqrt{225}}{\sqrt{4}} = \dfrac{15}{2}$

e. $\sqrt{\dfrac{144}{100}} = \sqrt{\dfrac{36}{25}} = \dfrac{6}{5}$  f. $\sqrt{\dfrac{9}{81}} = \sqrt{\dfrac{1}{9}} = \dfrac{1}{3}$

$\dfrac{\sqrt{144}}{\sqrt{100}} = \dfrac{12}{10} = \dfrac{6}{5}$  $\dfrac{\sqrt{9}}{\sqrt{81}} = \dfrac{3}{9} = \dfrac{1}{3}$

**4.** If both $a$ and $b$ are positive, $\sqrt{\dfrac{a}{b}} = \dfrac{\sqrt{a}}{\sqrt{b}}$

### 9.2 Guided Practice (p. 514)

**1.** a. Yes; there are no perfect square factors other than 1 in the radicand, there are no fractions in the radicand, and no radicals appear in the denominator of the fraction.

b. No; there is a fraction in the radicand and the denominator of the fraction is a perfect square.

c. No; the radicand contains a perfect square factor.

**2.** $\sqrt{3} \cdot \sqrt{15} = \sqrt{3} \cdot \sqrt{3 \cdot 5} = \sqrt{3} \cdot \sqrt{3} \cdot \sqrt{5} = 3\sqrt{5}$

**3.** $\sqrt{\dfrac{4}{25}} = \dfrac{\sqrt{4}}{\sqrt{25}} = \dfrac{2}{5}$

**4.** D  **5.** A  **6.** B  **7.** C

# Chapter 9 *continued*

**8.** $s^2 = \dfrac{16}{9}x$

$s^2 = \dfrac{16}{9}(25)$

$s = \dfrac{\sqrt{16(25)}}{\sqrt{9}} = \dfrac{20}{3} \approx 6.7$ knots

**9.** The factor 5 cannot be taken outside the radicand. The first step should be $\sqrt{50} = \sqrt{25 \cdot 2}$. Then, by the product property, this equals $\sqrt{25} \cdot \sqrt{2}$, or $5\sqrt{2}$.

**10.** $\sqrt{44} = \sqrt{4 \cdot 11}$
$= \sqrt{4} \cdot \sqrt{11}$
$= 2\sqrt{11}$

**11.** $\sqrt{27} = \sqrt{9 \cdot 3}$
$= \sqrt{9} \cdot \sqrt{3}$
$= 3\sqrt{3}$

**12.** $\sqrt{48} = \sqrt{16 \cdot 3}$
$= \sqrt{16} \cdot \sqrt{3}$
$= 4\sqrt{3}$

**13.** $\sqrt{75} = \sqrt{25 \cdot 3}$
$= \sqrt{25} \cdot \sqrt{3}$
$= 5\sqrt{3}$

**14.** $\sqrt{90} = \sqrt{9 \cdot 10}$
$= \sqrt{9} \cdot \sqrt{10}$
$= 3\sqrt{10}$

**15.** $\sqrt{125} = \sqrt{25 \cdot 5}$
$= \sqrt{25} \cdot \sqrt{5}$
$= 5\sqrt{5}$

**16.** $\sqrt{200} = \sqrt{100 \cdot 2}$
$= \sqrt{100} \cdot \sqrt{2}$
$= 10\sqrt{2}$

**17.** $\sqrt{80} = \sqrt{16 \cdot 5}$
$= \sqrt{16} \cdot \sqrt{5}$
$= 4\sqrt{5}$

**18.** $\dfrac{1}{2}\sqrt{112} = \dfrac{1}{2}\sqrt{16 \cdot 7}$
$= \dfrac{1}{2}\sqrt{16} \cdot \sqrt{7}$
$= \dfrac{1}{2}(4) \cdot \sqrt{7}$
$= 2\sqrt{7}$

**19.** $\dfrac{1}{3}\sqrt{54} = \dfrac{1}{3}\sqrt{9 \cdot 6}$
$= \dfrac{1}{3}\sqrt{9} \cdot \sqrt{6}$
$= \dfrac{1}{3}(3) \cdot \sqrt{6}$
$= \sqrt{6}$

**20.** $\sqrt{2} \cdot \sqrt{8} = \sqrt{2} \cdot \sqrt{4} \cdot \sqrt{2}$
$= 2 \cdot 2$
$= 4$

**21.** $\sqrt{6} \cdot \sqrt{8} = \sqrt{2} \cdot \sqrt{3} \cdot \sqrt{4} \cdot \sqrt{2}$
$= 2 \cdot 2 \cdot \sqrt{3}$
$= 4\sqrt{3}$

**22.** $\sqrt{\dfrac{7}{9}} = \dfrac{\sqrt{7}}{\sqrt{9}}$
$= \dfrac{\sqrt{7}}{3}$

**23.** $\sqrt{\dfrac{11}{16}} = \dfrac{\sqrt{11}}{\sqrt{16}}$
$= \dfrac{\sqrt{11}}{4}$

**24.** $2\sqrt{\dfrac{5}{4}} = \dfrac{2\sqrt{5}}{\sqrt{4}}$
$= \dfrac{2\sqrt{5}}{2}$
$= \sqrt{5}$

**25.** $18\sqrt{\dfrac{5}{81}} = 18\dfrac{\sqrt{5}}{\sqrt{81}}$
$= 18\dfrac{\sqrt{5}}{9}$
$= 2\sqrt{5}$

**26.** $2\sqrt{\dfrac{10}{2}} = \dfrac{2\sqrt{10}}{\sqrt{2}}$
$= \dfrac{2\sqrt{5} \cdot \sqrt{2}}{\sqrt{2}}$
$= 2\sqrt{5}$

**27.** $3\sqrt{\dfrac{9}{3}} = \dfrac{3\sqrt{9}}{\sqrt{3}}$
$= \dfrac{3\sqrt{3} \cdot \sqrt{3}}{\sqrt{3}}$
$= 3\sqrt{3}$

**28.** $8\sqrt{\dfrac{13}{9}} = \dfrac{8\sqrt{13}}{\sqrt{9}}$
$= \dfrac{8\sqrt{13}}{3}$

**29.** $3\sqrt{\dfrac{8}{64}} = \dfrac{3\sqrt{8}}{\sqrt{64}}$
$= \dfrac{3\sqrt{4} \cdot \sqrt{2}}{8}$
$= \dfrac{3\sqrt{2}}{4}$

**30.** $4\sqrt{\dfrac{16}{4}} = 4\sqrt{4}$
$= 8$

**31.** $3\sqrt{\dfrac{3}{16}} = \dfrac{3\sqrt{3}}{\sqrt{16}}$
$= \dfrac{3\sqrt{3}}{4}$

**32.** $5\sqrt{\dfrac{6}{2}} = 5\sqrt{3}$

**33.** $8\sqrt{\dfrac{20}{4}} = 8\sqrt{5}$

**34.** $\dfrac{\sqrt{32}}{\sqrt{25}} = \dfrac{\sqrt{16} \cdot \sqrt{2}}{5}$
$= \dfrac{4\sqrt{2}}{5}$

**35.** $\sqrt{\dfrac{27}{36}} = \sqrt{\dfrac{3}{4}}$
$= \dfrac{\sqrt{3}}{\sqrt{4}}$
$= \dfrac{\sqrt{3}}{2}$

**36.** $\dfrac{\sqrt{49}}{\sqrt{4}} = \dfrac{7}{2}$

**37.** $\dfrac{\sqrt{36}}{\sqrt{9}} = \dfrac{6}{3}$
$= 2$

**38.** $\dfrac{\sqrt{9}}{\sqrt{49}} = \dfrac{3}{7}$

**39.** $\dfrac{\sqrt{48}}{\sqrt{81}} = \dfrac{\sqrt{16} \cdot \sqrt{3}}{9}$
$= \dfrac{4\sqrt{3}}{9}$

**40.** $\dfrac{\sqrt{64}}{\sqrt{16}} = \dfrac{8}{4}$
$= 2$

**41.** $\dfrac{\sqrt{120}}{\sqrt{4}} = \dfrac{\sqrt{4} \cdot \sqrt{30}}{2}$
$= \sqrt{30}$

**42.** $\dfrac{1}{2}\sqrt{32} \cdot \sqrt{2} = \dfrac{1}{2}\sqrt{64} = \dfrac{1}{2}(8) = 4$

**43.** $3\sqrt{63} \cdot \sqrt{4} = 3\sqrt{9} \cdot \sqrt{7} \cdot \sqrt{4}$
$= 3(3)(2)\sqrt{7}$
$= 18\sqrt{7}$

**44.** $\sqrt{9} \cdot 4\sqrt{25} = (3)(4)(5)$
$= 60$

**45.** $-2\sqrt{27} \cdot \sqrt{3} = -2\sqrt{9} \cdot \sqrt{3} \cdot \sqrt{3}$
$= -2(3)(3)$
$= -18$

**46.** $\sqrt{7} \cdot \dfrac{\sqrt{18}}{\sqrt{2}} = \sqrt{7} \cdot \sqrt{\dfrac{18}{2}}$
$= \sqrt{7} \cdot \sqrt{9}$
$= 3\sqrt{7}$

# Chapter 9 *continued*

**47.** $-\sqrt{4} \cdot \dfrac{\sqrt{81}}{\sqrt{36}} = \dfrac{-2(9)}{6}$

$\qquad = -3$

**48.** $\dfrac{\sqrt{10} \cdot \sqrt{16}}{\sqrt{5}} = \dfrac{\sqrt{2} \cdot \sqrt{5} \cdot 4}{\sqrt{5}}$

$\qquad = 4\sqrt{2}$

**49.** $\dfrac{-2\sqrt{20}}{\sqrt{100}} = \dfrac{-2\sqrt{4} \cdot \sqrt{5}}{10}$

$\qquad = \dfrac{-2\sqrt{5}}{5}$

**50.** $A = \sqrt{20} \cdot \sqrt{10}$

$\qquad = \sqrt{4} \cdot \sqrt{5} \cdot \sqrt{2} \cdot \sqrt{5}$

$\qquad = 2(5)\sqrt{2}$

$\qquad = 10\sqrt{2}$

$\qquad \approx 14.14$

**51.** $A = \dfrac{1}{2}\left(\dfrac{\sqrt{2}}{2}\right)\left(\dfrac{\sqrt{2}}{2}\right) = \dfrac{2}{8} = \dfrac{1}{4} = 0.25$

**52.** $A = (7\sqrt{2})(7\sqrt{2})$

$\qquad = 49(2)$

$\qquad = 98; \ 98.00$

**53.** $A = \sqrt{\dfrac{(180)(75)}{3600}}$

$\qquad = \sqrt{\dfrac{13,500}{3600}}$

$\qquad = \sqrt{\dfrac{15}{4}}$

$\qquad = \dfrac{\sqrt{15}}{2}$

$\qquad \approx 1.94 \ \text{m}^2$

**54.** $A = \sqrt{\dfrac{(160)(50)}{3600}}$

$\qquad = \sqrt{\dfrac{8000}{3600}}$

$\qquad = \sqrt{\dfrac{20}{9}}$

$\qquad = \dfrac{2\sqrt{5}}{3}$

$\qquad \approx 1.49 \ \text{m}^2$

**55.** $S = \sqrt{(9.8)(1000)}$

$\quad S = \sqrt{9800}$

$\quad S = \sqrt{4900 \cdot 2}$

$\quad S = 70\sqrt{2} \ \text{m/sec}$

**56.** $S = \sqrt{(9.8)(4000)}$

$\quad S = \sqrt{39,200}$

$\quad S = \sqrt{19,600 \cdot 2}$

$\quad S = 140\sqrt{2} \ \text{m/sec}$

**57.** No; the speed is twice the speed of a tsunami at 1000 m deep; the speed is a function of the square root of the depth, not of the depth.

**58.** $\dfrac{4\sqrt{125}}{\sqrt{25}} = \dfrac{4\sqrt{25} \cdot \sqrt{5}}{\sqrt{25}}$

$\qquad = 4\sqrt{5}$

B

**59.** $\sqrt{\dfrac{2}{121}} = \dfrac{33\sqrt{2}}{\sqrt{121}}$

$\qquad = \dfrac{33\sqrt{2}}{11}$

$\qquad = 3\sqrt{2}$

C

**60.** $\dfrac{6\sqrt{52}}{\sqrt{2} \cdot \sqrt{8}} = \dfrac{6\sqrt{13} \cdot \sqrt{14}}{4}$

$\qquad = 3\sqrt{13}$

B

**61. a.**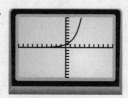

**b.** $y = 1.4142136$

**c.** The two values are the same.

**62.** $6x^{\frac{1}{2}} = 6\sqrt{x}$

**63.** $x^{\frac{1}{2}} \cdot 4\sqrt{2} = \sqrt{x} \cdot 4\sqrt{2}$

$\qquad = 4\sqrt{2x}$

**64.** $18^{\frac{1}{2}}x \cdot 9x^{\frac{1}{2}}x = \sqrt{18} \cdot x \cdot 9\sqrt{x} \cdot x$

$\qquad = 27x^2\sqrt{2x}$

**65.**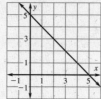

$y = -x + 5$

| $x$ | 0 | 1 | 2 | 3 |
|---|---|---|---|---|
| $y$ | 5 | 4 | 3 | 2 |

**66.**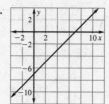

$y = x - 7$

| $x$ | 3 | 4 | 5 | 6 |
|---|---|---|---|---|
| $y$ | $-4$ | $-3$ | $-2$ | $-1$ |

**67.**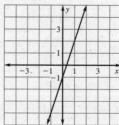

$y = 3x - 1$

| $x$ | 0 | 1 | 2 | 3 |
|---|---|---|---|---|
| $y$ | $-1$ | 2 | 5 | 8 |

**68.** $(a^3)^2 = a^6$

$\qquad = 1^6$

$\qquad = 1$

**69.** $b^6 \cdot b^2 = b^8$

$\qquad = 2^8$

$\qquad = 256$

**70.** $(a^3b)^4 = a^{12}b^4$

$\qquad = 1^{12} \cdot 2^4$

$\qquad = 16$

**71.** $\dfrac{1}{2x^{-5}} = \dfrac{x^5}{2}$

**72.** $\dfrac{1}{4x^{-7}} = \dfrac{x^7}{4}$

**73.** $x^{-4}y^3 = \dfrac{y^3}{x^4}$

**74.** $6x^{-2}y^{-6} = \dfrac{6}{x^2y^6}$

**75.** $\dfrac{455\left(\dfrac{13}{10}\right)^5}{455\left(\dfrac{13}{10}\right)^1} = \left(\dfrac{13}{10}\right)^4 \approx 2.86$

200 **Algebra 1**
Chapter 9 Worked-out Solution Key

# Chapter 9 *continued*

## Lesson 9.3

### Developing Concepts Activity 9.3 (p. 517)
### Exploring the Concept

1. As $|a|$ increases, the graph of $y = ax^2$ gets narrower, and as $|a|$ decreases, the graph of $y = ax^2$ gets wider. If $a$ is positive, the graph of $y = ax^2$ opens up, and if $a$ is negative, the graph of $y = ax^2$ opens down.

2. For negative values of $b$, the graph of $y = x^2 + bx$ is the graph of $y = x^2$ moved down and to the right. For positive values of $b$, the graph of $y = x^2 + bx$ is the graph of $y = x^2$ moved down and to the left.

3. For negative values of $c$, the graph of $y = x^2 + c$ is the graph of $y = x^2$ moved down. For positive values of $c$, the graph of $y = x^2 + c$ is the graph of $y = x^2$ moved up.

### Drawing Conclusions

1. $a = 1$ (or $a > 0$), $b = 0$, $c = 0$

2. $a = 1$ (or $a > 0$), $b = 0$, $c > 0$

3. $a < 0$, $b < 0$, $c < 0$

4.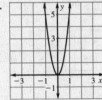

5.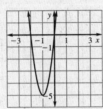

6.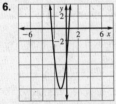

7.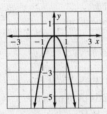

8.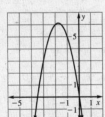

9.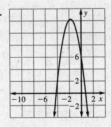

10. D   For the graph shown, $b \neq 0$.

### 9.3 Guided Practice (p. 521)

1. $a = -5$, $b = 7$, $c = -4$

2. The axis of symmetry divides the parabola into two parts that are symmetrical, that is, they are mirror images of each other.

3. The graph opens up because $a = 3$, which is positive.

4. $x = \dfrac{-4}{2(2)} = -1$

   $y = 2(-1)^2 + 4(-1) + 2$

   $y = -4$

   vertex: $(-1, -4)$

5. up

   $x = \dfrac{-4}{2(1)} = -2$

   axis of symmetry: $x = -2$

6. up

   $x = \dfrac{-8}{2(3)} = -\dfrac{4}{3}$

   axis of symmetry: $x = -\dfrac{4}{3}$

7. up

   $x = -\dfrac{7}{2(1)} = -\dfrac{7}{2}$

   axis of symmetry: $x = -\dfrac{7}{2}$

8. down

   $x = \dfrac{4}{2(-1)} = -2$

   axis of symmetry: $x = -2$

9. up

   $x = \dfrac{2}{2(5)} = \dfrac{1}{5}$

   axis of symmetry: $x = \dfrac{1}{5}$

10. down

    $x = \dfrac{-0}{2(-1)} = 0$

    axis of symmetry: $x = 0$

11. $y = -3x^2$

    $x = \dfrac{-0}{2(-3)} = 0$

    vertex: $(0, 0)$

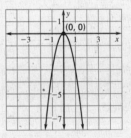

12. $y = -3x^2 + 6x + 2$

    $x = \dfrac{-6}{2(-3)} = 1$

    vertex: $(1, 5)$

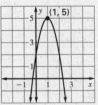

# Chapter 9 *continued*

**13.** $y = -5x^2 + 10$

$x = \dfrac{-0}{2(-5)} = 0$

vertex: $(0, 10)$

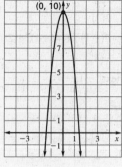

**14.** $y = x^2 + 4x + 7$

$x = \dfrac{-4}{2(1)} = -2$

vertex: $(-2, 3)$

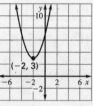

**15.** $y = x^2 - 6x + 8$

$x = \dfrac{+6}{2(1)} = 3$

vertex: $(3, -1)$

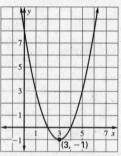

**16.** $y = 5x^2 + 5x - 2$

$x = \dfrac{-5}{5(2)} = -\dfrac{1}{2}$

vertex: $\left(-\dfrac{1}{2}, -\dfrac{13}{4}\right)$

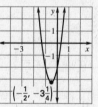

**17.** $y = -4x^2 - 4x + 12$

$x = \dfrac{+4}{2(-4)} = -\dfrac{1}{2}$

vertex: $\left(-\dfrac{1}{2}, 13\right)$

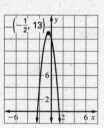

**18.** $y = 3x^2 - 6x + 1$

$x = \dfrac{6}{2(3)} = 1$

vertex: $(1, -2)$

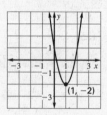

**19.** $y = 2x^2 - 8x + 3$

$x = \dfrac{8}{2(2)} = 2$

vertex: $(2, -5)$

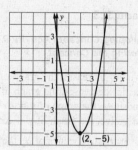

**20. a.** $y = -16x^2 + 15x + 6$

$x = \dfrac{-15}{2(-16)} = \dfrac{15}{32}$

$y = -16\left(\dfrac{15}{32}\right)^2 + 15\left(\dfrac{15}{32}\right) + 6$

$= \dfrac{-3600}{1024} + \dfrac{225}{32} + \dfrac{192}{32}$

$y = \dfrac{-3600 + 7200 + 6144}{1024}$

$y \approx 9.5$ ft

**b.**

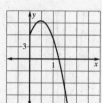

vertex: $\left(\dfrac{15}{32}, 9\dfrac{1}{2}\right)$

| $x$ | 0 | 1 | 1.5 |
|-----|---|---|-----|
| $y$ | 6 | 5 | $-7.5$ |

about 1.2 sec.

**21.** $y = 2x^2$

  **a.** opens up

  **b.** $x = \dfrac{-0}{2(2)} = 0$

    vertex: $(0, 0)$

  **c.** axis of symmetry:
    $x = 0$

**22.** $y = -7x^2$

  **a.** opens down

  **b.** $x = \dfrac{-0}{2(-7)} = 0$

    vertex: $(0, 0)$

  **c.** axis of symmetry:
    $x = 0$

**23.** $y = 6x^2$

  **a.** opens up

  **b.** $x = \dfrac{-0}{2(6)} = 0$

    vertex: $(0, 0)$

  **c.** axis of symmetry:
    $x = 0$

**24.** $y = \dfrac{1}{2}x^2$

  **a.** opens up

  **b.** $x = \dfrac{-0}{2\left(\dfrac{1}{2}\right)} = 0$

    vertex: $(0, 0)$

  **c.** axis of symmetry:
    $x = 0$

**25.** $y = -5x^2$

  **a.** opens down

  **b.** $x = \dfrac{-0}{2(-5)} = 0$

    vertex: $(0, 0)$

  **c.** axis of symmetry:
    $x = 0$

**26.** $y = -4x^2$

  **a.** opens down

  **b.** $x = \dfrac{-0}{2(-4)} = 0$

    vertex: $(0, 0)$

  **c.** axis of symmetry:
    $x = 0$

# Chapter 9 *continued*

**27.** $y = -16x^2$

  **a.** opens down

  **b.** $x = \dfrac{-0}{-16(2)} = 0$

    vertex: $(0, 0)$

  **c.** axis of symmetry:
    $x = 0$

**28.** $y = 5x^2 - x$

  **a.** opens up

  **b.** $x = \dfrac{+1}{2(5)} = \dfrac{1}{10}$

    vertex: $\left(\dfrac{1}{10}, -\dfrac{1}{20}\right)$

  **c.** axis of symmetry: $x = \dfrac{1}{10}$

**29.** $y = 2x^2 - 10x$

  **a.** opens up

  **b.** $x = \dfrac{10}{2(2)} = \dfrac{5}{2}$

    vertex: $\left(\dfrac{5}{2}, -\dfrac{25}{2}\right)$

  **c.** axis of symmetry:

    $x = \dfrac{5}{2}$

**30.** $y = -7x^2 + 2x$

  **a.** opens down

  **b.** $x = \dfrac{-2}{2(-7)} = \dfrac{1}{7}$

    vertex: $\left(\dfrac{1}{7}, \dfrac{1}{7}\right)$

  **c.** axis of symmetry: $x = \dfrac{1}{7}$

**31.** $y = -10x^2 + 12x$

  **a.** opens down

  **b.** $x = \dfrac{-12}{2(-10)} = \dfrac{3}{5}$

    vertex: $\left(\dfrac{3}{5}, \dfrac{18}{5}\right)$

  **c.** axis of symmetry:

    $x = \dfrac{3}{5}$

**32.** $y = 6x^2 + 2x + 4$

  **a.** opens up

  **b.** $x = \dfrac{-2}{2(6)} = -\dfrac{1}{6}$

    vertex: $\left(-\dfrac{1}{6}, \dfrac{23}{6}\right)$

  **c.** axis of symmetry: $x = -\dfrac{1}{6}$

**33.** $y = 5x^2 + 10x + 7$

  **a.** opens up

  **b.** $x = \dfrac{-10}{2(5)} = -1$

    vertex: $(-1, 2)$

  **c.** axis of symmetry:
    $x = -1$

**34.** $y = -4x^2 - 4x + 8$

  **a.** opens down

  **b.** $x = \dfrac{4}{2(-4)} = -\dfrac{1}{2}$

    vertex: $\left(-\dfrac{1}{2}, 9\right)$

  **c.** axis of symmetry: $x = -\dfrac{1}{2}$

**35.** $y = 2x^2 - 7x - 8$

  **a.** opens up

  **b.** $x = \dfrac{7}{2(2)} = \dfrac{7}{4}$

    vertex: $\left(\dfrac{7}{4}, -\dfrac{113}{8}\right)$

  **c.** axis of symmetry:

    $x = \dfrac{7}{4}$

**36.** $y = 2x^2 + 7x - 21$

  **a.** opens up

  **b.** $x = \dfrac{-7}{2(2)} = -\dfrac{7}{4}$

    vertex: $\left(-\dfrac{7}{4}, -\dfrac{217}{8}\right)$

  **c.** axis of symmetry: $x = -\dfrac{7}{4}$

**37.** $y = -x^2 + 8x + 32$

  **a.** opens down

  **b.** $x = \dfrac{-8}{2(-1)} = 4$

    vertex: $(4, 48)$

  **c.** axis of symmetry:
    $x = 4$

**39.** $y = 4x^2 + \dfrac{1}{4}x - 8$

  **a.** opens up

  **b.** $\dfrac{-\dfrac{1}{4}}{2(4)} = \dfrac{-1}{32}$

    vertex: $\left(-\dfrac{1}{32}, -\dfrac{2049}{256}\right)$

  **c.** axis of symmetry:

    $x = -\dfrac{1}{32}$

**41.** $y = 0.78x^2 - 4x - 8$

  **a.** opens up

  **b.** $x = \dfrac{4}{2(0.78)} = \dfrac{2}{0.78} \approx 2.56$

    vertex: $(2.56, -13.1)$

  **c.** axis of symmetry: $x \approx 2.56$

**42.** $y = 3.5x^2 + 2x - 8$

  **a.** opens up

  **b.** $x = \dfrac{-2}{2(3.5)} = -\dfrac{2}{7}$

    vertex: $\left(-\dfrac{2}{7}, -\dfrac{58}{7}\right)$

  **c.** axis of symmetry: $x = -\dfrac{2}{7}$

**43.** $y = -10x^2 - 7x + 2.66$

  **a.** opens down

  **b.** $x = \dfrac{7}{2(-10)} = -0.35$

    vertex: $(-0.35, 3.885)$

  **c.** axis of symmetry: $x = -0.35$

**44.** $y = x^2$

  $x = \dfrac{0}{2(1)} = 0$

  vertex: $(0, 0)$

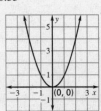

**38.** $y = \dfrac{1}{2}x^2 + 3x - 7$

  **a.** opens up

  **b.** $x = \dfrac{-3}{2\left(\dfrac{1}{2}\right)} = -3$

    vertex: $\left(-3, -\dfrac{23}{2}\right)$

  **c.** axis of symmetry:
    $x = -3$

**40.** $y = -10x^2 + 5x - 3$

  **a.** opens down

  **b.** $x = \dfrac{-5}{2(-10)} = \dfrac{1}{4}$

    vertex: $\left(\dfrac{1}{4}, -\dfrac{19}{8}\right)$

  **c.** axis of symmetry:

    $x = \dfrac{1}{4}$

# Chapter 9 *continued*

**45.** $y = -2x^2$

$x = \dfrac{0}{2(-2)} = 0$

vertex: $(0, 0)$

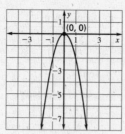

**46.** $y = 4x^2$

$x = \dfrac{0}{2(4)} = 0$

vertex: $(0, 0)$

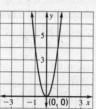

**47.** $y = x^2 + 4x - 1$

$x = \dfrac{-4}{2(1)} = -2$

vertex: $(-2, -5)$

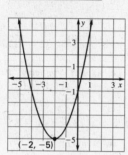

$(1, -6)$

**48.** $y = -3x^2 + 6x - 9$

$x = \dfrac{-6}{2(-3)} = 1$

vertex: $(1, -6)$

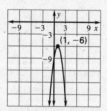

**49.** $y = 4x^2 + 8x - 3$

$x = \dfrac{-8}{2(4)} = -1$

vertex: $(-1, -7)$

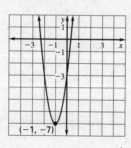

**50.** $y = 2x^2 - x$

$x = \dfrac{1}{2(2)} = \dfrac{1}{4}$

vertex: $\left(\dfrac{1}{4}, -\dfrac{1}{8}\right)$

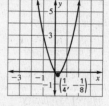

**51.** $y = 6x^2 - 4x$

$x = \dfrac{4}{2(6)} = \dfrac{1}{3}$

vertex: $\left(\dfrac{1}{3}, -\dfrac{2}{3}\right)$

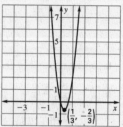

**52.** $y = 3x^2 - 2x$

$x = \dfrac{2}{2(3)} = \dfrac{1}{3}$

vertex: $\left(\dfrac{1}{3}, -\dfrac{1}{3}\right)$

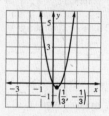

**53.** $y = x^2 + x + 4$

$x = \dfrac{-1}{2(1)} = \dfrac{-1}{2}$

vertex: $\left(-\dfrac{1}{2}, \dfrac{15}{4}\right)$

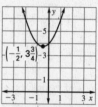

**54.** $y = x^2 + x + \dfrac{1}{4}$

$x = \dfrac{-1}{2(1)} = \dfrac{-1}{2}$

vertex: $\left(-\dfrac{1}{2}, 0\right)$

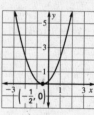

**55.** $y = 3x^2 - 2x - 1$

$x = \dfrac{2}{2(3)} = \dfrac{1}{3}$

vertex: $\left(\dfrac{1}{3}, -\dfrac{4}{3}\right)$

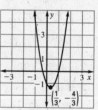

**56.** $y = 2x^2 + 6x - 5$

$x = \dfrac{-6}{2(2)} = -\dfrac{3}{2}$

vertex: $\left(-\dfrac{3}{2}, -\dfrac{19}{2}\right)$

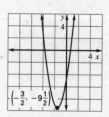

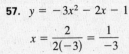

**57.** $y = -3x^2 - 2x - 1$

$x = \dfrac{2}{2(-3)} = \dfrac{1}{-3}$

vertex: $\left(-\dfrac{1}{3}, -\dfrac{2}{3}\right)$

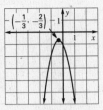

**58.** $y = -4x^2 + 32x - 20$

$x = \dfrac{-32}{2(-4)} = 4$

vertex: $(4, 44)$

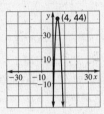

**59.** $y = -4x^2 + 4x + 7$

$x = \dfrac{-4}{2(-4)} = \dfrac{1}{2}$

vertex: $\left(\dfrac{1}{2}, 8\right)$

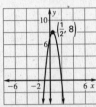

**60.** $y = -3x^2 - 3x + 4$

$x = \dfrac{3}{2(-3)} = \dfrac{-1}{2}$

vertex: $\left(-\dfrac{1}{2}, \dfrac{19}{4}\right)$

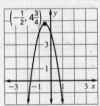

**61.** $y = -2x^2 + 6x - 5$

$x = \dfrac{-6}{2(-2)} = \dfrac{3}{2}$

vertex: $\left(\dfrac{3}{2}, -\dfrac{1}{2}\right)$

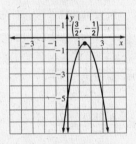

**62.** $y = -\dfrac{1}{3}x^2 + 2x - 3$

$x = \dfrac{-2}{2\left(-\dfrac{1}{3}\right)}$

vertex: $(3, 0)$

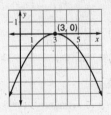

**63.** $y = -\dfrac{1}{2}x^2 - 4x + 6$

$x = \dfrac{4}{2\left(-\dfrac{1}{2}\right)} = -4$

vertex: $(-4, 14)$

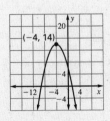

**64.** $y = -\dfrac{1}{4}x^2 - x - 1$

$x = \dfrac{1}{2\left(-\dfrac{1}{4}\right)} = -2$

vertex: $(-2, 0)$

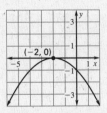

**65.** 5 ft    **66.** 10 ft    **67.** 70 ft

**68.** about 208 ft    **69.** 1980 to 1984    **70.** 1984 to 1995

**71.** The vertex is the point at which the graph changes direction, that is, when the values of $G$ stop decreasing and begin increasing.

**72.** about 0.22 m    **73.** about 0.21 sec

**74.** about 0.42 sec

**75.** The force with which the ball is hit, the angle at which it is hit, and the spin placed on the ball; a harder hit, a steeper angle, and a favorable spin; a softer hit, a flatter angle, and an unfavorable spin.

**76. a.** $35°$: $x = \dfrac{-0.70}{2(-0.06)} = 5.83$

$y = -0.06(5.83)^2 + 0.70(5.83) + 0.5$

$y = 2.5$ ft

$60°$: $x = \dfrac{-1.73}{2(-0.16)} = 5.41$

$y = -0.16(5.41)^2 + 1.73(5.41) + 0.5$

$y = 5.2$ ft

$75°$: $x = \dfrac{-3.73}{2(-0.6)} = 3.11$

$y = -0.60(3.11)^2 + 3.73(3.11) + 0.5$

$y = 6.3$ ft

**b.** $35°$: 12.3 ft
$60°$: 11.1 ft
$75°$: 6.3 ft

**c.** yes; the angle setting is between $35°$ and $60°$ and the reach is greater than that of any of the graphs shown; $90°$

**77. a.**

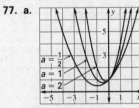

The functions differ in the value of $a$, and vary in the width of the parabola. When $a$ changes from 1 to $\dfrac{1}{2}$, the graph becomes wider. When $a$ changes from 1 to 2, the graph becomes narrower.

**b.**

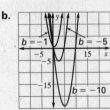

The functions differ in the value of $b$, and vary in the position of the parabola. When $b$ changes from $-1$ to $-5$, the vertex of the graph moves to the right and down. When $b$ changes from $-5$ to $-10$, the vertex of the graph moves farther to the right and down.

# Chapter 9 *continued*

**c.**
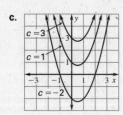

The functions differ in the value of *c*, and vary in the position of the parabola. When *c* changes from 1 to 3, the vertex of the graph moves up 2 units. When *c* changes from 1 to $-2$, the vertex of the graph moves down 3 units.

**78.** A change in the value of *a* makes the graph of $y = ax^2 + bx + c$ narrower or wider, and affects whether the graph opens up or down.

**79.** A change in the value of *b* moves the graph of $y = ax^2 + bx + c$ horizontally and vertically.

**80.** A change in the value of *c* moves the graph of $y = ax^2 + bx + c$ up or down.

## 9.3 Mixed Review (p. 524)

**81.** $-3x + y + 6 = 0$

$y = 3x - 6$

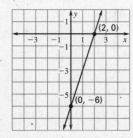

**82.** $-x + y - 7 = 0$

$y = x + 7$

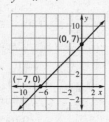

**83.** $4x + 2y - 12 = 0$

$2y = -4x + 12$

$y = -2x + 6$

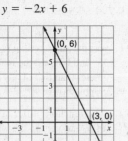

**84.** $x + 2y - 7 = 5x + 1$

$2y = 4x + 8$

$y = 2x + 4$

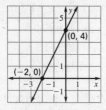

**85.**
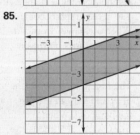

**86.**
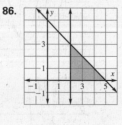

**87.**
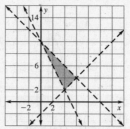

**88.** $4^5 \cdot 4^8 = 4^{13}$, or 67,108,864

**89.** $(3^3)^2 = 3^6$, or 729

**90.** $(3^6)^3 = 3^{18}$, or 387,420,489

**91.** $a \cdot a^5 = a^6$

**92.** $(3b^4)^2 = 9b^8$
**93.** $6x \cdot (6x)^2 = 216x^3$

**94.** $(3t)^3 \cdot (-t^4) = -27t^7$
**95.** $(-3a^2b^2)^3 = -27a^6b^6$

**96.** $1.2 \times 10^{-3}$
**97.** $9.87 \times 10^5$

**98.** $3.984328 \times 10^6$
**99.** $1.229 \times 10^9$

**100.** $4.32 \times 10^{-4}$
**101.** $9.99 \times 10^{-3}$

## Quiz 1 (p. 524)

**1.** 12 **2.** $-14$ **3.** $-26$ **4.** $-5.20$

**5.** 2.45 **6.** 1.22 **7.** 0.4 **8.** 1.5

**9.** $x^2 = 169$
$x = 13, -13$

**10.** $4x^2 = 64$
$x^2 = 16$
$x = 4, -4$

**11.** $12x^2 = 120$
$x^2 = 10$
$x = -\sqrt{10}, \sqrt{10}$

**12.** $-6x^2 = -48$
$x^2 = 8$
$x = 2\sqrt{2}, -2\sqrt{2}$

**13.** $\sqrt{18} = \sqrt{9 \cdot 2}$
$= 3\sqrt{2}$

**14.** $\sqrt{5} \cdot \sqrt{20} = \sqrt{100}$
$= 10$

**15.** $\dfrac{2\sqrt{121}}{\sqrt{4}} = \dfrac{2 \cdot 11}{2}$
$= 11$

**16.** $\sqrt{\dfrac{45}{36}} = \dfrac{\sqrt{45}}{\sqrt{36}}$
$= \dfrac{3\sqrt{5}}{6}$
$= \dfrac{\sqrt{5}}{2}$

**17.** opens up

$x = \dfrac{-2}{2(1)} = -1$

vertex: $(-1, -12)$

axis of symmetry: $x = -1$

**18.** opens up

$x = \dfrac{8}{2(2)} = 2$

vertex: $(2, -14)$

axis of symmetry: $x = 2$

# Chapter 9 *continued*

**19.** opens up

$$x = \frac{-6}{2(3)} = -1$$

vertex: $(-1, -13)$

axis of symmetry: $x = -1$

**20.** opens up

$$\frac{-5}{2\left(\frac{1}{2}\right)} = -5$$

vertex: $\left(-5, -\frac{31}{2}\right)$

axis of symmetry:
$x = -5$

**21.** opens up

$$x = \frac{7}{2(7)} = \frac{1}{2}$$

vertex: $\left(\frac{1}{2}, \frac{21}{4}\right)$

axis of symmetry: $x = \frac{1}{2}$

**22.** opens up

$$x - \frac{-9}{2(1)} - \frac{9}{2}$$

vertex: $\left(-\frac{9}{2}, -\frac{81}{4}\right)$

axis of symmetry:
$x = -\frac{9}{2}$

**23.** $y = -x^2 + 5x - 5$

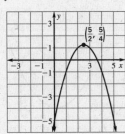

$$x = \frac{-5}{2(-1)} = \frac{5}{2}$$

vertex: $\left(\frac{5}{2}, \frac{5}{4}\right)$

**24.** $y = 3x^2 + 3x + 1$

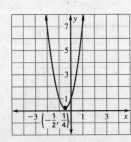

$$x = \frac{-3}{2(3)} = -\frac{1}{2}$$

vertex: $\left(-\frac{1}{2}, \frac{1}{4}\right)$

**25.** $y = 3x^2 + 3x + 1$

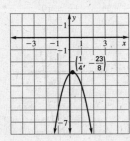

$$x = \frac{-1}{2(-2)} = \frac{1}{4}$$

vertex: $\left(\frac{1}{4}, -\frac{23}{8}\right)$

## Technology Activity 9.3 (p. 525)

**1.** $y = 5.25x^2 + 67.21x - 81.46$

**2.** $y = -0.86x^2 + 2.42x + 2.12$

**3.** Answers may vary.

## Lesson 9.4

### 9.4 Guided Practice (p. 529)

**1.** The roots of a quadratic equation are the solutions, which correspond to the $x$-intercepts of its graph.

**2.** Write the equation in the form $ax^2 + bx + c = 0$ and then write the related function $y = ax^2 + bx + c$. Sketch the graph of the function. The solutions of the original equation are the $x$-intercepts of the graph.

**3.** C  **4.** A  **5.** B

**6.** $3x^2 - 12$

$x^2 = 4$

$x = \pm 2$

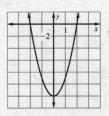

**7.** $4x^2 = 16$

$x^2 = 4$

$x = \pm 2$

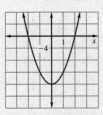

**8.** $5x^2 = 125$

$x^2 = 25$

$x = \pm 5$

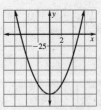

**9.** $3x^2 = 27$

$x^2 = 9$

$x = \pm 3$

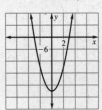

**10.** $8x^2 = 32$

$x^2 = 4$

$x = \pm 2$

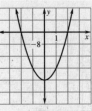

**11.** $-2x^2 = -18$

$x^2 = 9$

$x = \pm 3$

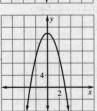

# Chapter 9 *continued*

**12.**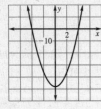

$x = 0$

vertex: $(0, -48)$

Check: $3x^2 = 48$

$\qquad x^2 = 16$

$\qquad x = \pm 4$

**13.**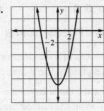

$x = 0$

vertex: $(0, -9)$

Check: $x^2 - 4 = 5$

$\qquad x^2 = 9$

$\qquad x = \pm 3$

**14.**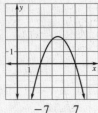

$x = \dfrac{-7}{2(-1)} = \dfrac{7}{2}$

vertex: $\left(\dfrac{7}{2}, \dfrac{9}{4}\right)$

Check: $x^2 - 7x + 10 = 0$

$\qquad (x - 2)(x - 5) = 0$

$\qquad x = 2 \quad x = 5$

**15.**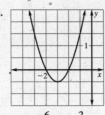

$x = \dfrac{-6}{2(2)} = -\dfrac{3}{2}$

vertex: $\left(-\dfrac{3}{2}, -\dfrac{1}{2}\right)$

Check: $2x^2 + 6x + 4 = 0$

$\qquad x^2 + 3x + 2 = 0$

$\qquad (x + 2)(x + 1) = 0$

$\qquad x = -2 \quad x = -1$

$x = \dfrac{-1}{2\left(\frac{1}{3}\right)} = -\dfrac{3}{2}$

vertex: $\left(-\dfrac{3}{2}, -\dfrac{27}{4}\right)$

Check: $\dfrac{1}{3}x^2 + x - 6 = 0$

$\qquad x^2 + 3x - 18 = 0$

$\qquad (x + 6)(x - 3) = 0$

$\qquad x = -6 \quad x = 3$

**16.**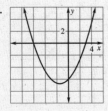

**17.**

$x = \dfrac{-1}{2(-1)} = \dfrac{1}{2}$

vertex: $\left(\dfrac{1}{2}, 20\dfrac{1}{4}\right)$

Check: $x - x^2 = -20$

$\qquad -x^2 + x + 20 = 0$

$\qquad x^2 - x - 20 = 0$

$\qquad (x - 5)(x + 4) = 0$

$\qquad x = 5 \quad x = -4$

## 9.4 Practice and Applications (pp. 529–531)

**18.** $1, 2$

**19.** $0.2, -4.2$

**20.** $-1, \dfrac{1}{3}$

**21.** $2x^2 = 32$

$\quad x^2 = 16$

$\quad x = \pm 4$

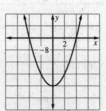

**22.** $4x^2 = 16$

$\quad x^2 = 4$

$\quad x = \pm 2$

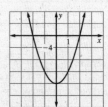

**23.** $4x^2 = 100$

$\quad x^2 = 25$

$\quad x = \pm 5$

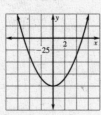

**24.** $\dfrac{1}{3}x^2 = 3$

$\quad x^2 = 9$

$\quad x = \pm 3$

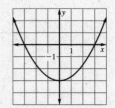

**25.** $\dfrac{1}{4}x^2 = 36$

$\quad x^2 = 144$

$\quad x = \pm 12$

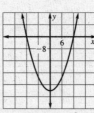

**26.** $\dfrac{1}{2}x^2 = 18$

$\quad x^2 = 36$

$\quad x = \pm 6$

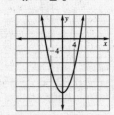

**27.** $x^2 - 11 = 14$

$\quad x^2 = 25$

$\quad x = \pm 5$

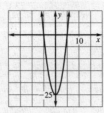

**28.** $x^2 - 13 = 36$

$\quad x^2 = 49$

$\quad x = \pm 7$

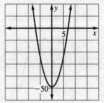

**29.** $x^2 - 4 = 12$

$\quad x^2 = 16$

$\quad x = \pm 4$

**30.** $x^2 - 53 = 11$

$\quad x^2 = 64$

$\quad x = \pm 8$

**Algebra 1**
Chapter 9 Worked-out Solution Key

# Chapter 9 *continued*

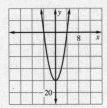

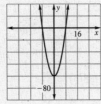

**31.** $x^2 + 37 = 118$
$x^2 = 81$
$x = \pm 9$

**32.** $2x^2 - 89 = 9$
$2x^2 = 98$
$x^2 = 49$
$x = \pm 7$

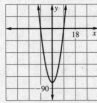

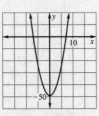

**33.**

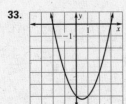

$x = \dfrac{1}{2(1)} = \dfrac{1}{2}$

vertex: $\left(\dfrac{1}{2}, -\dfrac{25}{4}\right)$

Check: $x^2 - x - 6 = 0$
$(x - 3)(x + 2) = 0$
$x = 3 \quad x = -2$

**34.**

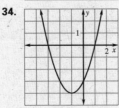

$x = \dfrac{-2}{2(1)} = -1$

vertex: $(-1, -4)$

Check: $x^2 + 2x - 3 = 0$
$(x + 3)(x - 1) = 0$
$x = -3 \quad x = 1$

**35.**

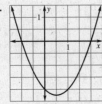

$x = \dfrac{1}{2(1)} = \dfrac{1}{2}$

vertex: $\left(\dfrac{1}{2}, -\dfrac{9}{4}\right)$

Check: $x^2 - x - 2 = 0$
$(x + 1)(x - 2) = 0$
$x = -1 \quad x = 2$

**36.**

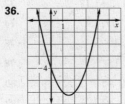

$x = \dfrac{3}{2(1)} = \dfrac{3}{2}$

vertex: $\left(\dfrac{3}{2}, -\dfrac{25}{4}\right)$

Check: $x^2 - 3x - 4 = 0$
$(x + 1)(x - 4) = 0$
$x = -1 \quad x = 4$

**37.**

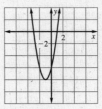

$x = \dfrac{-4}{2(2)} = -1$

vertex: $(-1, -8)$

Check: $2x^2 + 4x - 6 = 0$
$(2x + 6)(x - 1) = 0$
$2x + 6 = 0 \quad x - 1 = 0$
$x = -3 \qquad x = 1$

**38.**

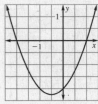

$\dfrac{-3}{2(3)} = \dfrac{-1}{2}$

vertex: $\left(-\dfrac{1}{2}, -\dfrac{9}{4}\right)$

Check: $3x^2 + 3x - 6 = 0$
$x^2 + x - 2 = 0$
$(x - 1)(x + 2) = 0$
$x = 1 \quad x = -2$

**39.**

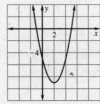

$x = \dfrac{4}{2(1)} = 2$

vertex: $(2, -9)$

Check: $x^2 - 4x - 5 = 0$
$(x + 1)(x - 5) = 0$
$x = -1, \quad x = 5$

**40.**

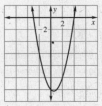

$x = \dfrac{1}{2(1)} = \dfrac{1}{2}$

vertex: $\left(\dfrac{1}{2}, -\dfrac{49}{4}\right)$

Check: $x^2 - x - 12 = 0$
$(x + 3)(x - 4) = 0$
$x = -3 \quad x = 4$

**41.**

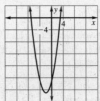

$x = \dfrac{-4}{2(1)} = -2$

vertex: $(-2, -25)$

Check: $x^2 + 4x - 21 = 0$
$(x + 7)(x - 3) = 0$
$x = -7 \quad x = 3$

**42.**

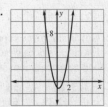

$x = \dfrac{4}{2(8)} = \dfrac{1}{4}$

vertex: $\left(\dfrac{1}{4}, -1\dfrac{1}{8}\right)$

Check: $8x^2 - 4x - 4 = 0$
$2x^2 - x - 1 = 0$
$(2x + 1)(x - 1) = 0$
$2x + 1 = 0 \quad x - 1 = 0$
$x = -\dfrac{1}{2} \qquad x = 1$

# Chapter 9 *continued*

**43.**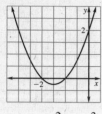

**44.**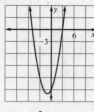

$$x = \frac{-3}{2(1)} = -\frac{3}{2}$$

$$x = \frac{-2}{2(1)} = -1$$

vertex: $\left(-\frac{3}{2}, -\frac{1}{4}\right)$

vertex: $(-1, -16)$

Check: $-2x^2 - 4x + 30 = 0$

Check: $-7x^2 - 21x - 14 = 0$

$x^2 + 2x - 15 = 0$

$x^2 + 3x + 2 = 0$

$(x + 5)(x - 3) = 0$

$(x + 2)(x + 1) = 0$

$x = -5 \quad x = 3$

$x = -2 \quad x = -1$

**45.** $-1, 4$      **46.** $-7, 1$

**47.** $-1.17, 0.17$      **48.** $-4, 1$

**49.** $-8, 4$      **50.** $-8, -4$

**51.** 69.19 ft; the distance was greater than that for an initial angle of 65°, but less than that for an initial angle of 35°.

**52.** 2009; no

**53.**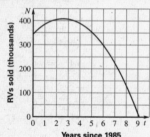

Years since 1985

$$t = -\frac{48.9}{2(-9.5)} \approx 2.57$$

Vertex: (2.57, 406.43)

**54.** 9.11      **55.** 1994

**56.** No; an improved economy, lower vehicle prices, and lower gas prices would all tend to increase sales, while opposite conditions would decrease sales.

**57.** $-1, 3$ D    **58.** B      **59.** $-1, 1.67$

**60.** $-0.82, 1.39$    **61.** $-0.26, 0.76$    **62.** $-1.35, 0.15$

**63.** $-0.66, 1.14$    **64.** $-0.45, 0$

### 9.4 Mixed Review (p. 531)

**65.**
$-2x + 8y = 11 \qquad x + 6\left(\frac{3}{4}\right) = 2$

$x + 6y = 2 \qquad\qquad x + \frac{9}{2} = 2$

$x = -6y + 2 \qquad\qquad x = -\frac{5}{2}$

$-2(-6y + 2) + 8y = 11$

$12y - 4 + 8y = 11$

$20y = 15$

$y = \frac{3}{4}$

one solution: $\left(-\frac{5}{2}, \frac{3}{4}\right)$

**66.**
$(4x + 5y = 37)(5) \rightarrow \quad 20x + 25y = 185$

$(-5x + 3y = 0)(4) \rightarrow \underline{-20x + 12y = \quad 0}$

$37y = 185$

$y = 5$

$4x + 5(5) = 37$

$4x = 12$

$x = 3$

one solution: (3, 5)

**67.** $-2x + 2y = \quad 4 \qquad -2x + 2y = \quad 4$

$(x - y = -2)(2) \rightarrow \underline{2x - 2y = -4}$

$0 = \quad 0$

infinitely many solutions

**68.** $8x + 4y = \quad -4 \qquad 8x + 4y = \quad -4$

$(4x - y = -20)(4) \rightarrow \underline{16x - 4y = -80}$

$24x \qquad = -84$

$x = -\frac{7}{2}$

$8\left(-\frac{7}{2}\right) + 4y = -4$

$-28 + 4y = -4$

$4y = 24$

$y = 6$

one solution: $\left(-\frac{7}{2}, 6\right)$

**69.** $10x - 6y = -5 \qquad 10x - 6y = -5$

$(-5x + 3y = 2)(2) \rightarrow \underline{-10x + 6y = \quad 4}$

$0 \neq -1$

no solution

**70.** $(5x + 4y = -3)(-3) \rightarrow -15x - 12y = \quad 9$

$\underline{15x + 12y = \quad 9}$

$0 \neq 18$

no solution

**71.** $6.17, -1.17$      **72.** $4.13, -2.80$

**73.** $-1.47, 7.47$      **74.** $-1.83, 3.83$

**75.** $2\sqrt{10}$      **76.** $2\sqrt{6}$

**77.** $2\sqrt{15}$      **78.** $10\sqrt{2}$

**79.** $\frac{1}{2}(4)\sqrt{5} = 2\sqrt{5}$      **80.** $\frac{1}{3}(3)\sqrt{3} = \sqrt{3}$

**81.** $\frac{1}{8}(4)\sqrt{2} = \frac{1}{2}\sqrt{2}$      **82.** $\frac{2}{3}(10)\sqrt{3} = \frac{20}{3}\sqrt{3}$

**83.** $2x + y = 13.85$

$x + 2y = 9.85$

$x = -2y + 9.85$

$2(-2y + 9.85) + y = 13.85 \qquad x + 2(1.95) = 9.85$

$-4y + 19.7 + y = 13.85 \qquad\qquad x = 5.95$

$-3y = -5.85$

$y = 1.95$

$5.95: pasta dish

$1.95: salad

# Chapter 9 *continued*

**Technology Activity 9.4 (p. 532)**

1. 0.26, 6.41
2. −0.77, 2.27
3. 0.21, 4.79
4. −1.31, 0.65
5. 1.71, 2
6. −0.43, 3.71
7. 3.04, 6.40
8. −2.38, 0.68

## Lesson 9.5

**9.5 Guided Practice (p. 536)**

1. the quadratic formula

2. Rewrite the equation in standard form; substitute $a = -2$, $b = 5$, and $c = 7$ into the quadratic formula.

The solutions are $\dfrac{-5 + \sqrt{5^2 - 4(-2)(7)}}{2(-2)}$ or

$x = -1$, and $\dfrac{-5 - \sqrt{5^2 - 4(-2)(7)}}{2(-2)}$ or $x = 3.5$.

3. In one model, the object is dropped, that is, its initial velocity is 0. In the other, the object is thrown, and, therefore, has a nonzero initial velocity.

4. $x = \dfrac{-6 \pm \sqrt{36 - 4(1)(-7)}}{2(1)}$

$= \dfrac{-6 \pm 8}{2}$

$= 1, -7$

5. $x = \dfrac{2 \pm \sqrt{4 - 4(1)(-15)}}{2(1)}$

$= \dfrac{2 \pm 8}{2}$

$= 5, -3$

6. $x = \dfrac{-12 \pm \sqrt{144 - 4(1)(36)}}{2(1)}$

$= \dfrac{-12}{2}$

$= -6$

7. $x = \dfrac{8 \pm \sqrt{64 - 4(4)(3)}}{2(4)}$

$= \dfrac{8 \pm 4}{8}$

$= 1.5, 0.5$

8. $x = \dfrac{-1 \pm \sqrt{1 - 4(3)(-1)}}{2(3)}$

$= \dfrac{-1 + \sqrt{13}}{6}, \dfrac{-1 - \sqrt{13}}{6}$

9. $x = \dfrac{-6 \pm \sqrt{36 - 4(1)(-3)}}{2(1)}$

$= \dfrac{-6 \pm \sqrt{48}}{2}$

$= \dfrac{-6 \pm 4\sqrt{3}}{2}$

$= -3 + 2\sqrt{3}, -3 - 2\sqrt{3}$

10. $2x^2 = -x + 6$

$2x^2 + x - 6 = 0$

$x = \dfrac{-1 \pm \sqrt{1 - 4(2)(-6)}}{2(2)}$

$= \dfrac{-1 \pm 7}{4}$

$= 1.5, -2$

11. $6x = -8x^2 + 2$

$8x^2 + 6x - 2 = 0$

$x = \dfrac{-6 \pm \sqrt{36 - 4(8)(-2)}}{2(8)}$

$= \dfrac{-6 \pm 10}{16}$

$= 0.25, -1$

12. $3 = 3x^2 + 8x$

$3x^2 + 8x - 3 = 0$

$x = \dfrac{-8 \pm \sqrt{64 - 4(3)(-3)}}{2(3)}$

$= \dfrac{-8 \pm 10}{6}$

$= \tfrac{1}{3}, -3$

13. $-14x = -2x^2 + 36$

$2x^2 - 14x - 36 = 0$

$x = \dfrac{14 \pm \sqrt{196 - 4(2)(-36)}}{2(2)}$

$x = \dfrac{14 \pm 22}{4}$

$= 9, -2$

14. $-x^2 + 4x = 3$

$-x^2 + 4x - 3 = 0$

$x = \dfrac{-4 \pm \sqrt{16 - 4(-1)(-3)}}{2(-1)}$

$= \dfrac{-4 \pm 2}{-2}$

$= 3, 1$

**15.** $4x^2 + 4x = -1$

$4x^2 + 4x + 1 = 0$

$x = \dfrac{-4 \pm \sqrt{16 - 4(4)(1)}}{2(4)}$

$= \dfrac{-4}{8}$

$= -\dfrac{1}{2}$

**16.** $0 = x^2 - 11x + 24$

$x = \dfrac{11 \pm \sqrt{121 - 4(1)(24)}}{2(1)}$

$= \dfrac{11 \pm 5}{2}$

$= 8, 3$

**17.** $0 = x^2 + 10x + 16$

$x = \dfrac{-10 \pm \sqrt{100 - 4(1)(16)}}{2(1)}$

$= \dfrac{-10 \pm 6}{2}$

$= -2, -8$

**18.** $0 = 2x^2 + 4x - 30$

$x = \dfrac{-4 \pm \sqrt{16 - 4(2)(-30)}}{2(2)}$

$= \dfrac{-4 \pm 16}{4}$

$= 3, -5$

**19.** $0 = 2x^2 + 6x - 9$

$x = \dfrac{-6 \pm \sqrt{36 - 4(2)(-9)}}{2(2)}$

$= \dfrac{-6 \pm 6\sqrt{3}}{4}$

$= \dfrac{-3 + 3\sqrt{3}}{2}, \dfrac{-3 - 3\sqrt{3}}{2}$

$\approx 1.10, -4.10$

**20.** $0 = 5x^2 + 8x - 8$

$x = \dfrac{-8 \pm \sqrt{64 - 4(5)(-8)}}{2(5)}$

$= \dfrac{-8 \pm 4\sqrt{14}}{10}$

$= \dfrac{-4 + 2\sqrt{14}}{5}, \dfrac{-4 - 2\sqrt{14}}{5}$

$\approx 0.70, -2.30$

**21.** $0 = 4x^2 + 8x - 1$

$x = \dfrac{-8 \pm \sqrt{64 - 4(4)(-1)}}{2(4)}$

$= \dfrac{-8 \pm 4\sqrt{5}}{8}$

$= \dfrac{-2 + \sqrt{5}}{2}, \dfrac{-2 - \sqrt{5}}{2}$

$\approx 0.12, -2.12$

**22.** $h = -16t^2 - 40t + 180$

$t = \dfrac{40 \pm \sqrt{1600 - 4(-16)(180)}}{2(-16)}$

$t \approx -4.83, 2.33$

about 2.33 sec

### 9.5 Practice and Applications (pp. 536–538)

**23.** $(-3)^2 - 4(1)(-4) = 25$

**24.** $(5)^2 - 4(4)(1) = 9$

**25.** $(-11)^2 - 4(1)(30) = 1$

**26.** $(-3)^2 - 4(5)(-2) = 49$

**27.** $(-13)^2 - 4(1)(42) = 1$

**28.** $(-5)^2 - 4(3)(-12) = 169$

**29.** $(5)^2 - 4(5)\left(\frac{1}{5}\right) = 21$

**30.** $(5)^2 - 4\left(\frac{1}{2}\right)(-8) = 41$

**31.** $(-6)^2 - 4\left(\frac{1}{4}\right)(-3) = 39$

**32.** $x = \dfrac{13 \pm \sqrt{169 - 4(4)(3)}}{2(4)}$

$= \dfrac{13 \pm 11}{8}$

$= 3, 0.25$

**33.** $x = \dfrac{-11 \pm \sqrt{121 - 4(1)(10)}}{2(1)}$

$= \dfrac{-11 \pm 9}{2}$

$= -1, -10$

**34.** $x = \dfrac{-8 \pm \sqrt{64 - 4(7)(1)}}{2(7)}$

$= \dfrac{-8 \pm 6}{14}$

$\approx -\dfrac{1}{7}, -1$

**35.** $x = \dfrac{-2 \pm \sqrt{4 - 4(-3)(8)}}{2(-3)}$

$= \dfrac{-2 \pm 10}{-6}$

$= -\dfrac{4}{3}, 2$

# Chapter 9 *continued*

**36.** $x = \dfrac{10 \pm \sqrt{100 - 4(6)(3)}}{2(6)}$

$\phantom{x} = \dfrac{10 \pm 2\sqrt{7}}{12}$

$\phantom{x} = \dfrac{5 \pm \sqrt{7}}{6}$

$\phantom{x} \approx 1.27, 0.39$

**37.** $x = \dfrac{-14 \pm \sqrt{196 - 4(9)(3)}}{2(9)}$

$\phantom{x} = \dfrac{-14 + 2\sqrt{22}}{18}$

$\phantom{x} = \dfrac{-7 \pm \sqrt{22}}{9}$

$\phantom{x} \approx -0.26, -1.30$

**38.** $x = \dfrac{-6 \pm \sqrt{36 - 4(8)(-1)}}{2(8)}$

$\phantom{x} = \dfrac{-6 \pm 2\sqrt{17}}{16}$

$\phantom{x} = \dfrac{-3 \pm \sqrt{17}}{8}$

$\phantom{x} \approx 0.14, -0.89$

**39.** $x = \dfrac{-2 \pm \sqrt{4 - 4(7)(-1)}}{2(7)}$

$\phantom{x} = \dfrac{-2 \pm 4\sqrt{2}}{14}$

$\phantom{x} = \dfrac{-1 \pm 2\sqrt{2}}{7}$

$\phantom{x} \approx 0.26, -0.55$

**40.** $x = \dfrac{3 \pm \sqrt{9 - 4(-6)(2)}}{2(-6)}$

$\phantom{x} = \dfrac{3 \pm \sqrt{57}}{-12}$

$\phantom{x} = -0.88, 0.38$

**41.** $x = \dfrac{-6 \pm \sqrt{36 - 4\left(-\frac{1}{2}\right)(13)}}{2\left(-\frac{1}{2}\right)}$

$\phantom{x} = \dfrac{-6 \pm \sqrt{62}}{-1}$

$\phantom{x} = 6 \pm \sqrt{62}$

$\phantom{x} \approx 13.87, -1.87$

**42.** $x = \dfrac{-3 \pm \sqrt{9 - 4(-10)(2)}}{2(-10)}$

$\phantom{x} = \dfrac{-3 \pm \sqrt{89}}{-20}$

$\phantom{x} \approx -0.32, 0.62$

**43.** $x = \dfrac{5 \pm \sqrt{25 - 4(-2)(19)}}{2(-2)}$

$\phantom{x} = \dfrac{5 \pm \sqrt{177}}{-4}$

$\phantom{x} \approx -4.58, 2.08$

**44.** $2x^2 = 4x + 30$

$2x^2 - 4x - 30 = 0$

$x = \dfrac{4 \pm \sqrt{16 - 4(2)(-30)}}{2(2)}$

$\phantom{x} = \dfrac{4 \pm 16}{4}$

$\phantom{x} = 5, -3$

**45.** $2 - 3x + x^2 = 0$

$x^2 - 3x + 2 = 0$

$x = \dfrac{3 \pm \sqrt{9 - 4(1)(2)}}{2(1)}$

$\phantom{x} = \dfrac{3 \pm 1}{2}$

$\phantom{x} = 2, 1$

**46.** $16 = -x^2 + 11x$

$x^2 - 11x + 16 = 0$

$x = \dfrac{11 \pm \sqrt{121 - 4(1)(16)}}{2(1)}$

$\phantom{x} = \dfrac{11 \pm \sqrt{57}}{2}$

$\phantom{x} \approx 9.27, 1.73$

**47.** $5x - 1 = -6x^2$

$6x^2 + 5x - 1 = 0$

$x = \dfrac{-5 \pm \sqrt{25 - 4(6)(-1)}}{2(6)}$

$\phantom{x} = \dfrac{-5 \pm 7}{12}$

$x = -1, \dfrac{1}{6}$

**48.** $2q^2 - 6 = -4q$

$2q^2 + 4q - 6 = 0$

$q = \dfrac{-4 \pm \sqrt{16 - 4(2)(-6)}}{2(2)}$

$\phantom{q} = \dfrac{-4 \pm 8}{4}$

$\phantom{q} = 1, -3$

# Chapter 9 *continued*

**49.** $5z - 2z^2 + 15 = 8$

$2z^2 - 5z - 7 = 0$

$z = \dfrac{5 \pm \sqrt{25 - 4(2)(-7)}}{2(2)}$

$= \dfrac{5 \pm 9}{4}$

$= -1, 3.5$

**50.** $-1 + 3x^2 = 2x$

$3x^2 - 2x - 1 = 0$

$x = \dfrac{2 \pm \sqrt{4 - 4(3)(-1)}}{2(3)}$

$= \dfrac{2 \pm 4}{6}$

$= 1, -\dfrac{1}{3}$

**51.** $-5c^2 + 9c = 4$

$5c^2 - 9c + 4 = 0$

$c = \dfrac{9 \pm \sqrt{81 - 4(5)(4)}}{2(5)}$

$= \dfrac{9 \pm 1}{10}$

$= 1, \dfrac{4}{5}$

**52.** $-16b = -8b^2 - 8$

$8b^2 - 16b + 8 = 0$

$b = \dfrac{16 \pm \sqrt{256 - 4(8)(8)}}{2(8)}$

$= \dfrac{16}{16}$

$= 1$

**53.** $0 = 3x^2 - 6x - 24$

$x = \dfrac{6 \pm \sqrt{36 - 4(3)(-24)}}{2(3)}$

$= \dfrac{6 \pm 18}{6}$

$= 4, -2$

**54.** $0 = 2x^2 - 6x - 8$

$x = \dfrac{6 \pm \sqrt{36 - 4(2)(-8)}}{2(2)}$

$= \dfrac{6 \pm 10}{4}$

$= 4, -1$

**55.** $0 = 2x^2 - 2x - 12$

$x = \dfrac{2 \pm \sqrt{4 - 4(2)(-12)}}{2(2)}$

$= \dfrac{2 \pm 10}{4}$

$= 3, -2$

**56.** $0 = -2x^2 + 6x + 9$

$x = \dfrac{-6 \pm \sqrt{36 - 4(-2)(9)}}{2(-2)}$

$= \dfrac{-6 \pm 6\sqrt{3}}{-4}$

$= \dfrac{3 \pm 3\sqrt{3}}{2}$

$= 4.10, -1.10$

**57.** $0 = x^2 + x - 4$

$x = \dfrac{-1 \pm \sqrt{1 - 4(1)(-4)}}{2(1)}$

$= \dfrac{-1 \pm \sqrt{17}}{2}$

$\approx 1.56, -2.56$

**58.** $0 = x^2 + 7x - 2$

$x = \dfrac{-7 \pm \sqrt{49 - 4(1)(-2)}}{2(1)}$

$= \dfrac{-7 \pm \sqrt{57}}{2}$

$\approx 0.27, -7.27$

**59.** $0 = -3x^2 - 2x + 1$

$x = \dfrac{2 \pm \sqrt{4 - 4(-3)(1)}}{2(-3)}$

$= \dfrac{2 \pm 4}{-6}$

$= -1, \dfrac{1}{3}$

**60.** $0 = -4x^2 + 8x - 2$

$x = \dfrac{-8 \pm \sqrt{64 - 4(-4)(-2)}}{2(-4)}$

$= \dfrac{-8 \pm 4\sqrt{2}}{-8}$

$= \dfrac{2 \pm \sqrt{2}}{2}$

$= 1.71, 0.29$

**Algebra 1**
Chapter 9 Worked-out Solution Key

# Chapter 9 *continued*

**61.** $0 = -5x^2 + 5x + 5$

$$x = \frac{-5 \pm \sqrt{25 - 4(-5)(5)}}{2(-5)}$$

$$= \frac{-5 \pm 5\sqrt{5}}{-10}$$

$$= \frac{1 \pm \sqrt{5}}{2}$$

$$\approx 1.62, -0.62$$

**62.** $x = \frac{-20 \pm \sqrt{400 - 4(6)(5)}}{2(6)}$

$$= \frac{-20 \pm 2\sqrt{70}}{12}$$

$$\approx -0.27, -3.06$$

*Sample answer:* quadratic formula; the expression on the left cannot be written as a perfect square.

**63.** $m^2 = 32$

$m = \pm 4\sqrt{2}$

$m \approx 5.66, -5.66$

*Sample answer:* finding square roots; simplest method

**64.** $x^2 = 625$

$x = \pm 25$

*Sample answer:* finding square roots; simplest method

**65.** $4y^2 - 49 = 0$

$4y^2 = 49$

$y^2 = 12.25$

$y = \pm 3.5$

*Sample answer:* finding square roots; simplest method

**66.** $x = \frac{-6 \pm \sqrt{36 - 4(-2)(1)}}{2(-2)}$

$$= \frac{-3 \pm \sqrt{11}}{-2}$$

$$\approx -0.16, 3.16$$

*Sample answer:* quadratic formula; the expression on the left cannot be written as a perfect square.

**67.** $(h + 9)^2 = 0$

$h = -9$

*Sample answer:* finding square roots; the expression on the left can be written as $(h + 9)^2$.

**68.** $5x^2 = 25$

$x^2 = 5$

$x = \pm 2.24$

*Sample answer:* finding square roots; simplest method

**69.** $9y^2 - 3y - 1 = 0$

$$y = \frac{3 \pm \sqrt{9 - 4(9)(-1)}}{2(9)}$$

$$= \frac{3 \pm \sqrt{9 - 4(9)(-1)}}{2(9)}$$

$$= \frac{3 \pm 3\sqrt{5}}{18}$$

$$\approx 0.54, -0.21$$

*Sample answer:* quadratic formula; when written in standard form, the expression on the left cannot be written as a perfect square.

**70.** $v^2 - 8v - 2 = 0$

$$v = \frac{8 \pm \sqrt{64 - 4(1)(-2)}}{2(1)}$$

$$= \frac{8 \pm 6\sqrt{2}}{2}$$

$$\approx 8.24, -0.24$$

*Sample answer:* quadratic formula; when written in standard form, the expression on the left cannot be written as a perfect square.

**71.** $h = -16t^2 - 50t + 200$

$$t = \frac{50 \pm \sqrt{2500 - 4(-16)(200)}}{2(-16)}$$

$$\approx -5.43, 2.30$$

2.30 sec

**72.** $h = -16t^2 - 25t + 150$

$$t = \frac{25 \pm \sqrt{625 - 4(-16)(150)}}{2(-16)}$$

$$\approx -3.94, 2.38$$

2.38 sec

**73.** $h = -16t^2 - 10t + 100$

$$t = \frac{10 \pm \sqrt{100 - 4(-16)(100)}}{2(-16)}$$

$$\approx -2.83, 2.21$$

2.21 sec

**74.** $h = -16t^2 - 33t + 150$

$$t = \frac{33 \pm \sqrt{1089 - 4(-16)(150)}}{2(-16)}$$

$$\approx 2.20 \text{ sec}$$

**75.** $h = -16t^2 - 40t + 80$

$$t = \frac{40 \pm \sqrt{1600 - 4(-16)(80)}}{2(-16)}$$

$$\approx 1.31 \text{sec}$$

**76.** $h = -16t^2 + 50$

$$t = \frac{0 \pm \sqrt{0 - 4(-16)(50)}}{2(-16)}$$

$$\approx 1.77 \text{sec}$$

**77.** $h = -16t^2 + 30$

$$t = \frac{0 \pm \sqrt{0 - 4(-16)(30)}}{2(-16)}$$

$$\approx 1.37 \text{ sec}$$

**78.** $h = -16t^2 + 45$

$$t = \frac{0 \pm \sqrt{0 - 4(-16)(45)}}{2(-16)}$$

$$\approx 1.68 \text{ sec}$$

**79.** $h = -16t^2 + 90t + 7$

$t = \dfrac{-90 \pm \sqrt{8100 - 4(-16)(7)}}{2(-16)}$

$\approx 5.70$ sec

**80.** $h = -16t^2 - 10t + 20$

$t = \dfrac{10 \pm \sqrt{100 - 4(-16)(20)}}{2(-16)}$

$\approx 0.85$ sec

**81.** $h = -16t^2 - 220t + 100$

$t = \dfrac{220 \pm \sqrt{48,400 - 4(-16)(100)}}{2(-16)}$

$\approx 0.44$ sec

**82.** $h = -16t^2 - 105t + 200$

$t = \dfrac{105 \pm \sqrt{11,025 - 4(-16)(200)}}{2(-16)}$

$\approx 1.54$ sec

**83.** $h = -16t^2 - 145t + 125$

$t = \dfrac{145 \pm \sqrt{21,025 - 4(-16)(125)}}{2(-16)}$

$\approx 0.79$ sec

**84. a.** $h = -16t^2 + 80t + 3$

$t = \dfrac{-80 \pm \sqrt{6400 - 4(-16)(3)}}{2(-16)}$

$\approx 5.04$ sec

**b.** $-16t^2 + 80t + 3 = 103$

$t = \dfrac{-80 \pm \sqrt{6400 - 4(-16)(-100)}}{2(-16)}$

$\approx 2.5$ sec

**c.** 2.5; this is the same as the value found in part (b).

**d.** The speed of the ball approaching the batter, the speed of the swinging bat, the angle of the bat and the wind and the weather conditions can change the path of the baseball. The dimensions of the park determine whether a hit is a homerun.

**85.** $x = 1\frac{1}{2}$; the $x$-intercepts are $-3$ and 6, and the midpoint of $(-3, 0)$ and $(6, 0)$ is $(1\frac{1}{2}, 0)$.

**86.** $x = \dfrac{-b}{2a}$ is the equation of the axis of symmetry of the graph of $y = ax^2 + bx + c$. The given values of $x$ are the $x$-intercepts of the equation, which are equidistant from the axis of symmetry. The distance of each point from the axis of symmetry is $\dfrac{\sqrt{b^2 - 4ac}}{2a}$.

### 9.5 Mixed Review (p. 538)

**87.** $(-5)^2 = 25$

**88.** $-(-1)^2 = -1$

**89.** $-4(-2)(-6) = -48$

**90.** $(-2)^2 + 2 = 6$

**91.** $2 \leq x < 5$

$x \geq 2$ and $x < 5$

**92.** $8 > 2x > -4$

$4 > x > -2$

$x < 4$ and $x > -2$

**93.** $-12 < 2x - 6 < 4$

$-6 < 2x < 10$

$-3 < x < 5$

**94.** $-3 < -x < 1$

$3 > x > -1$

**95.** $|x + 5| \geq 10$

$x + 5 \geq 10$ or $x + 5 \leq 10$

$x \geq 5$ or $x \leq -15$

**96.** $|2x + 9| \leq 15$

$2x + 9 \leq 15$ and $2x + 9 \geq -15$

$2x \leq 6$ and $2x \geq -24$

$x \leq 3$ and $x \geq -12$

**97.** $y = 6x^2 - 4x - 1$

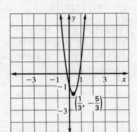

$x = \dfrac{4}{2(6)} = \dfrac{1}{3}$

vertex: $\left(\dfrac{1}{3}, -\dfrac{5}{3}\right)$

**98.** $y = -3x^2 - 5x + 3$

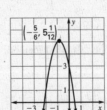

$x = \dfrac{5}{2(-3)} = -\dfrac{5}{6}$

vertex: $\left(-\dfrac{5}{6}, 5\dfrac{1}{12}\right)$

# Chapter 9 *continued*

**99.** $y = -2x^2 - 3x + 2$

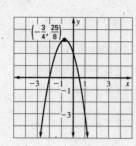

$$x = \frac{3}{2(-2)} = -\frac{3}{4}$$

vertex: $\left(-\frac{3}{4}, 3\frac{1}{8}\right)$

**100.** $y = \frac{1}{2}x^2 + 2x - 1$

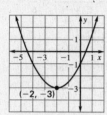

$$x = \frac{-2}{2\left(\frac{1}{2}\right)} = -2$$

vertex: $(-2, -3)$

**101.** $y = 4x^2 - \frac{1}{4}x + 4$

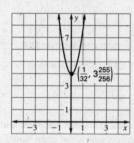

$$x = \frac{\frac{1}{4}}{2(4)} = \frac{1}{32}$$

vertex: $\left(\frac{1}{32}, 3\frac{255}{256}\right)$

**102.** $y = -5x^2 - 0.5x + 0.5$

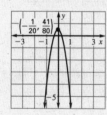

$$x = \frac{0.5}{2(-5)} = -\frac{1}{20}$$

vertex: $\left(-\frac{1}{20}, \frac{41}{80}\right)$

**103.** $\dfrac{1.4 \times 10^7}{365} \approx 3.84 \times 10^4$ people per day

$\dfrac{1.4 \times 10^7}{12} \approx 1.17 \times 10^6$ people per month

## Technology Activity 9.5 (p. 539)

**1.** $x = -6, 5$

**2.** $x = -3.5, 3$

**3.** $x = -2$

**4.** no solution

**5.** no solution

**6.** no solution

**7.** $x = -2.4, 1$

**8.** $x = -1.5, 2.88$

**9.** $x = -6.35, -2.16$

**10.** $x = -31.25, 9$

---

In the solutions that follow, the words *no solution* mean that the equation has no real solution.

## Lesson 9.6

### Developing Concepts Activity 9.6 (p. 540)
**Exploring the Concept**

**1.** $y = x^2 + 3x - 2$

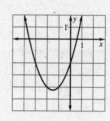

$$x = \frac{-3}{2(1)} = -\frac{3}{2}$$

vertex: $\left(-\frac{3}{2}, -\frac{17}{4}\right)$

**2.** two

**3.** two

**4.** $y = x^2 - 6x + 9$

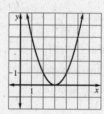

$$x = \frac{6}{2(1)} = 3$$

vertex: $(3, 0)$

one $x$-intercept

one solution

$y = x^2 + 3$

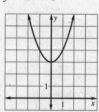

$x = 0$

vertex: $(0, 3)$

no $x$-intercept

no solutions

**5.** Each graph has as many $x$-intercepts as the related equation has solutions.

### Drawing Conclusions

**1.** (1) $y = 9x^2 - 24x + 16$

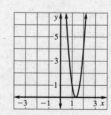

$$x = \frac{24}{2(9)} = \frac{24}{18} = \frac{4}{3}$$

vertex: $\left(\frac{4}{3}, 0\right)$

(2) one $x$-intercept

(3) one solution

---

# Chapter 9 continued

**2.** (1) $y = 2x^2 - 5x - 4$

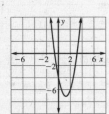

$$x = \frac{5}{2(2)} = \frac{5}{4}$$

vertex: $\left(\frac{5}{4}, -7\frac{1}{8}\right)$

(2) two $x$-intercepts

(3) two solutions

**3.** (1) $y = 3x^2 - x + 2$

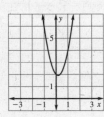

$$x = \frac{1}{2(3)} = \frac{1}{6}$$

vertex: $\left(\frac{1}{6}, \frac{23}{12}\right)$

(2) no $x$-intercepts

(3) no solutions

**4.** (1) $y = -x^2 - 4x + 3$

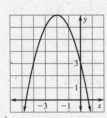

$$x = \frac{4}{2(-1)} = -2$$

vertex: $(-2, 7)$

(2) two $x$-intercepts

(3) two solutions

**5.** (1) $y = -5x^2 - 5x - 12$

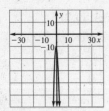

$$x = \frac{5}{2(-5)} = -\frac{1}{2}$$

vertex: $\left(-\frac{1}{2}, -\frac{43}{4}\right)$

(2) no $x$-intercepts

(3) no solutions

**6.** (1) $y = -2x^2 - 4x - 2$

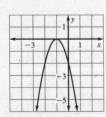

$$x = \frac{4}{2(-2)} = -1$$

vertex: $(-1, 0)$

(2) one $x$-intercept

(3) one solution

**7.** (1) $y = 4x^2 + 8x - 2$

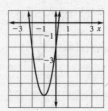

$$x = \frac{-8}{2(4)} = -1$$

vertex: $(-1, -6)$

(2) two $x$-intercepts

(3) two solutions

**8.** (1) $y = -6x^2 + 2x - \frac{1}{2}$

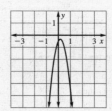

$$x = \frac{-2}{(2)(-6)} = \frac{1}{6}$$

vertex: $\left(\frac{1}{6}, -\frac{1}{3}\right)$

(2) no $x$-intercepts

(3) no solutions

**9.** (1) $y = x^2 + 2x - 1$

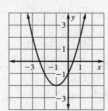

$$x = \frac{-2}{2(1)} = -1$$

vertex: $(-1, -2)$

(2) two $x$-intercepts

(3) two solutions

**10.** All of the graphs are parabolas, but some are wide and some are narrow, some open up and some open down. Some intersect the $x$-axis in one point, some in two points, and some do not intersect the $x$-axis at all.

**11.** (1) 0  (2) 57  (3) −23  (4) 28  (5) −215

(6) 0  (7) 96  (8) −8  (9) 8

**12.** The quadratic equation $ax^2 + bx + c = 0$ has two solutions if $b^2 - 4ac > 0$, one solution if $b^2 - 4ac = 0$, and no solution if $b^2 - 4ac < 0$.

### 9.6 Guided Practice (p. 544)

**1.** $x = \dfrac{-b \pm \sqrt{b^2 - 4ac}}{2a}$

**2.** The quadratic equation $ax^2 + bx + c = 0$ has two solutions if $b^2 - 4ac > 0$, one solution if $b^2 - 4ac = 0$, and no solution if $b^2 - 4ac < 0$. Similarly, the related function $y = ax^2 + bx + c$ has two $x$-intercepts if $b^2 - 4ac > 0$, one $x$-intercept if $b^2 - 4ac = 0$, and no $x$-intercept if $b^2 - 4ac < 0$.

**3.** $4 - 4(3)(5) = \overset{-}{}56$

no solution

**4.** $36 - 4(-3)(-3) = 0$

one solution

**5.** $25 - 4(1)(-10) = 65$

two solutions

**6.** C      **7.** A      **8.** B

### *9.6 Practice and Applications (pp. 544–546)*

**9.** $9 - 4(1)(2) = 1$

two solutions

**10.** $16 - 4(2)(3) = -8$

no solution

**11.** $25 - 4(-3)(-1) = 13$

two solutions

**12.** $1 - 4\left(-1\frac{1}{3}\right)(4) = \frac{19}{3}$

two solutions

**13.** $4 - 4(6)(4) = -92$

no solution

**14.** $36 - 4(3)(3) = 0$

one solution

**15.** $0 = 2x^2 + 4x + 2$

$16 - 4(2)(2) = 0$

one $x$-intercept

B

**16.** $0 = 3x^2 - 5x + 1$

$25 - 4(3)(1) = 13$

two $x$-intercepts

C

**17.** $0 = x^2 + 2x + 5$

$4 - 4(1)(5) = -16$

no $x$-intercept

A

**18.** $\left(\frac{2}{3}\right)^2 - 4\left(\frac{1}{2}\right)(-3) = \frac{4}{9} + 6$

$\qquad = 6\frac{4}{9}$

**19.** two solutions

**20.** The graph of $y = \frac{1}{2}x^2 + \frac{2}{3}x - 3$ crosses the $x$-axis at two different points.

**21.** $16 - 4(1)(c) < 0$

$\qquad -4c < -16$

$\qquad\quad c > 4$

$16 - 4(1)c = 0$

$\qquad -4c = -16$

$\qquad\quad c = 4$

$16 - 4(1)c > 0$

$\qquad -4c > -16$

$\qquad\quad c < 4$

For all values $c > 4$, the equation has no solution; for all values $c < 4$, the equation has 2 solutions. For $c = 4$, the equation has 1 solution.

$c = 5$

$y = x^2 + 4x + 5$

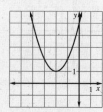

$c = 3$

$y = x^2 + 4x + 3$

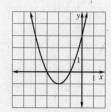

**21. (CONTINUED)**

$c = 4$

$y = x^2 + 4x + 4$

**22.** $x^2 - 2x + c = 0$

$4 - 4(1)c > 0$

$\qquad -4c > -4$

$\qquad\quad c < 1$

$4 - 4(1)c = 0$

$\qquad -4c = -4$

$\qquad\quad c = 1$

$4 - 4(1)c < 0$

$\qquad -4c < -4$

$\qquad\quad c > 1$

For all values $c < 1$, the equation has two solutions; for all values $c > 1$, the equation has no solution; for $c = 1$, the equation has 1 solution.

$c = 0$

$y = x^2 - 2x$

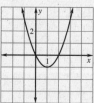

$c = 2$

$y = x^2 - 2x + 2$

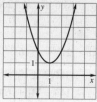

$c = 1$

$y = x^2 - 2x + 1$

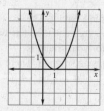

**23.** $2x^2 + 3x + c = 0$

$9 - 4(2)c > 0$

$\qquad -8c > -9$

$\qquad\quad c < \frac{9}{8}$

$9 - 4(2)c = 0$

$\qquad -8c = -9$

$\qquad\quad c = \frac{9}{8}$

$9 - 4(2)c < 0$

$\qquad -8c < -9$

$\qquad\quad c > \frac{9}{8}$

For all values $c < \frac{9}{8}$, the equation has 2 solutions; for all values $c > \frac{9}{8}$, the equation has no solution. For $c = \frac{9}{8}$, the equation has 1 solution.

$c = 1$
$y = 2x^2 + 3x + 1$

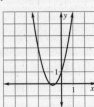

$c = 2$
$y = 2x^2 + 3x + 2$

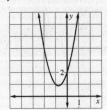

$c = \frac{9}{8}$
$y = 2x^2 + 3x + \frac{9}{8}$

**24.** $26 = -0.01d^2 + 1.06d + 4$

$0 = -0.01d^2 + 1.06d - 22$

$d = \dfrac{-1.06 \pm \sqrt{1.1236 - 4(-0.01)(-22)}}{2(-0.01)}$

$\approx 28$ ft

**25.** You: $2.2 = -16t^2 + 12t + 0$

$0 = -16t^2 + 12t - 2.2$

$144 - 4(-16)(-2.2) = 3.2$

Friend: $3.4 = -16t^2 + 14t + 0$

$0 = -16t^2 + 14t - 3.4$

$196 - 4(-16)(-3.4) = -21.6$

You can dunk the ball but your friend cannot. The discriminant for you is positive; so your equation has two solutions, since the discriminant for your friend is negative, your friend's equation has no solution. There is no time when your friend's jump height will be 3.4 ft.

**26.** You: $0 = -16t^2 + 11.5t - 2.2$

$132.25 - 4(-16)(-2.2) = -8.55$

Friend: $0 = -16t^2 + 15.5 + 3.4$

$240.25 - 4(-16)(3.4) = 22.65$

In this case, you would not be able to dunk but your friend would be able to dunk.

**27.** $80,000 = 26t^2 + 1629t + 19,958$

$0 = 26t^2 + 1629t - 60,042$

$(1629)^2 - 4(26)(-60,042) = 8,898,009$

The discriminant is positive therefore there are two solutions to the equation. The payroll will reach 80 billion dollars.

**28.** about 26 years

**29.** $650 = 6.84t^2 - 3.76t + 9.29$

$0 = 6.84t^2 - 3.76t - 640.71$

$14.1376 - 4(6.84)(-640.71) \approx 17,544$

Yes, the profit will reach 650 million dollars.

**30.** about 8.5 years

**31. a.** $220 = 0.039x^2 - 0.331x + 1.850$

$0 = 0.039x^2 - 0.331x - 218.15$

$0.109561 - 4(0.039)(-218.15) = 34.14$

The hill will reach 220 ft.

**b.** $x = \dfrac{0.331 \pm \sqrt{34.1}}{2(0.039)}$

$\approx 79.2$ ft

**c.** No; solving the equation $0.039x^2 - 0.331x - 218.5 = 0$ gives a solution of approximately 79.2 horizontal feet required. A horizontal distance of 75 feet will allow a height of only 196.4 ft.

**d.** No. The horizontal and vertical scales are not the same.

**32.** If $b^2 - 4ac > 0$, then $\sqrt{b^2 - 4ac}$ is defined and is a positive number, so the values $\dfrac{-b \pm \sqrt{b^2 - 4ac}}{2a}$ are distinct solutions of the equation.

**33.** If $b^2 - 4ac = 0$, then $\sqrt{b^2 - 4ac}$ is 0 so $\dfrac{-b - \sqrt{b^2 - 4ac}}{2a}$ and $\dfrac{-b + \sqrt{b^2 - 4ac}}{2a}$ are both equal to $-\dfrac{b}{2a}$ which is the only solution of the equation.

**34.** If $b^2 - 4ac < 0$, then $\sqrt{b^2 - 4ac}$ is not defined, so $\dfrac{-b + \sqrt{b^2 - 4ac}}{2a}$ and $\dfrac{-b - \sqrt{b^2 - 4ac}}{2a}$ are not defined, and there is no solution to the equation.

### 9.6 Mixed Review (p. 546)

**35.** $f(x) = -x + 1$

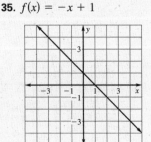

**36.** $f(x) = -6x + 1$

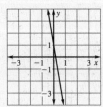

# Chapter 9 *continued*

**37.** $f(x) = 3x - 9$

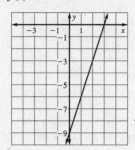

**38.** $\dfrac{x}{6} \le -2$

$x \le -12$

**39.** $-\dfrac{x}{3} \ge 15$

$x \le -45$

**40.** $-12.3x > 86.1$

$x < -7$

**41.** $11.2x \le 134.4$

$x \le 12$

**42.**

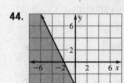

**43.**

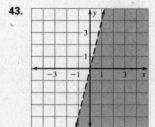

**44.**

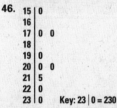

**45.**

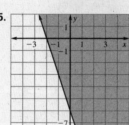

**46.**

```
15 | 0
16 |
17 | 0 0
18 |
19 | 0
20 | 0 0
21 | 5
22 | 0
23 | 0    Key: 23 | 0 = 230
```

$150, $170, $170, $190,
$200, $200, $215, $220,
$230

**47.**

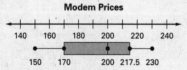

Modem Prices

**48.** The prices range from $150 to $230, with half the prices between $170 and $217.50. The median price is $200.

---

## Quiz 2 (p. 547)

**1.** $y = x^2 - 3x - 10$

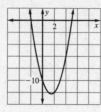

$x = \dfrac{3}{2(1)} = \dfrac{3}{2}$

vertex: $\left(\dfrac{3}{2}, -\dfrac{49}{4}\right)$

$x^2 - 3x - 10 = 0$

$x = \dfrac{-3 \pm \sqrt{9 - 4(1)(-10)}}{2(1)}$

$= \dfrac{3 \pm 7}{2}$

$= 5, -2$

**2.** $y = x^2 - 12x + 36$

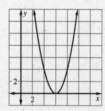

$x = \dfrac{12}{2(1)} = 6$

vertex: $(6, 0)$

$x^2 - 12x + 36 = 0$

$x = \dfrac{12 \pm \sqrt{144 - 4(1)(36)}}{2(1)}$

$= 6$

**3.** $y = 3x^2 + 12x + 9$

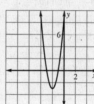

$x = \dfrac{-12}{2(3)} = -2$

vertex: $(-2, -3)$

$x = \dfrac{-12 \pm \sqrt{144 - 4(3)(9)}}{2(3)}$

$= \dfrac{-12 \pm 6}{6}$

$x = -1, -3$

**4.** $x^2 + 6x + 9 = 0$

$x = \dfrac{-6 \pm \sqrt{36 - 4(1)(9)}}{2(1)}$

$= -3$

**5.** $2x^2 + 13x + 6 = 0$

$x = \dfrac{-13 \pm \sqrt{169 - 4(2)(6)}}{2(2)}$

$= \dfrac{-13 \pm 11}{4}$

$= -0.5, -6$

**6.** $-x^2 + 6x + 16 = 0$

$x = \dfrac{-6 \pm \sqrt{36 - 4(-1)(16)}}{2(-1)}$

$= \dfrac{-6 \pm 10}{-2}$

$= -2, 8$

# Chapter 9 *continued*

**7.** $-2x^2 + 7x - 6 = 0$

$$x = \frac{-7 \pm \sqrt{49 - 4(-2)(-6)}}{2(-2)}$$

$$= \frac{-7 \pm 1}{-4}$$

$$= 1.5, 2$$

**8.** $-3x^2 - 5x + 12 = 0$

$$x = \frac{5 \pm \sqrt{25 - 4(-3)(12)}}{2(-3)}$$

$$= \frac{5 \pm 13}{-6}$$

$$= -3, 1\frac{1}{3}$$

**9.** $5x^2 + 8x + 3 = 0$

$$x = \frac{-8 \pm \sqrt{64 - 4(5)(3)}}{2(5)}$$

$$= \frac{-8 \pm 2}{10}$$

$$= -0.6, -1$$

**10.** $x^2 - 15x + 56 = 0$

$225 - 4(1)(56) = 1$

two solutions

**11.** $x^2 + 8x + 16 = 0$

$64 - 4(1)(16) = 0$

one solution

**12.** $x^2 - 3x + 4 = 0$

$9 - 4(1)(4) = -7$

no solution

**13.** $-16t^2 + 50t + 6 = 45$

$2500 - 4(-16)(6) = 2884$

Yes. The discriminant of the equation
$-16t^2 + 50t + 6 = 45$ is positive, so the equation has
two real solutions, meaning the ball would reach a height
of 45 ft.

*Math & History (p. 547)*

**1.** $x^2 + y^2 = 1000$

$y = \frac{2}{3}x - 10$

$x^2 + \left(\frac{2}{3}x - 10\right)^2 - 1000 = 0$

$x^2 + \frac{4}{9}x^2 - \frac{40}{3}x + 100 - 1000 = 0$

$\frac{13}{9}x^2 - \frac{40}{3}x - 900 = 0$ or $13x^2 - 120x - 8100 = 0$

**2.** $14,400 - 4(13)(-8100) = 435,600$

two solutions; one reasonable solution

**3.** $x = \frac{120 \pm \sqrt{435,600}}{2(13)}$

$= 30, -20.8$

The larger square is 30 ft by 30 ft, the smaller is 10 ft by
10 ft.

## Lesson 9.7

*9.7 Guided Practice (p. 551)*

**1.** $y < 2x^2 + 3x - 1$

$y \le -2x^2 + 7x - 3$

$y > x^2 + x + 10$

$y \ge -5x^2 - 3x - 2$

**2.** Write the corresponding equation by replacing the
inequality symbol with an equal sign. Graph the related
function. Use a dashed curve for $<$ and $>$ inequalities
and a solid curve for $\le$ and $\ge$ inequalities. Test a
point in one of the two regions of the plane determined
by the curve. If the point is a solution of the inequality,
shade the region containing the point. If not, shade the
other region.

**3.** inside

**4.** $y < -x^2$

$(1, 1): 1 \not< -1$

not a solution

$(0, -4): -4 < 0$

solution

**5.** $y \ge x^2 - 2$

$(0, 0): 0 \ge -2$

solution

$(1, -2): -2 \ge 1 - 2$

$-2 \not\ge -1$

not a solution

**6.** $y \le 2x^2 + 5x$

$(-1, 1): 1 \overset{?}{\le} 2(1) - 5$

$1 \not\le -3$

not a solution

$(1, 1): 1 \overset{?}{\le} 2(1) + 5(1)$

$1 \le 7$

solution

**7.**

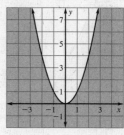

$(0, 1)$

$y \le x^2$

$1 \not\le 0$

**8.**

$(0, 0): y > -x^2 + 3$

$0 \not> 3$

**9.**

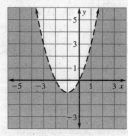

$(-1, 0): y < x^2 + 2x$

$0 \overset{?}{<} 1 - 2$

$0 \not< -1$

# Chapter 9 *continued*

**10.**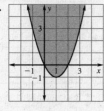

$(1, 0): y \geq x^2 - 2x$

$0 \overset{?}{\geq} 1 - 2$

$0 \geq -1$

**12.**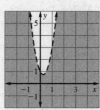

$(0, 0): y < 4x^2 - 2x + 1$

$0 < 1$

### 9.7 Practice and Applications (pp. 551–553)

**13.** $y \geq 2x^2 + x$

$6 \overset{?}{\geq} 2(2)^2 - 2$

$6 \geq 6$

solution

**14.** $y < x^2 + 9x$

$10 \overset{?}{<} (-3)^2 + 9(-3)$

$10 \overset{?}{<} 9 - 27$

$10 \not< -18$

not a solution

**15.** $y > 4x^2 - 7x$

$-10 \overset{?}{>} 4(4) - 7(2)$

$-10 \overset{?}{>} 8 - 14$

$-10 \not> -6$

not a solution

**16.** $y \geq x^2 - 13x$

$14 \overset{?}{\geq} (-1)^2 - 13(-1)$

$14 \overset{?}{\geq} 1 + 13$

$14 \geq 14$

solution

**17.** C    **18.** F    **19.** E

**20.** A    **21.** D    **22.** B

**23.**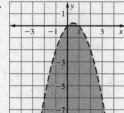

vertex: $\left(\frac{1}{2}, \frac{1}{4}\right)$

$(0, 1): y < -x^2 + x$

$1 \not< 0$

**24.**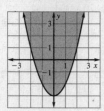

vertex: $(0, -3)$

$(0, 0): y \geq x^2 - 3$

$0 \geq -3$

**11.**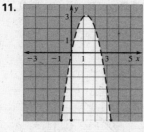

$(1, 0): y > -2x^2 + 5x$

$0 \overset{?}{>} -2(1) + 5$

$0 \not> 3$

**25.**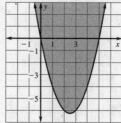

vertex: $\left(\frac{5}{2}, -6\frac{1}{4}\right)$

$(2, 0): y \geq x^2 - 5x$

$0 \geq -6$

**26.**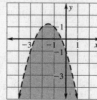

vertex: $\left(-\frac{3}{2}, \frac{5}{4}\right)$

$(-1, 0): y < -x^2 - 3x - 1$

$0 \overset{?}{<} -1 + 3 - 1$

$0 < 1$

**27.**

vertex: $\left(-\frac{1}{2}, 7\frac{3}{4}\right)$

$(0, 0): y < x^2 + x + 8$

$0 < 8$

**28.**

vertex: $\left(\frac{3}{2}, \frac{17}{4}\right)$

$(0, 0): y \leq -x^2 + 3x + 2$

$0 \leq 2$

**29.**

vertex: $(-1, 0)$

$(0, 0): y > -4x^2 - 8x - 4$

$0 > -4$

**30.**

vertex: $\left(-\frac{3}{8}, 8\frac{9}{16}\right)$

$(0, 0): y \geq -4x^2 - 3x + 8$

$0 \not\geq 8$

**31.**

vertex: $\left(-\frac{5}{4}, -\frac{1}{8}\right)$

$(0, 0): y \geq 2x^2 + 5x + 3$

$0 \not\geq 3$

# Chapter 9 *continued*

**32.** $y = 0.000112x^2 + 5$

$y = 0.000112(2100)^2 + 5$

$y \approx 499$ ft

**33.**

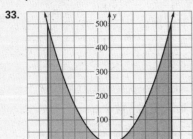

vertex: $(0, 5)$

| x | y |
|------|-------|
| 400 | 22.9 |
| 800 | 76.7 |
| 1600 | 291.7 |
| 2100 | 498.9 |

**34.** IF; for every value of $x$, the value of $y$ (that is, the cost of the diamond) is greater for that grade than for either of the others.

**35.** I3; for every value of $x$, the value of $y$ (that is, the cost of the diamond) is lower for that grade than for either of the others.

**36.** The region described by $y > 82.12x^2 + 119.13x - 48.15$; yes; the region is bounded below by the equation for an IF diamond.

**37.** $y > 1.87x^2 + 14.83x - 2.3$

$12,000 \overset{?}{>} 1.87 + 14.83 - 2.3$

$12,000 > 14.4$

A; VS2

**38.** no    **39.** c    **40.** C

**41.**

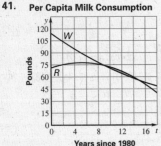

Per Capita Milk Consumption

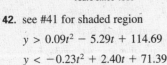

**42.** see #41 for shaded region

$y > 0.09t^2 - 5.29t + 114.69$

$y < -0.23t^2 + 2.40t + 71.39$

**43.** about $(15, 55.5)$ and $(9, 74.4)$; a year in which per capita consumption of whole milk and reduced fat milk was the same.

**9.7 Mixed Review (p. 553)**

**44.** $y = x + 4$

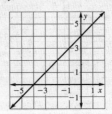

**45.** $3x + 6y = -18$

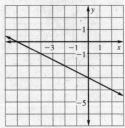

**46.** $2x - 4y = 24$

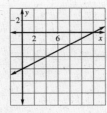

**47.** $y = x^2 + x + 2$

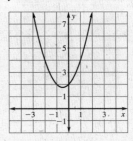

vertex: $\left(-\frac{1}{2}, \frac{7}{4}\right)$

**48.** $y = -x^2 + 4x + 1$

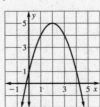

vertex: $(2, 5)$

**49.** $y = 3x^2 - 2x + 6$

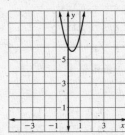

vertex: $\left(\frac{1}{3}, 5\frac{2}{3}\right)$

**50.** $6k = 42$

$k = 7$

$7x = y$

**51.** $54k = -9$

$k = -\frac{1}{6}$

$-\frac{1}{6}x = y$

**52.** $7k = 5$

$k = \frac{5}{7}$

$\frac{5}{7}x = y$

**53.** $-13k = -52$

$k = 4$

$4x = y$

**54.** $\frac{3}{4}k = 3$

$k = 4$

$4x = y$

**55.** $4 \cdot 6k = 1.2$

$k = \frac{6}{23}$

$\frac{6}{23}x = y$

**56.**

**57.**

**Algebra 1**
Chapter 9 Worked-out Solution Key

# Chapter 9 *continued*

**58.**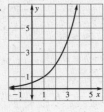

**59.**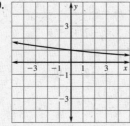

**60.**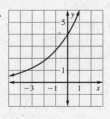

**61.**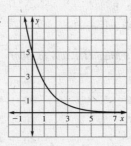

**62.** $x^2 - 2x - 3 = 0$

$$x = \frac{2 \pm \sqrt{4 - 4(1)(-3)}}{2(1)}$$

$$= \frac{2 \pm 4}{2}$$

$$= 3, -1$$

**63.** $2x^2 - 6x + 4 = 0$

$$x = \frac{6 \pm \sqrt{36 - 4(2)(4)}}{2(2)}$$

$$= \frac{6 \pm 2}{4}$$

$$= 2, 1$$

**64.** $2x^2 - 2x - 12 = 0$

$$x = \frac{2 \pm \sqrt{4 - 4(2)(-12)}}{2(2)}$$

$$= \frac{2 \pm 10}{4}$$

$$= 3, -2$$

**65.** $-\frac{2}{3}x^2 - 3x + 1 = 0$

$$x = \frac{3 \pm \sqrt{9 - 4\left(-\frac{2}{3}\right)(1)}}{2\left(-\frac{2}{3}\right)}$$

$$= \frac{3 \pm \sqrt{\frac{35}{3}}}{-\frac{4}{3}}$$

$$\approx -4.81, 0.31$$

**66.** $-7x^2 - 2.5x + 3 = 0$

$$x = \frac{2.5 \pm \sqrt{6.25 - 4(-7)(3)}}{2(-7)}$$

$$= \frac{2.5 \pm \sqrt{90.25}}{-14}$$

$$= -0.86, 0.5$$

**67.** $2x^2 + 4x - 3 = 0$

$$x = \frac{-4 \pm \sqrt{16 - 4(2)(-3)}}{2(2)}$$

$$= \frac{-4 \pm 2\sqrt{10}}{4}$$

$$\approx 0.58, -2.58$$

## Lesson 9.8

### 9.8 Guided Practice (p. 557)

**1.** linear: $y = x$   exponential: $y = 2^x$

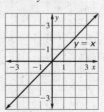 

quadratic: $y = x^2$

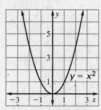

**2.** Make a scatter plot. If you cannot determine whether the best fit is an exponential or quadratic model, compare the ratios of consecutive $y$-coordinates.

**3.** quadratic   **4.** exponential   **5.** linear

**6.**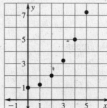

**7.** Quadratic; the graph is a curve but the ratios of consecutive $y$-coordinates are not the same.

**8.** $y = \frac{1}{4}x^2 + 1$

### 9.8 Practice and Applications (pp. 557–559)

**9.**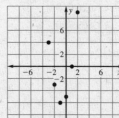

quadratic

**10.**

quadratic

# Chapter 9 continued

**11.**

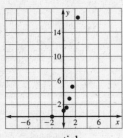

exponential

**12.**

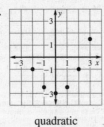

quadratic

**23.** $m = \dfrac{3-7}{3-1} = \dfrac{-4}{2} = -2$

$3 = -2(3) + b$

$9 = b$

$y = -2x + 9$

**24.** $m = \dfrac{5-7}{6-7} = \dfrac{-2}{-1} = 2$

$5 = 6(2) + b$

$-7 = b$

$y = 2x - 7$

**25.** $(4, 1)$

**26.** $a = 2, b = 4, c = 1$

**27.** A

**28.** D

**29.** B

**30.** linear

**31.** exponential

**32.** quadratic

**13.**

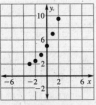

exponential

**14.**

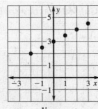

linear

### 9.8 Mixed Review (p. 560)

**33.** $-3, x$

**34.** $a, -5$

**35.** $-5, -8x$

**36.** $-4r, s, -1$

**37.** $-5y^2$

**38.** $-27x^2$

**39.** $2r^3$

**40.** $-4n^2$

**41.** $-y^3$

**42.** $-x^4$

**15.**

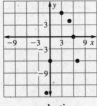

quadratic

**16.**

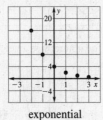

exponential

**43.** $15.67x + 23.61 = 1.56 + 45.8x$

$22.05 = 30.13x$

$x \approx 0.73$

**44.** $17.87 - 2.87x = 1.87 - 4.92x$

$2.05x = -16$

$x \approx -7.80$

**45.** $6.35x - 9.94 = 3.88 + 40.34x$

$-13.82 = 33.99x$

$x \approx -0.41$

**46.** $5.6(1.2 + 1.9x) = 20.4x + 6.8$

$6.72 + 10.64x = 20.4x + 6.8$

$-0.08 = 9.76x$

$x \approx -0.01$

**17.**

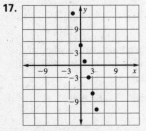

linear

**18.** $S = 4\pi(r^2)$

**47.** $m = \dfrac{4-(-2)}{5-3} = \dfrac{6}{2} = 3$

$4 = 5(3) + b$

$-11 = b$

$y = 3x - 11$

**48.** $m = \dfrac{7-(-9)}{5-(-3)} = \dfrac{16}{8} = 2$

$7 = 5(2) + b$

$-3 = b$

$y = 2x - 3$

**19.** linear

$19.55 = 45.40k$

$0.43 \approx k$

$B = 0.43w$

**20.** linear

$2.6 = 1k$

$2.6 = k$

$d = 2.6M$

**49.** $m = \dfrac{6-3}{-4-2} = \dfrac{3}{-6} = \dfrac{-1}{2}$

$6 = -\dfrac{1}{2}(-4) + b$

$4 = b$

$y = -\dfrac{1}{2}x + 4$

**21.**

| Year t | Attendance, A | Change in A from previous year |
|---|---|---|
| **0** | 12,580,660 | |
| 1 | 12,338,980 | −241,680 |
| 2 | 12,781,160 | 442,180 |
| 3 | 13,907,200 | 1,126,040 |
| 4 | 15,717,100 | 1,809,900 |

**22.** The change, $C$, fits a linear model,

$C = 683,860t - 925,540.$

**Algebra 1**
Chapter 9 Worked-out Solution Key

# Chapter 9 *continued*

**Quiz 3 (p. 560)**

**1.** $10 \overset{?}{\geq} 2(4) - 2 + 9$
$10 \overset{?}{\geq} 8 - 2 + 9$
$10 \not\geq 15$
not a solution

**2.** $-60 \overset{?}{>} 4(144) - 64(12) + 115$
$-60 > -77$
solution

**3.** $6 \overset{?}{<} 1 - 6 + 12$
$6 < 7$
solution

**4.** $16 \leq 1 + 7 + 9$
$16 \leq 17$
solution

**5.** B   **6.** C   **7.** A

**8.**

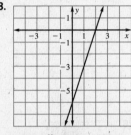

linear

**9.**

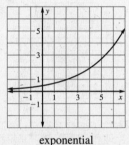

exponential

**10.**

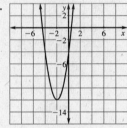

quadratic

**11.**

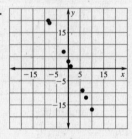

linear
$y = -2x + 3$

**12.**

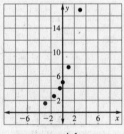

exponential
$y = 5(1.5)^x$

**13.**

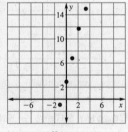

linear
$y = 3.72x + 3.16$

## Chapter 9 Review (pp. 562–564)

**1.** $x^2 - 144 = 0$
$x^2 = 144$
$x = \pm 12$

**2.** $8y^2 = 968$
$y^2 = 121$
$y = \pm 11$

**3.** $4t^2 + 19 = 19$
$4t^2 = 0$
$t^2 = 0$
$t = 0$

**4.** $16y^2 - 80 = 0$
$16y^2 = 80$
$y^2 = 5$
$y = \pm \sqrt{5}$

**5.** $\frac{1}{5}a^2 = 5$
$a^2 = 25$
$a = \pm 5$

**6.** $\frac{1}{3}x^2 - 7 = -4$
$\frac{1}{3}x^2 = 3$
$x^2 = 9$
$x = \pm 3$

**7.** $\sqrt{45} = \sqrt{9 \cdot 5}$
$= \sqrt{9} \cdot \sqrt{5}$
$= 3\sqrt{5}$

**8.** $\sqrt{441} = \sqrt{9 \cdot 49}$
$= \sqrt{9} \cdot \sqrt{49}$
$= 3 \cdot 7$
$= 21$

**9.** $\sqrt{\frac{36}{64}} = \frac{\sqrt{36}}{\sqrt{64}}$
$= \frac{6}{8}$
$= \frac{3}{4}$

**10.** $\sqrt{\frac{99}{25}} = \frac{\sqrt{9 \cdot 11}}{\sqrt{25}} = \frac{3\sqrt{11}}{5}$

**11.**

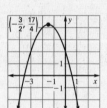

$x = \frac{1}{2(1)} = \frac{1}{2}$
vertex: $\left(\frac{1}{2}, -\frac{21}{4}\right)$

| $x$ | $-1$ | $0$ | $1$ | $2$ |
|-----|------|-----|-----|-----|
| $y$ | $-3$ | $-5$ | $-5$ | $-3$ |

**12.**

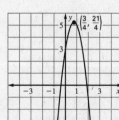

$x = \frac{3}{2(-1)} = \frac{-3}{2}$
vertex: $\left(-\frac{3}{2}, \frac{17}{4}\right)$

| $x$ | $-3$ | $-2$ | $-1$ | $0$ |
|-----|------|------|------|-----|
| $y$ | $2$ | $4$ | $4$ | $2$ |

**13.**

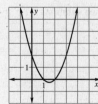

$x = \frac{-6}{2(-4)} = \frac{3}{4}$
vertex: $\left(\frac{3}{4}, \frac{21}{4}\right)$

| $x$ | $-1$ | $0$ | $1$ | $2$ |
|-----|------|-----|-----|-----|
| $y$ | $-7$ | $3$ | $5$ | $-1$ |

**14.**

Check: $1^2 - 3(1) + 2 \overset{?}{=} 0$
$0 = 0$
$4 - 3(2) + 2 \overset{?}{=} 0$
$0 = 0$

# Chapter 9 *continued*

**15.**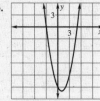

Check: $0 \stackrel{?}{=} (-3)^2 - 2(-3) - 15$

$0 = 0$

$0 \stackrel{?}{=} 5^2 - 2(5) - 15$

$0 = 0$

**16.**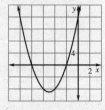

Check: $x^2 + 10x + 16 = 0$

$(x + 8)(x + 2) = 0$

$x = -8 \quad x = -2$

**17.**

Check: $-x^2 - 2x + 24 = 0$

$(x + 6)(4 - x) = 0$

$x = -6 \quad x = 4$

**18.** $3x^2 - 4x + 1 = 0$

$x = \dfrac{4 \pm \sqrt{16 - 4(3)(1)}}{2(3)}$

$= \dfrac{4 \pm 2}{6}$

$= 1, \dfrac{1}{3}$

**19.** $-2x^2 + x + 6 = 0$

$x = \dfrac{-1 \pm \sqrt{1 - 4(-2)(6)}}{2(-2)}$

$= \dfrac{-1 \pm 7}{-4}$

$= -\dfrac{3}{2}, 2$

**20.** $10x^2 - 11x + 3 = 0$

$x = \dfrac{11 \pm \sqrt{121 - 4(10)(3)}}{2(10)}$

$= \dfrac{11 \pm 1}{20}$

$= \dfrac{3}{5}, \dfrac{1}{2}$

**21.** $144 - 4(3)(12)$

$144 - 144 = 0$

one solution

**22.** $100 - 4(2)(6)$

$100 - 48 = 52$

two solutions

**23.** $9 - 4(-1)(-5)$

$9 - 20 = -11$

no solution

**24.** $x^2 - 3 \geq y$

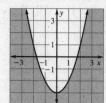

**25.** $-x^2 - 2x + 3 \leq y$

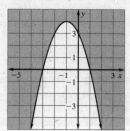

**26.** $\frac{1}{2}x^2 + 3x - 4 < y$

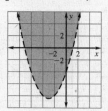

**27.**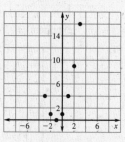

quadratic

**28.**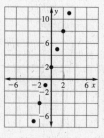

linear

**29.**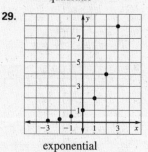

exponential

## Chapter 9 Test *(p. 565)*

**1.** $-3$

**2.** $0.08$

**3.** $\pm 11$

**4.** $0$

**5.** $\sqrt{192} = \sqrt{64 \cdot 3}$

$= 8\sqrt{3}$

**6.** $\sqrt{5} \cdot \sqrt{30} = \sqrt{5 \cdot 5 \cdot 6}$

$= 5\sqrt{6}$

**7.** $\sqrt{\dfrac{27}{147}} = \sqrt{\dfrac{9}{49}}$

$= \dfrac{3}{7}$

**8.** $\dfrac{10\sqrt{8}}{\sqrt{16}} = \dfrac{20\sqrt{2}}{4}$

$= 5\sqrt{2}$

**9.** $8x^2 = 800$

$x^2 = 100$

$x = \pm 10$

**10.** $4x^2 + 11 = 12$

$4x^2 = 1$

$x^2 = \dfrac{1}{4}$

$x = \pm\dfrac{1}{2}$

# Chapter 9 *continued*

**11.** $2x^2 + 5x - 7 = 0$

$$x = \frac{-5 \pm \sqrt{25 - 4(2)(-7)}}{2(2)}$$

$$= \frac{-5 \pm 9}{4}$$

$$= 1, -\frac{7}{2}$$

**12.** $-2x^2 + 4x + 6 = 0$

$$x = \frac{-4 \pm \sqrt{16 - 4(-2)(6)}}{2(-2)}$$

$$= \frac{-4 \pm 8}{-4}$$

$$= -1, 3$$

**13.** $64x^2 - 5 = 11$

$64x^2 = 16$

$x^2 = \frac{1}{4}$

$x = \pm\frac{1}{2}$

**14.** $10x^2 + 17x - 11 = 0$

$$x = \frac{-17 \pm \sqrt{289 - 4(10)(-11)}}{2(10)}$$

$$= \frac{-17 \pm 27}{20}$$

$$= \frac{1}{2}, -\frac{11}{5}$$

**15.** C      **16.** A

**17.** D      **18.** B

**19.** $400 - 4(5)(-60) = 1600$

two solutions

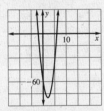

**20.** $16 - 4(-3)(-2) = -8$

no solution

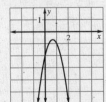

**21.** $16 - 4(1)(4) = 0$

one solution

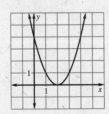

**22.**

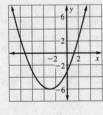

vertex: $(-3, -6)$

**23.**

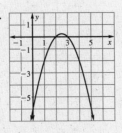

vertex: $\left(2\frac{1}{2}, \frac{1}{4}\right)$

**24.**

vertex: $\left(-\frac{7}{2}, -\frac{25}{4}\right)$

**25.**

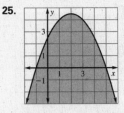

vertex: $\left(2, 4\frac{1}{2}\right)$

**26.** exponential      **27.** linear

**28.** quadratic      **29.** $0 = -16t^2 + 20$

$$t^2 = \frac{5}{4}$$

$$t \approx 1.12 \text{ sec}$$

**30.** $0 = -16t^2 + 30t + 20$

$$t = \frac{-30 \pm \sqrt{900 - (4)(-16)(20)}}{2(-16)}$$

$$= \frac{-30 \pm \sqrt{2180}}{-32}$$

$$\approx 2.40 \text{ sec}$$

**31.** $0 = -16t^2 + 50t - 60$

$2500 - 4(-16)(-60) = -1340$

no

## Chapter 9 Standardized Test *(pp. 566–567)*

**1.** D      **2.** A

**3.** $h = -16t^2 + 320$

$$t = \frac{0 \pm \sqrt{0 - 4(-16)(320)}}{2(-16)}$$

$$t \approx 4.5 \text{ sec.}$$

D

**4.** $150 = 6s^2$

$25 = s^2$

$5 = s$

C

**5.** $A = \sqrt{12} \cdot \sqrt{20}$

$= 4\sqrt{15}$

E

**6.** $3x^2 - 78 = 114$

$3x^2 = 192$

$x^2 = 64$

$x = \pm 8$

D

**7.** Column A: $\sqrt{144}$

Column B: $\sqrt{125}$

A

**8.** Column A: $\sqrt{9}$

Column B: $\sqrt{15}$

B

**9.** $x = \frac{1}{2\left(-\frac{1}{2}\right)} = -1$

B

**10.** $0 = -x^2 - 6x + 40$

$0 = x^2 + 6x - 40$

$(x - 4)(x + 10) = 0$

$x = 4 \quad x = -10$

C

**11.** D

# Chapter 9 *continued*

**12.** E

**13.** $4 - 4(-7)(5) = 144$

B

**14.** E

**15.** B

**16.** a.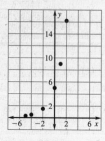

b. exponential

c. $y = 5(1.8)^x$

**17.** a. $x = \dfrac{12}{2(1)} = 6$ m

b. $y = 36 - 12(6) + 32$

$= -4$

4 m

c. $0 = x^2 - 12x + 32$

$x = \dfrac{12 \pm \sqrt{144 - 4(1)(32)}}{2(1)}$

$= 8$ m, $4$ m

d. $y = 32$ ft

e. The graph of $y < x^2 - 12x + 32$ represents the parabola below which the fish can swim out of reach of the bird.

**Algebra 1**
Chapter 9 Worked-out Solution Key

# Chapter 9 *continued*

## Chapter 9 Cumulative Practice
### Chs. 1–9 *(pp. 568–569)*

1. Yes; for each input, there is exactly one output.

2. No; the input value 5 has two different output values.

3. Yes; for each input, there is exactly one output.

4. probability: $\frac{30}{84} = \frac{5}{14} \approx 36\%$

   odds: 5 to 9

5. probability: $\frac{16}{72} = \frac{2}{9} \approx 22\%$

   odds: 2 to 7

6. *Sample equation:*

   $\dfrac{x}{8} = \dfrac{12}{16}$

   $16x = 96$

   $x = 6$ ft

7. *Sample equation:*

   $\dfrac{15}{35} = \dfrac{6}{x}$

   $15x = 210$

   $x = 14$ in.

8. $\dfrac{1}{3} = \dfrac{8 - y}{11 - 2}$

   $9 = 24 - 3y$

   $-15 = -3y$

   $5 = y$

9. $\dfrac{-1}{1} = \dfrac{3 - y}{7 - 6}$

   $-1 = 3 - y$

   $-4 = -y$

   $4 = y$

10. $\dfrac{4}{1} = \dfrac{-10 - y}{5 - 8}$

    $-12 = -10 - y$

    $-2 = -y$

    $y = 2$

11. $-\dfrac{1}{2} = \dfrac{y - 7}{5 + 5}$

    $-10 = 2y - 14$

    $y = 2y$

    $2 = y$

12. $\dfrac{5}{1} = \dfrac{y + 12}{3 - 0}$

    $15 = y + 12$

    $3 = y$

13. $\dfrac{-2}{1} = \dfrac{y + 8}{-4 - 3}$

    $14 = y + 8$

    $6 = y$

14. $3x - 5y = -6$

15. $2x - 6y = -4$

16. $-2x + 7y = 15$

17. 5, 7, 8

18. 7, 9, 12

19. 4, 4, 8

20. $x + y = 8$ $\qquad 2 + y = 8$

    $2x + y = 10$ $\qquad y = 6$

    $y = -x + 8$

    $2x - x + 8 = 10$

    $\qquad x = 2$

    solution: $(2, 6)$

21. $\left(\frac{1}{4}x - y = 7\right)$ Mult. by 4 → $x - 4y = 28$

    $x + 4y = 0$ $\qquad \underline{x + 4y = \;\;0}$

    $\qquad\qquad\qquad 2x \qquad = 28$

    $\qquad\qquad\qquad\quad x = 14$

    $\qquad\qquad 14 + 4y = 0$

    $\qquad\qquad\qquad 4y = -14$

    $\qquad\qquad\qquad\quad y = -\frac{7}{2}$

    solution: $\left(14, -\frac{7}{2}\right)$

22. $-2x + 20y = 10$ $\qquad\qquad -2x + 20y = \;\;\;10$

    $(x - 5y = -5)$ Mult. by 2 → $\underline{2x - 10y = -10}$

    $\qquad\qquad\qquad\qquad\qquad\qquad 10y = 0$

    $\qquad\qquad\qquad\qquad\qquad\qquad\quad y = 0$

    $\qquad\qquad\qquad\qquad x - 5(0) = -5$

    $\qquad\qquad\qquad\qquad\qquad\quad x = -5$

    solution: $(-5, 0)$

23.

    $(3.2x + 1.1y = -19.3)$ Mult. by 10 → $32x + 11y = -193$

    $-32x + 4y = 148$ $\qquad\qquad \underline{-32x + 4y = \;\;\;148}$

    $\qquad\qquad\qquad\qquad\qquad\qquad\qquad 15y = -45$

    $\qquad\qquad\qquad\qquad\qquad\qquad\quad y = -3$

    $\qquad\qquad -32x + 4(-3) = 148$

    $\qquad\qquad\qquad\qquad\qquad x = -5$

    solution: $(-5, -3)$

24. $\left(\frac{1}{10}x - \frac{3}{2}y = -1\right)$ Mult. by 100 → $10x - 150y = -100$

    $-10x + 3y = 2$ $\qquad\qquad \underline{-10x + \;\;3y = \;\;\;\;\;2}$

    $\qquad\qquad\qquad\qquad\qquad\qquad -147 = -98$

    $\qquad\qquad\qquad\qquad\qquad\qquad\quad y = \;\;\;\frac{2}{3}$

    $\qquad\qquad \frac{1}{10}x - \frac{3}{2}\left(\frac{2}{3}\right) = 1$

    $\qquad\qquad\qquad\qquad \frac{1}{10}x = 0$

    $\qquad\qquad\qquad\qquad\quad x = 0$

    solution: $\left(0, \frac{2}{3}\right)$

25.

    $(1.4x + 2.1y = 1.75)$ Mult. by $-2$ → $-2.8x - 4.2y = -3.5$

    $2.8x - 4.2y = 34.58$ $\qquad\qquad \underline{2.8x - 4.2y = 34.58}$

    $\qquad\qquad\qquad\qquad\qquad\qquad\qquad -8.4y = 31.08$

    $\qquad\qquad\qquad\qquad\qquad\qquad\qquad\quad y = -3.7$

    $\qquad\qquad 1.4x + 2.1(-3.7) = 1.75$

    $\qquad\qquad\qquad\qquad\qquad 1.4x = 9.52$

    $\qquad\qquad\qquad\qquad\qquad\quad x = 6.8$

    solution: $(6.8, -3.7)$

26.

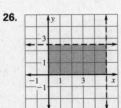

27.

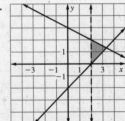

# Chapter 9 *continued*

**28.**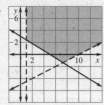

**29.**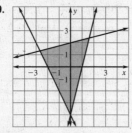

**30.** $\dfrac{1}{x^{14}}$

**31.** $\dfrac{1}{x^2}$

**32.** $\dfrac{y}{(2x)^{-3}} = 32x^3$

**33.** $\left(\dfrac{-8x^3}{4xy^5}\right)^2 = \left(\dfrac{-2x^2}{y^5}\right)^2$

$$= \dfrac{4x^4}{y^{10}}$$

**34.** $5x \cdot (x \cdot x^{-4})^2 = 5x(x^{-6})$

$$= \dfrac{5}{x^5}$$

**35.** $(6a^3)^2\left(\tfrac{1}{2}a^3\right)^2 = (36a^6)\left(\tfrac{1}{4}a^6\right)$

$$= 9a^{12}$$

**36.** $(r^2st^5)^0(s^4t^2)^3 = s^{12}t^6$

**37.** $\dfrac{6x^4y^4}{3xy} \cdot \dfrac{5x^2y^3}{2y^2} = (2x^3y^3)\left(\tfrac{5}{2}x^2y\right)$

$$= 5x^5y^4$$

**38.** $(5 \times 10^{-2})(3 \times 10^4) = 1.5 \times 10^3 = 1500$

**39.** $(6 \times 10^{-4})(7 \times 10^{-5}) = 4.2 \times 10^{-8} = 0.000000042$

**40.** $\dfrac{20 \times 10^{-4}}{2.5 \times 10^{-8}} = 8 \times 10^4 = 80{,}000$

**41.** $(7 \times 10^3)^{-3} = 7000^{-3}$

$$= 2.92 \times 10^{-12}$$

$$= 0.00000000000292$$

**42.** $\dfrac{8.8 \times 10^{-1}}{11 \times 10^{-1}} = 8 \times 10^{-1} = 0.8$

**43.** $(2.8 \times 10^{-2}) = 2.1952 \times 10^{-5} = 0.000021952$

**44.** D

**45.** B

**46.** C

**47.** A

**48.** $4\sqrt{3}$

**49.** $\sqrt{\dfrac{28}{36}} = \dfrac{2\sqrt{7}}{6}$

$$= \dfrac{\sqrt{7}}{3}$$

**50.** $\dfrac{1}{4}\left(2\sqrt{21}\right) = \dfrac{\sqrt{21}}{2}$

**51.** $\dfrac{\sqrt{112}}{\sqrt{49}} = \dfrac{4\sqrt{7}}{7}$

**52.** $\sqrt{12} \cdot \sqrt{63} = 2\sqrt{3}\left(3\sqrt{7}\right)$

$$= 6\sqrt{21}$$

**53.** $\sqrt{9} \cdot \dfrac{\sqrt{18}}{\sqrt{54}} = \dfrac{3(3\sqrt{2})}{3\sqrt{6}}$

$$= \dfrac{3\sqrt{2}}{\sqrt{6}}$$

$$= \dfrac{6\sqrt{3}}{6}$$

$$= \sqrt{3}$$

**54.** $\dfrac{-2\sqrt{98}}{\sqrt{7}} = -2\sqrt{14}$

**55.** $\dfrac{\sqrt{33} \cdot \sqrt{75}}{\sqrt{11}} = \sqrt{3} \cdot \left(5\sqrt{3}\right)$

$$= 15$$

**56.**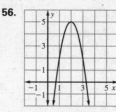

vertex: (2, 5)

**57.**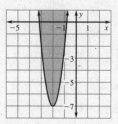

vertex: $(-2, -7)$

**58.**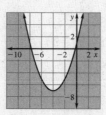

vertex: $(-4, -7)$

**59.** $x = \dfrac{-10 \pm \sqrt{100 - 4(1)(9)}}{2(1)}$

$$= \dfrac{-10 \pm 8}{2}$$

$$= -1, -9$$

**60.** $x = \dfrac{-5 \pm \sqrt{25 - 4(-1)(-6)}}{2(-1)}$

$$= \dfrac{-5 \pm 1}{-2}$$

$$= 2, 3$$

**61.** $x = \dfrac{-8 \pm \sqrt{64 - 4(3)(-5)}}{2(3)}$

$$= \dfrac{-8 \pm 2\sqrt{31}}{6}$$

$$\approx 0.52, -3.19$$

**Algebra 1**
Chapter 9 Worked-out Solution Key

# Chapter 9 *continued*

**62.** $x = \dfrac{-5 \pm \sqrt{25 - 4(-2)(12)}}{2(-2)}$

$= \dfrac{-5 \pm 11}{-4}$

$= -1.5, 4$

**63.** $x = \dfrac{-3 \pm \sqrt{9 - 4\left(-\frac{1}{2}\right)\left(-\frac{5}{2}\right)}}{2\left(-\frac{1}{2}\right)}$

$= \dfrac{3 \pm 2}{-1}$

$= 1, 5$

**64.** $x = \dfrac{-12 \pm \sqrt{144 - 4(7)(-2)}}{2(7)}$

$= \dfrac{-12 \pm 10\sqrt{2}}{14}$

$\approx 0.15, -1.87$

**65.** $A = 567(1.048)^5$

$= \$716.79$

$A = 567(1.051)^5$

$= \$727.10$

Amt. extra: $\$10.31$

$A = 567(1.048)^{10}$

$= \$906.14$

$A = 567(1.051)^{10}$

$= \$932.42$

Amt. extra: $\$26.28$

**66.** $150 = -16t^2 + 100t + 0$

$0 = -16t^2 + 100t - 150$

$t = \dfrac{-100 \pm \sqrt{10,000 - 4(-16)(-150)}}{2(-16)}$

$= \dfrac{-100 \pm 20}{-32}$

$\approx 2.5 \text{ sec}$

**67.** $180 = -16t^2 + 100t + 0$

$0 = -16t^2 + 100t - 180$

$10,000 - 4(-16)(-180) = -1520$

No; the discriminant is negative, therefore there are no *x*-intercepts (in other words, no solution) so the flare will never reach 180 ft.

**68.** Yes; the area of the rectangle is $x(26 - x)$. The discriminant of the equation $x(26 - x) = 148.75$ is positive, so there are two solutions. The rectangle has dimensions 8.5 cm and 17.5 cm.

$8.5 + 8.5 + 17.5 + 17.5 = 52$

$8.5(17.5) = 148.75$

## Project Chs. 7–9 *(pp. 570–571)*

**1.** Pizza Business Sales Report

| Year | Number of Pizzas Sold | Total Sales ($9 per pizza) | |
|---|---|---|---|
| 1 | 1200 | $9 \times 1200$ | $= 10{,}800$ |
| 2 | 2040 | $9 \times 2040$ | $= 18{,}360$ |
| 3 | 3470 | $9 \times 3470$ | $= 31{,}230$ |
| 4 | 5900 | $9 \times 5900$ | $= 53{,}100$ |
| 5 | 10,025 | $9 \times 10{,}025$ | $= 90{,}225$ |
| 6 | 17,040 | $9 \times 17{,}040$ | $= 153{,}360$ |
| 7. | 28,970 | $9 \times 28{,}970$ | $= 260{,}730$ |
| 8 | 49,240 | $9 \times 49{,}240$ | $= 443{,}160$ |

**2.**

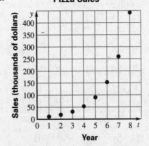

Pizza Sales

The more that time increases, the steeper the graph for number of pizzas sold climbs.

**3.** An exponential model, such as $y = 6353(1.7)^x$, fits the data best. From the graph it is clear that a linear model would not fit the data well. An exponential model will fit the data well because if you calculate the ratio of each *y*-coordinate to the preceding *y*-coordinate, the ratios are approximately equal.

**4.** $y = 706(1.7)^{10}$

$= 142{,}329$

Sales $= 142.329 \times 9$

$\approx \$1{,}280{,}000$

**5–7.** Answers may vary.

# CHAPTER 10

## Think & Discuss *(p. 573)*

1. −500, 500; the diameter is the distance between the *x*-intercepts, 1000 ft.

2. Estimates will vary; actual depth is 167 feet.

## Skill Review *(p. 574)*

1. $3x(x + 6) = 3x^2 + 18x$  2. $(2 − 4x)7 = 14 − 28x$

3. $−4(x + 5) = −4x − 20$  4. $(8 − 2x)(−8) = −64 + 16x$

5. $2(x − 3) + x = 2x − 6 + x = 3x − 6$

6. $5x + 3(1 − x) = 5x + 3 − 3x = 2x + 3$

7. $−(x + 4) − x = −x − 4 − x = −2x − 4$

8. $2x^2 − x(1 + x) = 2x^2 − x − x^2 = x^2 − x$

9. $x^5 \cdot x^4 \cdot x = x^{5+4+1} = x^{10}$  10. $(x^6)^2 = x^{6 \cdot 2} = x^{12}$

11. $(−2ab)^3 = −2^3a^3b^3 = −8a^3b^3$

12. $(3xy^4)^2 \cdot x^2 = 3^2x^2y^8 \cdot x^2 = 9x^4y^8$

13. $3x^2 − 4x + 6 = 0$   14. $5x^2 + 6x − 8 = 0$

   −56, no real solution    196, two solutions

15. $2x^2 − 12x + 18 = 0$

   0; one solution

## Lesson 10.1

### Activity 10.1 *(p. 575)*
### Drawing Conclusions

1.

   $(−x^2 + x − 1) + (4x^2 + 2x − 3)$

   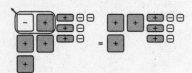

   $3x^2 + 3x − 4$

2.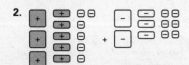

   $(3x^2 + 5x − 6) + (−2x^2 − 3x − 6)$

   $x^2 + 2x − 12$

3.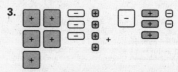

   $(5x^2 − 3x + 4) + (−x^2 + 3x − 2)$

   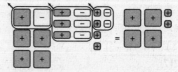

   $4x^2 + 2$

4.

   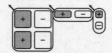

   $(2x^2 − x − 1) + (−2x^2 + x + 1) = 0$

5.

   $(−x^2 + 3x + 7) + (x^2 − 7) = 3x$

6.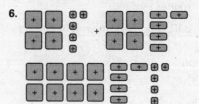

   $(4x^2 + 5) + (4x^2 + 5x) = 8x^2 + 5x + 5$

7. *Sample answer:* Subtracting a polynomial is the same as adding its opposite. Multiply each term of the second polynomial by −1 and add the resulting polynomial to the first one.

8. $(x^2 + 3x + 4) − (x^2 + 3) =$

   $(x^2 + 3x + 4) + (−x^2 − 3)$

   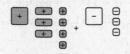

     $3x + 1$

# Chapter 10 *continued*

**9.** $(x^2 - 2x + 5) - (3 - 2x) =$
$(x^2 - 2x + 5) + (2x - 3)$

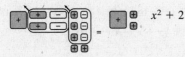

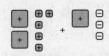

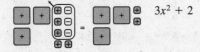 $x^2 + 2$

**10.** $(2x^2 + 5) - (-x^2 + 3) = (2x^2 + 5) + (x^2 - 3)$

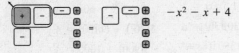 $3x^2 + 2$

**11.** $(x^2 + 4) - (2x^2 + x) = (x^2 + 4) + (-2x^2 - x)$

$-x^2 - x + 4$

## 10.1 Guided Practice (p. 579)

**1.** No; in standard form, the polynomial is
$-4x^3 + 9x^2 + 8x + 3$; the degree is 3.

**2.** $-3x^3 - 4x^2 + 5x + 6$; $-3x^3, -4x^2, 5x, 6$

**3.** $-3, -4, 5, 6; -3$   **4.** 3

**5.** $-3x^2$ and $-5x$ cannot be combined as they are not like terms.

**6.** $-9x - 3x \neq -9x + 3x$ and $0 - (-7) \neq -7$.

**7.** linear, binomial   **8.** quadratic, binomial

**9.** cubic, trinomial   **10.** quadratic, binomial

**11.** constant, monomial   **12.** quartic, polynomial

**13.** $(x^2 - 4x + 3) + (3x^2 - 3x - 5)$
$= x^2 - 4x + 3 + 3x^2 - 3x - 5 = 4x^2 - 7x - 2$

**14.** $(-x^2 + 3x - 4) - (2x^2 + x - 1)$
$= -x^2 + 3x - 4 - 2x^2 - x + 1$
$= -3x^2 + 2x - 3$

**15.** $(-3x^2 + x + 8) - (x^2 - 8x + 4)$
$= -3x^2 + x + 8 - x^2 + 8x - 4$
$= -4x^2 + 9x + 4$

**16.** $(5x^2 - 2x - 1) + (-3x^2 - 6x - 2)$
$= 5x^2 - 2x - 1 - 3x^2 - 6x - 2 = 2x^2 - 8x - 3$

**17.** $(4x^2 - 2x - 9) + (x - 7 - 5x^2)$
$= 4x^2 - 2x - 9 + x - 7 - 5x^2 = -x^2 - x - 16$

**18.** $(2x - 3 + 7x^2) - (3 - 9x^2 - 2x)$
$= 2x - 3 + 7x^2 - 3 + 9x^2 + 2x = 16x^2 + 4x - 6$

### 10.1 Practice and Applications (pp. 580–582)

**19.** $-3$; linear, binomial   **20.** $-4$; quadratic, trinomial

**21.** 1; cubic, polynomial   **22.** 5; quadratic, trinomial

**23.** $-6$; constant, monomial   **24.** 14; quartic, binomial

**25.** 5; quartic, trinomial   **26.** 7.4; cubic, binomial

**27.** $-4$; quartic, polynomial   **28.** 9; cubic, polynomial

**29.** $-16$; cubic, monomial   **30.** $-95$; quartic, polynomial

**31.**
$$\begin{array}{r} 12x^3 + 0x^2 + 0x + 10 \\ + -18x^3 + 3x^2 + 0x - 6 \\ \hline -6x^3 + 3x^2 + 0x + 4 \end{array}$$
$= -6x^3 + 3x^2 + 4$

**32.**
$$\begin{array}{r} 2a^3 + 3a^2 + \ a + 0 \\ + -a^4 + 1a^3 + 0a^2 + 0a + 0 \\ \hline -a^4 + 3a^3 + 3a^2 + \ a \end{array}$$

**33.**
$$\begin{array}{r} -8m^2 + 2m - 3 \\ + \ m^2 + 5m + 0 \\ \hline -7m^2 + 7m - 3 \end{array}$$

**34.**
$$\begin{array}{r} 8y^2 + 2 \\ + -3y^2 + 5 \\ \hline 5y^2 + 7 \end{array}$$

**35.**
$$\begin{array}{r} 3x^2 + 7x - 6 \\ - 3x^2 - 7x + 0 \\ \hline 0x^2 - 0x - 6 \end{array}$$
$= -6$

**36.**
$$\begin{array}{r} 4x^2 - 7x + 2 \\ + -x^2 + \ x - 2 \\ \hline 3x^2 - 6x \end{array}$$

**37.**
$$\begin{array}{r} 8y^3 + 4y^2 + 3y - 7 \\ - 0y^3 + 2y^2 - 6y + 4 \\ \hline 8y^3 + 6y^2 - 3y - 3 \end{array}$$

**38.**
$$\begin{array}{r} 7x^4 + \ 0x^3 - 1x^2 + 3x + 0 \\ + \ -x^3 - 6x^2 + 2x - 9 \\ \hline 7x^4 - \ x^3 - 7x^2 + 5x - 9 \end{array}$$

**39.** $(x^2 - 7) + (2x^2 + 2) = x^2 - 7 + 2x^2 + 2 = 3x^2 - 5$

**40.** $(-3a^2 + 5) + (-a^2 + 4a - 6)$
$= -3a^2 + 5 - a^2 + 4a - 6 = -4a^2 + 4a - 1$

**41.** $(x^3 + x^2 + 1) - x^2 = x^3 + x^2 + 1 - x^2 = x^3 + 1$

**42.** $12 - (y^3 + 4) = 12 - y^3 - 4 = 8 - y^3 = -y^3 + 8$

**43.** $(3n^3 + 2n - 7) - (n^3 - n - 2)$
$= 3n^3 + 2n - 7 - n^3 + n + 2 = 2n^3 + 3n - 5$

# Chapter 10 *continued*

**44.** $(3a^3 - 4a^2 + 3) - (a^3 + 3a^2 - a - 4)$

$= 3a^3 + 4a^2 + 3 - a^3 - 3a^2 + a + 4$

$= 2a^3 - 7a^2 + a + 7$

**45.** $(6b^4 - 3b^3 - 7b^2 + 9b + 3) + (4b^4 - 6b^2 + 11b - 7)$

$= 6b^4 - 3b^3 - 7b^2 + 9b + 3 + 4b^4 - 6b^2 + 11b - 7$

$= 10b^4 - 3b^3 - 13b^2 + 20b - 4$

**46.** $(x^3 - 6x) - (2x^3 + 9) - (4x^2 + x^3)$

$= x^3 - 6x - 2x^3 - 9 - 4x^2 - x^3$

$= -2x^3 - 4x^2 - 6x - 9$

**47.** $(9x^3 + 12) + (16x^3 - 4x + 2)$

$= 9x^3 + 12 + 16x^3 - 4x + 2$

$= 25x^3 - 4x + 14$

**48.** $(-2t^4 + 6t^2 + 5) - (-2t^4 + 5t^2 + 1)$

$= -2t^4 + 6t^2 + 5 + 2t^4 - 5t^2 - 1 = t^2 + 4$

**49.** $(3x + 2x^2 - 4) - (x^2 + x - 6)$

$= 3x + 2x^2 - 4 - x^2 - x + 6$

$= x^2 + 2x + 2$

**50.** $(u^3 - u) - (u^2 + 5) = u^3 - u - u^2 - 5$

$= u^3 - u^2 - u - 5$

**51.** $(-7x^2 + 12) - (6 - 4x^2) = -7x^2 + 12 - 6 + 4x^2$

$= -3x^2 + 6$

**52.** $(10x^3 + 2x^2 - 11) + (9x^2 + 2x - 1)$

$= 10x^3 + 2x^2 - 11 + 9x^2 + 2x - 1$

$= 10x^3 + 11x^2 + 2x - 12$

**53.** $(-9z^3 - 3z) + (13z - 8z^2)$

$= -9z^3 - 3z + 13z - 8z^2$

$= -9z^3 - 8z^2 + 10z$

**54.** $(21t^4 - 3t^2 + 43) - (19t^3 + 33t - 58)$

$= 21t^4 - 3t^2 + 43 - 19t^3 - 33t + 58$

$= 21t^4 - 19t^3 - 3t^2 - 33t + 101$

**55.** $(6t^2 - 19t) - (3 - 2t^2) - (8t^2 - 5)$

$= 6t^2 - 19t - 3 + 2t^2 - 8t^2 + 5$

$= -19t + 2$

**56.** $(7y^2 + 15y) + (5 - 15y + y^2) + (24 - 17y^2)$

$= 7y^2 + 15y + 5 - 15y + y^2 + 24 - 17y^2$

$= -9y^2 + 29$

**57.** $\left(x^4 - \frac{1}{2}x^2\right) + \left(x^3 + \frac{1}{3}x^2\right) + \left(\frac{1}{4}x^2 - 9\right)$

$= x^4 - \frac{1}{2}x^2 + x^3 + \frac{1}{3}x^2 + \frac{1}{4}x^2 - 9$

$= x^4 + x^3 + \frac{1}{12}x^2 - 9$

**58.** $(10w^3 + 20w^2 - 55w + 60) + (-25w^2 + 15w - 10) +$

$(-5w^2 + 10w - 20)$

$= 10w^3 + 20w^2 - 55w + 60 - 25w^2 + 15w - 10 -$

$5w^2 + 10w - 20$

$= 10w^3 - 10w^2 - 30w + 30$

**59.** $(9x^4 - x^2 + 7x) + (x^3 - 6x^2 + 2x - 9) - (4x^3 + 3x + 8)$

$= 9x^4 - x^2 + 7x + x^3 - 6x^2 + 2x - 9 - 4x^3 - 3x - 8$

$= 9x^4 - 3x^3 - 7x^2 + 6x - 17$

**60.** $(6.2b^4 - 3.1b + 8.5) + (-4.7 + 5.8b^2 - 2.4b^4)$

$= 6.2b^4 - 3.1b + 8.5 - 4.7 + 5.8b^2 - 2.4b^4$

$= 3.8b^4 + 5.8b^2 - 3.1b + 3.8$

**61.** $(-3.8y^3 + 6.9y^2 - y + 6.3) - (-3.1y^3 + 2.9y - 4.1)$

$= -3.8y^3 + 6.9y^2 - y + 6.3 + 3.1y^3 - 2.9y + 4.1$

$= -0.7y^3 + 6.9y^2 - 3.9y + 10.4$

**62.** $\left(\frac{2}{5}a^4 - 2a + 7\right) - \left(-\frac{3}{10}a^4 + 6a^3\right) - (2a^2 - 7)$

$= \frac{2}{5}a^4 - 2a + 7 + \frac{3}{10}a^4 - 6a^3 - 2a^2 + 7$

$= \frac{7}{10}a^4 - 6a^3 - 2a^2 - 2a + 14$

**63.** $3x(x + 20) - (1.5x)(x) = 1.5x^2 + 60x$

**64.**

$1.5x(x) = A \qquad 3x^2 + 60x = A$

$1.5(30)(30) = A \qquad 3(30^2) + 60(30) = A$

$1350 \text{ ft}^2 = A \qquad 3(900) + 1800 = A$

$2700 + 1800 = A$

$4500 \text{ ft}^2 = A$

**65.** $F = 1223.58t + 79,589.03$

**66.** $F = 298,475.86 - 145,472.03 = 153,003.83$

The population will be about 153,003,830.

**67.** $A = 1.381t^2 + 3.494t + 235.325$   **68.** natural gas

**69.**

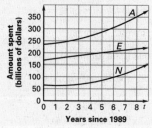

**Energy Spending**

Amount spent (billions of dollars) vs. Years since 1989

**70. a.** $S = 475g^3 + 50g^2 + 300g + 540$

**b.** $T = 725g^3 + 450g^2 + 470g + 1165$

**c.** $T = 725(1.025^3) + 450(1.025^2) + 470(1.025) + 1165$

$\approx 780.75 + 472.78 + 481.75 + 1165$

$\approx \$2900.28$

**d.** Yes

# Chapter 10 *continued*

**71.** An odd number can be written $2x + 1$, where $x$ is an integer.

$$(x) + (x + 1) = 2x + 1$$
$$2x + 1 = 2x + 1$$

Therefore, the sum of 2 consecutive integers is odd.

**72.** Four consecutive integers can be written $x, x + 1, x + 2, x + 3$. Their sum is $4x + 6 = 2(2x + 3)$. Since $x$ is an integer, $2x + 3$ is an integer and $2(2x + 3)$ is even. Therefore, the sum of 4 consecutive integers is even.

### 10.1 Mixed Review *(p. 582)*

**73.** $-3(x + 1) - 2 = -3x - 3 - 2 = -3x - 5$

**74.** $(2x - 1)(2) + x = 4x - 2 + x = 5x - 2$

**75.** $11x + 3(8 - x) = 11x + 24 - 3x = 8x + 24$

**76.** $(5x - 1)(-3) + 6 = -15x + 3 + 6 = -15x + 9$

**77.** $-4(1 - x) + 7 = -4 + 4x + 7 = 4x + 3$

**78.** $-12x - 5(11 - x) = -12x - 55 + 5x = -7x - 55$

**79.**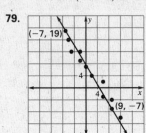

*Sample answer:*

$$m = \frac{19 - (-7)}{-7 - (9)}$$
$$= \frac{26}{-16} = -\frac{13}{8}$$
$$= -1.625$$

$$19 = -1.625 \cdot -7 + b$$
$$19 = 11.375 + b$$
$$7.625 = b$$
$$y = -1.625x + 7.625$$

**80.** $(4 \cdot 3^2 \cdot 2^3)^4 = 4^4 \cdot 3^8 \cdot 2^{12} = 6{,}879{,}707{,}136$

**81.** $(2^4 \cdot 2^4)^2 = 2^8 \cdot 2^8 = 65{,}536$

**82.** $(-6 \cdot 3^4)^3 = -6^3 \cdot 3^{12} = -216 \cdot 3^{12} = -114{,}791{,}256$

**83.** $(1.1 \cdot 3.3)^3 = 1.1^3 \cdot 3.3^3 = 1.331 \cdot 35.937 \approx 47.83$

**84.** $5.5^3 \cdot 5.5^4 = 5.5^7 \approx 152{,}243.52$

**85.** $(2.9^3)^5 = 2.9^{15} \approx 8{,}629{,}188.75$

**86.** $P = 4227(1.0104)^t$

2025: $P = 4227(1.0104)^{30}$    2000: $P = 4227(1.0104)^5$

Ratio: $\dfrac{4227(1.0104)^{30}}{4227(1.0104)^5} = (1.0104)^{25} \approx 1.30$

*Sample answer:* The ratio of the population in 2025 to the population in 2000 is about 1.23, or $1.0104^{20}$ times the ratio of the population in 2000 to the population in 1995.

### Technology Activity 10.1 *(p. 583)*

**1.** correct

**2.** incorrect

$$(-x^2 - 3x - 1) - (-2x^2 + 4x + 5)$$
$$= -x^2 - 3x - 1 + 2x^2 - 4x - 5 = x^2 - 7x - 6$$

**3.** $(2x^2 - 6x - 3) + (x^2 - 3x + 3)$
$$= 2x^2 - 6x - 3 + x^2 - 3x + 3 = 3x^2 - 9x$$

**4.** $(x^2 - 14x + 5) + (-2x^2 - 3x + 2)$
$$= x^2 - 14x + 5 - 2x^2 - 3x + 2 = -x^2 - 17x + 7$$

**5.** $(x^2 + 12x + 6) - (3x^2 - 2x + 2)$
$$= x^2 + 12x + 6 - 3x^2 + 2x - 2$$
$$= -2x^2 + 14x + 4$$

**6.** $(-3x^2 + 5x - 8) - (x^2 - 5x - 8)$
$$= -3x^2 + 5x - 8 - x^2 + 5x + 8 = -4x^2 + 10x$$

**7.** $(2x^2 + 10x + 3) + (4x^2 + 2x - 4)$
$$= 2x^2 + 10x + 3 + 4x^2 + 2x - 4$$
$$= 6x^2 + 12x - 1$$

**8.** $(x^2 - 4x + 5) - (-2x^2 - 3x + 7)$
$$= x^2 - 4x + 5 + 2x^2 + 3x - 7 = 3x^2 - x - 2$$

### Activity *(p. 584)*

**1.**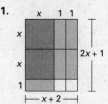

The area of the blue part is $2x^2$, the area of the green part is $5x$, and the area of the yellow part is 2.

**2.** $2x^2 + 5x + 2$

**3.** $(x + 2)(2x + 1) = 2x^2 + 5x + 2$

**4.**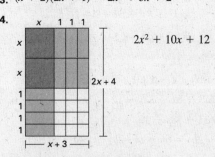

$2x^2 + 10x + 12$

### 10.2 Guided Practice *(p. 587)*

**1.** "F" stands for first: multiply the first two terms; "O" stands for outer: next, multiply the two outer terms. "I" stands for inner: next, multiply the two inner terms; "L" stands for last: multiply the two last terms. Finally, add the four products.

**2.** $(x + 1)(2x + 1) = 2x^2 + 3x + 1$

# Chapter 10 *continued*

**3.** Multiply $2x(x + 4)$, then $(-3)(x + 4)$, and add the products.

**4.** Multiply $x(x^2 - x + 1)$, then $(1)(x^2 - x + 1)$, and add the products.

**5.**

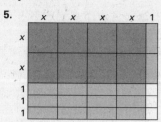

$8x^2 + 14x + 3$

**6.** $2x(4x + 1) + 3(4x + 1)$

$= 8x^2 + 2x + 12x + 3$

$= 8x^2 + 14x + 3$

**7.** $2x(4x) + 2x(1) + 3(4x) + 3(1)$

$= 8x^2 + 2x + 12x + 3$

$= 8x^2 + 14x + 3$

**8.** Answers may vary.  **9.** $(4x + 7)(-2x) = -8x^2 - 14x$

**10.** $-4x^2(3x^2 + 2x - 6) = -12x^4 - 8x^3 + 24x^2$

**11.** $(-y - 2)(y + 8)$

$= -y(y) + 8(-y) + (-2)(y) + (-2)(8)$

$= -y^2 - 8y - 2y - 16$

$= -y^2 - 10y - 16$

**12.** $(w - 3)(2w + 5)$

$= w(2w) + 5(w) + (-3)(2w) + (-3)(5)$

$= 2w^2 + 5w - 6w - 15$

$= 2w^2 - w - 15$

**13.** $(x + 6)(x + 9) = x(x) + 9(x) + 6(x) + 6(9)$

$= x^2 + 9x + 6x + 54$

$= x^2 + 15x + 54$

**14.** $(-2x - 4)(8x + 3)$

$= -2x(8x) + (-2x)(3) + 8x(-4) + (-4)(3)$

$= -16x^2 - 6x - 32x - 12$

$= -16x^2 - 38x - 12$

**15.** $(b + 8)(6 - 2b) = 6b + b(-2b) + 8(6) + 8(-2b)$

$= 6b - 2b^2 + 48 - 16b$

$= -2b^2 - 10b + 48$

**16.** $(-4y + 5)(-7 - 3y)$

$= -4y(-7) + (-3y)(-4y) + 5(-7) + 5(-3y)$

$= 28y + 12y^2 - 35 - 15y$

$= 12y^2 + 13y - 35$

**17.** $(3w^2 - 9)(5w - 1)$

$= 3w^2(5w) + (-1)(3w^2) + (-9)(5w) + (-9)(-1)$

$= 15w^3 - 3w^2 - 45w + 9$

### 10.2 Practice and Applications (pp. 587–589)

**18.** $(2x - 5)(-4x) = -8x^2 + 20x$

**19.** $3t^2(7t - t^3 - 3) = 21t^3 - 3t^5 - 9t^2$

**20.** $2x(x^2 - 8x + 1) = 2x^3 - 16x^2 + 2x$

**21.** $(-y)(6y^2 + 5y) = -6y^3 - 5y^2$

**22.** $4w^2(3w^3 - 2w^2 - w) = 12w^5 - 8w^4 - 4w^3$

**23.** $-b^2(6b^3 - 16b + 11) = -6b^5 + 16b^3 - 11b^2$

**24.** $(t + 8)(t + 5) = t(t + 5) + 8(t + 5)$

$= t^2 + 5t + 8t + 40$

$= t^2 + 13t + 40$

**25.** $(2d + 3)(3d + 1) = 2d(3d + 1) + 3(3d + 1)$

$= 6d^2 + 2d + 9d + 3$

$= 6d^2 + 11d + 3$

**26.** $(4y^2 + y - 7)(2y - 1)$

$= 2y(4y^2 + y - 7) - 1(4y^2 + y - 7)$

$= 8y^3 + 2y^2 - 14y - 4y^2 - y + 7$

$= 8y^3 - 2y^2 - 15y + 7$

**27.** $(3s^2 - s - 1)(s + 2)$

$= s(3s^2 - s - 1) + 2(3s^2 - s - 1)$

$= 3s^3 - s^2 - s + 6s^2 - 2s - 2$

$= 3s^3 + 5s^2 - 3s - 2$

**28.** $(a^2 + 8)(a^2 - a - 3)$

$= a^2(a^2 - a - 3) + 8(a^2 - a - 3)$

$= a^4 - a^3 - 3a^2 + 8a^2 - 8a - 24$

$= a^4 - a^3 + 5a^2 - 8a - 24$

**29.** $(x + 6)(x^2 - 6x - 2)$

$= x(x^2 - 6x - 2) + 6(x^2 - 6x - 2)$

$= x^3 - 6x^2 - 2x + 6x^2 - 36x - 12$

$= x^3 - 38x - 12$

**30.** $(4q - 1)(3q + 8)$

$= 4q(3q) + 8(4q) + (-1)(3q) + (-1)(8)$

$= 12q^2 + 32q - 3q - 8$

$= 12q^2 + 29q - 8$

**31.** $(2z + 7)(3z + 2)$

$= 2z(3z) + 2z(2) + 7(3z) + 7(2)$

$= 6z^2 + 4z + 21z + 14$

$= 6z^2 + 25z + 14$

# Chapter 10 *continued*

**32.** $(x + 6)(x - 6)$

$\quad = x(x) + (-6)(x) + 6x + 6(-6)$

$\quad = x^2 - 6x + 6x - 36$

$\quad = x^2 - 36$

**33.** $(2w - 5)(w + 5)$

$\quad = 2w(w) + 5(2w) + (-5)(w) + (-5)(5)$

$\quad = 2w^2 + 10w - 5w - 25$

$\quad = 2w^2 + 5w - 25$

**34.** $(x - 9)(2x + 15)$

$\quad = x(2x) + 15x + (-9)(2x) + (-9)(15)$

$\quad = 2x^2 + 15x - 18x - 135$

$\quad = 2x^2 - 3x - 135$

**35.** $(5t - 3)(2t + 3)$

$\quad = 5t(2t) + 3(5t) + (-3)(2t) + (-3)(3)$

$\quad = 10t^2 + 15t - 6t - 9$

$\quad = 10t^2 + 9t - 9$

**36.** $(d - 5)(d + 3) = d^2 + 3d - 5d - 15 = d^2 - 2d - 15$

**37.** $(4x + 1)(x - 8) = 4x^2 - 32x + x - 8 = 4x^2 - 31x - 8$

**38.** $(3b - 1)(b - 9)$

$\quad = 3b^2 - 27b - b + 9$

$\quad = 3b^2 - 28b + 9$

**39.** $(9w + 8)(11w - 10)$

$\quad = 99w^2 - 90w + 88w - 80$

$\quad = 99w^2 - 2w - 80$

**40.** $(11t - 30)(5t - 21)$

$\quad = 55t^2 - 231t - 150t + 630$

$\quad = 55t^2 - 381t + 630$

**41.** $(9.4y - 5.1)(7.3y - 12.2)$

$\quad = 68.62y^2 - 114.68y - 37.23y + 62.22$

$\quad = 68.62y^2 - 151.91y + 62.22$

**42.** $(3x + 4)\left(\frac{2}{3}x + 1\right)$

$\quad = 2x^2 + 3x + \frac{8}{3}x + 4$

$\quad = 2x^2 + \frac{17}{3}x + 4$

**43.** $\left(n + \frac{6}{5}\right)(4n - 10)$

$\quad = 4n^2 - 10n + \frac{24}{5}n - 12$

$\quad = 4n^2 - \frac{26}{5}n - 12$

**44.** $\left(x + \frac{1}{8}\right)\left(x - \frac{9}{8}\right) = x^2 - \frac{9}{8}x + \frac{1}{8}x - \frac{9}{64} = x^2 - x - \frac{9}{64}$

**45.** $(2.5z - 6.1)(z + 4.3)$

$\quad = 2.5z^2 + 10.75z - 6.1z - 26.23$

$\quad = 2.5z^2 + 4.65z - 26.23$

**46.** $(t^2 + 6t - 8)(t - 6)$

$\quad = t^3 + 6t^2 - 8t - 6t^2 - 36t + 48$

$\quad = t^3 - 44t + 48$

**47.** $(-4s^2 + s - 1)(s + 4)$

$\quad = -4s^3 + s^2 - s - 16s^2 + 4s - 4$

$\quad = -4s^3 - 15s^2 + 3s - 4$

**48.** $(x + 5)(x + 4) = x^2 + 4x + 5x + 20 = x^2 + 9x + 20$

**49.** $(2x + 1)(x - 6) = 2x^2 - 12x + x - 6 = 2x^2 - 11x - 6$

**50.** $(3x^2 - 8x - 1)(9x + 4)$

$\quad = 27x^3 - 72x^2 - 9x + 12x^2 - 32x - 4$

$\quad = 27x^3 - 60x^2 - 41x - 4$

**51.** $(x + 7)(-x^2 - 6x + 2)$

$\quad = -x^3 - 6x^2 + 2x - 7x^2 - 42x + 14$

$\quad = -x^3 - 13x^2 - 40x + 14$

**52.** $A = \frac{1}{2}h(b_1 + b_2)$

$\quad = \frac{1}{2}(x + 1)(3x + 4 + 5x + 7)$

$\quad = \frac{1}{2}(x + 1)(8x + 11)$

$\quad = \frac{1}{2}(8x^2 + 11x + 8x + 11)$

$\quad = \frac{1}{2}(8x^2 + 19x + 11)$

$\quad = 4x^2 + \frac{19}{2}x + \frac{11}{2}$

**53.** $A = \frac{1}{2}bh = \frac{1}{2}(6x - 5)(x + 4)$

$\quad = \frac{1}{2}(6x^2 + 24x - 5x - 20)$

$\quad = \frac{1}{2}(6x^2 + 19x - 20)$

$\quad = 3x^2 + \frac{19}{2}x - 10$

**54.** $A = l \cdot w = \left(\frac{5}{4}x - 15\right)\left(\frac{1}{2}x + 10\right)$

$\quad = \frac{5}{8}x^2 + \frac{25}{2}x - \frac{15}{2}x - 150$

$\quad = \frac{5}{8}x^2 + 5x - 150$

**55.** $\left(\frac{5}{4}x - 15\right) = 360$

$\quad \frac{5}{4}x = 375$

$\quad x = 375\left(\frac{4}{5}\right)$

$\quad x = 300$

**56.** $R = -3.15t^2 - 6.21t + 989.12$, where $R$ is measured in millions.

**57.** The revenue will decrease from \$989.12 million to \$678.08 million, because the value of the model $-3.15t^2 - 6.21t + 989.12$ will decrease as $t$ increases.

**58.** $R = 1.5125t^2 + 156.1846t + 3066.344$, where $R$ is measured in millions.

**59.** As time goes on, the revenue will increase, because the value of the model $1.5125t^2 + 156.1846t + 3066.344$ will increase as $t$ increases.

# Chapter 10 *continued*

**60. a.**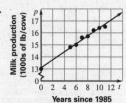

**b.**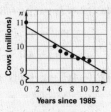

*Sample answer:*

**c.** $m = \dfrac{16.5 - 15}{11 - 6}$

$= \dfrac{1.5}{5} = 0.3$

$15 = 0.3 \cdot (6) + b$

$15 = 1.8 + b$

$13.7 = b$

$p = 0.3t + 13.2$

*Sample answer:*

$m = \dfrac{9.5 - 9.8}{10 - 6}$

$= -\dfrac{0.3}{4} = -0.075$

$9.8 = -0.075(6) + b$

$9.8 = -0.45 + b$

$10.25 = b$

$n = -0.075t + 10.25$

**d.** $A = p \cdot n = (0.3t + 13.2)(-0.075t + 10.25)$

$= -0.0225t^2 + 2.085t + 135.3$

**e.** Even though the number of cows is decreasing, the actual milk production is increasing because the milk production per cow is increasing.

**61. a.** $(x - 1)(x + 1) = x^2 - 1$

$(x - 1)(x^2 + x + 1)$

$= x^3 + x^2 + x - x^2 - x - 1$

$= x^3 - 1$

$(x - 1)(x^3 + x^2 + x + 1)$

$= x^4 + x^3 + x^2 + x - x^3 - x^2 - 1$

$= x^4 - 1$

Pattern: $(x - 1)(x^n + x^{n-1} + x^{n-2} + \ldots + x + 1)$

$= x^{n+1} - 1$

**b.** $x^5 - 1$

## 10.2 Mixed Review (p. 589)

**62.** $(7x)^2 = 49x^2$    **63.** $\left(\frac{1}{3}m\right)^2 = \frac{1}{9}m^2$    **64.** $\left(\frac{2}{5}y\right)^2 = \frac{4}{25}y^2$

**65.** $(0.5w)^2 = 0.25w^2$    **66.** $9^3 \cdot 9^5 = 9^8$, or $43,046,721$

**67.** $(4^2)^4 = 4^8$, or $65,536$

**68.** $b^2 \cdot b^5 = b^7$    **69.** $(4c^2)^4 = 256c^8$

**70.** $(2t)^4 \cdot 3^3 = 16t^4 \cdot 27 = 432t^4$    **71.** $(-w^4)^3 = -w^{12}$

**72.** $(-3xy)^3(2y)^2 = -3^3x^3y^32^2y^2$

$= -27x^3y^54$

$= -108x^3y^5$

**73.** $(8x^2y^8)^3 = 8^3x^6y^{24} = 512x^6y^{24}$

**74.** two solutions    **75.** one solution    **76.** no solution

**77.** two solutions    **78.** two solutions    **79.** two solutions

**80.** $y \geq 4x^2 - 7x$

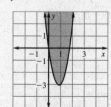

**81.** $y < x^2 - 3x - 10$

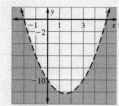

**82.** $y > -2x^2 + 4x + 16$

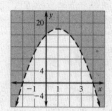

**83.** $y < 6x^2 - 1$

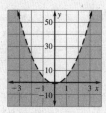

**84.** $y \geq x^2 - 3x + 1$

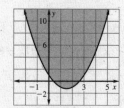

**85.** $y \leq 8x^2 - 3$

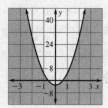

## Lesson 10.3

### Activity (p. 590)

**Step 1**

**1.** $(x - 2)(x + 2) = x^2 - 2x + 2x - 4 = x^2 - 4$

$(2n + 3)(2n - 3) = 4n^2 + 6n - 6n - 9 = 4n^2 - 9$

$(4t - 1)(4t + 1) = 16t^2 - 4t + 4t - 1 = 16t^2 - 1$

$(x + y)(x - y) = x^2 + xy - xy - y^2 = x^2 - y^2$

**2.** $(x + 3)(x + 3) = x^2 + 3x + 3x + 9 = x^2 + 6x + 9$

$(3m + 1)^2 = (3m + 1)(3m + 1)$

$= 9m^2 + 3m + 3m + 1$

$= 9m^2 + 6m + 1$

$(5s + 2)^2 = (5s + 2)(5s + 2)$

$= 25s^2 + 10s + 10s + 4$

$= 25s^2 + 20s + 4$

$(x + y)^2 = (x + y)(x + y)$

$= x^2 + xy + xy + y^2$

$= x^2 + 2xy + y^2$

# Chapter 10 *continued*

**3.** $(z - 2)^2 = (z - 2)(z - 2)$

$$= z^2 - 2z - 2z + 4$$

$$= z^2 - 4z + 4$$

$(6x - 4)^2 = (6x - 4)(6x - 4)$

$$= 36x^2 - 24x - 24x + 16$$

$$= 36x^2 - 48x + 16$$

$(5p - 7)^2 = (5p - 7)(5p - 7)$

$$= 25p^2 - 35p - 35p + 49$$

$$= 25p^2 - 70p + 49$$

$(x - y)^2 = (x - y)(x - y)$

$$= x^2 - xy - xy + y^2$$

$$= x^2 - 2xy + y^2$$

### Step 2

**1.** The factors in the expression are a sum and difference of two terms. The product is the difference of the squares of the two terms.

**2.** The expression is the square of a sum of two terms. The product is the square of the first plus twice the product of the terms plus the square of the second.

**3.** The expression is the square of a difference of two terms. The product is the square of the first minus twice the product of the terms plus the square of the second.

### 10.3 Guided Practice (p. 593)

**1.** The square of the sum and difference of two terms is the difference of the squares of the terms.

**2.** $(2x + 2)(2x + 2)$, $4x^2 + 8x + 4$

**3.** False; $(a + b)(a + b) = a^2 + 2ab + b^2$

**4.** $(a - b)^2 = (a - b)(a - b) = a^2 - 2ab + b^2$

**5.**  $x^2 + 4x + 4$

**6.** $(x - 6)^2 = x^2 - 2(6)(x) + 6^2 = x^2 - 12x + 36$

**7.** $(w + 11)(w - 11) = w^2 - 121$

**8.** $(6 + p)^2 = 6^2 + 2(6)p + p^2 = 36 + 12p + p^2$

**9.** $(2y - 3)^2 = (2y)^2 - 2(2y)(3) + 3^2 = 4y^2 - 12y + 9$

**10.** $(3z + 2)^2 = (3z)^2 + 2(3z)(2) + 2^2 = 9z^2 + 12z + 4$

**11.** $(t - 6)(t + 6) = t^2 - 36$

**12.** $(x - 3)(x - 3) = x^2 - 2x(3) + 3^2 = x^2 - 6x + 9$

**13.** $(2y + 5)(2y - 5) = (2y)^2 - 5^2 = 4y^2 - 25$

**14.** $(4n + 3)^2 = (4n)^2 + 2(4n)(3) + 3^2 = 16n^2 + 24n + 9$

### 10.3 Practice and Applications (pp. 593–595)

**15.** $(x + 3)(x - 3) = x^2 - 3x + 3x - 9 = x^2 - 9$

**16.** $(y - 1)(y + 1) = y^2 + y - y - 1 = y^2 - 1$

**17.** $(2m + 2)(2m - 2) = 4m^2 - 4m + 4m - 4 = 4m^2 - 4$

**18.** $(3b - 1)(3b + 1) = 9b^2 + 3b - 3b - 1 = 9b^2 - 1$

**19.** $(3 + 2x)(3 - 2x) = 9 - 6x + 6x - 4x^2 = 9 - 4x^2$

**20.** $(6 - 5n)(6 + 5n) = 36 + 30n - 30n - 25n^2$

$$= 36 - 25n^2$$

**21.** $(x + 5)^2 = (x + 5)(x + 5)$

$$= x^2 + 5x + 5x + 25$$

$$= x^2 + 10x + 25$$

**22.** $(a + 8)^2 = (a + 8)(a + 8)$

$$= a^2 + 8a + 8a + 64$$

$$= a^2 + 16a + 64$$

**23.** $(3t + 1)^2 = (3t + 1)(3t + 1)$

$$= 9t^2 + 3t + 3t + 1$$

$$= 9t^2 + 6t + 1$$

**24.** $(2s - 4)^2 = (2s - 4)(2s - 4)$

$$= 4s^2 - 8s - 8s + 16$$

$$= 4s^2 - 16s + 16$$

**25.** $(4b - 3)^2 = (4b - 3)(4b - 3)$

$$= 16b^2 - 12b - 12b + 9$$

$$= 16b^2 - 24b + 9$$

**26.** $(x - 7)^2 = (x - 7)(x - 7)$

$$= x^2 - 7x - 7x + 49$$

$$= x^2 - 14x + 49$$

**27.** $(x + 4)(x - 4) = x^2 - 4^2 = x^2 - 16$

**28.** $(x - 3)(x + 3) = x^2 - 3^2 = x^2 - 9$

**29.** $(3x + 1)(3x - 1) = (3x)^2 - 1^2 = 9x^2 - 1$

**30.** $(6x + 5)(6x - 5) = (6x)^2 - 5^2 = 36x^2 - 25$

**31.** $(a + 2b)(a - 2b) = a^2 - (2b)^2 = a^2 - 4b^2$

**32.** $(4n - 8m)(4n + 8m) = (4n)^2 - (8m)^2 = 16n^2 - 64m^2$

**33.** $(3y + 8)^2 = (3y)^2 + 2(3y)(8) + 8^2 = 9y^2 + 48y + 64$

**34.** $(9 - 4t)(9 + 4t) = 9^2 - (4t)^2 = 81 - 16t^2$

**35.** $\left(2x + \frac{1}{2}\right)\left(2x - \frac{1}{2}\right) = (2x)^2 - \left(\frac{1}{2}\right)^2 = 4x^2 - \frac{1}{4}$

**36.** $(-5 - 4x)^2 = (-5)^2 - 2(-5)(4x) + (4x)^2$

$$= 25 + 40x + 16x^2$$

**37.** $(3s + 4t)(3s - 4t) = (3s)^2 - (4t)^2 = 9s^2 - 16t^2$

# Chapter 10 *continued*

**38.** $(-a - 2b)^2 = -a^2 - 2(-a)(2b) + (2b)^2$
$= a^2 + 4ab + 4b^2$

**39.** true

**40.** false; $36y^2 - 2(6y)(7w) + 49w^2$    **41.** true
$= 36y^2 - 84yw + 49w^2$

**42.** false; $\frac{4}{49}n^2 - 2(\frac{2}{7}n)(3m) + 9m^2 = \frac{4}{49}n^2 - \frac{12}{7}nm + 9m^2$

**43.** $26 \cdot 34 = 884$    **44.** $45 \cdot 55 = 2475$    **45.** $16^2 = 256$

**46.** $41^2 = 1681$

**47.** $(x + 3)^2, x^2 + 2(x)(3) + 9 = x^2 + 6x + 9$; square of a binomial

**48.** $(2y^2) - 3^2, (2y + 3)(2y - 3)$; sum and difference

**49.** $(0.5C + 0.5c)^2 = 0.25C^2 + 0.5Cc + 0.25c^2$

**50.** 75%, 25%

**51.** $(0.5F + 0.5f)^2 = 0.25F^2 + 0.5Ff + 0.25f^2$

**52.** 25%, 50%, 25%    **53.** $(P(1 - r))(1 + r) = P(1 - r^2)$

**54.** decrease; $Pr^2$ dollars    **55.** decrease of $Pr^2$ dollars

**56.** A $(3x - 7)(3x + 7) = 9x^2 - 49 = 9(4^2) - 49$
$= 144 - 49 = 95$
$(3x - 7)^2 = 9x^2 - 2(3x)(7) + 49$
$= 9x^2 - 2(3x)(7) + 499x^2 - 42x + 49$
$= 9(4^2) - 42(4) + 49$
$= 144 - 168 + 49$
$= 73$

**57.** A $a^2 + b^2 = 1^2 + (-2)^2 = 1 + 4 = 5$
$(a + b)^2 = a^2 + 2ab + b^2$
$= 1^2 + 2(1)(-2) + (-2^2)$
$= 1 - 4 + 4 = 1$

**58.** D $(3x + 2)^2 - (x + 5)(x - 5)$
$= 9x^2 + 2(3x)(2) + 4 - (x^2 - 5x + 5x - 25)$
$= 9x^2 + 12x + 4 - x^2 + 25$
$= 8x^2 + 12x + 29$

**59.** $(x + 6)(x + 6)$    **60.** $(x - 2)(x + 2)$

**61.** $(5x - 3)(5x - 3)$    **62.** $(7x + 13)(7x - 13)$

## 10.3 Mixed Review (p. 595)

**63.** $\left(\frac{x}{4}\right)^3 = \frac{x^3}{64}$    **64.** $\frac{x^3}{x^2} = x^{3-2} = x$    **65.** $\left(\frac{4x}{y^3}\right)^3 = \frac{64x^3}{y^9}$

**66.** $x^7 \cdot \frac{1}{x^4} = \frac{x^7}{x^4} = x^{7-4} = x^3$

**67.** $\frac{3x^2y}{2x} \cdot \frac{6xy^2}{y^3} = \frac{3}{2}x^{2-1}y \cdot 6xy^{2-3}\frac{3}{2}xy \cdot 6xy^{-1}$
$= \frac{3xy \cdot 6x}{2y} = \frac{18x^2}{2} = 9x^2$

**68.** $\frac{5x^4y}{3xy^2} \cdot \frac{9xy}{x^2y} = \frac{5}{3}x^{4-1}y^{1-2} \cdot 9x^{1-2}y^{1-1}$
$= \frac{5x^39}{3yx} = \frac{5x^{3-19}}{3y} = \frac{45x^2}{3y} = \frac{15x^2}{y}$

**69.** $y = 2x^2 + 3x + 6$

Vertex: $x = -\frac{3}{2(2)} = -\frac{3}{4}$

$y = 2\left(-\frac{3}{4}\right)^2 + 3\left(-\frac{3}{4}\right) + 6 = \frac{9}{8} - \frac{9}{4} + 6 = 4\frac{7}{8}$

The vertex is $\left(-\frac{3}{4}, 4\frac{7}{8}\right)$.

Axis of symmetry: $x = -\frac{3}{2(2)} = -\frac{3}{4}$

The axis of symmetry is $x = -\frac{3}{4}$.

**70.** $y = 3x^2 - 9x - 12$

Vertex: $x = -\frac{-9}{2(3)} = 1\frac{1}{2}$

$y = 3\left(1\frac{1}{2}\right)^2 - 9\left(1\frac{1}{2}\right) - 12 = \frac{27}{4} - \frac{27}{2} - 12$
$= -18\frac{3}{4}$

The vertex is $\left(1\frac{1}{2}, -18\frac{3}{4}\right)$.

Axis of symmetry: $x = -\frac{-9}{2(3)} = 1\frac{1}{2}$

Axis of symmetry is $x = 1\frac{1}{2}$.

**71.** $y = -x^2 + 4x + 16$

Vertex: $x = -\frac{4}{2(-1)} = 2$

$y = -(2)^2 + 4(2) + 16 = -4 + 8 + 16 = 20$

The vertex is $(2, 20)$

Axis of symmetry: $x = -\frac{4}{2(-1)} = 2$

The axis of symmetry is $x = 2$.

# Chapter 10 continued

**72.** $y = -4x^2 - 2x + 5$

Vertex: $x = -\dfrac{-2}{2(-4)} = -\dfrac{1}{4}$

$y = -4\left(-\dfrac{1}{4}\right)^2 - 2\left(-\dfrac{1}{4}\right) + 5 = -\dfrac{1}{4} + \dfrac{1}{2} + 5 = 5\dfrac{1}{4}$

The vertex is $\left(-\dfrac{1}{4}, 5\dfrac{1}{4}\right)$.

Axis of symmetry: $x = -\dfrac{-2}{2(-4)} = -\dfrac{1}{4}$

The axis of symmetry is $x = -\dfrac{1}{4}$.

**73.** $y = -\dfrac{1}{2}x^2 + 6x - 4$

Vertex: $x = -\dfrac{6}{2\left(-\dfrac{1}{2}\right)} = 6$

$y = -\dfrac{1}{2}(6)^2 + 6(6) - 4 = -18 + 36 - 4 = 14$

The vertex is $(6, 14)$.

Axis of symmetry: $x = -\dfrac{6}{2\left(-\dfrac{1}{2}\right)} = 6$

The axis of symmetry is $x = 6$.

**74.** $y = \dfrac{1}{6}x^2 - \dfrac{1}{3}x + 2$

Vertex: $x = -\dfrac{\left(-\dfrac{1}{3}\right)}{2\left(\dfrac{1}{6}\right)}$

$x = 1$

$y = \dfrac{1}{6}(1)^2 - \dfrac{1}{3}(1) + 2 = \dfrac{1}{6} - \dfrac{1}{3} + 2 = 1\dfrac{5}{6}$

The vertex is $\left(1, 1\dfrac{5}{6}\right)$

Axis of symmetry: $x = -\dfrac{\left(-\dfrac{1}{3}\right)}{2\left(\dfrac{1}{6}\right)}$

$x = 1$

The axis of symmetry is $x = 1$.

**75.**
$x^2 - 10 = 6$
$x^2 - 16 = 0$
$(x + 4)(x - 4) = 0$
$x = 4, -4$

**76.**
$x^2 + 12 = 48$
$x^2 - 36 = 0$
$(x + 6)(x - 6) = 0$
$x = 6, -6$

**77.** $\dfrac{1}{5}x^2 = 5$
$x^2 = 25$
$x = 5, -5$

**78.** $3x^2 = 192$
$x^2 = 64$
$x = 8, -8$

**79.** $\dfrac{2}{3}x^2 = 6$
$x^2 = 9$
$x = 3, -3$

**80.** $2x^2 - 66 = 96$
$2x^2 = 162$
$x^2 = 81$
$x = 9, -9$

**81.** Yes; let $x$ be the length of the rectangle and $y$ be the width. Then $2x + 2y = 52$ and $xy = 148.75$. Substituting, $x(26 - x) = 148.75$, which is equivalent to $x^2 - 26x + 148.75 = 0$. Since the discriminant of the equation is 81, the equation has two solutions. The rectangle is 17.5 by 8.5.

*Quiz 1 (p. 596)*

**1.** $(2x^2 + 7x + 1) + (x^2 - 2x + 8)$
$= 2x^2 + 7x + 1 + x^2 - 2x + 8$
$= 3x^2 + 5x + 9$

**2.** $(-4x^3 - 5x^2 + 2x) - (2x^3 + 9x^2 + 2)$
$= -4x^3 - 5x^2 + 2x - 2x^3 - 9x^2 - 2$
$= -6x^3 - 14x^2 + 2x - 2$

**3.** $(7t^2 - 3t + 5) - (4t^2 + 10t - 9)$
$- 7t^2 - 3t + 5 - = 3t^2 - 13t + 14$
$= 3t^2 - 13t + 14$

**4.** $(5x^3 - x^2 + 3x + 3) + (x^3 - 4x^2 + x)$
$= 5x^3 - x^2 + 3x + 3 + x^3 - 4x^2 + x$
$= 6x^3 - 5x^2 + 4x + 3$

**5.** $(x + 8)(x - 1) = x^2 - x + 8x - 8 = x^2 + 7x - 8$

**6.** $(4n + 7)(4n - 7) = 16n^2 - 49$

**7.** $(-2x^2 + x - 4)(x - 2)$
$= -2x^3 + x^2 - 4x + 4x^2 - 2x + 8$
$= -2x^3 + 5x^2 - 6x + 8$

**8.** $(9m - 4)(3m + 1) = 27m^2 + 9m - 12m - 4$
$= 27m^2 - 3m - 4$

**9.** $\left(\dfrac{1}{2}x - 3\right)\left(\dfrac{1}{2}x + 5\right) = \dfrac{1}{4}x^2 + \dfrac{5}{2}x - \dfrac{3}{2}x - 15$
$= \dfrac{1}{4}x^2 + x - 15$

**10.** $-x^2(12x^3 - 11x^2 + 3) = -12x^5 + 11x^4 - 3x^2$

**11.** $(x - 6)(x + 6) = x^2 - 36$

**12.** $(4x + 3)(4x - 3) = 16x^2 - 9$

**13.** $(5 + 3b)(5 - 3b) = 25 - 9b^2$

**14.** $(2x - 7y)(2x + 7y) = 4x^2 - 49y^2$

**15.** $(3x + 6)^2 = 9x^2 + 2(3x)(6) + 36 = 9x^2 + 36x + 36$

**16.** $(-6 - 8x)^2 = 36 + 2(-6)(-8x) + 64x^2$
$= 36 + 96x + 64x^2$

**17.**

$(2x - 16)^2 = 4x^2 - 2(2x)(-16) + 256$
$= (4x^2 - 64x + 256) \text{ in.}^2$

# Chapter 10 *continued*

**Math & History (p. 597)**

**1.** $\frac{121}{160} \approx 0.76$    **2.** $\frac{81}{316} \approx 0.26$

**Activity (p. 597)**

**1.**

| Expression | $-3$ | $-2$ | $-1$ | 0 | 1 | 2 | 3 |
|---|---|---|---|---|---|---|---|
| $(x-3)(x+2)$ | 6 | 0 | $-4$ | $-6$ | $-6$ | $-4$ | 0 |
| $(x+1)(x+3)$ | 0 | $-1$ | 0 | 3 | 8 | 15 | 24 |
| $(x-2)(x-2)$ | 25 | 16 | 9 | 4 | 1 | 0 | 1 |
| $(x-1)(x+2)$ | 4 | 0 | $-2$ | $-2$ | 0 | 4 | 10 |

**2. a.** $3, -2$    **b.** $-1, -3$

   **c.** $2$      **d.** $1, -2$

**3.** *Sample answer:* To solve an equation in factored form, set each factor equal to zero and solve the resulting linear equations. The solution to each of the linear equations is a solution of the original equation.

## Lesson 10.4

### 10.4 Guided Practice (p. 600)

**1.** If the product of two numbers is zero, at least one of the numbers is zero. (If $ab = 0$, then $a = 0$ or $b = 0$).

**2.** No; the solutions are solutions of $x + 1 = 0$ and $x - 4 = 0$, or $x = -1$, and $x = 4$.

**3.** No; the equation has two linear factors, so it has only two solutions, the solutions of $x - 2 = 0$ and $x + 5 = 0$, or $x = 2$, and $x = -5$.

**4.** true   **5.** true   **6.** true   **7.** true   **8.** no   **9.** no

**10.** yes   **11.** yes

**12.** $(b + 1)(b + 3) = 0$      **13.** $(t - 3)(t - 5) = 0$

   $b + 1 = 0$    $b + 3 = 0$      $t - 3 = 0$   $t - 5 = 0$

     $b = -1$      $b = -3$        $t = 3$     $t = 5$

**14.** $(x - 7)^2 = 0$

     $x - 7 = 0$

        $x = 7$

**15.** $(y + 9)(y - 2) = 0$

   $y + 9 = 0$    $y - 2 = 0$

     $y = -9$      $y = 2$

**16.** $(3d + 6)(2d + 5) = 0$     **17.** $\left(2w + \frac{1}{2}\right)^2 = 0$

   $3d + 6 = 0$    $2d + 5 = 0$      $2w + \frac{1}{2} = 0$

     $3d = -6$      $2d = -5$       $2w = -\frac{1}{2}$

       $d = -2$        $d = -\frac{5}{2}$        $w = -\frac{1}{4}$

**18.**

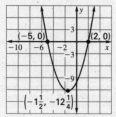

$\dfrac{2 + -5}{2} = -\dfrac{3}{2} = -1\dfrac{1}{2}$

$y = (x - 2)(x + 5)$

$\phantom{y} = \left(-1\dfrac{1}{2} - 2\right)\left(-1\dfrac{1}{2} + 5\right)$

$\phantom{y} = \left(-3\dfrac{1}{2}\right)\left(3\dfrac{1}{2}\right) = -12\dfrac{1}{4}$

Vertex: $\left(-1\dfrac{1}{2}, -12\dfrac{1}{4}\right)$

### 10.4 Practice and Applications (pp. 600–602)

**19.** $(x + 4)(x + 1) = 0$      **20.** $(y + 3)^2 = 0$

     $x = -4, -1$              $y = -3$

**21.** $(t + 8)(t - 6) = 0$      **22.** $(w - 17)^2 = 0$

     $t = -8, 6$              $w = 17$

**23.** $(b - 9)(b + 8) = 0$      **24.** $(d + 7)^2 = 0$

     $b = 9, -8$             $d = -7$

**25.** $(y - 2)(y + 1) = 0$      **26.** $(z + 2)(z + 3) = 0$

     $y = 2, -1$            $z = -2, -3$

**27.** $(v - 7)(v - 5) = 0$      **28.** $\left(t + \frac{1}{2}\right)(t - 4) = 0$

     $v = 7, 5$             $t = -\frac{1}{2}, 4$

**29.** $4(c + 9)^2 = 0$         **30.** $(u - 3)\left(u - \frac{2}{3}\right) = 0$

     $c = -9$             $u = 3, \frac{2}{3}$

**31.** $(y - 5.6)^2 = 0$       **32.** $(a - 40)(a + 12) = 0$

     $y = 5.6$             $a = 40, -12$

**33.** $7(b - 5)^3 = 0$        **34.** $(4x - 8)(7x + 21) = 0$

     $b = 5$             $4x - 8 = 0$    $7x + 21 = 0$

                           $4x = 8$      $7x = -21$

                            $x = 2$       $x = -3$

**35.** $(2d + 8)(3d + 12) = 0$

     $2d + 8 = 0$     $3d + 12 = 0$

       $2d = -8$        $3d = -12$

        $d = -4$         $d = -4$

**36.** $5(3m + 9)(5m - 15) = 0$

     $3m + 9 = 0$     $5m - 15 = 0$

       $3m = -9$        $5m = 15$

        $m = -3$         $m = 3$

# Chapter 10 *continued*

**37.** $8(9n + 27)(6n - 9) = 0$

$9n + 27 = 0 \qquad 6n - 9 = 0$

$9n = -27 \qquad\quad 6n = 9$

$n = -3 \qquad\qquad n = \frac{3}{2}$

**38.** $(6b - 18)(2b + 2)(2b + 2) = 0$

$6b - 18 = 0 \qquad 2b + 2 = 0$

$6b = 18 \qquad\quad 2b = -2$

$b = 3 \qquad\qquad b = -1$

**39.** $(4y - 5)(2y - 6)(3y - 4) = 0$

$4y - 5 = 0 \qquad 2y - 6 = 0 \qquad 3y - 4 = 0$

$4y = 5 \qquad\quad 2y = 6 \qquad\quad 3y = 4$

$y = \frac{5}{4} \qquad\qquad y = 3 \qquad\qquad y = \frac{4}{3}$

**40.** $(x + 44)(3x - 2)^2 = 0$

$x + 44 = 0 \qquad 3x - 2 = 0$

$x = -44 \qquad\quad 3x = 2$

$\qquad\qquad\qquad x = \frac{2}{3}$

**41.** $(5x - 9.5)^2(3x + 6.3) = 0$

$5x - 9.5 = 0 \qquad 3x + 6.3 = 0$

$5x = 9.5 \qquad\quad 3x = -6.3$

$x = 1.9 \qquad\qquad x = -2.1$

**42.** $\left(\frac{1}{2}x + 2\right)\left(\frac{2}{3}x + 6\right)\left(\frac{1}{6}x - 1\right) = 0$

$\frac{1}{2}x + 2 = 0 \qquad \frac{2}{3}x + 6 = 0 \qquad \frac{1}{6}x - 1 = 0$

$\frac{1}{2}x = -2 \qquad\quad \frac{2}{3}x = -6 \qquad\quad \frac{1}{6}x = 1$

$x = -4 \qquad\qquad x = -9 \qquad\qquad x = 6$

**43.** $\left(2n - \frac{1}{4}\right)\left(5n + \frac{3}{10}\right)\left(3n - \frac{2}{3}\right) = 0$

$2n - \frac{1}{4} = 0 \qquad 5n + \frac{3}{10} = 0 \qquad 3n - \frac{2}{3} = 0$

$2n = \frac{1}{4} \qquad\quad 5n = -\frac{3}{10} \qquad\quad 3n = \frac{2}{3}$

$n = \frac{1}{8} \qquad\qquad n = -\frac{3}{50} \qquad\qquad n = \frac{2}{9}$

**44.** C  **45.** B  **46.** A

**47.** $y = (x - 4)(x + 2)$

$x = 4, -2$

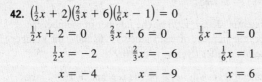

$x = \dfrac{4 + (-2)}{2} = \dfrac{2}{2} = 1$

$y = (1 - 4)(1 + 2)$

$\quad = (-3)(3) = -9$

Vertex: $(1, -9)$

**48.** $y = (x + 5)(x + 3)$

$x = -5, -3$

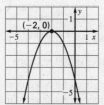

$x = \dfrac{-5 + (-3)}{2} = -4$

$y = (-4 + 5)(-4 + 3)$

$\quad = 1(-1) = -1$

Vertex: $(-4, 1)$

**49.** $y = (-x - 2)(x + 2)$

$x = -2$

Vertex: $(-2, 0)$

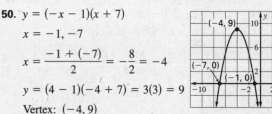

**50.** $y = (-x - 1)(x + 7)$

$x = -1, -7$

$x = \dfrac{-1 + (-7)}{2} = -\dfrac{8}{2} = -4$

$y = (4 - 1)(-4 + 7) = 3(3) = 9$

Vertex: $(-4, 9)$

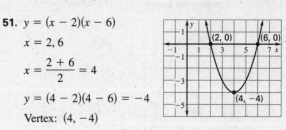

**51.** $y = (x - 2)(x - 6)$

$x = 2, 6$

$x = \dfrac{2 + 6}{2} = 4$

$y = (4 - 2)(4 - 6) = -4$

Vertex: $(4, -4)$

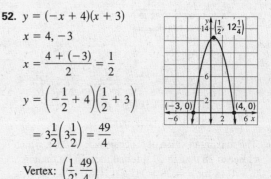

**52.** $y = (-x + 4)(x + 3)$

$x = 4, -3$

$x = \dfrac{4 + (-3)}{2} = \dfrac{1}{2}$

$y = \left(-\dfrac{1}{2} + 4\right)\left(\dfrac{1}{2} + 3\right)$

$\quad = 3\dfrac{1}{2}\left(3\dfrac{1}{2}\right) = \dfrac{49}{4}$

Vertex: $\left(\dfrac{1}{2}, \dfrac{49}{4}\right)$

**53.** Use the distance between the *x*-intercepts at $(-500, 0)$ and $(500, 0)$.

**54.** $x = \dfrac{-500 + 500}{2} = \dfrac{0}{2} = 0$

$y = \dfrac{167}{500^2}(0 + 500)(0 - 500) = \dfrac{167}{500^2}(500)(-500) = -167$

$(0, -167)$

# Chapter 10 *continued*

**55.** $y = -\frac{7}{1000}(x + 300)(x - 300)$

$x = 300, -300$

The legs of the arch are 600 ft apart.

**56.** Vertex: $(0, 630)$

$y = -\frac{7}{100}(300)(-300) = 630$ ft high

**57.** $x = 600, -600$

$600 - (-600) = 1200$ m wide

**58.** $x = \dfrac{600 + (-600)}{2} = \dfrac{0}{2} = 0$

$y = \dfrac{1}{1800}(-600)(600) = -200$

The crater is 200 m deep.

**59. a.** The price of a hot dog after $n$ \$0.10 increases; the number of hot dogs sold after $n$ \$0.10 increases.

**b.** $R = (1 + 0.1n)(200 - 3n)$

$1 + 0.1n = 0 \qquad 200 - 3n = 0$

$0.1n = -1 \qquad -3n = -200$

$n = -10 \qquad n = \frac{200}{3}$

$\qquad\qquad\qquad n = 66.\overline{6}$

**Selling Hot Dogs**          67 times

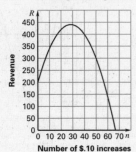

Vertex: $(28.\overline{3}, 440.8\overline{3})$

$n = \dfrac{-10 + \frac{200}{3}}{2} = \dfrac{-\frac{30}{3} + \frac{200}{3}}{2} = \dfrac{\frac{170}{3}}{2} = \dfrac{170}{6}$

$R = (1 + 0.1(28.3))(200 - 3(28.3))$

$\quad = (3.8\overline{3})(115.1) = 480.\overline{3}$

**c.** The maximum value of $R$ occurs at $n = 28.\overline{3}$, which is closer to 28 than to 29. I found the cost by finding $1 + 0.1n$ for $n = 28 = (1 + 0.1(28)) = \$3.80$

**60.** Answers may vary.

### *10.4 Mixed Review (p. 602)*

**61.** 210,000 **62.** 0.04443 **63.** 857,000,000 **64.** 1,250,000

**65.** 0.00371 **66.** 9,960,000 **67.** 0.000722

**68.** 81,700,000

**69.** $(x - 2)(x - 7) = x^2 - 7x - 2x + 14 = x^2 - 9x + 14$

**70.** $(x + 8)(x - 8) = x^2 - 64$

**71.** $(x - 4)(x + 5) = x^2 + 5x - 4x - 20 = x^2 + x - 20$

**72.** $(x + 6)(x - 7) = x^2 - 7x + 6x - 42 = x^2 - x - 42$

**73.** $\left(x + \frac{2}{3}\right)\left(x - \frac{1}{3}\right) = x^2 - \frac{1}{3}x + \frac{2}{3}x - \frac{2}{9} = x^2 + \frac{1}{3}x - \frac{2}{9}$

**74.** $\left(x - 3\right)\left(x - \frac{1}{6}\right) = x^2 - \frac{1}{6}x - 3x + \frac{1}{2} = x^2 - \frac{19}{6}x + \frac{1}{2}$

**75.** $(2x + 7)(3x - 1) = 6x^2 - 2x + 21x - 7$

$\qquad\qquad\qquad\qquad = 6x^2 + 19x - 7$

**76.** $(5x - 1)(5x + 2) = 25x^2 + 10x - 5x - 2$

$\qquad\qquad\qquad\qquad = 25x^2 + 5x - 2$

**77.** $(3x + 1)(8x - 3) = 24x^2 - 9x + 8x - 3$

$\qquad\qquad\qquad\qquad = 24x^2 - x - 3$

**78.** $(2x - 4)\left(\frac{1}{4}x - 2\right) = \frac{1}{2}x^2 - 4x - x + 8 = \frac{1}{2}x^2 - 5x + 8$

**79.** $(x + 10)(x + 10) = x^2 + 2(x)(10) + 100$

$\qquad\qquad\qquad\qquad = x^2 + 20x + 100$

**80.** $(3x + 5)\left(\frac{2}{3}x - 3\right) = 2x^2 - 9x + \frac{10}{3}x - 15$

$\qquad\qquad\qquad\qquad = 2x^2 - \frac{17}{3}x - 15$

**81.** Exponential decay; let $P$ be the price in 1996, $t$ the number of years since 1996 and $y$ the price after $t$ years; $y = P(0.84)^t$

**82.** Exponential growth; let $n$ be the number sold in 1995, $t$ the number of years since 1995, and $y$ the price after $t$ years; $y = n(1.23)^t$

**83.** Exponential decay; let $n$ be the number of members in 1996, $t$ the number of years since 1996, and $y$ the number of members after $t$ years; $y = n(0.97)^t$

**84.** Exponential growth; let $R$ be the revenue in 1993, $t$ the number of years since 1993, and $y$ the revenue after $t$ years; $y = R(2.37)^t$

## Lesson 10.5

### *Developing Concepts Activity 10.5 (p. 603)*

**1.** $(x + 2)(x + 6)$ **2.** $(x + 3)(x + 7)$

**3.** $x^2 + 7x + 6$

$(x + 1)(x + 6)$

**4.** $x^2 + 6x + 8$

$(x + 2)(x + 4)$

**5.** $x^2 + 8x + 15$

$(x + 3)(x + 5)$

# Chapter 10 *continued*

**6.** $x^2 + 6x + 9$

$(x + 3)(x + 3) = (x + 3)^2$

**7.** $x^2 + 4x + 4$

$(x + 2)(x + 2) = (x + 2)^2$

**8.** $x^2 + 7x + 10$

$(x + 2)(x + 5)$

**9.** $x^2 + 3x + 4$

The tiles cannot be formed into a rectangle.

## 10.5 Guided Practice  (p. 607)

**1.** to write it as the product of two linear expressions

**2.** $x^2 - 4x + 3$

$(x - 3)(x - 1)$; the product of the constant terms is positive, so both must be positive or both must be negative.

**3.** $x^2 + 2x - 3$

$(x + 3)(x - 1)$; the product of the constant terms is negative, so one must be positive and the other negative.

**4.** No; 35 is not a perfect square.

**5.** D  **6.** A  **7.** B  **8.** C

**9.** $b^2 - 4ac$

$-4^2 - 4(1)(4)$

$16 - 16 = 0$

Yes, 0 is a perfect square.

**10.** $b^2 - 4ac$

$-4^2 - 4(1)(-5)$

$16 + 20 = 36$

Yes; 36 is a perfect square.

**11.** $b^2 - 4ac$

$-4^2 - 4(1)(-6)$

$16 + 24 = 40$

No; 40 is not a perfect square.

## 10.5 Practice and Applications  (pp. 607–609)

**12.** B  **13.** B  **14.** $x^2 + 11x - 216$  **15.** $x^2 + 8x - 9$

$(x + 13)(x - 2)$        $(x + 9)(x - 1)$

**16.** $t^2 - 10t + 21$  **17.** $b^2 + 5b - 24$

$(t - 7)(t - 3)$        $(b + 8)(b - 3)$

**18.** $w^2 + 13w + 36$

$(w + 4)(w + 9)$

**19.** $y^2 - 3y - 18$

$(y - 6)(y + 3)$

**20.** $c^2 + 14c + 40$

$(c + 10)(c + 4)$

**21.** $m^2 - 7m - 30$

$(m - 10)(m + 3)$

**22.** $32 + 12n + n^2$

$(8 + n)(4 + n)$

**23.** $44 - 15s + s^2$

$(11 - s)(4 - s)$

**24.** $z^2 + 65z + 1000$

$(z + 40)(z + 25)$

**25.** $x^2 - 45x + 450$

$(x - 15)(x - 30)$

**26.** $d^2 - 33d - 280$

$(d + 7)(d - 40)$

**27.** $x^2 + 7x + 10 = 0$

$(x + 5)(x + 2) = 0$

$x = -5, -2$

**28.** $x^2 + 5x - 14 = 0$

$(x + 7)(x - 2) = 0$

$x = -7, 2$

**29.** $x^2 - 9x = -14$

$x^2 - 9x + 14 = 0$

$(x - 7)(x - 2) = 0$

$x = 7, 2$

**30.** $x^2 + 32x = -220$

$x^2 + 32x + 220 = 0$

$(x + 22)(x + 10) = 0$

$x = -22, -10$

**31.** $x^2 + 16x = -15$

$x^2 + 16x + 15 = 0$

$(x + 15)(x + 1) = 0$

$x = -15, -1$

**32.** $x^2 + 3x = 54$

$x^2 + 3x - 54 = 0$

$(x + 9)(x - 6) = 0$

$x = -9, 6$

**33.** $x^2 + 8x = 65$

$x^2 + 8x - 65 = 0$

$(x + 13)(x - 5) = 0$

$x = -13, 5$

**34.** $-x + x^2 = 56$

$x^2 - x - 56 = 0$

$(x - 8)(x + 7) = 0$

$x = 8, -7$

**35.** $x^2 - 20x = -51$

$x^2 - 20x + 51 = 0$

$(x - 17)(x - 3) = 0$

$x = 17, 3$

**36.** $x^2 - 5x = 84$

$x^2 - 5x - 84 = 0$

$(x + 7)(x - 12) = 0$

$x = -7, 12$

**37.** $x^2 + 3x - 31 = -3$

$x^2 + 3x - 28 = 0$

$(x + 7)(x - 4) = 0$

$x = -7, 4$

**38.** $x^2 - 2x - 19 = -4$

$x^2 - 2x - 15 = 0$

$(x - 5)(x + 3) = 0$

$x = 5, -3$

**39.** $x^2 - x - 8 = 82$

$x^2 - x - 90 = 0$

$(x + 9)(x - 10) = 0$

$x = -9, 10$

**40.** $x^2 + 42 = 13x$

$x^2 - 13x + 42 = 0$

$(x - 7)(x - 6) = 0$

$x = 7, 6$

**41.** $x^2 - 9x + 18 = 2x$

$x^2 - 11x + 18 = 0$

$(x - 9)(x - 2) = 0$

$x = 9, 2$

**42.** $b^2 - 4ac = 49 - 4(1)(-144) = 49 + 576 = 625$

yes; $(x + 16)(x - 9)$

**43.** $b^2 - 4ac = 19^2 - 4(1)(60) = 361 - 240 = 121$

yes; $(y + 15)(y + 4)$

**44.** $b^2 - 4ac = -11^2 - 4(1)(24) = 121 - 96 = 25$

yes; $(x - 3)(x - 8)$

**45.** $b^2 - 4ac = -6^2 - 4(1)(16) = 36 - 64 = -28$

no

**46.** $b^2 - 4ac = -26^2 - 4(1)(-87) = 676 + 348 = 1024$

yes; $(z + 3)(z - 29)$

**47.** $b^2 - 4ac = 14^2 - 4(1)(35) = 196 - 140 = 56$

no

**48.** $x^2 - 17x + 30 = 0$

$(x - 15)(x - 2) = 0$

$x = 15, 2$

**49.** $x^2 + 8x = 105$

$x^2 + 8x - 105 = 0$

$(x + 15)(x - 7) = 0$

$x = -15, 7$

**50.** $x^2 - 20x + 21 = 2$

$x^2 - 20x + 19 = 0$

$(x - 19)(x - 1) = 0$

$x = 19, 1$

**51.** $x^2 + 52x + 680 = 40$

$x^2 + 52x + 640 = 0$

$(x + 32)(x + 20) = 0$

$x = -32, -20$

**52–55** *Sample equations:*

**52.** $x^2 + 9x - 252 = 0$  **53.** $x^2 + 11x + 30 = 0$

**54.** $x^2 + 36x - 205 = 0$  **55.** $x^2 - 427x = 0$

**56.** $x^2 - 17x + 66$

$(x - 6)(x - 11)$

$(x - 6)$ and $(x - 11)$

**57.** $x^2 - 17x + 66 = 84$

$x^2 - 17x - 18 = 0$

$(x + 1)(x - 18) = 0$

$x = 18$

**58.** $(x - 6)(x - 11)$

$18 - 6 = 12$

$18 - 11 = 7$

12 ft by 7 ft

**59.** $A = \pi(x^2 - 18x + 81)$

$(x - 9)^2$

radius $= x - 9$

**60.** $12.56 = \pi(x - 9)^2$

$\sqrt{\dfrac{12.56}{\pi}} = \sqrt{(x - 9)^2}$

$\sqrt{\dfrac{12.56}{\pi}} = x - 9$

$\sqrt{\dfrac{12.56}{\pi}} + 9 = x$

$11 = x$

**61.**

$20 \geq \frac{1}{2}(2h - 2)(h)$

$20 \geq \frac{1}{2}(2h^2 - 2h)$

$20 \geq h^2 - h$

**62.**
$h^2 - h \leq 20$

$h^2 - h - 20 \leq 0$

$(h - 5)(h + 4) \leq 0$

$h = 5$ ft

height $= 5$ ft

base $= 2h - 2 = 2(5) - 2 = 8$ ft

**63.**

Platform $= 9025$

$\qquad = 95 \text{ m} \times 95 \text{ m}$

Building $= 95 - 38 = 57$ m

$\qquad = 57 \text{ m} \times 57 \text{ m}$

**64.** $x = $ width

$x + 245 = $ length

Area $= x(x + 245)$

$167,750 = x(x + 245)$

$167,750 = x^2 + 245x$

$x^2 + 245x - 167,750 = 0$

$(x + 550)(x - 305) = 0$

$x = 305$ m

305 m $\times$ 550 m

**65.** A  **66.** B  **67.** D  **68.** $a^{2n} - b^{2n} = (a^n + b^n)(a^n - b^n)$

**69.** $a^{2n} + 2a^nb^n + b^{2n} = (a^n + b^n)^2$

**70.** $a^{2n} + 18a^nb^n + 81b^{2n} = (a^n + 9b^n)^2$

**71.** $5a^2 - 9a^nb^n - 2b^{2n} = (5a^n + b^n)(a^n - 2b^n)$

## 10.5 Mixed Review (p. 609)

**72.** 15  **73.** 1  **74.** 2  **75.** 7  **76.** 4  **77.** 18

**78.** $3q(q^3 - 5q^2 + 6) = 3q^4 - 15q^3 + 18q$

**79.** $(y + 9)(y - 4) = y^2 - 4y + 9y - 36 = y^2 + 5y - 36$

**80.** $(7x - 11)^2 = 49x^2 - 2(7x)(11) + 121$

$\qquad = 49x^2 - 154x + 121$

**81.** $(5 - w)(12 + 3w) = 60 + 15w - 12w - 3w^2$

$\qquad = 60 + 3w - 3w^2$

**82.** $(3a - 2)(4a + 6) = 12a^2 + 18a - 8a - 12$

$\qquad = 12a^2 + 10a - 12$

**83.** $(2b - 4)(b^3 + 4b^2 + 5b)$

$\qquad = 2b^4 + 8b^3 + 10b^2 - 4b^3 - 16b^2 - 20b$

$\qquad = 2b^4 + 4b^3 - 6b^2 - 20b$

**84.** $(9x + 8)(9x - 8) = 81x^2 - 64$

**85.** $\left(6z + \frac{1}{3}\right)^2 = 36z^2 + 2(6z)\left(\frac{1}{3}\right) + \frac{1}{9} = 36z^2 + 4z + \frac{1}{9}$

**86.** $(5t - 3)(4t - 10) = 20t^2 - 50t - 12t + 30$

$\qquad = 20t^2 - 62t + 30$

# Chapter 10 *continued*

**87.** $(x + 12)(x + 7) = 0$    **88.** $(z + 2)(z + 3) = 0$
     $x = -12, -7$            $z = -2, -3$

**89.** $(t - 19)^2 = 0$      **90.** $\left(b - \frac{2}{5}\right)\left(b - \frac{5}{6}\right) = 0$
     $t = 19$                $b = \frac{2}{5}, \frac{5}{6}$

**91.** $(x - 9)(x - 6) = 0$    **92.** $(y + 47)(y - 27) = 0$
     $x = 9, 6$             $y = -47, 27$

**93.** $(z - 1)(4z + 2) = 0$    **94.** $(3a - 8)(a + 5) = 0$
   $z = 1$    $4z + 2 = 0$     $3a - 8 = 0$    $a + 5 = 0$
           $4z = -2$       $3a - 8$       $a = -5$
            $z = -\frac{1}{2}$        $a = \frac{8}{3}$

**95.** $(4n - 6)^3 = 0$      **96.** $h = -16t^2 + s$
     $4n - 6 = 0$           $0 = -16t^2 + 40$
       $4n = 6$            $16t^2 = 40$
         $n = \frac{6}{4}$           $t^2 = 2.5$
           $= \frac{3}{2}$         $t \approx 1.58$ sec

## Lesson 10.6

### Activity 10.6 (p. 610)

### Developing Concepts

**3.** Width: $x + 1$, length:
$2x + 3, 2x^2 + 5x + 3 = (x + 1)(2x + 3)$

### Drawing Conclusions

**1.** $2x^2 + 9x + 9 = (x + 3)(2x + 3)$

**2.** $2x^2 + 7x + 3 = (x + 3)(2x + 1)$

**3.** $3x^2 + 4x + 1 = (x + 1)(3x + 1)$

**4.** $3x^2 + 10 + 3 = (x + 3)(3x + 1)$

**5.** $3x^2 + 10x + 8 = (x + 2)(3x + 4)$

**6.** $4x^2 + 5x + 1 = (x + 1)(4x + 1)$

**7.** $2x^2 + 3x + 1 = (x + 1)(2x + 1)$

**8.** $2x^2 + 4x + 2 - (x + 1)(2x + 2)$

**9.** $4x^2 + 4x + 1 = (2x + 1)(2x + 1)$

### 10.6 Guided Practice (p. 614)

**1.** False; There is no product property of 1 similar to the zero product property.

**2.** Before factoring, the equation should have been rewritten with one side equal to zero.

     $2x^2 - 3x + 1 = 10$
     $2x^2 - 3x - 9 = 0$
     $(2x + 3)(x - 3) = 0$
     $2x + 3 = 0$     $x - 3 = 0$
        $2x = -3$         $x = 3$
          $x = -\frac{3}{2}$

**3.** Yes; 1 is a perfect square.

**4.** $x + x^2 = 0$            **5.** B   **6.** D   **7.** A   **8.** C
    $x(1 + x) = 0$
    $x = 0$    $1 + x = 0$
             $x = -1$

**9.** $b^2 - 4ac = -3^2 - 4(2)(-2) = 9 + 16 = 25$
    $2x^2 - 3x - 2 = (2x + 1)(x - 2)$

**10.** $b^2 - 4ac = (-4)^2 - 4(-3)(7) = 16 + 84 = 100$
    $-(3y^2 + 4y - 7) = (-3y - 7)(y - 1)$

# Chapter 10 *continued*

**11.** $6t^2 - 4t - 5$

$b^2 - 4ac$

$= -4^2 - 4(6)(-5)$

$= 16 + 120 = 136$

not factorable

**12.** $3b^2 + 26b + 35 = 0$

$(3b + 5)(b + 7) = 0$

$3b + 5 = 0 \qquad b + 7 = 0$

$3b = -5 \qquad b = -7$

$b = -\frac{5}{3}$

**13.** $\qquad 2z^2 + 15z = 8$

$2z^2 + 15z - 8 = 0$

$(2z - 1)(z + 8) = 0$

$2z - 1 = 0 \qquad z + 8 = 0$

$2z = 1 \qquad z = -8$

$z = \frac{1}{2}$

**14.** $\qquad -7n^2 - 40n = -12$

$-7n^2 - 40n + 12 = 0$

$-(7n^2 + 40n - 12) = 0$

$-(7n - 2)(n + 6) = 0$

$7n - 2 = 0 \qquad n + 6 = 0$

$7n = 2 \qquad n = -6$

$n = \frac{2}{7}$

## 10.6 Practice and Applications (pp. 614–616)

**15.** A

**16.** $6y^2 - 29y - 5$

$(6y + 1)(y - 5)$

**17.** A

**18.** $3t^2 + 16t + 5 = (3t + 1)(t + 5)$

**19.** $6b^2 - 11b - 2 = (6b + 1)(b - 2)$

**20.** $4n^2 - 26n - 42$  not factorable

**21.** $5w^2 - 9w - 2 = (5w + 1)(w - 2)$

**22.** $4x^2 + 27x + 35 = (4x + 7)(x + 5)$

**23.** $6y^2 - 11y - 10 = (2y - 5)(3y + 2)$

**24.** $6x^2 - 21x - 9$  not factorable

**25.** $3c^2 - 37c + 44 = (3c - 4)(c - 11)$

**26.** $10x^2 + 17x + 6 = (2x + 1)(5x + 6)$

**27.** $14y^2 - 15y + 4 = (7y - 4)(2y - 1)$

**28.** $4z^2 + 32z + 63 = (2z + 9)(2z + 7)$

**29.** $6t^2 + t - 70 = (2t + 7)(3t - 10)$

**30.** $8b^2 + 2b - 3 = (2b - 1)(4b + 3)$

**31.** $2z^2 + 19z - 10(2z - 1)(z + 10)$

**32.** $12m^2 + 48m + 96$  not factorable

**33.** $2x^2 - 9x - 35 = 0$

$(2x + 5)(x - 7) = 0$

$2x + 5 = 0 \qquad x - 7 = 0$

$x = -\frac{5}{2} \qquad x = 7$

**34.** $7x^2 - 10x + 3 = 0$

$(7x - 3)(x - 1) = 0$

$7x - 3 = 0 \qquad x - 1 = 0$

$7x = 3 \qquad x = 1$

$x = \frac{3}{7}$

**35.** $3x^2 + 34x + 11 = 0$

$(3x + 1)(x + 11) = 0$

$3x + 1 = 0 \qquad x + 11 = 0$

$3x = -1 \qquad x = -11$

$x = -\frac{1}{3}$

**36.** $4x^2 - 21x + 5 = 0$

$(4x - 1)(x - 5) = 0$

$4x - 1 = 0 \qquad x - 5 = 0$

$4x = 1 \qquad x = 5$

$x = \frac{1}{4}$

**37.** $2x^2 - 17x - 19 = 0$

$(2x - 19)(x + 1) = 0$

$2x - 19 = 0 \qquad x + 1 = 0$

$2x = 19 \qquad x = -1$

$x = \frac{19}{2}$

**38.** $5x^2 - 3x - 26 = 0$

$(5x - 13)(x + 2) = 0$

$5x - 13 = 0 \qquad x + 2 = 0$

$5x = 13 \qquad x = -2$

$x = \frac{13}{5}$

**39.** $\qquad 2x^2 + 19x = -24$

$2x^2 + 19x + 24 = 0$

$(2x + 3)(x + 8) = 0$

$2x + 3 = 0 \qquad x + 8 = 0$

$2x = -3 \qquad x = -8$

$x = -\frac{3}{2}$

**40.** $\qquad 4x^2 - 8x = -3$

$4x^2 - 8x + 3 = 0$

$(2x - 1)(2x - 3) = 0$

$2x - 1 = 0 \qquad 2x - 3 = 0$

$2x = 1 \qquad 2x = 3$

$x = \frac{1}{2} \qquad x = \frac{3}{2}$

**41.** $\qquad 6x^2 - 23x = 18$

$6x^2 - 23x - 18 = 0$

$(3x + 2)(2x - 9) = 0$

$3x + 2 = 0 \qquad 2x - 9 = 0$

$3x = -2 \qquad 2x = 9$

$x = -\frac{2}{3} \qquad x = \frac{9}{2}$

# Chapter 10 *continued*

**42.** $8x^2 - 34x + 24 = -11$

$8x^2 - 34x + 35 = 0$

$(4x - 7)(2x - 5) = 0$

$4x - 7 = 0 \quad 2x - 5 = 0$

$4x = 7 \qquad 2x = 5$

$x = \frac{7}{4} \qquad x = \frac{5}{2}$

**43.** $10x^2 + x - 10 = -2x + 8$

$10x^2 + 3x - 18 = 0$

$(5x - 6)(2x + 3) = 0$

$5x - 6 = 0 \quad 2x + 3 = 0$

$5x = 6 \qquad 2x = -3$

$x = \frac{6}{5} \qquad x = -\frac{3}{2}$

**44.** $28x^2 - 9x - 1 = -4x + 2$

$28x^2 - 5x - 3 = 0$

$(7x - 3)(4x + 1) = 0$

$7x - 3 = 0 \quad 4x + 1 = 0$

$7x = 3 \qquad 4x = -1$

$x = \frac{3}{7} \qquad x = -\frac{1}{4}$

**45.** $24x^2 + 39x + 15 = 0$

$(8x + 5)(3x + 3) = 0$

$8x + 5 = 0 \quad 3x + 3 = 0$

$8x = -5 \qquad 3x = -3$

$x = -\frac{5}{8} \qquad x = -1$

**46.** $30x^2 - 80x + 50 = 7x - 4$

$30x^2 - 87x + 54 = 0$

$(3x - 6)(10x - 9) = 0$

$3x - 6 = 0 \quad 10x - 9 = 0$

$3x = 6 \qquad 10x = 9$

$x = 2 \qquad x = \frac{9}{10}$

**47.** $18x^2 - 30x - 100 = 67x + 30$

$18x^2 - 97x - 130 = 0$

$(9x + 10)(2x - 13) = 0$

$9x + 10 = 0 \quad 2x - 13 = 0$

$9x = -10 \qquad 2x = 13$

$x = -1\frac{1}{9} \qquad x = 6\frac{1}{2}$

**48.** $\frac{1}{16}s^2 = 4$

$s^2 = 64$

$s = 8, -8$

**49.** $x^2 - 10x = -25$

$x^2 - 10x + 25 = 0$

$(x - 5)(x - 5) = 0$

$x = 5$

**50.** $y^2 - 7y + 6 = -6$

$y^2 - 7y + 12 = 0$

$(y - 4)(y - 3) = 0$

$y = 4 \quad y = 3$

**51.** $4n^2 + 2n = 0$

$2n(2n + 1) = 0$

$2n = 0 \quad 2n + 1 = 0$

$n = 0 \qquad 2n = -1$

$n = -\frac{1}{2}$

**52.** $12t^2 = 0$

$t^2 = 0$

$t = 0$

**53.** $35b^2 - 61b + 24 = 0$

$(7b - 8)(5b - 3) = 0$

$7b - 8 = 0 \quad 5b - 3 = 0$

$7b = 8 \qquad 5b = 3$

$b = 1\frac{1}{7} \qquad b = \frac{3}{5}$

**54.** $9x^2 - 19 = -3$

$9x^2 - 16 = 0$

$(3x - 4)(3x + 4) = 0$

$3x - 4 = 0 \quad 3x + 4 = 0$

$3x = 4 \qquad 3x = -4$

$x = 1\frac{1}{3} \qquad x = -1\frac{1}{3}$

**55.** $20d^2 - 10d = 100$

$20d^2 - 10d - 100 = 0$

$(5d + 10)(4d - 10) = 0$

$5d + 10 = 0 \quad 4d - 10 = 0$

$5d = -10 \qquad 4d = 10$

$d = -2 \qquad d = 2\frac{1}{2}$

**56.** $8c^2 - 27c - 24 = -6$

$8c^2 - 27c - 18 = 0$

$c = \dfrac{27 \pm \sqrt{27^2 - 4(8)(-18)}}{16}$

$= \dfrac{27 \pm \sqrt{1305}}{16}$

$= \dfrac{27 + 3\sqrt{145}}{16}, \dfrac{27 - 3\sqrt{145}}{16}$

**57.** $56w^2 - 61w - 22 = 0$

$(7w + 2)(8w - 11) = 0$

$7w + 2 = 0 \quad 8w - 11 = 0$

$7w = -2 \qquad 8w = 11$

$w = -\frac{2}{7} \qquad w = 1\frac{3}{8}$

**58.** $3n^2 + 12n + 9 = -1$

$3n^2 + 12n + 10 = 0$

$n = \dfrac{-12 \pm \sqrt{12^2 - 4(3)(10)}}{2(3)}$

$= \dfrac{-12 \pm \sqrt{24}}{6}$

$= \dfrac{-12 + 2\sqrt{6}}{6}, -\dfrac{12 - 2\sqrt{6}}{6}$

$= \dfrac{-6 + \sqrt{6}}{3}, \dfrac{-6 - \sqrt{6}}{3}$

**59.** $24z^2 + 46z - 55 = 10$

$24z^2 + 46z - 65 = 0$

$z = \dfrac{-46 \pm \sqrt{46^2 - 4(24)(-65)}}{48}$

$= \dfrac{-46 \pm \sqrt{8356}}{48}$

$= \dfrac{-46 \pm 2\sqrt{2089}}{48}$

$= \dfrac{-23 + \sqrt{2089}}{24}, \dfrac{-23 - \sqrt{2089}}{24}$

**60.** $0.8x^2 + 3.2x + 2.40 = 0$

$8x^2 + 32x + 24 = 0$

$8(x^2 + 4x + 3) = 0$

$8(x + 3)(x + 1) = 0$

$x = -3, -1$

**61.** $0.23t^2 - 0.54t + 0.16 = 0$

$23t^2 - 0.54t + 16 = 0$

$(23t - 8)(t - 2) = 0$

$23t - 8 = 0 \quad t - 2 = 0$

$23t = 8 \qquad t = 2$

$t = \frac{8}{23}$

**62.** $0.3n^2 - 2.2n + 8.4 = 0$

$3n^2 - 22n + 84 = 0$

no solution

**63.** $0.119y^2 - 0.162y + 0.055 = 0$

$119y^2 - 162y + 55 = 0$

$(17y - 11)(7y - 5) = 0$

$17y - 11 = 0 \quad 7y - 5 = 0$

$17y = 11 \qquad 7y = 5$

$y = \frac{11}{17} \qquad y = \frac{5}{7}$

**64.** $h = -16t^2 + vt + s$

$h = -16t^2 + 8t + 8$

$= -8(2t^2 - t - 1)$

$= -8(2t + 1)(t - 1)$

$2t + 1 = 0 \quad t - 1 = 0$

$t = \frac{-1}{2} \qquad t = 1$

Time: 1 sec; yes

**65.** $h = -16t^2 + 50t$

$10 = -16t^2 + 50t + 4$

$0 = -16t^2 + 50t - 6$

$= -2(8t^2 - 25t + 3)$

$= -2(8t - 1)(t - 3)$

$8t - 1 = 0 \quad t - 3 = 0$

$8t = 1 \qquad t = 3$

$t = \frac{1}{8}$

3 sec; $\frac{1}{8}$ is not reasonable, because at that time, the acrobat is still rising.

**66.** $30 = -16t^2 + 44t + 6$

$0 = -16t^2 + 44t - 24$

$= -4(4t^2 - 11t + 6)$

$= -4(4t - 3)(t - 2)$

$4t - 3 = 0 \quad t - 2 = 0$

$4t = 3 \qquad t = 2$

$t = \frac{3}{4}$

2 sec; $\frac{3}{4}$ is not reasonable, because at that time the T-shirt is leaving the cannon.

**67.** $2.5x^2 = 9000$

**68.** $2.5x^2 = 9000$

$x^2 = 3600$

$x = 60$ warp threads

Ratio is 5 to 2

(150 weft threads to 60 warp threads)

Total: 210 threads/in.$^2$

**69.** $x^2 - x - 12 = 0$ **70.** $x^2 - 64 = 0$

**71.** $6x^2 + x - 1 = 0$ **72.** $12x^2 + 47x + 40 = 0$

**69–72:** *Sample procedure:* If integers a and b are solutions of a quadratic equation, then the equation can be written in the form $(x - a)(x - b) = 0$. If fractions $\frac{a}{b}$ and $\frac{c}{d}$ are solutions of a quadratic equation, then the equation can be written in the form $(bx - a)(dx - c) = 0$. The equations given were written first in this form, then the factors were multiplied.

**73.** D **74.** C **75.** B $7x^2 - 11x - 6 = 0$

$(7x + 3)(x - 2) = 0$

$7x + 3 = 0 \quad x - 2 = 0$

$7x = -3 \qquad x = 2$

$x = -\frac{3}{7}$

# Chapter 10 *continued*

**76.** C $\quad 0 = -16t^2 + 4t + 132$

$\qquad = -4(4t^2 - t - 33)$

$\qquad = -4(4t + 11)(t - 3)$

$\quad 4t + 11 = 0 \qquad t - 3 = 0$

$\qquad 4t = -11 \qquad t = 3$

$\qquad t = -2\frac{3}{4}$

**77.** $1000(1 + r)^2 + 1800(1 + r) = 3600$

**78.** Yes $\quad 1000(1 + r)^2 + 1800(1 + r) - 3600 = 0$

$\qquad\qquad\qquad 1000r^2 + 3800r - 800 = 0$

$\qquad\qquad\qquad 200(5r^2 + 19r - 4) = 0$

$\qquad\qquad\qquad 200(5r - 1)(r + 4) = 0$

$\quad 5r - 1 = 0 \qquad r + 4 = 0$

$\qquad 5r = 1 \qquad\qquad r = -4$

$\qquad r = \frac{1}{5} \qquad\qquad \frac{1}{5} = 20\%$

## 10.6 Mixed Review (p. 617)

**79.** $\quad 4x + 5y = 7 \qquad\qquad -12x - 15y = -21$

$\qquad 6x - 2y = -18 \qquad \underline{\;\;12x - 4y = -36\;\;}$

$\qquad\qquad\qquad\qquad\qquad\qquad -19y = -57$

$\quad 4x + 5(3) = 7 \qquad\qquad\qquad y = 3$

$\quad 4x + 15 = 7$

$\qquad 4x = -8$

$\qquad x = -2 \qquad\qquad (-2, 3)$

**80.** $\quad 6x - 5y = 3 \qquad\qquad\quad 12 - 10y = 6$

$\quad -12x + 8y = 5 \qquad\qquad \underline{\;\;-12x + 8y = 5\;\;}$

$\qquad\qquad\qquad\qquad\qquad\qquad\quad -2y = 11$

$\quad 6x - 5\left(-\frac{11}{2}\right) = 3 \qquad\qquad\quad y = -\frac{11}{2}$

$\quad 6x + \frac{55}{2} = 3$

$\qquad 6x = -\frac{49}{2}$

$\qquad x = -\frac{49}{12} = -4\frac{1}{12} \quad \left(-4\frac{1}{12}, -5\frac{1}{2}\right)$

**81.** $\quad 2x + y = 120 \qquad\qquad 2x + y = 120$

$\qquad x + 2y = 120 \qquad\quad \underline{\;\;-2x - 4y = -240\;\;}$

$\qquad\qquad\qquad\qquad\qquad\qquad -3y = -120$

$\quad 2x + 40 = 120 \qquad\qquad\quad y = 40$

$\qquad 2x = 80$

$\qquad x = 40 \qquad\qquad (40, 40)$

**82.** $\quad y > 2x^2 - x + 7$

$\quad 15 \overset{?}{>} 2(2^2) - 2 + 7$

$\quad 15 \overset{?}{>} 2(4) - 2 + 7$

$\quad 15 \overset{?}{>} 8 - 2 + 7$

$\quad 15 > 13$

$\quad$ solution

**83.** $\quad y \geq 4x^2 - 64x + 92$

$\quad 30 \overset{?}{\geq} 4(1^2) - 64(1) + 92$

$\quad 30 \overset{?}{\geq} 4 - 64 + 92$

$\quad 30 \not\geq 32$

$\quad$ not a solution

**84.** $(4t - 1)^2 = 16t^2 - 2(4t)(1) + 1 = 16t^2 - 8t + 1$

**85.** $(b + 9)(b - 9) = b^2 - 81$

**86.** $(3x + 5)(3x + 5) = 9x^2 + 15x + 15x + 25$

$\qquad\qquad\qquad\qquad\quad = 9x^2 + 30x + 25$

**87.** $(2a - 7)(2a + 7) = 4a^2 - 49$

**88.** $(11 - 6x)^2 = 121 - 2(11)(6x) + 36x^2$

$\qquad\qquad\quad = 121 - (132x) + 36x^2$

$\qquad\qquad\quad = 121 - 132x + 36x^2$

**89.** $(100 + 27x)^2 = 100^2 + 2(100)(27x) + 27^2x^2$

$\qquad\qquad\qquad\quad = 10,000 + 5400x + 729x^2$

**90.** quadratic

## Quiz 2 (p. 617)

**1.** $(x + 5)(2x + 10) = 0$

$\quad x + 5 = 0 \qquad 2x + 10 = 0$

$\qquad x = -5 \qquad\quad 2x = -10$

$\qquad\qquad\qquad\qquad\quad x = -5$

**2.** $(4x - 6)(5x - 20) = 0$

$\quad 4x - 6 = 0 \quad 5x - 20 = 0$

$\qquad 4x = 6 \qquad\quad 5x = 20$

$\qquad x = \frac{3}{2} \qquad\qquad x = 4$

**3.** $(2x + 7)(3x - 12) = 0$

$\quad 2x + 7 = 0 \qquad 3x - 12 = 0$

$\qquad 2x = -7 \qquad\quad 3x = 12$

$\qquad x = -3\frac{1}{2} \qquad\quad x = 4$

**4.** $\left(4x + \frac{1}{3}\right)^2 = 0$

$\quad 4x + \frac{1}{3} = 0$

$\qquad 4x = -\frac{1}{3}$

$\qquad x = -\frac{1}{12}$

**5.** $(2x + 8)^2 = 0$

$\quad 2x + 8 = 0$

$\qquad 2x = -8$

$\qquad x = -4$

**6.** $(x - 4)(x + 7)(x + 1) = 0$

$\quad x - 4 = 0 \quad x + 7 = 0 \quad x + 1 = 0$

$\qquad x = 4 \qquad\quad x = -7 \qquad\quad x = -1$

# Chapter 10 *continued*

**7.** x-intercepts: $-2, \frac{1}{2}$

Vertex: $\left(-\frac{3}{4}, -\frac{25}{16}\right)$

$x = \frac{-2 + \frac{1}{2}}{2} = -\frac{1\frac{1}{2}}{2} = -\frac{3}{4}$

$y = \left(x - \frac{1}{2}\right)(x + 2)$

$= \left(-\frac{3}{4} - \frac{1}{2}\right)\left(-\frac{3}{4} + 2\right)$

$= -\frac{5}{4}\left(\frac{5}{4}\right) = -\frac{25}{16}$

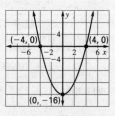

**8.** x-intercepts: $-4, 4$

Vertex: $(0, -16)$

$x = \frac{-4 + 4}{2} = \frac{0}{2} = 0$

$y = (x + 4)(x - 4)$

$= (0 + 4)(0 - 4)$

$= 4(-4) = -16$

**9.** x-intercepts: $\frac{1}{3}, -2$

Vertex: $\left(-\frac{5}{6}, -\frac{49}{12}\right)$

$x = \frac{\frac{1}{3} + (-2)}{2} = \frac{-\frac{5}{3}}{2} = -\frac{5}{6}$

$y = (3x - 1)(x + 2)$

$= \left(3\left(-\frac{5}{6}\right) - 1\right)\left(-\frac{5}{6} + 2\right)$

$= \left(-\frac{7}{2}\right)\left(\frac{7}{6}\right) = -\frac{49}{12}$

**10.** $y^2 + 3y - 4 = (y - 1)(y + 4)$

**11.** $w^2 + 13w + 22 = (w + 2)(w + 11)$

**12.** $n^2 + 16n - 57 = (n + 19)(n - 3)$

**13.** $x^2 + 17x + 66 = (x + 6)(x + 11)$

**14.** $t^2 - 41t - 86 = (t - 43)(t + 2)$

**15.** $-45 + 14z - z^2 = (9 - z)(-5 + z)$

**16.** $12b^2 - 17b - 99 = (4b + 9)(3b - 11)$

**17.** $2t^2 + 63t + 145 = (t + 29)(2t + 5)$

**18.** $18d^2 - 54d + 28$
$= 2(9d^2 - 27d + 14)$
$= 2(3d - 2)(3d - 7)$

**19.** $y^2 + 5y - 6 = 0$
$(y + 6)(y - 1) = 0$
$y = -6 \quad y = 1$

**20.** $n^2 + 26n + 25 = 0$
$(n + 1)(n + 25) = 0$
$n = -1 \quad n = -25$

**21.** $t^2 - 11t = -18$
$t^2 - 11t + 18 = 0$
$(t - 2)(t - 9) = 0$
$t = 2 \quad t = 9$

**22.** $w^2 - 29w = 170$
$w^2 - 29w - 170 = 0$
$(w - 34)(w + 5) = 0$
$w = 34 \quad w = -5$

**23.** $x^2 + 35x + 3 = 77$
$x^2 + 35x - 74 = 0$
$(x + 37)(x - 2) = 0$
$x = -37 \quad x = 2$

**24.** $2a^2 + 33a + 136 = 0$
$(2a + 17)(a + 8) = 0$
$2a + 17 = 0 \qquad a = -8$
$2a = -17$
$a = -\frac{17}{2}$

**25.** $15n^2 + 41n = -14$
$15n^2 + 41n + 14 = 0$
$(3n + 7)(5n + 2) = 0$
$3n + 7 = 0 \quad 5n + 2 = 0$
$3n = -7 \qquad 5n = -2$
$n = -\frac{7}{3} \qquad n = -\frac{2}{5}$

**26.** $18b^2 - 89b = -36$
$18b^2 - 89b + 36 = 0$
$(9b - 4)(2b - 9) = 0$
$9b - 4 = 0 \quad 2b - 9 = 0$
$9b = 4 \qquad 2b = 9$
$b = \frac{4}{9} \qquad b = \frac{9}{2}$

**27.** $6x^2 + x - 96 = 80$
$6x^2 + x - 176 = 0$
$(3x - 16)(2x + 11) = 0$
$3x - 16 = 0 \quad 2x + 11 = 0$
$3x = 16 \qquad 2x = -11$
$x = \frac{16}{3} \qquad x = -\frac{11}{2}$

### Activity 10.7 (p. 618)

**Developing Concepts**

**3.** $x + 3; x + 3; x + 3; x + 3$

**Drawing Conclusions**

**1.** $(x + 2)^2$

**2.** $(x + 4)^2$

**3.** $(2x + 1)^2$

**4.** $(2x + 3)^2$

**5.** $a^2 + 2ab + b^2 = (a + b)(a + b) = (a + b)^2$

**6.** $(3x + 1)^2$

**7.** $(4x + 7)^2$

**8.** $(7x + 2)^2$

**9.** $a^2 - 2ab + b^2 = (a - b)(a - b) = (a - b)^2$

# Chapter 10 *continued*

## Lesson 10.7

### 10.7 Guided Practice (p. 622)

**1.** The difference of the squares of two terms is equal to the product of the sum and the difference of the two terms.

$a^2 - b^2 = (a + b)(a - b)$

**2.** The wrong special products pattern is used.

$9x^2 - 6x + 1 = 0$

$(3x - 1)^2 = 0$

$3x - 1 = 0$

$3x = 1$

$x = \frac{1}{3}$

**3.** Each of the special products is the basis for one of the rules for factoring.

**4.** *Sample answers:*

1) $a^2 - b^2 = (a + b)(a - b)$

$9x^2 - 16 = (3x + 4)(3x - 4)$

2) $a^2 + 2ab + b^2 = (a + b)^2$

$x^2 + 8x + 16 = (x + 4)^2$

3) $a^2 - 2ab + b^2 = (a - b)^2$

$x^2 - 12x + 36 = (x - 6)^2$

**5.** $3x^2 - 6x + 9 = 3(x^2 - 2x + 3)$

**6.** $x^2 - 9 = (x + 3)(x - 3)$

**7.** $t^2 + 10t + 25 = (t + 5)^2$

**8.** $w^2 - 16w + 64 = (w - 8)^2$

**9.** $16 - t^2 = (4 - t)(4 + t)$

**10.** $6y^2 - 24 = 6(y^2 - 4) = 6(y + 2)(y - 2)$

**11.** $18 - 2z^2 = 2(9 - z^2) = 2(3 - z)(3 + z)$

**12.** $x^2 + 6x + 9 = 0$

$(x + 3)^2 = 0$

$x + 3 = 0$

$x = -3$

**13.** $144 - y^2 = 0$

$(12 - y)(12 + y) = 0$

$12 - y = 0 \qquad 12 + y = 0$

$-y = -12 \qquad y = -12$

$y = 12$

**14.** $s^2 - 14s + 49 = 0$

$(s - 7)^2 = 0$

$s - 7 = 0$

$s = 7$

**15.** $-25 + x^2 = 0$

$x^2 - 25 = 0$

$(x - 5)(x + 5) = 0$

$x = 5, x = -5$

**16.** $7x^2 + 28x + 28 = 0$

$7(x^2 + 4x + 4) = 0$

$7(x + 2)(x + 2) = 0$

$x + 2 = 0$

$x = -2$

**17.** $-4y^2 + 24y - 36 = 0$

$-4(y^2 - 6y + 9) = 0$

$-4(y - 3)(y - 3) = 0$

$y - 3 = 0$

$y = 3$

### 10.7 Practice and Applications (pp. 622–624)

**18.** $n^2 - 16 = (n + 4)(n - 4)$

**19.** $100x^2 - 121 = (10x - 11)(10x + 11)$

**20.** $6m^2 - 150 = 6(m^2 - 25) = 6(m + 5)(m - 5)$

**21.** $60y^2 - 540 = 60(y^2 - 9) = 60(y - 3)(y + 3)$

**22.** $16 - 81r^2 = (4 - 9r)(4 + 9r)$

**23.** $98 - 2t^2 = 2(49 - t^2) = 2(7 - t)(7 + t)$

**24.** $w^2 - y^2 = (w - y)(w + y)$

**25.** $9t^2 - 4q^2 = (3t - 2q)(3t + 2q)$

**26.** $-28y^2 + 7t^2 = -7(4y^2 - t^2) = -7(2y + t)(2y - t)$

**27.** $x^2 + 8x + 16 = (x + 4)^2$

**28.** $x^2 - 20x + 100 = (x - 10)^2$

**29.** $y^2 + 30y + 225 = (y + 15)^2$

**30.** $b^2 - 14b + 49 = (b - 7)^2$

**31.** $9x^2 + 6x + 1 = (3x + 1)^2$

**32.** $4r^2 + 12r + 9 = (2r + 3)^2$

**33.** $25n^2 - 20n + 4 = (5n - 2)^2$

**34.** $36m^2 - 84m + 49 = (6m - 7)^2$

**35.** $18x^2 + 12x + 2 = 2(9x^2 + 6x + 1) = 2(3x + 1)^2$

**36.** $48y^2 - 72xy + 27x^2 =$

$3(16y^2 - 24xy + 9x^2) =$

$3(4y - 3x)^2$

# Chapter 10 *continued*

**37.** $-16w^2 - 80w - 100 = -4(4w^2 + 20w + 25)$
$$= -4(2w + 5)^2$$

**38.** $3k^2 + 42k - 147 = -3(k^2 - 14k + 49) = -3(k - 7)^2$

**39.** $z^2 - 25 = (z + 5)(z - 5)$

difference of two squares

**40.** $y^2 + 12y + 36 = (y + 6)^2$

perfect square trinomial

**41.** $4n^2 - 36 = 4(n^2 - 9) = 4(n - 3)(n + 3)$

difference of two squares

**42.** $32 - 18x^2 = 2(16 - 9x^2) = 2(4 - 3x)(4 + 3x)$

difference of two squares

**43.** $4b^2 - 40b + 100 = 4(b^2 - 10b + 25) = 4(b - 5)^2$

perfect square trinomial

**44.** $-27t^2 - 18t - 3 = -3(9t^2 + 6t + 1) = -3(3t + 1)^2$

perfect square trinomial

**45.** $-2x^2 + 52x - 338 = -2(x^2 - 26x + 169)$
$$= -2(x - 13)^2$$

perfect square trinomial

**46.** $169 - x^2 = (13 - x)(13 + x)$

difference of two squares

**47.** $x^2 - 10{,}000w^2 = (x + 100w)(x - 100w)$

difference of two squares

**48.** $-108 + 147x^2 = -3(36 - 49x^2)$
$$= -3(6 - 7x)(6 + 7x)$$

difference of two squares

**49.** $x^2 + \frac{2}{3}x + \frac{1}{9} = \left(x + \frac{1}{3}\right)^2$

perfect square trinomial

**50.** $\frac{3}{4} - 12x^2 = \frac{3}{4}(1 - 16x^2) = \frac{3}{4}(1 - 4x)(1 + 4x)$

difference of two squares

**51.** $2x^2 - 72 = 0$

$2(x^2 - 36) = 0$

$2(x - 6)(x + 6) = 0$

$x = 6, x = -6$

**52.** $3x^2 - 24x + 48 = 0$

$3(x^2 - 8x + 16) = 0$

$3(x - 4)^2 = 0$

$x - 4 = 0$

$x = 4$

**53.** $25x^2 - 4 = 0$

$(5x + 2)(5x - 2) = 0$

$5x + 2 = 0 \qquad 5x - 2 = 0$

$5x = -2 \qquad 5x = 2$

$x = -\frac{2}{5} \qquad x = \frac{2}{5}$

**54.** $\frac{1}{5}x^2 - 2x + 5 = 0$

$x^2 - 10x + 25 = 0$

$(x - 5)^2 = 0$

$x = 5$

**55.** $27 - 12x^2 = 0$

$3(9 - 4x^2) = 0$

$3(3 - 2x)(3 + 2x) = 0$

$3 - 2x = 0 \qquad 3 + 2x = 0$

$-2x = -3 \qquad 2x = -3$

$x = \frac{3}{2} \qquad x = -\frac{3}{2}$

**56.** $50x^2 + 60x + 18 = 0$

$2(25x^2 + 30x + 9) = 0$

$2(5x + 3)^2 = 0$

$5x + 3 = 0$

$5x = -3$

$x = -\frac{3}{5}$

**57.** $\frac{1}{3}x^2 - 6x + 27 = 0$

$x^2 - 18x + 81 = 0$

$(x - 9)(x - 9) = 0$

$x - 9 = 0$

$x = 9$

**58.** $90x^2 - 120x + 40 = 0$

$10(9x^2 - 12x + 4) = 0$

$10(3x - 2)(3x - 2) = 0$

$3x - 2 = 0$

$3x = 2$

$x = \frac{2}{3}$

**59.** $x^2 - \frac{5}{3}x + \frac{25}{36} = 0$

$36x^2 - 60x + 25 = 0$

$(6x - 5)(6x - 5) = 0$

$6x - 5 = 0$

$6x = 5$

$x = \frac{5}{6}$

**60.** $112x^2 - 252 = 0$

$28(4x^2 - 9) = 0$

$28(2x - 3)(2x + 3) = 0$

$2x - 3 = 0 \qquad 2x + 3 = 0$

$2x = 3 \qquad 2x = -3$

$x = \frac{3}{2} \qquad x = -\frac{3}{2}$

**61.** $-16x^2 + 56x - 49 = 0$

$-(16x^2 - 56x + 49) = 0$

$-(4x - 7)(4x - 7) = 0$

$4x - 7 = 0$

$4x = 7$

$x = \frac{7}{4}$

**62.** $-\frac{4}{5}x^2 - \frac{4}{5}x - \frac{1}{5} = 0$

$-4x^2 - 4x - 1 = 0$

$-(4x^2 + 4x + 1) = 0$

$-(2x + 1)(2x + 1) = 0$

$2x + 1 = 0$

$2x = -1$

$x = -\frac{1}{2}$

# Chapter 10 *continued*

**63.** $4 \cdot D^2 = S$
$4 \cdot D^2 = 9$
$D^2 = \frac{9}{4}$
$D = \frac{3}{2} = 1.5$ in.

**64.** $2 \cdot D^2 = S$
$2 \cdot D^2 = 9$
$D^2 = \frac{9}{2}$
$D = \frac{3}{\sqrt{2}}$
$D = \frac{3\sqrt{2}}{2} \approx 2.12$ in.

**65.** $h = 4t^2$
$1 = 4t^2$
$\frac{1}{4} = t^2$
$\frac{1}{2} = t$
$t = \frac{1}{2}$ sec

**66.** $n = 4t^2$
$4 = 4t^2$
$1 = t^2$
$1 = t$
$t = 1$ sec

**67.** $-16t^2 + vt + s = h$
$-16t^2 + 224t + 784 = h$
$-(16t^2 - 224t + 784) = h$
$(4t - 28)(4t - 28) = h$

$4t - 28 = h$
$4t = 28$
$t = 7$
yes; 7 sec

**68.** $h = \frac{v^2}{2g}$
$9 = \frac{v^2}{2g}$
$9 = \frac{v^2}{2(32)}$
$9 = \frac{v^2}{64}$
$576 = v^2$
$v = 24$ ft/sec

**69.** $16 = \frac{v^2}{2(32)}$
$16 = \frac{v^2}{64}$
$1024 = v^2$
$v = 32$ ft/sec

**70.** D   **71.** C

**72.** D   $-4x^2 + 24x - 36 = 0$
$-4(x^2 - 6x + 9) = 0$
$-4(x - 3)(x - 3) = 0$
$x - 3 = 0$
$x = 3$

**73.** If a wire rope and a natural fiber rope have the same radius, the ratio of the safe load of the wire rope to the safe load of the natual fiber rope is

$$\frac{4}{150\pi^2} \approx \frac{1}{370} \approx 0.003.$$

That is, the natural fiber rope is about 370 times stronger.

### 10.7 Mixed Review (p. 624)

**74.** 9 and 12 = 3
9: 1, 3, 9
12: 1, 2, 3, 4, 6, 12

**75.** 15 and 45 = 5
15: 1, 3, 5, 15
45: 1, 3, 5, 9, 15, 45

**76.** 55 and 132 = 11
55: 1, 5, 11, 55
132: 1, 2, 3, 4, 6, 11, 12, 22, 33, 43, 66, 132

**77.** 14 and 18 = 2
14: 1, 2, 7, 14
18: 1, 2, 3, 6, 9, 18

**78.** $7 + 9(-2) \stackrel{?}{=} -11$
$7 - 18 \stackrel{?}{=} -11$
$-11 = -11$
solution

$-4(7) + (-2) \stackrel{?}{=} -30$
$-30 = -30$

**79.** $2(-5) + 6(-2) \stackrel{?}{=} 22$
$-10 - 12 \stackrel{?}{=} 22$
$-22 \neq 22$
not a solution

**80.** $3(-10) + 5(3) \stackrel{?}{=} 15$
$-30 + 15 \stackrel{?}{=} 15$
$-15 \neq 15$
not a solution

**81.** $9(4) - 2(-6) \stackrel{?}{=} 48$
$36 + 12 \stackrel{?}{=} 48$
$48 = 48$
solution

$-5(4) - 8(-6) \stackrel{?}{=} 28$
$-20 + 48 \stackrel{?}{=} 28$
$28 = 28$

**82.** $x - y = 2$
$2x + y = 1$

$1 - y = 2$
$y = -1$
$(1, -1)$

$y = x - 2$
$2x + (x - 2) = 1$
$3x - 2 = 1$
$3x = 3$
$x = 1$

**83.** $x - 2y = 10$
$3x - y = 0$

$3(-2) - y = 0$
$y = -6$
$(-2, -6)$

$y = 3x$
$x - 2(3x) = 10$
$-5x = 10$
$x = -2$

**84.** $-x + y = 0$
$2x + y = 0$

$0 + y = 0$
$y = 0$
$(0, 0)$

$y = x$
$2x + x = 0$
$3x = 0$
$x = 0$

**85.** $2x + 3y = -5$
$x - 2y = -6$

$-4 - 2y = -6$
$-2y = -2$
$y = 1$
$(-4, 1)$

$y = \frac{1}{2}x + 3$
$2x + 3\left(\frac{1}{2}x + 3\right) = -5$
$2x + \frac{3}{2}x + 9 = -5$
$\frac{7}{2}x = -14$
$x = -4$

**86.** $\sqrt{216} = \sqrt{36}\sqrt{6} = 6\sqrt{6}$

**87.** $\sqrt{5} \cdot \sqrt{15} = \sqrt{75} = \sqrt{25}\sqrt{3} = 5\sqrt{3}$

**88.** $\sqrt{10} \cdot \sqrt{20} = \sqrt{200} = \sqrt{100}\sqrt{2} = 10\sqrt{2}$

**89.** $\sqrt{4} \cdot 3\sqrt{9} = 3\sqrt{36} = 18$

**90.** $\sqrt{\dfrac{28}{49}} = \dfrac{\sqrt{28}}{7} = \dfrac{\sqrt{4}\sqrt{7}}{7} = \dfrac{2\sqrt{7}}{7}$

**91.** $\dfrac{10\sqrt{8}}{\sqrt{25}} = \dfrac{10\sqrt{4}\sqrt{2}}{5} = \dfrac{20\sqrt{2}}{5} = 4\sqrt{2}$

**92.** $\dfrac{12\sqrt{4}}{\sqrt{9}} = \dfrac{24}{3} = 8$

**93.** $-\dfrac{6\sqrt{12}}{\sqrt{4}} = -\dfrac{6\sqrt{4}\sqrt{3}}{2} = -\dfrac{12\sqrt{3}}{2} = -6\sqrt{3}$

**94.** $9x^2 - 14x - 7 = 0$

$$x = \dfrac{14 \pm \sqrt{(-14)^2 - 4(9)(-7)}}{2(9)}$$

$$= \dfrac{14 \pm \sqrt{448}}{18} = \dfrac{14 \pm \sqrt{64}\sqrt{7}}{18}$$

$$= \dfrac{2(7 \pm 4\sqrt{7})}{18}$$

$$= \dfrac{7 + 4\sqrt{7}}{9}, \dfrac{7 - 4\sqrt{7}}{9}$$

**95.** $9d^2 - 58d + 24 = 0$

$$d = \dfrac{58 \pm \sqrt{58^2 - 4(9)(24)}}{2(9)}$$

$$= \dfrac{58 \pm \sqrt{2500}}{18}$$

$$= \dfrac{108}{18}, \dfrac{8}{18}$$

$$= 6, \dfrac{4}{9}$$

**96.** $7y^2 - 9y - 17 = 0$

$$y = \dfrac{9 \pm \sqrt{9^2 - 4(7)(-17)}}{2(7)}$$

$$= \dfrac{9 \pm \sqrt{557}}{14}$$

$$= \dfrac{9 + \sqrt{557}}{14}, \dfrac{9 - \sqrt{557}}{14}$$

## Lesson 10.8

### 10.8 Guided Practice (p. 629)

**1.** It cannot be factored using integer coefficients.

**2.** monomial factors and prime factors with at least two terms

**3.** Graphing; the quadratic formula, factoring; *Sample answer:* Graphing may be very convenient, but may give only approximate solutions. The quadratic formula may be used for any equation, but may involve more calculation than is necessary if the polynomial expression is factorable. Factoring can be used only for equations involving factorable expressions.

**4.** $x^2 + 9$ does not factor to $(x + 3)(x - 3)$

$4x^3 + 36x =$

$4x(x^2 + 9)$

not factorable

**5.** $-2b^3 + 12b^2 - 14b = -2b(b^2 + 6b + 7)$

not $-2b(b^2 + 6b - 7)$

**6.** $5n^3 - 20n = 5n(n^2 - 4)$    **7.** $6x^2 + 3x^4 = 3x^2(2 + x^2)$

**8.** $6y^4 + 14y^3 - 10y^2 = 2y^2(3y^2 + 7y - 5)$

**9.** no; $x(7x^2 - 11)$    **10.** yes

**11.** no; $3w(3w + 4)(3w - 4)$

**12.**    $y^2 - 4y - 5 = 0$

$(y - 5)(y + 1) = 0$

$y - 5 = 0 \quad y + 1 = 0$

$y = 5 \qquad y = -1$

factoring

**13.**    $z^2 + 11z + 30 = 0$

$(z + 6)(z + 5) = 0$

$z + 6 = 0 \quad z + 5 = 0$

$z = -6 \qquad z = -5$

factoring

**14.**    $5a^2 + 11a + 2 = 0$

$(5a + 1)(a + 2) = 0$

$5a + 1 = 0 \quad a + 2 = 0$

$5a = -1 \qquad a = -2$

$a = -\frac{1}{5}$

factoring

### 10.8 Practice and Applications (pp. 629–631)

**15.** $6v^3 - 18v = 6v(v^2 - 3)$    **16.** $4q^4 + 12q = 4q(q^3 + 3)$

**17.** $3x - 9x^2 = 3x(1 - 3x)$

**18.** $24t^5 + 6t^3 = 6t^3(4t^2 + 1)$

**19.** $4a^5 + 8a^3 - 2a^2 = 2a^2(2a^3 + 4a - 1)$

**20.** $18d^6 - 6d^2 + 3d = 3d(6d^5 - 2d + 1)$

**21.** $24x^3 + 18x^2 = 6x^2(4x + 3)$

**22.** $-3w^4 + 21w^3 = -3w^3(w - 7)$

# Chapter 10 *continued*

**23.** $2y^3 - 10y^2 - 12y = 2y(y^2 - 5y - 6)$
$\qquad\qquad\qquad\quad = 2y(y - 6)(y + 1)$

**24.** $5s^3 + 30s^2 + 40s = 5s(s^2 + 6s + 8) = 5s(s + 4)(s + 2)$

**25.** $-7m^3 + 28m^2 - 21m = -7m(m^2 - 4m + 3)$
$\qquad\qquad\qquad\qquad\quad = -7m(m - 3)(m - 1)$

**26.** $2d^4 + 2d^3 - 60d^2 = 2d^2(d^2 + d - 30)$
$\qquad\qquad\qquad\qquad = 2d^2(d + 6)(d - 5)$

**27.** $4t^3 - 144t = 4t(t^2 - 36) = 4t(t - 6)(t + 6)$

**28.** $-12z^4 + 3z^2 = -3z^2(4z^2 - 1) = -3z^2(2z + 1)(2z - 1)$

**29.** $c^4 + c^3 - 12c - 12 = (c^4 + c^3) - (12c + 12)$
$\qquad\qquad\qquad\qquad\quad = c^3(c + 1) - 12(c + 1)$
$\qquad\qquad\qquad\qquad\quad = (c + 1)(c^3 - 12)$

**30.** $x^3 - 3x^2 + x - 3 = (x^3 - 3x^2) + (x - 3)$
$\qquad\qquad\qquad\qquad = x^2(x - 3) + (x - 3)$
$\qquad\qquad\qquad\qquad = (x - 3)(x^2 + 1)$

**31.** $6b^4 + 5b^3 - 24b - 20 = (6b^4 + 5b^3) - (24b + 20)$
$\qquad\qquad\qquad\qquad\qquad = b^3(6b + 5) - 4(6b + 5)$
$\qquad\qquad\qquad\qquad\qquad = (6b + 5)(b^3 - 4)$

**32.** $3y^3 - y^2 - 21y + 7 = (3y^3 - y^2) - (21y - 7)$
$\qquad\qquad\qquad\qquad\quad = y^2(3y - 1) - 7(3y - 1)$
$\qquad\qquad\qquad\qquad\quad = (3y - 1)(y^2 - 7)$

**33.** $a^3 + 6a^2 - 4a - 24 = (a^3 + 6a^2) - (4a + 24)$
$\qquad\qquad\qquad\qquad\quad = a^2(a + 6) - 4(a + 6)$
$\qquad\qquad\qquad\qquad\quad = (a + 6)(a^2 - 4)$
$\qquad\qquad\qquad\qquad\quad = (a + 6)(a - 2)(a + 2)$

**34.** $t^3 - t^2 - 16t + 16 = (t^3 - t^2) - (16t - 16)$
$\qquad\qquad\qquad\qquad = t^2(t - 1) - 16(t - 1)$
$\qquad\qquad\qquad\qquad = (t - 1)(t^2 - 16)$
$\qquad\qquad\qquad\qquad = (t - 1)(t + 4)(t - 4)$

**35.** $3m^3 - 15m^2 - 6m + 30 = (3m^3 - 15m^2) - (6m - 30)$
$\qquad\qquad\qquad\qquad\qquad = 3m^2(m - 5) - 6(m - 5)$
$\qquad\qquad\qquad\qquad\qquad = (m - 5)(3m^2 - 6)$
$\qquad\qquad\qquad\qquad\qquad = 3(m - 5)(m^2 - 2)$

**36.** $7n^5 + 7n^4 - 3n^2 - 6n - 3$
$\qquad = (7n^5 + 7n^4) - (3n^2 + 6n + 3)$
$\qquad = 7n^4(n + 1) - 3(n^2 + 2n + 1)$
$\qquad = 7n^4(n + 1) - 3(n + 1)(n + 1)$
$\qquad = (n + 1)(7n^4 - 3(n + 1))$
$\qquad = (n + 1)(7n^4 - 3n - 3)$

---

**37–50.** Sample solution methods are given.

**37.** $y^2 + 7y + 12 = 0$
$(y + 4)(y + 3) = 0$
$y = -4 \quad y = -3$
factoring

**38.** $x^2 - 3x - 4 = 0$
$(x - 4)(x + 1) = 0$
$x = 4 \quad x = -1$
factoring

**39.** $b^2 + 4b - 117 = 0$
$(b + 13)(b - 9) = 0$
$b = -13 \quad b = 9$
factoring

**40.** $t^2 - 16t + 65 = 0$
no solution

**41.** $27 + 6w - w^2 = 0$
$(9 - w)(3 + w) = 0$
$w = 9 \quad w = -3$
factoring

**42.** $x^2 - 21x + 84 = 0$
$x = \dfrac{21 \pm \sqrt{21^2 - 4(1)(84)}}{2}$
$x = \dfrac{21 + \sqrt{105}}{2}, \dfrac{21 - \sqrt{105}}{2}$
quadratic formula

**43.** $5x^4 - 80x^2 = 0$
$5x^2(x^2 - 16) = 0$
$5x^2(x - 4)(x + 4) = 0$
$x = 0 \quad x = 4 \quad x = -4$
factoring

**44.** $-16x^3 + 4x = 0$
$-4x(4x^2 - 1) = 0$
$-4x(2x - 1)(2x + 1) = 0$
$x = 0 \quad x = \frac{1}{2} \quad x = -\frac{1}{2}$
factoring

**45.** $10x^3 - 290x^2 + 620x = 0$
$10x(x^2 - 29x + 62) = 0$
$x = \dfrac{29 \pm \sqrt{29^2 - 4(1)(62)}}{2} \qquad 10x = 0$
$\qquad\qquad\qquad\qquad\qquad\qquad x = 0$
$= \dfrac{29 \pm \sqrt{593}}{2}$
$= \dfrac{29 + \sqrt{593}}{2}, \dfrac{29 - \sqrt{593}}{2}, 0$
factoring and quadratic formula

**46.** $34x^4 - 85x^3 + 51x^2 = 0$
$x^2(34x^2 - 85x + 51) = 0$
$x^2(17x - 17)(2x - 3) = 0$
$x = 0 \quad x = 1 \quad x = \frac{3}{2}$
factoring

**47.** $8x^2 + 9x - 7 = 0$
$x = \dfrac{-9 \pm \sqrt{9^2 - (4)(8)(-7)}}{2(8)}$
$= \dfrac{-9 + \sqrt{305}}{16}, \dfrac{-9 - \sqrt{305}}{16}$
quadratic formula

**48.** $18x^2 - 21x + 28 = 0$
no solution

# Chapter 10 *continued*

**49.** $24x^3 + 18x^2 - 168x = 0$

$2x(12x^2 + 9x - 84) = 0 \qquad 2x = 0$

$x = \dfrac{-9 \pm \sqrt{81 - 4(12)(84)}}{24} \qquad x = 0$

$= \dfrac{-9 \pm \sqrt{4113}}{24} = \dfrac{-9 \pm 3\sqrt{457}}{24}$

$= \dfrac{-3 + \sqrt{457}}{8}, \dfrac{-3 - \sqrt{457}}{8}, 0$

factoring and quadratic formula

**50.** $-14x^4 + 118x^3 + 72x^2 = 0$

$-2x^2(7x^2 - 59x - 36) = 0$

$-2x^2(7x + 4)(x - 9) = 0$

$-2x^2 = 0 \quad 7x + 4 = 0 \quad x - 9 = 0$

$x = 0 \qquad 7x = -4 \qquad x = 9$

$x = -\tfrac{4}{7}$

factoring

**51.** $96 = 16t^2 - 16t$

$0 = 16t^2 - 16t - 96$

$0 = (4t + 8)(4t - 12)$

$4t + 8 = 0 \quad 4t - 12 = 0$

$4t = -8 \qquad 4t = 12$

$t = -2 \qquad t = 3 \text{ sec}$

**52.** $96 = \dfrac{16}{6}t^2 - 16t$

$0 = \dfrac{8}{3}t^2 - 16t - 96$

$0 = 8t^2 - 48t - 288$

$0 = 8(t^2 - 6t - 36)$

$t = \dfrac{6 \pm \sqrt{6^2 - 4(1)(-36)}}{2}$

$= \dfrac{6 + \sqrt{180}}{2}, \dfrac{6 - \sqrt{180}}{2}$

$= \dfrac{6 + 6\sqrt{5}}{2}, \dfrac{6 - 6\sqrt{5}}{2} = 3 + 3\sqrt{5}, 3 - 3\sqrt{5}$

$(3 + 3\sqrt{5})$ sec (about 9.7 sec)

**53.** Earth; the coefficient of the $t^2$-term in the vertical motion model for Earth is 6 times greater than that of the moon.

**54.** $h = 2.4(16t^2) - vt$ or $38.4t^2 - vt$

**55.** $l = h - 3, w = h - 9$

**56.** $h(h - 3)(h - 9) = 324$ or $h^3 - 12h^2 + 27h - 324 = 0$

**57.** $h(h - 3)(h - 9) = 324$

$h^3 - 12h^2 + 27h - 324 = 0$

$(h^3 - 12h^2) + (27h - 324) = 0$

$h^2(h - 12) + 27(h - 12) = 0$

$(h - 12)(h^2 + 27) = 0$

$h - 12 = 0 \quad h^2 + 27 = 0$

$h = 12 \qquad h^2 = -27$

$l = h - 3 = 12 - 3 = 9$ in.

$w = h - 9 = 12 - 9 = 3$ in.

$h = 12$ in.

**58.** $8x^2 - 2x - 3$

$8(-3) = -24$

$-24 = -6(4)$

$8x^2 - 2x - 3 = 8x^2 - 6x + 4x - 3$

$= 2x(4x - 3) + (4x - 3)$

$= (2x + 1)(4x - 3)$

**59.** $3x^2 + 13x + 14$

$3(14) = 42$

$42 = 6(7)$

$3x^2 + 13x + 14 = 3x^2 + 6x + 7x + 14$

$= 3x(x + 2) + 7(x + 2)$

$= (3x + 7)(x + 2)$

**60.** $5x^2 + 27x - 18$

$5(-18) = -90$

$-90 = 30(-3)$

$5x^2 + 27x - 18 = 5x^2 + 30x - 3x - 18$

$= 5x(x + 6) - 3(x + 6)$

$= (5x - 3)(x + 6)$

**61.** C  **62.** D  **63.** A

**64.** $y^3 - 125 = (y - 5)(y^2 + 5y + 25)$

**65.** $b^3 + 27 = (b + 3)(b^2 - 3b + 9)$

**66.** $\tfrac{1}{8} + 8x^3 = \left(\tfrac{1}{2} + 2x\right)\left(\tfrac{1}{4} - x + 4x^2\right)$

**67.** $216 - 343t^3 = (6 - 7t)(36 + 42t + 49t^2)$

# Chapter 10 *continued*

**68.** Ex. 64: $(y - 5)(y^2 + 5y + 25) = 0$

$$y = 5$$

Ex. 65: $(b + 3)(b^2 - 3b + 9) = 0$

$$b + 3 = 0$$

$$b = -3$$

Ex. 66: $(\frac{1}{2} + 2x)(\frac{1}{4} - x + 4x^2) = 0$

$$(\frac{1}{2} + 2x)(4x^2 - x + \frac{1}{4}) = 0$$

$$\frac{1}{2} + 2x = 0$$

$$2x = -\frac{1}{2}$$

$$x = -\frac{1}{4}$$

Ex. 67: $(6 - 7t)(36 + 42t + 49t^2) = 0$

$$6 - 7t = 0$$

$$-7t = -6$$

$$t = \frac{6}{7}$$

### 10.8 Mixed Review (p. 632)

**69.** $7 + x \leq -9$

$$x \leq -16$$

**70.** $-3 > 2x - 5$

$$2 > 2x$$

$$1 > x$$

**71.** $-x + 6 \leq 13$

$$-x \leq 7$$

$$x \geq -7$$

**72.** $|x| = 3$

$$3, -3$$

**73.** $|x - 5| = 7$

$$x - 5 = 7 \quad x - 5 = -7$$

$$x = 12 \quad\quad x = -2$$

**74.** $|x + 6| = 13$

$$x + 6 = 13 \quad x + 6 = -13$$

$$x = 7 \quad\quad x = -19$$

**75.** $|x + 8| - 2 = -12$

no solution

**76.** $|2x - 5| + 7 = 16$

$$|2x - 5| = 9$$

$$2x - 5 = 9 \quad 2x - 5 = -9$$

$$2x = 14 \quad 2x = -4$$

$$x = 7 \quad\quad x = -2$$

**77.** $\left|x + \frac{3}{4}\right| = \frac{9}{4}$

$$x + \frac{3}{4} = \frac{9}{4} \quad x + \frac{3}{4} = -\frac{9}{4}$$

$$x = \frac{6}{4} \quad\quad x = -\frac{12}{4}$$

$$x = \frac{3}{2} \quad\quad x = -3$$

**78.** $x + y < 9$

$$y < 9 - x$$

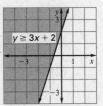

**79.** $y - 3x \geq 2$

$$y \geq 2 - 3x$$

**80.** $y - 4x \leq 10$

$$y \leq 10 + 4x$$

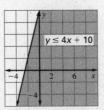

**81.** $\frac{93}{69} - \frac{31}{23} \approx 134.78\%$

**82.** Bedroom: $10 \times 12 = 120 \text{ ft}^2$

Living Room: $24 \times 15 = 360 \text{ ft}^2$

$$\frac{120 \text{ ft}^2}{360 \text{ ft}^2} = \frac{1}{3}$$

### Quiz 3 (p. 632)

**1.** $49x^2 - 64 = (7x + 8)(7x - 8)$

difference of two squares

**2.** $121 - 9x^2 = (11 + 3x)(11 - 3x)$

difference of two squares

**3.** $4t^2 + 20t + 25 = (2t + 5)(2t + 5)$

perfect square trinomial

**4.** $72 - 50y^2 = 2(6 + 5y)(6 - 5y)$

difference of two squares

**5.** $9y^2 + 42y + 49 = (3y + 7)(3y + 7) = (3y + 7)^2$

perfect square trinomial

**6.** $3n^2 - 36n + 108 = 3(n - 6)(n - 6) = 3(n - 6)^2$

perfect square trinomial

**7.** $3x^2 - 192 = 0$

$$3(x^2 - 64) = 0$$

$$3(x - 8)(x + 8) = 0$$

$$x = 8 \quad x = -8$$

**8.** $4x^2 + 32x + 64 = 0$

$$4(x^2 + 8x + 16) = 0$$

$$4(x + 4)(x + 4) = 0$$

$$x = -4$$

**9.** $9x^2 + 96x + 256 = 0$

$(3x + 16)(3x + 16) = 0$

$3x + 16 = 0$

$3x = -16$

$x = -\frac{16}{3}$

**10.** $\frac{1}{2}x - 4x + 8 = 0$

$x - 8x + 16 = 0$

$(x - 4)(x - 4) = 0$

$x = 4$

**11.** $216x^2 - 96 = 0$

$24(9x^2 - 4) = 0$

$24(3x + 2)(3x - 2) = 0$

$3x + 2 = 0 \qquad 3x - 2 = 0$

$3x = -2 \qquad\quad 3x = 2$

$x = -\frac{2}{3} \qquad\quad x = \frac{2}{3}$

**12.** $-3x^2 - 2x - \frac{1}{3} = 0$

$-9x^2 - 6x - 1 = 0$

$-(9x^2 + 6x + 1) = 0$

$-(3x + 1)(3x + 1) = 0$

$3x + 1 = 0$

$3x = -1$

$x = -\frac{1}{3}$

**13.** $3x^3 + 12x^2 = 3x^2(x + 4)$

**14.** $6x^2 + 3x = 3x(2x + 1)$

**15.** $18x^4 - 9x^3 = 9x^3(2x - 1)$

**16.** $8x^5 + 4x^2 - 2x = 2x(4x^4 + 2x - 1)$

**17.** $2x^3 - 6x^2 + 4x = 2x(x^2 - 3x + 2) = 2x(x - 2)(x - 1)$

**18.** $48x^3 - 75x = 3x(16x^2 - 25) = 3x(4x - 5)(4x + 5)$

**19.** $x^3 + 3x^2 + 4x + 12 = x^2(x + 3) + 4(x + 3)$

$= (x^2 + 4)(x + 3)$

**20.** $12x^4 - 27x^2 = 0$

$3x^2(4x^2 - 9) = 0$

$3x^2(2x - 3)(2x - 3) = 0$

$3x^2 = 0$

$x^2 = 0$

$x = 0 \qquad 2x - 3 = 0 \qquad 2x + 3 = 0$

$\qquad\qquad\qquad 2x = 3 \qquad\qquad 2x = -3$

$\qquad\qquad\qquad x = \frac{3}{2} \qquad\qquad x = -\frac{3}{2}$

factoring special products

**21.** $-3x^2 - 4x + 15 = 0$

$-(3x^2 + 4x - 15) = 0$

$-(3x - 5)(x + 3) = 0$

$3x - 5 = 0 \quad x + 3 = 0$

$3x = 5 \qquad x = -3$

$x = \frac{5}{3}$

factoring

**22.** $3x^3 - 6x^2 + 5x - 10 = 0$

$3x^2(x - 2) + 5(x - 2) = 0$

$(3x^2 + 5)(x - 2) = 0$

$3x^2 + 5 = 0 \qquad x - 2 = 0$

$3x^2 = -5 \qquad\quad x = 2$

$x^2 = -\frac{5}{3}$

no solution

factoring

**23.** $56x^3 + 98x = 170x^2$

$56x^3 - 170x^2 + 98x = 0$

$2x(28x^2 - 85x + 49) = 0$

$2x = 0$

$x = 0$

$x = \dfrac{85 \pm \sqrt{85^2 - 4(28)(49)}}{2(28)}$

$= \dfrac{85 + \sqrt{1737}}{56}, \dfrac{85 - \sqrt{1737}}{56}$

$\approx 2.26, 0.77, 0$

quadratic formula

**24.** $17x^6 = 17x^5 + 204x^4$

$17x^6 - 17x^5 - 204x^4 = 0$

$17x^4(x^2 - x - 12) = 0$

$17x^4(x - 4)(x + 3) = 0$

$17x^4 = 0 \quad x - 4 = 0 \quad x + 3 = 0$

$x = 0 \qquad\quad x = 4 \qquad\quad x = -3$

factoring

**25.** $6x^3 + 45x^2 + 15x = 0$

$3x(2x^2 + 15x + 5) = 0$

$3x = 0$

$x = 0$

$\dfrac{-15 \pm \sqrt{15^2 - 4(2)(5)}}{2(2)}$

$\dfrac{-15 + \sqrt{185}}{4}, \dfrac{-15 - \sqrt{185}}{4} \approx -0.35, -7.15, 0$

quadratic formula

## Chapter 10 Review *(pp. 634–636)*

**1.** $(-x^2 + x + 2) + (3x^2 + 4x + 5)$

$= -x^2 + x + 2 + 3x^2 + 4x + 5 = 2x^2 + 5x + 7$

**2.** $(x^2 + 3x - 1) - (4x^2 - 5x + 6)$

$= x^2 + 3x - 1 - 4x^2 + 5x - 6 = -3x^2 + 8x - 7$

**3.** $(x^3 + 5x^2 - 4x) - (3x^2 - 6x + 2)$

$= x^3 + 5x^2 - 4x - 3x^2 + 6x - 2$

$= x^3 + 2x^2 + 2x - 2$

**4.** $(4x^3 + x^2 - 1) + (2 - x - x^2)$

$\quad = 4x^3 + x^2 - 1 + 2 - x - x^2 = 4x^3 - x + 1$

**5.** $(-x)(8x^3 - 12x^2) = -8x^4 + 12x^3$

**6.** $4x^3(-x^2 + 2x - 7) = -4x^5 + 8x^4 - 28x^3$

**7.** $(x + 3)(x + 11) = x^2 + 11x + 3x + 33$

$\quad\quad\quad\quad\quad = x^2 + 14x + 33$

**8.** $(7x - 1)(5x + 2) = 35x^2 + 14x - 5x - 2$

$\quad\quad\quad\quad\quad\quad = 35x^2 + 9x - 2$

**9.** $(x + 5)(2x^2 + x - 10)$

$\quad\quad = 2x^3 + x^2 - 10x + 10x^2 + 5x - 50$

$\quad\quad = 2x^3 + 11x^2 - 5x - 50$

**10.** $(4x^2 + 9x)(-x^2 - 3)$

$\quad\quad = -4x^4 - 12x^2 - 9x^3 - 27x$

$\quad\quad = -4x^4 - 9x^3 - 12x^2 - 27x$

**11.** $(x + 15)(x - 15) = x^2 - 15x + 15x - 225 = x^2 - 225$

**12.** $(5x - 2)(5x + 2) = 25x^2 + 10x - 10x - 4 = 25x^2 - 4$

**13.** $(x + 2)^2 = (x + 2)(x + 2)$

$\quad\quad\quad = x^2 + 2x + 2x + 2$

$\quad\quad\quad = x^2 + 4x + 2$

**14.** $(7m - 6)^2 = (7m - 6)(7m - 6)$

$\quad\quad\quad = 49m^2 - 42m - 42m + 36$

$\quad\quad\quad = 49m^2 - 84m + 36$

**15.** $(x + 1)(x + 10) = 0$

$\quad x + 1 = 0 \quad\quad x + 10 = 0$

$\quad\quad x = -1 \quad\quad\quad x = -10$

**16.** $(x - 3)(x - 2) = 0$

$\quad x - 3 = 0 \quad x - 2 = 0$

$\quad\quad x = 3 \quad\quad x = 2$

**17.** $\quad\quad (x + 9)^2 = 0$

$\quad (x + 9)(x + 9) = 0$

$\quad x + 9 = 0$

$\quad\quad x = -9$

**18.** $(3x + 6)(4x - 1)(x - 4) = 0$

$\quad 3x + 6 = 0 \quad\quad 4x - 1 = 0 \quad x - 4 = 0$

$\quad\quad 3x = -6 \quad\quad\quad 4x = 1 \quad\quad x = 4$

$\quad\quad\quad x = -2 \quad\quad\quad x = \frac{1}{4}$

**19.** $x^2 - 21x + 108 = 0$

$\quad (x - 9)(x - 12) = 0$

$\quad x - 9 = 0 \quad x - 12 = 0$

$\quad\quad x = 9 \quad\quad x = 12$

**20.** $x^2 + 10x - 200 = 0$

$\quad (x + 20)(x - 10) = 0$

$\quad x + 20 = 0 \quad\quad x - 10 = 0$

$\quad\quad x = -20 \quad\quad\quad x = 10$

**21.** $\quad\quad\quad x^2 + 26x = -169$

$\quad x^2 + 26x + 169 = 0$

$\quad (x + 13)(x + 13) = 0$

$\quad x + 13 = 0$

$\quad\quad x = -13$

**22.** $12x^2 + 7x + 1$

$\quad (3x + 1)(4x + 1)$

**23.** $2x^2 + 5x - 12 = (2x - 3)(x + 4)$

**24.** $6x^2 + 4x - 10 = (3x + 5)(2x - 2)$

**25.** $4x^2 - 12x + 9 = (2x - 3)(2x - 3)$

**26.** $\quad\quad 100x^2 - 121 = 0$

$\quad (10x - 11)(10x + 11) = 0$

$\quad 10x - 11 = 0 \quad\quad 10x + 11 = 0$

$\quad\quad 10x = 11 \quad\quad\quad 10x = -11$

$\quad\quad x = \frac{11}{10} \quad\quad\quad\quad x = -\frac{11}{10}$

**27.** $16x^2 + 24x + 9 = 0$

$\quad (4x + 3)(4x + 3) = 0$

$\quad 4x + 3 = 0$

$\quad\quad 4x = -3$

$\quad\quad x = -\frac{3}{4}$

**28.** $9x^2 - 18x + 9 = 0$

$\quad 9(x^2 - 2x + 1) = 0$

$\quad 9(x - 1)(x - 1) = 0$

$\quad\quad x - 1 = 0$

$\quad\quad\quad x = 1$

**29.** $-2x^5 - 2x^4 + 4x^3 = -2x^3(x^2 + x - 2)$

$\quad\quad\quad\quad\quad\quad = -2x^3(x + 2)(x - 1)$

**30.** $2x^4 - 32x^2 = 2x^2(x^2 - 16)$

$\quad\quad\quad\quad\quad = 2x^2(x - 4)(x + 4)$

**31.** $3x^3 + x^2 + 15x + 5 = x^2(3x + 1) + 5(3x + 1)$

$\quad\quad\quad\quad\quad\quad\quad = (x^2 + 5)(3x + 1)$

**32.** $x^3 + 3x^2 - 4x - 12 = x^2(x + 3) - 4(x + 3)$

$\quad\quad\quad\quad\quad\quad\quad = (x^2 - 4)(x + 3)$

$\quad\quad\quad\quad\quad\quad\quad = (x - 2)(x + 2)(x + 3)$

## Chapter 10 Test *(p. 637)*

**1.** $(x^2 + 4x - 1) + (5x^2 + 2) = x^2 + 4x - 1 + 5x^2 + 2$

$\quad\quad\quad\quad\quad\quad\quad\quad = 6x^2 + 4x + 1$

**2.** $(5t^2 - 9t + 1) - (8t + 13) = 5t^2 - 9t + 1 - 8t - 13$

$\quad\quad\quad\quad\quad\quad\quad\quad = 5t^2 - 17t - 12$

**3.** $(7n^3 + 2n^2 - n - 4) - (4n^3 - 3n^2 + 8)$

$\quad\quad = 7n^3 + 2n^2 - n - 4 - 4n^3 + 3n^2 - 8$

$\quad\quad = 3n^3 + 5n^2 - n - 12$

# Chapter 10 *continued*

**4.** $(x^4 + 6x^2 + 7) + (2x^4 - 3x^2 + 1)$

$\quad = x^4 + 6x^2 + 7 + 2x^4 - 3x^2 + 1$

$\quad = 3x^4 + 3x^2 + 8$

**5.** $(x + 3)(2x + 3) = 2x^2 + 6x + 3x + 9 = 2x^2 + 9x + 9$

**6.** $(9x - 1)(7x + 4) = 63x^2 - 7x + 36x - 4$

$\qquad\qquad\qquad = 63x^2 + 29x - 4$

**7.** $(w - 6)(4w^2 + w - 7)$

$\quad = 4w^3 + w^2 - 7w - 24w^2 - 6w + 42$

$\quad = 4w^3 - 23w^2 - 13w + 42$

**8.** $(5t^3 + 2)(4t^2 + 8t - 7)$

$\quad = 20t^5 + 40t^4 - 35t^3 + 8t^2 + 16t - 14$

**9.** $(3z^3 - 5z^2 + 8)(z + 2)$

$\quad = 3z^4 - 5z^3 + 8z + 6z^3 - 10z^2 + 16$

$\quad = 3z^4 + z^3 - 10z^2 + 8z + 16$

**10.** $(4x + 1)(\frac{1}{2}x - \frac{3}{8}) = 2x^2 - x - \frac{3}{8}$

**11.** $(x - 12)^2 = (x - 12)(x - 12)$

$\qquad\qquad = x^2 - 12x - 12x + 144$

$\qquad\qquad = x^2 - 24x + 144$

**12.** $(7x + 2)^2 = (7x + 2)(7x + 2)$

$\qquad\qquad = (7x + 2)(7x + 2) = 49x^2 + 14x + 14x + 4$

$\qquad\qquad = 49x^2 + 28x + 4$

**13.** $(8x + 3)(8x - 3) = 64x^2 + 24x - 24x - 9 = 64x^2 - 9$

**14.** $(6x - 5)(x + 2) = 0$

$\quad 6x - 5 = 0 \quad x + 2 = 0$

$\qquad 6x = 5 \qquad\quad x = -2$

$\qquad\ x = \frac{5}{6}$

**15.** $\qquad (x + 8)^2 = 0$

$\quad (x + 8)(x + 8) = 0$

$\quad x + 8 = 0$

$\qquad\ x = -8$

**16.** $(4x + 3)(x - 1)(3x + 9) = 0$

$\quad 4x + 3 = 0 \quad\ x - 1 = 0 \quad 3x + 9 = 0$

$\qquad 4x = -3 \qquad\ x = 1 \qquad\ 3x = 9$

$\qquad\ x = -\frac{3}{4} \qquad\qquad\qquad\quad x = -3$

**17.** $y = (x + 1)(x - 5)$

$\quad 0 = (x + 1)(x - 5)$

$\quad x + 1 = 0 \quad\ x - 5 = 0$

$\qquad\ x = -1 \qquad\ x = 5$

Vertex: $(2, -9)$

$\qquad x = \frac{-1 + 5}{2} = \frac{4}{2} = 2$

$\qquad y = (2 + 1)(2 - 5) = 3(-3) = -9$

**18.** $y = (x + 2)(x + 6)$

$\quad 0 = (x + 2)(x + 6)$

$\quad x + 2 = 0 \qquad x + 6 = 0$

$\qquad\ x = -2 \qquad\quad x = -6$

Vertex: $(-4, -4)$

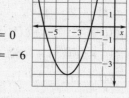

$\qquad x = \frac{-2 + -6}{2} = -\frac{8}{2} = -4$

$\qquad y = (-4 + 2)(-4 + 6) = -2(2) = -4$

**19.** $y = (-x - 4)(x + 7)$

$\quad 0 = (-x - 4)(x + 7)$

$\quad -x - 4 = 0 \qquad x + 7 = 0$

$\qquad -x = 4 \qquad\qquad x = -7$

$\qquad\ x = -4$

Vertex: $\left(-\dfrac{11}{2}, \dfrac{9}{4}\right)$

$\quad x = \dfrac{-4 + -7}{2} = -\dfrac{11}{2}$

$\quad y = \left(\dfrac{11}{2} - 4\right)\left(-\dfrac{11}{2} + 7\right)$

$\quad y = \dfrac{3}{2}\left(\dfrac{3}{2}\right) = \dfrac{9}{4}$

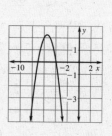

**20.** $x^2 + 13x + 30 = 0$

$\quad (x + 3)(x + 10) = 0$

$\quad x + 3 = 0 \qquad x + 10 = 0$

$\qquad\ x = -3 \qquad\qquad x = -10$

**21.** $x^2 - 19x + 84 = 0$

$\quad (x - 12)(x - 7) = 0$

$\quad x - 12 = 0 \quad\ x - 7 = 0$

$\qquad\ x = 12 \qquad\quad x = 7$

**22.** $x^2 - 34x - 240 = 0$

$\quad (x + 6)(x - 40) = 0$

$\quad x + 6 = 0 \qquad x - 40 = 0$

$\qquad\ x = -6 \qquad\quad x = 40$

**23.** $2x^2 + 15x - 108 = 0$

$\quad (2x - 9)(x + 12) = 0$

$\quad 2x - 9 = 0 \quad x + 12 = 0$

$\qquad 2x = 9 \qquad\quad x = -12$

$\qquad\ x = \frac{9}{2}$

# Chapter 10 *continued*

**24.**
$$9x^2 - 9x = 28$$
$$9x^2 - 9x - 28 = 0$$
$$(3x - 7)(3x + 4) = 0$$
$$3x - 7 = 0 \quad 3x + 4 = 0$$
$$3x = 7 \qquad 3x = -4$$
$$x = \frac{7}{3} \qquad x = -\frac{4}{3}$$

**25.**
$$18x^2 - 57x = -35$$
$$18x^2 - 57x + 35 = 0$$
$$(6x - 5)(3x - 7) = 0$$
$$6x - 5 = 0 \quad 3x - 7 = 0$$
$$6x = 5 \qquad 3x = 7$$
$$x = \frac{5}{6} \qquad x = \frac{7}{3}$$

**26.** $x^2 - 196 = (x - 14)(x + 14)$

**27.** $16x^2 - 36 = 4(4x^2 - 9) = 4(2x - 3)(2x + 3)$

**28.** $128 - 50x^2 = 2(64 - 25x^2) = 2(8 - 5x)(8 + 5x)$

**29.** $x^2 - 6x + 9 = (x - 3)(x - 3) = (x - 3)^2$

**30.** $4x^2 + 44x + 121 = (2x + 11)(2x - 11) = (2x + 11)^2$

**31.** $2x^2 + 28x + 98 = 2(x^2 + 14x + 49)$
$$= 2(x + 7)(x + 7)$$
$$= 2(x + 7)^2$$

**32.** $9t^2 - 54 = 9(t^2 - 6)$

**33.** $4x^2 + 38x + 34 = 2(2x^2 + 19x + 17) = 2(2x + 17)(x + 1)$

**34.** $x^3 + 2x^2 - 16x - 32 = (x^3 + 2x^2) - (16x + 32)$
$$= x^2(x + 2) - 16(x + 2)$$
$$= (x^2 - 16)(x + 2)$$
$$= (x + 4)(x - 4)(x + 2)$$

**35.** $x^2 - 60 = -11$
$$x^2 = 49$$
$$x = -7, 7$$

**36.** $2x^2 + 15x - 8 = 0$
$$(2x - 1)(x + 8) = 0$$
$$2x - 1 = 0 \quad x + 8 = 0$$
$$2x = 1 \qquad x = -8$$
$$x = \frac{1}{2}$$

**37.**
$$x^2 - 13x = -40$$
$$x^2 - 13x + 40 = 0$$
$$(x - 8)(x - 5) = 0$$
$$x - 8 = 0 \quad x - 5 = 0$$
$$x = 8 \qquad x = 5$$

**38.** $x(x - 16) = 0$
$$x = 0 \quad x - 16 = 0$$
$$x = 16$$

**39.** $12x^2 + 3x = 0$
$$3x(4x + 1) = 0$$
$$3x = 0 \quad 4x + 1 = 0$$
$$x = 0 \qquad 4x = -1$$
$$x = -\frac{1}{4}$$

**40.**
$$8x^2 + 6x = 5$$
$$8x^2 + 6x - 5 = 0$$
$$(4x + 5)(2x - 1) = 0$$
$$4x + 5 = 0 \quad 2x - 1 = 0$$
$$4x = -5 \qquad 2x = 1$$
$$x = -\frac{5}{4} \qquad x = \frac{1}{2}$$

**41.**
$$12x^2 + 17x = 7$$
$$12x^2 + 17x - 7 = 0$$
$$(4x + 7)(3x - 1) = 0$$
$$4x + 7 = 0 \quad 3x - 1 = 0$$
$$4x = -7 \qquad 3x = 1$$
$$x = -\frac{7}{4} \qquad x = \frac{1}{3}$$

**42.** $81x^2 - 6 = 30$
$$81x^2 = 36$$
$$x^2 = \frac{36}{81}$$
$$x = \frac{6}{9}, -\frac{6}{9}$$
$$x = \frac{2}{3}, -\frac{2}{3}$$

**43.** $16x^2 - 34x - 15 = 0$
$$(8x + 3)(2x - 5) = 0$$
$$8x + 3 = 0 \quad 2x - 5 = 0$$
$$8x = -3 \qquad 2x = 5$$
$$x = -\frac{3}{8} \qquad x = \frac{5}{2}$$

**44.**

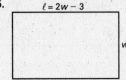

$$A = (3x + 2)(3x + 2)$$
$$= (3x + 2)^2$$
$$= 9x^2 + 6x + 6x + 4$$
$$= 9x^2 + 12x + 4$$

**45.**

$\ell = 2w - 3$, $w$

$$A = (2w - 3)w$$
$$135 = 2w^2 - 3w$$
$$0 = 2w^2 - 3w - 135 = (2w + 15)(w - 9)$$
$$2w + 15 = 0 \qquad w - 9 = 0$$
$$2w = -15 \qquad w = 9$$
$$w = -\frac{15}{2}$$
$$L = 2w - 3 = 2(9) - 3 = 18 - 3 = 15$$
width = 9 ft    length = 15 ft

**46.** $h = -16t^2 + vt$
$$0 = -16t^2 + 31t + 2$$
$$0 = -(16t^2 - 31t - 2) = -(16t + 1)(t - 2)$$
$$16t + 1 = 0 \qquad t - 2 = 0$$
$$16t = -1 \qquad t = 2 \text{ sec}$$
$$t = -\frac{1}{16}$$

## Chapter 10 Standardized Test *(pp. 638–639)*

**1.** B cubic polynomial

**2.** E $(-x^2 - 5x + 7) + (-7x^2 + 5x - 2)$
$$= -x^2 - 5x + 7 - 7x^2 + 5x - 2$$
$$= -8x^2 + 5$$

# Chapter 10 *continued*

**3.** D $(5x^3 + 3x^2 - x + 1) - (2x^3 + x - 5)$

$\quad = 5x^3 + 3x^2 - x + 1 - 2x^3 - x + 5$

$\quad = 3x^3 + 3x^2 - 2x + 6$

**4.** B

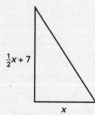

$A = \frac{1}{2}bh = \frac{1}{2}x\left(\frac{1}{2}x + 7\right)$

$\quad = \frac{1}{4}x^2 + \frac{7}{2}x$

**5.** A $(4x - 9)(7x - 2) = 28x^2 - 8x - 63x + 18$

$\quad = 28x^2 - 71x + 18$

**6.** C

$A = \frac{1}{2}(x + 3)(2x + 1 + 3x + 3)$

$\quad = \frac{1}{2}(x + 3)(5x + 4)$

$\quad = \frac{1}{2}(5x^2 + 4x + 15x + 12)$

$\quad = \frac{1}{2}(5x^2 + 19x + 12)$

**7.** E $(2x - 9)^2 = (2x - 9)(2x - 9)$

$\quad = 4x^2 - 18x - 18x + 81$

$\quad = 4x^2 - 36x + 81$

**8.** D $y = (x - 6)(x + 5)$

$\quad x - 6 = 0 \quad x + 5 = 0$

$\quad x = 6 \qquad x = -5$

Vertex: $\left(\frac{1}{2}, -30\frac{1}{4}\right)$

$x = \dfrac{6 + -5}{2} = \dfrac{1}{2}$

$y = \left(\frac{1}{2} - 6\right)\left(\frac{1}{2} + 5\right) = -5\frac{1}{2}\left(5\frac{1}{2}\right) = -30\frac{1}{4}$

**9.** B $\qquad x^2 - 2x = 120$

$\quad -10^2 - 2(-10) \overset{?}{=} 120$

$\qquad 100 + 20 \overset{?}{=} 120$

$\qquad\qquad 120 = 120$

**10.** A $A = \pi(9x^2 + 30x + 25)$

$\quad = \pi(3x + 5)(3r + 5)$

$\quad = \pi(3x + 5)^2$

$r = |3x + 5|$

**11.** C $\qquad x^2 - bx - 16 = 0$

$\quad (-2)^2 - b(-2) - 16 = 0$

$\qquad 4 + 2b - 16 = 0$

$\qquad\qquad 2b - 12 = 0$

$\qquad\qquad\quad 2b = 12$

$\qquad\qquad\quad\ b = 6$

**12.** C $\quad h = 16t^2 - vt$

$\quad 10 = 16t^2 - 12t$

$\quad 0 = 16t^2 + 12t - 10$

$\quad 0 = 2(8t^2 - 6t - 5)$

$\quad 0 = 2(4t - 5)(2t + 1)$

$\quad 4t - 5 = 0 \qquad 2t + 1 = 0$

$\qquad 4t = 5 \qquad\qquad 2t = -1$

$\qquad\ t = \frac{5}{4} \qquad\qquad\ t = -\frac{1}{2}$

$\qquad\ t = 1\frac{1}{4}$ sec.

**13.** D $\ -45x^2 + 150x - 125 = -5(9x^2 - 30x + 25)$

$\qquad\qquad\qquad\qquad\quad = -5(3x - 5)(3x - 5)$

$\qquad\qquad\qquad\qquad\quad = -5(3x - 5)^2$

**14.** B $(a + b)^2 = (17 - 8)^2 = (9)^2 = 81$

$\quad (a - b)^2 = (17 + 8)^2 = (25)^2 = 625$

**15.** B $(a^2 - b^2) = (3^2 - (-4)^2) = (9 - 16) = -7$

$\quad (a - b)^2 = (3 + 4)^2 = 7^2 = 49$

**16.** C $x^3 - 2x^2 - 11x + 22 = (x^3 - 2x^2) - (11x - 22)$

$\qquad\qquad\qquad\qquad\ = x^2(x - 2) - 11(x - 2)$

$\qquad\qquad\qquad\qquad\ = (x^2 - 11)(x - 2)$

**17. a.**

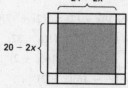

$A = (24 - 2x)(20 - 2x)$

$\quad = 480 - 40x - 48x + 4x^2$

$\quad = 480 - 88x + 4x^2$

**b.** $V = $ Area of Base $\cdot$ Height $= (480 - 88x + 4x^2)x$

**c.** No; the combined length of two squares would be longer than the shorter side and the same length as the longer side.

**d.** Yes; Example: The squares could be $6\frac{1}{2}$ in. on a side. The box would then be 7 in. wide, 11 in. long, and $6\frac{1}{2}$ in. high. The animal would fit inside the box and its head would show at the top.

# CHAPTER 11

## Think & Discuss (p. 641)

**1.** $\dfrac{7}{1} = \dfrac{x}{48}$

$336 = x$

$336 \div 12 = 28 \text{ ft}$

**2.** $\dfrac{x}{1} = \dfrac{12}{48}$

$48x = 12$

$x = \dfrac{1}{4} \text{ in.}$

## Skill Review (p. 642)

**1.** $\dfrac{7}{14} = \dfrac{1}{2}$

**2.** $\dfrac{9}{15} + \dfrac{7}{10} = \dfrac{18}{30} + \dfrac{21}{30}$

$= \dfrac{39}{30}$

$= 1\dfrac{3}{10}$

**3.** $\dfrac{6}{11} - \dfrac{3}{5} = \dfrac{30}{55} - \dfrac{33}{55}$

$= -\dfrac{3}{55}$

**4.** $\dfrac{8}{12} + \dfrac{11}{18} = \dfrac{24}{36} + \dfrac{22}{36}$

$= \dfrac{46}{36}$

$= 1\dfrac{5}{18}$

**5.** $\dfrac{36x}{15} \cdot \dfrac{5}{9} = \dfrac{4x}{3}$

**6.** $49x^2 \cdot \dfrac{3}{-7x} = -21x$

**7.** $\dfrac{15x + 25}{5} = 3x + 5$

**8.** $\dfrac{16 - 4x}{8} = \dfrac{4 - x}{2}$

**9.** $3x^2 - 65 = 178$

$3x^2 = 243$

$x^2 = 81; \ x = 9 \text{ or } x = -9$

**10.** $4x^2 - 10x + 6 = 0$

$(2x - 2)(2x - 3) = 0$

$2x - 2 = 0 \quad 2x - 3 = 0$

$x = 1 \qquad x = 1\dfrac{1}{2}$

**11.** $10x^2 - 30x - 40 = 0$

$x^2 - 3x - 4 = 0$

$(x - 4)(x + 1) = 0$

$x - 4 = 0 \quad\quad x + 1 = 0$

$x = 4 \qquad\quad x = -1$

## Lesson 11.1

### 11.1 Guided Practice (p. 646)

**1.** extremes: 3, 12

means: 4, 9

**2.** no  **3.** no  **4.** yes  **5.** no  **6.** yes  **7.** no

**8.** $\dfrac{4}{x + 1} = \dfrac{7}{2}$  reciprocal property: $\dfrac{x + 1}{4} = \dfrac{2}{7}$

$x + 1 = 4 \cdot \dfrac{2}{7}$

$x + 1 = \dfrac{8}{7}$

$x = \dfrac{1}{7}$

cross-product: $\dfrac{4}{x + 1} = \dfrac{7}{2}$

$4(2) = 7(x + 1)$

$\dfrac{8}{7} = x + 1$

$\dfrac{1}{7} = x$

Answers may vary.

**9.** $\dfrac{x}{3} = \dfrac{2}{7}$

$x = 3 \cdot \dfrac{2}{7}$

$x = \dfrac{6}{7}$

solution: $\dfrac{6}{7}$

check: $\dfrac{\frac{6}{7}}{3} \overset{?}{=} \dfrac{2}{7}$

$\dfrac{6}{7} \overset{?}{=} 3 \cdot \dfrac{2}{7}$

$\dfrac{6}{7} = \dfrac{6}{7}$

**10.** $\dfrac{6}{x} = \dfrac{5}{3}$

$\dfrac{x}{6} = \dfrac{3}{5}$

$x = 6 \cdot \dfrac{3}{5}$

$x = \dfrac{18}{5}$

solution: $\dfrac{18}{5}$

check: $\dfrac{\frac{18}{5}}{6} \overset{?}{=} \dfrac{3}{5}$

$\dfrac{18}{5} \overset{?}{=} 6 \cdot \dfrac{3}{5}$

$\dfrac{18}{5} = \dfrac{18}{5}$

**11.** $\dfrac{2}{2x + 1} = \dfrac{1}{5}$

$10 = 2x + 1$

$9 = 2x$

$\dfrac{9}{2} = x$

solution: $\dfrac{9}{2}$

check: $\dfrac{2}{2\left(\frac{9}{2}\right) + 1} \overset{?}{=} \dfrac{1}{5}$

$\dfrac{2}{9 + 1} \overset{?}{=} \dfrac{1}{5}$

$\dfrac{2}{10} = \dfrac{1}{5}$

**12.**

$$\frac{3}{x} = \frac{x+1}{4}$$

$$12 = x^2 + x$$

$$x^2 + x - 12 = 0$$

$$(x-3)(x+4) = 0$$

$$x - 3 = 0 \qquad x + 4 = 0$$

$$x = 3 \qquad x = -4$$

solutions: 3, −4

check: $\dfrac{3}{3} \overset{?}{=} \dfrac{3+1}{4}$

$$\frac{3}{3} = \frac{4}{4}$$

$$\frac{3}{-4} \overset{?}{=} \frac{-4+1}{4}$$

$$\frac{3}{-4} = \frac{-3}{4}$$

**13.**

$$\frac{t-2}{t} = \frac{2}{t+3}$$

$$(t-2)(t+3) = 2t$$

$$t^2 + t - 6 = 2t$$

$$t^2 - t - 6 = 0$$

$$(t-3)(t+2) = 0$$

$$t - 3 = 0 \qquad t + 2 = 0$$

$$t = 3 \qquad t = -2$$

solutions: 3, −2

check: $\dfrac{3-2}{3} \overset{?}{=} \dfrac{2}{3+3}$

$$\frac{1}{3} = \frac{2}{6}$$

$$\frac{-2-2}{-2} \overset{?}{=} \frac{2}{-2+3}$$

$$\frac{-4}{-2} \overset{?}{=} \frac{2}{1}$$

$$2 = 2$$

**14.**

$$\frac{2u-3}{4u} = \frac{u-1}{u}$$

$$(2u-3)u = 4u(u-1)$$

$$2u^2 - 3u = 4u^2 - 4u$$

$$2u^2 - u = 0$$

$$u(2u-1) = 0$$

$$u = 0 \qquad 2u - 1 = 0$$

$$u = \frac{1}{2}$$

solution: $\dfrac{1}{2}$

check: $\dfrac{2(0)-3}{4(0)} \overset{?}{=} \dfrac{(0)-1}{0}$

$$\frac{-3}{0} \neq \frac{-1}{0}$$

$$\frac{2\left(\frac{1}{2}\right)-3}{4\left(\frac{1}{2}\right)} \overset{?}{=} \frac{\frac{1}{2}-1}{\frac{1}{2}}$$

$$\frac{-2}{2} \overset{?}{=} \frac{-\frac{1}{2}}{\frac{1}{2}}$$

$$-1 = -1$$

**15.**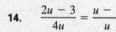

$$\frac{\text{height of actual statue}}{\text{height of model}} = \frac{\text{length of actual statue}}{\text{length of model}}$$

Answers may vary.

**16.**

$$\frac{1.5}{h} = \frac{2}{18}$$

$$27 = 2h$$

$$13\frac{1}{2} = h$$

same solution

## 11.1 Practice and Applications (pp. 646–648)

**17.**

$$\frac{16}{4} = \frac{12}{x}$$

$$16x = 48$$

$$x = 3$$

solution: 3

check: $\dfrac{16}{4} \overset{?}{=} \dfrac{12}{3}$

$$4 = 4$$

**18.**

$$\frac{4}{2x} = \frac{7}{3}$$

$$12 = 14x$$

$$\frac{6}{7} = x$$

solution: $\dfrac{6}{7}$

check: $\dfrac{4}{2\left(\frac{6}{7}\right)} \overset{?}{=} \dfrac{7}{3}$

$$\frac{4}{\frac{12}{7}} \overset{?}{=} \frac{7}{3}$$

$$\frac{28}{12} = \frac{7}{3}$$

**19.**

$$\frac{5}{8} = \frac{c}{9}$$

$$45 = 8c$$

$$\frac{45}{8} = c$$

solution: $\dfrac{45}{8}$

check: $\dfrac{5}{8} \overset{?}{=} \dfrac{\frac{45}{8}}{9}$

$$\frac{5}{8} = \frac{5}{8}$$

**20.**

$$\frac{x}{3} = \frac{2}{5}$$

$$5x = 6$$

$$x = \frac{6}{5}$$

solution: $\dfrac{6}{5}$

check: $\dfrac{\frac{6}{5}}{3} \overset{?}{=} \dfrac{2}{5}$

$$\frac{2}{5} = \frac{2}{5}$$

**21.**

$$\frac{5}{3c} = \frac{2}{3}$$

$$15 = 6c$$

$$\frac{5}{2} = c$$

solution: $\dfrac{5}{2}$

check: $\dfrac{5}{3\left(\frac{5}{2}\right)} \overset{?}{=} \dfrac{2}{3}$

$$\frac{10}{15} = \frac{2}{3}$$

**22.**

$$\frac{24}{5} = \frac{9}{y+2}$$

$$24y + 48 = 45$$

$$24y = -3$$

$$y = -\frac{1}{8}$$

solution: $-\dfrac{1}{8}$

check: $\dfrac{24}{5} \overset{?}{=} \dfrac{9}{-\frac{1}{8}+2}$

$$\frac{24}{5} \overset{?}{=} \frac{9}{\frac{15}{8}}$$

$$\frac{24}{5} = \frac{24}{5}$$

**23.**

$$\frac{6}{3} = \frac{x+8}{-1}$$

$$-6 = 3x + 24$$

$$-30 = 3x$$

$$-10 = x$$

solution: −10

check: $\dfrac{6}{3} \overset{?}{=} \dfrac{-10+8}{-1}$

$$\frac{6}{3} = \frac{-2}{-1}$$

**24.**

$$\frac{r+4}{3} = \frac{r}{5}$$

$$5r + 20 = 3r$$

$$20 = -2r$$

$$-10 = r$$

solution: −10

check: $\dfrac{-10+4}{3} \overset{?}{=} \dfrac{-10}{5}$

$$\frac{-6}{3} = \frac{-10}{5}$$

# Chapter 11 *continued*

**25.** $\dfrac{w + 4}{2w} = \dfrac{-5}{6}$

$6w + 24 = -10w$

$24 = -16w$

$-\dfrac{3}{2} = w$

solution: $-\dfrac{3}{2}$

check: $\dfrac{\frac{-3}{2} + 4}{2\left(-\frac{3}{2}\right)} \stackrel{?}{=} \dfrac{-5}{6}$

$\dfrac{\frac{5}{2}}{-\frac{6}{2}} = \dfrac{-5}{6}$

**26.** $\dfrac{5}{2y} = \dfrac{7}{y - 3}$

$5y - 15 = 14y$

$-15 = 9y$

$-\dfrac{5}{3} = y$

solution: $-\dfrac{5}{3}$

check: $\dfrac{5}{2\left(-\frac{5}{3}\right)} \stackrel{?}{=} \dfrac{7}{-\frac{5}{3} - 3}$

$\dfrac{5}{-\frac{10}{3}} \stackrel{?}{=} \dfrac{7}{-\frac{14}{3}}$

$\dfrac{15}{-10} \stackrel{?}{=} \dfrac{21}{-14}$

$-\dfrac{3}{2} = -\dfrac{3}{2}$

**27.** $\dfrac{x + 6}{3} = \dfrac{x - 5}{2}$

$2x + 12 = 3x - 15$

$27 = x$

solution: 27

check: $\dfrac{27 + 6}{3} \stackrel{?}{=} \dfrac{27 - 5}{2}$

$\dfrac{33}{3} \stackrel{?}{=} \dfrac{22}{2}$

$11 = 11$

**28.** $\dfrac{x - 2}{4} = \dfrac{x + 10}{10}$

$10x - 20 = 4x + 40$

$6x = 60$

$x = 10$

solution: 10

check: $\dfrac{10 - 2}{4} \stackrel{?}{=} \dfrac{10 + 10}{10}$

$2 = 2$

**29.** $\dfrac{8}{x + 2} = \dfrac{3}{x - 1}$

$8x - 8 = 3x + 6$

$5x = 14$

$x = \dfrac{14}{5}$

solution: $\dfrac{14}{5}$

check: $\dfrac{8}{\frac{14}{5} + 2} \stackrel{?}{=} \dfrac{3}{\frac{14}{5} - 1}$

$\dfrac{8}{\frac{24}{5}} \stackrel{?}{=} \dfrac{3}{\frac{9}{5}}$

$\dfrac{40}{24} \stackrel{?}{=} \dfrac{15}{9}$

$\dfrac{5}{3} = \dfrac{5}{3}$

**30.** $\dfrac{x - 3}{18} = \dfrac{3}{x}$

$x^2 - 3x = 54$

$x^2 - 3x - 54 = 0$

$(x - 9)(x + 6) = 0$

$x - 9 = 0 \qquad x + 6 = 0$

$x = 9 \qquad x = -6$

solutions: 9, $-6$

check: $\dfrac{9 - 3}{18} \stackrel{?}{=} \dfrac{3}{9}$

$\dfrac{1}{3} = \dfrac{1}{3}$

$\dfrac{-6 - 3}{18} \stackrel{?}{=} \dfrac{3}{-6}$

$-\dfrac{1}{2} = -\dfrac{1}{2}$

**31.** $\dfrac{-2}{a - 7} = \dfrac{a}{5}$

$-10 = a^2 - 7a$

$a^2 - 7a + 10 = 0$

$(a - 5)(0 - 2) = 0$

$a - 5 = 0 \qquad a - 2 = 0$

$a = 5 \qquad a = 2$

solutions: 5, 2

check: $\dfrac{-2}{5 - 7} \stackrel{?}{=} \dfrac{5}{5}$

$1 = 1$

$\dfrac{-2}{2 - 7} \stackrel{?}{=} \dfrac{2}{5}$

$\dfrac{-2}{-5} = \dfrac{2}{5}$

**32.** $\dfrac{u}{3} = \dfrac{1}{2u - 1}$

$2u^2 - u = 3$

$2u^2 - u - 3 = 0$

$(2u - 3)(u + 1) = 0$

$2u - 3 = 0 \qquad u + 1 = 0$

$u = \dfrac{3}{2} \qquad u = -1$

solutions: $\dfrac{3}{2}$, $-1$

check: $\dfrac{\frac{3}{2}}{3} \stackrel{?}{=} \dfrac{1}{2\left(\frac{3}{2}\right) - 1}$

$\dfrac{1}{2} = \dfrac{1}{2}$

$\dfrac{-1}{3} \stackrel{?}{=} \dfrac{1}{2(-1) - 1}$

$-\dfrac{1}{3} = -\dfrac{1}{3}$

**33.** $\dfrac{d}{d + 4} = \dfrac{d - 2}{d}$

$d^2 = d^2 + 2d - 8$

$0 = 2d - 8$

$d = 4$

solution: 4

check: $\dfrac{4}{4 + 4} \stackrel{?}{=} \dfrac{4 - 2}{4}$

$\dfrac{4}{8} = \dfrac{2}{4}$

**34.** $\dfrac{3x}{4x - 1} = \dfrac{1}{x}$

$3x^2 = 4x - 1$

$3x^2 - 4x + 1 = 0$

$(3x - 1)(x - 1) = 0$

$3x - 1 = 0 \qquad x - 1 = 0$

$x = \dfrac{1}{3} \qquad x = 1$

solutions: $\dfrac{1}{3}$, 1

check: $\dfrac{3\left(\frac{1}{3}\right)}{4\left(\frac{1}{3}\right) - 1} \stackrel{?}{=} \dfrac{1}{\frac{1}{3}}$

$\dfrac{1}{\frac{1}{3}} = \dfrac{1}{\frac{1}{3}}$

$\dfrac{3(1)}{4(1) - 1} \stackrel{?}{=} \dfrac{1}{1}$

$\dfrac{3}{3} = \dfrac{1}{1}$

**35.** $\dfrac{x - 3}{x} = \dfrac{x}{x + 6}$

$x^2 = x^2 + 3x - 18$

$0 = 3x - 18$

$x = 6$

solution: 6

check: $\dfrac{6 - 3}{6} \stackrel{?}{=} \dfrac{6}{6 + 6}$

$\dfrac{3}{6} = \dfrac{6}{12}$

**36.** $\dfrac{5}{m + 1} = \dfrac{4m}{m}$

$\dfrac{5}{m + 1} = \dfrac{4}{1}$

$5 = 4m + 4$

$1 = 4m$

$\dfrac{1}{4} = m$

solution: $\dfrac{1}{4}$

check: $\dfrac{5}{\frac{1}{4} + 1} \stackrel{?}{=} \dfrac{4\left(\frac{1}{4}\right)}{\frac{1}{4}}$

$\dfrac{5}{\frac{5}{4}} \stackrel{?}{=} \dfrac{1}{\frac{1}{4}}$

$4 = 4$

**37.** $\dfrac{2}{3t} = \dfrac{t-1}{t}$

$2t = 3t^2 - 3t$

$0 = 3t^2 - 5t$

$0 = t(3t - 5)$

$t = 0 \quad 3t - 5 = 0$

$t = \dfrac{5}{3}$

solution: $\dfrac{5}{3}$

check: $\dfrac{2}{3\left(\frac{5}{3}\right)} \overset{?}{=} \dfrac{\frac{5}{3} - 1}{\frac{5}{3}}$

$\dfrac{2}{5} = \dfrac{\frac{2}{3}}{\frac{5}{3}}$

$\dfrac{2}{3(0)} \overset{?}{=} \dfrac{0 - 1}{0}$

extraneous root

**38.** $\dfrac{2}{6x+1} = \dfrac{2x}{1}$

$2 = 12x^2 + 2x$

$12x^2 + 2x - 2 = 0$

$6x^2 + x - 1 = 0$

$(3x - 1)(2x + 1) = 0$

$3x - 1 = 0 \quad 2x + 1 = 0$

$x = \dfrac{1}{3} \qquad x = -\dfrac{1}{2}$

solutions: $\dfrac{1}{3}, -\dfrac{1}{2}$

check: $\dfrac{2}{6\left(\frac{1}{3}\right) + 1} \overset{?}{=} \dfrac{2\left(\frac{1}{3}\right)}{1}$

$\dfrac{2}{3} = \dfrac{2}{3}$

$\dfrac{2}{6\left(-\frac{1}{2}\right) + 1} = \dfrac{2^{-\frac{1}{2}}}{1}$

$\dfrac{2}{-2} = -1$

**39.** $\dfrac{-2}{q} = \dfrac{q+1}{q^2}$

$-2q^2 = q^2 + q$

$3q^2 + q = 0$

$q(3q + 1) = 0$

$q = 0 \qquad 3q + 1 = 0$

$q = -\dfrac{1}{3}$

solution: $-\dfrac{1}{3}$

check: $\dfrac{-2}{0} \overset{?}{=} \dfrac{0+1}{0^2}$

extraneous root

$\dfrac{-2}{-\frac{1}{3}} \overset{?}{=} \dfrac{-\frac{1}{3} + 1}{\left(-\frac{1}{3}\right)^2}$

$6 \overset{?}{=} \dfrac{\frac{2}{3}}{\frac{1}{9}}$

$6 \overset{?}{=} \dfrac{2}{3} \cdot \dfrac{9}{1}$

$6 = 6$

**40.** $\dfrac{6}{19n} = \dfrac{-2}{n^2 + 2}$

$6n^2 + 12 = -38n$

$6n^2 + 38n + 12 = 0$

$3n^2 + 19n + 6 = 0$

$(3n + 1)(n + 6) = 0$

$3n + 1 = 0 \quad n + 6 = 0$

$n = -\dfrac{1}{3} \qquad n = -6$

solutions: $-\dfrac{1}{3}, -6$

check: $\dfrac{6}{19\left(-\frac{1}{3}\right)} \overset{?}{=} \dfrac{-2}{\left(-\frac{1}{3}\right)^2 + 2}$

$-\dfrac{18}{19} \overset{?}{=} -\dfrac{2}{\frac{19}{9}}$

$-\dfrac{18}{19} = -\dfrac{18}{19}$

$\dfrac{6}{19(-6)} \overset{?}{=} \dfrac{-2}{(-6)^2 + 2}$

$\dfrac{-6}{114} = \dfrac{-2}{38}$

**41.** $\dfrac{4 \text{ ft}}{8 \text{ ft}} = \dfrac{11 \text{ in.}}{x \text{ in.}}$

$4x = 88$

$x = 22 \text{ in.}$

**42.** $\dfrac{2\frac{1}{2} \text{ ft}}{4 \text{ ft}} = \dfrac{x \text{ in.}}{11 \text{ in.}}$

$27\frac{1}{2} = 4x$

$x = 6\frac{7}{8} \text{ in.}$

**43.** Rewrite 1 foot as 12 inches. Then use the proportion $\dfrac{\frac{1}{16}}{12} = \dfrac{1}{192}$ and cross-multiply. The cross-products should be equal. $\dfrac{1}{16}(192) \overset{?}{=} 12;\ 12 = 12$

**44.** $\dfrac{20}{100} = \dfrac{x}{500}$

$10{,}000 = 100x$

$100 = x$

100 students

**45.** *Sample answer:*

Food: $\dfrac{78}{100} = \dfrac{x}{20}$

$100x = 1560$

$x = 15.6$

about 16 students

Movie tickets: $\dfrac{14}{100} = \dfrac{x}{20}$

$100x = 280$

$x = 2.8$

about 3 students

Music: $\dfrac{12}{100} = \dfrac{x}{20}$

$100x = 240$

$x = 2.4$

about 2 students

**46.** Answers may vary.

**47.** $\dfrac{16}{329} = \dfrac{232}{x}$

$16x = 76{,}328$

$x = 4770.5$

approximately 4771

**48. a.** high school: $\dfrac{x}{200} = \dfrac{33.8}{100}$

$100x = 6760$

$x = 67.6$

about 68 people

at least 4 yrs. of college: $\dfrac{x}{200} = \dfrac{23.8}{100}$

$100x = 4760$

$x = 47.6$

about 48 people

**b.** $\dfrac{23.8}{100} = \dfrac{x}{170{,}581}$

$x = 40598.278$

about 40,598,000

**c.** $\dfrac{23.8}{100} = \dfrac{x}{20{,}000}$

$x = 4760$ people

**d.** Use the ratio $\dfrac{7581}{15{,}860}$ to find the percent and compare the result (47.8%) to the typical percent (23.8%).

**49.** $\dfrac{a}{b} = \dfrac{c}{d}$

$\dfrac{ad}{c} = \dfrac{bc}{c}$

$\dfrac{ad}{cd} = \dfrac{b}{d}$

therefore, $\dfrac{a}{c} = \dfrac{b}{d}$

# Chapter 11 *continued*

## 11.1 Mixed Review (p. 648)

**50.**

| Decimal | Percent | | Decimal | Percent |
|---------|---------|---|---------|---------|
| 0.78 | 78% | | 0.666... | 66.7% |
| 0.2 | 20% | | 1.76 | 176% |
| 0.03 | 3% | | 1.1 | 110% |
| 0.073 | 7.3% | | 2 | 200% |

**51.** $\sqrt{64} = 8$ or $-8$  **52.** $\sqrt{-9}$ no square roots

**53.** $\sqrt{12} = \sqrt{4 \cdot 3} = 2\sqrt{3}$ or $-2\sqrt{3}$

**54.** $\sqrt{169} = 13$ or $-13$  **55.** $\sqrt{-20}$ no square roots

**56.** $\sqrt{50} = \sqrt{25 \cdot 2}$  **57.** $\sqrt{\dfrac{9}{25}} = \dfrac{3}{5}$ or $-\dfrac{3}{5}$
$= 5\sqrt{2}$ or $-5\sqrt{2}$

**58.** $\sqrt{0.04} = 0.2$ or $-0.2$  **59.** $\sqrt{18} = \sqrt{9 \cdot 2}$
$= 3\sqrt{2}$

**60.** $\sqrt{20} = \sqrt{4 \cdot 5}$  **61.** $\sqrt{80} = \sqrt{16 \cdot 5}$
$= 2\sqrt{5}$      $= 4\sqrt{5}$

**62.** $\sqrt{162} = \sqrt{81 \cdot 2}$  **63.** $9\sqrt{36} = 9 \cdot 6$  **64.** $\sqrt{\dfrac{11}{9}} = \dfrac{\sqrt{11}}{3}$
$= 9\sqrt{2}$      $= 54$

**65.** $\dfrac{1}{2}\sqrt{28} = \dfrac{1}{2}(2)\sqrt{7}$  **66.** $4\sqrt{\dfrac{5}{4}} = 4\dfrac{\sqrt{5}}{2} = 2\sqrt{5}$
$= \sqrt{7}$

## Lesson 11.2

### 11.2 Guided Practice (p. 652)

**1.** $0.1(160) = 16$; base number: 160  **2.** 75%

**3.** regular price  **4.** $0.75b = 17.25$; $b = \$23$

**5.** $35 = p(20)$  **6.** $0.12(5) = a$
$1.75 = p$       $0.6 = a$
$175\% = p$

**7.** $18 = 0.375b$  **8.** $13.2 = 1.2b$
$48 = b$       $11 = b$

**9.** Method 1: $a = 0.06(5.99)$
$a \approx 0.36$
Method 2: $\dfrac{6}{100} = \dfrac{x}{5.99}$
$x \approx 0.36$
The right side of the second equation is the rate expressed as a fraction instead of a decimal. Both methods involve multiplying the base by the rate.

### 11.2 Practice and Applications (pp. 653–655)

**10.** C  **11.** A  **12.** B

**13.** $a = 0.25(80)$  **14.** $0.85(300) = a$
$a = 20$       $255 = a$

**15.** $18 = p(60)$  **16.** $52 = 0.125b$
$0.3 = p$; 30%      $416 = b$

---

**17.** $0.14(220) = a$  **18.** $a = 0.35(750)$
$30.8 \text{ ft} = a$       $a = \$262.50$

**19.** $42 = 0.5(b)$  **20.** $a = 0.24(710)$
$84 \text{ ft} = b$       $a = 170.4 \text{ mi}$

**21.** $0.16b = 8$  **22.** $4 = 0.025b$
$b = 50$       $\$160 = b$

**23.** $33 = 0.22b$  **24.** $55 = p(20)$
$150 \text{ grams} = b$      $275\% = p$

**25.** $a = 0.082(800)$  **26.** $9 = p(60)$
$a = 65.6 \text{ tons}$      $15\% = p$

**27.** $62 = p(72)$  **28.** $30 = p(480)$
$86.1\% = p$       $6.25\% = p$

**29.** $240 = p(50)$  **30.** $0.02b = 200$
$480\% = p$       $b = \$10,000$

**31.** blue region: $(50)(60) - (20)(20) = 2600$
yellow region: $(20)(20) = 400$
entire region: $(50)(60) = 3000$
percent shaded blue: $2600 = p(3000)$
$86.7\% = p$
percent shaded yellow: $400 = p(3000)$
$13.3\% = p$

**32.** entire region: $(10)(17) = 170$
yellow region: $2(5)(5) = 50$
blue region: $170 - 50 = 120$
percent shaded blue: $120 = p(170)$
$70.6\% = p$
percent shaded yellow: $50 = p(170)$
$29.4\% = p$

**33.** $a = 0.288(861)$  **34.** $a = 0.104(861)$
$a \approx 248 \text{ people}$      $a \approx 90 \text{ people}$

**35.** $a = 0.445(2500)$  **36.** $735 = p(3500)$
$a \approx 1113 \text{ people}$      $21\% = p$

**37.** $630 = p(3500)$  **38.** $1120 = p(3500)$
$18\% = p$       $32\% = p$

**39.** $\dfrac{735}{3500} = \dfrac{x}{2000}$
$x = 420 \text{ students}$

**40.** The correct tax is $2.41, so the total is $32.59.

**41.** before tax: $4.75 = p(30.18)$
$15.7\% \approx p$
yes
after tax: $4.75 = p(32.59)$
$14.6\% = p$
no

**42.** $572 = 0.55b$  **43.** $a = 0.38(1040)$
$1040 \text{ people} = b$      $a \approx 395 \text{ people}$

**44.** No; *Sample answer:* Scientists are likely to be better informed than those in a representative sample of people 18 and over, and their answers are more likely to be based on information rather than on personal opinion.

**45.** First sale: $a = 0.7(80)$

$a = \$56$

Second sale: $a = 0.8(80)$

$a = \$64$

$a = 0.9(64)$

$a = \$57.60$

No. The coat is $56 at the first store and $57.60 at the second store.

**46.** $a = 0.2(11.50)$

$a = 2.3$

$a$ represents the tip.

$\$11.50 + \$2.30 = \$13.80$

**47.** $a = 1.2(11.50)$

$a = \$13.80$

$a$ represents the total cost of the ride.

**48.** C

**49.** $0.24x = 450 \qquad 0.12x = 225$

$\quad x = 1875 \qquad\quad x = 1875$

C

**50.** $0.16x = 28 \qquad \dfrac{16(28)}{100} = x$

$\quad x = 175 \qquad\quad 4.48 = x$

A

**51.** No; let $x$ be the amount your sister earns. Then the amount you earn is $1.1x$ and $\dfrac{x}{1.1x} \approx 91\%$. So your sister earns about 9% less.

**52.** Yes; let $x$ be the original price; a price that is 220% greater is $x + 2.2x = 3.2x$ or 320% of $x$; also, 31.25% of 320% of $x = 0.3125(3.2)(x) = x$.

## 11.2 Mixed Review (p. 655)

**53.** $4k = 8$

$k = 2$

$2x = y$

**54.** $33k = 9$

$k = \dfrac{9}{33}$

$k = \dfrac{3}{11}$

$y = \dfrac{3}{11}x$

**55.** $-2k = -1$

$k = \dfrac{1}{2}$

$\dfrac{1}{2}x = y$

**56.** $6.3k = 1.5$

$k = \dfrac{1.5}{6.3}$

$k = \dfrac{5}{21}$

$y = \dfrac{5}{21}x$

**57.** $5\dfrac{1}{3}k = 8$

$k = \dfrac{8}{\frac{16}{3}}$

$k = \dfrac{3}{2}$

$y = \dfrac{3}{2}x$

**58.** $9.8k = 3.6$

$k = \dfrac{3.6}{9.8}$

$k = \dfrac{18}{49}$

$y = \dfrac{18}{49}x$

**59.** $4 \overset{?}{<} (-1)^2 + 6(-1) + 12$

$4 < 7$

solution

**60.** $2 \overset{?}{\leq} (-1)^2 - 7(-1) + 9$

$2 \leq 17$

solution

**61.** $5 \overset{?}{>} 2(2)^2 - 7(2) - 15$

$5 > -21$

solution

**62.** $-4 \overset{?}{\geq} 1^2 + 6(1) + 12$

$-4 \not\geq 19$

not a solution

**63.** $x^2 + 5x - 14 = (x + 7)(x - 2)$

**64.** $7x^2 + 8x + 1 = (7x + 1)(x + 1)$

**65.** $5x^2 - 51x + 54 = (5x - 6)(x - 9)$

**66.** $4x^2 - 28x + 49 = (2x - 7)(2x - 7)$

**67.** $6x^2 + 16x = 2x(3x + 8)$

**68.** $36x^5 - 90x^3 = 18x^3(2x^2 - 5)$

**69.** $3x^3 + 21x^2 + 30x = 3x(x^2 + 7x + 10)$

$\qquad\qquad\qquad\qquad = 3x(x + 5)(x + 2)$

**70.** $36x^3 - 9x = 9x(4x^2 - 1) = 9x(2x - 1)(2x + 1)$

**71.** $15x^4 - 50x^3 - 40x^2 = 5x^2(3x^2 - 10x - 8)$

$\qquad\qquad\qquad\qquad\qquad = 5x^2(3x + 2)(x - 4)$

**72.** $P = 18,870(1.0124)^t$

$\dfrac{18,870(1.0124)^{30}}{18,870(1.0124)^5} \approx 1.36$

## Lesson 11.3

### Activity (p. 657)

**1.**

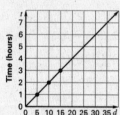

| t | d |
|---|---|
| 1 | 5 |
| 2 | 10 |
| 3 | 15 |

$d = 5t$

The distance and time are related directly.

**2.**

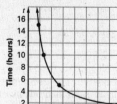

| t | r |
|---|---|
| 5 | 1 |
| 10 | $\frac{1}{2}$ |
| 15 | $\frac{1}{3}$ |

$5 = r \cdot t$

$\dfrac{5}{t} = r$

Rate and time are related inversely.

# Chapter 11 *continued*

**3.**

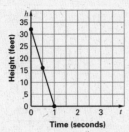

| $h$ | $t$ |
|-----|-----|
| 32 | 0 |
| 16 | $\frac{1}{2}$ |
| 0 | 1 |

$h = 32(1 - t)$

Height and time are not related directly or inversely.

**4.**

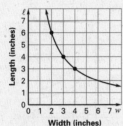

| $w$ | $l$ |
|-----|-----|
| 2 | 6 |
| 3 | 4 |
| 4 | 3 |

$l \cdot w = 12$

$l = \dfrac{12}{w}$

Length and width are related inversely.

## 11.3 Guided Practice (p. 659)

**1.** Two numbers $x$ and $y$ vary directly if there is a constant $k$, $k \neq 0$, such that $y = kx$. Two numbers $x$ and $y$ vary inversely if there is a constant $k$, $k \neq 0$, such that $y = \dfrac{k}{x}$.

**2.** Direct variation; the equation of the graph is $y = \frac{3}{2}x$.

**3.** Neither; the equation of the graph is $y = -x^2 + 2$.

**4.** Inverse variation; the equation of the graph is $y = \dfrac{4}{x}$.

**5.** Direct variation; the equation is of the form $y = kx$.

**6.** inverse variation

**7.** neither

**8.** direct variation

**9.** inverse variation

**10.** $6 = 4k$

$\dfrac{3}{2} = k$

$y = \dfrac{3}{2}x$

$y = \dfrac{3}{2}(8)$

$y = 12$

**11.** $4(6) = k$

$24 = k$

$y = \dfrac{24}{x}$

$y = \dfrac{24}{8}$

$y = 3$

## 11.3 Practice and Applications (pp. 659–661)

**12.** $9 = 3k$

$3 = k$

$y = 3x$

**13.** $8 = 2k$

$4 = k$

$y = 4x$

**14.** $6 = 18k$

$\dfrac{1}{3} = k$

$y = \frac{1}{3}x$

**15.** $24 = 8k$

$3 = k$

$y = 3x$

**16.** $12 = 36k$

$\dfrac{1}{3} = k$

$y = \frac{1}{3}x$

**17.** $3 = 27k$

$\dfrac{1}{9} = k$

$y = \frac{1}{9}x$

**18.** $16 = 24k$

$\dfrac{2}{3} = k$

$y = \frac{2}{3}x$

**19.** $81 = 45k$

$\dfrac{9}{5} = k$

$y = \frac{9}{5}x$

**20.** $27 = 54k$

$\dfrac{1}{2} = k$

$y = \frac{1}{2}x$

**21.** $2(5) = k$

$10 = k$

$y = \dfrac{10}{x}$

**22.** $3(7) = k$

$21 = k$

$y = \dfrac{21}{x}$

**23.** $16(1) = k$

$16 = k$

$y = \dfrac{16}{x}$

**24.** $11(2) = k$

$22 = k$

$y = \dfrac{22}{x}$

**25.** $\left(\dfrac{1}{2}\right)(8) = k$

$4 = k$

$y = \dfrac{4}{x}$

**26.** $\left(\dfrac{13}{5}\right)(5) = k$

$13 = k$

$y = \dfrac{13}{x}$

**27.** $12\left(\dfrac{3}{4}\right) = k$

$9 = k$

$y = \dfrac{9}{x}$

**28.** $5\left(\dfrac{1}{3}\right) = k$

$\dfrac{5}{3} = k$

$y = \dfrac{5}{3x}$

**29.** $30(7.5) = k$

$225 = k$

$y = \dfrac{225}{x}$

**30.** $(1.5)(50) = k$

$75 = k$

$y = \dfrac{75}{x}$

**31.** $45\left(\dfrac{3}{5}\right) = k$

$27 = k$

$y = \dfrac{27}{x}$

**32.** $10.5(7) = k$

$73.5 = k$

$y = \dfrac{73.5}{x}$

**33.**

| $x$ | $y$ |
|-----|-----|
| 1 | 4 |
| 2 | 2 |
| 3 | $\frac{4}{3}$ |
| 4 | 1 |

$y = \dfrac{4}{x}$

inversely

**34.**

| $x$ | $y$ |
|-----|-----|
| 1 | $\frac{3}{2}$ |
| 2 | $\frac{3}{4}$ |
| 3 | $\frac{1}{2}$ |
| 4 | $\frac{3}{8}$ |

$y = \dfrac{3}{2x}$

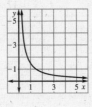

inversely

**35.**

| $x$ | $y$ |
|-----|-----|
| 1 | 3 |
| 2 | 6 |
| 3 | 9 |
| 4 | 12 |

$y = 3x$

directly

**36.**

| $x$ | $y$ |
|-----|-----|
| 1 | 6 |
| 2 | 3 |
| 3 | 2 |
| 4 | $\frac{3}{2}$ |

$y = \dfrac{6}{x}$

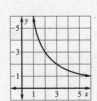

inversely

**37.** direct variation

$y = 5x$

**38.** inverse variation

$30(1) = k$

$30 = k$

$y = \dfrac{30}{x}$

**39.** inverse variation

**40.** direct variation

**41.** neither

**42.** inverse variation

**43.** $PA = W$

$4(29) = W$

$116 = W$

She weighs 116 lb.

**44.** $P = \dfrac{W}{A}$

$P = \dfrac{116}{29(11)}$

$P = \dfrac{4}{11}$ lb/in.$^2$

**45.** $3700(1.2) = k$

$4440 = k$

$T = \dfrac{4440}{d}$

**46.** $T = \dfrac{4440}{5000}$

$T = 0.9°C$

**47.** $26(12) = 312$ miles

$\dfrac{12 \text{ gal}}{312 \text{ mi}} = \dfrac{1}{26}$ gal/mi

$\approx 0.04$ gal/mi

**48.** $g = 12 - \dfrac{m}{26}$

**49.** Neither; the equation cannot be written in the form

$g = km$ or $g = \dfrac{k}{m}$ for any constant, $k$.

**50. a.** Let $x$ = width

Let $y$ = length

| $x$ | $y$ |
|-----|-----|
| 2 | 6 |
| 3 | 4 |
| 4 | 3 |
| 6 | 2 |

$12 = xy$

$\dfrac{12}{x} = y$

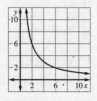

**b.** Inversely; an equation of the graph is $y = \dfrac{12}{x}$.

**c.**

| $x$ | $y$ |
|-----|-----|
| 4 | 6 |
| 6 | 4 |
| 8 | 3 |

$\dfrac{24}{x} = y$

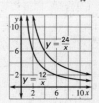

**d.** The area of the first rectangle is half the area of the second rectangle; the value of $y$ for the second rectangle is twice the value of $y$ for the first; the value of $x$ for the first rectangle is half the value of $x$ for the second rectangle.

**51.** $\dfrac{3}{2} = \dfrac{6}{x_2}$

$3x_2 = 12$

$x_2 = 4$

**52.** $\dfrac{8}{-4} = \dfrac{y_2}{-1}$

$-8 = -4y_2$

$2 = y_2$

**53.** $4(5) = 8(y_2)$

$20 = 8y_2$

$\dfrac{5}{2} = y_2$

**54.** $9(-3) = x_2(12)$

$-27 = 12x_2$

$-\dfrac{9}{4} = x_2$

**11.3 Mixed Review (p. 661)**

**55.** $\dfrac{36 \div 12}{48 \div 12} = \dfrac{3}{4}$

**56.** $\dfrac{27 \div 27}{108 \div 27} = \dfrac{1}{4}$

**57.** $\dfrac{96 \div 12}{180 \div 12} = \dfrac{8}{15}$

**58.** $\dfrac{-15 \div 5}{125 \div 5} = -\dfrac{3}{25}$

**59.** $\dfrac{1}{6}$

**60.** $\dfrac{3}{6} = \dfrac{1}{2}$

**61.** $\dfrac{y^4 \cdot y^7}{y^5} = y^6$

**62.** $\dfrac{5xy}{5x^2} = \dfrac{y}{x}$

**63.** $\dfrac{-3xy^3}{3x^3y} = -\dfrac{y^2}{x^2}$

**64.** $\dfrac{56x^2y^5}{64x^2y} = \dfrac{7y^4}{8}$

**65.** leading coefficient: $-5$

linear; binomial

**66.** leading coefficient: 8

quartic; binomial

**67.** leading coefficient: $-1$

cubic; binomial

**68.** leading coefficient: 3

quadratic; trinomial

**Algebra 1**
Chapter 11 Worked-out Solution Key

# Chapter 11 *continued*

## Quiz 1 (p. 662)

**1.** $\dfrac{x}{10} = \dfrac{4}{5}$

$5x = 40$

$x = 8$

**2.** $\dfrac{3}{x} = \dfrac{7}{9}$

$7x = 27$

$x = \dfrac{27}{7}$

**3.** $\dfrac{x}{4x - 8} = \dfrac{2}{x}$

$x^2 = 8x - 16$

$x^2 - 8x + 16 = 0$

$(x - 4)(x - 4) = 0$

$x = 4$

**4.** $\dfrac{6x + 4}{5} = \dfrac{2}{x}$

$6x^2 + 4x = 10$

$6x^2 + 4x - 10 = 0$

$3x^2 + 2x - 5 = 0$

$(3x + 5)(x - 1) = 0$

$3x + 5 = 0 \qquad x - 1 = 0$

$x = -\dfrac{5}{3} \qquad x = 1$

**5.** $4 = \dfrac{k}{3}$

$12 = k$

$y = \dfrac{12}{x}$

**6.** $1 = \dfrac{k}{12}$

$12 = k$

$y = \dfrac{12}{x}$

**7.** $\dfrac{3}{4} = \dfrac{k}{24}$

$18 = k$

$y = \dfrac{18}{x}$

**8.** $413 = 0.457x$

$x \approx 904$ people

**9.** $x = 0.226(904)$

$x \approx 204$ people

## Math & History (p. 662)

**1.** candidate 1: $0.37(2{,}075{,}280) \approx 767{,}854$ votes

candidate 2: $0.34(2{,}075{,}280) \approx 705{,}595$ votes

candidate 3: $0.28(2{,}075{,}280) \approx 581{,}078$ votes

**2.** no

## Technology Activity 11.3 (p. 663)

**1.** directly; $k = 0.825$; $y = 0.825x$

**2.** inversely; $k = 25$; $y = \dfrac{25}{x}$

## Lesson 11.4

### 11.4 Guided Practice (p. 667)

**1.** A rational expression is a fraction whose numerator and denominator are nonzero polynomials. Example: $\dfrac{x}{x - 1}$

**2.** There are no common factors in the numerator and denominator that can be canceled. The expression is in simplest form.

**3.** The numerator is equal to $1(x + 4)$, so when $x + 4$ is canceled in the numerator and denominator, the resulting numerator is 1, not 0; $\dfrac{x + 4}{2(x + 4)} = \dfrac{1}{2}$.

**4.** $8x = 0$

$x = 0$

**5.** $x - 5 = 0$

$x = 5$

**6.** $x^2 - x - 2 = 0$

$(x - 2)(x + 1) = 0$

$x = 2 \qquad x = -1$

**7.** $\dfrac{6 + 2x}{x^2 + 5x + 6} = \dfrac{2(3 + x)}{(x + 3)(x + 2)}$

$= \dfrac{2}{x + 2}$    C

**8.** $\dfrac{x(x + 4)}{2x(2x^2 + 8x)} = \dfrac{x(x + 4)}{4x^2(x + 4)}$

$= \dfrac{1}{4x}$    B

### 11.4 Practice and Applications (pp. 667–669)

**9.** $\dfrac{4x}{20} = \dfrac{x}{5}$

**10.** $\dfrac{15x}{45} = \dfrac{x}{3}$

**11.** $\dfrac{-18x^2}{12x} = -\dfrac{3x}{2}$

**12.** $\dfrac{14x^2}{50x^4} = \dfrac{7}{25x^2}$

**13.** $\dfrac{3x^2 - 18x}{-9x^2} = \dfrac{3x(x - 6)}{-9x^2}$

$= \dfrac{x - 6}{-3x}$

$= \dfrac{6 - x}{3x}$

**14.** $\dfrac{42x - 6x^3}{36x} = \dfrac{6x(7 - x^2)}{36x}$

$= \dfrac{7 - x^2}{6}$

**15.** $\dfrac{7x}{12x + x^2} = \dfrac{7}{12 + x}$

**16.** $\dfrac{x + 2x^2}{x + 2} = \dfrac{x(1 + 2x)}{x + 2}$

cannot be simplified

**17.** $\dfrac{12 - 5x}{10x^2 - 24x} = \dfrac{12 - 5x}{2x(5x - 12)}$

$= -\dfrac{1}{2x}$

**18.** $\dfrac{x^2 + 25}{2x + 10} = \dfrac{x^2 + 25}{2(x + 5)}$

cannot be simplified

**19.** $\dfrac{5 - x}{x^2 - 8x + 15} = \dfrac{-1(x - 5)}{(x - 5)(x - 3)}$

$= \dfrac{-1}{x - 3} = \dfrac{1}{3 - x}$

**20.** $\dfrac{2x^2 + 11x - 6}{x + 6} = \dfrac{(x + 6)(2x - 1)}{x + 6}$

$= 2x - 1$

**21.** $\dfrac{x^2 + x - 20}{x^2 + 2x - 15} = \dfrac{(x + 5)(x - 4)}{(x + 5)(x - 3)}$

$= \dfrac{x - 4}{x - 3}$

# Chapter 11 *continued*

**22.** $\dfrac{x^3 + 9x^2 + 14x}{x^2 - 4} = \dfrac{x(x^2 + 9x + 14)}{(x-2)(x+2)}$

$= \dfrac{x(x+2)(x+7)}{(x-2)(x+2)}$

$= \dfrac{x(x+7)}{x-2}$

**23.** $\dfrac{x^3 - x}{x^3 + 5x^2 - 6x} = \dfrac{x(x-1)(x+1)}{x(x^2 + 5x - 6)}$

$= \dfrac{\cancel{x}(x-1)(x+1)}{\cancel{x}(x+6)(x-1)}$

$= \dfrac{x+1}{x+6}$

**24.** $x - 3 = 0$
$x = 3$

**25.** $x - 8 = 0$
$x = 8$

**26.** $x^2 - 1 = 0$
$(x-1)(x+1) = 0$
$x = 1, -1$

**27.** $x^2 - 9 = 0$
$(x-3)(x+3) = 0$
$x = 3, -3$

**28.** $x^2 + x - 12 = 0$
$(x+4)(x-3) = 0$
$x = -4, 3$

**29.** $x^2 + 5x - 6 = 0$
$(x-1)(x+6) = 0$
$x = 1, -6$

**30.** $\dfrac{(x-1)(x+2)}{(2x+2)(3x+6)} = \dfrac{(x-1)(x+2)}{6(x+1)(x+2)}$

$= \dfrac{x-1}{6(x+1)}$

$\dfrac{3-1}{6(3+1)} = \dfrac{2}{24}$

$= \dfrac{1}{12}$

**31.** $\dfrac{\frac{1}{2}(5x+3)(4x)}{(10x+6)(5x+3)} = \dfrac{(5x+3)(4x)}{4(5x+3)(5x+3)}$

$= \dfrac{x}{5x+3}$

$\dfrac{3}{5(3)+3} = \dfrac{3}{18}$

$= \dfrac{1}{6}$

**32.** $x$ and $2x - 2y = 2(x-y)$

**33.** $\dfrac{2x(x-y)}{x(2x)} = \dfrac{x-y}{x}$

**34.** $72 \le 2x^2 \le 120$
$36 \le x^2 \le 60$
$6 \le x \le 2\sqrt{15}$
Correct values will have
$6 \le x \le 2\sqrt{15}(\approx 7.75)$.
Choose $y$ so that $y = \frac{1}{2}x$.

**35.** *Sample answer:* To create such problems, write the numerator and denominator in factored form, then multiply. For example, $\dfrac{(x+1)(x+2)}{(x+1)(x-2)} = \dfrac{x^2 + 3x + 2}{x^2 - x - 2}$.

**36.** $A = 3250(2\pi R)$
$= 6500\pi R$

**37.** $P = \dfrac{\overset{1625}{\cancel{6500}}\cancel{\pi R}}{\cancel{4\pi R^2}}$

$= \dfrac{1625}{R}$

**38.** $P = \dfrac{1625}{3963}$
$\approx 0.41$

**39. a.**

| x-values | $-2$ | $-1$ | 0 | 1 | 2 | 3 | 4 |
|---|---|---|---|---|---|---|---|
| $\dfrac{x^2 - x - 6}{x - 3}$ | 0 | 1 | 2 | 3 | 4 | undef. | 6 |
| $x + 2$ | 0 | 1 | 2 | 3 | 4 | 5 | 6 |

**b.** The second expression is the simplified form of the first. The two expressions have the same value unless $x = 3$, in which case the first expression is not defined.

**c.** *Sample answer:* Any pair of expressions of the form $\dfrac{(ax+b)(cx+d)}{ax+b}$ and $cx + d$; first I wrote an expression that had a common term in the numerator and denominator, then let the other expression be the other factor of numerator of the first expression. The two expressions will be equal except when $x = -\dfrac{b}{a}$.

**40.** $\dfrac{\frac{x}{5}\left(\frac{x+2}{3}\right)}{x\left(\frac{x+5}{3}\right)} = \dfrac{3(x)(x+2)}{15x(x+5)}$

$= \dfrac{x+2}{5(x+5)}$

**41.** $\dfrac{x^2 + 5x + 6}{x^2 - 2x - 8} = \dfrac{x^2 - 4x - 5}{x^2 - 8x + 15}$

$\dfrac{(x+3)(x+2)}{(x+2)(x-4)} = \dfrac{(x+1)(x-5)}{(x-3)(x-5)}$

$x^2 - 9 = x^2 - 3x - 4$

$3x - 5 = 0$

$x = \dfrac{5}{3}$

### 11.4 Mixed Review (p. 669)

**42.** $-\dfrac{1}{3}$

**43.** $\dfrac{25}{2}$

**44.** $\dfrac{2}{7} \cdot \dfrac{24}{14} = \dfrac{24}{49}$

**45.** $\dfrac{4}{9} \cdot \dfrac{-1}{36} = -\dfrac{1}{81}$

**46.** $\dfrac{9y}{20}$

**47.** $4m^3$

**48.** $\dfrac{36}{45a} \cdot \dfrac{5}{-9a} = -\dfrac{4}{9a^2}$

**49.** $-18c^3 \cdot \dfrac{4}{27c} = -\dfrac{8c^2}{3}$

**276** **Algebra 1**
Chapter 11 Worked-out Solution Key

Copyright © McDougal Littell Inc.
All rights reserved.

# Chapter 11 *continued*

**50.** $192 = \frac{1}{2}(6x)(4x)$     $P = 5x + 5x + 6x$

        $192 = 12x^2$            $= 16x$

          $16 = x^2$          $P = 16(4)$

       $4 \text{ m} = x$           $= 64 \text{ m}$

**51.**

| x | y |
|----|----|
| −2 | 4 |
| 2 | 4 |
| 0 | 0 |

$y - x^2$

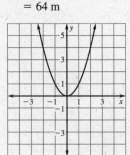

**52.**

| x | y |
|----|----|
| 0 | 4 |
| −2 | 0 |
| 2 | 0 |

$y = 4 - x^2$

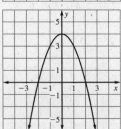

**53.**

| x | y |
|----|----|
| −2 | 2 |
| 2 | 2 |
| 0 | 0 |

$y = \frac{1}{2}x^2$

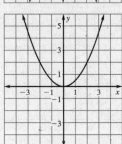

**54.**

| x | y |
|----|----|
| −1 | −4 |
| 0 | −5 |
| 1 | 4 |

$y = 5x^2 + 4x - 5$

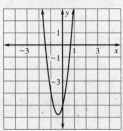

**55.**

| x | y |
|----|----|
| −1 | 11 |
| 0 | 6 |
| 1 | 9 |

$y = 4x^2 - x + 6$

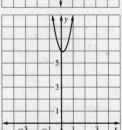

**56.**

| x | y |
|----|----|
| 0 | 7 |
| 1 | 3 |
| −1 | 5 |

$y = -3x^2 - x + 7$

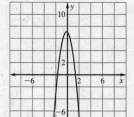

**57.** $y = 2500(0.92)^x$

| Distance... (miles) | 2 | 3 | 4 | 5 | 6 |
|---|---|---|---|---|---|
| Population... (people) | 2116 | 1947 | 1791 | 1648 | 1516 |

$y = 2500(0.92)^2$    $y = 2500(0.92)^3$    $y = 2500(0.92)^4$

   $= 2116$            $= 1946.72$         $= 1790.9824$

$y = 2500(0.92)^5$    $y = 2500(0.92)^6$

   $\approx 1647.70$        $\approx 1515.89$

## Lesson 11.5

### 11.5 Guided Practice (p. 673)

1. Multiply the numerators and denominators, then factor and divide out common factors.

2. Multiply the first expression by the reciprocal of the divisor, multiplying numerators and denominators, then divide out common factors.

3. In the first step, the expression on the left should have been multiplied by the reciprocal of the expression on the right.

   $\dfrac{x + 3}{x - 3} \cdot \dfrac{x^2 - 9}{4x} = \dfrac{(x + 3)(x + 3)(x - 3)}{(x - 3)(4x)}$

                  $= \dfrac{(x + 3)^2}{4x}$

4. $\dfrac{3x}{8x^2} \cdot \dfrac{4x^3}{3x^4} = \dfrac{3 \cdot 4x^2}{8 \cdot 3x^6}$

                $= \dfrac{1}{2x^2}$

5. $\dfrac{x^2 - 1}{x} \cdot \dfrac{2x}{3x - 3} = \dfrac{(x - 1)(x + 1)(2x)}{x(3)(x - 1)}$

                $= \dfrac{2(x + 1)}{3}$

6. $\dfrac{x}{x^2 - 25} \cdot \dfrac{x - 5}{x + 5} = \dfrac{x(x - 5)}{(x - 5)(x + 5)^2}$

                $= \dfrac{x}{(x + 5)^2}$

7. $\dfrac{3x}{x^2 - 2x - 15} \cdot \dfrac{(x + 3)}{1} = \dfrac{3x(x + 3)}{(x + 3)(x - 5)}$

                   $= \dfrac{3x}{x - 5}$

# Chapter 11 *continued*

**8.** $\dfrac{x}{8-2x} \cdot \dfrac{4-x}{2x} = \dfrac{x(4-x)}{4x(4-x)}$

$\qquad = \dfrac{1}{4}$

**9.** $\dfrac{4x^2-25}{4x} \cdot \dfrac{1}{2x-5} = \dfrac{(2x-5)(2x+5)}{4x(2x-5)}$

$\qquad = \dfrac{2x+5}{4x}$

**10.** $\dfrac{x^2-4x+3}{2x} \cdot \dfrac{2}{x-1} = \dfrac{2(x-3)(x-1)}{2x(x-1)}$

$\qquad = \dfrac{x-3}{x}$

**11.** $\dfrac{9x^2+6x+1}{x+5} \cdot \dfrac{x^2+5x}{3x+1} = \dfrac{(3x+1)(3x+1)(x)(x+5)}{(x+5)(3x+1)}$

$\qquad = x(3x+1)$

## 11.5 Practice and Applications (pp. 673–675)

**12.** $\dfrac{4x}{3} \cdot \dfrac{1}{x} = \dfrac{4x}{3x}$

$\qquad = \dfrac{4}{3}$

**13.** $\dfrac{9x^2}{4} \cdot \dfrac{8}{18x} = \dfrac{3^2 \cdot 2^3 x^2}{2^3 \cdot 3^2 x}$

$\qquad = x$

**14.** $\dfrac{7x^2}{6x} \cdot \dfrac{12x^2}{2x} = \dfrac{7 \cdot 2^2 \cdot 3 x^{4}}{2^2 \cdot 3x^2}$

$\qquad = 7x^2$

**15.** $\dfrac{16x^2}{8x} \cdot \dfrac{16x}{4x^2} = \dfrac{2^8 x^3}{2^5 x^3}$

$\qquad = 8$

**16.** $\dfrac{25x^2}{10x} \cdot \dfrac{10x}{5x} = \dfrac{5^3 \cdot 2x^3}{5^2 \cdot 2x^2}$

$\qquad = 5x$

**17.** $\dfrac{13x^4}{7x} \cdot \dfrac{7x}{x^3} = \dfrac{13 \cdot 7x^5}{7x^4}$

$\qquad = 13x$

**18.** $\dfrac{5-2x}{-2} \cdot \dfrac{24}{10-4x} = \dfrac{^{6}24(5-2x)}{-4(5-2x)}$

$\qquad = -6$

**19.** $\dfrac{4x}{x^2-9} \cdot \dfrac{x-3}{8x^2+12x} = \dfrac{4x(x-3)}{(x-3)(x+3)(4x)(2x+3)}$

$\qquad = \dfrac{1}{(x+3)(2x+3)}$

**20.** $\dfrac{-3}{x-4} \cdot \dfrac{x-4}{12(x-7)} = \dfrac{-3(x-4)}{_{4}12(x-4)(x-7)}$

$\qquad = \dfrac{-1}{4(x-7)}$

**21.** $\dfrac{3x^2}{10} \cdot \dfrac{25}{9x^3} = \dfrac{(3x^2)5^{2}}{(2)(5)3^2 x^3}$

$\qquad = \dfrac{5}{6x}$

**22.** $\dfrac{x}{x+2} \cdot \dfrac{x+2}{x+5} = \dfrac{x(x+2)}{(x+2)(x+5)}$

$\qquad = \dfrac{x}{x+5}$

**23.** $\dfrac{5x+15}{3x} \cdot \dfrac{9x}{x+3} = \dfrac{^{15}45x(x+3)}{3x(x+3)}$

$\qquad = 15$

**24.** $\dfrac{2(x+2)}{5(x-3)} \cdot \dfrac{5x-15}{4(x-2)} = \dfrac{10(x+2)(x-3)}{_{2}20(x-3)(x-2)}$

$\qquad = \dfrac{x+2}{2(x-2)}$

**25.** $\dfrac{x^2-36}{-5x^2} \cdot \dfrac{1}{x-6} = \dfrac{(x-6)(x+6)}{-5x^2(x-6)}$

$\qquad = -\dfrac{x+6}{5x^2}$

**26.** $\dfrac{8}{2+3x} \cdot \dfrac{8+12x}{1} = \dfrac{32(2+3x)}{2+3x}$

$\qquad = 32$

**27.** $\dfrac{3x}{x^2-2x-24} \cdot \dfrac{x-6}{6x^2+9x} = \dfrac{3x(x-6)}{3x(x-6)(x+4)(2x+3)}$

$\qquad = \dfrac{1}{(x+4)(2x+3)}$

**28.** $\dfrac{x}{3x^2+2x-8} \cdot \dfrac{3x-4}{1} = \dfrac{x(3x-4)}{(3x-4)(x+2)}$

$\qquad = \dfrac{x}{x+2}$

**29.** $\dfrac{x+1}{x^3(3-x)} \cdot \dfrac{x(x-3)}{5} = \dfrac{x(x+1)(x-3)}{-5x^3(x-3)}$

$\qquad = -\dfrac{x+1}{5x^2}$

**30.** $\dfrac{4x^2+x-3}{1} \cdot \dfrac{1}{(4x+3)(x-1)} = \dfrac{(4x-3)(x+1)}{(4x+3)(x-1)}$

**31.** $\dfrac{x^2-8x+15}{x^2-3x} \cdot \dfrac{1}{3x-15} = \dfrac{(x-5)(x-3)}{3x(x-3)(x-5)}$

$\qquad = \dfrac{1}{3x}$

**32.** $\dfrac{6x^2+7x-33}{x+4} \cdot \dfrac{1}{6x-11} = \dfrac{(6x-11)(x+3)}{(x+4)(6x-11)}$

$\qquad = \dfrac{x+3}{x+4}$

**33.** $\left(\dfrac{x^2}{5} \cdot \dfrac{x+2}{2}\right) \cdot \dfrac{30}{x} = \dfrac{x^2(x+2)(30)^{3}}{10x}$

$\qquad = 3x(x+2)$

**34.** $\left(\dfrac{2x^2}{3} \cdot \dfrac{5}{x}\right) \cdot \dfrac{25}{6x^2} = \dfrac{^{5}10x(25)}{3(6x^2)_{3}}$

$\qquad = \dfrac{125}{9x}$

**35.** $A = \dfrac{6300-800t}{1-0.12t} \cdot \dfrac{10-t}{222-24t}$

$\qquad = \dfrac{100(63-8t)(10-t)}{6(1-0.12t)(37-4t)}$

$\qquad = \dfrac{100(63-8t)(10-t)}{3t(1-0.12t)(37-4t)}$

$\quad A = \dfrac{50(63-8t)(10-t)}{3(1-0.12t)(37-4t)}$

# Chapter 11 continued

**36.** $\dfrac{50(63-40)(5)}{3(1-0.6)(37-20)} = \dfrac{5750}{20.4}$

$\qquad\qquad\qquad\qquad \approx 282 \text{ mi}$

**37.** $\dfrac{50(63-120)(10-15)}{3(1-1.8)(37-60)} = \dfrac{14,250}{55.2} \approx 258 \text{ mi}$

**38.** Total sales in billions of dollars:

services: $\dfrac{1055}{1} = \$1{,}055$

hotel services: $\dfrac{46}{1} = \$46$

auto repair services : $\dfrac{48}{1} = \$48$

**39.** $R = \dfrac{46+0.7t}{1-0.04t} \cdot \dfrac{1-0.04t}{1055+23t}$

$\quad R = \dfrac{46+0.7t}{1055+23t}$

| | | | |
|---|---|---|---|
| 1990: | 0.04360 | 1994: | 0.04255 |
| 1991: | 0.04332 | 1995: | 0.04231 |
| 1992: | 0.04305 | 1996: | 0.04208 |
| 1993: | 0.04279 | 1997: | 0.04186 |

decreasing

**40.** $R = \dfrac{48-t}{1-0.06t} \cdot \dfrac{1-0.04t}{1055+23t}$

$\quad R = \dfrac{(48-t)(1-0.04t)}{(1-0.06t)(1055+23t)}$

| | | | |
|---|---|---|---|
| 1990: | 0.04550 | 1994: | 0.04240 |
| 1991: | 0.04453 | 1995: | 0.04200 |
| 1992: | 0.04368 | 1996: | 0.04181 |
| 1993: | 0.04297 | 1997: | 0.04186 |

The ratio was decreasing from 1990 to 1996 then increased in 1997.

**41.** In general, sales of hotel services and auto repair services were not increasing as rapidly as the total sales of service.

**42.** 

| Solution Steps | Explanation |
|---|---|
| 1. $\dfrac{ac}{bc} = \dfrac{a}{b} \cdot \dfrac{c}{c}$ | 1. Apply the rule for multiplying rational expressions. |
| 2. $\dfrac{ac}{bc} = \dfrac{a}{b} \cdot 1$ | 2. Any nonzero number divided by itself is 1. |
| 3. $\dfrac{ac}{bc} = \dfrac{a}{b}$ | 3. Any nonzero number multiplied by 1 is itself. |

**43.** $\dfrac{2x-4}{x^2-4} = \dfrac{2(x-2)}{(x+2)(x-2)}$

$\qquad = \dfrac{2}{x+2} \cdot \dfrac{x-2}{x-2}$

$\qquad = \dfrac{2}{x+2} \cdot 1$

$\qquad = \dfrac{2}{x+2}$

**44.** $\dfrac{x^2-3x}{x^2-5x+6} \cdot \dfrac{(x-2)^2}{2x} = \dfrac{x(x-3)(x-2)^2}{2x(x-3)(x-2)}$

$\qquad\qquad\qquad\qquad = \dfrac{x-2}{2} \qquad \text{C}$

**45.** D $\qquad$ **46.** $\dfrac{n-r}{n} \cdot \dfrac{r}{n} = \dfrac{r(n-r)}{n^2}$ $\qquad$ **47.** $\dfrac{(x-y)^2}{x^2}$

### 11.5 Mixed Review (p. 675)

**48.** $\dfrac{3}{4}, \dfrac{2}{5}$

$\quad 4: 2^2$

$\quad 5: 5$

$\quad \text{LCD: } 2^2 \cdot 5 = 20$

**49.** $\dfrac{2}{9}, \dfrac{3}{18}$

$\quad 9: 3^2$

$\quad 18: 3^2 \cdot 2$

$\quad \text{LCD: } 3^2 \cdot 2 = 18$

**50.** $\dfrac{1}{16}, \dfrac{9}{20}$

$\quad 16: 2^4$

$\quad 20: 2^2 \cdot 5$

$\quad \text{LCD: } 2^4 \cdot 5 = 80$

**51.** $\dfrac{14}{54}, \dfrac{31}{81}$

$\quad 54: 3^3 \cdot 2$

$\quad 81: 3^4$

$\quad \text{LCD: } 3^4 \cdot 2 = 162$

**52.** $x = \dfrac{-b \pm \sqrt{b^2-4ac}}{2a}$

$\quad x = \dfrac{-12 \pm \sqrt{144-4(2)(-6)}}{2(2)}$

$\quad = \dfrac{-12 \pm \sqrt{192}}{4}$

$\quad = \dfrac{-12 \pm 8\sqrt{3}}{4}$

$\quad = -3 + 2\sqrt{3}, -3 - 2\sqrt{3}$

**53.** $x = \dfrac{6 \pm \sqrt{36-4(1)(7)}}{2(1)}$

$\quad = \dfrac{6 \pm \sqrt{8}}{2}$

$\quad = \dfrac{6 \pm 2\sqrt{2}}{2}$

$\quad = 3 + \sqrt{2}, 3 - \sqrt{2}$

**54.** $x = \dfrac{-11 \pm \sqrt{121-4(3)(10)}}{2(3)}$

$\quad = \dfrac{-11 \pm 1}{6}$

$\quad = -\dfrac{5}{3}, -2$

**55.** $4t^2 + 5t + 2 - t^2 + 3t + 8 = 3t^2 + 8t + 10$

**56.** $16p^3 - p^2 + 24 + 12p^2 - 8p - 16$

$\qquad = 16p^3 + 11p^2 - 8p + 8$

**57.** $a^4 - 12a + 4a^3 + 11a - 1 = a^4 + 4a^3 - a - 1$

**58.** $-5x^2 + 2x - 12 - 6 + 9x + 7x^2 = 2x^2 + 11x - 18$

**59.** $1000(1+r)^2 = 1000(1 + 2r + r^2)$

$\qquad\qquad\qquad = 1000r^2 + 2000r + 1000$

# Chapter 11 *continued*

## Lesson 11.6

### 11.6 Guided Practice (p. 679)

**1.** The least common multiple of the denominators of the expressions is the least number that is a multiple of both denominators.

**2.** When $2x - 2$ is subtracted from $3x$, the result is $3x - 2x + 2$; the correct answer is $\dfrac{x + 2}{x(x - 1)}$.

**3.** $-1$

**4.** $\dfrac{1}{3x} + \dfrac{5}{3x} = \dfrac{6}{3x} = \dfrac{2}{x}$

**5.** $\dfrac{5x}{x + 4} + \dfrac{20}{4 + x} = \dfrac{5(x + 4)}{x + 4} = 5$

**6.** $\dfrac{x}{x^2 - 9} - \dfrac{3x + 1}{x^2 - 9} = \dfrac{x - 3x - 1}{x^2 - 9} = \dfrac{-2x - 1}{x^2 - 9}$

**7.** $\dfrac{3}{10x} - \dfrac{1}{4x^2}$    $10x: 2 \cdot 5 \cdot x$

$\qquad\qquad\qquad 4x^2: 2^2 \cdot x^2$

$\qquad\qquad$ LCD: $2^2 \cdot 5 \cdot x^2 = 20x^2$

$\dfrac{6x}{20x^2} - \dfrac{5}{20x^2} = \dfrac{6x - 5}{20x^2}$

**8.** $\dfrac{x + 6}{x + 1} - \dfrac{4}{2x + 3} = \dfrac{(x + 6)(2x + 3) - 4(x + 1)}{(x + 1)(2x + 3)}$

$\qquad = \dfrac{2x^2 + 15x + 18 - 4x - 4}{(x + 1)(2x + 3)}$

$\qquad = \dfrac{2x^2 + 11x + 14}{(x + 1)(2x + 3)}$

$\qquad = \dfrac{(2x + 7)(x + 2)}{(x + 1)(2x + 3)}$

**9.** $\dfrac{x - 2}{2x - 10} + \dfrac{x + 3}{x - 5} = \dfrac{x - 2}{2x - 10} + \dfrac{2(x + 3)}{2x - 10}$

$\qquad = \dfrac{x - 2 + 2x + 6}{2x - 10}$

$\qquad = \dfrac{3x + 4}{2(x - 5)}$

### 11.6 Practice and Applications (pp. 679–681)

**10.** $\dfrac{7}{2x} + \dfrac{x + 2}{2x} = \dfrac{7 + x + 2}{2x} = \dfrac{9 + x}{2x}$

**11.** $\dfrac{4}{x + 1} + \dfrac{2x - 2}{x + 1} = \dfrac{4 + 2x - 2}{x + 1}$

$\qquad = \dfrac{2x + 2}{x + 1}$

$\qquad = \dfrac{2(x + 1)}{x + 1}$

$\qquad = 2$

**12.** $\dfrac{7x}{x^3} - \dfrac{6x}{x^3} = \dfrac{x}{x^3}$

$\qquad = \dfrac{1}{x^2}$

**13.** $\dfrac{2}{3x - 1} - \dfrac{5x}{3x - 1} = \dfrac{2 - 5x}{3x - 1}$

**14.** $\dfrac{3x}{4x + 1} + \dfrac{5x}{4x + 1} = \dfrac{8x}{4x + 1}$

**15.** $\dfrac{-8}{3x^2} + \dfrac{11}{3x^2} = \dfrac{3}{3x^2}$

$\qquad = \dfrac{1}{x^2}$

**16.** $\dfrac{9}{4x} + \dfrac{7}{-5x} = \dfrac{-45 + 28}{-20x}$

$\qquad = \dfrac{17}{20x}$

**17.** $\dfrac{11}{6x} + \dfrac{2}{13x} = \dfrac{143 + 12}{78x}$

$\qquad = \dfrac{155}{78x}$

**18.** $\dfrac{9}{5x} - \dfrac{2}{x^2} = \dfrac{9x - 10}{5x^2}$

**19.** $\dfrac{4}{x + 4} - \dfrac{7}{x - 2} = \dfrac{4(x - 2) - 7(x + 4)}{(x + 4)(x - 2)}$

$\qquad = \dfrac{4x - 8 - 7x - 28}{(x + 4)(x - 2)}$

$\qquad = \dfrac{-3x - 36}{(x + 4)(x - 2)}$

$\qquad = -\dfrac{3(x + 12)}{(x + 4)(x - 2)}$

**20.** $\dfrac{3}{x + 3} + \dfrac{4x}{2x + 6} = \dfrac{6 + 4x}{2x + 6}$

$\qquad = \dfrac{3 + 2x}{x + 3}$

**21.** $\dfrac{9}{x^2 - 3x} + \dfrac{3}{x - 3} = \dfrac{9 + 3x}{x^2 - 3x}$

$\qquad = \dfrac{3(x + 3)}{x(x - 3)}$

**22.** $\dfrac{2x}{x - 1} - \dfrac{7x}{x + 4} = \dfrac{2x(x + 4) - 7x(x - 1)}{(x - 1)(x + 4)}$

$\qquad = \dfrac{2x^2 + 8x - 7x^2 + 7x}{(x - 1)(x + 4)}$

$\qquad = \dfrac{-5x^2 + 15x}{(x - 1)(x + 4)}$

$\qquad = -\dfrac{5x(x - 3)}{(x - 1)(x + 4)}$

**23.** $\dfrac{x}{x - 10} + \dfrac{x + 4}{x + 6} = \dfrac{x(x + 6) + (x + 4)(x - 10)}{(x - 10)(x + 6)}$

$\qquad = \dfrac{x^2 + 6x + x^2 - 6x - 40}{(x - 10)(x + 6)}$

$\qquad = \dfrac{2x^2 - 40}{(x - 10)(x + 6)}$

$\qquad = \dfrac{2(x^2 - 20)}{(x - 10)(x + 6)}$

# Chapter 11 *continued*

**24.** $\dfrac{4x}{5x-2} - \dfrac{2x}{5x+1} = \dfrac{4x(5x+1) - 2x(5x-2)}{(5x-2)(5x+1)}$

$\qquad = \dfrac{20x^2 + 4x - 10x^2 + 4x}{(5x-2)(5x+1)}$

$\qquad = \dfrac{10x^2 + 8x}{(5x-2)(5x+1)}$

$\qquad = \dfrac{2x(5x+4)}{(5x-2)(5x+1)}$

**25.** $\dfrac{x+8}{3x-1} + \dfrac{x+3}{x+1} = \dfrac{(x+8)(x+1) + (3x-1)(x+3)}{(3x-1)(x+1)}$

$\qquad = \dfrac{x^2 + 9x + 8 + 3x^2 + 8x - 3}{(3x-1)(x+1)}$

$\qquad = \dfrac{4x^2 + 17x + 5}{(3x-1)(x+1)}$

**26.** $\dfrac{3x+10}{7x-4} - \dfrac{x}{4x+3} = \dfrac{(3x+10)(4x+3) - x(7x-4)}{(7x-4)(4x+3)}$

$\qquad = \dfrac{12x^2 + 49x + 30 - 7x^2 + 4x}{(7x-4)(4x+3)}$

$\qquad = \dfrac{5x^2 + 53x + 30}{(7x-4)(4x+3)}$

**27.** $\dfrac{2x+1}{3x-1} - \dfrac{x+4}{x-2} = \dfrac{(2x+1)(x-2) - (3x-1)(x+4)}{(3x-1)(x-2)}$

$\qquad = \dfrac{2x^2 - 3x - 2 - 3x^2 - 11x + 4}{(3x-1)(x-2)}$

$\qquad = \dfrac{-x^2 - 14x + 2}{(3x-1)(x-2)}$

$\qquad = -\dfrac{x^2 + 14x - 2}{(3x-1)(x-2)}$

**28.** $\dfrac{x}{x^2+5x-24} + \dfrac{8}{x^2+5x-24} = \dfrac{x+8}{(x+8)(x-3)}$

$\qquad = \dfrac{1}{x-3}$

**29.** $\dfrac{x^2+1}{x^2-4} + \dfrac{5x}{x^2-4} - \dfrac{2x+11}{x^2-4} = \dfrac{x^2 + 3x - 10}{(x-2)(x+2)}$

$\qquad = \dfrac{(x+5)(x-2)}{(x-2)(x+2)}$

$\qquad = \dfrac{x+5}{x+2}$

**30.** $\dfrac{x^2-9}{x+3} + \dfrac{x^2+9}{x-3} = \dfrac{(x^2-9)(x-3) + (x+3)(x^2+9)}{(x+3)(x-3)}$

$\qquad = \dfrac{(x-3)^2 + (x^2+9)}{x-3}$

$\qquad = \dfrac{x^2 - 6x + 9 + x^2 + 9}{x-3}$

$\qquad = \dfrac{2x^2 - 6x + 18}{x-3}$

$\qquad = \dfrac{2(x^2 - 3x + 9)}{x-3}$

**31.** $\dfrac{2}{x+1} + \dfrac{3}{x-2} + \dfrac{3}{x+4}$

$\qquad = \dfrac{2(x-2)(x+4) + 3(x+1)(x+4) + 3(x+1)(x-2)}{(x+1)(x-2)(x+4)}$

$\qquad = \dfrac{2x^2 + 4x - 16 + 3x^2 + 15x + 12 + 3x^2 - 3x - 6}{(x+1)(x-2)(x+4)}$

$\qquad = \dfrac{8x^2 + 16x - 10}{(x+1)(x-2)(x+4)}$

$\qquad = \dfrac{2(4x^2 + 8x - 5)}{(x+1)(x-2)(x+4)}$

**32.** $2\left(\dfrac{2x-5}{x-2}\right) + 2\left(\dfrac{x^2-3}{x-2}\right) = \dfrac{4x - 10 + 2x^2 - 6}{x-2}$

$\qquad = \dfrac{2x^2 + 4x - 16}{x-2}$

$\qquad = \dfrac{2(x^2 + 2x - 8)}{x-2}$

$\qquad = \dfrac{2(x+4)(x-2)}{x-2}$

$\qquad = 2(x+4)$

**33.** $\dfrac{6x+10}{2x-1} + \dfrac{4x-2}{x-3}$

$\qquad = \dfrac{(x-3)(6x+10) + (2x-1)(4x-2)}{(2x-1)(x-3)}$

$\qquad = \dfrac{6x^2 - 8x - 30 + 8x^2 - 8x + 2}{(2x-1)(x-3)}$

$\qquad = \dfrac{14x^2 - 16x - 28}{(2x-1)(x-3)}$

$\qquad = \dfrac{2(7x^2 - 8x - 14)}{(2x-1)(x-3)}$

**34.**

| $x$ | 10 | 100 | 1000 |
|---|---|---|---|
| $2x-5$ | 15 | 195 | 1995 |
| $x-2$ | 8 | 98 | 998 |
| $\dfrac{2x-5}{x-2}$ | 1.875 | 1.9898 | 1.999 |

| $x$ | 10,000 | 100,000 | 1,000,000 |
|---|---|---|---|
| $2x-5$ | 19,995 | 199,995 | 1,999,995 |
| $x-2$ | 9998 | 99,998 | 999,998 |
| $\dfrac{2x-5}{x-2}$ | 1.9999 | 1.9999 | 1.9999 |

**35.** The numerator values get large; the denominator values get large; the rational expression values approach 2. As $x$ gets very large, the numbers subtracted in the numerator and denominator become insignificantly small compared to $2x$ and $x$, so the rational expression gets closer and closer to $\dfrac{2x}{x} = 2$.

# Chapter 11 *continued*

**36.** $\dfrac{24}{x-2} + \dfrac{24}{x+2}$

**37.** $\dfrac{24x + 48 + 24x - 48}{(x-2)(x+2)} = \dfrac{48x}{(x-2)(x+2)}$

**38.** $\dfrac{48(10)}{8(12)} = 5\text{ h}$

**39.** $\dfrac{1}{x} = $ partner's rate

$\dfrac{1}{x} + \dfrac{1}{35} = $ combined rate

**40.** $\dfrac{35 + x}{35x}$

**41. a.** partner A: $\dfrac{35 + 40}{35(40)} = \dfrac{75}{1400}$  **b.** partner A or B

$= \dfrac{3}{56}$

partner B: $\dfrac{35 + 45}{35(45)} = \dfrac{80}{1575}$

$= \dfrac{16}{315}$

partner C: $\dfrac{35 + 55}{35(55)} = \dfrac{90}{1925}$

$= \dfrac{18}{385}$

**42.** $\dfrac{2x}{x+5} - \dfrac{3x+2}{x+5} - \dfrac{4}{x+5} = \dfrac{2x - 3x - 2 - 4}{x+5}$

$= \dfrac{-(x+6)}{x+5}$

**43.** $\left(\dfrac{3x^2}{56}\right)\left(\dfrac{3}{x} + \dfrac{5}{x}\right) = \dfrac{3x^2}{56} \cdot \dfrac{8}{x}$

$= \dfrac{24x^2}{56x}$

$= \dfrac{3x}{7}$

**44.** $\left(\dfrac{3x-5}{x} + \dfrac{1}{x}\right) \cdot \left(\dfrac{6x-8}{x}\right) = \dfrac{3x-4}{x} \cdot \dfrac{6x-8}{x}$

$= \dfrac{18x^2 - 48x - 32}{x^2}$

$= \dfrac{2(9x^2 - 24x - 16)}{x^2}$

$= \dfrac{2(3x-4)^2}{x^2}$

**45.** $\dfrac{(x-2)}{x+6} \cdot \dfrac{4x-24}{x+8} \cdot \dfrac{x-8}{(x-2)} = \dfrac{4(x-6)(x-8)}{(x+6)(x+8)}$

**46.** $\dfrac{2}{x-3} + \dfrac{x}{x^2 + 3x - 18} = \dfrac{2}{x-3} + \dfrac{x}{(x-3)(x+6)}$

$= \dfrac{2(x+6) + x}{(x-3)(x+6)}$

$= \dfrac{3(x+4)}{(x-3)(x+6)}$

**47.** $\dfrac{2}{(x-2)(x+2)} + \dfrac{3}{(x+3)(x-2)} = \dfrac{2(x+3) + 3(x+2)}{(x-2)(x+2)(x+3)}$

$= \dfrac{5x + 12}{(x-2)(x+2)(x+3)}$

**48.** $\dfrac{7x+2}{-1(x-4)(4+x)} + \dfrac{7}{x-4} = \dfrac{7x + 2 + (-1)(7)(x+4)}{-(x-4)(x+4)}$

$= \dfrac{26}{(x-4)(x+4)}$

**49.** $\dfrac{5x-1}{(2x+3)(x-5)} - \dfrac{-3x+4}{(2x+3)(x+1)}$

$= \dfrac{(5x-1)(x+1) - (-3x+4)(x-5)}{(2x+3)(x-5)(x+1)}$

$= \dfrac{5x^2 + 4x - 1 + 3x^2 - 19x + 20}{(2x+3)(x-5)(x+1)}$

$= \dfrac{8x^2 - 15x + 19}{(2x+3)(x-5)(x+1)}$

**50.** B

**51.** $\dfrac{x(2x+1) - (x-1)}{(x-1)(2x+1)} = \dfrac{2x^2 + 1}{(x-1)(2x+1)}$  C

**52.** $\dfrac{350}{s} - \dfrac{350}{s+5}$  C

**53.** $\dfrac{d}{x} + \dfrac{d}{y}; \dfrac{d(y+x)}{xy}$

**54.** $2d \div \dfrac{d(y+x)}{xy} = \dfrac{2dxy}{d(y+x)} = \dfrac{2xy}{y+x}$

**55.** The variable $d$ that represents the distance is not involved in the expression, which indicates that the average speed is not dependent on the distance traveled; the average speed is not changed.

### 11.6 Mixed Review (p. 682)

**56.** $\dfrac{1}{2x}$  **57.** $\dfrac{2m}{3}$  **58.** $\dfrac{1}{2x^4}$  **59.** $\dfrac{7x}{y^6}$

**60.** $6x^2 = 5x - 7$

$6x^2 - 5x + 7 = 0$

**61.** $9 - 6x = 2x^2$

$2x^2 + 6x - 9 = 0$

**62.** $-4 + 3y^2 = y$

$3y^2 - y - 4 = 0$

**63.**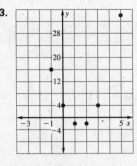

quadratic

**64.**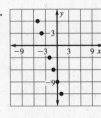

linear

# Chapter 11 *continued*

**65.** $|x - 8500| \le 1000$

## Quiz 2 (p. 682)

**1.** $\dfrac{3x}{2}$

**2.** $\dfrac{5}{11 + x}$

**3.** $\dfrac{3 - x}{(x - 3)(x - 2)} = \dfrac{-1}{x - 2}$

**4.** $\dfrac{(x - 4)(x - 3)}{x^2 + 3x + 18}$     cannot be simplified

**5.** $\dfrac{\cancel{8} \cdot 2 \cdot 7x^{\cancel{4}2}}{2^{\cancel{2}} \cdot \cancel{8}x^2} = \dfrac{7x^2}{2}$

**6.** $\dfrac{\cancel{10}20(5 + 2x)}{2(5 + 2x)} = 10$

**7.** $\dfrac{3(x + 4)}{4x} \cdot \dfrac{2x}{x + 4} = \dfrac{3}{2}$

**8.** $\dfrac{5(x^2 - 6x + 9)}{x + 2} \cdot \dfrac{1}{5(x - 3)} = \dfrac{\cancel{5}(x - 3)(x - 3)}{\cancel{5}(x + 2)(x - 3)}$

$$= \dfrac{x - 3}{x + 2}$$

**9.** $\dfrac{x + 5}{(x - 7)(x + 5)} = \dfrac{1}{x - 7}$

**10.** $\dfrac{(4x - 1) - (x - 6)}{(3x + 5)(x + 1)} = \dfrac{3x + 5}{(3x + 5)(x + 1)}$

$$= \dfrac{1}{x + 1}$$

**11.** $\dfrac{15}{x + 2}$ : downstream

$\dfrac{15}{x - 2}$ : upstream

**12.** $\dfrac{15(x - 2) + 15(x + 2)}{(x - 2)(x + 2)} = \dfrac{15(2x)}{(x - 2)(x + 2)}$

$$= \dfrac{30x}{(x - 2)(x + 2)}$$

**13.** $\dfrac{30(4)}{2(6)} = 10$ h

## Lesson 11.7

### Developing Concepts Activity 11.7 (p. 683)
### Exploring the Concept

**1.**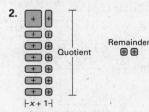

**2.**
├─ x + 3 ─┤

**3.**
├─ x + 3 ─┤
 x + 3 does not divide evenly into $x^2 + 4x + 4$

**4.**
├─ x + 3 ─┤
quotient = x + 1, remainder = 1

## Drawing Conclusions

**1.**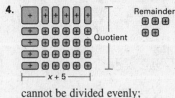
├ x + 2 ┤

cannot be divided evenly;

quotient = x + 2

remainder = 4

**2.**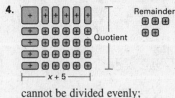
├ x + 1 ┤

cannot be divided evenly;

quotient = x + 6

remainder = 2

**3.**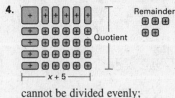
├─── x + 4 ───┤

cannot be divided evenly;

quotient = x + 2

remainder = 4

**4.**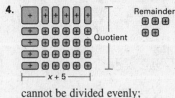
├───── x + 5 ─────┤

cannot be divided evenly;

quotient = x + 4

remainder = 5

**5.** $x^2 + 7x + 18$; $x + 3$; 6

**6.** The model without the remainder represents the product $(x + 4)(x + 3)$, so the whole model represents the product of the quotient and the divisor plus the remainder; $(x + 4)(x + 3) = x^2 + 7x + 12$; $x^2 + 7x + 12 + 6 = x^2 + 7x + 18$.

### Activity (p. 684)

**1.** *Sample answer:* Write the division as a rational expression; divide by dividing each term by the denominator:

$$\dfrac{2x^2 - 3}{6x} = \dfrac{2x^2}{6x} - \dfrac{3}{6x} = \dfrac{x}{3} - \dfrac{1}{2x}.$$

**2.** *Sample answer:* Multiply $\dfrac{x}{3} - \dfrac{1}{2x}$ by $6x$:

$$6x\left(\dfrac{x}{3} - \dfrac{1}{2x}\right) = 2x^2 - 3.$$

# Chapter 11 *continued*

**1.** The method is the same; the quantities used are polynomials instead of digits.

**2.** Exs. 10, 11, 13; In Exs. 10 and 13 you are dividing by a monomial. In Ex. 11, you can divide by factoring the dividend and then canceling.

**3.** $x + 1 - \dfrac{4}{x - 1}$

**4.** Multiply the quotient by the divisor and add the remainder; the result should be the dividend.

**5.** $y + 2 \overline{)y^2 + 0y + 8}$

**6.** $-x + 3 \overline{)-x^2 - 4x + 21}$

**7.** $3y + 5 \overline{)8y^2 - 2y + 0}$

**8.** $x - 6 \overline{)x^2 - 18x + 72}$

**9.**
$$29\tfrac{15}{29}$$
$$29 \overline{)856}$$
$$\underline{-58}$$
$$276$$
$$\underline{-261}$$
$$15$$

**10.** $\dfrac{18x^2 + 45x - 36}{9x}$

$$= \dfrac{18x^2}{9x} + \dfrac{45x}{9x} - \dfrac{36}{9x}$$

$$= 2x + 5 - \dfrac{4}{x}$$

**11.** $\dfrac{x^2 - 8x + 15}{x - 3} = \dfrac{(x - 3)(x - 5)}{x - 3}$

$$= x - 5$$

**12.**
$$y + 3$$
$$y + 3 \overline{)y^2 + 6y + 2}$$
$$\underline{y^2 + 3y}$$
$$3y + 2$$
$$\underline{3y + 9}$$
$$-7$$

$$y + 3 - \dfrac{7}{y + 3}$$

**13.** $\dfrac{10b^3 - 8b^2 - 5b}{-2b} = \dfrac{10b^3}{-2b} + \dfrac{-8b^2}{-2b} + \dfrac{-5b}{-2b}$

$$= -5b^2 + 4b + \dfrac{5}{2}$$

**14.**
$$\tfrac{2}{3}x + 1$$
$$3x - 6 \overline{)2x^2 - x + 4}$$
$$\underline{2x^2 - 4x}$$
$$3x + 4$$
$$\underline{3x - 6}$$
$$10$$

$$\tfrac{2}{3}x + 1 + \dfrac{10}{3x - 6}$$

**15.** $\dfrac{8x + 13}{2} = \dfrac{8x}{2} + \dfrac{13}{2}$

$$= 4x + \dfrac{13}{2}$$

**16.** $\dfrac{16y - 9}{4} = \dfrac{16y}{4} - \dfrac{9}{4}$

$$= 4y - \dfrac{9}{4}$$

**17.** $\dfrac{9c^2 + 3c}{c} = 9c + 3$

**18.** $\dfrac{9m^3 + 4m^2 - 8m}{m} = 9m^2 + 4m - 8$

**19.** $\dfrac{-2x^2 - 12x}{-2x} = x + 6$

**20.** $\dfrac{7p^3 + 18p^2}{p^2} = 7p + 18$

**21.** $\dfrac{9a^2 - 54a - 36}{3a} = 3a - 18 - \dfrac{12}{a}$

**22.** $\dfrac{16y^3 - 36y^2 - 64}{-4y^2} = -4y + 9 + \dfrac{16}{y^2}$

**23.** D
$$5x - 8$$
$$x + 2 \overline{)5x^2 + 2x + 3}$$
$$\underline{5x^2 + 10x}$$
$$-8x + 3$$
$$\underline{-8x - 16}$$
$$19$$

$$5x - 8 + \dfrac{19}{x + 2}$$

**24.** A
$$6$$
$$2x - 3 \overline{)12x - 5}$$
$$\underline{12x - 18}$$
$$13$$

$$6 + \dfrac{13}{2x - 3}$$

**25.** B $\dfrac{10x^2 - 7x - 12}{2x - 3} = \dfrac{(2x - 3)(5x + 4)}{2x - 3}$

**26.** C $\dfrac{2x^2 + 5x + 2}{x + 2} = \dfrac{(x + 2)(2x + 1)}{x + 2}$

**27.** $\dfrac{a^2 - 3a + 2}{a - 1} = \dfrac{(a - 1)(a - 2)}{a - 1}$

$$= a - 2$$

**28.**
$$y + 3$$
$$y + 2 \overline{)y^2 + 5y + 7}$$
$$\underline{y^2 + 2y}$$
$$3y + 7$$
$$\underline{3y + 6}$$
$$1$$

$$y + 3 + \dfrac{1}{y + 2}$$

**29.**
$$2b + 1$$
$$b - 2 \overline{)2b^2 - 3b - 4}$$
$$\underline{2b^2 - 4b}$$
$$b - 4$$
$$\underline{b - 2}$$
$$-2$$

$$2b + 1 - \dfrac{2}{b - 2}$$

**30.** $\dfrac{3p^2 + 10p + 3}{p + 3} = \dfrac{(p + 3)(3p + 1)}{p + 3}$

$$= 3p + 1$$

# Chapter 11 *continued*

**31.**
$$\begin{array}{r} 5g - 1 \phantom{{}-2} \\ g + 3 \overline{\smash{\big)}\, 5g^2 + 14g - 2} \\ \underline{5g^2 + 15g} \phantom{{}-2} \\ -1g - 2 \\ \underline{-1g - 3} \\ 1 \end{array}$$

$$5g - 1 + \frac{1}{g + 3}$$

**32.** $\dfrac{c^2 - 25}{c - 5} = \dfrac{(c + 5)(c - 5)}{c - 5}$

$$= c + 5$$

**33.**
$$\begin{array}{r} x + 6 \phantom{{}-59} \\ x - 9 \overline{\smash{\big)}\, x^2 - 3x - 59} \\ \underline{x^2 - 9x} \phantom{{}-59} \\ 6x - 59 \\ \underline{6x - 54} \\ -5 \end{array}$$

$$x + 6 - \frac{5}{x - 9}$$

**34.**
$$\begin{array}{r} d + 10 \phantom{{}+45} \\ d + 5 \overline{\smash{\big)}\, d^2 + 15d + 45} \\ \underline{d^2 + 5d} \phantom{{}+45} \\ 10d + 45 \\ \underline{10d + 50} \\ -5 \end{array}$$

$$d + 10 - \frac{5}{d + 5}$$

**35.**
$$\begin{array}{r} -x - 4 \phantom{{}-16} \\ x + 2 \overline{\smash{\big)}\, -x^2 - 6x - 16} \\ \underline{-x^2 - 2x} \phantom{{}-16} \\ -4x - 16 \\ \underline{-4x - 8} \\ -8 \end{array}$$

$$-x - 4 - \frac{8}{x + 2}$$

**36.**
$$\begin{array}{r} x - 11 \phantom{{}-12} \\ -x - 2 \overline{\smash{\big)}\, -x^2 + 9x - 12} \\ \underline{-x^2 - 2x} \phantom{{}-12} \\ 11x - 12 \\ \underline{11x + 22} \\ -34 \end{array}$$

$$x - 11 + \frac{34}{x + 2}$$

**37.**
$$\begin{array}{r} b - 10 \phantom{{}+4} \\ b + 3 \overline{\smash{\big)}\, b^2 - 7b + 4} \\ \underline{b^2 + 3b} \phantom{{}+4} \\ -10b + 4 \\ \underline{-10b - 30} \\ 34 \end{array}$$

$$b - 10 + \frac{34}{b + 3}$$

**38.**
$$\begin{array}{r} 3m + 2 \phantom{{}+5} \\ m - 3 \overline{\smash{\big)}\, 3m^2 - 7m + 5} \\ \underline{3m^2 - 9m} \phantom{{}+5} \\ 2m + 5 \\ \underline{2m - 6} \\ 11 \end{array}$$

$$3m + 2 + \frac{11}{m - 3}$$

**39.**
$$\begin{array}{r} -x + 4 \phantom{{}+9} \\ -x - 4 \overline{\smash{\big)}\, x^2 + 0x + 9} \\ \underline{x^2 + 4x} \phantom{{}+9} \\ -4x + 9 \\ \underline{-4x - 16} \\ 25 \end{array}$$

$$-x + 4 - \frac{25}{x + 4}$$

**40.**
$$\begin{array}{r} 4x + 20 \phantom{{}-10} \\ x - 2 \overline{\smash{\big)}\, 4x^2 + 12x - 10} \\ \underline{4x^2 - 8x} \phantom{{}-10} \\ 20x - 10 \\ \underline{20x - 40} \\ 30 \end{array}$$

$$4x + 20 + \frac{30}{x - 2}$$

**41.**
$$\begin{array}{r} -5m - 5 \phantom{{}+2} \\ m - 1 \overline{\smash{\big)}\, -5m^2 + 0m + 2} \\ \underline{-5m^2 + 5m} \phantom{{}+2} \\ -5m + 2 \\ \underline{-5m + 5} \\ -3 \end{array}$$

$$-5m - 5 - \frac{3}{m - 1}$$

**42.** $\dfrac{6q^2 + 11q + 4}{2q + 1} = \dfrac{(2q + 1)(3q + 4)}{2q + 1}$

$$= 3q + 4$$

**43.**
$$\begin{array}{r} -s + 5 \phantom{{}+4} \\ s + 5 \overline{\smash{\big)}\, -s^2 + 0s + 4} \\ \underline{-s^2 - 5s} \phantom{{}+4} \\ 5s + 4 \\ \underline{5s + 25} \\ -21 \end{array}$$

$$-s + 5 - \frac{21}{s + 5}$$

**44.**
$$\begin{array}{r} 4a - 3 \phantom{{}-25} \\ 4a + 3 \overline{\smash{\big)}\, 16a^2 + 0a - 25} \\ \underline{16a^2 + 12a} \phantom{{}-25} \\ -12a - 25 \\ \underline{-12a - 9} \\ -16 \end{array}$$

$$4a - 3 - \frac{16}{4a + 3}$$

# Chapter 11 *continued*

**45.**

$$\begin{array}{r} \frac{1}{2}c - 2 \\ 2c - 6 \overline{)\, c^2 - 7c + 21} \\ \underline{c^2 - 3c} \\ -4c + 21 \\ \underline{-4c + 12} \\ 9 \end{array}$$

$$\frac{1}{2}c - 2 + \frac{9}{2c - 6}$$

**46.**

$$\begin{array}{r} \frac{1}{4}b - 2 \\ 4b + 4 \overline{)\, b^2 - 7b - 12} \\ \underline{b^2 + b} \\ -8b - 12 \\ \underline{-8b - 8} \\ -4 \end{array}$$

$$\frac{1}{4}b - 2 - \frac{1}{b + 1}$$

**47.**

$$\begin{array}{r} 5x + 3 \\ x - 4 \overline{)\, 5x^2 - 17x - 12} \\ \underline{5x^2 - 20x} \\ 3x - 12 \\ \underline{3x - 12} \\ 0 \end{array}$$

$$5x + 3$$

**48.**

$$\begin{array}{r} 7x + 3 \\ x + 2 \overline{)\, 7x^2 + 17x + 6} \\ \underline{7x^2 + 14x} \\ 3x + 6 \\ \underline{3x + 6} \\ 0 \end{array}$$

$$7x + 3$$

**49.**

$$\begin{array}{r} 3x + 2 \\ 4x - 1 \overline{)\, 12x^2 + 5x - 2} \\ \underline{12x^2 - 3x} \\ 8x - 2 \\ \underline{8x + 2} \\ 0 \end{array}$$

$$3x + 2$$

**50.** $\dfrac{200t + 1400}{900t + 9900} = \dfrac{200(t + 7)}{900(t + 11)} = \dfrac{2t + 14}{9t + 99}$

**51.**

$$\begin{array}{r} \frac{2}{9} \\ 9t + 99 \overline{)\, 2t + 14} \\ \underline{2t + 22} \\ -8 \end{array}$$

$$\frac{2}{9} - \frac{8}{9t + 99}$$

**52.**

| Year $t$ | 0 | 1 | 2 | 3 | 4 |
|---|---|---|---|---|---|
| Ratio of $E$ to $S$ | 0.141 | 0.148 | 0.154 | 0.159 | 0.163 |

| Year $t$ | 5 | 6 | 7 | 8 | 9 |
|---|---|---|---|---|---|
| Ratio of $E$ to $S$ | 0.167 | 0.170 | 0.173 | 0.175 | 0.178 |

$R = \dfrac{2}{9} - \dfrac{8}{9(0) + 99}$  　　$R = \dfrac{2}{9} - \dfrac{8}{9(1) + 99}$

$R \approx 0.141$  　　　　$R \approx 0.148$

$R = \dfrac{2}{9} - \dfrac{8}{9(2) + 99}$  　　$R = \dfrac{2}{9} - \dfrac{8}{9(3) + 99}$

$R \approx 0.154$  　　　　$R \approx 0.159$

$R = \dfrac{2}{9} - \dfrac{8}{9(4) + 99}$  　　$R = \dfrac{2}{9} - \dfrac{8}{9(5) + 99}$

$R \approx 0.163$  　　　　$R \approx 0.167$

$R = \dfrac{2}{9} - \dfrac{8}{9(6) + 99}$  　　$R = \dfrac{2}{9} - \dfrac{8}{9(7) + 99}$

$R \approx 0.170$  　　　　$R \approx 0.173$

$R = \dfrac{2}{9} - \dfrac{8}{9(8) + 99}$  　　$R = \dfrac{2}{9} - \dfrac{8}{9(9) + 99}$

$R \approx 0.175$  　　　　$R \approx 0.178$

**53.** The ratio was increasing.

**54.** The ratio is increasing because as $t$ increases, $9t + 99$ increases, so $\dfrac{8}{9t + 99}$ decreases. The amount subtracted from $\dfrac{2}{9}$ decreases, so the value of the rational expression increases.

**55. a.** $4 + t$

**b.** $\dfrac{14 + t}{4 + t} \Rightarrow t + 4 \overline{\smash{)}\, \begin{array}{l}1 \\ t + 14\end{array}} \Rightarrow 1 + \dfrac{10}{t + 4}$
$$\phantom{xxxxxxxxxxx}\underline{t + 4}$$
$$\phantom{xxxxxxxxxxx}10$$

**c.** now: $1 + \frac{10}{4} = 3.5$

5 years: $1 + \frac{10}{5 + 4} = 1 + \frac{10}{9} \approx 2.111$

10 years: $1 + \frac{10}{10 + 4} = 1 + \frac{10}{14} \approx 1.714$

25 years: $1 + \frac{10}{25 + 4} \approx 1.345$

50 years: $1 + \frac{10}{50 + 4} \approx 1.185$

80 years: $1 + \frac{10}{80 + 4} \approx 1.119$

**d.** smaller; 1

**e.** *Sample answer:* Given the form $\dfrac{14 + t}{4 + t}$, you can see that as $t$ gets larger and larger, the numbers added in the numerator and denominator become insignificant in comparison and that the ratios approach $\dfrac{t}{t}$ or 1. Given the form $1 + \dfrac{10}{t + 4}$, you can see that as $t$ gets larger and larger, $\dfrac{10}{t + 4}$ approaches 0, so that $1 + \dfrac{10}{t + 4}$ approaches 1.

**56.** quotient = 6

remainder = 1

**57.** $6^2 \div 5$

quotient = 7

remainder = 1

# Chapter 11 *continued*

**58.**

**59.** The quotient is one more than the number being squared, and the remainder is 1; $x^2 \div (x - 1) = x + 1 + \dfrac{1}{x - 1}$

**60.**

$$\begin{array}{r} x + 1 \\ x - 1 \overline{)\, x^2 + 0x + 0} \\ \underline{x^2 - x} \\ x + 0 \\ \underline{x - 1} \\ 1 \end{array}$$

$x + 1 + \dfrac{1}{x - 1}$; yes

## 11.7 Mixed Review (p. 689)

**61.** $\dfrac{x}{5} = 3$

$x = 15$

**62.** $\dfrac{a}{-3} = 7$

$a = -21$

**63.** $\dfrac{c}{4} = \dfrac{6}{8}$

$c = 3$

**64.** $\dfrac{y}{-2} = \dfrac{5}{4}$

$4y = -10$

$y = -\dfrac{5}{2}$

**65.** $\dfrac{7}{5} = \dfrac{2}{x}$

$7x = 10$

$x = \dfrac{10}{7}$

**66.** $\dfrac{2}{x} = \dfrac{x - 1}{6}$

$12 = x^2 - x$

$x^2 - x - 12 = 0$

$(x + 3)(x - 4) = 0$

$x = -3 \quad x = 4$

**67.** $(6x - 7)x = 20$

$6x^2 - 7x - 20 = 0$

$(3x + 4)(2x - 5) = 0$

$3x + 4 = 0 \quad 2x - 5 = 0$

$x = -\dfrac{4}{3} \qquad x = \dfrac{5}{2}$

**68.** $3(8b^2 + 4b) = 4b(2b - 5)$

$24b^2 + 12b = 8b^2 - 20b$

$16b^2 + 32b = 0$

$16b(b + 2) = 0$

$16b = 0 \quad b + 2 = 0$

$b = 0 \qquad b = -2$

extraneous root

solution: $-2$

**69.** $2(5p^2 - 9) = 5(2p^2 + 3p)$

$10p^2 - 18 = 10p^2 + 15p$

$15p + 18 = 0$

$p = -\dfrac{6}{5}$

**70.** $10(a^2 - 4) = (a - 2)(a + 2)$

$10a^2 - 40 = a^2 - 4$

$9a^2 - 36 = 0$

$(3a - 6)(3a + 6) = 0$

$3a - 6 = 0 \qquad 3a + 6 = 0$

$a = 2 \qquad\qquad a = -2$

extraneous root    solution: $-2$

**71.**

| x | y |
|---|---|
| 1 | 12 |
| 2 | 6 |
| 3 | 4 |
| 4 | 3 |
| 5 | $\dfrac{12}{5}$ |

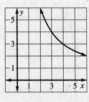

$y = \dfrac{12}{x}$

**72.** $\dfrac{1.3 \times 10^5}{2 \times 10^{-3}} = 6.5 \times 10^7$

## 11.8 Guided Practice (p. 694)

**1.** A hyperbola consists of two branches, each of which approaches, but does not intersect, two lines called the asymptotes of the hyperbola. The asymptotes intersect at the center of the hyperbola.

**2.** $3x$

**3.** $2 \cdot 2 \cdot x$          LCD $= 2^3 \cdot 3x^2$

$2 \cdot 3 \cdot x^2$          $= 24x^2$

$2 \cdot 2 \cdot 2x^2$

**4.** $x^2 + 6x + 9 = (x + 3)(x + 3)$

$x + 3$

LCD $= (x + 3)(x + 3)$

**5.** The equations in Exs. 6, 7, 9 and 10; each side of the equation is a single fraction.

**6.** $\qquad 36 = x^2$

$\pm\sqrt{36} = x$

$x = 6; \quad x = -6$

**7.** $-4x(x - 2) = 2(x + 1)$

$-4x^2 + 8x = 2x + 2$

$-4x^2 + 6x - 2 = 0$

$2x^2 - 3x + 1 = 0$

$(2x - 1)(x - 1) = 0$

$2x - 1 = 0 \quad x - 1 = 0$

$x = \dfrac{1}{2} \qquad x = 1$

# Chapter 11 *continued*

**8.** $x^2 + 90 = 5x$

$x^2 - 5x + 90 = 0$

no solution

**9.** $4(3x - 6) = 4(x^2 - 2x)$

$12x - 24 = 4x^2 - 8x$

$4x^2 - 20x + 24 = 0$

$x^2 - 5x + 6 = 0$

$(x - 3)(x - 2) = 0$

$x - 3 = 0 \quad x - 2 = 0$

$x = 3 \qquad x = 2$    extraneous solution

solution: 3

**10.** $4x = 3x(x + 1)$

$4x = 3x^2 + 3x$

$3x^2 - x = 0$

$x(3x - 1) = 0$

$x = 0 \qquad\qquad 3x - 1 = 0$

extraneous solution $\qquad x = \frac{1}{3}$

solution: $\frac{1}{3}$

**11.** $3x + 4(x + 4) = -5$

$3x + 4x + 16 = -5$

$7x + 21 = 0$

$x = -3$

**12.** center: $(-5, -3)$

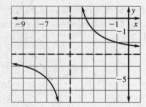

**13.** center: $(2, 3)$

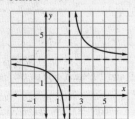

## 11.8 Practice and Applications (pp. 694–696)

**14.** $3x = 35$

$x = \frac{35}{3}$

**15.** $5x = 140$

$x = 28$

**16.** $20(x + 2) = 12x$

$20x + 40 = 12x$

$8x + 40 = 0$

$x + 5 = 0$

$x = -5$

**17.** $15(x + 1) = 5(x + 4)$

$15x + 15 = 5x + 20$

$10x - 5 = 0$

$2x - 1 = 0$

$x = \frac{1}{2}$

**18.** $24 = x(x + 2)$

$24 = x^2 + 2x$

$x^2 + 2x - 24 = 0$

$(x - 4)(x + 6) = 0$

$x - 4 = 0 \quad x + 6 = 0$

$x = 4 \qquad x = -6$

**19.** $7x - 21 = 5x + 5$

$2x - 26 = 0$

$x - 13 = 0$

$x = 13$

**20.** $112 = 9x - x^2 \qquad\qquad$ LCD $= 2x$

$x^2 - 9x + 112 = 0$

no solution

**21.** $x = 9 + 2x + 18 \quad$ LCD $= x + 9$

$x + 27 = 0$

$x = -27$

**22.** $7 - 3 = 2(x - 4) \qquad$ LCD $= 3(x - 4)$

$4 = 2x - 8$

$2x = 12$

$x = 6$

**23.** LCD $= (x + 4)(x - 4)$

$x + 4 + x - 4 = 22$

$2x - 22 = 0$

$x = 11$

**24.** LCD $= (x - 4)(x + 5)$

$x + 5 + (x + 5)(x - 4) = -7$

$x + 5 + x^2 + x - 20 = -7$

$x^2 + 2x - 8 = 0$

$(x + 4)(x - 2) = 0$

$x + 4 = 0 \qquad x - 2 = 0$

$x = -4 \qquad x = 2$

**25.** LCD $= (x - 5)(x + 5)$

$2(x^2 - 25) + 8(x + 5) = x + 5$

$2x^2 - 50 + 8x + 40 = x + 5$

$2x^2 + 7x - 15 = 0$

$(2x - 3)(x + 5) = 0$

$2x - 3 = 0 \quad x + 5 = 0$

$x = \frac{3}{2} \qquad x = -5$   extraneous solution

solution: $1\frac{1}{2}$

**26.** LCD $= 4x$

$x + 16 = 4$

$x = -12$

# Chapter 11 *continued*

**27.** $-3x(x-1) = -2(x+1)$

$\quad -3x^2 + 3x = -2x - 2$

$\quad 3x^2 - 5x - 2 = 0$

$\quad (3x+1)(x-2) = 0$

$\quad 3x+1 = 0 \qquad x - 2 = 0$

$\qquad x = -\frac{1}{3} \qquad x = 2$

**28.** LCD $= 5x$

$\quad x - 2 = 5$

$\qquad x = 7$

**29.** LCD $= 9x$

$\quad x^2 - 72 = x$

$\quad x^2 - x - 72 = 0$

$\quad (x-9)(x+8) = 0$

$\quad x - 9 = 0 \quad x + 8 = 0$

$\qquad x = 9 \qquad x = -8$

**30.** $\quad x + 42 = x^2$

$\quad x^2 - x - 42 = 0$

$\quad (x-7)(x+6) = 0$

$\quad x - 7 = 0 \quad x + 6 = 0$

$\qquad x = 7 \qquad x = -6$

**31.** LCD $= 8x$

$\quad 16 - x^2 = 6x$

$\quad x^2 + 6x - 16 = 0$

$\quad (x-2)(x+8) = 0$

$\quad x - 2 = 0 \quad x + 8 = 0$

$\qquad x = 2 \qquad x = -8$

**32.** $-3(x+2) = 2(x+7)$

$\quad -3x - 6 = 2x + 14$

$\quad 5x + 20 = 0$

$\qquad x = -4$

**33.** LCD $= 3x(x+3)$

$\quad 6x + 3(x+3) = 4(x+3)$

$\quad 6x + 3x + 9 = 4x + 12$

$\quad 5x - 3 = 0$

$\qquad x = \frac{3}{5}$

**34.** LCD $= 15(x+3)$

$\quad 150 - 3(3x+9) = 50x + 5$

$\quad 150 - 9x - 27 = 50x + 5$

$\quad 59x - 118 = 0$

$\qquad x = 2$

**35.** LCD $= (x-5)(x-8)$

$\quad (x+3)(x-8) = 56 - 3x$

$\quad x^2 - 5x - 24 = 56 - 3x$

$\quad x^2 - 2x - 80 = 0$

$\quad (x-10)(x+8) = 0$

$\quad x - 10 = 0 \quad x + 8 = 0$

$\qquad x = 10 \qquad x = -8$

**36.** LCD $= (x+4)(x-6)$

$\quad 8(x-6) + (x+4)(x-6) = 5x$

$\quad 8x - 48 + x^2 - 2x - 24 = 5x$

$\qquad x^2 + x - 72 = 0$

$\qquad (x+9)(x-8) = 0$

$\quad x + 9 = 0 \quad x - 8 = 0$

$\qquad x = -9 \qquad x = 8$

**37.** LCD $= (x-11)(x+6)$

$\quad x(x+6) - (x-11)(x+6) = 22$

$\quad x^2 + 6x - (x^2 - 5x - 66) = 22$

$\quad x^2 + 6x - x^2 + 5x + 66 = 22$

$\qquad 11x + 44 = 0$

$\qquad x = -4$

**38.** LCD $= (x+3)(x+7)$

$\quad 2x(x+7) - x(x+3) = x^2 - 1$

$\quad 2x^2 + 14x - x^2 - 3x = x^2 - 1$

$\qquad 11x = -1$

$\qquad x = -\frac{1}{11}$

**39.** B  **40.** C  **41.** A

center $(0, 4)$

**42.**

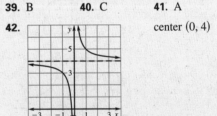

domain: all real numbers except 0

**43.** center $(3, -8)$  **44.** center $(4, 6)$

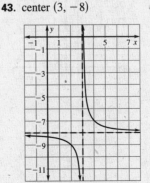

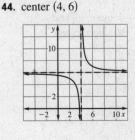

domain: all real numbers except 3.

domain: all real numbers except 4.

# Chapter 11 *continued*

**45.** center $(-1, 8)$

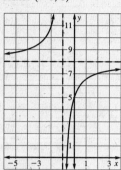

domain: all real numbers except $-1$.

**46.** center $(-9, -7)$

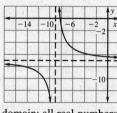

domain: all real numbers except $-9$

**47.**

$$\begin{array}{r} 3 \\ x+3\overline{\smash{\big)}\,3x+11} \\ \underline{3x+\phantom{0}9} \\ 2 \end{array}$$

$$y = \frac{2}{x+3} + 3$$

center $(-3, 3)$

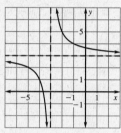

domain: all real numbers except $-3$.

**48.**

$$\begin{array}{r} -2 \\ x-5\overline{\smash{\big)}\,-2x+11} \\ \underline{-2x+10} \\ 1 \end{array}$$

$$y = \frac{1}{x-5} - 2$$

domain: all real numbers except 5.

center $(5, -2)$

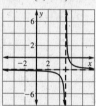

**49.**

$$\begin{array}{r} 9 \\ x-1\overline{\smash{\big)}\,9x-6} \\ \underline{9x-9} \\ 3 \end{array}$$

$$y = \frac{3}{x-1} + 9$$

domain: all real numbers except 1.

center $(1, 9)$

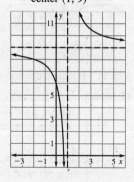

**50.**

$$\begin{array}{r} -5 \\ x-3\overline{\smash{\big)}\,-5x+19} \\ \underline{-5x+15} \\ 4 \end{array}$$

$$y = \frac{4}{x-3} - 5$$

domain: all real numbers except 3.

**51.**

$$0.250 = \frac{8+x}{50+x}$$

$$0.250(50+x) = 8+x$$

$$12.5 + 0.250x = 8+x$$

$$4.5 = 0.750x$$

$$x = 6 \text{ hits}$$

**52.**

$$88 = \frac{216+100x}{3+x}$$

$$88(3+x) = 216+100x$$

$$264 + 88x = 216+100x$$

$$48 = 12x$$

$$x = 4 \text{ tests}$$

**53.** $y = \dfrac{5000+10x}{x+1}$

$$\begin{array}{r} 10 \\ x+1\overline{\smash{\big)}\,10x+5000} \\ \underline{10x+\phantom{00}10} \\ 4990 \end{array}$$

$$y = \frac{4990}{x+1} + 10$$

center $(3, -5)$

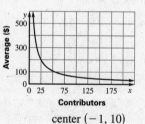

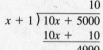

center $(-1, 10)$

**54.** The smaller the number of sponsors, the more the average donation is inflated by the single donation of \$5000. The fewer the sponsors, the less representative the average is of a typical donation.

**55.** $\dfrac{1}{2000} \cdot 60 = \dfrac{60}{2000} = \dfrac{3}{100}$

**56.** Your friend can fold $\dfrac{1}{1000x}$ cranes per minute.

$$60\left(\frac{1}{1000x}\right) = \frac{60}{1000x}$$

**57.** $\dfrac{3}{100} + \dfrac{3}{50x} = \dfrac{3x+6}{100x}$

**58.**

$$\frac{3x+6}{100x} = \frac{1}{20}$$

$$20(3x+6) = 100x$$

$$60x + 120 = 100x$$

$$120 = 40x$$

$$x = 3$$

# Chapter 11 *continued*

**59.**

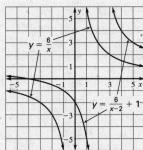

graph of $f(x) = \dfrac{6}{x}$

**60.** The graphs seem to have the same shape but they have different asymptotes; yes; in the first equation the asymptotes are $x = h$ and $y = k$. In the second they are $x = 0$ and $y = 0$. The effect of the $h$ and $k$ in the equation is to shift the position of the branches.

**61.** For $a > 0$, as $a$ increases, the branches of the hyperbola (which are in the first and third quadrants) become wider and more open. For $a < 0$, as $|a|$ increases, the branches of the hyperbola (which are in the second and fourth quadrants) become wider and more open.

**62.**
$$\frac{9}{x + 5} = \frac{7}{x - 5}$$
$$9x - 45 = 7x + 35$$
$$2x = 80$$
$$x = 40 \qquad \text{D}$$

**63.** B

**64.**

| X | 0 | 20 | 40 | 60 | 80 | 100 |
|---|---|----|----|----|----|-----|
| $\dfrac{x^2 + 6}{x + 2}$ | 3 | 18.45 | 38.24 | 58.16 | 78.12 | 98.10 |
| $x - 2$ | $-2$ | 18 | 38 | 58 | 78 | 98 |
| $\dfrac{10}{x + 2}$ | 5 | 0.45 | 0.24 | 0.16 | 0.12 | 0.10 |

**65.** The values of the first two expressions increase and the value of the third decreases.

**66.** B; A

**67.** The graph of $y = x - 2$ is an asymptote of the graph of $y = \dfrac{x^2 + 6}{x + 2}$. Notice that for each column in the table, the value of the first expression is equal to the sum of the other two. $\dfrac{x^2 + 6}{x + 2} = x - 2 + \dfrac{10}{x + 2}$. As $x$ gets very large, $\dfrac{10}{x + 2}$ approaches, but never equals, 0, and $\dfrac{x^2 + 6}{x + 2}$ approaches, but never equals $x - 2$.

## 11.8 Mixed Review (p. 697)

**68.** $f(x) = 4x$
$f(0) = 0$
$f(1) = 4$
$f(2) = 8$
$f(3) = 12$
$f(4) = 16$

**69.** $f(x) = -x + 9$
$f(0) = 9$
$f(1) = 8$
$f(2) = 7$
$f(3) = 6$
$f(4) = 5$

**70.** $f(x) = 3x + 1$
$f(0) = 1$
$f(1) = 4$
$f(2) = 7$
$f(3) = 10$
$f(4) = 13$

**71.** $f(x) = -x^2$
$f(0) = 0$
$f(1) = -1$
$f(2) = -4$
$f(3) = -9$
$f(4) = -16$

**72.** $f(x) = x^2 - 1$
$f(0) = -1$
$f(1) = 0$
$f(2) = 3$
$f(3) = 8$
$f(4) = 15$

**73.** $f(x) = \dfrac{x^2}{2}$
$f(0) = 0$
$f(1) = \dfrac{1}{2}$
$f(2) = 2$
$f(3) = \dfrac{9}{2}$
$f(4) = 8$

**74.** $2^4 \cdot 2^3 = 2^7$
$= 128$

**75.** $6^3 \cdot 6^{-1} = 6^2$
$= 36$

**76.** $(3^3)^2 = 3^6$
$= 729$

**77.** $(-4^{-2})^{-1} = -4^2$
$= -16$

**78.** $\sqrt{50} = 5\sqrt{2}$

**79.** $\sqrt{72} = \sqrt{9 \cdot 4 \cdot 2}$
$= 6\sqrt{2}$

**80.** $\dfrac{1}{4}\sqrt{112} = \dfrac{1}{4}(4)\sqrt{7}$
$= \sqrt{7}$

**81.** $\dfrac{1}{2}\sqrt{52} = \dfrac{1}{2}(2)\sqrt{13}$
$= \sqrt{13}$

**82.** $\dfrac{1}{4}\sqrt{64} = \dfrac{1}{4}(8)$
$= 2$

**83.** $\sqrt{256} = 16$

**84.** $\dfrac{1}{5}\sqrt{625} = \dfrac{1}{5}(25)$
$= 5$

**85.** $\sqrt{396} = \sqrt{36 \cdot 11}$
$= 6\sqrt{11}$

**86.** $A = 500(1.04)^6$
$\approx \$632.66$

## Quiz 3 (p. 697)

**1.** $\dfrac{x^2 - 8}{6x} = \dfrac{x^2}{6x} - \dfrac{8}{6x}$
$= \dfrac{x}{6} - \dfrac{4}{3x}$

**2.** $\dfrac{6a^3 + 5a^2}{10a^2} = \dfrac{6a^3}{10a^2} + \dfrac{5a^2}{10a^2}$
$= \dfrac{3a}{5} + \dfrac{1}{2}$

**3.**

$$x + 4 \overline{\smash{\big)}\, x^2 + 0x + 16} \quad \underset{\textstyle x - 4}{\phantom{)}}$$

$$\underline{x^2 + 4x}$$
$$-4x + 16$$
$$\underline{-4x - 16}$$
$$32$$

$$x - 4 + \frac{32}{x + 4}$$

**4.**

$$2y - 5 \overline{\smash{\big)}\, 2y^2 + 8y - 5} \quad \underset{\textstyle y + \frac{13}{2}}{\phantom{)}}$$

$$\underline{2y^2 - 5y}$$
$$13y - 5$$
$$\underline{13y - \frac{65}{2}}$$
$$\frac{55}{2}$$

$$y + \frac{13}{2} + \frac{27.5}{2y - 5}$$

**5.**

$$3z - 6 \overline{\smash{\big)}\, z^2 + 4z + 8} \quad \underset{\textstyle \frac{1}{3}z + 2}{\phantom{)}}$$

$$\underline{z^2 - 2z}$$
$$6x + 8$$
$$\underline{6z - 12}$$
$$20$$

$$\frac{1}{3}z + 2 + \frac{20}{3z - 6}$$

**6.**

$$3x + 2 \overline{\smash{\big)}\, 12x^2 + 17x - 5} \quad \underset{\textstyle 4x + 3}{\phantom{)}}$$

$$\underline{12x^2 + 8x}$$
$$9x - 5$$
$$\underline{9x + 6}$$
$$-11$$

$$4x + 3 - \frac{11}{3x + 2}$$

**7.** $\dfrac{1}{2} + \dfrac{2}{t} = \dfrac{1}{t}$

$$t + 4 = 2$$
$$t = -2$$

**8.** $6(x + 2) = 9x$

$$6x + 12 = 9x$$
$$12 = 3x$$
$$x = 4$$

**9.** $\dfrac{1}{x - 5} + \dfrac{1}{x + 5} = \dfrac{x + 3}{x^2 - 25}$

$$x + 5 + x - 5 = x + 3$$
$$2x = x + 3$$
$$x = 3$$

**10.** $\dfrac{7}{8} - \dfrac{16}{s - 2} = \dfrac{3}{4}$

$$\frac{-16}{s - 2} = -\frac{1}{8}$$
$$-128 = -s + 2$$
$$-130 = -s$$
$$s = 130$$

**11.**

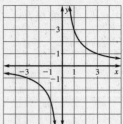

$$f(x) = \frac{3}{x}$$
center $(0, 0)$

**12.**

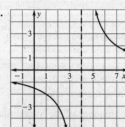

$$f(x) = \frac{6}{x - 4}$$
center $(4, 0)$

**13.**

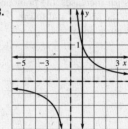

$$f(x) = \frac{3}{x + 1} - 2$$
center $(-1, -2)$

**14.** $\dfrac{14t + 505}{16t + 1048}$

$$16t + 1048 \overline{\smash{\big)}\, 14t + 505} \quad \underset{\textstyle \frac{14}{16}}{\phantom{)}}$$

$$\underline{14t + 917}$$
$$-412$$

$$\frac{14}{16} - \frac{412}{16t + 1048}$$

$$\frac{U}{T} = \frac{7}{8} - \frac{103}{4t + 262}$$

**Technology Activity 11.8 (p. 698)**
**Exploring the Concept**

**1.** $y = \dfrac{4x^2 + 9x + 2}{x^2 - 4}$

$$y = \frac{(4x + 1)(x + 2)}{(x + 2)(x - 2)}$$

$$y = \frac{4x + 1}{x - 2}$$

**2.** Check graph.

**3.** Check graph.

**4.** original: $x = 2$; $x = -2$

   simplified: $x = 2$

**5.** No; the original function has the restriction that $x \neq -2$, since $x^2 - 4$ cannot equal zero.

# Chapter 11 *continued*

### Drawing Conclusions

**1.** $y = \dfrac{3x^2 - 13x - 10}{x^2 - 25}$

$y = \dfrac{(3x + 2)(x - 5)}{(x - 5)(x + 5)}$

$y = \dfrac{3x + 2}{x + 5}$

Domain: all real numbers except $-5$ and $5$; all real numbers except $-5$

**2.** $y = \dfrac{x^2 - 6x + 8}{x^2 - 2x - 8}$

$= \dfrac{(x - 2)(x - 4)}{(x + 2)(x - 4)}$

$y = \dfrac{x - 2}{x + 2}$

Domain: all real numbers except $-2$ and $4$; all real numbers except $-2$

**3.** $y = \dfrac{2x^2 + 9x + 4}{x^2 + x - 12}$

$y = \dfrac{(2x + 1)(x + 4)}{(x + 4)(x - 3)}$

$y = \dfrac{2x + 1}{x - 3}$

Domain: all real numbers except $-4$ and $3$; all real numbers except $3$

**4.** $y = \dfrac{x^2 - x - 20}{x^2 - 16}$

$y = \dfrac{(x - 5)(x + 4)}{(x - 4)(x + 4)}$

$y = \dfrac{x - 5}{x - 4}$

Domain: all real numbers except $4$ and $-4$; all real numbers except $4$

**5.** $y = \dfrac{2x^2 - 3x - 9}{x^2 - 2x - 3}$

$y = \dfrac{(2x + 3)(x - 3)}{(x - 3)(x + 1)}$

$y = \dfrac{2x + 3}{x + 1}$

Domain: all real numbers except $3$ and $-1$; all real numbers except $-1$

**6.** $y = \dfrac{2x^2 - 5x - 3}{x^2 - 8x + 15}$

$y = \dfrac{(2x + 1)(x - 3)}{(x - 3)(x - 5)}$

$y = \dfrac{2x + 1}{x - 5}$

Domain: all real numbers except $3$ and $5$; all real numbers except $5$

**7.** No; they do not have the same domain. The domain of the first is all real numbers except $-3$ and $1$; the domain of the second is all real numbers except $-3$.

## Chapter 11 Review (pp. 700–702)

**1.** $\dfrac{x}{2} = \dfrac{4}{7}$

$7x = 8$

$x = \dfrac{8}{7}$

**2.** $\dfrac{7}{10} = \dfrac{9 + x}{x}$

$7x = 90 + 10x$

$-3x = 90$

$x = -30$

**3.** $\dfrac{x^2 - 16}{x + 4} = \dfrac{x - 4}{3}$

$\dfrac{x - 4}{1} = \dfrac{x - 4}{3}$

$3x - 12 = x - 4$

$2x = 8$

$x = 4$

**4.** $5x = x^2 - 36$

$x^2 - 5x - 36 = 0$

$(x - 9)(x + 4) = 0$

$x - 9 = 0 \quad x + 4 = 0$

$x = 9 \qquad x = -4$

**5.** $x = 0.80(95)$

$x = \$76$

**6.** $24 = 2.5x$

$9.6 \text{ inches} = x$

**7.** $90 = 0.75x$

$\$120 = x$

**8.** $35 = x(175)$

$20\% = x$

**9.** $51 = 17k$

$3 = k$

$y = 3x$

**10.** $51 = \dfrac{k}{17}$

$867 = k$

$y = \dfrac{867}{x}$

**11.** $\dfrac{3x}{9x^2 + 3} = \dfrac{x}{3x^2 + 1}$

**12.** $\dfrac{6x^2}{12x^4 + 18x^2} = \dfrac{6x^2}{6x^2(2x^2 + 3)}$

$= \dfrac{1}{2x^2 + 3}$

**13.** $\dfrac{7x^3 - 21x}{-14x^2} = \dfrac{7x(x^2 - 3)}{-14x^2}$

$= \dfrac{x^2 - 3}{-2x} = \dfrac{-x^2 + 3}{2x}$

**14.** $\dfrac{5x^2 + 21x + 4}{25x + 100} = \dfrac{(5x + 1)(x + 4)}{25(x + 4)}$

$= \dfrac{5x + 1}{25}$

**15.** $\dfrac{\overset{4}{\cancel{12x^2}}}{15x^3} \cdot \dfrac{\overset{5}{\cancel{25x^4}}}{13x} = 20x^2$

**16.** $\dfrac{9x^3}{x^3 - x^2} \cdot \dfrac{x^2 - 9x + 8}{x - 8} = \dfrac{9x^3}{x^2(x - 1)} \cdot \dfrac{(x - 1)(x - 8)}{(x - 8)}$

$= 9x$

**17.** $\dfrac{x^2 + 3x + 2}{x^2 + 7x + 12} \cdot \dfrac{x^2 + 5x + 6}{x^2 + 5x + 4}$

$= \dfrac{(x + 2)(x + 1)}{(x + 3)(x + 4)} \cdot \dfrac{(x + 3)(x + 2)}{(x + 4)(x + 1)} = \dfrac{(x + 2)^2}{(x + 4)^2}$

# Chapter 11 *continued*

**18.**
$$\frac{6x}{x+4} - \frac{5x-4}{x+4} = \frac{6x-5x+4}{x+4}$$
$$= \frac{x+4}{x+4}$$
$$= 1$$

**19.**
$$\frac{2x+1}{8x} - \frac{x}{12x} = \frac{6(2x+1)-4x}{48x}$$
$$= \frac{12x+6-4x}{48x}$$
$$= \frac{8x+6}{48x} = \frac{4x+3}{24x}$$

**20.**
$$\frac{x+3}{3x-1} + \frac{4}{x-3} = \frac{(x+3)(x-3)+4(3x-1)}{(3x-1)(x-3)}$$
$$= \frac{x^2-9+12x-4}{(3x-1)(x-3)}$$
$$= \frac{x^2+12x-13}{(3x-1)(x-3)}$$
$$= \frac{(x+13)(x-1)}{(3x-1)(x-3)}$$

**21.**
$$\frac{-5x-10}{x^2-4} + \frac{4x}{x-2} = \frac{-5x-10}{(x+2)(x-2)} + \frac{4x(x+2)}{(x+2)(x-2)}$$
$$= \frac{-5x-10+4x^2+8x}{(x+2)(x-2)}$$
$$= \frac{4x^2+3x-10}{(x+2)(x-2)}$$
$$= \frac{(4x-5)(x+2)}{(x+2)(x-2)}$$
$$= \frac{4x-5}{x-2}$$

**22.**
$$x-2\overline{)3x^2-x-1}$$ quotient $3x+5$
$$\underline{3x^2-6x}$$
$$5x-1$$
$$\underline{5x-10}$$
$$9$$
$$3x+5+\frac{9}{x-2}$$

**23.**
$$\frac{6x^2}{6x} - \frac{36x}{6x} + \frac{5}{6x} = x-6+\frac{5}{6x}$$

**24.**
$$2x-1\overline{)4x^2+6x-5}$$ quotient $2x+4$
$$\underline{4x^2-2x}$$
$$8x-5$$
$$\underline{8x-4}$$
$$-1$$
$$2x+4-\frac{1}{2x-1}$$

**25.**
$$\frac{5x^2+13x-6}{5x-2} = \frac{(5x-2)(x+3)}{5x-2}$$
$$= x+3$$

**26.** $x-24=12$
$x=36$

**27.**
$$x(x+2)=8$$
$$x^2+2x-8=0$$
$$(x-2)(x+4)=0$$
$$x-2=0 \quad x+4=0$$
$$x=2 \quad\quad x=-4$$

**28.**
$$\frac{6}{x+4} + \frac{3}{4} = \frac{2x+1}{3(x+4)}$$
$$72+9(x+4)=4(2x+1)$$
$$72+9x+36=8x+4$$
$$x+104=0$$
$$x=-104$$

## Chapter 11 Test *(p. 703)*

**1.**
$$\frac{6}{x} = \frac{17}{5}$$
$$30=17x$$
$$\frac{30}{17}=x$$

**2.**
$$\frac{x}{4} = \frac{x+8}{x}$$
$$x^2=4x+32$$
$$x^2-4x-32=0$$
$$(x-8)(x+4)=0$$
$$x-8=0 \quad x+4=0$$
$$x=8 \quad\quad x=-4$$

**3.**
$$\frac{x}{-3} = \frac{7}{x-10}$$
$$x^2-10x=-21$$
$$x^2-10x+21=0$$
$$(x-7)(x-3)=0$$
$$x-7=0 \quad x-3=0$$
$$x=7 \quad\quad x=3$$

**4.**
$$\frac{x^2-64}{x+8} = \frac{x-8}{2}$$
$$\frac{x-8}{1} = \frac{x-8}{2}$$
$$2(x-8)=x-8$$
$$2x-16=x-8$$
$$x=8$$

**5.** $x=0.34(100)$
$x=34$ L

**6.** $x=0.86(350)$
$x=\$301$

**7.** $24=0.12x$
$200$ yd $=x$

**8.** $36=x(900)$
$4\%=x$

**9.**

| x | y |
|---|---|
| 1 | 4 |
| 2 | 8 |
| 3 | 12 |
| 4 | 16 |

$y=4x$

directly

**10.**

| x | y |
|---|---|
| 1 | 50 |
| 2 | 25 |
| 3 | $16\frac{2}{3}$ |
| 4 | $12\frac{1}{2}$ |

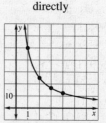

$y=\frac{50}{x}$

inversely

**294** **Algebra 1**
Chapter 11 Worked-out Solution Key

Copyright © McDougal Littell Inc.
All rights reserved.

# Chapter 11 *continued*

**11.**

| x | y |
|---|---|
| 1 | $4\frac{1}{2}$ |
| 2 | 9 |
| 3 | $13\frac{1}{2}$ |
| 4 | 18 |

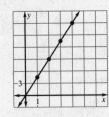

$$y = \frac{9}{2}x$$

directly

**12.**

| x | y |
|---|---|
| 1 | $7\frac{1}{2}$ |
| 2 | $3\frac{3}{4}$ |
| 3 | $2\frac{1}{2}$ |
| 4 | $1\frac{7}{8}$ |

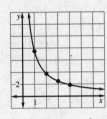

$$y = \frac{15}{2x}$$

inversely

**13.** $\dfrac{56x^6}{4x^4} = 14x^2$

**14.** $\dfrac{5x^2 - 15x}{15x^4} = \dfrac{5x(x - 3)}{3\,15x^4\;^3}$

$$= \dfrac{x - 3}{3x^3}$$

**15.** $\dfrac{x^2 - x - 6}{x^2 - 4} = \dfrac{(x - 3)(x + 2)}{(x - 2)(x + 2)}$

$$= \dfrac{x - 3}{x - 2}$$

**16.** $\dfrac{^3 6x^2}{_2 8x} \cdot \dfrac{-4x^3\,^2}{2x^2} = \dfrac{-3x^2}{2}$

**17.** $\dfrac{x + 3}{x^3 - x^2 - 6x} \cdot \dfrac{x^2 + x - 12}{x^2 - 9}$

$$= \dfrac{x + 3}{x(x - 3)(x + 2)} \cdot \dfrac{(x + 4)(x - 3)}{(x - 3)(x + 3)}$$

$$= \dfrac{x + 4}{x(x - 3)(x + 2)}$$

**18.** $\dfrac{x^3 + x^2}{x^2 - 16} \cdot \dfrac{x + 4}{3x^4 + x^3 - 2x^2}$

$$= \dfrac{x^2(x + 1)}{(x - 4)(x + 4)} \cdot \dfrac{x + 4}{x^2(3x - 2)(x + 1)}$$

$$= \dfrac{1}{(x - 4)(3x - 2)}$$

**19.** $\dfrac{3x^2 + 6x}{4x} \cdot \dfrac{8x^2}{15} = \dfrac{3x(x + 2)}{4x} \cdot \dfrac{^2 8x^2}{15_5} = \dfrac{2x^2(x + 2)}{5}$

**20.** $\dfrac{12x - 4}{x - 1} + \dfrac{4x}{x - 1} = \dfrac{16x - 4}{x - 1} = \dfrac{4(4x - 1)}{x - 1}$

**21.** $\dfrac{5}{2x^2} + \dfrac{4}{3x} = \dfrac{15 + 8x}{6x^2}$

**22.** $\dfrac{8}{5x} - \dfrac{4}{x^2} = \dfrac{8x - 20}{5x^2}$

**23.** $\dfrac{4}{x + 3} + \dfrac{3x}{x - 2} = \dfrac{4x - 8 + 3x^2 + 9x}{(x + 3)(x - 2)} = \dfrac{3x^2 + 13x - 8}{(x + 3)(x - 2)}$

**24.** $\dfrac{5x + 1}{x - 3} - \dfrac{2x}{x - 1} = \dfrac{(5x + 1)(x - 1) - 2x(x - 3)}{(x - 3)(x - 1)}$

$$= \dfrac{5x^2 - 4x - 1 - 2x^2 + 6x}{(x - 3)(x - 1)}$$

$$= \dfrac{3x^2 + 2x - 1}{(x - 3)(x - 1)}$$

$$= \dfrac{(3x - 1)(x + 1)}{(x - 3)(x - 1)}$$

**25.** $\dfrac{4x^3}{3x} - \dfrac{15x^2}{3x} - \dfrac{6x}{3x} = \dfrac{4x^2}{3} - 5x - 2$

**26.** $\dfrac{81x^2 - 25}{9x - 5} = \dfrac{(9x + 5)(9x - 5)}{9x - 5} = 9x + 5$

**27.** $\dfrac{2x^2 + 11x + 12}{x + 4} = \dfrac{(2x + 3)(x + 4)}{x + 4} = 2x + 3$

**28.**

$$
\begin{array}{r}
5x - 6 \phantom{00} \\
x + 2 \overline{)\;5x^2 + 4x - 7} \\
\underline{5x^2 + 10x} \phantom{00} \\
-6x - 7 \\
\underline{-6x - 12} \\
5
\end{array}
$$

$$5x - 6 + \dfrac{5}{x + 2}$$

**29.** $\dfrac{3}{4x - 9} = \dfrac{x}{3}$

$$9 = 4x^2 - 9x$$

$$4x^2 - 9x - 9 = 0$$

$$(4x + 3)(x - 3) = 0;\; x = -\dfrac{3}{4}\quad x = 3$$

**30.** $5x + 2 = 27$

$$5x = 25$$

$$x = 5$$

**31.** $\dfrac{5}{x + 3} - \dfrac{3}{x - 2} = \dfrac{5}{3(x - 2)}$

$$15(x - 2) - 9(x + 3) = 5(x + 3)$$

$$15x - 30 - 9x - 27 = 5x + 15$$

$$x = 72$$

**32.**

center (4, 3)

$$y = \dfrac{1}{x - 4} + 3$$

**33.**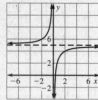

center (0, 5)

$$y = \frac{-2}{x} + 5$$

**34.**

$$x + 2 \overline{)\, x - 5}$$
$$\underline{x + 2}$$
$$-7$$

$$y = \frac{-7}{x + 2} + 1$$

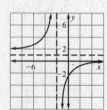

center (−2, 1)

**35.** $30{,}000 + 1.20x = 1.80x$

$30{,}000 = 0.60x$

$50{,}000 \text{ dozen} = x$

**36.** $\dfrac{2\overset{x}{\cancel{x}}(x + 1)}{\underset{6}{\cancel{12x}}(x + 3)} = \dfrac{x + 1}{6(x + 3)}$

$x = 5: \dfrac{6}{48} = 0.125$

## Chapter 11 Standardized Test *(pp. 704–705)*

**1.** $\dfrac{4}{y + 9} = \dfrac{6}{y - 7}$

$4y - 28 = 6y + 54$

$-2y = 82$

$y = -41$

B

**2.**

| Column A | Column B |
|---|---|
| $3x^2 - 12 = 2x^2 - 2x - 4$ | $5y + 30 = 2y + 24$ |
| $x^2 + 2x - 8 = 0$ | $3y = -6$ |
| $(x + 4)(x - 2) = 0$ | $y = -2$ |
| $x = -4 \quad x = 2$ | |
| extraneous | |

B $(-2 > -4)$

**3.** $x = 3.8(52)$

$x = 197.6$

E

**4.** $18.6 = 0.15x$

$124 = x$

D

**5.** $614 = x(840)$

$x = 73.1\%$

C

**6.** D

**7.** $36 = \dfrac{k}{9}$

$k = 324$

$y = \dfrac{324}{3}$

$y = 108$ E

**8.** $\dfrac{x(x^2 - 10x + 9)}{(x + 6)(x - 1)} = \dfrac{x(x - 9)(\cancel{x - 1})}{(x + 6)(\cancel{x - 1})}$

D

**9.** $\dfrac{3\cancel{x}(x + 6)}{\underset{5x}{\cancel{15x^2}}(2x - 1)} = \dfrac{x + 6}{5x(2x - 1)} = \dfrac{3 + 6}{5(3)(6 - 1)} = \dfrac{9}{75}$ B

**10.** $\dfrac{9x^2}{4x} \cdot \dfrac{\overset{4}{\cancel{16x^3}}}{x^{\cancel{3}}} = \dfrac{36}{x}$ C

**11.** $\dfrac{(\cancel{x - 8})(x + 8)}{3x^2} \cdot \dfrac{1}{(\cancel{x - 8})} = \dfrac{x + 8}{3x^2}$ A

**12.** $x^2 - 9{:}(x - 3)(x + 3)$

$x^2 + x - 6{:}(x + 3)(x - 2)$

LCD: $(x - 3)(x + 3)(x - 2)$ D

**13.** $\dfrac{2x + 9}{x + 5} - \dfrac{x - 4}{x - 2} = \dfrac{(2x + 9)(x - 2) - (x - 4)(x + 5)}{(x + 5)(x - 2)}$

$= \dfrac{2x^2 + 5x - 18 - x^2 - x + 20}{(x + 5)(x - 2)}$

$= \dfrac{x^2 + 4x + 2}{(x + 5)(x - 2)}$ D

**14.**

$$x - 4 \overline{)\, x^2 + 24x - 3} \quad\quad x + 28$$
$$\underline{x^2 - 4x}$$
$$28x - 3$$
$$\underline{28x - 112}$$
$$109$$

$x + 28 + \dfrac{109}{x - 4}$ A

**15.** A

**16. a.** $\dfrac{20{,}000(1.20)}{x} = \dfrac{24{,}000}{x}$

**b.** $\dfrac{24{,}000}{x + 5}$

**c.** $\dfrac{24{,}000}{x} - 400 = \dfrac{24{,}000}{x + 5}$

$24{,}000(x + 5) - 400x(x + 5) = 24{,}000x$

$60(x + 5) - x(x + 5) = 60x$

$60x + 300 - x^2 - 5x = 60x$

$x^2 + 5x - 300 = 0$

$(x + 20)(x - 15) = 0$

$x = -20 \quad x = 15$

old car: 15 mi/gal

new car: 20 mi/gal

**d.** old car: $\dfrac{24{,}000}{15} = \$1{,}600$

new car: $\dfrac{24{,}000}{20} = \$1{,}200$

# CHAPTER 12

## Think & Discuss (p. 707)

**1.** No

**2.** Somewhere between 12 and 17

## Skill Review (p. 708)

**1.** $\dfrac{4}{x} = \dfrac{7}{9.1}$

$7x = 36.4$

$x = 5.2$

**2.** $\dfrac{5}{1.5} = \dfrac{13}{x}$

$5x = 19.5$

$x = 3.9$

**3.** $\sqrt{98} = \sqrt{49}\sqrt{2}$

$\quad = 7\sqrt{2}$

**4.** $\sqrt{140} = \sqrt{4}\sqrt{35}$

$\quad = 2\sqrt{35}$

**5.** $\sqrt{\dfrac{7}{4}} = \dfrac{\sqrt{7}}{\sqrt{4}}$

$\quad = \dfrac{\sqrt{7}}{2}$

**6.** $\dfrac{\sqrt{144}}{\sqrt{16}} = \dfrac{12}{4}$

$\quad = 3$

**7.** $x^2 - 3x - 18 = (x - 6)(x + 3)$

**8.** $x^2 + 2x - 8 = (x + 4)(x - 2)$

**9.** $4x^2 + 20x + 25 = (2x + 5)(2x + 5) = (2x + 5)^2$

## Lesson 12.1

### Developing Concepts Activity 12.1 (p. 709)

**1.**

| x | 0 | 1 | 2 | 3 | 4 | 5 |
|---|---|---|-----|-----|---|-----|
| y | 0 | 1 | 1.4 | 1.7 | 2 | 2.2 |

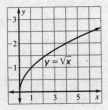

$y = \sqrt{x}$

**2.** Both the domain and range are all nonnegative numbers.

**3.** Square roots are not defined for negative values of $x$.

### 12.1 Guided Practice (p. 712)

**1.** The square root function is $y = \sqrt{x}$, where the domain and range are all nonnegative numbers.

**2.** Yes; the domain includes all numbers greater than or equal to $-3$. For those values, $\sqrt{x + 3}$ is defined.

**3.** $\begin{aligned} y &= 4\sqrt{0} \\ &= 4(0) \\ &= 0 \end{aligned}$  $\begin{aligned} y &= 4\sqrt{1} \\ &= 4(1) \\ &= 4 \end{aligned}$  $\begin{aligned} y &= 4\sqrt{2} \\ &= 4(1.41) \\ &= 5.7 \end{aligned}$

$\begin{aligned} y &= 4\sqrt{3} \\ &= 4(1.73) \\ &= 6.9 \end{aligned}$  $\begin{aligned} y &= 4\sqrt{4} \\ &= 4(2) \\ &= 8 \end{aligned}$

**4.** $\begin{aligned} y &= \tfrac{1}{2}\sqrt{0} \\ &= \tfrac{1}{2}(0) \\ &= 0 \end{aligned}$  $\begin{aligned} y &= \tfrac{1}{2}\sqrt{1} \\ &= \tfrac{1}{2}(1) \\ &= 0.5 \end{aligned}$  $\begin{aligned} y &= \tfrac{1}{2}\sqrt{2} \\ &= 0.5(1.41) \\ &= 0.7 \end{aligned}$

$\begin{aligned} y &= \tfrac{1}{2}\sqrt{3} \\ &= 0.5(1.73) \\ &= 0.9 \end{aligned}$  $\begin{aligned} y &= \tfrac{1}{2}\sqrt{4} \\ &= \tfrac{1}{2}(2) \\ &= 1 \end{aligned}$

**5.** $\begin{aligned} y &= 3\sqrt{0} + 4 \\ &= 3(0) + 4 \\ &= 4 \end{aligned}$  $\begin{aligned} y &= 3\sqrt{1} + 4 \\ &= 3(1) + 4 \\ &= 7 \end{aligned}$  $\begin{aligned} y &= 3\sqrt{2} + 4 \\ &= 3(1.41) + 4 \\ &= 8.2 \end{aligned}$

$\begin{aligned} y &= 3\sqrt{3} + 4 \\ &= 4(1.73) + 4 \\ &= 9.2 \end{aligned}$  $\begin{aligned} y &= 3\sqrt{4} + 4 \\ &= 3(2) + 4 \\ &= 10 \end{aligned}$

**6.** $\begin{aligned} y &= 6\sqrt{0} - 3 \\ &= 6(0) - 3 \\ &= -3 \end{aligned}$  $\begin{aligned} y &= 6\sqrt{1} - 3 \\ &= 6(1) - 3 \\ &= 3 \end{aligned}$  $\begin{aligned} y &= 6\sqrt{2} - 3 \\ &= 6(1.41) - 3 \\ &= 5.5 \end{aligned}$

$\begin{aligned} y &= 6\sqrt{3} - 3 \\ &= 6(1.73) - 3 \\ &= 7.4 \end{aligned}$  $\begin{aligned} y &= 6\sqrt{4} - 3 \\ &= 6(2) - 3 \\ &= 9 \end{aligned}$

**7.** $\begin{aligned} y &= \sqrt{0 + 2} \\ &= \sqrt{2} \\ &= 1.4 \end{aligned}$  $\begin{aligned} y &= \sqrt{1 + 2} \\ &= \sqrt{3} \\ &= 1.7 \end{aligned}$  $\begin{aligned} y &= \sqrt{2 + 2} \\ &= \sqrt{4} \\ &= 2 \end{aligned}$

$\begin{aligned} y &= \sqrt{3 + 2} \\ &= \sqrt{5} \\ &= 2.2 \end{aligned}$  $\begin{aligned} y &= \sqrt{4 + 2} \\ &= \sqrt{6} \\ &= 2.4 \end{aligned}$

**8.** $\begin{aligned} y &= \sqrt{4(0) - 1} \\ &= \sqrt{0 - 1} \\ &= \sqrt{-1} \\ &\text{undefined} \end{aligned}$  $\begin{aligned} y &= \sqrt{4(1) - 1} \\ &= \sqrt{4 - 1} \\ &= \sqrt{3} \\ &= 1.7 \end{aligned}$  $\begin{aligned} y &= \sqrt{4(2) - 1} \\ &= \sqrt{8 - 1} \\ &= \sqrt{7} \\ &= 2.6 \end{aligned}$

$\begin{aligned} y &= \sqrt{4(3) - 1} \\ &= \sqrt{12 - 1} \\ &= \sqrt{11} \\ &= 3.3 \end{aligned}$  $\begin{aligned} y &= \sqrt{4(4) - 1} \\ &= \sqrt{16 - 1} \\ &= \sqrt{15} \\ &= 3.9 \end{aligned}$

**9.** $y = 3\sqrt{x}$

Domain: all nonnegative numbers

Range: all nonnegative numbers

**10.** $y = \sqrt{x}$

Domain: all nonnegative numbers

Range: all nonnegative numbers

**11.** $y = \sqrt{x} - 10$

Domain: all nonnegative numbers.

Range: all numbers greater than or equal to $-10$

**12.** $y = \sqrt{x} + 6$

Domain: all nonnegative numbers

Range: all numbers greater than or equal to 6

**13.** $y = \sqrt{x + 5}$

Domain: all numbers greater than or equal to $-5$

Range: all nonnegative numbers.

**14.** $y = \sqrt{x - 10}$

Domain: all numbers greater than or equal to 10

Range: all nonnegative numbers

**15.** $y = 3\sqrt{x}$

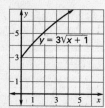

**16.** $y = \sqrt{x} + 5$

**17.** $y = 3\sqrt{x + 1}$

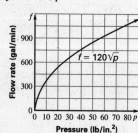

**18.** $f = 120\sqrt{p}$

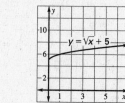

Pressure (lb/in.²)

**19.**
$$f = 120\sqrt{p}$$
$$1200 = 120\sqrt{p}$$
$$10 = \sqrt{p}$$
$$100 = p$$
$$p = 100 \text{ lb/in.}^2$$

## 12.1 Practice and Applications (pp. 712–714)

**20.** $y = 3\sqrt{9}$

$= 3(3)$

$= 9$

**21.** $y = \frac{1}{2}\sqrt{16} - 1$

$= \frac{1}{2}(4) - 1$

$= 1$

**22.** $y = \sqrt{15 - 7}$

$= \sqrt{8}$

$= 2.8$

**23.** $y = \sqrt{3(7) - 5}$

$= \sqrt{21 - 5}$

$= \sqrt{16}$

$= 4$

**24.** $y = 6\sqrt{15 + 1}$

$= 6\sqrt{16}$

$= 6(4)$

$= 24$

**25.** $y = \sqrt{21 - 2(-2)}$

$= \sqrt{21 + 4}$

$= \sqrt{25}$

$= 5$

**26.** $y = \sqrt{\frac{22}{2} - 2}$

$= \sqrt{11 - 2}$

$= \sqrt{9}$

$= 3$

**27.** $y = \sqrt{8\left(\frac{1}{4}\right)^2 + \frac{3}{2}}$

$= \sqrt{8\left(\frac{1}{16}\right) + \frac{3}{2}}$

$= \sqrt{\frac{8}{16} + \frac{3}{2}}$

$= \sqrt{\frac{32}{16}}$

$= \frac{\sqrt{32}}{4}$

$= 1.4$

**28.** $y = \sqrt{\frac{2(6)}{3} + 5}$

$= \sqrt{\frac{12}{3} + 5}$

$= \sqrt{4 + 5}$

$= \sqrt{9}$

$= 3$

**29.** $y = \sqrt{36\left(\frac{1}{2}\right) - 2}$

$= \sqrt{18 - 2}$

$= \sqrt{16}$

$= 4$

**30.** $y = 6\sqrt{x}$

all nonnegative numbers

**31.** $y = \sqrt{x - 17}$

$x \geq 17$

**32.** $y = \sqrt{3x - 10}$

$x \geq 3\frac{1}{3}$

**33.** $y = \sqrt{x + 5}$

$x \geq -5$

**34.** $y = 4 + \sqrt{x}$

all nonnegative numbers

**35.** $y = \sqrt{x} - 3$

all nonnegative numbers

**36.** $y = 5 - \sqrt{x}$

all nonnegative numbers

**37.** $y = 4\sqrt{x}$

all nonnegative numbers

**38.** $y = 2\sqrt{4x}$

all nonnegative numbers

**39.** $y = 0.2\sqrt{x}$

all nonnegative numbers

**40.** $y = x\sqrt{x}$

all nonnegative numbers

**41.** $y = \sqrt{x + 9}$

$x \geq -9$

**42.** $y = 8\sqrt{\frac{5}{2}x}$

all nonnegative numbers

**43.** $y = \frac{\sqrt{x}}{5}$

all nonnegative numbers

**44.** $y = \frac{\sqrt{4 - x}}{x}$

$x \leq 4, x \neq 0$

**45.** $y = 7\sqrt{x}$

Domain: all nonnegative numbers

Range: all nonnegative numbers

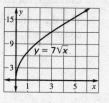

## Chapter 12 *continued*

**46.** $y = 4\sqrt{x}$

Domain: all nonnegative numbers

Range: all nonnegative numbers

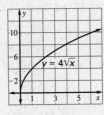

**47.** $y = 5\sqrt{x}$

Domain: all nonnegative numbers

Range: all nonnegative numbers

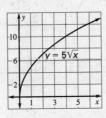

**48.** $y = \sqrt{x} - 2$

Domain: all nonnegative numbers

Range: $y \geq -2$

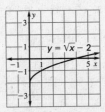

**49.** $y = \sqrt{x} + 4$

Domain: all nonnegative numbers

Range: $y \geq 4$

**50.** $y = \sqrt{x} - 3$

Domain: all nonnegative numbers

Range: $y \geq -3$

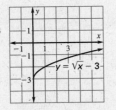

**51.** $y = \sqrt{x - 4}$

Domain: $x \geq 4$

Range: all nonnegative numbers

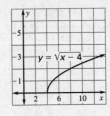

**52.** $y = \sqrt{x + 1}$

Domain: $x \geq -1$

Range: all nonnegative numbers

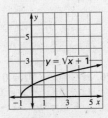

**53.** $y = \sqrt{x - 6}$

Domain: $x \geq 6$

Range: all nonnegative numbers

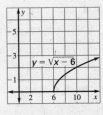

**54.** $y = 2\sqrt{4x + 10}$

Domain: $x \geq -2\frac{1}{2}$

Range: all nonnegative numbers

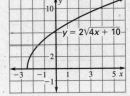

**55.** $y = \sqrt{2x + 5}$

Domain: $x \geq -2\frac{1}{2}$

Range: all nonnegative numbers

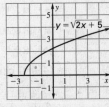

**56.** $y = x\sqrt{8x}$

Domain: all nonnegative numbers

Range: all nonnegative numbers

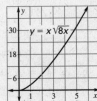

**57.** $A = \sqrt{\dfrac{h \cdot w}{3131}}$

$= \sqrt{\dfrac{62 \cdot 100}{3131}}$

$= \sqrt{\dfrac{6200}{3131}}$

$= 1.41 \text{ m}^2$

**58.**

$T = 2\pi\sqrt{\dfrac{L}{384}}$

$8 = 2\pi\sqrt{\dfrac{L}{384}}$

$\dfrac{8}{2\pi} = \sqrt{\dfrac{L}{384}}$

$\dfrac{8}{2\pi} = \dfrac{\sqrt{L}}{\sqrt{384}}$

$\dfrac{156.77}{2\pi} = \sqrt{L}$

$\left(\dfrac{156.77}{2\pi}\right)^2 = L$

$622.5 \text{ in.} \approx L$

**59. a.** $cm^2 = cm \cdot \sqrt{cm^2 + cm^2}$

$\qquad = cm \cdot \sqrt{cm^2}$

$\qquad = cm \cdot cm$

$\qquad = cm^2$

**b.** $S = \pi \cdot 14\sqrt{14^2 + h^2}$

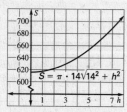

**c.** $S = \pi \cdot 14\sqrt{14^2 + 30^2}$

$\qquad = \pi \cdot 14\sqrt{196 + 900}$

$\qquad = \pi \cdot 14\sqrt{1096}$

$\qquad \approx 463.48\pi$

$\qquad \approx 1456.07 \ cm^2$

**60. a.** $S = \sqrt{30df}$

$\qquad = \sqrt{30(74)(0.5)}$

$\qquad = \sqrt{1110}$

$\qquad \approx 33.3 \ mi/h$

No; the speed was 33.3 mi/h.

**b.** $\quad 45 = \sqrt{30df}$

$\quad 45 = \sqrt{30d(0.5)}$

$\quad 45 = \sqrt{15d}$

$\ 2025 = 15d$

$\quad 135 = d$

The skid marks would have to be longer than 135 ft.

**61.** Answers may vary.

**62.** Answers may vary.

**63. a.** $v = \sqrt{2(32)h}$

$\qquad = \sqrt{64h}$

$\qquad = 8\sqrt{h}$

**b.** $v = 8\sqrt{h}$

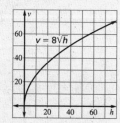

**c.** When $h$ is doubled, $v = 8\sqrt{2h}$. The velocity is multiplied by a factor of $\sqrt{2} \approx 1.4$, not 2. The graph is not linear. If doubling the height doubled the drop, the graph would be linear.

**64.** Domain: all nonnegative numbers except 4

---

**65.** $\sqrt{24} = \sqrt{4}\sqrt{6}$

$\qquad = 2\sqrt{6}$

**66.** $\sqrt{60} = \sqrt{4}\sqrt{15}$

$\qquad = 2\sqrt{15}$

**67.** $\sqrt{175} = \sqrt{25}\sqrt{7}$

$\qquad = 5\sqrt{7}$

**68.** $\sqrt{9900} = \sqrt{900}\sqrt{11}$

$\qquad = 30\sqrt{11}$

**69.** $\sqrt{\dfrac{20}{25}} = \dfrac{\sqrt{4}\sqrt{5}}{\sqrt{25}}$

$\qquad = \dfrac{2\sqrt{5}}{5}$

**70.** $\dfrac{1}{2}\sqrt{80} = \dfrac{1}{2}\sqrt{16}\sqrt{5}$

$\qquad = 2\sqrt{5}$

**71.** $\dfrac{3\sqrt{7}}{\sqrt{9}} = \dfrac{3\sqrt{7}}{3}$

$\qquad = \sqrt{7}$

**72.** $4\sqrt{\dfrac{11}{16}} = 4\dfrac{\sqrt{11}}{\sqrt{16}}$

$\qquad = \dfrac{4\sqrt{11}}{4} = \sqrt{11}$

**73.** $x^2 + 4x - 8 = 0 \qquad x = \dfrac{-4 \pm \sqrt{4^2 - 4(1)(-8)}}{2(1)}$

$\qquad\qquad\qquad = \dfrac{-4 \pm \sqrt{16 + 32}}{2}$

$\qquad\qquad\qquad = \dfrac{-4 \pm \sqrt{48}}{2} = \dfrac{-4 \pm \sqrt{16}\sqrt{3}}{2}$

$\qquad\qquad\qquad = \dfrac{-4 + 4\sqrt{3}}{2}, \dfrac{-4 - 4\sqrt{3}}{2}$

$\qquad\qquad\qquad = -2 + 2\sqrt{3}, -2 - 2\sqrt{3}$

**74.** $x^2 - 2x - 4 = 0 \qquad x = \dfrac{2 \pm \sqrt{(-2)^2 - 4(1)(-4)}}{2(1)}$

$\qquad\qquad\qquad = \dfrac{2 \pm \sqrt{4 + 16}}{2}$

$\qquad\qquad\qquad = \dfrac{2 \pm \sqrt{20}}{2} = \dfrac{2 \pm \sqrt{4}\sqrt{5}}{2}$

$\qquad\qquad\qquad = \dfrac{2 + 2\sqrt{5}}{2}, \dfrac{2 - 2\sqrt{5}}{2}$

$\qquad\qquad\qquad = 1 + \sqrt{5}, 1 - \sqrt{5}$

**75.** $x^2 - 6x + 1 = 0 \qquad x = \dfrac{6 \pm \sqrt{(-6)^2 - 4(1)(1)}}{2(1)}$

$\qquad\qquad\qquad = \dfrac{6 \pm \sqrt{36 - 4}}{2}$

$\qquad\qquad\qquad = \dfrac{6 \pm \sqrt{32}}{2} = \dfrac{6 \pm \sqrt{16}\sqrt{2}}{2}$

$\qquad\qquad\qquad = \dfrac{6 + 4\sqrt{2}}{2}, \dfrac{6 - 4\sqrt{2}}{2}$

$\qquad\qquad\qquad = 3 + 2\sqrt{2}, 3 - 2\sqrt{2}$

**76.** $x^2 + 3x - 10 = 0$

$\quad (x + 5)(x - 2) = 0$

$\quad x + 5 = 0 \qquad x - 2 = 0$

$\qquad x = -5 \qquad\quad x = 2$

---

# Chapter 12 *continued*

**77.**
$$2x^2 + x = 3$$
$$2x^2 + x - 3 = 0$$
$$(2x + 3)(x - 1) = 0$$
$$2x + 3 = 0 \quad x - 1 = 0$$
$$2x = -3 \qquad x = 1$$
$$x = -\frac{3}{2}$$

**78.** $4x^2 - 6x + 1 = 0$
$$\frac{6 \pm \sqrt{(-6)^2 - 4(4)(1)}}{2(4)} = \frac{6 \pm \sqrt{36 - 16}}{8}$$
$$= \frac{6 \pm \sqrt{20}}{8} = \frac{6 \pm \sqrt{4}\sqrt{5}}{8}$$
$$= \frac{6 + 2\sqrt{5}}{8}, \frac{6 - 2\sqrt{5}}{8}$$
$$= \frac{3 + \sqrt{5}}{4}, \frac{3 - \sqrt{5}}{4}$$

**79.** $(x - 2)(x + 11) = x^2 + 11x - 2x - 22$
$$= x^2 + 9x - 22$$

**80.** $(x + 4)(3x - 7) = 3x^2 - 7x + 12x - 28$
$$= 3x^2 + 5x - 28$$

**81.** $(2x - 3)(5x - 9) = 10x^2 - 18x - 15x + 27$
$$= 10x^2 - 33x + 27$$

**82.** $(x - 5)(x - 4) = x^2 - 4x - 5x + 20$
$$= x^2 - 9x + 20$$

**83.** $(6x + 2)(x^2 - x - 1)$
$$= 6x^3 - 6x^2 - 6x + 2x^2 - 2x - 2$$
$$= 6x^3 - 4x^2 - 8x - 2$$

**84.** $(2x - 1)(x^2 + x + 1)$
$$= 2x^3 + 2x^2 + 2x - x^2 - x - 1$$
$$= 2x^3 + x^2 + x - 1$$

**85.** $\dfrac{8x}{3} \cdot \dfrac{1}{x} = \dfrac{8x}{3x} = \dfrac{8}{3}$

**86.** $\dfrac{\overset{1}{8x^2}}{\underset{1}{13}} \cdot \dfrac{\overset{3}{9}}{\underset{2}{16x}} = \dfrac{3x}{2}$

**87.** $\dfrac{x}{x + 6} \div \dfrac{x + 1}{x + 6} = \dfrac{x}{x + 6} \cdot \dfrac{x + 6}{x + 1}$
$$= \dfrac{x}{x + 1}$$

**88.** 12 to 1
$$\frac{12}{1} = \frac{n}{5}$$
$$n = 60 \text{ ft}$$

***Technology Activity 12.1 (p. 715)***
***Exploring the Concept***

**1.** Check graphs.

**2.** Check graphs.

**3.** For $a > 0$, as $a$ increases, the graph of $y = a\sqrt{x}$ gets steeper.

**4.** Check graphs.

**5.** Check graphs.

**6.** Check graphs.

**7.** For $a < 0$, the graph of $y = a\sqrt{x}$ is in the third quadrant and gets steeper as $|a|$ increases.

***Drawing Conclusions***

**1.** Check graphs; $k$ moves the graph of $y = \sqrt{x}$ up $k$ units if $k$ is positive and down $|k|$ units if $k$ is negative.

**2.** Check graphs; $k$ moves the graph of $y = \sqrt{x}$ to the left $k$ units if $k$ is positive and $|k|$ units to the right if $k$ is negative.

**3.** $-2$

**4.** 4

**5.** 3

## Lesson 12.2

*12.2 Guided Practice (p. 719)*

**1.** like radicals

**2.** Any expressions of the form $a + \sqrt{b}$ and $a - \sqrt{b}$ where $b > 0$

**3.** Multiply the numerator and denominator of the expression by the conjugate of the denominator, $\sqrt{3} + 1$.

**4.** $4\sqrt{5} + 5\sqrt{5} = 9\sqrt{5}$

**5.** $3\sqrt{7} - 2\sqrt{7} = \sqrt{7}$

**6.** $3\sqrt{6} + \sqrt{24} = 3\sqrt{6} + \sqrt{4}\sqrt{6}$
$$= 3\sqrt{6} + 2\sqrt{6}$$
$$= 5\sqrt{6}$$

**7.** $\sqrt{3} \cdot \sqrt{8} = \sqrt{24}$
$$= \sqrt{4}\sqrt{6}$$
$$= 2\sqrt{6}$$

**8.** $(2 + \sqrt{3})^2 = (2 + \sqrt{3})(2 + \sqrt{3})$
$$= 4 + 2\sqrt{3} + 2\sqrt{3} + 3$$
$$= 7 + 4\sqrt{3}$$

**9.** $\sqrt{3}(5\sqrt{3} - 2\sqrt{6}) = 5\sqrt{9} - 2\sqrt{18}$
$$= 15 - 2\sqrt{9}\sqrt{2}$$
$$= 15 - 2(3)\sqrt{2}$$
$$= 15 - 6\sqrt{2}$$

**10.** $\dfrac{4}{\sqrt{13}} \cdot \dfrac{\sqrt{13}}{\sqrt{13}} = \dfrac{4\sqrt{13}}{13}$

**11.** $\dfrac{3}{8 - \sqrt{10}} \cdot \dfrac{8 + \sqrt{10}}{8 + \sqrt{10}} = \dfrac{24 + 3\sqrt{10}}{64 - 10}$
$$= \dfrac{24 + 3\sqrt{10}}{54}$$
$$= \dfrac{8 + \sqrt{10}}{18}$$

**12.** $\dfrac{6}{\sqrt{10}} \cdot \dfrac{\sqrt{10}}{\sqrt{10}} = \dfrac{6\sqrt{10}}{10} = \dfrac{3\sqrt{10}}{5}$

# Chapter 12 *continued*

**13.** $6(\sqrt{26})^2 - 156 \stackrel{?}{=} 0$

$6(26) - 156 \stackrel{?}{=} 0$

$156 - 156 \stackrel{?}{=} 0$

$0 = 0$

solution

**14.** $(-4\sqrt{3})^2 - 48 \stackrel{?}{=} 0$

$16(3) - 48 \stackrel{?}{=} 0$

$48 - 48 \stackrel{?}{=} 0$

$0 = 0$

solution

**15.** $\left(6 + \sqrt{31}\right)^2 - 12\left(6 + \sqrt{31}\right) + 5 \stackrel{?}{=} 0$

$\left(6 + \sqrt{31}\right)\left(6 + \sqrt{31}\right) - 12\left(6 + \sqrt{31}\right) + 5 \stackrel{?}{=} 0$

$36 + 12\sqrt{31} + 31 - 72 - 12\sqrt{31} + 5 \stackrel{?}{=} 0$

$0 = 0$

solution

**16.** $\left(4 + 2\sqrt{2}\right)^2 - 8\left(4 + 2\sqrt{2}\right) + 8 \stackrel{?}{=} 0$

$\left(4 + 2\sqrt{2}\right)\left(4 + 2\sqrt{2}\right) - 32 - 16\sqrt{2} + 8 \stackrel{?}{=} 0$

$16 + 16\sqrt{2} + 8 - 32 - 16\sqrt{2} + 8 \stackrel{?}{=} 0$

$0 = 0$

solution

**17.** $D = \sqrt{\dfrac{3(24)}{2}} - \sqrt{\dfrac{3(12)}{2}}$

$= \sqrt{\dfrac{72}{2}} - \sqrt{\dfrac{36}{2}}$

$= \sqrt{36} - \sqrt{18}$

$= 6 - \sqrt{9}\sqrt{2}$

$= 6 - 3\sqrt{2}$

$\approx 1.8$ miles

### 12.2 Practice and Applications (pp. 719–721)

**18.** $5\sqrt{7} + 2\sqrt{7} = 7\sqrt{7}$

**19.** $\sqrt{3} + 5\sqrt{3} = 6\sqrt{3}$

**20.** $11\sqrt{3} - 12\sqrt{3} = -1\sqrt{3}$

$= -\sqrt{3}$

**21.** $2\sqrt{6} - \sqrt{6} = \sqrt{6}$

**22.** $\sqrt{32} + \sqrt{2} = \sqrt{16}\sqrt{2} + \sqrt{2}$

$= 4\sqrt{2} + \sqrt{2}$

$= 5\sqrt{2}$

**23.** $\sqrt{75} + \sqrt{3} = \sqrt{25}\sqrt{3} + \sqrt{3}$

$= 5\sqrt{3} + \sqrt{3}$

$= 6\sqrt{3}$

**24.** $\sqrt{80} - \sqrt{45} = \sqrt{16}\sqrt{5} - \sqrt{9}\sqrt{5}$

$= 4\sqrt{5} - 3\sqrt{5}$

$= \sqrt{5}$

**25.** $\sqrt{72} - \sqrt{18} = \sqrt{36}\sqrt{2} - \sqrt{9}\sqrt{2}$

$= 6\sqrt{2} - 3\sqrt{2}$

$= 3\sqrt{2}$

**26.** $\sqrt{147} - 7\sqrt{3} = \sqrt{49}\sqrt{3} - 7\sqrt{3}$

$= 7\sqrt{3} - 7\sqrt{3}$

$= 0$

**27.** $4\sqrt{5} + \sqrt{125} + \sqrt{45}$

$= 4\sqrt{5} + \sqrt{25}\sqrt{5} + \sqrt{9}\sqrt{5}$

$= 4\sqrt{5} + 5\sqrt{5} + 3\sqrt{5}$

$= 12\sqrt{5}$

**28.** $3\sqrt{11} + \sqrt{176} + \sqrt{11} = 3\sqrt{11} + \sqrt{16}\sqrt{11} + \sqrt{11}$

$= 3\sqrt{11} + 4\sqrt{11} + \sqrt{11}$

$= 8\sqrt{11}$

**29.** $\sqrt{24} - \sqrt{96} + \sqrt{6} = \sqrt{4}\sqrt{6} - \sqrt{16}\sqrt{6} + \sqrt{6}$

$= 2\sqrt{6} - 4\sqrt{6} + \sqrt{6}$

$= -\sqrt{6}$

**30.** $\sqrt{243} - \sqrt{75} + \sqrt{300} = \sqrt{81}\sqrt{3} - \sqrt{25}\sqrt{3} + \sqrt{100}\sqrt{3}$

$= 9\sqrt{3} - 5\sqrt{3} + 10\sqrt{3}$

$= 14\sqrt{3}$

**31.** $\sqrt{3} \cdot \sqrt{12} = \sqrt{36} = 6$

**32.** $\sqrt{5} \cdot \sqrt{8} = \sqrt{40} = \sqrt{4}\sqrt{10} = 2\sqrt{10}$

**33.** $\left(1 + \sqrt{13}\right)\left(1 - \sqrt{13}\right) = 1 - \sqrt{13} + \sqrt{13} - 13$

$= -12$

**34.** $\sqrt{3}\left(5\sqrt{2} + \sqrt{3}\right) = 5\sqrt{6} + \sqrt{9} = 5\sqrt{6} + 3$

**35.** $\sqrt{6}\left(7\sqrt{3} + 6\right) = 7\sqrt{18} + 6\sqrt{6}$

$= 7\sqrt{9}\sqrt{2} + 6\sqrt{6}$

$= 21\sqrt{2} + 6\sqrt{6}$

**36.** $\left(\sqrt{6} + 5\right)^2 = \left(\sqrt{6} + 5\right)\left(\sqrt{6} + 5\right)$

$= 6 + 5\sqrt{6} + 5\sqrt{6} + 25$

$= 31 + 10\sqrt{6}$

**37.** $\left(\sqrt{a} - b\right)^2 = \left(\sqrt{a} - b\right)\left(\sqrt{a} - b\right)$

$= a - b\sqrt{a} - b\sqrt{a} + b^2$

$= a - 2b\sqrt{a} + b^2$

**38.** $\left(\sqrt{c} + d\right)\left(3 + \sqrt{5}\right) = 3\sqrt{c} + \sqrt{c}\sqrt{5} + 3d + \sqrt{5}d$

$= 3\sqrt{c} + \sqrt{5c} + 3d + d\sqrt{5}$

**39.** $\left(2\sqrt{3} - 5\right)^2 = \left(2\sqrt{3} - 5\right)\left(2\sqrt{3} - 5\right)$

$= 4(3) - 5(2\sqrt{3}) - 5(2\sqrt{3}) + 25$

$= 12 - 10\sqrt{3} - 10\sqrt{3} + 25$

$= 12 - 20\sqrt{3} + 25$

$= 37 - 20\sqrt{3}$

**40.** $\dfrac{5}{\sqrt{7}} \cdot \dfrac{\sqrt{7}}{\sqrt{7}} = \dfrac{5\sqrt{7}}{7}$

# Chapter 12 *continued*

**41.** $\dfrac{2}{\sqrt{2}} \cdot \dfrac{\sqrt{2}}{\sqrt{2}} = \dfrac{2\sqrt{2}}{2} = \sqrt{2}$

**42.** $\dfrac{9}{5 - \sqrt{7}} \cdot \dfrac{5 + \sqrt{7}}{5 + \sqrt{7}} = \dfrac{45 + 9\sqrt{7}}{25 - 7}$

$\qquad\qquad\qquad = \dfrac{45 + 9\sqrt{7}}{18}$

$\qquad\qquad\qquad = \dfrac{5 + \sqrt{7}}{2}$

**43.** $\dfrac{3}{\sqrt{48}} \cdot \dfrac{\sqrt{48}}{\sqrt{48}} = \dfrac{3\sqrt{48}}{48}$

$\qquad\qquad\qquad = \dfrac{3\sqrt{16}\sqrt{3}}{48}$

$\qquad\qquad\qquad = \dfrac{12\sqrt{3}}{48}$

$\qquad\qquad\qquad = \dfrac{\sqrt{3}}{4}$

**44.** $\dfrac{6}{10 + \sqrt{2}} \cdot \dfrac{10 - \sqrt{2}}{10 - \sqrt{2}} = \dfrac{60 - 6\sqrt{2}}{100 - 2}$

$\qquad\qquad\qquad = \dfrac{60 - 6\sqrt{2}}{98}$

$\qquad\qquad\qquad = \dfrac{30 - 3\sqrt{2}}{49}$

**45.** $\dfrac{\sqrt{3}}{\sqrt{3} - 1} \cdot \dfrac{\sqrt{3} + 1}{\sqrt{3} + 1} = \dfrac{3 + \sqrt{3}}{3 + \sqrt{3} - \sqrt{3} - 1}$

$\qquad\qquad\qquad = \dfrac{3 + \sqrt{3}}{2}$

**46.** $\dfrac{14}{60 - \sqrt{578}} \cdot \dfrac{60 + \sqrt{578}}{60 + \sqrt{578}} = \dfrac{840 + 14\sqrt{578}}{3600 - 578}$

$\qquad\qquad\qquad = \dfrac{840 + 238\sqrt{2}}{3022}$

$\qquad\qquad\qquad = \dfrac{420 + 119\sqrt{2}}{1511}$

**47.** $\dfrac{12}{7 - \sqrt{3}} \cdot \dfrac{7 + \sqrt{3}}{7 + \sqrt{3}} = \dfrac{84 + 12\sqrt{3}}{49 - 3}$

$\qquad\qquad\qquad = \dfrac{84 + 12\sqrt{3}}{46}$

$\qquad\qquad\qquad = \dfrac{42 + 6\sqrt{3}}{23}$

**48.** $\dfrac{4 + \sqrt{3}}{a - \sqrt{b}} \cdot \dfrac{a + \sqrt{b}}{a + \sqrt{b}} = \dfrac{4a + 4\sqrt{b} + a\sqrt{3} + \sqrt{3b}}{a^2 - b}$

**49.** $A = \sqrt{68} \cdot \left(\sqrt{17} + 9\right)$

$\qquad = \sqrt{1156} + 9\sqrt{68}$

$\qquad = 34 + 9\sqrt{4}\sqrt{17}$

$\qquad = 34 + 9(2)\sqrt{17}$

$\qquad = 34 + 18\sqrt{17}$

**50.** $A = \frac{1}{2}\left(\sqrt{99} + 2\right)\left(\sqrt{44} + 12\right)$

$\qquad = \frac{1}{2}\left(66 + 2\sqrt{44} + 12\sqrt{99} + 24\right)$

$\qquad = 33 + \sqrt{44} + 6\sqrt{99} + 12$

$\qquad = 45 + \sqrt{44} + 6\sqrt{99}$

$\qquad = 45 + \sqrt{4}\sqrt{11} + 6\sqrt{9}\sqrt{11}$

$\qquad = 45 + 2\sqrt{11} + 18\sqrt{11}$

$\qquad = 45 + 20\sqrt{11}$

**51.** $A = \frac{1}{2}\left(\sqrt{25} + 4\right)\left(\sqrt{75} + 10\right)$

$\qquad = \frac{1}{2}(9)\left(\sqrt{75} + 10\right)$

$\qquad = \frac{9}{2}\left(\sqrt{75} + 10\right)$

$\qquad = \frac{9}{2}\sqrt{75} + 45$

$\qquad = \frac{9}{2}\sqrt{25}\sqrt{3} + 45$

$\qquad = \frac{45}{2}\sqrt{3} + 45$

$\qquad = \dfrac{45\sqrt{3} + 90}{2}$

**52.** $x^2 + 10x + 13 = 0$

$\qquad \dfrac{-10 \pm \sqrt{10^2 - 4(1)(13)}}{2}$

$\qquad = \dfrac{-10 \pm \sqrt{100 - 52}}{2}$

$\qquad = \dfrac{-10 \pm \sqrt{48}}{2}$

$\qquad = \dfrac{-10 + \sqrt{16}\sqrt{3}}{2}, \dfrac{-10 - \sqrt{16}\sqrt{3}}{2}$

$\qquad = \dfrac{-10 + 4\sqrt{3}}{2}, \dfrac{-10 - 4\sqrt{3}}{2}$

$\qquad = -5 + 2\sqrt{3}, -5 - 2\sqrt{3}$

**53.** $x^2 - 4x - 6 = 0$

$\qquad \dfrac{4 \pm \sqrt{(-4)^2 - 4(1)(-6)}}{2}$

$\qquad = \dfrac{4 \pm \sqrt{16 + 24}}{2}$

$\qquad = \dfrac{4 + \sqrt{40}}{2}$

$\qquad = \dfrac{4 + \sqrt{4}\sqrt{10}}{2}, \dfrac{4 - \sqrt{4}\sqrt{10}}{2}$

$\qquad = \dfrac{4 + 2\sqrt{10}}{2}, \dfrac{4 - 2\sqrt{10}}{2}$

$\qquad = 2 + \sqrt{10}, 2 - \sqrt{10}$

# Chapter 12 *continued*

**54.** $x^2 - 4x - 15 = 0$

$$\frac{4 \pm \sqrt{(-4)^2 - 4(1)(-15)}}{2} = \frac{4 \pm \sqrt{16 + 60}}{2}$$

$$= \frac{4 \pm \sqrt{76}}{2}$$

$$= \frac{4 \pm \sqrt{4}\sqrt{19}}{2}$$

$$= \frac{4 + 2\sqrt{19}}{2}, \frac{4 - 2\sqrt{19}}{2}$$

$$= 2 + \sqrt{19}, 2 - \sqrt{19}$$

**55.** $4x^2 - 2x - 1 = 0$

$$\frac{2 \pm \sqrt{(-2)^2 - 4(4)(-1)}}{8} = \frac{2 \pm \sqrt{4 + 16}}{8}$$

$$= \frac{2 \pm \sqrt{20}}{8}$$

$$= \frac{2 \pm \sqrt{4}\sqrt{5}}{8}$$

$$= \frac{2 + 2\sqrt{5}}{8}, \frac{2 - 2\sqrt{5}}{8}$$

$$= \frac{1 + \sqrt{5}}{4}, \frac{1 - \sqrt{5}}{4}$$

**56.** $x^2 - 6x - 1 = 0$

$$\frac{6 \pm \sqrt{(-6)^2 - 4(1)(-1)}}{2} = \frac{6 \pm \sqrt{36 + 4}}{2}$$

$$= \frac{6 \pm \sqrt{40}}{2}$$

$$= \frac{6 \pm \sqrt{4}\sqrt{10}}{2}$$

$$= \frac{6 + 2\sqrt{10}}{2}, \frac{6 - 2\sqrt{10}}{2}$$

$$= 3 + \sqrt{10}, 3 - \sqrt{10}$$

**57.** $a^2 - 6a - 13 = 0$

$$\frac{6 \pm \sqrt{(-6)^2 - 4(1)(-13)}}{2} = \frac{6 \pm \sqrt{36 + 52}}{2}$$

$$= \frac{6 \pm \sqrt{88}}{2}$$

$$= \frac{6 \pm \sqrt{4}\sqrt{22}}{2}$$

$$= \frac{6 + 2\sqrt{22}}{2}, \frac{6 - 2\sqrt{22}}{2}$$

$$= 3 + \sqrt{22}, 3 - \sqrt{22}$$

**58.** $V = 8\sqrt{20} - 8\sqrt{16}$

$$= 8\sqrt{4}\sqrt{5} - 8(4)$$

$$= 16\sqrt{5} - 32$$

$$\approx 3.78 \text{ ft/sec}$$

**59.** $T = \dfrac{\sqrt{50}}{4} - \dfrac{\sqrt{32}}{4}$

$$= \frac{\sqrt{25 \cdot 2} - \sqrt{16 \cdot 2}}{4}$$

$$= \frac{5\sqrt{2} - 4\sqrt{2}}{4}$$

$$= \frac{\sqrt{2}}{4} \approx 0.35 \text{ sec}$$

**60.** $S = \dfrac{d}{\dfrac{\sqrt{d}}{4}}$

$$= \frac{d}{1} \cdot \frac{4}{\sqrt{d}}$$

$$= \frac{4d}{\sqrt{d}} \cdot \frac{\sqrt{d}}{\sqrt{d}}$$

$$= \frac{4d\sqrt{d}}{d}$$

$$= 4\sqrt{d}$$

**61.** $S = 4\sqrt{d}$

$$= 4\sqrt{400}$$

$$= 80 \text{ ft/sec}$$

**62.** Graphs to be done on graphing utility

**63.** C; $\sqrt{5}(6 + \sqrt{5})^2 = \sqrt{5}(6 + \sqrt{5})(6 + \sqrt{5})$

$$= \sqrt{5}(36 + 12\sqrt{5} + 5)$$

$$= \sqrt{5}(41 + 12\sqrt{5})$$

$$= 41\sqrt{5} + 12(5)$$

$$= 41\sqrt{5} + 60$$

**64.** B; $\sqrt{3} - 5\sqrt{9} = \sqrt{3} - 5(3)$

$$= \sqrt{3} - 15$$

**65.** A; $\dfrac{3}{5 - \sqrt{5}} = \dfrac{3}{5 - \sqrt{5}} \cdot \dfrac{5 + \sqrt{5}}{5 + \sqrt{5}}$

$$= \frac{15 + 3\sqrt{5}}{25 - 5}$$

$$= \frac{15 + 3\sqrt{5}}{20}$$

**66.** $t = \dfrac{\sqrt{d}}{4} = \dfrac{\sqrt{d}}{\sqrt{16}} = \sqrt{\dfrac{d}{16}}$ where $d$ is in feet. Changing $d$

from feet to inches gives $\sqrt{\dfrac{d}{16 \cdot 12}} = \sqrt{\dfrac{d}{192}}$. So

$$t = \sqrt{\frac{d}{192}}.$$

**67.** $t = \sqrt{\dfrac{d}{192}}$

$$t = \sqrt{\frac{4}{192}} \qquad t = \sqrt{\frac{1}{192}} \qquad t = \sqrt{\frac{3}{192}}$$

$$= \frac{\sqrt{4} + \sqrt{1} + \sqrt{3}}{\sqrt{192}}$$

$$= \frac{3 + \sqrt{3}}{\sqrt{192}} \cdot \frac{\sqrt{192}}{\sqrt{192}} = \frac{3\sqrt{192} + \sqrt{576}}{192} \div 3$$

**—CONTINUED—**

# Chapter 12 *continued*

**67.** —CONTINUED—

$$= \frac{3\sqrt{192} + \sqrt{576}}{192} \cdot \frac{1}{3} = \frac{3\sqrt{192} + \sqrt{576}}{576}$$

$$= \frac{3\sqrt{64}\sqrt{3} + 24}{576} = \frac{24\sqrt{3} + 24}{576}$$

$$= \frac{\sqrt{3} + 1}{24} \approx 0.11 \text{ sec}$$

**68.** Answers may vary

### 12.2 Mixed Review (p. 721)

**69.** $N = 0.30 \cdot 160$

$= 48$

**70.** $\frac{105}{240} = \frac{N \cdot 240}{240}$

$0.4375 = N$

$43.75\% = N$

**71.** $y = \frac{1}{x - 6} - 1$

**72.** $y = \frac{1}{x - 5} + 2$

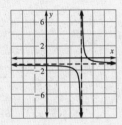

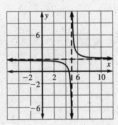

**73.** $y = \frac{2}{x - 6} + 9$

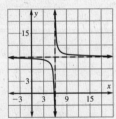

**74.** $f(x) = \sqrt{x} - 3$

Domain: all nonnegative numbers

Range: all numbers greater than or equal to $-3$

**75.** $f(x) = \sqrt{x - 8}$

Domain: all numbers greater than or equal to 8

Range: all nonnegative numbers

**76.** $f(x) = \sqrt{\frac{1}{2}x^2}$

Domain: all real numbers

Range: all nonnegative numbers

**77.** $f(x) = \sqrt{x} + 4$

Domain: all nonnegative numbers

Range: all numbers greater than or equal to 4

**78.** $f(x) = 6x$

Domain: all Real numbers

Range: all Real numbers

**79.** $f(x) = \sqrt{x + 3}$

Domain: all numbers greater than or equal to $-3$

Range: all nonnegative numbers

## Lesson 12.3

### 12.3 Guided Practice (p. 725)

**1.** A solution of the equation produced by the solution method that is not a solution of the original equation

**2.** To check for extraneous solutions

**3.** No; $\sqrt{x}$ is the positive square root of $x$, so $\sqrt{36} = 6$, not $-6$. ($\sqrt{x} = -6$ has no solution.)

**4.** $\sqrt{x} - 20 = 0$

$\left(\sqrt{x}\right)^2 = (20)^2$

$x = 400$

**5.** $\sqrt{5x + 1} + 8 = 12$

$\sqrt{5x + 1} = 4$

$5x + 1 = 16$

$5x = 15$

$x = 3$

**6.** $\sqrt{4x} - 1 = 3$

$\sqrt{4x} = 4$

$4x = 16$

$x = 4$

**7.** $\sqrt{x} + 6 = 0$

$\sqrt{x} = -6$

No solution

**8.** $\sqrt{4x + 5} = x$

$4x + 5 = x^2$

$x^2 - 4x - 5 = 0$

$(x - 5)(x + 1) = 0$

$x = 5$

$x - 5 = 0 \quad x + 1 = 0$

$x = 5 \qquad x = -1$

(extraneous)

**9.** $\sqrt{x + 6} - x = 0$

$\sqrt{x + 6} = x$

$x + 6 = x^2$

$x^2 - x - 6 = 0$

$(x - 3)(x + 2) = 0$

$x = 3$

**10.** $x = \sqrt{x + 12}$

$x^2 = x + 12$

$x^2 - x - 12 = 0$

$(x - 4)(x + 3) = 0$

$x - 4 = 0 \quad x + 3 = 0$

$x = 4 \qquad x = -3$

(extraneous)

**11.** $-5 + \sqrt{x} = 0$

$\sqrt{x} = 5$

$x = 25$

**12.** $x = \sqrt{5x + 24}$

$x^2 = 5x + 24$

$x^2 - 5x - 24 = 0$

$(x - 8)(x + 3) = 0$

$x - 8 = 0 \quad x + 3 = 0$

$x = 8 \qquad x = -3$

(extraneous)

**13.** $6 = \sqrt{12a}$

$36 = 12a$

$3 = a$

# Chapter 12 *continued*

**14.**
$$r = 30d^2\sqrt{P}$$
$$250 = 30(2)^2\sqrt{P}$$
$$250 = 30(4)\sqrt{P}$$
$$250 = 120\sqrt{P}$$
$$\frac{250}{120} = \sqrt{P}$$
$$\frac{62500}{14400} = P$$
$$4.3 \text{ lb/in.}^2 = P$$

## 12.3 Practice and Applications (pp. 725–727)

**15.**
$$\sqrt{x} - 9 = 0$$
$$\sqrt{x} = 9$$
$$x = 81$$

**16.**
$$\sqrt{x} - 1 = 0$$
$$\sqrt{x} = 1$$
$$x = 1$$

**17.**
$$\sqrt{x} + 5 = 0$$
$$\sqrt{x} = -5$$
No solution

**18.**
$$\sqrt{x} - 10 = 0$$
$$\sqrt{x} = 10$$
$$x = 100$$

**19.**
$$\sqrt{x} - 15 = 0$$
$$\sqrt{x} = 15$$
$$x = 225$$

**20.**
$$\sqrt{x} - 0 = 0$$
$$\sqrt{x} = 0$$
$$x = 0$$

**21.**
$$\sqrt{6x} - 13 = 23$$
$$\sqrt{6x} = 36$$
$$6x = 1296$$
$$x = 216$$

**22.**
$$\sqrt{4x + 1} + 5 = 10$$
$$\sqrt{4x + 1} = 5$$
$$4x + 1 = 25$$
$$4x = 24$$
$$x = 6$$

**23.**
$$\sqrt{9 - x} - 10 = 14$$
$$\sqrt{9 - x} = 24$$
$$9 - x = 576$$
$$x = -567$$

**24.**
$$\sqrt{5x + 1} + 2 = 6$$
$$\sqrt{5x + 1} = 4$$
$$5x + 1 = 16$$
$$5x = 15$$
$$x = 3$$

**25.**
$$\sqrt{6x - 2} - 3 = 7$$
$$\sqrt{6x - 2} = 10$$
$$6x - 2 = 100$$
$$6x = 102$$
$$x = 17$$

**26.**
$$4 = 7 - \sqrt{33x - 2}$$
$$\sqrt{33x - 2} = 7 - 4$$
$$33x - 2 = 9$$
$$x = \frac{11}{33}$$
$$x = \frac{1}{3}$$

**27.**
$$10 = 4 + \sqrt{5x + 11}$$
$$6 = \sqrt{5x + 11}$$
$$36 = 5x + 11$$
$$25 = 5x$$
$$5 = x$$

**28.**
$$-5 - \sqrt{10x - 2} = 5$$
$$-10 = \sqrt{10x - 2}$$
No solution

**29.**
$$\sqrt{-x} - \frac{3}{2} = \frac{3}{2}$$
$$\sqrt{-x} = 3$$
$$-x = 9$$
$$x = -9$$

**30.**
$$\sqrt{x} + \frac{1}{3} = \frac{13}{3}$$
$$\sqrt{x} = 4$$
$$x = 16$$

**31.**
$$\sqrt{\tfrac{1}{5}x - 2} - \tfrac{1}{10} = \tfrac{7}{10}$$
$$\sqrt{\tfrac{1}{5}x - 2} = \tfrac{8}{10}$$
$$\tfrac{1}{5}x - 2 = \tfrac{16}{25}$$
$$\tfrac{1}{5}x = \tfrac{66}{25}$$
$$x = \tfrac{66}{5}$$
$$x = 13\tfrac{2}{10}$$
$$x = 13\tfrac{1}{5}$$

**32.**
$$x = \sqrt{\tfrac{3}{2}x + \tfrac{5}{2}}$$
$$x^2 = \tfrac{3}{2}x + \tfrac{5}{2}$$
$$2x^2 = 3x + 5$$
$$2x^2 - 3x - 5 = 0$$
$$(2x - 5)(x + 1) = 0$$
$$2x - 5 = 0 \qquad x + 1 = 0$$
$$2x = 5 \qquad x = -1 \text{(extraneous)}$$
$$x = \tfrac{5}{2}$$

**33.**
$$\sqrt{\tfrac{1}{4}x - 4} - 3 = 5$$
$$\sqrt{\tfrac{1}{4}x - 4} = 8$$
$$\tfrac{1}{4}x - 4 = 64$$
$$\tfrac{1}{4}x = 68$$
$$x = 272$$

**34.**
$$6 - \sqrt{7x - 9} = 3$$
$$6 = 3 + \sqrt{7x - 9}$$
$$3 = \sqrt{7x - 9}$$
$$9 = 7x - 9$$
$$18 = 7x$$
$$\tfrac{18}{7} = x$$
$$2\tfrac{4}{7} = x$$

**35.**
$$\sqrt{\tfrac{1}{9}x + 1} - \tfrac{2}{3} = \tfrac{5}{3}$$
$$\sqrt{\tfrac{1}{9}x + 1} = \tfrac{7}{3}$$
$$\tfrac{1}{9}x + 1 = \tfrac{49}{9}$$
$$\tfrac{1}{9}x = \tfrac{40}{9}$$
$$x = 40$$

**36.**
$$x = \sqrt{35 + 2x}$$
$$x^2 = 35 + 2x$$
$$x^2 - 2x - 35 = 0$$
$$(x - 7)(x + 5) = 0$$
$$x - 7 = 0 \qquad x + 5 = 0$$
$$x = 7 \qquad x = -5 \text{ (extraneous)}$$

**Algebra 1**
Chapter 12 Worked-out Solution Key

# Chapter 12 *continued*

**37.**
$$x = \sqrt{-4x - 4}$$
$$x^2 = -4x - 4$$
$$x^2 + 4x + 4 = 0$$
$$(x + 2)(x + 2) = 0$$
$$x + 2 = 0$$
$$x = -2 \text{ (extraneous)}$$
No solution

**38.**
$$x = \sqrt{6x - 9}$$
$$x^2 = 6x - 9$$
$$x^2 - 6x + 9 = 0$$
$$(x - 3)(x - 3) = 0$$
$$x - 3 = 0$$
$$x = 3$$

**39.**
$$x = \sqrt{30 - x}$$
$$x^2 = 30 - x$$
$$x^2 + x - 30 = 0$$
$$(x + 6)(x - 5) = 0$$
$$x - 5 = 0$$
$$x = 5$$

**40.**
$$x = \sqrt{11x - 28}$$
$$x^2 = 11x - 28$$
$$x^2 - 11x + 28 = 0$$
$$(x - 7)(x - 4) = 0$$
$$x - 7 = 0 \quad x - 4 = 0$$
$$x = 7 \quad\quad x = 4$$

**41.**
$$\sqrt{-10x - 4} = 2x$$
$$-10x - 4 = 4x^2$$
$$0 = 4x^2 + 10x + 4$$
$$0 = (2x + 4)(2x + 1)$$
$$2x + 4 = 0 \quad\quad 2x + 1 = 0$$
$$2x = -4 \quad\quad 2x = -1$$
$$x = -\tfrac{4}{2} \text{ (extraneous)} \quad x = -\tfrac{1}{2} \text{ (extraneous)}$$
No solution

**42.**
$$x = \sqrt{200 - 35x}$$
$$x^2 = 200 - 35x$$
$$x^2 + 35x - 200 = 0$$
$$(x + 40)(x - 5) = 0$$
$$x + 40 = 0 \quad\quad x - 5 = 0$$
$$x = -40 \text{ (extraneous)} \quad x = 5$$

**43.**
$$\sqrt{110 - x} = x$$
$$110 - x = x^2$$
$$x^2 + x - 110 = 0$$
$$(x - 10)(x + 11) = 0$$
$$x - 10 = 0 \quad x + 11 = 0$$
$$x = 10 \quad\quad x = -11 \text{(extraneous)}$$

**44.**
$$2x = \sqrt{-13x - 10}$$
$$4x^2 = -13x - 10$$
$$4x^2 + 13x + 10 = 0$$
$$(4x + 5)(x + 2) = 0$$
$$4x + 5 = 0 \quad\quad x + 2 = 0$$
$$4x = -5 \quad\quad x = -2 \text{ (extraneous)}$$
$$x = -\tfrac{5}{4} \text{ (extraneous)}$$
No solution

**45.**
$$x = \sqrt{4x + 45}$$
$$x^2 = 4x + 45$$
$$x^2 - 4x - 45 = 0$$
$$(x - 9)(x + 5) = 0$$
$$x - 9 = 0 \quad x + 5 = 0$$
$$x = 9 \quad\quad x = -5 \text{ (extraneous)}$$

**46.**
$$\tfrac{1}{5}x = \sqrt{x - 6}$$
$$\tfrac{1}{25}x^2 = x - 6$$
$$\tfrac{1}{25}x^2 - x + 6 = 0$$
$$x^2 - 25x + 150 = 0$$
$$(x - 10)(x - 15) = 0$$
$$x - 10 = 0 \quad x - 15 = 0$$
$$x = 10 \quad\quad x = 15$$

**47.**
$$\tfrac{2}{3}x = \sqrt{24x - 128}$$
$$\tfrac{4}{9}x^2 = 24x - 128$$
$$\tfrac{4}{9}x^2 - 24x + 128 = 0$$
$$x^2 - 54x + 288 = 0$$
$$(x - 6)(x - 48) = 0$$
$$x - 6 = 0 \quad x - 48 = 0$$
$$x = 6 \quad\quad x = 48$$

**48.**
$$27 = \sqrt{12a}$$
$$729 = 12a$$
$$60.75 = a$$

**49.**
$$16 = \sqrt{4a}$$
$$256 = 4a$$
$$64 = a$$

**50.**
$$27 = \sqrt{6a}$$
$$729 = 6a$$
$$121.5 = a$$

**51.**
$$14 = \sqrt{4a}$$
$$196 = 4a$$
$$49 = a$$

**52.**
$$72 = \sqrt{6a}$$
$$5184 = 6a$$
$$864 = a$$

**53.**
$$104 = \sqrt{8a}$$
$$10816 = 8a$$
$$1352 = a$$

# Chapter 12 *continued*

**54.** $\sqrt{x} = 7$

$(\sqrt{x})^2 = 7^2$

$x = 49$

**55.** $\sqrt{x} - 15 = 0$

$\sqrt{x} = 15$

$(\sqrt{x})^2 = 15^2$

$x = 225$

**56.–64. Check graphs on graphing calculator.**

**56.** $\sqrt{x + 4} = 3$

$x + 4 = 9$

$x = 5$

**57.** $\sqrt{x - 5.6} = 2.5$

$x - 5.6 = 6.25$

$x = 11.85$

**58.** $\sqrt{9.2 - x} = 1.8$

$9.2 - x = 3.24$

$-x = -5.96$

$x = 5.96$

**59.** $\sqrt{6x - 2} - 3 = 7$

$\sqrt{6x - 2} = 10$

$6x - 2 = 100$

$6x = 102$

$x = 17$

**60.** $4 + \sqrt{x} = 9$

$\sqrt{x} = 5$

$x = 25$

**61.** $\sqrt{7x - 12} = x$

$7x - 12 = x^2$

$x^2 - 7x + 12 = 0$

$(x - 4)(x - 3) = 0$

$x - 4 = 0 \quad x - 3 = 0$

$x = 4 \qquad x = 3$

**62.** $\sqrt{2x + 7} = x + 2$

$2x + 7 = x^2 + 4x + 4$

$0 = x^2 + 2x - 3$

$(x - 1)(x + 3) = 0$

$x - 1 = 0 \quad x + 3 = 0$

$x = 1 \qquad x = -3 \text{ (extraneous)}$

**63.** $\sqrt{3x - 2} = 4 - x$

$3x - 2 = 16 - 8x + x^2$

$0 = x^2 - 11x + 18$

$(x - 2)(x - 9) = 0$

$x - 2 = 0 \quad x - 9 = 0$

$x = 2 \qquad x = 9 \text{ (extranous)}$

**64.** $\sqrt{15 - 4x} = 2x$

$15 - 4x = 4x^2$

$0 = 4x^2 + 4x - 15$

$0 = (2x + 5)(2x - 3)$

$2x + 5 = 0 \quad 2x - 3 = 0$

$2x = -5 \qquad 2x = 3$

$x = -\frac{5}{2} \qquad x = \frac{3}{2}$

(extraneous)

**65.** *Sample answer:* The graphs do not intersect; the graph of $y = \sqrt{11x - 30}$ is entirely in the first quadrant, while the graph of $y = -x$ is entirely in the second and fourth quadrants.

**66.** $2(\sqrt{x + 6}) + 2(6) = 30$

$2(\sqrt{x + 6}) + 12 = 30$

$2\sqrt{x + 6} = 18$

$\sqrt{x + 6} = 9$

$x + 6 = 81$

$x = 75$

**67.** $82\frac{1}{2} = \frac{1}{2}(\sqrt{2x - 1})(11)$

$82\frac{1}{2} = \frac{11}{2}(\sqrt{2x - 1})$

$15 = \sqrt{2x - 1}$

$225 = 2x - 1$

$226 = 2x$

$113 = x$

**68.** *Sample answer:* $\sqrt{x + 18} = 6$

**69.** $t = \sqrt{\dfrac{4\pi^2 mr}{F}}$

$10 = \sqrt{\dfrac{4\pi^2 67.5(6)}{F}}$

$10 = \sqrt{\dfrac{1620\pi^2}{F}}$

$100 = \dfrac{1620\pi^2}{F}$

$100F = 1620\pi^2$

$F = \dfrac{1620\pi^2}{100}$

$F \approx 159.9$

$F \approx 160$

**70.** Yes; if you solve the equation for $F$,

$F = \dfrac{4\pi^2 mr}{t^2}$,

so, if $r$ and $t$ do not vary, $F$ is directly proportional to $m$.

**71.** $d = 0.444\sqrt{t}$

$28 = 0.444\sqrt{t}$

$63.063 = \sqrt{t}$

$3976.9 = t$

$66.3 \text{ min}$

**72.** $d = 0.444\sqrt{33\frac{1}{3}}$

$\approx 2.6 \text{ cm}$

**73.** $\dfrac{a}{b} = \dfrac{b}{d}$

$ad = b^2$

$\sqrt{ad} = b$

**74. a.** $\dfrac{a}{4} = \dfrac{4}{d}$

**b.** $x + 6; \dfrac{x}{4} = \dfrac{4}{x + 6}$

**c.** $16 = x^2 + 6x$

$0 = x^2 + 6x - 16$

$0 = (x + 8)(x - 2)$

$x + 8 = 0 \quad x - 2 = 0$

$x = -8 \qquad \boxed{x = 2}$

The numbers are 2 and $2 + 6 = 8$.

# Chapter 12 *continued*

**75.** $\dfrac{x}{10} = \dfrac{10}{x - 21}$

$100 = x^2 - 21x$

$0 = x^2 - 21x - 100$

$0 = (x - 25)(x + 4)$

$x - 25 = 0 \qquad x + 4 = 0$

$\boxed{x = 25} \qquad x = -4$

The numbers are 25 and $25 - 21 = 4$.

**76.** $\dfrac{x}{12} = \dfrac{12}{x + 32}$

$144 = x^2 + 32x$

$0 = x^2 + 32x - 144$

$0 = (x + 36)(x - 4)$

$x + 36 = 0 \qquad x - 4 = 0$

$x = -36 \qquad \boxed{x = 4}$

The numbers are 4 and $4 + 32 = 36$.

**77.** C; $n = \sqrt{3 \cdot 12}$

$n = \sqrt{36}$

$n = 6$

$n = \sqrt{1 \cdot 36}$

$n = \sqrt{36}$

$n = 6$

**78.** B; $\sqrt{x} + 4 = 5 \qquad \sqrt{x} - 4 = 5$

$\sqrt{x} = 1 \qquad \sqrt{x} = 9$

$x = 1 \qquad x = 81$

**79.** A; $\sqrt{x} - 3 = 5 \qquad \sqrt{x - 3} = 5$

$\sqrt{x} = 8 \qquad x - 3 = 25$

$x = 64 \qquad x = 28$

**80.** C; $n = \sqrt{-1(-64)} \qquad n = \sqrt{-2(-32)}$

$n = \sqrt{64} \qquad n = \sqrt{64}$

$n = 8 \qquad n = 8$

**81.** The equation has two solutions when $k < 2$, one solution when $k = 2$, and no solution when $k > 2$; squaring both sides of the equation yields the quadratic equation $4x^2 + (4k - 16)x + k^2 = 0$, which has determinant $-128k + 256$. The equation has two solutions if the determinant is positive, one solution if the determinant is zero, and no solution if the determinant is negative.

### 12.3 Mixed Review (p. 728)

**82.** $x^2 = 36$

$x = 6, -6$

**83.** $x^2 = 11$

$x = \sqrt{11}, -\sqrt{11}$

**84.** $7x^2 = 700$

$x^2 = 100$

$x = -10, 10$

**85.** $25x^2 - 9 = -5$

$25x^2 - 4 = 0$

$(5x - 2)(5x + 2) = 0$

$5x - 2 = 0 \qquad 5x + 2 = 0$

$5x = 2 \qquad 5x = -2$

$x = \dfrac{2}{5} \qquad x = -\dfrac{2}{5}$

**86.** $\dfrac{1}{7}x^2 - 7 = -7$

$\dfrac{1}{7}x^2 = 0$

$x^2 = 0$

$x = 0$

**87.** $-16t^2 + 48 = 0$

$\dfrac{0 \pm \sqrt{0^2 - 4(-16)(48)}}{2(-16)}$

$= \dfrac{0 \pm \sqrt{3072}}{-32}$

$= \dfrac{\pm \sqrt{1024}\sqrt{3}}{-32} = \dfrac{\pm 32\sqrt{3}}{-32}$

$= -\sqrt{3}, \sqrt{3}$

**88.** $(x + 5)^2 = (x + 5)(x + 5)$

$= x^2 + 10x + 25$

**89.** $(2x - 3)^2 = (2x - 3)(2x - 3)$

$= 4x^2 - 12x + 9$

**90.** $(3x + 5y)(3x - 5y) = 9x^2 + 15xy - 15xy - 25y^2$

$= 9x^2 - 25y^2$

**91.** $(6y - 4)(6y + 4) = 36y^2 - 24y + 24y - 16$

$= 36y^2 - 16$

**92.** $(x + 7y)^2 = (x + 7y)(x + 7y)$

$= x^2 + 7xy + 7xy + 49y^2$

$= x^2 + 14xy + 49y^2$

**93.** $(2a - 9b)^2 = (2a - 9b)(2a - 9b)$

$= 4a^2 - 18ab - 18ab + 81b^2$

$= 4a^2 - 36ab + 81b^2$

**94.** $x^2 + 18x + 81 = (x + 9)(x + 9) = (x + 9)^2$

**95.** $x^2 - 12x + 36 = (x - 6)(x - 6) = (x - 6)^2$

**96.** $4x^2 + 28x + 49 = (2x + 7)(2x + 7) = (2x + 7)^2$

**97.** $\dfrac{5}{x - 4}$

undefined when $x = 4$

**98.** $\dfrac{x + 2}{x^2 - 4}$

undefined when $x = 2, -2$

**99.** $\dfrac{x + 4}{x^2 + x - 6}$

undefined when $x = -3$ and 2

**100.**

| | | |
|---|---|---|
| $25 | $100 | Cost of Membership |
| $750 | $750 | Cost of groceries |
| $0 | $-75 | Discount |
| $775 | $775 | |

# Chapter 12 *continued*

**Quiz 1 (p. 728)**

**1.** $y = 10\sqrt{x}$

Domain: all nonnegative numbers

Range: all nonnegative numbers

**2.** $y = \sqrt{x - 9}$

Domain: all numbers greater than or equal to 9

Range: all nonnegative numbers

**3.** $y = \sqrt{2x - 1}$

Domain: all numbers greater than or equal to 1/2

Range: all nonnegative numbers

**4.** $y = \dfrac{\sqrt{x} - 2}{3}$

Domain: all nonnegative numbers

Range: all numbers greater than or equal to $-\frac{2}{3}$

**5.** $7\sqrt{10} + 11\sqrt{10} = 18\sqrt{10}$

**6.** $2\sqrt{7} - 5\sqrt{28} = 2\sqrt{7} - 5\sqrt{4}\sqrt{7}$
$= 2\sqrt{7} - 10\sqrt{7}$
$= -8\sqrt{7}$

**7.** $4\sqrt{5} + \sqrt{125} - \sqrt{80} = 4\sqrt{5} + \sqrt{25}\sqrt{5} - \sqrt{16}\sqrt{5}$
$= 4\sqrt{5} + 5\sqrt{5} - 4\sqrt{5}$
$= 5\sqrt{5}$

**8.** $\sqrt{3}(3\sqrt{2} + \sqrt{3}) = 3\sqrt{6} + \sqrt{9}$
$= 3\sqrt{6} + 3$

**9.** $(\sqrt{7} + 5)^2 = (\sqrt{7} + 5)(\sqrt{7} + 5)$
$= \sqrt{49} + 5\sqrt{7} + 5\sqrt{7} + 25$
$= 7 + 10\sqrt{7} + 25$
$= 32 + 10\sqrt{7}$

**10.** $\dfrac{15}{8 + \sqrt{7}}\left(\dfrac{8 - \sqrt{7}}{8 - \sqrt{7}}\right) = \dfrac{120 - 15\sqrt{7}}{64 - 7}$
$= \dfrac{120 - 15\sqrt{7}}{57}$
$= \dfrac{3(40 - 5\sqrt{7})}{57}$
$= \dfrac{40 - 5\sqrt{7}}{19}$

**11.** $\sqrt{x} - 12 = 0$
$\sqrt{x} = 12$
$x = 144$

**12.** $\sqrt{x} - 8 = 0$
$\sqrt{x} = 8$
$x = 64$

**13.** $\sqrt{3x + 2} + 2 = 3$
$\sqrt{3x + 2} = 1$
$3x + 2 = 1$
$3x = -1$
$x = -\frac{1}{3}$

**14.** $\sqrt{3x - 2} + 3 = 7$
$\sqrt{3x - 2} = 4$
$3x - 2 = 16$
$3x = 18$
$x = 6$

**15.** $\sqrt{77 - 4x} = x$
$77 - 4x = x^2$
$0 = x^2 + 4x - 77$
$0 = (x + 11)(x - 7)$
$x + 11 = 0 \qquad\qquad x - 7 = 0$
$x = -11 \text{ (extraneous)} \quad x = 7$

**16.** $x = \sqrt{2x + 3}$
$x^2 = 2x + 3$
$x^2 - 2x - 3 = 0$
$(x - 3)(x + 1) = 0$
$x - 3 = 0 \quad x + 1 = 0$
$x = 3 \qquad x = -1 \text{ (extraneous)}$

**17.** $r = 30d^2\sqrt{P}$
$250 = 30(2.5)^2\sqrt{P}$
$250 = 187.5\sqrt{P}$
$1.\overline{3} = \sqrt{P}$
$1.\overline{7} = P$
$1\frac{7}{9} \text{ lb/in.}^2 = P$

**Developing Concepts Activity 12.4 (p. 729)**

**1.**

| $x^2 + 8x$ | 16 | $4^2$ |
|---|---|---|
| $x^2 + 4x$ | 4 | $2^2$ |
| $x^2 + 2x$ | 1 | $1^2$ |

$x^2 + 8x + 16 \qquad x^2 + 4x + 4 \qquad x^2 + 2x + 1$
$= (x + 4)^2 \qquad = (x + 2)^2 \qquad = (x + 1)^2$

**2.** The base of the expression is half the coefficient of $x$.

**3.** $x^2 + 14x \rightarrow x^2 + 14x + 49$

Half of 14 is 7, so $7^2 = 49$ tiles are needed.

## Lesson 12.4

**Activity (p. 732)**

**1.** $ax^2 + bx = -c$

**2.** $x^2 + \dfrac{bx}{a} = -\dfrac{c}{a}$

**3.** $x^2 + \dfrac{bx}{a} + \dfrac{b^2}{4a^2} = -\dfrac{c}{a} + \dfrac{b^2}{4a^2}$

**4.** $\left(x + \dfrac{b}{2a^2}\right)^2 = -\dfrac{c}{a} + \dfrac{b^2}{4a^2}$

**Algebra 1**
Chapter 12 Worked-out Solution Key

# Chapter 12 *continued*

**5.** $\left(x + \dfrac{b}{2a}\right)^2 = \dfrac{-4ac + b^2}{4a^2}$

**6.** $x + \dfrac{b}{2a} = \pm\dfrac{\sqrt{b^2 - 4ac}}{2a}$

**7.** $x = -\dfrac{b}{2a} \pm \dfrac{\sqrt{b^2 - 4ac}}{2a}$

**8.** $x = \dfrac{-b \pm \sqrt{b^2 - 4ac}}{2a}$

### 12.4 Guided Practice (p. 734)

**1.** 3

**2.** To complete the square, you need to square, $\dfrac{b}{2}$, which is an integer if $b$ is even.

**3.** $\left(\dfrac{20}{2}\right)^2 = \dfrac{400}{4}$
$= 100$

**4.** $\left(\dfrac{50}{2}\right)^2 = \dfrac{2500}{4}$
$= 625$

**5.** $\left(\dfrac{10}{2}\right)^2 = \dfrac{100}{4}$
$= 25$

**6.** $\left(\dfrac{14}{2}\right)^2 = \dfrac{196}{4}$
$= 49$

**7.** $\left(\dfrac{22}{2}\right)^2 = \dfrac{484}{4}$
$= 121$

**8.** $\left(\dfrac{100}{2}\right)^2 = \dfrac{10{,}000}{4}$
$= 2500$

**9.**
$x^2 - 3x = 8$
$x^2 - 3x + 2.25 = 8 + 2.25$
$x^2 - 3x + 2.25 = 10.25$
$(x - 1.5)^2 = 10.25$
$x - 1.5 = \pm 3.2016$
$x = 4.7016, -1.7016$

$x^2 - 3x = 8$
$x^2 - 3x - 8 = 0$
$\dfrac{3 \pm \sqrt{(-3)^2 - 4(1)(-8)}}{2} = \dfrac{3 \pm \sqrt{9 + 32}}{2}$
$= \dfrac{3 + \sqrt{41}}{2}, \dfrac{3 - \sqrt{41}}{2}$

**10.** $x^2 - 2x - 18 = 0$
$x^2 - 2x + 1 = 18 + 1$
$(x - 1)^2 = 19$
$x - 1 = \pm\sqrt{19}$
$x = 1 + \sqrt{19}$
$x = 1 - \sqrt{19}$

**11.** $x^2 + 14x + 13 = 0$
$x^2 + 14x + 49 = -13 + 49$
$(x + 7)^2 = 36$
$x + 7 = \pm 6$
$x = 6 - 7 \quad x = -6 - 7$
$x = -1 \quad\quad x = -13$

**12.** $3x^2 + 4x - 1 = 0$
$x^2 + \dfrac{4}{3}x = \dfrac{1}{3}$
$x^2 + \dfrac{4}{3}x + \dfrac{4}{9} = \dfrac{1}{3} + \dfrac{4}{9}$
$\left(x + \dfrac{2}{3}\right)^2 = \dfrac{7}{9}$
$x + \dfrac{2}{3} = \pm\sqrt{\dfrac{7}{9}}$
$x = \pm\dfrac{\sqrt{7}}{3} - \dfrac{2}{3}$
$x = \dfrac{-2 + \sqrt{7}}{3}, \dfrac{-2 - \sqrt{7}}{3}$

**13.** $3x^2 - 7x + 6 = 0$
$x^2 - \dfrac{7}{3}x + 2 = 0$
$x^2 - \dfrac{7}{3}x = -2$
no solution

**14.** $x^2 - x - 2 = 0$
$(x - 2)(x + 1) = 0$
$x - 2 = 0 \quad x + 1 = 0$
$x = 2 \quad\quad x = -1$
can be factored

**15.** $3x^2 + 17x + 10 = 0$
$(3x + 2)(x + 5) = 0$
$3x + 2 = 0 \quad x + 5 = 0$
$3x = -2 \quad\quad x = -5$
$x = -\dfrac{2}{3}$
can be factored

**16.** $x^2 - 9 = 0$
$x^2 = 9$
$x = 3, -3$
Find square roots.

**17.** $-3x^2 + 5x + 5 = 0$
Quadratic formula
$\dfrac{-5 \pm \sqrt{5^2 - 4(-3)(5)}}{2(-3)} = \dfrac{-5 \pm \sqrt{25 + 60}}{-6}$
$= \dfrac{-5 \pm \sqrt{85}}{-6}$
$= \dfrac{5 + \sqrt{85}}{6}, \dfrac{5 - \sqrt{85}}{6}$

**18.** $x^2 + 2x - 14 = 0$
Complete the square.
$x^2 + 2x + 1 = 14 + 1$
$(x + 1)^2 = 15$
$x + 1 = \sqrt{15}$
$x = 1 - \sqrt{15}$
$x = 1 + \sqrt{15}$

**19.** $3x^2 - 2 = 0$
$3x^2 = 2$
$x^2 = \dfrac{2}{3}$
$x = \pm\sqrt{\dfrac{2}{3}\left(\dfrac{\sqrt{3}}{\sqrt{3}}\right)}$
$x = \dfrac{\sqrt{6}}{3}, \dfrac{-\sqrt{6}}{3}$

Find square roots.

# Chapter 12 *continued*

**20.** $\left(\frac{12}{2}\right)^2 = \frac{144}{4}$
$= 36$

**21.** $\left(\frac{8}{2}\right)^2 = \frac{64}{4}$
$= 16$

**22.** $\left(\frac{21}{2}\right)^2 = \frac{441}{4}$

**23.** $\left(\frac{22}{2}\right)^2 = \frac{484}{4}$
$= 121$

**24.** $\left(\frac{11}{2}\right)^2 = \frac{121}{4}$

**25.** $\left(\frac{40}{2}\right)^2 = \frac{1600}{4}$
$= 400$

**26.** $\left(\frac{0.4}{2}\right)^2 = \frac{0.16}{4}$
$= 0.04$

**27.** $\left(\frac{3}{4} \cdot \frac{1}{2}\right)^2 = \frac{9}{64}$

**28.** $\left(\frac{4}{5} \cdot \frac{1}{2}\right)^2 = \frac{16}{100}$
$= \frac{4}{25}$

**29.** $\left(\frac{5.2}{2}\right)^2 = \frac{27.04}{4}$
$= 6.76$

**30.** $\left(\frac{0.3}{2}\right)^2 = \frac{0.09}{4}$
$= 0.0225$

**31.** $\left(\frac{2}{3} \cdot \frac{1}{2}\right)^2 = \frac{4}{36}$
$= \frac{1}{9}$

**32.**
$$x^2 + 10x = 39$$
$$x^2 + 10x + 25 = 39 + 25$$
$$(x + 5)^2 = 64$$
$$x + 5 = \pm 8$$
$$x = 3, -13$$

**33.**
$$x^2 + 16x = 17$$
$$x^2 + 16x + 64 = 17 + 64$$
$$(x + 8)^2 = 81$$
$$x + 8 = \pm 9$$
$$x = 1, -17$$

**34.**
$$x^2 - 24x = -44$$
$$x^2 - 24x + 144 = -44 + 144$$
$$(x - 12)^2 = 100$$
$$x - 12 = \pm 10$$
$$x = 22, 2$$

**35.** $x^2 - 8x + 12 = 0$
$$x^2 - 8x + 16 = -12 + 16$$
$$(x - 4)^2 = 4$$
$$x - 4 = \pm 2$$
$$x = 6, 2$$

**36.** $x^2 + 5x - \frac{11}{4} = 0$
$$x^2 + 5x + \left(\frac{5}{2}\right)^2 = \frac{11}{4} + \frac{25}{4}$$
$$\left(x + \frac{5}{2}\right)^2 = \frac{36}{4}$$
$$x + \frac{5}{2} = \pm 3$$
$$x = \frac{1}{2}, -\frac{11}{2}$$

**37.** $x^2 + 11x + \frac{21}{4} = 0$
$$x^2 + 11x + \left(\frac{11}{2}\right)^2 = -\frac{21}{4} + \frac{121}{4}$$
$$\left(x + \frac{11}{2}\right)^2 = \frac{100}{4}$$
$$x + \frac{11}{2} = \pm \frac{10}{2}$$
$$x = -\frac{1}{2}, -\frac{21}{2}$$

**38.**
$$x^2 - \frac{2}{3}x - 3 = 0$$
$$x^2 - \frac{2}{3}x + \left(\frac{2}{3} \cdot \frac{1}{2}\right)^2 = 3 + \frac{4}{36}$$
$$\left(x - \frac{2}{6}\right)^2 = \frac{28}{9}$$
$$x - \frac{2}{6} = \pm \frac{\sqrt{28}}{\sqrt{9}}$$
$$x = \frac{\pm 2\sqrt{7}}{3} + \frac{1}{3}$$
$$x = \frac{1 - 2\sqrt{7}}{3}, \frac{1 + 2\sqrt{7}}{3}$$

**39.**
$$x^2 + \frac{3}{5}x - 1 = 0$$
$$x^2 + \frac{3}{5}x + \left(\frac{3}{5} \cdot \frac{1}{2}\right)^2 = 1 + \frac{9}{100}$$
$$\left(x + \frac{3}{10}\right)^2 = \frac{109}{100}$$
$$x + \frac{3}{10} = \pm \frac{\sqrt{109}}{\sqrt{100}}$$
$$x = \frac{-3 - \sqrt{109}}{10}, \frac{-3 + \sqrt{109}}{10}$$

**40.**
$$x^2 + x - 1 = 0$$
$$x^2 + x + \left(\frac{1}{2}\right)^2 = 1 + \frac{1}{4}$$
$$\left(x + \frac{1}{2}\right)^2 = \frac{5}{4}$$
$$x + \frac{1}{2} = \pm \sqrt{\frac{5}{4}}$$
$$x = \frac{-1 - \sqrt{5}}{2}, \frac{-1 + \sqrt{5}}{2}$$

**41.** $4x^2 + 4x - 11 = 0$
$$x^2 + x - \frac{11}{4} = 0$$
$$x^2 + x + \left(\frac{1}{2}\right)^2 = \frac{11}{4} + \frac{1}{4}$$
$$\left(x + \frac{1}{2}\right)^2 = \frac{12}{4}$$
$$x + \frac{1}{2} = \pm \frac{\sqrt{12}}{2}$$
$$x = \pm \frac{\sqrt{4}\sqrt{3} - 1}{2}$$
$$x = \frac{-1 + 2\sqrt{3}}{2}, \frac{-1 - 2\sqrt{3}}{2}$$

# Chapter 12 *continued*

**42.** $3x^2 - 24x - 1 = 0$

$$x^2 - 8x - \frac{1}{3} = 0$$

$$x^2 - 8x + 16 = \frac{1}{3} + 16$$

$$(x - 4)^2 = \frac{49}{3}$$

$$x - 4 = \pm\sqrt{\frac{49}{3}}$$

$$x - 4 = \pm\left(\frac{7}{\sqrt{3}} \cdot \frac{\sqrt{3}}{\sqrt{3}}\right)$$

$$x - 4 = \frac{\pm 7\sqrt{3}}{3}$$

$$x = \frac{12 + 7\sqrt{3}}{3}, \frac{12 - 7\sqrt{3}}{3}$$

**43.** $4x^2 - 40x - 7 = 0$

$$x^2 - 10x - \frac{7}{4} = 0$$

$$x^2 - 10x + 25 = \frac{7}{4} + 25$$

$$(x - 5)^2 = \frac{107}{4}$$

$$x - 5 = \pm\sqrt{\frac{107}{4}}$$

$$x = \pm\frac{\sqrt{107}}{2} + 5$$

$$x = \frac{10 + \sqrt{107}}{2}, \frac{10 - \sqrt{107}}{2}$$

**44.** $2x^2 - 8x - 13 = 7$

$$2x^2 - 8x = 20$$

$$x^2 - 4x = 10$$

$$x^2 - 4x + 4 = 10 + 4$$

$$(x - 2)^2 = 14$$

$$x - 2 = \pm\sqrt{14}$$

$$x = 2 + \sqrt{14}, 2 - \sqrt{14}$$

**45.** $5x^2 - 20x - 20 = 5$

$$5x^2 - 20x = 25$$

$$x^2 - 4x = 5$$

$$x^2 - 4x + 4 = 5 + 4$$

$$(x - 2)^2 = 9$$

$$x - 2 = \pm 3$$

$$x = 5, -1$$

**46.** $3x^2 + 4x + 4 = 3$

$$3x^2 + 4x = -1$$

$$x^2 + \frac{4}{3}x = -\frac{1}{3}$$

$$x^2 + \frac{4}{3}x + \left(\frac{4}{3} \cdot \frac{1}{2}\right)^2 = -\frac{1}{3} + \frac{4}{9}$$

$$\left(x + \frac{4}{6}\right)^2 = \frac{1}{9}$$

$$x + \frac{4}{6} = \pm\sqrt{\frac{1}{9}}$$

$$x = \pm\frac{1}{3} - \frac{4}{6}$$

$$x = -\frac{2}{6}, -\frac{6}{6} = -\frac{1}{3}, -1$$

**47.** $4x^2 + 6x - 6 = 2$

$$4x^2 + 6x = 8$$

$$x^2 + \frac{6}{4}x = 2$$

$$x^2 + \frac{3}{2}x + \left(\frac{3}{2} \cdot \frac{1}{2}\right)^2 = 2 + \frac{9}{16}$$

$$\left(x + \frac{3}{4}\right)^2 = \frac{41}{16}$$

$$x + \frac{3}{4} = \pm\sqrt{\frac{41}{16}}$$

$$x = \pm\frac{\sqrt{41}}{4} - \frac{3}{4}$$

$$x = \frac{-3 - \sqrt{41}}{4}, \frac{-3 + \sqrt{41}}{4}$$

**48.** $6x^2 + 24x - 41 = 0$

$$6x^2 + 24x = 41$$

$$x^2 + 4x = \frac{41}{6}$$

$$x^2 + 4x + 4 = \frac{41}{6} + 4$$

$$(x + 2)^2 = \frac{65}{6}$$

$$x + 2 = \pm\sqrt{\frac{65}{6} \cdot \frac{\sqrt{6}}{\sqrt{6}}}$$

$$x + 2 = \pm\frac{\sqrt{390}}{6}$$

$$x = \frac{-12 + \sqrt{390}}{6}, \frac{-12 - \sqrt{390}}{6}$$

# Chapter 12 *continued*

**49.** $20x^2 - 120x - 109 = 0$

$$20x^2 - 120x = 109$$

$$x^2 - 6x = \frac{109}{20}$$

$$x^2 - 6x + 9 = \frac{109}{20} + 9$$

$$(x - 3)^2 = \frac{289}{20}$$

$$x - 3 = \pm\sqrt{\frac{289}{20}} \cdot \frac{\sqrt{20}}{\sqrt{20}}$$

$$x - 3 = \pm\frac{\sqrt{5780}}{20}$$

$$x - 3 = \pm\frac{\sqrt{1156}\sqrt{5}}{20}$$

$$x - 3 = \pm\frac{34\sqrt{5}}{20}$$

$$x = \pm\frac{34\sqrt{5}}{20} + 3$$

$$x = \frac{60 + 34\sqrt{5}}{20}, \frac{60 - 34\sqrt{5}}{20}$$

$$x = \frac{30 + 17\sqrt{5}}{10}, \frac{30 - 17\sqrt{5}}{10}$$

**50.** $x^2 - 5x - 1 = 0$

Quadratic formula

$$\frac{5 \pm \sqrt{(-5)^2 - 4(1)(-1)}}{2(1)}$$

$$= \frac{5 \pm \sqrt{25 + 4}}{2}$$

$$= \frac{5 + \sqrt{29}}{2}, \frac{5 - \sqrt{29}}{2}$$

**51.** $4x^2 - 12 = 0$

$$4x^2 = 12$$

$$x^2 = 3$$

$$x = \sqrt{3}, -\sqrt{3}$$

Find square roots.

**52.** $n^2 + 5n - 24 = 0$

$(n + 8)(n - 3) = 0$

$n + 8 = 0 \quad n - 3 = 0$

$n = -8 \quad\quad n = 3$

can be factored

**53.** $9a^2 - 25 = 0$

$9a^2 = 25$

$a^2 = \frac{25}{9}$

$a = \frac{5}{3}, -\frac{5}{3}$

Find square roots.

**54.** $x^2 - x - 20 = 0$

$(x - 5)(x + 4) = 0$

$x - 5 = 0 \quad x + 4 = 0$

$x = 5 \quad\quad x = -4$

can be factored

**55.** $x^2 + 6x - 55 = 0$

Complete the square.

$x^2 + 6x + 9 = 55 + 9$

$(x + 3)^2 = 64$

$x + 3 = \pm 8$

$x = 5, -11$

**56.** $x^2 - 10x = 0$

$x(x - 10) = 0$

$x - 10 = 0$

$x = 0, 10$

**57.** $c^2 + 2c - 26 = 0$

$c^2 + 2c + 1 = 26 + 1$

$(c + 1)^2 = 27$

$c + 1 = \pm\sqrt{27}$

$c = \sqrt{27} - 1, -\sqrt{27} - 1$

$c = -1 - 3\sqrt{3}, -1 + 3\sqrt{3}$

**58.** $8x^2 + 14x = -5$

$8x^2 + 14x + 5 = 0$

$(4x + 5)(2x + 1) = 0$

$4x + 5 = 0 \quad 2x + 1 = 0$

$4x = -5 \quad\quad 2x = -1$

$x = -\frac{5}{4} \quad\quad x = -\frac{1}{2}$

**59.** $x^2 - 16 = 0$

$x^2 = 16$

$x = 4, -4$

**60.** $x^2 + 12x + 20 = 0$

$(x + 10)(x + 2) = 0$

$x + 10 = 0 \quad x + 2 = 0$

$x = -10 \quad\quad x = -2$

**61.** $x^2 - 4x = \frac{5}{6}$

$$x^2 - 4x + 4 = \frac{5}{6} + 4$$

$$(x - 2)^2 = \frac{29}{6}$$

$$x - 2 = \pm\sqrt{\frac{29}{6}}$$

$$x - 2 = \pm\frac{\sqrt{29}}{\sqrt{6}} \cdot \frac{\sqrt{6}}{\sqrt{6}}$$

$$x = \pm\frac{\sqrt{174}}{6} + 2$$

$$x = \frac{12 + \sqrt{174}}{6}, \frac{12 - \sqrt{174}}{6}$$

**62.** $4x^2 + 4x + 1 = 0$

$(2x + 1)(2x + 1) = 0$

$2x + 1 = 0$

$2x = -1$

$x = -\frac{1}{2}$

# Chapter 12 *continued*

**63.** $13x^2 - 26x = 0$
$x^2 - 2x = 0$
$x^2 - 2x + 1 = 1$
$(x - 1)^2 = 1$
$x - 1 = \pm 1$
$x = 2, 0$

**64.** $4p^2 - 12p + 5 = 0$
$(2p - 5)(2p - 1) = 0$
$2p - 5 = 0 \quad 2p - 1 = 0$
$2p = 5 \qquad 2p = 1$
$p = \frac{5}{2} \qquad p = \frac{1}{2}$

**65.** $7z^2 - 46z = 21$
$(7z + 3)(z - 7) = 0$
$7z + 3 = 0 \quad z - 7 = 0$
$7z = -3 \qquad z = 7$
$z = -\frac{3}{7}$

**66.** $11x^2 - 22 = 0$
$11x^2 = 22$
$x^2 = 2$
$x = \sqrt{2}, -\sqrt{2}$

**67.** $x^2 + 20x + 10 = 0$
$x^2 + 20x + 100 = -10 + 100$
$(x + 10)^2 = 90$
$x + 10 = \pm\sqrt{90}$
$x + 10 = \pm 3\sqrt{10}$
$x = -10 + 3\sqrt{10}, -10 - 3\sqrt{10}$

**68.** $x \cdot x = 150$
$x^2 = 150$
$x = \sqrt{150}$
$x = 5\sqrt{6} \text{ ft}$
$5\sqrt{6} \text{ ft} \times 5\sqrt{6} \text{ ft} \approx 12.25 \text{ ft} \times 12.25 \text{ ft}$

**69.** $2x(x + 5) = 600 \qquad 2x = 2(15)$
$2x^2 + 10x = 600 \qquad\quad = 30$
$x^2 + 5x = 300 \qquad x + 5 = 15 + 5$
$x^2 + 5x - 300 = 0 \qquad\quad = 20$
$(x - 15)(x + 20) = 0 \qquad 30 \text{ ft} \times 20 \text{ ft}$
$x = 15$

**70.** $\frac{1}{2}b(4 + 2b) = A$
$2b + b^2 = 60$
$b^2 + 2b = 60$
$b^2 + 2b + 1 = 60 + 1$
$(b + 1)^2 = 61$
$b + 1 = \pm\sqrt{61}$
$b = -1 + \sqrt{61}$
base $(b) = -1 + \sqrt{61} \approx 6.8 \text{ cm}$
height $(4 + 2b) = 4 + 2(-1 + \sqrt{61})$
$= 4 - 2 + 2\sqrt{61}$
$= 2 + 2\sqrt{61} \approx 17.6 \text{ cm}$

(triangle with height labeled $4 + 2b$ and base labeled $b$)

**71.** $0 = -0.44x^2 + 2.61x + 10$
$0.44x^2 - 2.61x - 10 = 0$
$x = \dfrac{2.61 \pm \sqrt{(2.61)^2 - 4(0.44)(-10)}}{2(0.44)}$
$x = 8.58 \text{ or } -2.65 \text{ (extraneous)}$
about 8.58 feet

**72.** $h = -0.05x^2 + 1.178x$

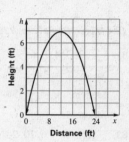

**73.** about 11.8 ft

**74.** B; $2x^2 + 8x - 25 = 5$
$2x^2 + 8x - 30 = 0$
$x^2 + 4x - 15 = 0$
$x^2 + 4x + 4 = 15 + 4$
$(x + 2)^2 = 19$
$x + 2 = \pm\sqrt{19}$
$x = -\sqrt{19} - 2, \sqrt{19} - 2$

**75.** C; $\left(\frac{1}{2}\right)^2 = \frac{1}{4}$

**76.** A; $x^2 + 8x - 2 = 0$
$x^2 + 8x + 16 = 2 + 16$
$(x + 4)^2 = 18$
$x + 4 = \pm\sqrt{18}$
$x = \pm 3\sqrt{2} - 4$
$x = -4 \pm 3\sqrt{2}$

**77.** $y = x^2 + 10x + 25$
$y = (x + 5)^2$
$(-5, 0)$

**78.** $y = 2x^2 + 12x + 13$
$18 - 13 = 2(x^2 + 6x + 9)$
$5 = 2(x + 3)^2$
$y = 2(x + 3)^2 - 5$
$(-3, -5)$

**79.** $y = -x^2 - 5x + 6$
$x^2 + 5x - 6 = 0$
$x^2 + 5x + \left(\frac{5}{2}\right)^2 = 6 + \frac{25}{4}$
$\left(x + \frac{5}{2}\right)^2 = \frac{49}{4}$
$y = -\left(x + \frac{5}{2}\right)^2 + \frac{49}{4}$
$\left(-\frac{5}{2}, \frac{49}{4}\right)$

# Chapter 12 continued

**80.** If $a = 0$, the rational expression $\dfrac{-b \pm \sqrt{b^2 - 4ac}}{2a}$ is not

defined because the denominator is zero. If

$b^2 - 4ac < 0$, the rational expression is not defined

because the numerator is the square root of a negative

number.

### 12.4 Mixed Review (p. 736)

**81.** 1, 1, 2, 3, 4, 5, 6

Mean: $\frac{22}{7} = 3\frac{1}{7}$

Median: 3

Mode: 1

**82.** 3, 6, 9, 10, 10, 14

Mean: $\frac{52}{6} = 8\frac{2}{3}$

Median: $9\frac{1}{2}$

Mode: 10

**83.** $-18, -8, -6, 10, 20$

Mean: $-\frac{2}{5}$

Median: $-6$

Mode: None

**84.** 3, 4, 8, 9, 11, 11, 15, 15, 17

Mean: $\frac{93}{9} = 10\frac{1}{3}$

Median: 11

Mode: 11, 15

**85.** $y = 4x$     $x + 4x = 10$     $y = 4(2)$

$x + y = 10$     $5x = 10$     $y = 8$

           $x = 2$     $(2, 8)$

**86.**   $3x + y = 12$   Multiply by $-3 \rightarrow -9x - 3y = -36$

$9x - y = 36$                $\underline{9x - y = \phantom{-}36}$

                                      $-4y = \phantom{-}0$

$3x + 0 = 12$                     $y = \phantom{-}0$

    $x = 4$

   $(4, 0)$

**87.**   $2x - y = 8$   Multiply by $-1 \rightarrow -2x + y = -8$

$2x + 2y = 2$                $\underline{2x + 2y = \phantom{-}2}$

                                   $3y = -6$

$2x - (-2) = 8$                $y = -2$

     $2x = 6$

       $x = 3$

  $(3, -2)$

**88.** $16 + x^2 = 64$

$x^2 = 48$

$x = 4\sqrt{3}, -4\sqrt{3}$

**89.** $x^2 + 81 = 144$

$x^2 = 63$

$x = 3\sqrt{7}, -3\sqrt{7}$

**90.** $x^2 + 25 = 81$

$x^2 = 56$

$x = -2\sqrt{14}, 2\sqrt{14}$

**91.** $4x^2 - 144 = 0$

$4x^2 = 144$

$x^2 = 36$

$x = 6, -6$

**92.** $x^2 - 30 = -3$

$x^2 = 27$

$x = -3\sqrt{3}, 3\sqrt{3}$

**93.** $x^2 = \frac{9}{25}$

$x = \pm\sqrt{\frac{9}{25}}$

$x = \frac{3}{5}, -\frac{3}{5}$

**94.**

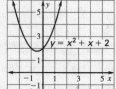

$y = x^2 + x + 2$

**95.**

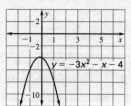

$y = -3x^2 - x - 4$

**96.**

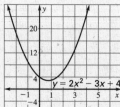

$y = 2x^2 - 3x + 4$

**97.** $(x + 4)(x - 8) = 0$

$x + 4 = 0$    $x - 8 = 0$

     $x = -4$      $x = 8$

**98.** $(x - 3)(x - 2) = 0$

$x - 3 = 0$    $x - 2 = 0$

    $x = 3$      $x = 2$

**99.** $(x + 5)(x + 6) = 0$

$x + 5 = 0$    $x + 6 = 0$

    $x = -5$     $x = -6$

**100.** $(x + 4)^2 = 0$

$x + 4 = 0$

   $x = -4$

**101.** $(x - 3)^2 = 0$

$x - 3 = 0$

   $x = 3$

**102.** $6(x - 14)^2 = 0$

$x - 14 = 0$

    $x = 14$

**103.** $x^2 + x - 20 = (x + 5)(x - 4)$

**104.** $x^2 - 10x + 24 = (x - 6)(x - 4)$

**105.** $x^2 + 2x + 4$   cannot be factored

**106.** $3x^2 - 15x + 18 = 3(x^2 - 5x + 6) = 3(x - 2)(x - 3)$

**107.** $2x^2 - x - 3 = (2x - 3)(x + 1)$

**108.** $14x^2 - 19x - 3 = (7x + 1)(2x - 3)$

## Lesson 12.5

### Developing Concepts Activity 12.5 (p. 737)

**1.** Check results.

**2.** Check results.

**3.** $<$      **4.** $=$      **5.** $>$

### 12.5 Guided Practice (p. 741)

**1.** legs

**2.** Hypothesis: $x$ is an even number.

Conclusion: $x^2$ is an even number.

# Chapter 12 *continued*

**3.** Check to see that the sum of the squares of the two shortest lengths is equal to the square of the longest length.

**4.** $a^2 + b^2 = c^2$
$7^2 + 24^2 = c^2$
$49 + 576 = c^2$
$625 = c^2$
$25 = c$

**5.** $5^2 + b^2 = 13^2$
$25 + b^2 = 169$
$b^2 = 144$
$b = 12$

**6.** $a^2 + 15^2 = 17^2$
$a^2 + 225 = 289$
$a^2 = 64$
$a = 8$

**7.** $9^2 + b^2 = 41^2$
$81 + b^2 = 1681$
$b^2 = 1600$
$b = 40$

**8.** $a^2 + 11^2 = 61^2$
$a^2 + 121 = 3721$
$a^2 = 3600$
$a = 60$

**9.** $12^2 + 35^2 = c^2$
$144 + 1225 = c^2$
$1369 = c^2$
$37 = c$

**10.** $12^2 + b^2 = 20^2$
$144 + b^2 = 400$
$b^2 = 256$
$b = 16$

**11.** $20^2 + 21^2 = c^2$
$400 + 441 = c^2$
$841 = c^2$
$c = 29$

**12.** $x^2 + (x + 2)^2 = 10^2$
$x^2 + x^2 + 4x + 4 = 100$
$2x^2 + 4x + 4 = 100$
$x^2 + 2x + 2 = 50$
$x^2 + 2x - 48 = 0$
$(x + 8)(x - 6) = 0$
$x = -8 \quad \boxed{x = 6}$
$x = 6 \quad x + 2 = 8$

## 12.5 Practice and Applications (pp. 741–743)

**13.** $3^2 + 4^2 = c^2$
$9 + 16 = c^2$
$25 = c^2$
$5 = c$

**14.** $5^2 + b^2 = 10^2$
$25 + b^2 = 100$
$b^2 = 75$
$b = 5\sqrt{3}$

**15.** $a^2 + 3^2 = 7^2$
$a^2 + 9 = 49$
$a^2 = 40$
$a = 2\sqrt{10}$

**16.** $10^2 + 24^2 = c^2$
$100 + 576 = c^2$
$676 = c^2$
$26 = c$

**17.** $a^2 + 9^2 = 16^2$
$a^2 + 81 = 256$
$a^2 = 175$
$a = 5\sqrt{7}$

**18.** $14^2 + b^2 = 21^2$
$196 + b^2 = 441$
$b^2 = 245$
$b = 7\sqrt{5}$

**19.** $(x - 6)^2 + x^2 = 30^2$
$x^2 - 12x + 36 + x^2 = 900$
$2x^2 - 12x + 36 = 900$
$x^2 - 6x + 18 = 450$
$x^2 - 6x - 432 = 0$
$(x - 24)(x + 18) = 0$
$\boxed{x = 24} \quad x = -18$
$x = 24 \quad x - 6 = 18$

**20.** $x^2 + (x + 1)^2 = \left(\sqrt{61}\right)^2$
$x^2 + x^2 + 2x + 1 = 61$
$2x^2 + 2x + 1 = 61$
$2x^2 + 2x - 60 = 0$
$x^2 + x - 30 = 0$
$(x + 6)(x - 5) = 0$
$x = -6 \quad \boxed{x = 5}$
$x = 5 \quad x + 1 = 6$

**21.** $(x + 5)^2 + x^2 = \left(5\sqrt{5}\right)^2$
$x^2 + 10x + 25 + x^2 = 25(5)$
$2x^2 + 10x + 25 = 125$
$2x^2 + 10x - 100 = 0$
$x^2 + 5x - 50 = 0$
$(x + 10)(x - 5) = 0$
$x = -10 \quad \boxed{x = 5}$
$x = 5 \quad x + 5 = 10$

**22.** $x^2 + (2x - 1)^2 = (2x + 1)^2$
$x^2 + 4x^2 - 4x + 1 = 4x^2 + 4x + 1$
$5x^2 - 4x + 1 = 4x^2 + 4x + 1$
$x^2 - 8x = 0$
$x(x - 8) = 0$
$x = 0 \quad \boxed{x = 8}$
$x = 8 \quad 2x + 1 = 17 \quad 2x - 1 = 15$

**23.** $1^2 + x^2 = \left(\sqrt{2x}\right)^2$
$1 + x^2 = 2x$
$x^2 - 2x + 1 = 0$
$(x - 1)^2 = 0$
$x = 1 \quad \sqrt{2x} = \sqrt{2(1)} = \sqrt{2}$

**24.** $x^2 + (x + 6)^2 = (2\sqrt{17})^2$
$x^2 + x^2 + 12x + 36 = 4(17)$
$2x^2 + 12x + 36 = 68$
$x^2 + 6x + 18 = 34$
$x^2 + 6x - 16 = 0$
$(x + 8)(x - 2) = 0$
$x = -8 \quad \boxed{x = 2}$
$x = 2 \quad x + 6 = 8$

**25.** $9^2 + 12^2 \overset{?}{=} 15^2$

$81 + 144 \overset{?}{=} 225$

$225 = 225$

yes

**26.** $6^2 + 9^2 \overset{?}{=} 11^2$

$36 + 81 \overset{?}{=} 121$

$117 \neq 121$

no

**27.** $10^2 + 8^2 \overset{?}{=} 13^2$

$100 + 64 \overset{?}{=} 169$

$164 \neq 169$

no

**28.** $2^2 + 10^2 \overset{?}{=} 11^2$

$4 + 100 \overset{?}{=} 121$

$104 \neq 121$

no

**29.** $15^2 + 20^2 \overset{?}{=} 25^2$

$225 + 400 \overset{?}{=} 625$

$625 = 625$

yes

**30.** $5^2 + 12^2 \overset{?}{=} 13^2$

$25 + 144 \overset{?}{=} 169$

$169 = 169$

yes

**31.** $11^2 + 60^2 \overset{?}{=} 61^2$

$121 + 3600 \overset{?}{=} 3721$

$3721 = 3721$

yes

**32.** $7^2 + 24^2 \overset{?}{=} 26^2$

$49 + 576 \overset{?}{=} 676$

$625 \neq 676$

no

**33.** $9.9^2 + 2^2 \overset{?}{=} 10.1^2$

$98.01 + 4 \overset{?}{=} 102.01$

$102.01 = 102.01$

yes

**34.** Hypothesis: Today is Tuesday.

Conclusion: Yesterday was Monday.

**35.** Hypothesis: A polygon is a square.

Conclusion: It is a parallelogram.

**36.** Hypothesis: $\frac{x}{3} = -15$

Conclusion: $x = -45$

**37.** Hypothesis: The area of a square is 25 ft$^2$.

Conclusion: The length of a side is 5 ft.

**38.** Hypothesis: A triangle has sides of 8 in. and 9 in.

Conclusion: The third side is greater than 1 in. and less than 17 in.

**39.**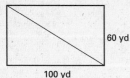

$100^2 + 60^2 = c^2$

$10{,}000 + 3600 = c^2$

$13{,}600 = c^2$

$c \approx 116.6$ ft

**40.** Use the rope to form a triangle with a hypotenuse of length 13 and leg lengths 5 and 12. The rope forms a right triangle since $5^2 + 12^2 = 169 = 13^2$. Then one of the angles of the triangle is a right angle.

**41.** $a^2 + 200^2 = 230^2$

$a^2 + 40{,}000 = 52{,}900$

$a^2 = 12{,}900$

$a = 113.58$ ft

**42.** $12^2 + 7^2 = d^2$

$144 + 49 = d^2$

$193 = d^2$

$d \approx 13.9$ in.

**43.** $13.9(4) = 55.6$ in.

**44.**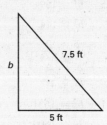

$5^2 + b^2 = 7.5^2$

$25 + b^2 = 56.25$

$b^2 = 31.25$

$b \approx 5.6$ ft

**45. a.** Amalia: $r$ mi

Cindy: $r + 2$ mi

**b.**

**c.** $r^2 + (r + 2)^2 = 10^2$

$r^2 + r^2 + 4r + 4 = 100$

$2r^2 + 4r - 96 = 0$

$r^2 + 2r - 48 = 0$

$(r + 8)(r - 6) = 0$

Amalia:        Cindy:

$r = 6$ mi/h        $r + 2 = 8$ mi/h

**d.** *Sample answer:* I used factoring because the simplified quadratic equation that results is $r^2 + 2r - 48 = 0$ and $r^2 + 2r - 48 = (r + 8)(r - 6)$.

**46.** *Sample answers:*   5, 12, 13; 8, 15, 17

**47.** Yes; Yes; Yes

**48.** If $a$, $b$, and $c$ are a Pythagorean triple, then $a \cdot b \cdot c = 60k$ for some whole number $k$. Each of the numbers 3, 4, and 5 is a factor of at least one of the three numbers in a Pythagorean triple.

### 12.5 Mixed Review (p. 743)

**49.**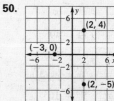

**50.**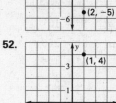

**51.**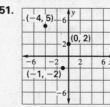

**52.**

**53.**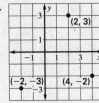

**54.**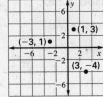

**55.** $0 = x^2 + 2x + 15$

No $x$-intercept

**56.** $0 = x^2 + 8x + 12$

$0 = (x + 6)(x + 2)$

$x = -6, -2$

**57.** $0 = x^2 + x - 10$

$x = \dfrac{-1 \pm \sqrt{1^2 - 4(1)(-10)}}{2(1)}$

$= \dfrac{-1 + \sqrt{41}}{2}, \dfrac{-1 - \sqrt{41}}{2}$

**58.** $0 = x^2 + 8x + 16$

$0 = (x + 4)^2$

$x = -4$

**59.** $0 = x^2 + 3x + 1$

$x = \dfrac{-3 \pm \sqrt{9 - 4(1)(1)}}{2(1)}$

$= \dfrac{-3 + \sqrt{5}}{2}, \dfrac{-3 - \sqrt{5}}{2}$

**60.** $0 = x^2 - 8x - 11$

$x = \dfrac{8 \pm \sqrt{8^2 - 4(1)(-11)}}{2(1)}$

$= \dfrac{8 \pm \sqrt{108}}{2} = \dfrac{8 \pm 6\sqrt{3}}{2}$

$= 4 + 3\sqrt{3}, 4 - 3\sqrt{3}$

**61.** $0 = x^2 + 8x - 10$

$x = \dfrac{-8 \pm \sqrt{64 - 4(1)(-10)}}{2(1)}$

$= \dfrac{-8 \pm \sqrt{104}}{2} = \dfrac{-8 \pm 2\sqrt{26}}{2}$

$= -4 + \sqrt{26}, -4 - \sqrt{26}$

**62.** $0 = 3x^2 + 20x + 1$

$x = \dfrac{-20 \pm \sqrt{20^2 - 4(3)(1)}}{2(3)}$

$= \dfrac{-20 \pm \sqrt{388}}{6} = \dfrac{-20 \pm 2\sqrt{97}}{6}$

$= \dfrac{-10 + \sqrt{97}}{3}, \dfrac{-10 - \sqrt{97}}{3}$

**63.** $0 = -x^2 + 4x + 1$

$x = \dfrac{-4 \pm \sqrt{16 - 4(-1)(1)}}{-2}$

$= \dfrac{-4 \pm \sqrt{20}}{-2} = \dfrac{-4 \pm 2\sqrt{5}}{-2}$

$= \dfrac{-2 + \sqrt{5}}{-1}, \dfrac{-2 - \sqrt{5}}{-1}$

$= 2 - \sqrt{5}, 2 + \sqrt{5}$

**64.** $x^2 - 64 = (x - 8)(x + 8)$

**65.** $16x^2 - 25 = (4x - 5)(4x + 5)$

**66.** $x^2 + 18x + 81 = (x + 9)(x + 9) = (x + 9)^2$

**67.** $7x^2 - 28x + 28 = 7(x^2 - 4x + 4) = 7(x - 2)^2$

**68.** $45x^2 - 60x + 20 = 5(9x^2 - 12x + 4) = 5(3x - 2)^2$

**69.** $-48x^2 + 216x - 243 = -3(16x^2 - 72x + 81)$

$\qquad\qquad\qquad\qquad\quad = -3(4x - 9)^2$

**70.** $\quad 32 = \sqrt{16a}$

$\quad 32^2 = 16a$

$\quad 1024 = 16a$

$\quad\quad 64 = a$

**71.** $0.04 \leq x \leq 0.33$

### Quiz 2 (p. 744)

**1.** $2x^2 - 6x - 15 = 5$

$2x^2 - 6x = 20$

$x^2 - 3x = 10$

$x^2 - 3x + 2.25 = 12.25$

$(x - 1.5)^2 = 12.25$

$x - 1.5 = \pm 3.5$

$x = 5, -2$

**2.** $4x^2 + 4x - 9 = 0$

$x^2 + x - \dfrac{9}{4} = 0$

$x^2 + x + \dfrac{1}{4} = \dfrac{9}{4} + \dfrac{1}{4}$

$\left(x + \dfrac{1}{2}\right)^2 = \dfrac{10}{4}$

$x + \dfrac{1}{2} = \pm \sqrt{\dfrac{10}{4}}$

$x + \dfrac{1}{2} = \pm \dfrac{\sqrt{10}}{2}$

$x = \pm \dfrac{\sqrt{10}}{2} - \dfrac{1}{2}$

$x = \dfrac{-1 + \sqrt{10}}{2}, \dfrac{-1 - \sqrt{10}}{2}$

# Chapter 12 *continued*

**3.**
$$x^2 + 2x = 2$$
$$x^2 + 2x + 1 = 3$$
$$(x + 1)^2 = 3$$
$$x + 1 = \pm\sqrt{3}$$
$$x = -1 + \sqrt{3}$$
$$x = -1 - \sqrt{3}$$

**4.**
$$6^2 + 7^2 \overset{?}{=} 8^2$$
$$36 + 49 \overset{?}{=} 64$$
$$85 \neq 64$$
no

**5.**
$$9^2 + 40^2 \overset{?}{=} 41^2$$
$$81 + 1600 \overset{?}{=} 1681$$
$$1681 = 1681$$
yes

**6.**
$$12^2 + 35^2 \overset{?}{=} 37^2$$
$$144 + 1225 \overset{?}{=} 1369$$
$$1369 = 1369$$
yes

**7.**
$$a^2 + b^2 = c^2$$
$$5^2 + 12^2 = c^2$$
$$25 + 144 = c^2$$
$$169 = c^2$$
$$13 = c$$

**8.**
$$a^2 + b^2 = c^2$$
$$11^2 + b^2 = 61^2$$
$$121 + b^2 = 3721$$
$$b^2 = 3600$$
$$b = 60$$

**9.**
$$a^2 + b^2 = c^2$$
$$a^2 + 63^2 = 65^2$$
$$a^2 + 3969 = 4225$$
$$a^2 = 256$$
$$a = 16$$

**10.**
$$1500^2 + b^2 = 2500^2$$
$$2,250,000 + b^2 = 6,250,000$$
$$b^2 = 4,000,000$$
$$b = 2000 \text{ ft}$$

## Math & History (p. 744)

| Leg | Leg | Hypotenuse |
|-----|-----|------------|
| 119 | 120 | 169 |
| 56 | 90 | 106 |
| 161 | 240 | 289 |
| 65 | 72 | 97 |
| 2700 | 3600 | 4500 |
| 1771 | 2700 | 3229 |

$$65^2 + b^2 = 97^2$$
$$4225 + b^2 = 9409$$
$$b^2 = 5184$$
$$b = 72$$

$$119^2 + b^2 = 169^2$$
$$14,161 + b^2 = 28,561$$
$$b^2 = 14,400$$
$$b = 120$$

$$2700^2 + b^2 = 4500^2$$
$$7,290,000 + b^2 = 20,250,000$$
$$b^2 = 12,960,000$$
$$b = 3600$$

$$56^2 + b^2 = 106^2$$
$$3136 + b^2 = 11,236$$
$$b^2 = 8100$$
$$b = 90$$

$$1771^2 + b^2 = 3229^2$$
$$3,136,441 + b^2 = 10,426,441$$
$$b^2 = 7,290,000$$
$$b = 2700$$

$$161^2 + b^2 = 289^2$$
$$25,921 + b^2 = 83,521$$
$$b^2 = 57,600$$
$$b = 240$$

## Lesson 12.6

### Activity (p. 745)

**1. and 2.**

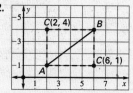

**3.**
$$AC = \sqrt{(1 - 1)^2 + (6 - 2)^2}$$
$$= \sqrt{0^2 + 4^2}$$
$$= \sqrt{16}$$
$$= 4$$

$$AC = \sqrt{(1 - 4)^2 + (2 - 2)^2}$$
$$= \sqrt{(-3)^2 + 0^2}$$
$$= \sqrt{9}$$
$$= 3$$

$$BC = \sqrt{(1 - 4)^2 + (6 - 6)^2}$$
$$= \sqrt{(-3)^2 + 0^2}$$
$$= \sqrt{9}$$
$$= 3$$

$$BC = \sqrt{(4 - 4)^2 + (6 - 2)^2}$$
$$= \sqrt{0 + 4^2}$$
$$= \sqrt{16}$$
$$= 4$$

**4.**
$$AB = 4^2 + 3^2 = c^2$$
$$16 + 9 = c^2$$
$$25 = c^2$$
$$5 = c$$

**5.** Check results.

### 12.6 Guided Practice (p. 748)

**1.** The point that lies on the line connecting two points and is halfway between them

**2.** Draw a right triangle that has a horizontal leg and a vertical leg and whose hypotenuse has the two points as endpoints. Then use the lengths of the legs and the Pythagorean theorem to determine the length of the hypotenuse.

**3.** $(1, 5)(-3, 1)$
$$d = \sqrt{(-3 - 1)^2 + (1 - 5)^2}$$
$$= \sqrt{(-4)^2 + (-4)^2}$$
$$= \sqrt{16 + 16}$$
$$= \sqrt{32}$$
$$\approx 5.66$$

**Algebra 1**
Chapter 12 Worked-out Solution Key

# Chapter 12 *continued*

**4.** $(-3, -2)(4, 1)$

$$d = \sqrt{[4 - (-3)]^2 + [1 - (-2)]^2}$$
$$= \sqrt{7^2 + 3^2}$$
$$= \sqrt{49 + 9}$$
$$= \sqrt{58}$$
$$\approx 7.62$$

**5.** $(5, -2)(-1, 1)$

$$d = \sqrt{(-1 - 5)^2 + [1 - (-2)]^2}$$
$$= \sqrt{(-6)^2 + 3^2}$$
$$= \sqrt{36 + 9}$$
$$= \sqrt{45}$$
$$\approx 6.71$$

**6.**

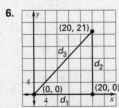

$$20^2 + 21^2 \overset{?}{=} 29^2$$
$$400 + 441 \overset{?}{=} 841$$
$$841 = 841 \quad \text{yes}$$

$$d_1 = \sqrt{(20 - 0)^2 + (0 - 0)^2}$$
$$= \sqrt{20^2 + 0^2}$$
$$= \sqrt{400}$$
$$= 20$$
$$d_2 = \sqrt{(20 - 20)^2 + (0 - 21)^2}$$
$$= \sqrt{0^2 + (-21)^2}$$
$$= \sqrt{441}$$
$$= 21$$
$$d_3 = \sqrt{(20 - 0)^2 + (21 - 0)^2}$$
$$= \sqrt{20^2 + 21^2}$$
$$= \sqrt{400 + 441}$$
$$= \sqrt{841}$$
$$= 29$$

**7.**

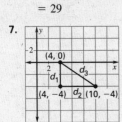

$$a^2 + b^2 = c^2$$
$$4^2 + 6^2 = 7.21^2$$
$$16 + 36 = 52$$
$$52 = 52 \quad \text{yes}$$

$$d_1 = \sqrt{(4 - 4)^2 + (0 + 4)^2}$$
$$= \sqrt{0^2 + 4^2}$$
$$= \sqrt{16}$$
$$= 4$$

$$d_2 = \sqrt{(10 - 4)^2 + (-4 + 4)^2}$$
$$= \sqrt{6^2 + 0^2}$$
$$= \sqrt{36}$$
$$= 6$$
$$d_3 = \sqrt{(10 - 4)^2 + (-4 - 0)^2}$$
$$= \sqrt{6^2 + -4^2}$$
$$= \sqrt{36 + 16}$$
$$= \sqrt{52}$$
$$\approx 7.21$$

**8.**

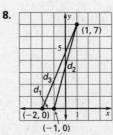

$$a^2 + b^2 = c^2$$
$$1^2 + 7.3^2 \overset{?}{=} 7.6^2$$
$$1 + 53.29 \overset{?}{=} 57.76$$
$$54.29 \neq 57.76$$
$$\text{no}$$

$$d_1 = \sqrt{(-1 + 2)^2 + (0 - 0)^2}$$
$$= \sqrt{1^2 + 0^2}$$
$$= \sqrt{1}$$
$$= 1$$
$$d_2 = \sqrt{(-1 - 1)^2 + (0 - 7)^2}$$
$$= \sqrt{4 + 49}$$
$$= \sqrt{53}$$
$$\approx 7.3$$
$$d_3 = \sqrt{(-2 - 1)^2 + (0 - 7)^2}$$
$$= \sqrt{-3^2 + -7^2}$$
$$= \sqrt{9 + 49}$$
$$= \sqrt{58}$$
$$\approx 7.6$$

**9.**

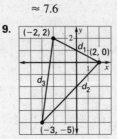

$$a^2 + b^2 = c^2$$
$$20 + 50 \overset{?}{=} 50$$
$$70 \neq 50 \quad \text{no}$$

$$d_1 = \sqrt{(2 + 2)^2 + (-2)^2}$$
$$= \sqrt{16 + 4}$$
$$= \sqrt{20}$$
$$d_2 = \sqrt{(-3 - 2)^2 + (-5 - 0)^2}$$
$$= \sqrt{(-5)^2 + (-5)^2}$$
$$= \sqrt{25 + 25}$$
$$= \sqrt{50}$$

# Chapter 12 *continued*

$$d_3 = \sqrt{(-3 + 2)^2 + (-5 - 2)^2}$$
$$= \sqrt{(-1)^2 + (-7)^2}$$
$$= \sqrt{1 + 49}$$
$$= \sqrt{50}$$

**10.** $(4, 4), (-1, 2)$

$$\left(\frac{4 + -1}{2}, \frac{4 + 2}{2}\right)$$

$$\left(\frac{3}{2}, \frac{6}{2}\right)$$

$$\left(\frac{3}{2}, 3\right)$$

**11.** $(6, 2), (2, -3)$

$$\left(\frac{6 + 2}{2}, \frac{2 + (-3)}{2}\right)$$

$$\left(\frac{8}{2}, -\frac{1}{2}\right)$$

$$\left(4, -\frac{1}{2}\right)$$

**12.** $(-5, 3), (-3, -3)$

$$\left(\frac{-5 + (-3)}{2}, \frac{3 + (-3)}{2}\right)$$

$$\left(-\frac{8}{2}, \frac{0}{2}\right)$$

$$(-4, 0)$$

**13.**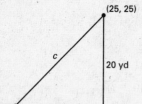

$$15^2 + 20^2 = c^2$$
$$225 + 400 = c^2$$
$$625 = c^2$$
$$25 \text{ yd} = c$$

(Figure is not drawn to scale.)

### 12.6 Practice and Applications (pp. 748–750)

**14.** $(2, 0), (8, -3)$

$$d = \sqrt{(8 - 2)^2 + (-3 - 0)^2}$$
$$= \sqrt{6^2 + (-3)^2}$$
$$= \sqrt{36 + 9}$$
$$= \sqrt{45}$$
$$\approx 6.71$$

**15.** $(2, -8), (-3, 3)$

$$d = \sqrt{(-3 - 2)^2 + (3 + 8)^2}$$
$$= \sqrt{(-5)^2 + 11^2}$$
$$= \sqrt{25 + 121}$$
$$= \sqrt{146}$$
$$\approx 12.08$$

**16.** $(3, -1), (0, 3)$

$$d = \sqrt{(0 - 3)^2 + (3 + 1)^2}$$
$$= \sqrt{(-3)^2 + 4^2}$$
$$= \sqrt{9 + 16}$$
$$= \sqrt{25}$$
$$= 5$$

**17.** $(5, 8), (-2, 3)$

$$d = \sqrt{(-2 - 5)^2 + (3 - 8)^2}$$
$$= \sqrt{(-7)^2 + (-5)^2}$$
$$= \sqrt{49 + 25}$$
$$= \sqrt{74}$$
$$\approx 8.60$$

**18.** $(-3, 1), (2, 6)$

$$d = \sqrt{(2 + 3)^2 + (6 - 1)^2}$$
$$= \sqrt{5^2 + 5^2}$$
$$= \sqrt{25 + 25}$$
$$= \sqrt{50}$$
$$\approx 7.07$$

**19.** $(-6, -2), (-3, -5)$

$$d = \sqrt{(-3 + 6)^2 + (-5 + 2)^2}$$
$$= \sqrt{3^2 + (-3)^2}$$
$$= \sqrt{9 + 9}$$
$$= \sqrt{18}$$
$$\approx 4.24$$

**20.** $(4, 5), (-1, 3)$

$$d = \sqrt{(-1 - 4)^2 + (3 - 5)^2}$$
$$= \sqrt{(-5)^2 + (-2)^2}$$
$$= \sqrt{25 + 4}$$
$$= \sqrt{29}$$
$$\approx 5.39$$

**21.** $(-6, 1), (3, 1)$

$$d = \sqrt{(3 + 6)^2 + (1 - 1)^2}$$
$$= \sqrt{9^2 + 0^2}$$
$$= \sqrt{81 + 0}$$
$$= \sqrt{81}$$
$$= 9$$

**22.** $(-2, -1), (3, -3)$

$$d = \sqrt{(3 + 2)^2 + (-3 + 1)^2}$$
$$= \sqrt{5^2 + (-2)^2}$$
$$= \sqrt{25 + 4}$$
$$= \sqrt{29}$$
$$\approx 5.39$$

**23.** $(3.5, 6), (-3.5, -2)$

$$d = \sqrt{(-3.5 - 3.5)^2 + (-2 - 6)^2}$$
$$= \sqrt{(-7)^2 + (-8)^2}$$
$$= \sqrt{49 + 64}$$
$$= \sqrt{113}$$
$$\approx 10.63$$

**Algebra 1**
Chapter 12 Worked-out Solution Key

# Chapter 12 *continued*

**24.** $\left(\frac{1}{2}, \frac{1}{4}\right), (2, 1)$

$$d = \sqrt{\left(2 - \frac{1}{2}\right)^2 + \left(1 - \frac{1}{4}\right)^2}$$

$$= \sqrt{\left(1\frac{1}{2}\right)^2 + \left(\frac{3}{4}\right)^2}$$

$$= \sqrt{\frac{9}{4} + \frac{9}{16}}$$

$$= \sqrt{\frac{45}{16}}$$

$$\approx 1.68$$

**25.** $\left(\frac{1}{3}, \frac{1}{6}\right), \left(-\frac{2}{3}, \frac{8}{3}\right)$

$$d = \sqrt{\left(-\frac{2}{3} - \frac{1}{3}\right)^2 + \left(\frac{8}{3} - \frac{1}{6}\right)^2}$$

$$= \sqrt{(-1)^2 + \left(\frac{15}{6}\right)^2}$$

$$= \sqrt{1 + \frac{225}{36}}$$

$$= \sqrt{\frac{261}{36}}$$

$$\approx 2.69$$

**26.**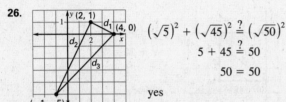

$$\left(\sqrt{5}\right)^2 + \left(\sqrt{45}\right)^2 \overset{?}{=} \left(\sqrt{50}\right)^2$$

$$5 + 45 \overset{?}{=} 50$$

$$50 = 50$$

yes

$$d_1 = \sqrt{(4 - 2)^2 + (0 - 1)^2}$$

$$= \sqrt{2^2 + (-1)^2}$$

$$= \sqrt{4 + 1}$$

$$= \sqrt{5}$$

$$d_2 = \sqrt{(2 + 1)^2 + (1 + 5)^2}$$

$$= \sqrt{3^2 + 6^2}$$

$$= \sqrt{9 + 36}$$

$$= \sqrt{45}$$

$$d_3 = \sqrt{(4 + 1)^2 + (0 + 5)^2}$$

$$= \sqrt{5^2 + 5^2}$$

$$= \sqrt{25 + 25}$$

$$= \sqrt{50}$$

**27.**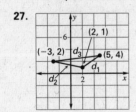

$$\left(\sqrt{18}\right)^2 + \left(\sqrt{26}\right)^2 \overset{?}{=} \left(\sqrt{68}\right)^2$$

$$18 + 26 \overset{?}{=} 68$$

$$44 \neq 68$$

no

$$d_1 = \sqrt{(2 - 5)^2 + (1 - 4)^2}$$

$$= \sqrt{(-3)^2 + (-3)^2}$$

$$= \sqrt{9 + 9}$$

$$= \sqrt{18}$$

$$d_2 = \sqrt{(2 + 3)^2 + (1 - 2)^2}$$

$$= \sqrt{5^2 + (-1)^2}$$

$$= \sqrt{25 + 1}$$

$$= \sqrt{26}$$

$$d_3 = \sqrt{(5 + 3)^2 + (4 - 2)^2}$$

$$= \sqrt{8^2 + 2^2}$$

$$= \sqrt{64 + 4}$$

$$= \sqrt{68}$$

**28.**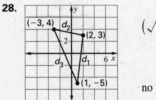

$$\left(\sqrt{65}\right)^2 + \left(\sqrt{26}\right)^2 \overset{?}{=} \left(\sqrt{97}\right)^2$$

$$65 + 26 \overset{?}{=} 97$$

$$91 \neq 97$$

no

$$d_1 = \sqrt{(2 - 1)^2 + (3 + 5)^2}$$

$$= \sqrt{1^2 + 8^2}$$

$$= \sqrt{1 + 64}$$

$$= \sqrt{65}$$

$$d_2 = \sqrt{(2 + 3)^2 + (3 - 4)^2}$$

$$= \sqrt{5^2 + (-1)^2}$$

$$= \sqrt{25 + 1}$$

$$= \sqrt{26}$$

$$d_3 = \sqrt{(-3 - 1)^2 + (4 + 5)^2}$$

$$= \sqrt{(-4)^2 + 9^2}$$

$$= \sqrt{16 + 81}$$

$$= \sqrt{97}$$

**29.**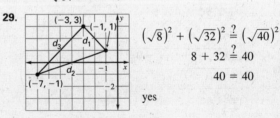

$$\left(\sqrt{8}\right)^2 + \left(\sqrt{32}\right)^2 \overset{?}{=} \left(\sqrt{40}\right)^2$$

$$8 + 32 \overset{?}{=} 40$$

$$40 = 40$$

yes

$$d_1 = \sqrt{(-3 + 1)^2 + (3 - 1)^2}$$

$$= \sqrt{(-2)^2 + 2^2}$$

$$= \sqrt{4 + 4}$$

$$= \sqrt{8}$$

$$d_2 = \sqrt{(-1 + 7)^2 + (1 + 1)^2}$$

$$= \sqrt{6^2 + 2^2}$$

$$= \sqrt{36 + 4}$$

$$= \sqrt{40}$$

$$d_3 = \sqrt{(-3 + 7)^2 + (3 + 1)^2}$$

$$= \sqrt{4^2 + 4^2}$$

$$= \sqrt{16 + 16}$$

$$= \sqrt{32}$$

**30.**

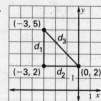

$$3^2 + 3^2 \stackrel{?}{=} \left(\sqrt{18}\right)^2$$
$$9 + 9 \stackrel{?}{=} 18$$
$$18 = 18$$

yes

$$d_1 = \sqrt{(-3 + 3)^2 + (5 - 2)^2}$$
$$= \sqrt{0^2 + 3^2}$$
$$= \sqrt{9}$$
$$= 3$$
$$d_2 = \sqrt{(-3 - 0)^2 + (2 - 2)^2}$$
$$= \sqrt{(-3)^2 + 0^2}$$
$$= \sqrt{9}$$
$$= 3$$
$$d_3 = \sqrt{(-3 - 0)^2 + (5 - 2)^2}$$
$$= \sqrt{(-3)^2 + 3^2}$$
$$= \sqrt{9 + 9}$$
$$= \sqrt{18}$$

**31.**

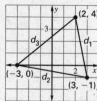

$$\left(\sqrt{26}\right)^2 + \left(\sqrt{37}\right)^2 \stackrel{?}{=} \left(\sqrt{41}\right)^2$$
$$26 + 37 \stackrel{?}{=} 41$$
$$63 \neq 41$$

no

$$d_1 = \sqrt{(3 - 2)^2 + (-1 - 4)^2}$$
$$= \sqrt{1^2 + (-5)^2}$$
$$= \sqrt{1 + 25}$$
$$= \sqrt{26}$$
$$d_2 = \sqrt{(3 + 3)^2 + (-1 - 0)^2}$$
$$= \sqrt{6^2 + (-1)^2}$$
$$= \sqrt{36 + 1}$$
$$= \sqrt{37}$$
$$d_3 = \sqrt{(2 + 3)^2 + (4 - 0)^2}$$
$$= \sqrt{5^2 + 4^2}$$
$$= \sqrt{25 + 16}$$
$$= \sqrt{41}$$

**32.**

$$\left(\sqrt{5}\right)^2 + 6^2 \stackrel{?}{=} \left(\sqrt{29}\right)^2$$
$$5 + 36 \stackrel{?}{=} 29$$
$$41 \neq 29$$

no

$$d_1 = \sqrt{(4 - 3)^2 + (2 - 4)^2}$$
$$= \sqrt{1^2 + (-2)^2}$$
$$= \sqrt{1 + 4}$$
$$= \sqrt{5}$$
$$d_2 = \sqrt{(4 + 2)^2 + (2 - 2)^2}$$
$$= \sqrt{6^2 + 0^2}$$
$$= \sqrt{36}$$
$$= 6$$
$$d_3 = \sqrt{(3 + 2)^2 + (4 - 2)^2}$$
$$= \sqrt{5^2 + 2^2}$$
$$= \sqrt{25 + 4}$$
$$= \sqrt{29}$$

**33.**

$$3^2 + 4^2 \stackrel{?}{=} 5^2$$
$$9 + 16 \stackrel{?}{=} 25$$
$$25 = 25$$

yes

$$d_1 = \sqrt{(4 - 0)^2 + (-4 + 4)^2}$$
$$= \sqrt{4^2 + 0^2}$$
$$= \sqrt{16}$$
$$= 4$$
$$d_2 = \sqrt{(4 - 4)^2 + (-4 + 1)^2}$$
$$= \sqrt{0^2 + (-3)^2}$$
$$= \sqrt{9}$$
$$= 3$$
$$d_3 = \sqrt{(4 - 0)^2 + (-1 + 4)^2}$$
$$= \sqrt{4^2 + 3^2}$$
$$= \sqrt{16 + 9}$$
$$= \sqrt{25}$$
$$= 5$$

**34.** $(3, 0), (-5, 4)$

$$\left(\frac{3 + -5}{2}, \frac{0 + 4}{2}\right)$$
$$\left(\frac{-2}{2}, \frac{4}{2}\right)$$
$$(-1, 2)$$

**35.** $(0, 0), (0, 8)$

$$\left(\frac{0 + 0}{2}, \frac{0 + 8}{2}\right)$$
$$\left(\frac{0}{2}, \frac{8}{2}\right)$$
$$(0, 4)$$

**36.** $(1, 2), (5, 4)$

$$\left(\frac{1 + 5}{2}, \frac{2 + 4}{2}\right)$$
$$\left(\frac{6}{2}, \frac{6}{2}\right)$$
$$(3, 3)$$

**37.** $(-1, 2), (7, 4)$

$$\left(\frac{-1 + 7}{2}, \frac{2 + 4}{2}\right)$$
$$\left(\frac{6}{2}, \frac{6}{2}\right)$$
$$(3, 3)$$

**Algebra 1**
Chapter 12 Worked-out Solution Key

# Chapter 12 *continued*

**38.** $(-3, 3), (2, -2)$

$\left(\dfrac{-3 + 2}{2}, \dfrac{3 + (-2)}{2}\right)$

$\left(-\dfrac{1}{2}, \dfrac{1}{2}\right)$

**39.** $(2, 7), (4, 3)$

$\left(\dfrac{2 + 4}{2}, \dfrac{7 + 3}{2}\right)$

$\left(\dfrac{6}{2}, \dfrac{10}{2}\right)$

$(3, 5)$

**40.** $(-1, 1), (-4, -4)$

$\left(\dfrac{-1 + (-4)}{2}, \dfrac{1 + (-4)}{2}\right)$

$\left(-\dfrac{5}{2}, -\dfrac{3}{2}\right)$

**41.** $(-4, 0), (-1, -5)$

$\left(\dfrac{-4 + (-1)}{2}, \dfrac{0 + (-5)}{2}\right)$

$\left(-\dfrac{5}{2}, -\dfrac{5}{2}\right)$

**42.** $(5, 1), (1, -5)$

$\left(\dfrac{5 + 1}{2}, \dfrac{1 + (-5)}{2}\right)$

$\left(\dfrac{6}{2}, -\dfrac{4}{2}\right)$

$(3, -2)$

**43.** $(0, -3), (-4, 2)$

$\left(\dfrac{0 + (-4)}{2}, \dfrac{-3 + 2}{2}\right)$

$\left(-\dfrac{4}{2}, -\dfrac{1}{2}\right)$

$\left(-2, -\dfrac{1}{2}\right)$

**44.** $(5, -5), (-5, 1)$

$\left(\dfrac{5 + (-5)}{2}, \dfrac{-5 + 1}{2}\right)$

$\left(\dfrac{0}{2}, -\dfrac{4}{2}\right)$

$(0, -2)$

**45.** $(-4, -3), (-1, -5)$

$\left(\dfrac{-4 + (-1)}{2}, \dfrac{-3 + (-5)}{2}\right)$

$\left(-\dfrac{5}{2}, -\dfrac{8}{2}\right)$

$\left(-\dfrac{5}{2}, -4\right)$

**46.–48. Estimates may vary.**

**46.** $(0.3, 4.5), (-3.2, -1.7)$   $1 = 95$ miles

$(28.5, 427.5), (-304, -161.5)$

$d = \sqrt{(-304 - 28.5)^2 + (-161.5 - 427.5)^2}$

$= \sqrt{110{,}556.25 + 346{,}921}$

$= \sqrt{457{,}477.25}$

$\approx 676.37$ miles

**47.** $(1.8, -0.3), (-2.2, 2.8)$   $1 = 95$ miles

$(171, -28.5), (-209, 266)$

$d = \sqrt{(-209 - 171)^2 + (266 + 28.5)^2}$

$= \sqrt{144{,}400 + 86{,}730.25}$

$= \sqrt{231{,}130.25}$

$\approx 480.76$ miles

**48.** $(2, -2.5), (4, 2.5)$   $1 = 95$ miles

$(190, -237.5), (380, 237.5)$

$d = \sqrt{(380 - 190)^2 + (237.5 + 237.5)^2}$

$= \sqrt{36{,}100 + 225{,}625}$

$= \sqrt{261{,}725}$

$\approx 511.59$ miles

**49.**

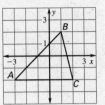

**50.** $AB = \sqrt{(-3 - 1)^2 + (-2 - 2)^2}$

$= \sqrt{(-4)^2 + (-4)^2}$

$= \sqrt{16 + 16}$

$= \sqrt{32}$

$= 4\sqrt{2}$

$BC = \sqrt{(1 - 2)^2 + (2 + 2)^2}$

$= \sqrt{(-1)^2 + 4^2}$

$= \sqrt{1 + 16}$

$= \sqrt{17}$

$CA = \sqrt{(2 + 3)^2 + (-2 + 2)^2}$

$= \sqrt{5^2 + 0^2}$

$= \sqrt{25}$

$= 5$

**51.** $\overline{BC}: \left(\dfrac{1 + 2}{2}, \dfrac{2 + (-2)}{2}\right) = \left(\dfrac{3}{2}, \dfrac{0}{2}\right) = \left(\dfrac{3}{2}, 0\right)$

$\overline{AB}: \left(\dfrac{(-3) + 1}{2}, \dfrac{-2 + 2}{2}\right) = \left(-\dfrac{2}{2}, \dfrac{0}{2}\right) = (-1, 0)$

$\overline{AC}: \left(\dfrac{-3 + 2}{2}, \dfrac{-2 + (-2)}{2}\right) = \left(-\dfrac{1}{2}, -\dfrac{4}{2}\right) = \left(-\dfrac{1}{2}, -2\right)$

**52.** Let midpoint of $\overline{BC}$ be $D$.
Let midpoint of $\overline{AB}$ be $E$.
Let midpoint of $\overline{AC}$ be $F$.

$DE = \sqrt{\left(\dfrac{3}{2} + 1\right)^2 + (0 - 2)^2}$

$= \sqrt{\left(\dfrac{5}{2}\right)^2}$

$= \dfrac{5}{2} = 2\dfrac{1}{2}$

$EF = \sqrt{\left(-1 + \dfrac{1}{2}\right)^2 + (0 + 2)^2}$

$= \sqrt{\left(-\dfrac{1}{2}\right)^2 + 2^2}$

$= \sqrt{\dfrac{1}{4} + 4}$

$= \sqrt{\dfrac{17}{4}} = \dfrac{\sqrt{17}}{2}$

$FD = \sqrt{\left(-\dfrac{1}{2} - \dfrac{3}{2}\right)^2 + (-2 - 0)^2}$

$= \sqrt{(-2)^2 + (-2)^2}$

$= \sqrt{4 + 4}$

$= \sqrt{8} = 2\sqrt{2}$

# Chapter 12 *continued*

**53.** The Perimeter of $\triangle ABC$ is $4\sqrt{2} + \sqrt{17} + 5 \approx 14.78$.

The Perimeter of $\triangle EDF$ is $2\frac{1}{2} + \frac{\sqrt{17}}{2} + 2\sqrt{2} \approx 7.39$.

The Perimeter of $\triangle ABC$ is twice that of $\triangle EDF$.

**54.**

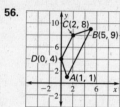

$$AB = \sqrt{(4 + 2)^2 + (-1 - 3)^2}$$
$$= \sqrt{6^2 + (-4)^2}$$
$$= \sqrt{36 + 16}$$
$$= \sqrt{52}$$
$$= 2\sqrt{13} \approx 7.2 \text{ miles}$$

**55.** You could meet at a point 1 mi east and 1 mi north of the starting point. You would walk 3 mi east and 2 mi south and your friend would walk 2 mi north and 3 mi west. You would each have to walk $\sqrt{13} = 3.6$ miles.

**56.**

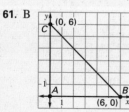

**57.** $\overline{DC}$ is parallel to $\overline{AB}$ because both have slope 2. $\overline{CB}$ and $\overline{DA}$ are not parallel because $\overline{DA}$ has slope $-3$ and $\overline{CB}$ has slope $\frac{1}{3}$.

**58.**
$$CB = \sqrt{(2 - 5)^2 + (8 - 9)^2}$$
$$= \sqrt{(-3)^2 + (-1)^2}$$
$$= \sqrt{9 + 1}$$
$$= \sqrt{10}$$
$$AD = \sqrt{(1 - 0)^2 + (1 - 4)^2}$$
$$= \sqrt{1^2 + (-3)^2}$$
$$= \sqrt{1 + 9}$$
$$= \sqrt{10}$$

Because the lengths of $\overline{AD}$ and $\overline{CB}$ are both $\sqrt{10}$, the trapezoid is isosceles.

**59.** C
$$d = \sqrt{(-6 - 2)^2 + (-2 - 4)^2}$$
$$= \sqrt{(-8)^2 + (-6)^2}$$
$$= \sqrt{64 + 36}$$
$$= \sqrt{100}$$
$$= 10$$

**60.** D $\left(\dfrac{-2 + 1}{2}, \dfrac{-3 + \frac{1}{2}}{2}\right) = \left(\dfrac{-1}{2}, \dfrac{-2\frac{1}{2}}{2}\right) = \left(-\dfrac{1}{2}, -1\dfrac{1}{4}\right)$

**61.** B

$$CB = \sqrt{(6 - 0)^2 + (0 - 6)^2}$$
$$= \sqrt{6^2 + (-6)^2}$$
$$= \sqrt{36 + 36}$$
$$= \sqrt{72}$$
$$= 6\sqrt{2}$$

**62.**
$$d = \sqrt{(100 - 0)^2 + (250 - 0)^2}$$
$$= \sqrt{100^2 + 250^2}$$
$$= \sqrt{10{,}000 + 62{,}500}$$
$$= \sqrt{72{,}500}$$
$$\approx 269.3 \text{ miles}$$

**63.** Home to Amusement Park = 269.3 mi

Amusement Park to Beach
$$d = \sqrt{(100 - 450)^2 + (250 - 450)^2}$$
$$= \sqrt{(-350)^2 + (-200)^2}$$
$$= \sqrt{122{,}500 + 40{,}000}$$
$$= \sqrt{162{,}500}$$
$$\approx 403.1 \text{ miles}$$

Beach to Home
$$d = \sqrt{(450 - 0)^2 + (450 - 0)^2}$$
$$= \sqrt{450^2 + 450^2}$$
$$= \sqrt{202{,}500 + 202{,}500}$$
$$= \sqrt{405{,}000}$$
$$\approx 636.4 \text{ miles}$$

$$269.3 + 403.1 + 636.4 = 1308.8 \text{ miles}$$

**64.** Visit the sites in one of the following two ways:

    (1) amusement park, beach, campground, zoo

    (2) zoo, campground, beach, amusement park

**12.6 Mixed Review (p. 750)**

**65.** $3x(x^2 + 4x - 5) = 3x(x + 5)(x - 1)$

**66.** $(x^4 - 3x^2) - (25x^2 - 75) = x^2(x^2 - 3) - 25(x^2 - 3)$
$$= (x^2 - 25)(x^2 - 3)$$
$$= (x + 5)(x - 5)(x^2 - 3)$$

**67.** inverse variation      **68.** direct variation

**69.** neither      **70.** $(6z + 10) \div 2 = 3z + 5$

**71.** $(7x^3 - 2x^2) \div 14x = \dfrac{7x^3 - 2x^2}{14x}$
$$= \dfrac{x(7x^2 - 2x)}{14x}$$
$$= \dfrac{7x^2 - 2x}{14}$$

**72.** $\dfrac{4}{8} = \dfrac{x}{6}$

$24 = 8x$

$3 \text{ ft} = x$

**73.** $\dfrac{14}{21} = \dfrac{x}{30}$

$21x = 420$

$x = 20 \text{ in.}$

# Chapter 12 *continued*

**Developing Concepts Activity 12.7 (p. 751)**
**Exploring the Concept**

1.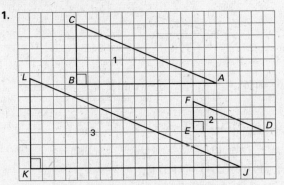

2. $AB = 12, BC = 5, AC = 13, DE = 6, EF = 2.5,$
$FD = 6.5, KL = 7.5, JL = 18, KJ = 19.5$

3.

| | Shorter Leg | Longer Leg | Hypotenuse |
|---|---|---|---|
| △ 1 | 5 | 12 | 13 |
| △ 2 | 2.5 | 6 | 6.5 |
| △ 3 | 7.5 | 18 | 19.5 |

| | $\frac{\text{Shorter Leg}}{\text{Hypotenuse}}$ | $\frac{\text{Longer Leg}}{\text{Hypotenuse}}$ | $\frac{\text{Shorter Leg}}{\text{Longer Leg}}$ |
|---|---|---|---|
| △ 1 | 0.38 | 0.92 | 0.42 |
| △ 2 | 0.38 | 0.92 | 0.42 |
| △ 3 | 0.38 | 0.92 | 0.42 |

**Drawing Conclusions**

1. Each ratio is the same for all three triangles.
2. Check results.
3. **a.** 0.38  **b.** 0.92  **c.** 0.42

## Lesson 12.7

**12.7 Guided Practice (p. 755)**

1. a ratio of the lengths of two sides of a right triangle
2. True; the value of a trigonometric ratio depends only on the measure of the angle, not on the size of the triangle.
3. $\frac{7}{25}$  4. $\frac{24}{25}$  5. $\frac{7}{24}$
6. $\sin 60° = \frac{4}{c}$     $\tan 60° = \frac{4}{b}$

$c = \frac{4}{\sin 60°}$     $b = \frac{4}{\tan 60°}$

$c \approx 4.62$      $b \approx 2.3$

Check: $4^2 + 2.3^2 \stackrel{?}{=} 4.62^2$
$16 + 5.29 \stackrel{?}{=} 21.34$
$21 = 21$

7. $\sin 61° = \frac{49}{f}$     $\tan 61° = \frac{49}{e}$

$f = \frac{49}{\sin 61°}$     $e = \frac{49}{\tan 61°}$

$f \approx 56.024$      $e \approx 27.161$

Check: $49^2 + 27.161^2 \stackrel{?}{=} 56.024^2$
$2401 + 738 \stackrel{?}{=} 3139$
$3139 = 3139$

8. $\sin 70° = \frac{16}{h}$     $\tan 70° = \frac{16}{g}$

$h = \frac{16}{\sin 70°}$     $g = \frac{16}{\tan 70°}$

$h = 17.03$       $g = 5.82$

Check: $16^2 + 5.82^2 \stackrel{?}{=} 17.03^2$
$256 + 34 \stackrel{?}{=} 290$
$290 = 290$

9. $\sin 35° = \frac{a}{800}$     $\cos 35° = \frac{c}{800}$

$0.574(800) = a$     $0.8191(800) = c$

$458.86 = a$       $655.3 = c$

$459$ ft        $655$ ft

**12.7 Practice and Applications (pp. 755–757)**

10. $\sin R = \frac{4}{5}, \cos R = \frac{3}{5}, \tan R = \frac{4}{3}$
$\sin S = \frac{3}{5}, \cos S = \frac{4}{5}, \tan S = \frac{3}{4}$
11. $\sin R = \frac{5}{13}, \cos R = \frac{12}{13}, \tan R = \frac{5}{12}$
$\sin S = \frac{12}{13}, \cos S = \frac{5}{13}, \tan S = \frac{12}{5}$
12. $\sin R = \frac{11}{61}, \cos R = \frac{60}{61}, \tan R = \frac{11}{60}$
$\sin S = \frac{60}{61}, \cos S = \frac{11}{61}, \tan S = \frac{60}{11}$
13. $\sin 53° = \frac{17}{c}$     $\tan 53° = \frac{17}{b}$

$c = \frac{17}{\sin 53°}$     $b = \frac{17}{\tan 53°}$

$c \approx 21.286$      $b \approx 12.810$

Check: $17^2 + 12.810^2 \stackrel{?}{=} 21.286^2$
$289 + 164.1 \stackrel{?}{=} 453$
$453 = 453$

14. $\cos 75° = \frac{5}{p}$     $\tan 75° = \frac{n}{5}$

$p = \frac{5}{\cos 75°}$     $3.73(5) = n$

$p \approx 19.32$      $18.66 \approx n$

Check: $5^2 + 18.66^2 \stackrel{?}{=} 19.32^2$
$25 + 348.2 \stackrel{?}{=} 373.26$
$373 = 373$

# Chapter 12 *continued*

**15.**
$$\cos 34° = \frac{y}{24} \qquad\qquad \sin 34° = \frac{x}{24}$$
$$0.8290(24) \approx y \qquad\qquad 0.5592(24) \approx x$$
$$y \approx 19.90 \qquad\qquad\qquad x \approx 13.42$$
Check: $19.9^2 + 13.42^2 \overset{?}{=} 24^2$
$$396.01 + 180.0964 \overset{?}{=} 576$$
$$576.1064 \approx 576$$

**16.**
$$\sin 7° = \frac{h}{10} \qquad\qquad \cos 7° = \frac{g}{10}$$
$$0.1219(10) \approx h \qquad\qquad 0.9925(10) \approx g$$
$$1.22 \approx h \qquad\qquad\qquad 9.93 \approx g$$
Check: $1.22^2 + 9.93^2 \overset{?}{=} 10^2$
$$1.4884 + 98.6049 \overset{?}{=} 100$$
$$100.0933 \approx 100$$

**17.**
$$\cos 60° = \frac{30}{w} \qquad\qquad \tan 60° = \frac{v}{30}$$
$$w = \frac{30}{\cos 60°} \qquad\qquad 1.732(30) \approx v$$
$$\qquad\qquad\qquad\qquad\qquad v \approx 51.96$$
$$w = 60$$
Check: $30^2 + 51.96^2 \overset{?}{=} 60^2$
$$3599.8416 \approx 3600$$

**18.**
$$\sin 39° = \frac{3}{d} \qquad\qquad \tan 39° = \frac{3}{f}$$
$$d = \frac{3}{\sin 39°} \qquad\qquad f = \frac{3}{\tan 39°}$$
$$d \approx 4.77 \qquad\qquad\qquad f \approx 3.70$$
Check: $3^2 + 3.7^2 \overset{?}{=} 4.77^2$
$$22.69 \approx 22.7529$$

**19.**
$$\cos 35° = \frac{a}{210}$$
$$0.8192(210) \approx a$$
$$a \approx 172 \text{ ft}$$

**20.**
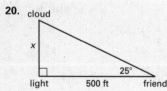
cloud, x, light, 500 ft, friend, 25°
$$\tan 25° = \frac{x}{500}$$
$$0.4663(500) \approx x$$
$$x \approx 233 \text{ ft}$$

**21.**
$$\tan 40° = \frac{x}{197}$$
$$0.8391(197) \approx x$$
$$165.3 \approx x$$
$$165.3 + 5 \approx 170 \text{ ft}$$

**22.**
$$\sin 20° = \frac{1.5}{x}$$
$$x = \frac{1.5}{\sin 20°}$$
$$x \approx 4.39 \text{ ft}$$

**23.**
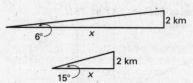
2 km, x, 6°
2 km, x, 15°

$$\tan 6° = \frac{2}{x} \qquad\qquad \tan 15° = \frac{2}{x}$$
$$x = \frac{2}{\tan 6°} \qquad\qquad x = \frac{2}{\tan 15°}$$
$$x \approx 19.03 \qquad\qquad x \approx 7.46$$
$$19.03 - 7.46 = 11.57 \text{ km}$$

**24. a.** $\tan 30° = \dfrac{3}{q}$
$$q = \frac{3}{\tan 30°}$$
$$q \approx 5.20$$

**b.** Use $\sin 30° = \dfrac{3}{r}$ or since you know the value of $q$, you could use the Pythagorean theorem.

**c.** $\sin 30° = \dfrac{3}{r}$
$$r = \frac{3}{\sin 30°}$$
$$r = 6$$

**d.** I used $\sin 30° = \dfrac{3}{r}$ because the calculations are simpler than if you used the Pythagorean theorem.

**25.** $\dfrac{2-0}{3-1} = \dfrac{2}{2} = 1$
slope $= \tan A$

**26.** $\dfrac{2-0}{5-2} = \dfrac{2}{3}$
slope $= \tan A$

**27.** $y = x \tan 60° \approx 1.73x$

## 12.7 Mixed Review (p. 757)

**28.** False; the absolute value of zero is not positive.

**29.** False; the opposite value of a positive number is negative.

**30.** $6(w - 3) = 6w - 18$

**31.** $-p(p + 1) = -p^2 - p$

**32.** $-(x - 8) = -x + 8$

**33.** $(x + 3)x = x^2 + 3x$

**34.** $(x - 2)5x = 5x^2 - 10x$

**35.** $(4 + x)(-6x) = -24x - 6x^2$

**36.**
$$\begin{array}{r} 6x^2 - 3x + 2 \\ 2x^2 + \phantom{0}x + 7 \\ \hline 4x^2 - 4x - 5 \end{array}$$

**37.**
$$\begin{array}{r} 4x^3 + 3x^2 + \phantom{0}8x + 6 \\ 2x^3 - 3x^2 - \phantom{0}7x \\ \hline 2x^3 + 6x^2 + 15x + 6 \end{array}$$

**38.**
$$\begin{array}{r} 10x^3 + 0x^2 + 0x + 15 \\ 17x^3 + 0x^2 - 4x + \phantom{0}5 \\ \hline -7x^3 + 0x^2 + 4x + 10 \end{array}$$
$$= -7x^3 + 4x + 10$$

**39.**
$$\begin{array}{r} -2x^3 + 5x^2 - \phantom{0}x + \phantom{0}8 \\ -2x^3 + 0x^2 + 3x - \phantom{0}4 \\ \hline 0x^3 + 5x^2 - 4x + 12 \end{array}$$
$$= 5x^2 - 4x + 12$$

**40.** $\dfrac{3}{x} + \dfrac{x+9}{x} = \dfrac{x+12}{x}$

**41.** $\dfrac{8}{4a+1} - \dfrac{5}{4a+1} = \dfrac{3}{4a+1}$

# Chapter 12 *continued*

**42.** $\dfrac{2}{2x} + \dfrac{12}{x} = \dfrac{2}{2x} + \dfrac{24}{2x}$

$\qquad\qquad = \dfrac{26}{2x}$

$\qquad\qquad = \dfrac{13}{x}$

**43.** $\dfrac{5}{4x} - \dfrac{7}{3x} = \dfrac{15}{12x} - \dfrac{28}{12x}$

$\qquad\qquad = -\dfrac{13}{12x}$

**44.** $\dfrac{2x}{x+1} + \dfrac{5}{x+3} = \dfrac{2x(x+3)}{(x+1)(x+3)} + \dfrac{5(x+1)}{(x+1)(x+3)}$

$\qquad\qquad = \dfrac{2x^2 + 6x + 5x + 5}{(x+1)(x+3)}$

$\qquad\qquad = \dfrac{2x^2 + 11x + 5}{(x+1)(x+3)}$

$\qquad\qquad = \dfrac{(2x+1)(x+5)}{(x+1)(x+3)}$

**45.** $\dfrac{6x}{x+1} - \dfrac{2x+4}{x+1} = \dfrac{6x - 2x - 4}{x+1}$

$\qquad\qquad = \dfrac{4x-4}{x+1}$

$\qquad\qquad = \dfrac{4(x-1)}{x+1}$

## Lesson 12.8

### 12.8 Guided Practice (p. 761)

1. An axiom is a rule accepted without proof; a theorem must be proved.

2. Assume that the opposite of what you want to prove is true.

3. Identity property of multiplication

4. Commutative property of addition

5. Distributive property

6. Associative property of multiplication

7. Identity property of addition

8. Inverse property of addition

9. Inverse property of multiplication

10. Commutative property of multiplication

11. Associative property of addition

12. No, not necessarily; if $b \geq a$, then $a - b$ is not positive.

13.
| | |
|---|---|
| $a = b$ | Given |
| $ac = bc$ | Multiplication axiom of equality |
| $c = d$ | Given |
| $bc = bd$ | Multiplication axiom of equality |
| $ac = bd$ | Substitution Property of equality |

### 12.8 Practice and Applications (pp. 762–763)

14.
| | |
|---|---|
| $(a+b) - b = (a+b) + (-b)$ | Definition of subtraction |
| $(a+b) - b = a + [b + (-b)]$ | Associative property of addition |
| $(a+b) - b = a + 0$ | Inverse property of addition |
| $(a+b) - b = a$ | Identity property of addition |

15.
| | |
|---|---|
| $a - b = a + (-b)$ | Definition of subtraction |
| $a - b = -b + a$ | Commutative property of addition |

16.
| | |
|---|---|
| $(a - b) - c = [a + (-b)] + (-c)$ | Definition of subtraction |
| $\quad = a + [(-b) + (-c)]$ | Associative property of addition |
| $\quad = a + (-1)(b + c)$ | Distributive property |
| $\quad = a - (b + c)$ | Definition of subtraction |

17.
| | |
|---|---|
| $a + (-1)(a) = 1(a) + (-1)(a)$ | Identity property of multiplication |
| $\quad = [1 + (-1)](a)$ | Distributive property |
| $\quad = (0)(a)$ | Inverse property of addition |
| $\quad = 0$ | Multiplication property of 0 |

Since $a + (-1)(a) = 0, (-1)(a) = -a$ by definition

18. No; no number of examples can prove a conjecture is true for all real numbers; the conjecture is true.

19. $(1 + 2)^2 \neq 1^2 + 2^2$

$\quad 3^2 \neq 1 + 4$

$\quad 9 \neq 5$

20. $(12 \div 6) \div 2 \neq 12 \div (6 \div 2)$

$\quad 2 \div 2 \neq 12 \div 3$

$\quad 1 \neq 4$

21. 2 and 3 are integers, but $\frac{2}{3}$ is not an integer.

22. Each block represents a sum of odd integers. $1 = 1^2$, $1 + 3 = 2^2$, $1 + 3 + 5 = 3^2$, $1 + 3 + 5 + 7 = 4^2$, $1 + 3 + 5 + 7 + 9 = 5^2$

23. Yes; this map shows that the proposal is false.

24. Assume that the bus will get you home in time for dinner at 5 P.M. The bus made the 45 mile trip home in half an hour. That means the bus was traveling 90 mi/h, which is impossible because the bus does not travel more than 60 mi/h.

25. Assume that $p$ is an integer, $p^2$ is odd, and $p$ is even. Let $p = 2n$. Then $p^2 = (2n)^2 = 4n^2 = 2(2n^2)$, an even number, which is impossible, since $p^2$ is odd. Then $p$ must be odd.

26. Assume that $a < b$ and $a + c \geq b + c$. Then $(a + c) - (b + c) \geq 0$. But $(a + c) - (b + c) = a - b$, so $a - b \geq 0$, and, since $a < b$, it is impossible for $a - b$ to be greater than or equal to 0. Then $a + c < b + c$.

**Algebra 1**

Chapter 12 Worked-out Solution Key

# Chapter 12 *continued*

**27.** Assume that $ac > bc$, $c > 0$, and $a \le b$. By the Multiplication property of inequality, $ac \le bc$, which is impossible since it was given that $ac > bc$. Then $a > b$.

**28.** $BC = \sqrt{x^2 + y^2}$ from the distance formula. Therefore, $BD = CD = \frac{1}{2}\sqrt{x^2 + y^2}$. Point $D$ would be located at $\left(\frac{1}{2}x, \frac{1}{2}y\right)$. Using the distance formula,

$$AD = \sqrt{\left(-\frac{x}{2}\right)^2 + \left(-\frac{y}{2}\right)^2} = \sqrt{\frac{x^2}{4} + \frac{y^2}{4}} = \frac{1}{2}\sqrt{x^2 + y^2}$$

which is the same as both $BD$ and $CD$.

**29. a.**

| Vertices | 2 | 3 | 4 | 5 | 6 |
|----------|---|---|---|---|----|
| Edges | 1 | 3 | 6 | 10 | 15 |

**b.** A complete graph with $n$ vertices has $\dfrac{n(n-1)}{2}$ edges.

**c.** 45 edges

**30.** *Sample answer:* In the figure, $c$ is the length of the hypotenuse of a right triangle and $a$ and $b$ are the lengths of the legs. The next two figures are squares having the same area. The area of the square on the left is $(a+b)^2$ or $a^2 + 2ab + b^2$. The area of the square on the right can be written as the sum of its parts, $4\left(\frac{1}{2}ab\right) + c^2$ or $2ab + c^2$. So, $a^2 + 2ab + b^2 = 2ab + c^2$, and $a^2 + b^2 = c^2$.

### 12.8 Mixed Review (p. 764)

**31.** $(-2)^2 - 4(4) = -12$
no real solution

**32.** $4^2 - 4(-2)(-2) = 0$
one solution

**33.** $(-8)^2 - 4(8)(2) = 0$
one solution

**34.** $(-14)^2 - 4(49) = 0$
one solution

**35.** $(-5)^2 - 4(-3)(1) = 37$
two solutions

**36.** $(-1)^2 - 4(6)(5) = -119$
no real solution

**37.** $(-2)^2 - 4(-15) = 64$
two solutions

**38.** $16^2 - 4(64) = 0$
one solution

**39.** $(11)^2 - 4(30) = 1$
two solutions

**40.** $1 \overset{?}{<} 1^2 - 2(1) - 5$
$1 \overset{?}{<} 1 - 2 - 5$
$1 \not< -6$
not a solution

**41.** $-2 \overset{?}{\ge} 2(3)^2 - 8(3) + 8$
$-2 \overset{?}{\ge} 2(9) - 24 + 8$
$-2 \overset{?}{\ge} 18 - 24 + 8$
$-2 \overset{?}{\ge} -6 + 8$
$-2 \not\ge 2$
not a solution

**42.** $20 \overset{?}{\le} 2(-2)^2 - 3(-2) + 10$
$20 \overset{?}{\le} 2(4) + 6 + 10$
$20 \overset{?}{\le} 8 + 6 + 10$
$20 \le 24$
solution

**43.** $17 \overset{?}{>} 4(1)^2 - 48(1) + 61$
$17 \overset{?}{>} 4(1) - 48 + 61$
$17 \overset{?}{>} 4 - 48 + 61$
$17 \not> 17$
not a solution

**44.** $-4 \overset{?}{\ge} (-2)^2 + 4(-2)$
$-4 \overset{?}{\ge} 4 - 8$
$-4 \ge -4$
solution

**45.** $10 \overset{?}{<} 3(5)^2 - 2(5)$
$10 \overset{?}{<} 3(25) - 10$
$10 \overset{?}{<} 75 - 10$
$10 < 65$
solution

**46.** $100 \overset{?}{>} 3(-6)^2 + 50(-6) + 500$
$100 \overset{?}{>} 3(36) - 300 + 500$
$100 \overset{?}{>} 108 - 300 + 500$
$100 \not> 308$
not a solution

**47.** $-3 \overset{?}{\ge} -2^2 + 3(2) - \frac{15}{4}$
$-3 \overset{?}{\ge} 4 + 6 - \frac{15}{4}$
$-3 \overset{?}{\ge} 10 - \frac{15}{4}$
$-3 \overset{?}{\ge} \frac{25}{4}$
$-3 \not\ge 6\frac{1}{4}$
not a solution

**48.** $a = 0.15 \cdot 15$
$a = \$2.25$

**49.** $41 = p \cdot 50$
$p = 0.82$
$p = 82\%$

**50.** $100 = 0.01 \cdot b$
$b = 10{,}000$

**51.** $6 = p \cdot 3$
$2 = p$
$p = 200\%$

**52.** $1240 = 0.80 \cdot b$
$b = 1550$

**53.** $5 = 0.33 \cdot b$
$b = \$15.15$

### Quiz 3 (p. 764)

**1.** $d = \sqrt{(1-7)^2 + (3+9)^2}$
$= \sqrt{(-6)^2 + 12^2}$
$= \sqrt{36 + 144}$
$= \sqrt{180}$
$\approx 13.42$

$\left(\dfrac{1+7}{2}, \dfrac{3+(-9)}{2}\right) = \left(\dfrac{8}{2}, \dfrac{-6}{2}\right) = (4, -3)$

# Chapter 12 *continued*

**2.** $d = \sqrt{(-2 - 6)^2 + (-5 + 11)^2}$

$= \sqrt{(-8)^2 + 6^2}$

$= \sqrt{64 + 36}$

$= \sqrt{100}$

$= 10$

$\left(\dfrac{-2 + 6}{2}, \dfrac{-5 + (-11)}{2}\right) = \left(\dfrac{4}{2}, \dfrac{-16}{2}\right) = (2, -8)$

**3.** $d = \sqrt{(0 - 8)^2 + (0 + 14)^2}$

$= \sqrt{(-8)^2 + 14^2}$

$= \sqrt{64 + 196}$

$= \sqrt{260}$

$\approx 16.12$

$\left(\dfrac{0 + 8}{2}, \dfrac{0 - 14}{2}\right) = \left(\dfrac{8}{2}, \dfrac{-14}{2}\right) = (4, -7)$

**4.** $d = \sqrt{(-8 + 8)^2 + (-8 - 8)^2}$

$= \sqrt{0^2 + (-16)^2}$

$= \sqrt{0 + 256}$

$= \sqrt{256}$

$= 16$

$\left(\dfrac{-8 + (-8)}{2}, \dfrac{-8 + 8}{2}\right) = \left(-\dfrac{16}{2}, \dfrac{0}{2}\right) = (-8, 0)$

**5.** $d = \sqrt{(3 + 3)^2 + (4 - 4)^2}$

$= \sqrt{6^2 + 0^2}$

$= \sqrt{36}$

$= 6$

$\left(\dfrac{3 + (-3)}{2}, \dfrac{4 + 4}{2}\right) = \left(\dfrac{0}{2}, \dfrac{8}{2}\right) = (0, 4)$

**6.** $d = \sqrt{(1 + 4)^2 + (7 + 2)^2}$

$= \sqrt{5^2 + 9^2}$

$= \sqrt{25 + 81}$

$= \sqrt{106}$

$\approx 10.30$

$\left(\dfrac{1 + (-4)}{2}, \dfrac{7 + 2}{2}\right) = \left(-\dfrac{3}{2}, \dfrac{5}{2}\right)$

**7.** $d = \sqrt{(2 + 2)^2 + (0 + 3)^2}$

$= \sqrt{4^2 + 3^2}$

$= \sqrt{16 + 9}$

$= \sqrt{25}$

$= 5$

$\left(\dfrac{2 + 2}{2}, \dfrac{0 + 3}{2}\right) = \left(\dfrac{0}{2}, -\dfrac{3}{2}\right) = \left(0, -\dfrac{3}{2}\right)$

**8.** $d = \sqrt{(-3 - 4)^2 + (3 - 1)^2}$

$= \sqrt{(-7)^2 + 2^2}$

$= \sqrt{49 + 4}$

$= \sqrt{53}$

$= 7.28$

$\left(\dfrac{-3 + 4}{2}, \dfrac{3 + 1}{2}\right) = \left(\dfrac{1}{2}, \dfrac{4}{2}\right) = \left(\dfrac{1}{2}, 2\right)$

**9.** $d = \sqrt{(3 - 2)^2 + (4 + 4)^2}$

$= \sqrt{1^2 + 8^2}$

$= \sqrt{1 + 64}$

$= \sqrt{65}$

$\approx 8.06$

$\left(\dfrac{3 + 2}{2}, \dfrac{4 + (-4)}{2}\right) = \left(\dfrac{5}{2}, \dfrac{0}{2}\right) = \left(\dfrac{5}{2}, 0\right)$

**10.** $\sin A = \dfrac{3}{5}$ $\qquad \sin B = \dfrac{4}{5}$

$\cos A = \dfrac{4}{5}$ $\qquad \cos B = \dfrac{3}{5}$

$\tan A = \dfrac{3}{4}$ $\qquad \tan B = \dfrac{4}{3}$

**11.** $\sin A = \dfrac{15}{17}$ $\qquad \sin B = \dfrac{8}{17}$

$\cos A = \dfrac{8}{17}$ $\qquad \cos B = \dfrac{15}{17}$

$\tan A = \dfrac{15}{8}$ $\qquad \tan B = \dfrac{8}{15}$

**12.** $(a - b)c = [a + (-b)]c$     Definition of subtraction

$\qquad\quad = a(c) + (-b)c$     Distributive property

$\qquad\quad = ac + (-bc)$     $(-b)c = -bc$

$\qquad\quad = ac - bc$     Definition of subtraction

**13.** Any real numbers $a$, $b$, and $c$ with $a < b$ and $c \leq 0$
*Sample answer:* $a = 1$, $b = 2$, and $c = -1$: $1 < 2$, and $1(-1) > 2(-1)$

**14.** Any real numbers $a$ and $b$ with $b \neq 0$. *Sample answer:*
$a = 1$ and $b = 2$. $-(1 + 2) \neq (-1) - (-2)$

$$-3 \neq -1 + 2$$

$$-3 \neq 1$$

## Chapter 12 Review *(pp. 766–768)*

**1.** Domain: All nonnegative numbers

Range: All nonnegative numbers

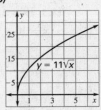

**2.** Domain: All numbers $\geq 5$

Range: All nonnegative numbers

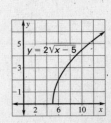

# Chapter 12 *continued*

**3.** Domain: All nonnegative numbers

Range: All numbers $\geq 3$

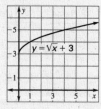

**4.** $\sqrt{5} + 2\sqrt{5} - \sqrt{3} = 3\sqrt{5} - \sqrt{3}$

**5.** $\sqrt{6}(2\sqrt{3} - 4\sqrt{2}) = 2\sqrt{18} - 4\sqrt{12} = 6\sqrt{2} - 8\sqrt{3}$

**6.** $(3 - \sqrt{10})^2 = (3 - \sqrt{10})(3 - \sqrt{10})$

$\qquad = 9 - 3\sqrt{10} - 3\sqrt{10} + 10$

$\qquad = 19 - 6\sqrt{10}$

**7.** $(\sqrt{8} + \sqrt{3})^2 = (\sqrt{8} + \sqrt{3})(\sqrt{8} + \sqrt{3})$

$\qquad = 8 + \sqrt{24} + \sqrt{24} + 3$

$\qquad = 11 + 2\sqrt{24}$

$\qquad = 11 + 4\sqrt{6}$

**8.** $\dfrac{21}{\sqrt{3}} \cdot \dfrac{\sqrt{3}}{\sqrt{3}} = \dfrac{21\sqrt{3}}{3} = 7\sqrt{3}$

**9.** $\dfrac{8}{6 - \sqrt{4}} = \dfrac{8}{6 - 2}$

$\qquad = \dfrac{8}{4}$

$\qquad = 2$

**10.** $2\sqrt{x} - 4 = 0$

$\qquad 2\sqrt{x} = 4$

$\qquad \sqrt{x} = 2$

$\qquad x = 4$

**11.** $x = \sqrt{-4x - 4}$

$x^2 = -4x - 4$

No solution

**12.** $\sqrt{x - 3} + 2 = 8$

$\qquad \sqrt{x - 3} = 6$

$\qquad x - 3 = 36$

$\qquad x = 39$

**13.** $x^2 - 4x - 1 = 7$

$x^2 - 4x + 4 = 8 + 4$

$\qquad (x - 2)^2 = 12$

$\qquad x - 2 = \pm\sqrt{12}$

$\qquad x = 2 + 2\sqrt{3}, 2 - 2\sqrt{3}$

**14.** $x^2 + 20x + 19 = 0$

$x^2 + 20x + 100 = -19 + 100$

$\qquad (x + 10)^2 = 81$

$\qquad x + 10 = \pm 9$

$\qquad x = -1, -19$

**15.** $2x^2 - x - 4 = 10$

$x^2 - \dfrac{1}{2}x - 2 = 5$

$x^2 - \dfrac{1}{2}x + \dfrac{1}{16} = 7 + \dfrac{1}{16}$

$\left(x - \dfrac{1}{4}\right)^2 = 7\dfrac{1}{16}$

$x - \dfrac{1}{4} = \pm\sqrt{\dfrac{113}{16}}$

—CONTINUED—

**15.** —CONTINUED—

$x = \pm\dfrac{\sqrt{113}}{4} + \dfrac{1}{4}$

$x = \dfrac{1 + \sqrt{113}}{4}, \dfrac{1 - \sqrt{113}}{4}$

**16.** $4^2 + 6^2 = c^2$

$16 + 36 = c^2$

$52 = c^2$

$\sqrt{52} = c$

$2\sqrt{13} = c$

**17.** $(\sqrt{2})^2 + b^2 = (\sqrt{3})^2$

$2 + b^2 = 3$

$b^2 = 1$

$b = 1$

**18.** $b^2 + (2b + 2)^2 = 13^2 \qquad b = 5$

$b^2 + 4b^2 + 8b + 4 = 169 \qquad 2b + 2 = 2(5) + 2$

$5b^2 + 8b + 4 = 169 \qquad\qquad = 10 + 2$

$5b^2 + 8b - 165 = 0 \qquad\qquad = 12$

$(5b + 33)(b - 5) = 0$

$5b = -33 \qquad b = 5$

$b = -\dfrac{33}{5}$

Solution: 5, 12

**19.** $d = \sqrt{(8 - 11)^2 + (5 + 4)^2}$

$\quad = \sqrt{(-3)^2 + 9^2}$

$\quad = \sqrt{9 + 81}$

$\quad = \sqrt{90}$

$\quad = 3\sqrt{10}$

$\left(\dfrac{8 + 11}{2}, \dfrac{5 + (-4)}{2}\right) = \left(\dfrac{19}{2}, \dfrac{1}{2}\right)$

**20.** $d = \sqrt{(-3 - 1)^2 + (6 - 7)^2}$

$\quad = \sqrt{(-4)^2 + -1^2}$

$\quad = \sqrt{16 + 1}$

$\quad = \sqrt{17}$

$\left(\dfrac{-3 + 1}{2}, \dfrac{6 + 7}{2}\right) = \left(-\dfrac{2}{2}, \dfrac{13}{2}\right) = \left(-1, \dfrac{13}{2}\right)$

**21.** $d = \sqrt{(-2 - 2)^2 + (-2 - 8)^2}$

$\quad = \sqrt{(-4)^2 + (-10)^2}$

$\quad = \sqrt{16 + 100}$

$\quad = \sqrt{116}$

$\quad = 2\sqrt{29}$

$\left(\dfrac{-2 + 2}{2}, \dfrac{-2 + 8}{2}\right) = \left(\dfrac{0}{2}, \dfrac{6}{2}\right) = (0, 3)$

# Chapter 12 *continued*

**22.** $\sin B = \frac{8}{10} = 0.8$; $\cos B = \frac{6}{10} = 0.6$; $\tan B = \frac{8}{6} = 1.\overline{3}$

**23.** Commutative property of multiplication

**24.** $c(-b) = c((-1) \cdot b)$     Property of opposites

$\phantom{c(-b)} = (c(-1))b$     Associative property of mult.

$\phantom{c(-b)} = -1(c)b$     Commutative property of mult.

$\phantom{c(-b)} = -cb$     Property of opposites

## Chapter 12 Test *(p. 769)*

**1.** Domain: all nonnegative numbers

Range: all nonnegative numbers

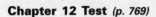

**2.** Domain: all numbers $\geq -\frac{7}{2}$

Range: all nonnegative numbers

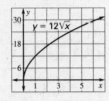

**3.** Domain: all nonnegative numbers

Range: all numbers $\geq -3$

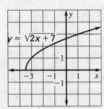

**4.** Domain: all numbers $\geq 5$

Range: all nonnegative numbers

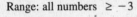
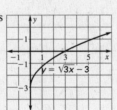

**5.** $3\sqrt{2} - \sqrt{2} = 2\sqrt{2}$

**6.** $\left(4 + \sqrt{7}\right)\left(4 - \sqrt{7}\right) = 16 + 4\sqrt{7} - 4\sqrt{7} - 7 = 9$

**7.** $\left(4\sqrt{5} + 1\right)^2 = \left(4\sqrt{5} + 1\right)\left(4\sqrt{5} + 1\right)$

$\phantom{\left(4\sqrt{5} + 1\right)^2} = 16(5) + 4\sqrt{5} + 4\sqrt{5} + 1$

$\phantom{\left(4\sqrt{5} + 1\right)^2} = 80 + 8\sqrt{5} + 1$

$\phantom{\left(4\sqrt{5} + 1\right)^2} = 81 + 8\sqrt{5}$

**8.** $\dfrac{8}{3 - \sqrt{5}} \cdot \dfrac{3 + \sqrt{5}}{3 + \sqrt{5}} = \dfrac{24 + 8\sqrt{5}}{9 - 5}$

$\phantom{\dfrac{8}{3 - \sqrt{5}}} = \dfrac{24 + 8\sqrt{5}}{4}$

$\phantom{\dfrac{8}{3 - \sqrt{5}}} = \dfrac{8\left(3 + \sqrt{5}\right)}{4}$

$\phantom{\dfrac{8}{3 - \sqrt{5}}} = 2\left(3 + \sqrt{5}\right)$

$\phantom{\dfrac{8}{3 - \sqrt{5}}} = 6 + 2\sqrt{5}$

**9.** 
$$\sqrt{y} + 6 = 10$$
$$\sqrt{y} = 4$$
$$y = 16$$

**10.** 
$$\sqrt{2m + 3} - 6 = 4$$
$$\sqrt{2m + 3} = 10$$
$$2m + 3 = 100$$
$$2m = 97$$
$$m = 48\tfrac{1}{2}$$

**11.** 
$$n = \sqrt{9n - 18}$$
$$n^2 = 9n - 18$$
$$n^2 - 9n + 18 = 0$$
$$(n - 6)(n - 3) = 0$$
$$n = 6, n = 3$$

**12.** 
$$p = \sqrt{-3p + 18}$$
$$p^2 = -3p + 18$$
$$p^2 + 3p - 18 = 0$$
$$(p + 6)(p - 3) = 0$$
$$p = -6 \text{ (extraneous)}, p = 3$$

**13.** 
$$x^2 - 6x = -5$$
$$x^2 - 6x + 9 = -5 + 9$$
$$(x - 3)^2 = 4$$
$$x - 3 = \pm 2$$
$$x = 5, 1$$

**14.** 
$$x^2 - 2x = 2$$
$$x^2 - 2x + 1 = 2 + 1$$
$$(x - 1)^2 = 3$$
$$x - 1 = \pm\sqrt{3}$$
$$x = 1 + \sqrt{3}, 1 - \sqrt{3}$$

**15.** 
$$x^2 + \frac{4}{5}x - 1 = 0$$
$$x^2 + \frac{4}{5}x + \frac{16}{100} = 1 + \frac{16}{100}$$
$$\left(x + \frac{4}{10}\right)^2 = \frac{116}{100}$$
$$x + \frac{4}{10} = \pm\sqrt{\frac{116}{100}}$$
$$x = \pm\frac{\sqrt{116}}{10} - \frac{4}{10}$$
$$x = \frac{-4 + \sqrt{116}}{10}, \frac{-4 - \sqrt{116}}{10}$$
$$x = \frac{-2 + \sqrt{29}}{5}, \frac{-2 - \sqrt{29}}{5}$$

**16.** 
$$6^2 + 18^2 \overset{?}{=} 36^2$$
$$36 + 324 \overset{?}{=} 1296$$
$$360 \neq 1296$$

no

**17.** 
$$9^2 + 40^2 \overset{?}{=} 41^2$$
$$81 + 1600 \overset{?}{=} 1681$$
$$1681 = 1681$$

yes

# Chapter 12 continued

**18.** $(1.5)^2 + (3.6)^2 \overset{?}{=} (3.9)^2$

$\qquad 2.25 + 12.96 \overset{?}{=} 15.21$

$\qquad\qquad\qquad 15.21 = 15.21$

yes

**19.** $d_1 = \sqrt{(-3+3)^2 + (1-5)^2}$

$\qquad = \sqrt{0^2 + (-4)^2}$

$\qquad = \sqrt{0 + 16}$

$\qquad = \sqrt{16}$

$\qquad = 4$

$d_2 = \sqrt{(-3-2)^2 + (5-7)^2}$

$\qquad = \sqrt{(-5)^2 + (-2)^2}$

$\qquad = \sqrt{25 + 4}$

$\qquad = \sqrt{29}$

$d_3 = \sqrt{(2-2)^2 + (7-3)^2}$

$\qquad = \sqrt{0^2 + 4^2}$

$\qquad = \sqrt{16}$

$\qquad = 4$

$d_4 = \sqrt{(2+3)^2 + (3-1)^2}$

$\qquad = \sqrt{5^2 + 2^2}$

$\qquad = \sqrt{25 + 4}$

$\qquad = \sqrt{29}$

$4 + \sqrt{29} + 4 + \sqrt{29} = 8 + 2\sqrt{29} \approx 18.77$

**20.** $\left(\dfrac{-3 + (-3)}{2}, \dfrac{1+5}{2}\right) = \left(\dfrac{-6}{2}, \dfrac{6}{2}\right) = (-3, 3)$

$\left(\dfrac{-3+2}{2}, \dfrac{5+7}{2}\right) = \left(\dfrac{-1}{2}, \dfrac{12}{2}\right) = \left(-\dfrac{1}{2}, 6\right)$

$\left(\dfrac{2+2}{2}, \dfrac{7+3}{2}\right) = \left(\dfrac{4}{2}, \dfrac{10}{2}\right) = (2, 5)$

$\left(\dfrac{2+(-3)}{2}, \dfrac{3+1}{2}\right) = \left(\dfrac{-1}{2}, \dfrac{4}{2}\right) = \left(-\dfrac{1}{2}, 2\right)$

**21.** $d_1 = \sqrt{\left(-3 + \dfrac{1}{2}\right)^2 + (3-6)^2}$

$\qquad = \sqrt{\left(-2\dfrac{1}{2}\right)^2 + (-3)^2}$

$\qquad = \sqrt{\dfrac{25}{4} + 9}$

$\qquad = \sqrt{\dfrac{61}{4}}$

$\qquad = \dfrac{\sqrt{61}}{2}$

$d_2 = \sqrt{\left(-\dfrac{1}{2} - 2\right)^2 + (6-5)^2}$

$\qquad = \sqrt{\left(-2\dfrac{1}{2}\right)^2 + 1^2}$

$\qquad = \sqrt{\dfrac{25}{4} + 1}$

$\qquad = \sqrt{\dfrac{29}{4}}$

$\qquad = \dfrac{\sqrt{29}}{2}$

$d_3 = \sqrt{\left(2 + \dfrac{1}{2}\right)^2 + (5-2)^2}$

$\qquad = \sqrt{\left(2\dfrac{1}{2}\right)^2 + 3^2}$

$\qquad = \sqrt{\dfrac{25}{4} + 9}$

$\qquad = \sqrt{\dfrac{61}{4}}$

$\qquad = \dfrac{\sqrt{61}}{2}$

$d_4 = \sqrt{\left(-\dfrac{1}{2} + 3\right)^2 + (2-3)^2}$

$\qquad = \sqrt{\left(-2\dfrac{1}{2}\right)^2 + (-1)^2}$

$\qquad = \sqrt{\dfrac{25}{4} + 1}$

$\qquad = \sqrt{\dfrac{29}{4}}$

$\qquad = \dfrac{\sqrt{29}}{2}$

$\qquad = \dfrac{2\sqrt{61} + 2\sqrt{29}}{2} = \sqrt{61} + \sqrt{29} \approx 13.20$

**22.** The perimeter of the original parallelogram is greater than the perimeter of the new parellelogram.

**23.** $\sin 45° = \dfrac{10}{x}$ $\qquad\qquad$ $\tan 45° = \dfrac{10}{b}$

$\qquad x = \dfrac{10}{\sin 45°}$ $\qquad\qquad$ $b = \dfrac{10}{\tan 45°}$

$\qquad x = 14.14$ $\qquad\qquad\qquad$ $b = 10$

**24.** $\sin 33° = \dfrac{x}{16}$ $\qquad\qquad$ $\cos 33° = \dfrac{y}{16}$

$\qquad 8.71 = x$ $\qquad\qquad\qquad$ $13.42 = y$

**25.** $\sin 15° = \dfrac{4}{x}$ $\qquad\qquad$ $\tan 15° = \dfrac{4}{y}$

$\qquad x = \dfrac{4}{\sin 15°}$ $\qquad\qquad$ $y = \dfrac{4}{\tan 15°}$

$\qquad x = 15.45$ $\qquad\qquad\qquad$ $y = 14.93$

**Algebra 1**
Chapter 12 Worked-out Solution Key

# Chapter 12 *continued*

**26.** $(a + c) + (-c) = (b + c) + (-c)$    Addition property
of equality

$a + [c + (-c)] = b + [c + (-c)]$    Associative property
of addition

$a + 0 = b + 0$    Inverse property of
addition

$a = b$    Identity property of
addition

**27.** $\left(\dfrac{39.2 + 39.2}{2}, \dfrac{119 + 118.5}{2}\right) = (39.2 \text{ N}, 118.75 \text{ W})$

**28.**

$\tan 78° = \dfrac{x}{134}$

$x \approx 630.42$ ft

## Chapter 12 Standardized Test *(pp. 770–771)*

**1.** C $f(x) = \dfrac{x\sqrt{x^2 - 1}}{x^2 + 8}$

$= \dfrac{8\sqrt{64 - 1}}{64 + 8}$

$= \dfrac{8\sqrt{63}}{72}$

$= \dfrac{\sqrt{63}}{9}$

$= \dfrac{3\sqrt{7}}{9}$

$= \dfrac{\sqrt{7}}{3}$

**2.** B

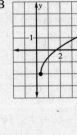

**3.** A $5\sqrt{7} + \sqrt{448} + \sqrt{175} - \sqrt{63}$

$= 5\sqrt{7} + 8\sqrt{7} + 5\sqrt{7} - 3\sqrt{7}$

$= 15\sqrt{7}$

**4.** D $\left(3 - \sqrt{6}\right)^2 = \left(3 - \sqrt{6}\right)\left(3 - \sqrt{6}\right)$

$= 9 - 3\sqrt{6} - 3\sqrt{6} + 6$

$= 15 - 6\sqrt{6}$

**5.** B   $6 = \sqrt{144a}$

$36 = 144a$

$\dfrac{1}{4} = a$

**6.** C    $x = \sqrt{880 - 18x}$

$x^2 = 880 - 18x$

$x^2 + 18x - 880 = 0$

$(x + 40)(x - 22) = 0$

$x = -40 \quad x = 22$

**7.** C $\dfrac{4}{3} \cdot \dfrac{1}{2} = \dfrac{4}{6}$

$\left(\dfrac{4}{6}\right)^2 = \dfrac{16}{36}$

$= \dfrac{4}{9}$

**8.** D $9^2 + x^2 = \left(5\sqrt{10}\right)^2$

$81 + x^2 = 25(10)$

$x^2 = 250 - 81$

$x^2 = 169$

$x = 13$

**9.** E $PQ = \sqrt{(-5 - 2)^2 + (6 - 2)^2}$

$= \sqrt{(-7)^2 + 4^2}$

$= \sqrt{49 + 16}$

$= \sqrt{65}$

**10.** A $\left(\dfrac{-5 + 2}{2}, \dfrac{6 + 2}{2}\right)$

$\left(-\dfrac{3}{2}, \dfrac{8}{2}\right)$

$\left(-\dfrac{3}{2}, 4\right)$

**11.** C $\tan A = \dfrac{18}{b}$

$\dfrac{9}{5} = \dfrac{18}{b}$

$b = \dfrac{18}{1} \cdot \dfrac{5}{9}$

$b = 10$

$c^2 = 18^2 + 10^2$

$c^2 = 324 + 100$

$c^2 = 424$

$c = \sqrt{424} = 2\sqrt{106}$

**12.** B $\sin D = \dfrac{3}{3\sqrt{5}}$

$= \dfrac{1}{\sqrt{5}}$

**13.** A $\cos D = \dfrac{6}{3\sqrt{5}}$    $\sin D = \dfrac{3}{3\sqrt{5}}$

$= \dfrac{2}{\sqrt{5}}$    $= \dfrac{1}{\sqrt{5}}$

**14.** A $\cos D = \dfrac{6}{3\sqrt{5}}$    $\cos E = \dfrac{3}{3\sqrt{5}}$

$= \dfrac{2}{\sqrt{5}}$    $= \dfrac{1}{\sqrt{5}}$

**15.** A $\sin E = \dfrac{6}{3\sqrt{5}}$    $\cos E = \dfrac{3}{3\sqrt{5}}$

$= \dfrac{2}{\sqrt{5}}$    $= \dfrac{1}{\sqrt{5}}$

**16.** D Distributive property

**17. a.** $w^2 + 20^2 = 25^2$    **b.** $15(2) = 30$

$w^2 = 625 - 400$    $20(2) = 40$

$w^2 = 225$    30 in. by 40 in.

$w = 15$ in.

**c.** $15 \times 20 = 300$ in.$^2$

$30 \times 40 = 1200$ in.$^2$

**d.** 4 times

**e.** 9 times

Copyright © McDougal Littell Inc.
All rights reserved.

**Algebra 1**
Chapter 12 Worked-out Solution Key

**335**

# Chapter 12 *continued*

## Cumulative Practice, Chs. 1–12 (pp. 772–773)

**1.** $\dfrac{m}{7} \geq 16$

$m \geq 112$

**2.** $4 + b^2 = 104$

$b^2 = 100$

$b = -10, 10$

**3.** $t = 3\frac{2}{3}(3) = \frac{11}{3}(3) = 11$ mi

**4.** $3 + x + (-4) = 3 + 5 + (-4) = 4$

**5.** $-x + 12 - 5 = -9 + 12 - 5 = -2$

**6.** $3.5 - (-x) = 3.5 - (-1.5) = 5$

**7.** $-(-3)^2(x) = -(9)(7) = -63$

**8.** $6x(x + 2) = 6(2)(2 + 2) = 12(4) = 48$

**9.** $(8x + 1)(-3) = \left[8\left(\frac{1}{2}\right) + 1\right](-3)$

$= (4 + 1)(-3)$

$= -15$

**10.** $\frac{1}{4}|(-x)(-x)(-x)| = \frac{1}{4}|(-4)(-4)(-4)|$

$= \frac{1}{4}|-64|$

$= \frac{1}{4}(64)$

$= 16$

**11.** $\dfrac{x^2 + 4}{6} = \dfrac{8^2 + 4}{6} = \dfrac{64 + 4}{6} = \dfrac{68}{6} = 11\frac{1}{3}$

**12.** $(-5)\left(-\dfrac{3}{4}x\right) = (-5)\left[-\dfrac{3}{4}(6)\right]$

$= (-5)\left(-\dfrac{18}{4}\right)$

$= \dfrac{90}{4}$

$= 22\frac{1}{2}$

**13.** $-\frac{2}{9}(x - 5) = 12$

$x - 5 = -54$

$x = -49$

**14.** $7x - (3x - 2) = 38$

$7x - 3x + 2 = 38$

$4x = 36$

$x = 9$

**15.** $\frac{1}{3}x + 7 = -7x - 5$

$7\frac{1}{3}x = -12$

$x = -1\frac{7}{11}$

**16.** $8(x + 3) - 2x = 4(x - 8)$

$8x + 24 - 2x = 4x - 32$

$6x + 24 = 4x - 32$

$2x = -56$

$x = -28$

**17.** $11.47 + 6.23x = 7.62 + 5.51x$

$0.72x = -3.85$

$x \approx -5.35$

**18.** $-3(2.98 - 4.1x) = 9.2x + 6.25$

$-8.94 + 12.3x = 9.2x + 6.25$

$3.1x = 15.19$

$x = 4.9$

**19.** $m = \dfrac{1 + 3}{3 + 3} = \dfrac{4}{6} = \dfrac{2}{3}$

$y = \dfrac{2}{3}x + 1$

**20.** $m = -\dfrac{3}{2}$

$y = -\dfrac{3}{2}x - 1$

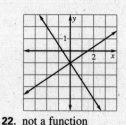

**21.** function

domain: $-1, 1, 3, 5, 7$

range: $-1, 1, 3, 5$

**22.** not a function

**23.** function

domain: $-2, -1, 0, 1, 2$

range: $-2, -1, 0, 1$

**24.** not a function

**25.** $y = \frac{4}{5}x - 3$

$5y - 4x = -15$

$4x - 5y = 15$

**26.** $y - 2 = \frac{1}{3}(x + 1)$

$y - 2 = \frac{1}{3}x + \frac{1}{3}$

$y = \frac{1}{3}x + 2\frac{1}{3}$

$3y = x + 7$

$x - 3y = -7$

**27.** $-3 < -4x + 9 \leq 14$

$-12 < -4x \leq 5$

$-\frac{5}{4} \leq x < 3$

```
        -5/4
   ●━━━━┫━━━━━━━━━━━○
   -1   0   1   2   3
```

**28.** $|3x + 16| + 2 < 10$

$|3x + 16| < 8$

| | |
|---|---|
| $3x + 16 < 8$ | $3x + 16 > -8$ |
| $3x < -8$ | $3x > -24$ |
| $x < -\frac{8}{3}$ | $x > -8$ |

$-2\frac{2}{3} > x > -8$

```
   ○━━━━┿━━━━┿━━━━○━┿━━┿
  -8   -6   -4   -2   0
```

# Chapter 12 *continued*

**29.** $3x - 4 > 5$ or $5x + 1 < 11$

$3x > 9$ or $5x < 10$

$x > 3$ or $x < 2$

```
    +--+--+--○--○--+--+--
   -1  0  1  2  3  4  5
```

**30.** $4y = -8x + 16$

$-2(2y = 11x - 7)$

$4y + 8x = 16$

$-4y + 22x = 14$

$\overline{\phantom{-4y + 2}30x = 30}$

$x = 1$

$4y + 8(1) = 16$

$4y = 8$

$y = 2$

$(1, 2)$

**31.** $5(-2x + 3y = 15)$

$10x - 11y = 9$

$-10x + 15y = 75$

$\underline{10x - 11y = \phantom{0}9}$

$4y = 84$

$y = 21$

$-10x + 15(21) = 75$

$-10x = -240$

$x = 24$

$(24, 21)$

**32.** $-7(y = 5x - 2)$

$3x + 7y = 5$

$-7y + 35x = 14$

$\underline{7y + 3x = \phantom{0}5}$

$38x = 19$

$x = \frac{1}{2}$

$y = 5\left(\frac{1}{2}\right) - 2$

$= \frac{5}{2} - 2$

$= \frac{5}{2} - \frac{4}{2}$

$= \frac{1}{2}$

$\left(\frac{1}{2}, \frac{1}{2}\right)$

**33.** $\dfrac{b^8}{b^2} = b^{8-2} = b^6$

$b^6 = 2^6 = 64$

**34.** $3a^4 \cdot a^{-3} = \dfrac{3a^4}{a^3}$

$= 3a$

$= 3(1)$

$= 3$

**35.** $(-a^3)(2b^2)^3 = (-a^3)(8b^6)$

$= -8a^3b^6$

$= -8(1^3)(2^6)$

$= -8(1)(64)$

$= -512$

**36.** $4b^3 \cdot (2 + b)^2 = 4b^3 \cdot (4 + 4b + b^2)$

$= 16b^3 + 16b^4 + 4b^5$

$= 16(2^3) + 16(2^4) + 4(2^5)$

$= 16(8) + 16(16) + 4(32)$

$= 128 + 256 + 128$

$= 512$

**37.** $\dfrac{4a^{-3}b^3}{ab^{-2}} = \dfrac{4b^3b^2}{a^3a}$

$= \dfrac{4b^5}{a^4}$

$= \dfrac{4(2^5)}{(1^4)}$

$= \dfrac{4(32)}{1}$

$= 128$

**38.** $\dfrac{(5ab^2)^{-2}}{a^{-3}b} = \dfrac{a^3}{25a^2b^4b}$

$= \dfrac{a}{25b^5}$

$= \dfrac{1}{25(2^5)}$

$= \dfrac{1}{25(32)}$

$= \dfrac{1}{800}$

**39.** two solutions

$6x^2 + 8 = 34$

$6x^2 = 26$

$x^2 = \frac{26}{6}$

$x = \pm \sqrt{\frac{26}{6}}$

$= \pm \dfrac{\sqrt{26}}{\sqrt{6}} \cdot \dfrac{\sqrt{6}}{\sqrt{6}}$

$= \pm \dfrac{\sqrt{156}}{6}$

$= \dfrac{2\sqrt{39}}{6}, -\dfrac{2\sqrt{39}}{6}$

$= \dfrac{\sqrt{39}}{3}, -\dfrac{\sqrt{39}}{3}$

**40.** two solutions

$4x^2 - 9x + 5 = 0$

$(x - 1)(4x - 5) = 0$

$x = 1$    $4x = 5$

$x = \frac{5}{4} = 1\frac{1}{4}$

**41.** one solution

$3x^2 + 6x + 3 = 0$

$x^2 + 2x + 1 = 0$

$(x + 1)^2 = 0$

$x = -1$

**42.** $x^2 + 6x + 8 = (x + 4)(x + 2)$

**43.** $x^2 - 24x - 112 = (x - 28)(x + 4)$

**44.** $3x^2 + 17x - 6 = (3x - 1)(x + 6)$

**45.** $4x^2 + 12x + 9 = (2x + 3)(2x + 3) = (2x + 3)^2$

**46.** $x^2 + 10x + 25 = (x + 5)^2$

**47.** $x^2 - 14x + 49 = (x - 7)^2$

**48.** $(3x + 1)(2x + 7) = 0$

$3x + 1 = 0$    $2x + 7 = 0$

$3x = -1$    $2x = -7$

$x = -\frac{1}{3}$    $x = -3\frac{1}{2}$

**49.** $6x^2 - x - 7 = 8$

$6x^2 - x - 15 = 0$

$(2x + 3)(3x - 5) = 0$

$2x + 3 = 0$    $3x - 5 = 0$

$2x = -3$    $3x = 5$

$x = -1\frac{1}{2}$    $x = 1\frac{2}{3}$

# Chapter 12 *continued*

**50.** $4x^2 + 16x + 16 = 0$

$\quad x^2 + 4x + 4 = 0$

$\quad\quad (x + 2)^2 = 0$

$x + 2 = 0$

$x = -2$

**51.** $\quad x^3 + 5x^2 - 4x - 20 = 0$

$\quad (x^3 + 5x^2) - (4x + 20) = 0$

$\quad\quad x^2(x + 5) - 4(x + 5) = 0$

$\quad\quad\quad (x^2 - 4)(x + 5) = 0$

$\quad (x + 2)(x - 2)(x + 5) = 0$

$\quad x = -2, 2, -5$

**52.** $\quad x^4 + 9x^3 + 18x^2 = 0$

$\quad x^2(x^2 + 9x + 18) = 0$

$\quad x^2(x + 6)(x + 3) = 0$

$\quad x = 0, -6, -3$

**53.** $x^2 - \dfrac{4}{3}x + \dfrac{4}{9} = 0$

$\left(x - \dfrac{2}{3}\right)^2 = 0$

$x = \dfrac{2}{3}$

**54.** $\dfrac{4x}{12x^2} = \dfrac{1}{3x}$

**55.** $\dfrac{2x + 6}{x^2 - 9} = \dfrac{2(x + 3)}{(x - 3)(x + 3)} = \dfrac{2}{x - 3}$

**56.** $\dfrac{3x}{x^2 - 2x - 24} \cdot \dfrac{x - 6}{6x^2 + 9x}$

$\quad = \dfrac{3x}{(x - 6)(x + 4)} \cdot \dfrac{x - 6}{3x(2x + 3)}$

$\quad = \dfrac{1}{(x + 4)(2x + 3)}$

**57.** $\dfrac{x^2 - 6x + 8}{x^2 - 2x} \div (3x - 12)$

$\quad = \dfrac{(x - 4)(x - 2)}{x(x - 2)} \cdot \dfrac{1}{3(x - 4)}$

$\quad = \dfrac{1}{3x}$

**58.** $\dfrac{4}{x + 2} + \dfrac{15x}{3x + 6}$

$\quad = \dfrac{4}{x + 2} + \dfrac{15x}{3(x + 2)}$

$\quad = \dfrac{12 + 15x}{3(x + 2)}$

$\quad = \dfrac{3(4 + 5x)}{3(x + 2)} = \dfrac{5x + 4}{x + 2}$

**59.** $\dfrac{3x}{x + 4} - \dfrac{x}{x - 1} = \dfrac{3x(x - 1) - x(x + 4)}{(x + 4)(x - 1)}$

$\quad = \dfrac{3x^2 - 3x - x^2 - 4x}{(x + 4)(x - 1)} = \dfrac{2x^2 - 7x}{(x + 4)(x - 1)}$

$\quad = \dfrac{x(2x - 7)}{(x + 4)(x - 1)}$

**60.** $4\sqrt{7} + 3\sqrt{7} = 7\sqrt{7}$

**61.** $9\sqrt{2} - 12\sqrt{8}$

$\quad = 9\sqrt{2} - 12\sqrt{4}\sqrt{2}$

$\quad = 9\sqrt{2} - 24\sqrt{2}$

$\quad = -15\sqrt{2}$

**62.** $\sqrt{6}(5\sqrt{3} + 6)$

$\quad = 5\sqrt{18} + 6\sqrt{6}$

$\quad = 15\sqrt{2} + 6\sqrt{6}$

**63.** $\dfrac{11}{7 - \sqrt{3}} \cdot \dfrac{7 + \sqrt{3}}{7 + \sqrt{3}} = \dfrac{77 + 11\sqrt{3}}{49 - 3}$

$\quad = \dfrac{77 + 11\sqrt{3}}{46}$

**64.** $\sin Q = \dfrac{12}{13} \approx 0.92 \quad \sin R = \dfrac{5}{13} \approx 0.38$

$\cos Q = \dfrac{5}{13} \approx 0.38 \quad \cos R = \dfrac{12}{13} \approx 0.92$

$\tan Q = \dfrac{12}{5} = 2.4 \quad\; \tan R = \dfrac{5}{12} \approx 0.42$

**65.** $\sin 22° = \dfrac{8}{c} \quad\quad\quad \tan 22° = \dfrac{8}{b}$

$\quad c = \dfrac{8}{\sin 22°} \quad\quad\; b = \dfrac{8}{\tan 22°}$

$\quad c \approx 21.36 \quad\quad\quad b \approx 19.80$

**66.**

| | Comp. A | Comp. B |
|---|---|---|
| PVR | 700 | 500 |
| 1 yr | 60 | 240 |
| 3yr | 180 | 720 |

Comp. A $760, $880

Comp. B $740, $1220

**67.** $5x + 700 = 20x + 500$

$\quad\quad 200 = 15x$

$\quad\quad 13\frac{1}{3} = x$

$x = 13\frac{1}{3}$ months

**Algebra 1**
Chapter 12 Worked-out Solution Key

# Chapter 12 *continued*

## Project, Chs. 10–12 (pp. 774–775)

**1.**

| length, b | 3 | 5 | 8 | 13 | 21 | 34 |
|-----------|---|---|---|----|----|----|
| width, a  | 2 | 3 | 5 | 8  | 13 | 21 |

**2.**

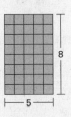

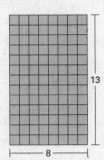

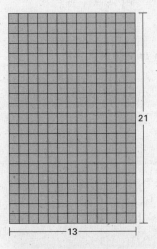

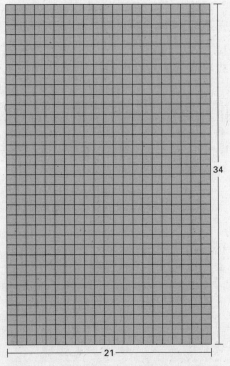

**3.** $\dfrac{\text{length}}{\text{width}}$: 1.5, 1.$\overline{6}$, 1.625, about 1.62, about 1.62

*Sample Answer:* the ratios get closer and closer to the golden ratio

**4.** $b = \dfrac{b + 1}{b}$ or $b^2 = b + 1$; using the quadratic formula and noting that $b$ (which is a length) must be positive, $b = \dfrac{1 + \sqrt{5}}{2}$.

# SKILLS REVIEW HANDBOOK

## Factors and Multiples *(p. 778)*

**1.** 18: 1, 2, 3, 6, 9, 18
**2.** 10: 1, 2, 5, 10
**3.** 77: 1, 7, 11, 77
**4.** 35: 1, 5, 7, 35
**5.** 27: 1, 3, 9, 27
**6.** 100: 1, 2, 4, 5, 10, 20, 25, 50, 100
**7.** 42: 1, 2, 3, 6, 7, 14, 21, 42
**8.** 49: 1, 7, 49
**9.** 52: 1, 2, 4, 13, 26, 52
**10.** 81: 1, 3, 9, 27, 81
**11.** 121: 1, 11, 121
**12.** 150: 1, 2, 3, 5, 6, 10, 15, 25, 30, 50, 75, 150
**13.** $3^3$
**14.** $2^3 \cdot 3$
**15.** $2^5$
**16.** Prime
**17.** $5 \cdot 11$
**18.** $2^2 \cdot 17$
**19.** $2^4 \cdot 3$
**20.** $3^2 \cdot 5^2$
**21.** $2 \cdot 3^2 \cdot 5$
**22.** $3 \cdot 5^2$
**23.** $3 \cdot 13$
**24.** $2^3 \cdot 5^3$
**25.** $2^3 \cdot 7 \cdot 13$
**26.** Prime
**27.** $2^9$
**28.** $2 \cdot 3 \cdot 5 \cdot 7$
**29.** 1, 3, 5, 15
**30.** 1, 2, 3, 6, 9, 18
**31.** 1, 5
**32.** 1, 7
**33.** 1, 3, 9
**34.** 1, 2, 4
**35.** 1, 5
**36.** 1, 2, 3, 6
**37.** 5
**38.** 8
**39.** 1
**40.** 5
**41.** 14
**42.** 13
**43.** 51
**44.** 8
**45.** 2
**46.** 45
**47.** 29
**48.** 1
**49.** 14
**50.** 11
**51.** 1
**52.** 31
**53.** 35
**54.** 12
**55.** 48
**56.** 84
**57.** 12
**58.** 45
**59.** 420
**60.** 42
**61.** 100
**62.** 120
**63.** 51
**64.** 120
**65.** 70
**66.** 900
**67.** 330

## Comparing and Ordering Numbers *(p. 780)*

**1.** 12,428 < 15,116
**2.** 905 < 961
**3.** 142,109 > 140,999
**4.** 16.82 > 14.09
**5.** 0.00456 < 0.40506
**6.** 23.03 < 23.3
**7.** 1005.2 < 1050.7
**8.** 932,778 < 934,112
**9.** 0.058 < 0.102
**10.** $\frac{7}{13} > \frac{3}{13}$
**11.** $17\frac{1}{4} > 15\frac{11}{12}$
**12.** $\frac{7}{10} < \frac{3}{4}$
**13.** $\frac{5}{9} = \frac{15}{27}$
**14.** $\frac{1}{2} > \frac{3}{8}$
**15.** $\frac{1}{8} > \frac{1}{9}$
**16.** $\frac{4}{5} > \frac{2}{3}$

**17.** $42\frac{1}{5} > 41\frac{7}{8}$
**18.** 508.881 > 508.793
**19.** 32,227 > 32,226.5
**20.** $\frac{5}{8} < \frac{2}{3}$
**21.** $17\frac{5}{6} > 17\frac{5}{7}$
**22.** 1207, 1220, 1702, 1772
**23.** 40,071, 40,099, 45,242, 45,617
**24.** 23.08, 23.12, 23.5, 24.0, 24.01
**25.** 9.003, 9.027, 9.10, 9.27, 9.3
**26.** 4.01, 4.07, 4.22, 4.5
**27.** $\frac{1}{3}, \frac{3}{8}, \frac{5}{6}, \frac{5}{4}$
**28.** $\frac{3}{10}, \frac{3}{7}, \frac{3}{5}, \frac{3}{4}, \frac{3}{2}$
**29.** $\frac{15}{16}, 1\frac{1}{8}, 1\frac{2}{5}, \frac{5}{3}, \frac{7}{4}$
**30.** $14\frac{7}{9}, 14\frac{5}{6}, 15\frac{1}{4}, 15\frac{1}{3}$
**31.** $\frac{5}{12}, \frac{7}{8}, \frac{5}{4}, 1\frac{1}{3}$
**32.** Yes, $6\frac{3}{4} > 6\frac{5}{8}$
**33.** Goran, 53.56 < 53.78

## Fraction Operations *(p. 783)*

**1.** $\frac{1}{6} + \frac{4}{6} = \frac{5}{6}$
**2.** $\frac{5}{8} - \frac{3}{8} = \frac{2}{8} = \frac{1}{4}$
**3.** $\frac{4}{9} - \frac{1}{9} = \frac{3}{9} = \frac{1}{3}$
**4.** $\frac{5}{12} + \frac{3}{12} = \frac{8}{12} = \frac{2}{3}$
**5.** $\frac{1}{2} + \frac{1}{8} = \frac{4}{8} + \frac{1}{8} = \frac{5}{8}$
**6.** $\frac{3}{5} - \frac{1}{10} = \frac{6}{10} - \frac{1}{10} = \frac{5}{10} = \frac{1}{2}$
**7.** $\frac{7}{10} + \frac{1}{3} = \frac{21}{30} + \frac{10}{30} = \frac{31}{30} = 1\frac{1}{30}$
**8.** $\frac{15}{24} - \frac{7}{12} = \frac{15}{24} - \frac{14}{24} = \frac{1}{24}$
**9.** $5\frac{1}{8} - 2\frac{3}{4} = 5\frac{1}{8} - 2\frac{6}{8} = 2\frac{3}{8}$
**10.** $1\frac{3}{7} + \frac{1}{2} = 1\frac{6}{14} + \frac{7}{14} = 1\frac{13}{14}$
**11.** $4\frac{3}{8} - 2\frac{5}{6} = 4\frac{9}{24} - 2\frac{20}{24} = 1\frac{13}{24}$
**12.** $\frac{3}{7} + \frac{3}{4} = \frac{12}{28} + \frac{21}{28} = \frac{33}{28} = 1\frac{5}{28}$
**13.** $7\frac{1}{2} + \frac{7}{10} = 7\frac{5}{10} + \frac{7}{10} = 7\frac{12}{10} = 8\frac{2}{10} = 8\frac{1}{5}$
**14.** $5\frac{5}{9} - 2\frac{1}{3} = 5\frac{5}{9} - 2\frac{3}{9} = 3\frac{2}{9}$
**15.** $4\frac{5}{8} - 1\frac{3}{16} = 4\frac{10}{16} - 1\frac{3}{16} = 3\frac{7}{16}$
**16.** $9\frac{2}{5} + 3\frac{1}{3} = 9\frac{6}{15} + 3\frac{5}{15} = 12\frac{11}{15}$
**17.** $\frac{1}{7}$
**18.** 14
**19.** $\frac{12}{7} = 1\frac{5}{7}$
**20.** $\frac{8}{5} = 1\frac{3}{5}$
**21.** 20
**22.** $\frac{1}{100}$
**23.** $\frac{13}{5} = 1\frac{8}{5} = 2\frac{3}{5}$
**24.** $\frac{7}{6} = 1\frac{1}{6}$
**25.** $\frac{5}{6}$
**26.** $\frac{5}{13}$
**27.** $\frac{9}{3} = 3$
**28.** $\frac{17}{12} = 1\frac{5}{12}$
**29.** $\frac{5}{32}$
**30.** $\frac{3}{31}$
**31.** $\frac{7}{2} = 3\frac{1}{2}$
**32.** $\frac{4}{19}$
**33.** $\frac{1}{2} \cdot \frac{1}{2} = \frac{1}{4}$
**34.** $\frac{2}{3} \cdot \frac{4}{5} = \frac{8}{15}$
**35.** $\frac{1\overset{6}{\cancel{6}}}{28} \cdot \frac{4\overset{1}{\cancel{4}}}{15\,\cancel{3}} = \frac{1}{6}$
**36.** $\frac{1\overset{6}{\cancel{6}}}{17} \cdot \frac{7\overset{1}{\cancel{1}}}{9\,\cancel{3}} = \frac{1}{3}$
**37.** $\frac{1\overset{6}{\cancel{6}}}{14} \cdot \frac{8\overset{2}{\cancel{2}}}{9\,\cancel{3}} = \frac{2}{3}$
**38.** $\frac{1\overset{6}{\cancel{6}}}{15} \cdot \frac{5\overset{1}{\cancel{1}}}{2} = \frac{1}{1} = 1$
**39.** $\frac{1\overset{6}{\cancel{6}}}{1} \cdot \frac{23}{9\,\cancel{3}} = \frac{23}{3} = 7\frac{2}{3}$
**40.** $\frac{3\overset{21}{\cancel{21}}}{14} \cdot \frac{8\overset{2}{\cancel{2}}}{7\,\cancel{1}} = \frac{6}{1} = 6$
**41.** $\frac{7}{2\,\cancel{8}} \cdot \frac{4\overset{1}{\cancel{1}}}{3} = \frac{7}{6} = 1\frac{1}{6}$
**42.** $\frac{5}{6\,\cancel{12}} \cdot \frac{2\overset{1}{\cancel{1}}}{1} = \frac{5}{6}$
**43.** $\frac{2\overset{4}{\cancel{4}}}{5} \cdot \frac{3}{2\,\cancel{1}} = \frac{6}{5} = 1\frac{1}{5}$
**44.** $\frac{11}{8\,\cancel{16}} \cdot \frac{2\overset{1}{\cancel{1}}}{3} = \frac{11}{24}$

# Skills Review Handbook *continued*

**45.** $\frac{3\cancel{9}}{1\cancel{2}} \cdot \frac{\cancel{2}}{\cancel{3}1} = \frac{6}{1} = 6$   **46.** $\frac{9}{4} \cdot \frac{3}{4} = \frac{27}{16} = 1\frac{11}{16}$

**47.** $\frac{17}{5} \cdot \frac{1}{4} = \frac{17}{20}$   **48.** $\frac{4\cancel{36}}{5} \cdot \frac{4}{\cancel{9}1} = \frac{16}{5} = 3\frac{1}{5}$

**49.** $\frac{15}{16} - \frac{1}{8} = \frac{15}{16} - \frac{2}{16} = \frac{13}{16}$   **50.** $\frac{5}{3\cancel{9}} \cdot \frac{\cancel{3}1}{2} = \frac{5}{6}$

**51.** $1\frac{\cancel{12}}{\cancel{13}} \cdot \frac{\cancel{13}1}{\cancel{12}1} = \frac{1}{1} = 1$

**52.** $\frac{24}{25} + \frac{1}{5} = \frac{24}{25} + \frac{5}{25} = \frac{29}{25} = 1\frac{4}{25}$

**53.** $5\frac{1}{2} - \frac{1}{8} = 5\frac{4}{8} - \frac{1}{8} = 5\frac{3}{8}$   **54.** $\frac{3}{2\cancel{10}} \cdot \frac{\cancel{5}1}{1} = \frac{3}{2} = 1\frac{1}{2}$

**55.** $\frac{7}{28} \cdot \frac{\cancel{4}1}{9} = \frac{7}{18}$   **56.** $\frac{1}{3} + \frac{1}{6} = \frac{2}{6} + \frac{1}{6} = \frac{3}{6} = \frac{1}{2}$

**57.** $\frac{17}{2\cancel{4}} \cdot \frac{\cancel{2}1}{3} = \frac{17}{6} = 2\frac{5}{6}$

**58.** $9\frac{2}{5} + 3\frac{1}{2} - 9\frac{4}{10} + 3\frac{5}{10} - 12\frac{9}{10}$

**59.** $\frac{4}{5} \cdot \frac{2}{1} = \frac{8}{5} = 1\frac{3}{5}$

**60.** $6\frac{5}{7} - 2\frac{1}{5} = 6\frac{25}{35} - 2\frac{7}{35} = 4\frac{18}{35}$

**61.** $\frac{9}{10} + \frac{3}{8} = \frac{36}{40} + \frac{15}{40} = \frac{51}{40} = 1\frac{11}{40}$

**62.** $\frac{17}{2} \cdot \frac{1}{4} = \frac{17}{8} = 2\frac{1}{8}$   **63.** $\frac{11}{5\cancel{15}} \cdot \frac{\cancel{3}1}{8} = \frac{11}{40}$

**64.** $\frac{1\cancel{4}}{7} \cdot \frac{5}{\cancel{4}1} = \frac{5}{7}$

**65.** $8\frac{15}{24} + 10\frac{12}{24} + 9\frac{6}{24} + 8\frac{18}{24} + 10\frac{16}{24} + 9\frac{20}{24} = 54\frac{87}{24}$

$$= 57\frac{15}{24} = 57\frac{5}{8}$$

$$57\frac{5}{8} \div 6 = \frac{461}{8} \cdot \frac{1}{6} = \frac{461}{48} = 9\frac{29}{48}\text{in.}$$

## Fractions, Decimals, and Percents (p. 785)

**1.** 0.63, $\frac{63}{100}$   **2.** 0.07, $\frac{7}{100}$

**3.** 0.24, $\frac{24}{100} = \frac{6}{25}$   **4.** 0.35, $\frac{35}{100} = \frac{7}{20}$

**5.** 0.17, $\frac{17}{100}$   **6.** 1.25, $1\frac{1}{4}$

**7.** 0.45, $\frac{45}{100} = \frac{9}{20}$   **8.** 2.5, $2\frac{1}{2}$

**9.** $0.\overline{3}, \frac{1}{3}$   **10.** 0.96, $\frac{96}{100} = \frac{24}{25}$

**11.** 0.625, $\frac{625}{1000} = \frac{5}{8}$   **12.** 7.25, $7\frac{1}{4}$

**13.** 0.052, $\frac{52}{1000} = \frac{13}{250}$   **14.** 0.008, $\frac{8}{1000} = \frac{1}{125}$

**15.** 0.0012, $\frac{12}{10,000} = \frac{3}{2500}$   **16.** 39%, $\frac{39}{100}$

**17.** 8%, $\frac{8}{100} = \frac{2}{25}$   **18.** 12%, $\frac{12}{100} = \frac{3}{25}$

**19.** 150%, $1\frac{1}{2}$   **20.** 72%, $\frac{72}{100} = \frac{18}{25}$

**21.** 5%, $\frac{5}{100} = \frac{1}{20}$   **22.** 208%, $2\frac{8}{100} = 2\frac{2}{25}$

**23.** 480%, $4\frac{80}{100} = 4\frac{4}{5}$   **24.** 2%, $\frac{2}{100} = \frac{1}{50}$

**25.** 375%, $3\frac{75}{100} = 3\frac{3}{4}$   **26.** 85%, $\frac{85}{100} = \frac{17}{20}$

**27.** 52%, $\frac{52}{100} = \frac{13}{25}$   **28.** 90%, $\frac{90}{100} = \frac{9}{10}$

**29.** 0.5%, $\frac{1}{200}$   **30.** 201%, $2\frac{1}{100}$

**31.** 0.7, 70%   **32.** 0.65, 65%

**33.** 0.44, 44%   **34.** 0.3, 30%

**35.** 0.375, 37.5%   **36.** 2.75, 275%

**37.** 5.125, 512.5%   **38.** 0.95, 95%

**39.** 0.875, 87.5%   **40.** 3.28, 328%

**41.** 0.833, 83.3%   **42.** 3.6, 360%

**43.** 0.533, 53.3%   **44.** 8.2, 820%

**45.** 1.417, 141.7%

## Using Percent (p. 786)

**1.** $N \cdot 90 = 15$
$N = 0.1\overline{6}$
$N = 16\frac{2}{3}\%$

**2.** $12 = N \cdot 60$
$0.2 = N$
$20\% = N$

**3.** $N \cdot 80 = 30$
$N = 0.375$
$N = 37.5\%$

**4.** $N \cdot 60 = 90$
$N = 1.5$
$= 150\%$

**5.** $N \cdot 90 = 60$
$N = 0.\overline{6}$
$N = 66\frac{2}{3}\%$

**6.** $6 = N \cdot 120$
$0.05 = N$
$5\% = N$

**7.** $15 = N \cdot 90$
$0.1\overline{6} = N$
$16\frac{2}{3}\% = N$

**8.** $N \cdot 18 = 4.5$
$N = 0.25$
$N = 25\%$

**9.** $18 = N \cdot 96$
$0.1875 = N$
$18.75\% = N$

**10.** $0.38 \cdot 250 = 95$

**11.** $0.12 \cdot 75 = 9$   **12.** $\frac{1}{4} \cdot 84 = 21$

**13.** $0.50 \cdot 96 = 48$   **14.** $0.42 \cdot 115 = 48.3$

**15.** $\frac{1}{3} \cdot 114 = 38$   **16.** $0.055 \cdot \$102.95 = \$5.66$

**17.** $0.70 \cdot 60 = 42$   **18.** $\frac{2}{3} \cdot 48 = 32$

**19.** $0.08 \cdot \$12.99 = \$1.04$   **20.** $1.5 \cdot 90 = 135$

**21.** $0.045 \cdot \$75 = \$3.38$   **22.** $0.065 \cdot \$42 = \$2.73$

## Ratio and Rate (p. 787)

**1.** 15 to 4   **2.** 9 : 2

**3.** $\frac{3}{5}$   **4.** 7 : 5

**5.** $\frac{12}{17}$   **6.** 7 to 12

**7.** 17 : 3   **8.** $\frac{9}{5}$

**9.** 2 to 3   **10.** 3 : 8

**11.** 15 to 7   **12.** 4 : 1

**13.** 1 to 4   **14.** 24 to 5

**15.** $22.50 per ticket   **16.** $8.50 per hour

**17.** 52 miles per hour   **18.** $175 per week

**19.** 28 miles per gallon   **20.** 8 ounces per person

**21.** $.10 per minute   **22.** $2.89 per notebook

**23.** 21.6 meters per second   **24.** 4.5 hours per day

**25.** $1.12 per mile   **26.** $10 per hour

# Skills Review Handbook *continued*

## Counting Methods *(p. 789)*

**1.**
    <u>Jeans</u>               <u>Shirts</u>

  Blue, Black, Gray       White, Blue, Yellow

Blue J-White S, Blue J-Blue S, Blue J-Yellow S
Black J-White S, Black J-Blue S, Black J-Yellow S } 9 outfits
Gray J-White S, Gray J-Blue S, Gray J- Yellow S

**2.**

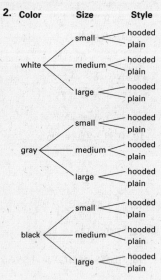

18 different sweatshirts

**3.** $270 \cdot 210 = 56{,}700$ pairs

**4.**

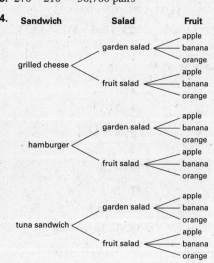

18 different lunches

**5.** $10 \cdot 10 \cdot 10 \cdot 10 = 10{,}000$ numbers

**6.** $12 \cdot 14 \cdot 13 = 2184$ combinations

**7.** $26 \cdot 26 \cdot 10 \cdot 10 = 67{,}000$ PINs

   $26 \cdot 25 \cdot 10 \cdot 9 = 58{,}500$ PINs

## Perimeter, Area and Volume *(p. 791)*

**1.** $10 + 7 + 10 + 7 = 34$ units

**2.** $0.5 + 0.75 + 0.5 + 0.75 = 2.5$ in.

**3.** $35 + 28 + 21 = 84$ ft

**4.** $3.5 + 3.5 + 3.5 + 3.5 = 14$ m

**5.** $4(18) = 72$ ft

**6.** $6 + 7 + 6 + 7 = 26$ m

**7.** $29(29) = 841$ yd$^2$     **8.** $7(4) = 28$ km$^2$

**9.** $3.5(3.5) = 12.25$ in.$^2$     **10.** $24(6) = 144$ ft$^2$

**11.** $7.2(7.2) = 51.84$ cm$^2$     **12.** $7.5(8) = 60$ in.$^2$

**13.** $45(45) = 2025$ km$^2$     **14.** $5.3(4) = 21.2$ m$^2$

**15.** $25(25)(25) = 15{,}625$ ft$^3$     **16.** $4.2(4.2)(4.2) = 74.088$ cm$^3$

**17.** $15(7)(4) = 420$ yd$^3$     **18.** $7.3(5)(3.2) = 116.8$ cm$^3$

**19.** $5.3(4)(10) = 212$ in.$^3$

## Data Displays *(p. 794)*

**1.** 0–25 with increases of 5

**2.**

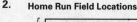

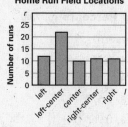

**3.** 0–20 with increases of 5

**4.**

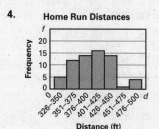

**5.** Expert $= \frac{51}{150} \cdot 360° = 122.4$

   Intermediate $= \frac{60}{150} \cdot 360° = 144$

   Beginner $= \frac{39}{150} \cdot 360° = 93.6$

# Skills Review Handbook *continued*

**6.**

**Patient's Temperature**

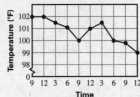

**7.**

**Stock Prices**

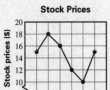

**8.**

**Passenger Car Stopping Distance (dry road)**

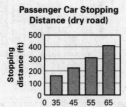

**9.**

**Fat in One Tablespoon of Canola Oil**

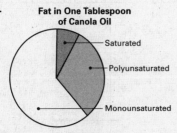

- Saturated
- Polyunsaturated
- Monounsaturated

**10.**

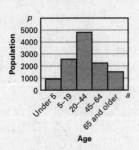

## Problem Solving (p. 796)

**1.** $2.75(5) + 0.80(3) \stackrel{?}{=} \$16.15$

$13.75 + 2.4 \stackrel{?}{=} 16.15$

$16.15 = 16.15$

5 sandwiches, 3 cartons of milk

**2.**

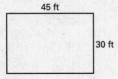

$150 - 2(45) = 2(w)$

$60 = 2w$

$30 = w$

$A = \ell w$

$= 45(30)$

$= 1350 \text{ ft}^2$

**3.** $\$18.75 \div 5 = \$3.75$

$\$3.75 \cdot 7 = \$26.25$

**4.**

9 diagonals

**5.** $25 + 15 + 20 = 60$ min

$7:25 - 1 \text{ hr} = 6:25$

No later than 6:25 A.M.

**6.**

31 games

**7.** People to Be Left Out

| | | | |
|---|---|---|---|
| Andrea, Betty | Andrea, Joyce | Andrea, Karen | Andrea, Paula |
| Betty, Joyce | Betty, Karen | Betty, Paula | |
| Joyce, Karen | Joyce, Paula | | |
| Karen, Paula | | | |

10 groups of 3 people

**8.** $135 + 5x = 90 + 10x$

$45 = 5x$

$9 = x$

It will take 9 weeks before they have the same amount.

**9.**

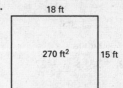

270 ft² , 18 ft, 15 ft

The problem cannot be solved as not enough information was given. We need to know how many tiles are in a carton.

# EXTRA PRACTICE

## Chapter 1 (p. 797)

**1.** $15(7) = 105$

**2.** $0.75 + 2.25 = 3$

**3.** $32 - 14 = 18$

**4.** $\frac{391}{23} = 17$

**5.** $\frac{1578}{3} = 526$

**6.** $\frac{3}{4} \cdot \frac{2}{3} = \frac{1}{2}$

**7.** $3(5)^2 = 75$

**8.** $(4(2))^3 = 512$

**9.** $6(4)^4 = 1536$

**10.** $3^4 - 5 = 76$

**11.** $(4 + 2)^3 = 216$

**12.** $(10 - 3)^2 = 49$

**13.** $3(5^2) + 8 = 83$

**14.** $44 - 4(7) = 16$

**15.** $(3^4 - 6) \div 5 = 15$

**16.** $3^3 - 12 \div 4 = 27 - 3 = 24$

**17.** $10^2 \div 4 + 6 = 25 + 6 = 31$

**18.** $10^2 \div (4 + 6) = 10$

**19.** $\frac{9 \cdot 7^2}{5 + 8^2 - 6} = \frac{441}{63} = 7$

**20.** $3 + 7(3.5 \div 5) = 3 + 4.9 = 7.9$

**21.** $2 + 21 \div 3 - 6 = 3$

**22.** $50 \div (6^2 - 11) - 2 = 50 \div 25 - 2 = 0$

**23.** $[(5 \cdot 2^3) + 8] \div 16 = 48 \div 16 = 3$

**24.** $x + 7 = 13$
$x = 6$

**25.** $3y = 21$
$y = 7$

**26.** $8t - 1 = 23$
$t = 3$

**27.** $\frac{m}{4} = 6$
$m = 24$

**28.** $12 + 10 \overset{?}{<} 22$
not a solution

**29.** $2(8) + 5 \overset{?}{>} 15$
$21 > 15$
solution

**30.** $7(2) \overset{?}{>} 15 - 2$
$14 > 13$
solution

**31.** $\frac{5 + 1}{2} \overset{?}{\le} 3$
$3 \le 3$
solution

**32.** $6(12 - 5) \overset{?}{<} 40$
$42 \not< 40$
not a solution

**33.** $12(3) \overset{?}{>} 3 + 27$
$36 > 30$
solution

**34.** $17 - 12 \overset{?}{\ge} 3$
$5 \ge 3$
solution

**35.** $9(2) \overset{?}{\le} 16$
$18 \not\le 16$
not a solution

**36.** $225 = 2x + 25$
$100 = x$

**37.** $25n - 13 = 37$

**38.** $5 + y < 12$

**39.**

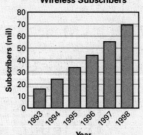
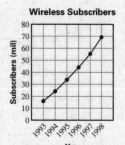

**40.**

| $x$ | 0 | 1 | 2 | 3 |
|---|---|---|---|---|
| $y$ | 8 | 6 | 4 | 2 |

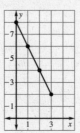

$y = 8 - 2x$

**41.**

| $x$ | 0 | 1 | 2 | 3 |
|---|---|---|---|---|
| $y$ | 4.5 | 5.5 | 6.5 | 7.5 |

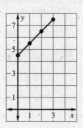

$y = 4.5 + x$

**42.**

| $x$ | 0 | 1 | 2 | 3 |
|---|---|---|---|---|
| $y$ | 1 | 8 | 15 | 22 |

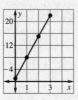

$y = 7x + 1$

**43.**

| $x$ | 0 | 1 | 2 | 3 |
|---|---|---|---|---|
| $y$ | 0 | 1 | 2 | 3 |

$y = x$

## Chapter 2 (p. 798)

**1.**

$-7 < 8, 8 > -7$

**2.**

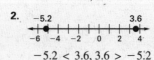

$-5.2 < 3.6, 3.6 > -5.2$

**3.**

$-2.5 < -2.4, -2.4 > -2.5$

# Extra Practice *continued*

**4.**

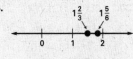

$$1\tfrac{2}{3} < 1\tfrac{5}{6}, 1\tfrac{5}{6} > 1\tfrac{2}{3}$$

**5.** $|8.5| = 8.5$

**6.** $|-3| = 3$

**7.** $|-4| + 3 = 4 + 3 = 7$

**8.** $7 - |-5| = 7 - 5 = 2$

**9.** $-3 + 8 = 5$

**10.** $18 + 27 = 45$

**11.** $5 + (-7) = -2$

**12.** $-4 + (-11) = -15$

**13.** $-4 + 13 + (-6) = 9 + (-6) = 3$

**14.** $15 + (-12) + (-4) = 3 + (-4) = -1$

**15.** $-2 + (-9) + 8 = -11 + 8 = -3$

**16.** $17 + (-5) + 15 = 12 + 15 = 27$

**17.** $-8 - 5 = -13$

**18.** $4.1 - 6.3 = -2.2$

**19.** $-3 - (-7) = 4$

**20.** $-6 + (-3) - 4 = -9 - 4 = -13$

**21.** $3.6 - 2.4 - (-6.1) = 1.2 - (-6.1) = 7.3$

**22.** $-15 + 4 - 12 = -11 - 12 = -23$

**23.** $-11 - (-6) - 7 = -5 - 7 = -12$

**24.** $\frac{9}{10} - \frac{1}{2} + \left(-\frac{1}{5}\right) = \frac{9}{10} - \frac{5}{10} + \left(-\frac{2}{10}\right)$
$$= \frac{4}{10} + \left(-\frac{2}{10}\right)$$
$$= \frac{2}{10} = \frac{1}{5}$$

**25.** $\begin{bmatrix} 8 & -4 \\ 9 & 3 \end{bmatrix} + \begin{bmatrix} -2 & 6 \\ -1 & 5 \end{bmatrix} = \begin{bmatrix} 8 + (-2) & -4 + 6 \\ 9 + (-1) & 3 + 5 \end{bmatrix}$
$$= \begin{bmatrix} 6 & 2 \\ 8 & 8 \end{bmatrix}$$

**26.** $\begin{bmatrix} 3 & -6 \\ 1 & 4 \end{bmatrix} - \begin{bmatrix} 5 & 9 \\ 2 & -2 \end{bmatrix} = \begin{bmatrix} 3 - 5 & -6 - 9 \\ 1 - 2 & 4 - (-2) \end{bmatrix}$
$$= \begin{bmatrix} -2 & -15 \\ -1 & 6 \end{bmatrix}$$

**27.** $\begin{bmatrix} -6 & 8 & 3 \\ 4 & 2 & 6 \end{bmatrix} + \begin{bmatrix} 3 & -6 & 7 \\ -4 & -5 & 8 \end{bmatrix}$
$$= \begin{bmatrix} -6 + 3 & 8 + (-6) & 3 + 7 \\ 4 + (-4) & 2 + (-5) & 6 + 8 \end{bmatrix}$$
$$= \begin{bmatrix} -3 & 2 & 10 \\ 0 & -3 & 14 \end{bmatrix}$$

**28.** $(-6)(-7) = 42$

**29.** $(-5)(9) = -45$

**30.** $(3)(-8)(-2) = 48$

**31.** $(-8)(-4x) = 32x$

**32.** $-3(-y)(-y) = -3y^2$

**33.** $(-c)^3(c) = -c^4$

**34.** $(-7)^2(b)(-b) = 49(b)(-b) = -49b^2$

**35.** $-4(-a^4) = 4a^4$

**36.** $6(y + 5) = 6y + 30$

**37.** $4(a - 6) = 4a - 24$

**38.** $(x + 3)(-5) = -5x - 15$

**39.** $-r(r - 5) = -r^2 + 5r$

**40.** $-k(7 + k) = -7k - k^2$

**41.** $(x + 4)6x = 6x^2 + 24x$

**42.** $s(s - s^2) = s^2 - s^3$

**43.** $(0.5z - 1.4)6 = 3z - 8.4$

**44.** $3x + 7x = 10x$

**45.** $5.4m - 2.3m = 3.1m$

**46.** $82p - (-29p) = 111p$

**47.** $6 - 4t - 4 = 2 - 4t$

**48.** $5 + 4(x - 2) = 5 + 4x - 8 = -3 + 4x$

**49.** $8x^2 + 5 - 2x^2 = 6x^2 + 5$

**50.** $2x(7 - x) + 3x^2 = 14x - 2x^2 + 3x^2 = x^2 + 14x$

**51.** $\frac{2}{3}x + \left(-\frac{1}{6}\right)x = \frac{2}{3}x - \frac{1}{6}x = \frac{4}{6}x - \frac{1}{6}x = \frac{3}{6}x = \frac{1}{2}x$

**52.** $18 \div (-2) = -9$

**53.** $-48 \div 12 = -4$

**54.** $16 \div \left(-\frac{4}{5}\right) = \frac{16}{1} \cdot \left(-\frac{5}{4}\right)$
$$= \frac{4}{1} \cdot -\frac{5}{1}$$
$$= -\frac{20}{1}$$
$$= -20$$

**55.** $\frac{3x}{8} \div \frac{1}{2} = \frac{3x}{8} \cdot \frac{2}{1} = \frac{3x}{4} \cdot \frac{1}{1} = \frac{3x}{4}$

**56.** $21x \div 7 = 3x$

**57.** $8x \div \left(-\frac{1}{4}\right) = \frac{8x}{1} \cdot \left(-\frac{4}{1}\right) = -\frac{32x}{1} = -32x$

**58.** $-24x \div \left(-\frac{2}{3}\right) = -\frac{24x}{1} \cdot \left(-\frac{3}{2}\right)$
$$= -\frac{12x}{1} \cdot -\frac{3}{1}$$
$$= \frac{36x}{1}$$
$$= 36x$$

**59.** $\frac{-22}{-\frac{1}{3}} = -\frac{22}{1} \div -\frac{1}{3} = -\frac{22}{1} \cdot -\frac{3}{1} = \frac{66}{1} = 66$

**60.** $\frac{9}{36} = \frac{1}{4}$ or 25%

**61.** $0.20 = \frac{1}{5}$ = Probability
odds = 1 to 4

**62.** Probability = $\frac{1}{3}$
odds = 1 to 2

## Chapter 3 *(p. 799)*

**1.** $y - 6 = 8$
$$y = 14$$

**2.** $n + 5 = -10$
$$n = -15$$

**3.** $a - (-6) = 22$
$$a = 16$$

**4.** $14 - r = 3$
$$-r = -11$$
$$r = 11$$

# Extra Practice *continued*

**5.** $|-7| + k = 4$
$k = -3$

**6.** $\frac{1}{5} + m = -\frac{2}{5}$
$m = -\frac{3}{5}$

**7.** $-t - (-3) = 0$
$-t = -3$
$t = 3$

**8.** $-b + 3 - 1 = 3 \cdot 4$
$-b = 10$
$b = -10$

**9.** $7x = 35$
$x = 5$

**10.** $6a = 3$
$a = \frac{1}{2}$

**11.** $\frac{x}{10} = -2$
$x = -20$

**12.** $-\frac{3}{8}t = 0$
$t = 0$

**13.** $|-5| = 15b$
$\frac{1}{3} = b$

**14.** $-\frac{7}{8}r = \frac{3}{4}$
$r = -\frac{6}{7}$

**15.** $\frac{y}{10} = -\frac{2}{5}$
$y = -4$

**16.** $-\frac{2}{3} = \frac{1}{9}k$
$k = -6$

**17.** $6x + 8 = 32$
$6x = 24$
$x = 4$

**18.** $2x - 1 = 11$
$2x = 12$
$x = 6$

**19.** $-x - 5 + 3x = 1$
$2x = 6$
$x = 3$

**20.** $4(x - 9) = 8$
$x - 9 = 2$
$x = 11$

**21.** $\frac{3}{5}x - 7 = 17$
$\frac{3}{5}x = 24$
$x = 40$

**22.** $x = 2(x - 1) + 6$
$x = 2x + 4$
$-4 = x$

**23.** $\frac{x}{4} = -\frac{x}{2} - 1$
$\frac{x}{4} + \frac{2x}{4} = -1$
$3x = -4$
$x = -\frac{4}{3}$

**24.** $-5 = \frac{3}{8}(x - 1)$
$-\frac{40}{3} = x - 1$
$-12\frac{1}{3} = x$

**25.** $-6 + 5x = 8x - 9$
$3 = 3x$
$1 = x$

**26.** $8x + 6 = 3(4 - x)$
$8x + 6 = 12 - 3x$
$11x = 6$
$x = \frac{6}{11}$

**27.** $4\left(\frac{1}{2}x + \frac{1}{2}\right) = 2x + 2$
$2x + 2 = 2x + 2$
$0 = 0$
all real numbers

**28.** $-3(-x - 4) = 2x + 1$
$3x + 12 = 2x + 1$
$x = -11$

**29.** $4x = -2(-2x + 3)$
$4x = 4x - 6$
$0 \neq -6$
no solution

**30.** $-4(x - 3) = 6(x + 5)$
$-4x + 12 = 6x + 30$
$-18 = 10x$
$-\frac{9}{5} = x$

**31.**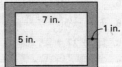
Length = 9 in.
Width = 7 in.

**32.** $4x = 82.50$
$x \approx 20.63$

**33.** $-26x - 59 = 135$
$-26x = 194$
$x \approx -7.46$

**34.** $2.3 - 4.8x = 8.2x + 5.6$
$-3.3 = 13x$
$x \approx -0.25$

**35.** $18.25x - 4.15 = 2.75x$
$15.5x = 4.15$
$x \approx 0.27$

**36.** $3(3.1x - 4.2) = 6.2x + 3.1$
$9.3x - 12.6 = 6.2x + 3.1$
$3.1x = 15.7$
$x \approx 5.06$

**37.** $8.4x - 3.2 = 4.1(3.4 - 2.1x)$
$8.4x - 3.2 = 13.94 - 8.61x$
$17.01x = 17.14$
$x \approx 1.01$

**38.** $37.5 = 2.5r$
$r = 15$ mi/h

**39.** $3 - y = x$
$y = -x + 3$

| $x$ | $-2$ | $-1$ | $0$ | $1$ |
|---|---|---|---|---|
| $y$ | 5 | 4 | 3 | 2 |

**40.** $6x - 2y = 10$
$-2y = -6x + 10$
$y = 3x - 5$

| $x$ | $-2$ | $-1$ | $0$ | $1$ |
|---|---|---|---|---|
| $y$ | $-11$ | $-8$ | $-5$ | $-2$ |

**41.** $-6x - 1 = -4y + 2$
$4y = 6x + 3$
$y = \frac{3}{2}x + \frac{3}{4}$

| $x$ | $-2$ | $-1$ | $0$ | $1$ |
|---|---|---|---|---|
| $y$ | $-2\frac{1}{4}$ | $-\frac{3}{4}$ | $\frac{3}{4}$ | $2\frac{1}{4}$ |

# Extra Practice *continued*

**42.** $-2y - 3x - 16 = 0$

$$-2y = 3x + 16$$
$$y = -\frac{3}{2}x - 8$$

| $x$ | $-2$ | $-1$ | $0$ | $1$ |
|---|---|---|---|---|
| $y$ | $-5$ | $-6\frac{1}{2}$ | $-8$ | $-9\frac{1}{2}$ |

**43.** $\frac{1}{2}x - \frac{1}{2}y = 6$

$$x - y = 12$$
$$-y = -x + 12$$
$$y = x - 12$$

| $x$ | $-2$ | $-1$ | $0$ | $1$ |
|---|---|---|---|---|
| $y$ | $-14$ | $-13$ | $-12$ | $-11$ |

**44.** $-2(x - 4) = 2(y + 5)$

$$-x + 4 = y + 5$$
$$y = -x - 1$$

| $x$ | $-2$ | $-1$ | $0$ | $1$ |
|---|---|---|---|---|
| $y$ | $1$ | $0$ | $-1$ | $-2$ |

**45.** $\dfrac{\$3.00}{5 \text{ snacks}} = \$0.60$ per yogurt snack

**46.** $\dfrac{\$50.75}{7 \text{ hr}} = \$7.25$ per hour   **47.** $\dfrac{122 \text{ mi}}{2.5 \text{ h}} = 48.8$ mi/h

**48.** $24(1.609) = 38.62$ km   **49.** $\frac{6}{8} = 0.75$ cup

**50.** $\dfrac{\$1.00}{\$19.99} \approx 5\%$   **51.** $\dfrac{9}{14} \approx 64\%$

**52.** $\dfrac{\$4.50}{\$22} \approx 20\%$   **53.** $\dfrac{127}{350} \approx 36\%$

## Chapter 4 *(p. 800)*

**1.**
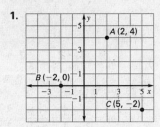

**2.**
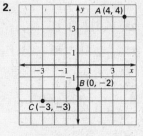

**3.**
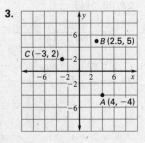

**4.**
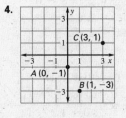

**5.**
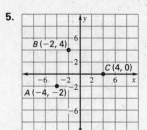

**6.**
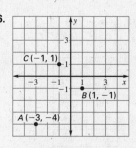

**7.** $y = 5x + 1$

| $x$ | $y$ |
|---|---|
| $-2$ | $-9$ |
| $-1$ | $-4$ |
| $0$ | $1$ |
| $1$ | $6$ |
| $2$ | $11$ |

**8.** $y = -2x + 4$

| $x$ | $y$ |
|---|---|
| $-2$ | $8$ |
| $-1$ | $6$ |
| $0$ | $4$ |
| $1$ | $2$ |
| $2$ | $0$ |

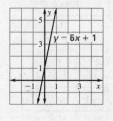

**9.** $y = -2$

| $x$ | $y$ |
|---|---|
| $-1$ | $-2$ |
| $0$ | $-2$ |
| $1$ | $-2$ |

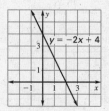

**10.** $4x + y = -8$

$$y = -4x - 8$$

| $x$ | $y$ |
|---|---|
| $-2$ | $0$ |
| $-1$ | $-4$ |
| $0$ | $-8$ |
| $1$ | $-12$ |
| $2$ | $-16$ |

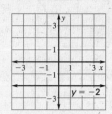

**11.** $x = 3$

| $x$ | $y$ |
|---|---|
| $3$ | $-2$ |
| $3$ | $0$ |
| $3$ | $2$ |

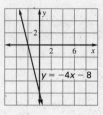

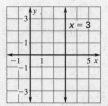

# Extra Practice *continued*

**12.** $y = -(6 + x)$

$y = -6 - x$

| $x$ | $y$ |
|-----|-----|
| $-2$ | $-4$ |
| $-1$ | $-5$ |
| $0$ | $-6$ |
| $1$ | $-7$ |
| $2$ | $-8$ |

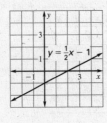

**13.** $y = \frac{1}{2}x - 1$

| $x$ | $y$ |
|-----|-----|
| $-2$ | $-2$ |
| $-1$ | $-1\frac{1}{2}$ |
| $0$ | $-1$ |
| $1$ | $-\frac{1}{2}$ |
| $2$ | $0$ |

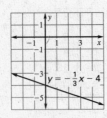

**14.** $y = -\frac{1}{3}x - 4$

| $x$ | $y$ |
|-----|-----|
| $-2$ | $-3\frac{1}{3}$ |
| $-1$ | $-3\frac{2}{3}$ |
| $0$ | $-4$ |
| $1$ | $-4\frac{1}{3}$ |
| $2$ | $-4\frac{2}{3}$ |

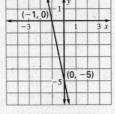

**15.** $5x + y = -5$    $5x + 0 = -5$

$5(0) + y = -5$    $5x = -5$

$y = -5$    $x = -1$

$(0, -5)$    $(-1, 0)$

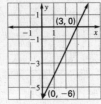

**16.** $2x - y = 6$

$2(0) - y = 6$    $2x - 0 = 6$

$0 - y = 6$    $2x = 6$

$y = -6$    $x = 3$

$(0, -6)$    $(3, 0)$

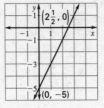

**17.** $y = 2x - 5$

$y = 2(0) - 5$    $0 = 2x - 5$

$y = -5$    $5 = 2x$

$(0, -5)$    $\frac{5}{2} = x$

$\left(\frac{5}{2}, 0\right)$

**18.** $6y + 2x = 12$

$6y + 2(0) = 12$    $6(0) + 2x = 12$

$6y = 12$    $2x = 12$

$y = 2$    $x = 6$

$(0, 2)$    $(6, 0)$

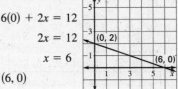

**19.** $14 = y - 2x$

$14 = y - 2(0)$    $14 = 0 - 2x$

$14 = y$    $14 = -2x$

   $-7 = x$

$(0, 14)$    $(-7, 0)$

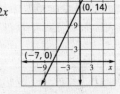

**20.** $8x + 2y = -16$

$8(0) + 2y = -16$    $8x + 2(0) = -16$

$2y = -16$    $8x = -16$

$y = -8$    $x = -2$

$(0, -8)$    $(-2, 0)$

**21.** $y = 6 - 3x$

$y = 6 - 3(0)$    $0 = 6 - 3x$

$y = 6$    $-6 = -3x$

   $2 = x$

$(0, 6)$    $(2, 0)$

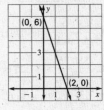

**22.** $1.5y = x - 3$

$1.5y = 0 - 3$    $1.5(0) = x - 3$

$1.5y = -3$    $0 = x - 3$

$y = -2$    $3 = x$

$(0, -2)$    $(3, 0)$

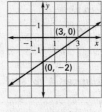

**23.** $(6, 1)\ (-4, 1)$

$m = \dfrac{1 - 1}{-4 - 6} = \dfrac{0}{-10} = 0$

**24.** $(2, 2)\ (-1, 4)$

$m = \dfrac{4 - 2}{-1 - 2} = -\dfrac{2}{3}$

**25.** $(0, 6)\ (-3, 0)$

$m = \dfrac{0 - 6}{-3 - 0} = \dfrac{-6}{-3} = 2$

**26.** $(4, 5)\ (2, 2)$

$m = \dfrac{2 - 5}{2 - 4} = \dfrac{-3}{-2} = \dfrac{3}{2}$

**27.** $(-4, 2)\ (-3, -5)$

$m = \dfrac{-5 - 2}{-3 + 4} = \dfrac{-7}{1} = -7$

**28.** $(3, 6)\ (3, -1)$

$m = \dfrac{-1 - 6}{3 - 3} = \dfrac{-7}{0}$

no slope

**29.** $(-1, 0)\ (0, -1)$

$m = \dfrac{-1 - 0}{0 + 1} = \dfrac{-1}{1} = -1$

**30.** $(-2, -3)\ (1, -2)$

$m = \dfrac{-2 + 3}{1 + 2} = \dfrac{1}{3}$

# Extra Practice *continued*

**31.**   $y = kx$
$18 = k(6)$
$3 = k$
$y = 3x$

**32.**   $y = kx$
$1 = k(4)$
$\frac{1}{4} = k$
$y = \frac{1}{4}x$

**33.**   $y = kx$
$-7 = k(8)$
$-\frac{7}{8} = k$
$y = -\frac{7}{8}x$

**34.**   $y = kx$
$-10 = k\left(-\frac{1}{2}\right)$
$20 = k$
$y = 20x$

**35.**   $y = kx$
$-2 = k(-2)$
$1 = k$
$y = x$

**36.**   $y = kx$
$-4 = k(8)$
$-\frac{1}{2} = k$
$y = -\frac{1}{2}x$

**37.**   $y = kx$
$2 = k(2.5)$
$0.8 = k$
$y = 0.8x$

**38.**   $y = kx$
$-6.3 = k(2.1)$
$-3 = k$
$y = -3x$

**39.**   $y = kx$    $y = 15x$
$60 = k(4)$  $y = 15(35)$
$15 = k$      $y = \$525$

**40.**  $x - y = 1$
$-y = 1 - x$
$y = x - 1$

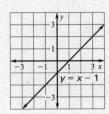

**41.**  $y = 3$

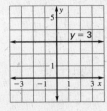

**42.**  $-3x + 2y = 6$
$2y = 6 + 3x$
$y = 3 + \frac{3}{2}x$
$y = \frac{3}{2}x + 3$

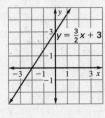

**43.**  $y + 4 = 0$
$y = -4$

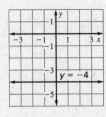

**44.**   $x = 2y + 1$
$-2y = 1 - x$
$y = -\frac{1}{2} + \frac{1}{2}x$
$y = \frac{1}{2}x - \frac{1}{2}$

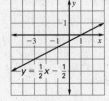

**45.**  $2(x + y + 1) = 4y$
$2x + 2y + 2 = 4y$
$-2y = -2x - 2$
$y = x + 1$

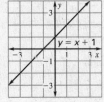

**46.**  $2x - 4y + 6 = 0$
$-4y = -2x - 6$
$y = \frac{1}{2}x + \frac{3}{2}$

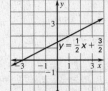

**47.**  $5x - 3y + 2 = 14 - 4x$
$-3y = 12 - 9x$
$y = -4 + 3x$
$y = 3x - 4$

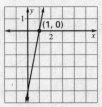

**48.**  $6 - 5x = 1$       $6 - 5(1) = 1$
$0 = -5 + 5x$     $6 - 5 = 1$
$y = 5x - 5$           $1 = 1$
Solution: 1

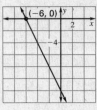

**49.**  $2x = -12$
$0 = -2x - 12$    $2(-6) = -12$
$y = -2x - 12$      $-12 = -12$
Solution: $-6$

**50.**  $6x + 8 = 2x$
$0 = -4x - 8$
$y = -4x - 8$
$6(-2) + 8 = 2(-2)$
$-12 + 8 = -4$
$-4 = -4$
Solution: $-2$

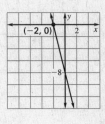

# Extra Practice *continued*

**51.** $8x + 3 = 19$ $\qquad$ $8(2) + 3 = 19$

$\quad 0 = -8x + 16$ $\qquad 16 + 3 = 19$

$\quad y = -8x + 16$ $\qquad\quad 19 = 19$

Solution: 2

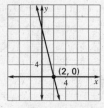

**52.** $\frac{4}{3}x - 2 = -6$ $\qquad \frac{4}{3}(-3) - 2 = -6$

$\quad 0 = -\frac{4}{3}x - 4$ $\qquad -4 - 2 = -6$

$\quad y = -\frac{4}{3}x - 4$ $\qquad\quad -6 = -6$

Solution: $-3$

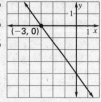

**53.** $15 - 2x = 3x$ $\qquad 15 - 2(3) = 3(3)$

$\quad 0 = 5x - 15$ $\qquad\quad 15 - 6 = 9$

$\quad y = 5x - 15$ $\qquad\qquad\quad 9 = 9$ Solution: 3

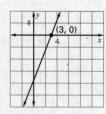

**54.** $\frac{2}{3}x + 5 = -2$ $\qquad \frac{2}{3}\left(-\frac{21}{2}\right) + 5 = -2$

$\quad 0 = -\frac{2}{3}x - 7$ $\qquad\quad -7 + 5 = -2$

$\quad y = -\frac{2}{3}x - 7$ $\qquad\qquad\quad -2 = -2$ $\qquad$ Solution: $-\frac{21}{2}$

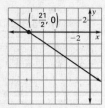

**55.** $7x - 4 = -3 + 4x$ $\qquad 7\left(\frac{1}{3}\right) - 4 = -3 + 4\left(\frac{1}{3}\right)$

$\quad 0 = -3x + 1$ $\qquad\qquad \frac{7}{3} - 4 = -3 + \frac{4}{3}$

$\quad y = -3x + 1$ $\qquad\qquad \frac{7}{3} - \frac{12}{3} = -\frac{9}{3} + \frac{4}{3}$

$\qquad\qquad\qquad\qquad\qquad\qquad -\frac{5}{3} = -\frac{5}{3}$

Solution: $\frac{1}{3}$

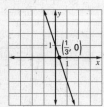

**56.** function

**57.** not a function

**58.** function

**59.** not a function

## Chapter 5 *(p. 801)*

**1.** $y = 2x + 1$ $\qquad\qquad$ **2.** $y = -3x - 2$

**3.** $y = x - 3$ $\qquad\qquad\quad$ **4.** $y = -4x$

**5.** $0 = 3(-1) + b$ $\qquad$ **6.** $\quad 2 = 5(-2) + b$

$\quad 3 = b$ $\qquad\qquad\qquad\qquad 12 = b$

$\quad y = 3x + 3$ $\qquad\qquad\quad y = -2x + 12$

**7.** $6 = 3(0) + b$ $\qquad$ **8.** $\quad 1 = -2(-5) + b$

$\quad 6 = b$ $\qquad\qquad\qquad\qquad -9 = b$

$\quad y = 6$ $\qquad\qquad\qquad\quad y = -5x - 9$

**9.** $-1 = -3(4) + b$ $\quad$ **10.** $\quad 5 = 1(8) + b$

$\quad 11 = b$ $\qquad\qquad\qquad\quad -3 = b$

$\quad y = 4x + 11$ $\qquad\qquad y = 8x - 3$

**11.** $-1 = 2\left(\frac{1}{2}\right) + b$ $\quad$ **12.** $3 = -4\left(-\frac{1}{3}\right) + b$

$\quad -2 = b$ $\qquad\qquad\qquad\quad \frac{5}{3} = b$

$\quad y = \frac{1}{2}x - 2$ $\qquad\qquad y = -\frac{1}{3}x + \frac{5}{3}$

**13.** $m = \dfrac{4 - (-2)}{5 - 3} = 3$ $\quad$ **14.** $y = \dfrac{17 - (-1)}{2 - (-1)} = 6$

$\quad 4 = 5(3) + b$ $\qquad\qquad -1 = -1(6) + b$

$\quad -11 = b$ $\qquad\qquad\qquad 5 = b$

$\quad y = 3x - 11$ $\qquad\qquad y = 6x + 5$

**15.** $m = \dfrac{-6 - 1}{0 - 5} = \dfrac{7}{5}$ $\quad$ **16.** $m = \dfrac{-4 - (-1)}{4 - (-2)} = -\dfrac{1}{2}$

$\quad -6 = \dfrac{7}{5}(0) + b$ $\qquad -1 = -2\left(-\dfrac{1}{2}\right) + b$

$\quad -6 = b$ $\qquad\qquad\qquad -2 = b$

$\quad y = \dfrac{7}{5}x - 6$ $\qquad\quad y = -\dfrac{1}{2}x - 2$

**17.** $m = \dfrac{7 - 7}{5 - (-1)} = 0$ $\quad$ **18.** $m = \dfrac{3 - 0}{2 - 0} = \dfrac{3}{2}$

$\quad 7 = -1(0) + b$ $\qquad\quad 0 = 0\left(\dfrac{3}{2}\right) + b$

$\quad y = 7$ $\qquad\qquad\qquad\quad 0 = b$

$\qquad\qquad\qquad\qquad\qquad\quad y = \dfrac{3}{2}x$

**19.** $m = \dfrac{8 - 5}{-6 + 3} = -1$ $\quad$ **20.** $m = \dfrac{14 - 2}{1 - 5} = -3$

$\quad 5 = -1(-3) + b$ $\qquad 2 = -3(5) + b$

$\quad 2 = b$ $\qquad\qquad\qquad\quad 17 = b$

$\quad y = -x + 2$ $\qquad\qquad y = -3x + 17$

# Extra Practice *continued*

*21–22. Sample answers are given.*

**21.**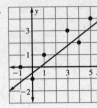

$(3, 2)$ and $(-1, -1)$

$m = \dfrac{2 - (-1)}{3 - (-1)} = \dfrac{3}{4}$

$2 = \dfrac{3}{4}(3) + b$

$-\dfrac{1}{4} = b$

$y = \dfrac{3}{4}x - \dfrac{1}{4}$

**22.**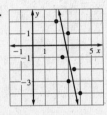

$(3, -1)$ and $(2, 4)$

$m = \dfrac{4 - (-1)}{2 - 3} = -5$

$-1 = 3(-5) + b$

$14 = b$

$y = -5x + 14$

**23.** $y - 3 = -2(x - 5)$
$y - 3 = -2x + 10$
$y = -2x + 13$

**24.** $y - 7 = -3(x - 3)$
$y - 7 = -3x + 9$
$y = -3x + 16$

**25.** $y - 4 = 5(x - 5)$
$y - 4 = 5x - 25$
$y = 5x - 21$

**26.** $y + 4 = \dfrac{3}{4}(x + 2)$
$y + 4 = \dfrac{3}{4}x + \dfrac{3}{2}$
$y = \dfrac{3}{4}x - \dfrac{5}{2}$

**27.** $y + 5 = -\dfrac{5}{3}(x + 3)$
$y + 5 = -\dfrac{5}{3}x - 5$
$y = -\dfrac{5}{3}x - 10$

**28.** $y - 8 = \dfrac{1}{4}(x - 0)$
$y - 8 = \dfrac{1}{4}x$
$y = \dfrac{1}{4}x + 8$

**29.** $y - 4 = -\dfrac{1}{2}(x - 2)$
$y - 4 = -\dfrac{1}{2}x + 1$
$y = -\dfrac{1}{2}x + 5$

**30.** $y + 7 = 4(x + 1)$
$y + 7 = 4x + 4$
$y = 4x - 3$

**31.** $-2 = 3(5) + b$
$-17 = b$
$y = 3x - 17$
$3x - y = 17$

**32.** $m = \dfrac{4 - 2}{3 - 1} = 1$
$2 = 1(1) + b$
$1 = b$
$y = x + 1$
$x - y = -1$

**33.** $m = \dfrac{-3 - 0}{-5 + 1} = \dfrac{3}{4}$
$0 = -1\left(\dfrac{3}{4}\right) + b$
$\dfrac{3}{4} = b$
$y = \dfrac{3}{4}x + \dfrac{3}{4}$
$4y = 3x + 3$
$3x - 4y = -3$

**34.** $3 = -\dfrac{5}{6}(-4) + b$
$-\dfrac{1}{3} = b$
$y = -\dfrac{5}{6}x - \dfrac{1}{3}$
$6y = -5x - 2$
$5x + 6y = -2$

**35.** $m = \dfrac{7 - 7}{-3 - 1} = 0$
$7 = 1(0) + b$
$7 = b$
$y = 7$

**36.** $7 = -\dfrac{3}{4}(5) + b$
$\dfrac{43}{4} = b$
$y = -\dfrac{3}{4}x + \dfrac{43}{4}$
$4y = -3x + 43$
$3x + 4y = 43$

**37.** $-5 = -2(5) + b$
$5 = b$
$y = 5x + 5$
$5x - y = -5$

**38.** $m = \dfrac{5 + 2}{4 - 4} = \dfrac{7}{0}$    no slope
$x = 4$

**39.**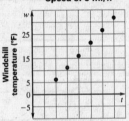

**Windchill for Wind Speed of 5 mi/h**

**40.** $w = t - 4$

**41.** $w = 5 - 4$
$w = 1°\text{F}$

## Chapter 6 *(p. 802)*

**1.** $x + 5 > -4$
$x > -9$
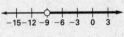

**2.** $m - 4 < -20$
$m < -16$

**3.** $3 \geq y - 4$
$7 \geq y$
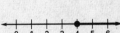

**4.** $9x \geq 36$
$x \geq 4$

# Extra Practice *continued*

**5.** $\dfrac{k}{9} \le 2$

$k \le 18$

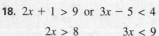

**6.** $-5a > 35$

$a < -7$

**7.** $-\dfrac{x}{10} < \dfrac{1}{5}$

$-x < 2$

$x > -2$

**8.** $0.5 \le \dfrac{b}{-6}$

$-3 \ge b$

**9.** $0 \le \frac{1}{2}x + 6$

$-6 \le \frac{1}{2}x$

$-12 \le x$

**10.** $\frac{3}{4}x + 5 \le 8$

$\frac{3}{4}x \le 3$

$x \le 4$

**11.** $3x + 8 \ge -2x + 3$

$5x \ge -5$

$x \ge -1$

**12.** $-3x - 7 < 2$

$-3x < 9$

$x > -3$

**13.** $-(x + 5) < -4x - 11$

$-x - 5 < -4x - 11$

$3x < -6$

$x < -2$

**14.** $-(4 + x) > 2(x - 5)$

$-4 - x > 2x - 10$

$-3x > -6$

$x < 2$

**15.** $8 \ge x + 4 > 3$

$4 \ge x > -1$

**16.** $-36 \le 6x < 12$

$-6 \le x < 2$

**17.** $-15 < -3x < 18$

$5 > x > -6$

**18.** $2x + 1 > 9$ or $3x - 5 < 4$

$2x > 8$ $\qquad$ $3x < 9$

$x > 4$ or $\qquad$ $x < 3$

**19.** $x + 1 > 4$ or $2x + 3 \le 5$

$x > 3$ or $\qquad$ $2x \le 2$

$x > 3$ or $\qquad$ $x \le 1$

**20.** $-4x + 1 \ge 17$ or $5x - 4 > 6$

$-4x \ge 16$ $\qquad$ $5x > 10$

$x \le -4$ or $\qquad$ $x > 2$

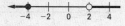

**21.** $|10 + x| = 4$

$10 + x = 4$ $\qquad$ $10 + x = -4$

$x = -6$ $\qquad$ $x = -14$

**22.** $|2x + 3| = 9$

$2x + 3 = 9$ $\qquad$ $2x + 3 = -9$

$2x = 6$ $\qquad$ $2x = -12$

$x = 3$ $\qquad$ $x = -6$

**23.** $|x - 4| + 4 = 7$

$x - 4 = 3$ $\qquad$ $x - 4 = -3$

$x = 7$ $\qquad$ $x = 1$

**24.** $|6x - 5| + 1 < 8$

$|6x - 5| < 7$

$6x - 5 < 7$ $\qquad$ $6x - 5 > -7$

$6x < 12$ and $\qquad$ $6x > -2$

$x < 2$ and $\qquad$ $x > -\frac{1}{3}$

$-\frac{1}{3} < x < 2$

**25.** $|3x + 4| - 6 \ge 14$

$|3x + 4| \ge 20$

$3x + 4 \ge 20$ $\qquad$ $3x + 4 \le -20$

$3x \ge 16$ or $\qquad$ $3x \le -24$

$x \ge 5\frac{1}{3}$ or $\qquad$ $x \le -8$

**26.** $|10 - 4x| \le 2$

$10 - 4x \le 2$ $\qquad$ $10 - 4x \ge -2$

$-4x \le -8$ $\qquad$ $-4x \ge -12$

$x \ge 2$ $\qquad$ $x \le 3$

$2 \le x \le 3$

**27.** $y \ge -2$

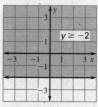

**28.** $x - y \le 0$

$-y \le -x$

$y \ge x$

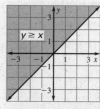

**29.** $x + y \ge 5$

$y \ge -x + 5$

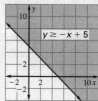

**Algebra 1**
Extra Practice  Worked-out Solution Key

# Extra Practice *continued*

**30.** $4y + x < 4$

$\quad 4y < 4 - x$

$\quad\quad y < 1 - \frac{1}{4}x$

$\quad\quad y < -\frac{1}{4}x + 1$

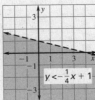

**31.** $x - 3y \le 0$

$\quad -3y \le -x$

$\quad\quad y \ge \frac{1}{3}x$

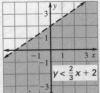

**32.** $3y - 2x < 6$

$\quad 3y < 6 + 2x$

$\quad\quad y < \frac{2}{3}x + 2$

**33.** $5x - 3y > 9$

$\quad -3y > -5x + 9$

$\quad\quad y < \frac{5}{3}x - 3$

**34.** $2y - x > 10$

$\quad 2y > x + 10$

$\quad\quad y > \frac{1}{2}x + 5$

**35.** $2x + 1.25y \le 20$

$\quad 1.25y \le 20 - 2x$

$\quad 1.25y \le -2x + 20$

$\quad\quad y \le -1.6x + 16$

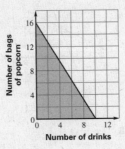

**36.**

| 1 | 3 4 8 8 |
|---|---|
| 2 | 0 2 |
| 3 | 1 3 |
| 4 | |
| 5 | |
| 6 | 6 |
| 7 | 1 6 |
| 8 | |
| 9 | 8 |

Key: 9 | 8 = 98

**37.**

| 1 | 3 6 |
|---|---|
| 2 | 1 2 |
| 3 | 1 3 5 |
| 4 | 0 1 5 |
| 5 | 1 |
| 6 | 5 6 |

Key: 6 | 5 = 65

**38.**

| 0 | 3 5 6 8 |
|---|---|
| 1 | 3 7 |
| 2 | 2 4 |
| 3 | 2 |
| 4 | 1 |
| 5 | 4 5 |
| 6 | 0 3 4 |
| 7 | 2 8 |
| 8 | |
| 9 | 1 |

Key: 9 | 1 = 91

**39.**

| 0 | 2 8 |
|---|---|
| 1 | 9 |
| 2 | 1 2 7 |
| 3 | 8 9 9 |
| 4 | 1 3 4 6 9 |
| 5 | 1 4 6 |
| 6 | 0 2 3 |

Key: 6 | 3 = 63

**40.** $\dfrac{8 + 3 + 2 + 8 + 1 + 4 + 8 + 5 + 6}{9} = \dfrac{45}{9} = 5$

8, 8, 8, 6, 5, 4, 3, 2, 1

Mean: 5

Mode: 8

Median: 5

**41.** $\dfrac{13 + 15 + 17 + 13 + 11 + 13 + 9 + 17 + 4 + 8}{10}$

$= \dfrac{120}{10} = 12$

17, 17, 15, 13, 13, 13, 11, 9, 8, 4

Mean: 12

Mode: 13

Median: 13

**42.** $\dfrac{3.2 + 5.4 + 2.3 + 1.2 + 3.2 + 5.1 + 4.2}{7} = \dfrac{24.6}{7} \approx 3.51$

5.4, 5.1, 4.2, 3.2, 3.2, 2.3, 1.2

Mean: 3.51

Median: 3.2

Mode: 3.2

**43.** $\dfrac{123 + 151 + 121 + 112 + 146 + 112 + 138}{7} = \dfrac{903}{7} = 129$

151, 146, 138, 123, 121, 112, 112

Mean: 129

Median: 123

Mode: 112

**44.** 4, 5, 6, 7, 7, 8, 9

1st Quartile- 5

2nd Quartile- 7

3rd Quartile- 8

# Extra Practice *continued*

**45.** 14, 16, 20, 31, 54, 76, 88

    1st Quartile- 16, 2nd Quartile- 31

    3rd Quartile- 76

**46.** 10, 11, 15, 18, 25, 31, 32, 33, 41, 45

    1st Quartile- 15

    2nd Quartile- 28

    3rd Quartile- 33

**47.** 14, 19, 21, 22, 31, 72, 82, 93

    1st Quartile- 20

    2nd Quartile- 26.5

    3rd Quartile- 77

**48.** 3, 5, 6, 8, 9, 11, 13, 14, 18

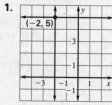

**49.** 2, 3, 3, 5, 7, 7, 8, 9

**50.** 4, 9, 13, 18, 22, 31, 40, 52

**51.** 1, 2, 3, 4, 5, 8, 9, 10, 12, 15, 21, 23, 31

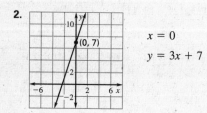

## Chapter 7 *(p. 803)*

**1.**

$y = 5$

$x = -2$

**2.**

$x = 0$

$y = 3x + 7$

**3.**

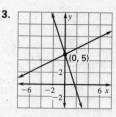

$y = \frac{1}{2}x + 5$

$y = -3x + 5$

**4.** $5x + 3y = 15$

    $4x - 3y = 12$

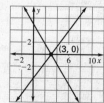

$3y = -5x + 15$

$y = -\frac{5}{3}x + 5$

$-3y = -4x + 12$

$y = \frac{4}{3}x - 4$

**5.**

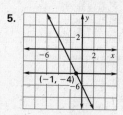

$y = -2x - 6$

$y = -4$

**6.** $x + y = 10$

    $x - y = -2$

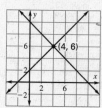

$y = -x + 10$

$y = x + 2$

**7.** $-2x + 4y = 12$

    $5x - 2y = 10$

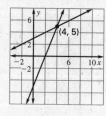

$4y = 2x + 12$

$y = \frac{1}{2}x + 3$

$-2y = -5x + 10$

$y = \frac{5}{2}x - 5$

**8.** $\frac{1}{8}(x + y) = 1$

    $x - y = 4$

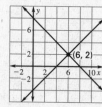

$\frac{1}{8}x + \frac{1}{8}y = 1$

$\frac{1}{8}y = -\frac{1}{8}x + 1$

$y = -x + 8$

$-y = -x + 4$

$y = x - 4$

# Extra Practice *continued*

**9.** $x = 5y$
$2x + 3y = -13$
$2(5y) + 3y = -13$
$10y + 3y = -13$
$13y = -13$
$y = -1$
$x = 5(-1)$
$x = -5$
$(-5, -1)$

**10.** $y = -2x$
$x + y = 7$
$x + (-2x) = 7$
$-x = 7$
$x = -7$
$-7 + y = 7$
$y = 14$
$(-7, 14)$

**11.** $x + y = 9$
$x - y = 3 \rightarrow x = 3 + y$
$(3 + y) + y = 9$
$3 + 2y = 9$      $x + 3 = 9$
$2y = 6$      $x = 6$
$y = 3$      $(6, 3)$

**12.** $2a + 3b = 3$
$a - 6b = -6 \rightarrow a = -6 + 6b$
$2(-6 + 6b) + 3b = 3$      $2a + 3(1) = 3$
$-12 + 12b + 3b = 3$      $2a + 3 = 3$
$-12 + 15b = 3$      $2a = 0$
$15b = 15$      $a = 0$
$b = 1$      $(0, 1)$

**13.** $s - t = 5 \rightarrow s = 5 + t$
$3s + 4t = 16$
$3(5 + t) + 4t = 16$      $s - \frac{1}{7} = 5$
$15 + 3t + 4t = 16$      $s = 5\frac{1}{7}$
$15 + 7t = 16$
$7t = 1$
$t = \frac{1}{7}$      $\left(5\frac{1}{7}, \frac{1}{7}\right)$

**14.** $5x - 8y = -17$
$3x = 5 + y \rightarrow y = 3x - 5$
$5x - 8(3x - 5) = -17$      $5(3) - 8y = -17$
$5x - 24x + 40 = -17$      $15 - 8y = -17$
$-19x + 40 = -17$      $-8y = -32$
$-19x = -57$      $y = 4$
$x = 3$      $(3, 4)$

**15.** $2m + n = 7 \rightarrow n = 7 - 2m$
$4m + 3n = -10$
$4m + 3(7 - 2m) = -10$      $2\left(15\frac{1}{2}\right) + n = 7$
$4m + 21 - 6m = -10$      $31 + n = 7$
$21 - 2m = -10$      $n = -24$
$-2m = -31$
$m = 15\frac{1}{2}$      $\left(15\frac{1}{2}, -24\right)$

**16.** $5a + b = 4 \rightarrow b = 4 - 5a$
$7a + 5b = 11$
$7a + 5(4 - 5a) = 11$      $5\left(\frac{1}{2}\right) + b = 4$
$7a + 20 - 25a = 11$      $\frac{5}{2} + b = 4$
$20 - 18a = 11$      $b = \frac{3}{2}$
$-18a = -9$
$a = \frac{1}{2}$      $\left(\frac{1}{2}, 1\frac{1}{2}\right)$

**17.** $3x + 3y = 6$
$\underline{2x - 3y = 4}$
$5x = 10$
$x = 2$
$3(2) + 3y = 6$
$6 + 3y = 6$
$3y = 0$
$y = 0$
$(2, 0)$

**18.** $3x + 7y = -1$
$7y = -6x$
$3x + 7y = -1$
$\underline{-6x - 7y = 0}$
$-3x = -1$
$x = \frac{1}{3}$
$7y = -6\left(\frac{1}{3}\right)$
$7y = -2$
$y = -\frac{2}{7}$    $\left(\frac{1}{3}, -\frac{2}{7}\right)$

**19.** $\frac{2}{5}x - \frac{1}{2}y = 1$
$\underline{\frac{1}{5}x + \frac{1}{2}y = -1}$
$\frac{3}{5}x = 0$
$x = 0$
$\frac{2}{5}(0) - \frac{1}{2}y = 1$
$0 - \frac{1}{2}y = 1$
$-\frac{1}{2}y = 1$
$y = -2$
$(0, -2)$

**20.** $x + 4y = \frac{9}{2}$      $x + 4\left(\frac{9}{10}\right) = \frac{9}{2}$
$\frac{1}{2}(x - y) = 0 \rightarrow \frac{1}{2}x - \frac{1}{2}y = 0$      $x + \frac{36}{10} = \frac{9}{2}$
$x + 4y = \frac{9}{2}$      $x = \frac{9}{10}$
$\underline{-x + y = 0}$      $\left(\frac{9}{10}, \frac{9}{10}\right)$
$5y = \frac{9}{2}$
$y = \frac{9}{10}$

# Extra Practice *continued*

**21.** $2x + y = 3(x - 5)$

$x + 5 = 4y + 2x$

$-x + y = -15$

$\underline{x + 4y = 5}$

$\quad 5y = -10$

$\quad\quad y = -2$

$\quad x + 5 = 4(-2) + 2x$

$\quad -x + 5 = -8$

$\quad\quad -x = -13$

$\quad\quad x = 13$

$(13, -2)$

**22.** $2x + 3y = 15$

$3y + 5x = 12$

$2x + 3y = 15$

$\underline{-5x - 3y = -12}$

$\quad -3x = 3$

$\quad\quad x = -1$

$2(-1) + 3y = 15$

$\quad -2 + 3y = 15$

$\quad\quad 3y = 17$

$\quad\quad y = 5\frac{2}{3}$

$\left(-1, 5\frac{2}{3}\right)$

**23.** $y = 2x - 36$

$3x - 0.5y = 26$

$0.5(-2x + y = -36)$

$3x - 0.5y = 26$

$-1x + 0.5y = -18$

$\underline{3x - 0.5y = \quad 26}$

$\quad\quad 2x = \quad 8$

$\quad\quad x = \quad 4$

$y = 2(4) - 36$

$y = 8 - 36$

$y = -28$

$(4, -28)$

**24.** $-4x - 15 = 5y$

$2y = 11 - 5x$

$5(-4x - 5y = 15)$

$4(5x + 2y = 11)$

$-20x - 25y = 75$

$\underline{20x + 8y = 44}$

$\quad -17y = 119$

$\quad\quad y = -7$

$-4x - 15 = 5(-7)$

$-4x - 15 = -35$

$\quad -4x = -20$

$\quad\quad x = 5$

$(5, -7)$

**25.** $10x + 15y = 105$

$-10(x + y = 8)$

$10x + 15y = 105$

$\underline{-10x - 10y = -80}$

$\quad\quad 5y = 25$

$\quad\quad y = 5$

$x + 5 = 8$

$\quad x = 3$

$x$ = student tickets

$y$ = adult tickets

$(3, 5)$

3 student, 5 adult

**26.** Company A: $20 + 8s$

Company B: $10 + 10s$

$20 + 8s = 10 + 10s$

$\quad 10 = 2s$

$\quad 5 = s$

$\quad s = 5$ shirts

**27.** Company A

$20 + 8(12) = 20 + 96$

$\quad\quad\quad = 116$

Company B

$10 + 10(12) = 10 + 120$

$\quad\quad\quad = 130$

Company A offers a better price than Company B.

**28.** $x + y = 4 \rightarrow x = 4 - y$

$2x + 3y = 9$

$2(4 - y) + 3y = 9$

$8 - 2y + 3y = 9$

$\quad 8 + y = 9$

$\quad\quad y = 1$

$x + 1 = 4$

$\quad x = 3$

one solution, $(3, 1)$

**29.** $x + y = 6 \rightarrow x = 6 - y$

$3x + 3y = 3$

$3(6 - y) + 3y = 3$

$18 - 3y + 3y = 3$

$\quad\quad 18 \neq 3$

no solution

**30.** $x + 2y = 5 \rightarrow x = 5 - 2y$

$3x - 15 = -6y$

$3(5 - 2y) - 15 = -6y$

$15 - 6y - 15 = -6y$

$\quad -6y = -6y$

$\quad\quad 0 = 0$

infinitely many solutions

**31.** $12x - y = 5$

$\underline{-8x + y = -5}$

$\quad 4x = 0$

$\quad\quad x = 0$

$12(0) - y = 5$

$\quad 0 - y = 5$

$\quad\quad -y = 5$

$\quad\quad y = -5$

one solution, $(0, -5)$

**32.** $y = -3x$

$6y - x = 38$

$6(-3x) - x = 38$

$-18x - x = 38$

$\quad -19x = 38$

$\quad\quad x = -2$

$y = -3(-2)$

$y = 6$

one solution, $(-2, 6)$

**33.** $(2x - 3y = 3)(-3)$

$6x - 9y = 9$

$-6x + 9y = -9$

$\underline{6x - 9y = 9}$

$\quad\quad 0 = 0$

infinitely many solutions

**34.** $3x + 6 = 7y$

$x + 2y = 11 \rightarrow x = 11 - 2y$

$3(11 - 2y) + 6 = 7y$

$33 - 6y + 6 = 7y$

$\quad\quad 39 = 13y$

$\quad\quad 3 = y$

$x + 2(3) = 11$

$\quad x + 6 = 11$

$\quad\quad x = 5$

one solution, $(5, 3)$

**35.** $3x - 8y = 4$

$6x - 42 = 16y$

$-2(3x - 8y = 4)$

$6x - 16y = 42$

$-6x + 16y = -8$

$\underline{6x - 16y = 42}$

$\quad\quad 0 \neq 34$

no solution

**36.** $y \geq 0$

  $x \leq 0$

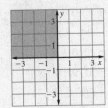

**37.** $y > x + 1$

  $y < x + 3$

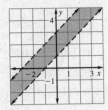

**38.** $x \geq 1$

  $y + x \leq 5$

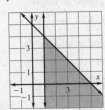

**39.** $y + 2 < -x$

  $2y - 4 > 3x$

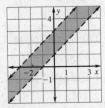

  $y < -x - 2$

  $y > \frac{3}{2}x + 2$

**40.** $x < 5$

  $x \geq 1$

  $y \geq -2$

  $y < 7$

**41.** $y > x - 4$

  $y \geq -x - 1$

  $y \leq 0$

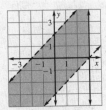

**42.** $y > x - 3$

  $x - y > -2$

  $x \leq 3$

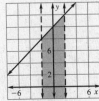

  $y > x - 3$

  $y < x + 2$

  $x \leq 3$

**43.** $3x - 1 < 5$

  $-x + y \leq 10$

  $-5x + 2 < 12$

  $3x < 6$

  $x < 2$

  $y \leq x + 10$

  $-5x < 10$

  $x > -2$

## Chapter 8 (p. 804)

**1.** $(7^2)(7^3) = 7^5$, or 16,807   **2.** $2^3 \cdot 2^4 = 2^7$, or 128

**3.** $(12x)^3 = 1728x^3$           **4.** $-(4x)^2 \cdot (5x^3) = -80x^5$

**5.** $(4r^2s)^2(-2s^2)^3 = 16r^4s^2(-8s^6)$

$\qquad = -128r^4s^8$

**6.** $(7x^3y) \cdot (-2x^4) = -14x^7y$

**7.** $(3x)^3(-5y)^2 = 27x^3(25y^2)$

$\qquad = 675x^3y^2$

**8.** $(-x^3)^2(x)^2(-x^4)^3 = x^6 \cdot x^2 \cdot -x^{12}$

$\qquad = -x^{20}$

**9.** $\dfrac{1}{m^4}$

**10.** $\dfrac{1}{\frac{x^2}{4}} = \dfrac{4}{x^2}$

**11.** $x^2y$

**12.** $6(2a)^3 = 6(8a^3)$

$\qquad = 48a^3$

**13.** $\dfrac{3x^4}{3y^3} = \dfrac{x^4}{y^3}$

**14.** $1 \cdot 2s^2 = 2s^2$

**15.** $6x \cdot x^3 = 6x^4$

**16.** $\dfrac{1}{\left(\frac{4}{2a^4b}\right)^2} = \dfrac{4a^8b^2}{16} = \dfrac{a^8b^2}{4}$

**17.** $y = 3^{-x}$

  $y = \dfrac{1}{3^x}$

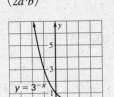

**18.** $y = -2^x$

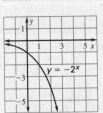

**19.** $y = \frac{1}{4} \cdot 2^x$

**20.** $y = \left(\frac{1}{2}\right)^x$

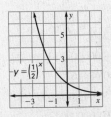

# Extra Practice *continued*

**21.** $\dfrac{2^{11}}{2^8} = 2^{11-8}$

$= 2^3$

$= 8$

**22.** $\dfrac{6^5}{6^7} = 6^{5-7}$

$= 6^{-2}$

$= \dfrac{1}{6^2}$

$= \dfrac{1}{36}$

**23.** $\left(\dfrac{3x^2z^4}{2xz}\right)^3 = \dfrac{27x^6z^{12}}{8x^3z^3}$

$= \dfrac{27x^{6-3}z^{12-3}}{8}$

$= \dfrac{27x^3z^9}{8}$

**24.** $\dfrac{(a^3)^5}{(a^4)^5} = \dfrac{a^{15}}{a^{20}}$

$= a^{15-20}$

$= a^{-5}$

$= \dfrac{1}{a^5}$

**25.** $\dfrac{18b^2c}{4bc^3} \cdot \dfrac{(3ab)^{-2}}{5a^2c^3} = \dfrac{9b^{2-1}}{2c^{3-1}} \cdot \dfrac{1}{9a^2b^25a^2c^3}$

$= \dfrac{9b}{2c^2} \cdot \dfrac{1}{45a^4b^2c^3}$

$= \dfrac{1}{2c^2} \cdot \dfrac{1}{5a^4bc^3}$

$= \dfrac{1}{10a^4bc^5}$

**26.** $\dfrac{(rst)^{-2}}{rs} \cdot \dfrac{(t^2)^3}{(s^{-3})^4} = \dfrac{1}{rsr^2s^2t^2} \cdot \dfrac{t^6}{s^{-12}}$

$= \dfrac{t^6s^{12}}{rsr^2s^2t^2}$

$= \dfrac{t^4s^9}{r^3}$

**27.** $5^0 \cdot \dfrac{(5xy)^2}{(x^3y^{-5})^3} = 1 \cdot \dfrac{25x^2y^2}{x^9y^{-15}}$

$= \dfrac{1 \cdot 25y^{17}}{x^7}$

$= \dfrac{25y^{17}}{x^7}$

**28.** $\left(\dfrac{3}{8}\right)^{-1} \cdot \dfrac{(2a^3x^5)^2}{(8a^{-3}x^{-1})^{-3}} = \dfrac{8}{3} \cdot \dfrac{4a^6x^{10}(8a^{-3}x^{-1})^3}{1}$

$= \dfrac{8 \cdot 4a^6x^{10}(512a^{-9}x^{-3})}{3}$

$= \dfrac{16{,}384a^6x^{10}}{3a^9x^3}$

$= \dfrac{16{,}384x^7}{3a^3}$

**29.** $\dfrac{1}{216} \approx 0.005$

**30.** $0.000004813$

**31.** $31{,}100$

**32.** $0.084162$

**33.** $9.43$

**34.** $50{,}645{,}000{,}000$

**35.** $0.0012468$

**36.** $0.0000000234$

**37.** $60{,}901{,}300{,}000$

**38.** $5.280 \times 10^3$

**39.** $3.78 \times 10^{-2}$

**40.** $1.138 \times 10^1$

**41.** $3.3 \times 10^7$

**42.** $8.2766 \times 10^2$

**43.** $2.08054 \times 10^{-1}$

**44.** $1.6354 \times 10^1$

**45.** $8.91 \times 10^{-4}$

**46.** $A = P(1 + r)^t$

$= 1100(1 + 0.05)^1$

$= \$1155$

**47.** $A = P(1 + r)^t$

$= 1100(1 + 0.05)^{10}$

$= \$1791.78$

**48.** $A = P(1 + r)^t$

$= 1100(1 + 0.05)^{15}$

$= \$2286.82$

**49.** $A = P(1 + r)^t$

$= 1100(1 + 0.05)^{25}$

$= \$3724.99$

**50.** Exponential decay; $0.99$; $1\%$

**51.** Exponential decay; $0.5$; $50\%$

**52.** Exponential growth; $1.12$; $12\%$

**53.** Exponential growth; $1.04$; $4\%$

**54.** Exponential growth; $\frac{7}{6}$; $16\frac{2}{3}\%$

**55.** Exponential decay; $\frac{2}{3}$; $33\frac{1}{3}\%$

**56.** Exponential growth; $1.5$; $50\%$

**57.** Exponential decay; $0.68$; $32\%$

**58.** $y = 120{,}000(1 - 0.10)^6$

$= \$63{,}772.92$

## Chapter 9 *(p. 805)*

**1.** $-10$

**2.** $\pm26$

**3.** $-0.5$

**4.** $\pm13$

**5.** $19.47$

**6.** $-6$

**7.** $14.83$

**8.** $0.1$

**9.** $x^2 = 25$

$x = 5, -5$

**10.** $4x^2 - 8 = 0$

$4x^2 = 8$

$x^2 = 2$

$x = \sqrt{2}, -\sqrt{2}$

**11.** $x^2 = -16$

no solution

**12.** $x^2 + 1 = 1$

$x^2 = 0$

$x = 0$

**13.** $3x^2 - 48 = 0$

$3x^2 = 48$

$x^2 = 16$

$x = 4, -4$

**14.** $6x^2 + 6 = 4$

$6x^2 = -2$

$x^2 = -\frac{1}{3}$

no solution

**15.** $2x^2 - 6 = 0$

$2x^2 = 6$

$x^2 = 3$

$x = \sqrt{3}, -\sqrt{3}$

**16.** $x^2 - 4 = -3$

$x^2 = 1$

$x = 1, -1$

**17.** $0 = -16t^2 + 150$

$t^2 = 9.375$

$t \approx 3.06$ sec.

**18.** $\sqrt{250} = 5\sqrt{10}$

# Extra Practice *continued*

**19.** $6\sqrt{8} \cdot 7\sqrt{2} = 12\sqrt{2} \cdot 7\sqrt{2} = 168$

**20.** $2 \cdot 3\sqrt{2} = 6\sqrt{2}$

**21.** $-2\sqrt{6} \cdot 7\sqrt{30} = -2\sqrt{6} \cdot 7\sqrt{6} \cdot \sqrt{5} = -84\sqrt{5}$

**22.** $\sqrt{\dfrac{11}{16}} = \dfrac{\sqrt{11}}{4}$

**23.** $\dfrac{\sqrt{20}}{\sqrt{5}} = \dfrac{2\sqrt{5}}{\sqrt{5}} = 2$

**24.** $\dfrac{1}{2}\sqrt{\dfrac{8}{50}} = \dfrac{1}{2}\sqrt{\dfrac{4}{25}} = \dfrac{1}{5}$

**25.** $\sqrt{\dfrac{2}{3}} \cdot \sqrt{\dfrac{5}{3}} = \dfrac{\sqrt{10}}{3}$

**26.** $y = 3x^2$

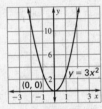

Vertex: $(0, 0)$

**27.** $y = x^2 - 4$

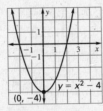

Vertex: $(0, -4)$

**28.** $y = -x^2 - 2x$

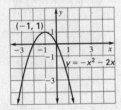

Vertex: $(-1, 1)$

**29.** $y = x^2 - 6x + 8$

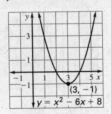

Vertex: $(3, -1)$

**30.** $y = 4x^2 + 4x - 5$

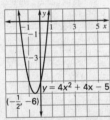

Vertex: $\left(-\dfrac{1}{2}, -6\right)$

**31.** $y = x^2 - 2x + 3$

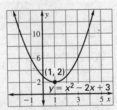

Vertex: $(1, 2)$

**32.** $y = -x^2 + 3x + 2$

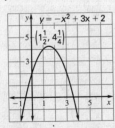

Vertex: $\left(\dfrac{3}{2}, \dfrac{17}{4}\right)$

**33.** $y = -3x^2 + 12x - 1$

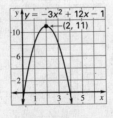

Vertex: $(2, 11)$

**34.** $y = x^2 - 6x + 5$

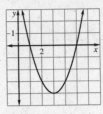

Vertex: $(3, -4)$

Check: $0 = x^2 - 6x + 5$
$0 = (x - 1)(x - 5)$
$x = 1 \quad x = 5$

**35.** $y = x^2 + 5x + 6$

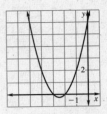

Vertex: $\left(-\dfrac{5}{2}, -\dfrac{1}{4}\right)$

Check: $0 = x^2 + 5x + 6$
$0 = (x + 3)(x + 2)$
$x = -3 \quad x = -2$

**36.** $y = x^2 - 3x - 4$

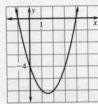

Vertex: $\left(\dfrac{3}{2}, -\dfrac{25}{4}\right)$

Check: $0 = x^2 - 3x - 4$
$0 = (x - 4)(x + 1)$
$x = 4 \quad x = -1$

**37.** $y = \dfrac{1}{4}x^2 - 9$

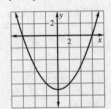

Vertex: $(0, -9)$

Check: $\dfrac{1}{4}x^2 - 9 = 0$
$x^2 - 36 = 0$
$(x - 6)(x + 6) = 0$
$x = 6 \quad x = -6$

**38.** $y = x^2 + 3x - 10$

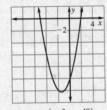

Vertex: $\left(-\dfrac{3}{2}, -\dfrac{49}{4}\right)$

Check: $0 = x^2 + 3x - 10$
$0 = (x + 5)(x - 2)$
$x = -5 \quad x = 2$

**39.** $y = x^2 - 9$

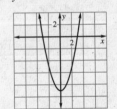

Vertex: $(0, -9)$

Check: $x^2 - 9 = 0$
$(x - 3)(x + 3) = 0$
$x = 3 \quad x = -3$

**40.** $y = \dfrac{1}{2}x^2 + 2x - 6$

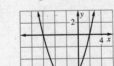

Vertex: $(-2, -8)$

Check: $0 = \dfrac{1}{2}x^2 + 2x - 6$
$0 = x^2 + 4x - 12$
$0 = (x + 6)(x - 2)$
$x = -6 \quad x = 2$

**41.** $y = -2x^2 + 4x + 6$

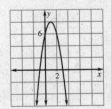

Vertex: $(1, 8)$

Check: $0 = -2x^2 + 4x + 6$
$0 = x^2 - 2x - 3$
$0 = (x + 1)(x - 3)$
$x = -1 \quad x = 3$

# Extra Practice *continued*

**42.** $x^2 + x - 12 = 0$

$$x = \frac{-1 \pm \sqrt{1 - 4(1)(-12)}}{2(1)}$$

$$= \frac{-1 \pm 7}{2}$$

$$= -4, 3$$

**43.** $3r^2 + 8r + 2 = 0$

$$r = \frac{-8 \pm \sqrt{64 - 4(3)(2)}}{2(3)}$$

$$= \frac{-8 \pm 2\sqrt{10}}{6}$$

$$= \frac{-4 \pm \sqrt{10}}{3}$$

**44.** $3k^2 + 11k - 4 = 0$

$$k = \frac{-11 \pm \sqrt{121 - 4(3)(-4)}}{2(3)}$$

$$= \frac{-11 \pm 13}{6}$$

$$= \frac{1}{3}, -4$$

**45.** $-x^2 + 5x - 4 = 0$

$$x = \frac{-5 \pm \sqrt{25 - 4(-1)(-4)}}{2(-1)}$$

$$= \frac{-5 \pm 3}{-2}$$

$$= 1, 4$$

**46.** $m^2 - 2m - 1 = 0$

$$m = \frac{2 \pm \sqrt{4 - 4(1)(-1)}}{2(1)}$$

$$= \frac{2 \pm 2\sqrt{2}}{2}$$

$$= 1 \pm \sqrt{2}$$

**47.** $2x^2 - 6x - 5 = 0$

$$x = \frac{6 \pm \sqrt{36 - 4(2)(-5)}}{2(2)}$$

$$= \frac{6 \pm 2\sqrt{19}}{4}$$

$$= \frac{3 \pm \sqrt{19}}{2}$$

**48.** $b^2 - 7b - 8 = 0$

$$b = \frac{7 \pm \sqrt{49 - 4(1)(-8)}}{2(1)}$$

$$= \frac{7 \pm 9}{2}$$

$$= 8, -1$$

**49.** $-2x^2 + x + 10 = 0$

$$x = \frac{-1 \pm \sqrt{1 - 4(-2)(10)}}{2(-2)}$$

$$= \frac{-1 \pm 9}{-4}$$

$$= -2, \frac{5}{2}$$

**50.** $196 - 4(3)(-5) = 256$

two solutions

**51.** $144 - 4(4)(9) = 0$

one solution

**52.** $100 - 4(1)(9) = 64$

two solutions

**53.** $64 - 4(2)(8) = 0$

one solution

**54.** $5x^2 + 125 = 0$

$0 - 4(5)(125) = -2500$

no solution

**55.** $4 - 4(1)(35) = -136$

no solution

**56.** $1 - 4(2)(-3) = 25$

two solutions

**57.** $25 - 4(-3)(-6) = -47$

no solution

**58.** $y > -x^2 + 4$

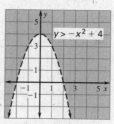

**59.** $y \le 4x^2$

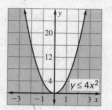

**60.** $y > 5x^2 + 10x$

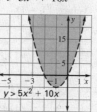

**61.**

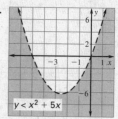

**62.**

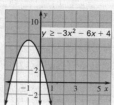

**63.**

**64.**

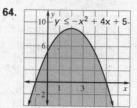

**65.**

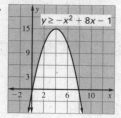

# Extra Practice *continued*

**66.**

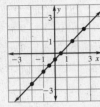

linear

**67.**

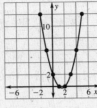

quadratic

## Chapter 10 (p. 806)

**1.** $(7x^2 - 4) + (x^2 + 5) = 7x^2 - 4 + x^2 + 5$
$$= 8x^2 + 1$$

**2.** $(3x^2 - 2) - (2x - 6x^2) = 3x^2 - 2 - 2x + 6x^2$
$$= 9x^2 - 2x - 2$$

**3.** $(8x^2 - 3x + 7) + (6x^2 - 4x + 1)$
$$= 8x^2 - 3x + 7 + 6x^2 - 4x + 1$$
$$= 14x^2 - 7x + 8$$

**4.** $(-z^3 + 3z) + (-z^2 - 4z - 6)$
$$= -z^3 + 3z - z^2 - 4z - 6$$
$$= -z^3 - z^2 - z - 6$$

**5.** $(5x^2 + 7x - 4) - (4x^2 - 2x)$
$$= 5x^2 + 7x - 4 - 4x^2 + 2x$$
$$= x^2 + 9x - 4$$

**6.** $(3a + 2a^4 - 5) - (a^3 + 2a^4 + 5a)$
$$= 3a + 2a^4 - 5 - a^3 - 2a^4 - 5a$$
$$= -a^3 - 2a - 5$$

**7.** $x(4x^2 - 8x + 7) = 4x^3 - 8x^2 + 7x$

**8.** $-3x(x^2 + 5x - 5) = -3x^3 - 15x^2 + 15x$

**9.** $5b^2(3b^3 - 2b^2 + 1) = 15b^5 - 10b^4 + 5b^2$

**10.** $(t + 9)(2t + 1) = 2t^2 + 18t + t + 9 = 2t^2 + 19t + 9$

**11.** $(d - 1)(d + 5) = d^2 - d + 5d - 5 = d^2 + 4d - 5$

**12.** $(3z + 4)(5z - 8) = 15z^2 + 20z - 24z - 32$
$$= 15z^2 - 4z - 32$$

**13.** $(x + 3)(x^2 - 2x + 6)$
$$= x^3 - 2x^2 + 6x + 3x^2 - 6x + 18$$
$$= x^3 + x^2 + 18$$

**14.** $(3 + 2s - s^2)(s - 1)$
$$= 3s + 2s^2 - s^3 - 3 - 2s + s^2$$
$$= -3 + s + 3s^2 - s^3$$

**15.** $(x + 9)^2 = (x + 9)(x + 9)$
$$= x^2 + 9x + 9x + 81$$
$$= x^2 + 18x + 81$$

**16.** $(-c - d)^2 = (-c - d)(-c - d)$
$$= c^2 + cd + cd + d^2$$
$$= c^2 + 2cd + d^2$$

**17.** $(a - 2)(a + 2) = a^2 - 2a + 2a - 4 = a^2 - 4$

**18.** $(-7 + m)(-7 - m) = 49 - 7m + 7m - m^2 = 49 - m^2$

**19.** $(4x + 5)^2 = (4x + 5)(4x + 5)$
$$= 16x^2 + 20x + 20x + 25$$
$$= 16x^2 + 40x + 25$$

**20.** $(5p - 6q)^2 = (5p - 6q)(5p - 6q)$
$$= 25p^2 - 30pq - 30pq + 36q^2$$
$$= 25p^2 - 60pq + 36q^2$$

**21.** $(2a + 3b)(2a - 3b)$
$$= 4a^2 + 6ab - 6ab - 9b^2$$
$$= 4a^2 - 9b^2$$

**22.** $(10x - 5y)(10x + 5y)$
$$= 100x^2 - 50xy + 50xy - 25y^2$$
$$= 100x^2 - 25y^2$$

**23.** $(x + 3)(x + 6) = 0$

$x + 3 = 0 \quad x + 6 = 0$

$x = -3 \quad\quad x = -6$

**24.** $(x - 11)^2 = 0$

$(x - 11)(x - 11) = 0$

$x - 11 = 0$

$x = 11$

**25.** $(z - 1)(z + 5) = 0$

$z - 1 = 0 \quad z + 5 = 0$

$z = 1 \quad\quad z = -5$

**26.** $-8(5w - 2)(w + 4) = 0$

$5w - 2 = 0 \quad w + 4 = 0$

$5w = 2 \quad\quad w = -4$

$w = \frac{2}{5}$

**27.** $(6n - 9)(n - 7) = 0$

$6n - 9 = 0 \quad n - 7 = 0$

$6n = 9 \quad\quad n = 7$

$n = \frac{9}{6}$

$n = \frac{3}{2}$

**28.** $3(x + 2)^2 = 0$

$x + 2 = 0$

$x = -2$

**29.** $(2d - 2)(4d - 8) = 0$

$2d - 2 = 0 \quad 4d - 8 = 0$

$2d = 2 \quad\quad 4d = 8$

$d = 1 \quad\quad d = 2$

**30.** $2(3x - 1)(2x + 5) = 0$

$3x - 1 = 0 \quad 2x + 5 = 0$

$3x = 1 \quad\quad 2x = -5$

$x = \frac{1}{3} \quad\quad x = -\frac{5}{2}$

# Extra Practice *continued*

**31.**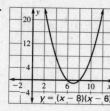

$(7, -1)$

$(x - 8)(x - 6) = 0$

$x - 8 = 0 \qquad x - 6 = 0$

$x = 8 \qquad\qquad x = 6$

$\dfrac{8 + 6}{2} = 7$

$y = (7 - 8)(7 - 6)$

$\quad = -1(1)$

$\quad = -1$

**32.**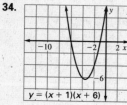

$(0, -16)$

$(x + 4)(x - 4) = 0$

$x + 4 = 0 \quad x - 4 = 0$

$x = -4 \qquad x = 4$

$\dfrac{-4 + 4}{2} = \dfrac{0}{2} = 0$

$y = (0 + 4)(0 - 4)$

$\quad = 4(-4)$

$\quad = -16$

**33.**

$(6, -1)$

$(x - 5)(x - 7) = 0$

$x - 5 = 0 \quad x - 7 = 0$

$x = 5 \qquad x = 7$

$\dfrac{5 + 7}{2} = 6$

$y = (x - 5)(x - 7)$

$\quad = (6 - 5)(6 - 7)$

$\quad = (1)(-1)$

$\quad = -1$

**34.**

$\left(-3\dfrac{1}{2}, -6\dfrac{1}{4}\right)$

$(x + 1)(x + 6) = 0$

$x + 1 = 0 \quad x + 6 = 0$

$x = -1 \qquad x = -6$

$\dfrac{-1 + (-6)}{2} = \dfrac{-7}{2}$

$\qquad\qquad = -3\dfrac{1}{2}$

$y = \left(-3\dfrac{1}{2} + 1\right)\left(-3\dfrac{1}{2} + 6\right)$

$\quad = \left(-2\dfrac{1}{2}\right)\left(2\dfrac{1}{2}\right)$

$\quad = -6\dfrac{1}{4}$

**35.**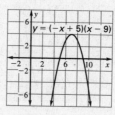

$(7, 4)$

$(-x + 5)(x - 9) = 0$

$-x + 5 = 0 \quad x - 9 = 0$

$-x = -5 \qquad x = 9$

$x = 5$

$\dfrac{5 + 9}{2} = 7$

$y = (-7 + 5)(7 - 9)$

$\quad = (-2)(-2)$

$\quad = 4$

**36.**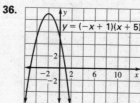

$(-2, 9)$

$(-x + 1)(x + 5) = 0$

$-x + 1 = 0 \quad x + 5 = 0$

$-x = -1 \qquad x = -5$

$x = 1$

$\dfrac{1 + (-5)}{2} = -2$

$y = (2 + 1)(-2 + 5)$

$\quad = 3(3)$

$\quad = 9$

**37.**

$(1, -4)$

$(x - 3)(x + 1) = 0$

$x - 3 = 0 \quad x + 1 = 0$

$x = 3 \qquad x = -1$

$\dfrac{3 + (-1)}{2} = 1$

$y = (1 - 3)(1 + 1)$

$\quad = (-2)(2)$

$\quad = -4$

**38.**

$(-5, 4)$

$(-x - 3)(x + 7) = 0$

$-x - 3 = 0 \quad x + 7 = 0$

$-x = 3 \qquad x = -7$

$x = -3$

$\dfrac{-3 + (-7)}{2} = -5$

$y = (5 - 3)(-5 + 7)$

$\quad = 2(2)$

$\quad = 4$

**39.** $x^2 + 6x + 9 = 0$

$(x + 3)(x + 3) = 0$

$x + 3 = 0$

$x = -3$

**40.** $x^2 + 2x - 35 = 0$

$(x + 7)(x - 5) = 0$

$x + 7 = 0 \quad x - 5 = 0$

$x = -7 \qquad x = 5$

**41.** $x^2 - 12x + 36 = 0$

$(x - 6)(x - 6) = 0$

$x - 6 = 0$

$x = 6$

**42.** $-x^2 - 4x - 3 = 0$

$0 = x^2 + 4x + 3$

$0 = (x + 3)(x + 1)$

$x + 3 = 0 \quad x + 1 = 0$

$x = -3 \qquad x = -1$

# Extra Practice *continued*

**43.** $x^2 - 15x + 54 = 0$
$(x - 6)(x - 9) = 0$
$x - 6 = 0 \quad x - 9 = 0$
$x = 6 \qquad x = 9$

**44.** $-x^2 + 14x - 48 = 0$
$0 = x^2 - 14x + 48$
$0 = (x - 6)(x - 8)$
$x - 6 = 0 \quad x - 8 = 0$
$x = 6 \qquad x = 8$

**45.** $x^2 - 2x - 24 = 0$
$(x - 6)(x + 4) = 0$
$x - 6 = 0 \quad x + 4 = 0$
$x = 6 \qquad x = -4$

**46.** $x^2 - 5x + 4 = 0$
$(x - 4)(x - 1) = 0$
$x - 4 = 0 \quad x - 1 = 0$
$x = 4 \qquad x = 1$

**47.** $2x^2 + x - 6 = 0$
$(2x - 3)(x + 2) = 0$
$2x - 3 = 0 \quad x + 2 = 0$
$2x = 3 \qquad x = -2$
$x = \frac{3}{2}$

**48.** $2x^2 + 7x + 3 = 0$
$(2x + 1)(x + 3) = 0$
$2x + 1 = 0 \quad x + 3 = 0$
$2x = -1 \qquad x = -3$
$x = -\frac{1}{2}$

**49.** $9x^2 + 24x + 16 = 0$
$(3x + 4)(3x + 4) = 0$
$3x + 4 = 0$
$3x = -4$
$x = -\frac{4}{3}$

**50.** $20x^2 + 23x + 6 = 0$
$(4x + 3)(5x + 2) = 0$
$4x + 3 = 0 \quad 5x + 2 = 0$
$4x = -3 \qquad 5x = -2$
$x = -\frac{3}{4} \qquad x = -\frac{2}{5}$

**51.** $4x^2 - 5x - 6 = 0$
$(4x + 3)(x - 2) = 0$
$4x + 3 = 0 \quad x - 2 = 0$
$4x = -3 \qquad x = 2$
$x = -\frac{3}{4}$

**52.** $3x^2 + 14x - 5 = 0$
$(3x - 1)(x + 5) = 0$
$3x - 1 = 0 \quad x + 5 = 0$
$3x = 1 \qquad x = -5$
$x = \frac{1}{3}$

**53.** $3x^2 - 17x - 56 = 0$
$(3x + 7)(x - 8) = 0$
$3x + 7 = 0 \quad x - 8 = 0$
$3x + 7 = 0 \quad x = 8$
$3x = -7$
$x = -\frac{7}{3}$

**54.** $12x^2 + 46x - 36 = 0$
$(4x + 18)(3x - 2) = 0$
$4x + 18 = 0 \quad 3x - 2 = 0$
$4x = -18 \qquad 3x = 2$
$x = -\frac{18}{4} \qquad x = \frac{2}{3}$

**55.** $x^2 - 1 = (x + 1)(x - 1)$

Difference of two squares

**56.** $9b^2 - 81 = (3b + 9)(3b - 9)$

Difference of two squares

**57.** $121 - x^2 = (11 - x)(11 + x)$

Difference of two squares

**58.** $12 - 27x^2 = 3(4 - 9x^2) = 3(2 + 3x)(2 - 3x)$

Difference of two squares

**59.** $t^2 + 2t + 1 = (t + 1)^2$

Perfect square trinomial

**60.** $x^2 + 20x + 100 = (x + 10)^2$

Perfect square trinomial

**61.** $64y^2 + 48y + 9 = (8y + 3)^2$

Perfect square trinomial

**62.** $20x^2 - 100x + 125 = 5(4x^2 - 20x + 25) = 5(2x - 5)^2$

Perfect square trinomial

**63.** $x^4 - 9x^2 = x^2(x^2 - 9) = x^2(x - 3)(x + 3)$

**64.** $m^3 + 11m^2 + 28m = m(m^2 + 11m + 28)$
$\qquad = m(m + 7)(m + 4)$

**65.** $x^4 + 4x^3 - 45x^2 = x^2(x^2 + 4x - 45)$
$\qquad = x^2(x + 9)(x - 5)$

**66.** $x^3 + 2x^2 - 4x - 8 = (x^3 + 2x^2) - (4x + 8)$
$\qquad = x^2(x + 2) - 4(x + 2)$
$\qquad = (x^2 - 4)(x + 2)$
$\qquad = (x - 2)(x + 2)^2$

**67.** $-3y^3 - 15y^2 - 12y = -3y(y^2 + 5y + 4)$
$\qquad = -3y(y + 1)(y + 4)$

**68.** $x^3 - x^2 + 4x - 4 = (x^3 - x^2) + (4x - 4)$
$\qquad = x^2(x - 1) + 4(x - 1)$
$\qquad = (x^2 + 4)(x - 1)$

**69.** $7x^6 - 21x^4 = 7x^4(x^2 - 3)$

**70.** $8t^3 - 3t^2 + 16t - 6 = (8t^3 - 3t^2) + (16t - 6)$
$\qquad = t^2(8t - 3) + 2(8t - 3)$
$\qquad = (t^2 + 2)(8t - 3)$

**71.**

Length = 6 in.
Width = 2 in.
Height = 15in.

$$\ell(\ell - 4)(\ell + 9) = 180$$
$$(\ell^2 - 4\ell)(\ell + 9) = 180$$
$$\ell^3 + 9\ell^2 - 4\ell^2 - 36\ell - 180 = 0$$
$$(\ell^3 + 5\ell^2) - (36\ell + 180) = 0$$
$$\ell^2(\ell + 5) - 36(\ell + 5) = 0$$
$$(\ell^2 - 36)(\ell + 5) = 0$$
$$(\ell + 6)(\ell - 6)(\ell + 5) = 0$$
$$\ell = 6$$

## Chapter 11 *(p. 807)*

**1.** $\dfrac{x}{2} = \dfrac{8}{x}$
$x^2 = 16$
$x = 4, -4$

**2.** $\dfrac{9}{m} = \dfrac{15}{10}$
$90 = 15m$
$6 = m$

# Extra Practice *continued*

**3.** $\dfrac{c^2 - 16}{c + 4} = \dfrac{c - 4}{3}$

$3(c^2 - 16) = (c^2 - 16)$

$3c^2 - 48 = c^2 - 16$

$2c^2 = 32$

$c^2 = 16$

$c = 4$

**4.** $\dfrac{3}{5} = \dfrac{x + 2}{6}$

$18 = 5x + 10$

$8 = 5x$

$\dfrac{8}{5} = x$

**5.** $\dfrac{14}{2} = \dfrac{7}{n - 5}$

$14n - 70 = 14$

$14n = 84$

$n = 6$

**6.** $\dfrac{12}{8} = \dfrac{5 + t}{t - 3}$

$12t - 36 = 40 + 8t$

$4t = 76$

$t = 19$

**7.** $\dfrac{x + 15}{16} = \dfrac{-9}{x - 10}$

$x^2 + 5x - 150 = -144$

$x^2 + 5x - 6 = 0$

$(x + 6)(x - 1) = 0$

$x = -6 \quad x = 1$

**8.** $\dfrac{x + 30}{-4} = \dfrac{143}{x - 18}$

$x^2 + 12x - 540 = -572$

$x^2 + 12x + 32 = 0$

$(x + 4)(x + 8) = 0$

$x = -4 \quad x = -8$

**9.** $x = 0.40(60)$

$x = 24$

**10.** $75 = x(250)$

$x = 0.3$ or $30\%$

**11.** $x = 0.04(525)$

$x = 21$ lb

**12.** $8 = 0.20x$

$40 = x$

**13.** $0.35(720) = x$

$252 = x$

**14.** $0.12x = \$1482$

$x = \$12,350$

**15.** inverse variation

$xy = 2$

**16.** direct variation

$3 = 1k \quad y = 3x$

$3 = k$

**17.** $\dfrac{12x^4}{42x} = \dfrac{2x^3}{7}$

**18.** $\dfrac{5x^2 - 15x^3}{10x} = \dfrac{5x^2(1 - 3x)}{10x}$

$= \dfrac{x(1 - 3x)}{2}$

$= \dfrac{x - 3x^2}{2}$

**19.** $\dfrac{x + 6}{(x^2 + 7x + 6)} = \dfrac{x + 6}{(x + 6)(x + 1)}$

$= \dfrac{1}{x + 1}$

**20.** $\dfrac{x^2 - 8x + 15}{x - 3} = \dfrac{(x - 3)(x - 5)}{x - 3}$

$= x - 5$

**21.** $\dfrac{x^2}{4x^2} = \dfrac{1}{4}$

**22.** $\dfrac{\pi\left(\dfrac{x}{2}\right)^2}{4x^2} = \dfrac{\dfrac{\pi x^2}{4}}{4x^2}$

$= \dfrac{\pi}{16}$

**23.** $\dfrac{3x}{5} \cdot \dfrac{15}{18x} = \dfrac{1}{2}$

**24.** $\dfrac{4x^2}{7} \cdot \dfrac{14}{8x} = x$

**25.** $\dfrac{2x^2}{3x} \cdot \dfrac{6x^3}{20x^2} = \dfrac{x^2}{5}$

**26.** $\dfrac{x - 3}{5(x + 4)} \cdot \dfrac{4(x + 4)}{x - 3} = \dfrac{4}{5}$

**27.** $\dfrac{1}{4x} \cdot \dfrac{15}{6x} = \dfrac{5}{8x^2}$

**28.** $\dfrac{10x^2}{x^2 - 25} \cdot (x - 5) = \dfrac{10x^2}{(x + 5)(x - 5)} \cdot (x - 5)$

$= \dfrac{10x^2}{x + 5}$

**29.** $\dfrac{5x}{x - 3} \cdot \dfrac{x - 3}{x - 8} = \dfrac{5x}{x - 8}$

**30.** $\dfrac{x^2 + 5x - 36}{x^2 - 81} \cdot \dfrac{1}{x^2 - 16}$

$= \dfrac{(x + 9)(x - 4)}{(x + 9)(x - 9)} \cdot \dfrac{1}{(x - 4)(x + 4)}$

$= \dfrac{1}{(x - 9)(x + 4)}$

**31.** $\dfrac{20}{5x} - \dfrac{15}{5x} = \dfrac{5}{5x} = \dfrac{1}{x}$

**32.** $\dfrac{4}{x + 2} + \dfrac{8}{x + 2} = \dfrac{12}{x + 2}$

**33.** $\dfrac{3}{x + 7} - \dfrac{7}{x - 4} = \dfrac{3x - 12 - 7x - 49}{(x + 7)(x - 4)}$

$= \dfrac{-4x - 61}{(x + 7)(x - 4)}$

**34.** $\dfrac{5x + 3}{x^2 - 25} + \dfrac{5}{x - 5} = \dfrac{5x + 3 + 5x + 25}{(x + 5)(x - 5)}$

$= \dfrac{10x + 28}{(x + 5)(x - 5)}$

**35.** $\dfrac{7}{2x} - \dfrac{1}{9x^2} = \dfrac{63x - 2}{18x^2}$

**36.** $\dfrac{3x}{6x^2 + 13x + 2} + \dfrac{x + 1}{x^2 + 5x + 6}$

$= \dfrac{3x}{(6x + 1)(x + 2)} + \dfrac{x + 1}{(x + 2)(x + 3)}$

$= \dfrac{3x(x + 3) + (x + 1)(6x + 1)}{(6x + 1)(x + 2)(x + 3)}$

$= \dfrac{3x^2 + 9x + 6x^2 + 7x + 1}{(6x + 1)(x + 2)(x + 3)}$

$= \dfrac{9x^2 + 16x + 1}{(6x + 1)(x + 2)(x + 3)}$

# Extra Practice *continued*

**37.** $\dfrac{14x^3 + 7x^2 + x}{7x} = 2x^2 + x + \dfrac{1}{7}$

**38.** $\dfrac{x^2 + 6x - 16}{x + 8} = \dfrac{(x + 8)(x - 2)}{x + 8} = x - 2$

**39.** $\dfrac{12x^2 - 2x - 2}{3x + 1} = \dfrac{(3x + 1)(4x - 2)}{3x + 1}$

$$= 4x - 2$$
$$= 2(2x - 1)$$

**40.**

$$2x + 1 \overline{)\,2x^3 + 9x^2 - 6x + 2} \quad x^2 + 4x - 5 + \dfrac{7}{2x + 1}$$

$$\underline{2x^3 + \phantom{0}x^2}$$
$$8x^2 - 6x$$
$$\underline{8x^2 + 4x}$$
$$-10x + 2$$
$$\underline{-10x - 5}$$
$$7$$

**41.**
$$\dfrac{4}{x - 6} = \dfrac{x}{10}$$
$$40 = x^2 - 6x$$
$$x^2 - 6x - 40 = 0$$
$$(x - 10)(x + 4) = 0 \quad x = 10, \; x = -4$$

**42.**
$$\dfrac{-2}{3x} = \dfrac{4 + x}{6}$$
$$-12 = 12x + 3x^2$$
$$3x^2 + 12x + 12 = 0$$
$$x^2 + 4x + 4 = 0$$
$$(x + 2)(x + 2) = 0$$
$$x = -2$$

**43.**
$$\dfrac{4}{x} + \dfrac{2}{3} = \dfrac{6}{x}$$
$$12 + 2x = 18$$
$$2x - 6 = 0$$
$$2x = 6$$
$$x = 3$$

**44.**
$$\dfrac{x}{x - 5} - \dfrac{11}{x - 5} = 7$$
$$x - 11 = 7x - 35$$
$$24 = 6x$$
$$4 = x$$

**45.**
$$\dfrac{1}{x - 3} = \dfrac{5}{x + 9}$$
$$x + 9 = 5x - 15$$
$$24 = 4x$$
$$6 = x$$

**46.**
$$\dfrac{5}{x - 1} + 1 = \dfrac{4}{x^2 + 3x - 4}$$
$$5(x + 4) + 1(x - 1)(x + 4) = 4$$
$$5x + 20 + x^2 + 3x - 4 = 4$$
$$x^2 + 8x + 12 = 0$$
$$(x + 2)(x + 6) = 0$$
$$x = -2 \quad x = -6$$

## Chapter 12 *(p. 808)*

**1.** Domain: all nonnegative numbers

Range: all nonnegative numbers

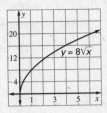

**2.** Domain: all nonnegative numbers

Range: all numbers greater than or equal to $-5$

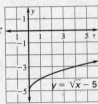

**3.** Domain: all numbers greater than or equal to $-3$

Range: all nonnegative numbers

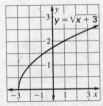

**4.** Domain: all nonnegative numbers

Range: all nonnegative numbers

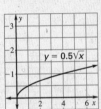

**5.** Domain: all numbers greater than or equal to 2

Range: all nonnegative numbers

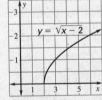

**6.** Domain: all nonnegative numbers

Range: all numbers greater than or equal to 1

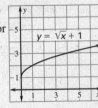

**7.** Domain: all numbers greater than or equal to $-\frac{2}{3}$

Range: all nonnegative numbers

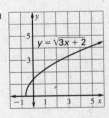

# Extra Practice *continued*

8. Domain: all numbers greater than or equal to $\frac{3}{4}$

   Range: all nonnegative numbers

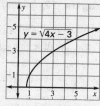

$y = \sqrt{4x - 3}$

9. $3\sqrt{5} + 2\sqrt{5} = 5\sqrt{5}$   10. $8\sqrt{7} - 15\sqrt{7} = -7\sqrt{7}$

11. $2\sqrt{8} + 3\sqrt{32} = 2\sqrt{4}\sqrt{2} + 3\sqrt{16}\sqrt{2}$
    $= 2(2)\sqrt{2} + 3(4)\sqrt{2}$
    $= 4\sqrt{2} + 12\sqrt{2}$
    $= 16\sqrt{2}$

12. $\sqrt{20} - \sqrt{45} + \sqrt{80} = \sqrt{4}\sqrt{5} - \sqrt{9}\sqrt{5} + \sqrt{16}\sqrt{5}$
    $= \left(\sqrt{4} - \sqrt{9} + \sqrt{16}\right)\sqrt{5}$
    $= (2 - 3 + 4)\sqrt{5}$
    $= 3\sqrt{5}$

13. $\sqrt{3}\left(7 - \sqrt{6}\right) = 7\sqrt{3} - \sqrt{18}$
    $= 7\sqrt{3} - \sqrt{9}\sqrt{2}$
    $= 7\sqrt{3} - 3\sqrt{2}$

14. $(4 + \sqrt{10})^2 = \left(4 + \sqrt{10}\right)\left(4 + \sqrt{10}\right)$
    $= 16 + 4\sqrt{10} + 4\sqrt{10} + 10$
    $= 26 + 8\sqrt{10}$

15. $\dfrac{4}{\sqrt{24}} \cdot \dfrac{\sqrt{24}}{\sqrt{24}} = \dfrac{4\sqrt{24}}{24} = \dfrac{\sqrt{6}}{3}$

16. $\dfrac{3}{5 - \sqrt{2}} \cdot \dfrac{5 + \sqrt{2}}{5 + \sqrt{2}} = \dfrac{15 + 3\sqrt{2}}{25 - 2} = \dfrac{15 + 3\sqrt{2}}{23}$

17. $\sqrt{x} - 11 = 0$
    $\sqrt{x} = 11$
    $x = 121$

18. $\sqrt{2x - 1} + 4 = 7$
    $\sqrt{2x - 1} = 3$
    $2x - 1 = 9$
    $2x = 10$
    $x = 5$

19. $\sqrt{x + 20} = x$
    $x + 20 = x^2$
    $x^2 - x - 20 = 0$
    $(x - 5)(x + 4) = 0$
    $x = 5$

20. $12 = \sqrt{3x + 1} + 7$
    $5 = \sqrt{3x + 1}$
    $25 = 3x + 1$
    $24 = 3x$
    $8 = x$

21. $\frac{1}{2}x = \sqrt{2x - 3}$
    $\frac{1}{4}x^2 = 2x - 3$
    $\frac{1}{4}x^2 - 2x + 3 = 0$
    $x^2 - 8x + 12 = 0$
    $(x - 6)(x - 2) = 0$
    $x = 6 \quad x = 2$

22. $\sqrt{18 - 2x} + 5 = x$
    $\sqrt{18 - 2x} = x - 5$
    $18 - 2x = x^2 - 10x + 25$
    $x^2 - 8x + 7 = 0$
    $(x - 7)(x - 1) = 0$
    $x = 7$

23. $x^2 + 10x + 25 = 56 + 25$
    $x^2 + 10x + 25 = 81$
    $(x + 5)^2 = 81$
    $x + 5 = \pm 9$
    $x = 4 \quad x = -14$

24. $x^2 - 6x + 9 = 16 + 9$
    $(x - 3)^2 = 25$
    $x - 3 = \pm 5$
    $x = 8 \quad x = -2$

25. $x^2 - 7x + \frac{49}{4} = -10 + \frac{49}{4}$
    $\left(x - \frac{7}{2}\right)^2 = \frac{9}{4}$
    $x - \frac{7}{2} = \pm\frac{3}{2}$
    $x = \frac{10}{2}, x = \frac{4}{2}$
    $x = 5, x = 2$

26. $x^2 + x + \frac{1}{4} = 3 + \frac{1}{4}$
    $\left(x + \frac{1}{2}\right)^2 = 3\frac{1}{4}$
    $x + \frac{1}{2} = \pm\sqrt{\frac{13}{4}}$
    $x + \frac{1}{2} = \pm\frac{\sqrt{13}}{2}$
    $x = \dfrac{\sqrt{13} - 1}{2}, \dfrac{-\sqrt{13} - 1}{2}$

27. $5x^2 - 12x - 15 = 0$
    $x^2 - \frac{12}{5}x - 3 = 0$
    $x^2 - \frac{12}{5}x + \frac{144}{100} = 3 + \frac{144}{100}$
    $\left(x - \frac{12}{10}\right)^2 = 3\frac{144}{100}$
    $x - \frac{6}{5} = \pm\sqrt{\frac{444}{100}}$
    $x - \frac{6}{5} = \pm\frac{2\sqrt{111}}{10}$
    $x = \dfrac{\sqrt{111} + 6}{5}, \dfrac{-\sqrt{111} + 6}{5}$

# Extra Practice *continued*

**28.** $3x^2 - 10x - 29 = 0$

$$x^2 - \frac{10}{3}x - \frac{29}{3} = 0$$

$$x^2 - \frac{10}{3}x + \frac{100}{36} = \frac{29}{3} + \frac{100}{36}$$

$$\left(x - \frac{10}{6}\right)^2 = \frac{448}{36}$$

$$x - \frac{10}{6} = \pm\sqrt{\frac{448}{36}}$$

$$x = \frac{10}{6} \pm \frac{8\sqrt{7}}{6}$$

$$x = \frac{10 \pm 8\sqrt{7}}{6}$$

$$x = \frac{5 + 4\sqrt{7}}{3}$$

$$x = \frac{5 + 4\sqrt{7}}{3}, \frac{5 - 4\sqrt{7}}{3}$$

**29.** $1^2 + 1^2 = c^2$

$1 + 1 = c^2$

$2 = c^2$

$\sqrt{2} = c$

**30.** $1^2 + b^2 = 2^2$

$1 + b^2 = 4$

$b^2 = 3$

$b = \sqrt{3}$

**31.** $a^2 + 6^2 = 10^2$

$a^2 + 36 = 100$

$a^2 = 64$

$a = 8$

**32.** $7^2 + 10^2 = c^2$

$49 + 100 = c^2$

$149 = c^2$

$c = \sqrt{149}$

**33.** $a^2 + 15^2 = 25^2$

$a^2 + 225 = 625$

$a^2 = 400$

$a = 20$

**34.** $30^2 + b^2 = 50^2$

$900 + b^2 = 2500$

$b^2 = 1600$

$b = 40$

**35.** $d = \sqrt{(5-4)^2 + (4-0)^2}$    $\left(\frac{(0+4)}{2}, \frac{(4+5)}{2}\right)$

$= \sqrt{1^2 + 4^2}$    $\left(\frac{4}{2}, \frac{9}{2}\right)$

$= \sqrt{1 + 16}$ 

$= \sqrt{17}$    $\left(2, 4\frac{1}{2}\right)$

$\approx 4.12$

**36.** $d = \sqrt{(-1-3)^2 + (6+3)^2}$    $\left(\frac{(-3+6)}{2}, \frac{[3+(-1)]}{2}\right)$

$= \sqrt{(-4)^2 + 9^2}$    $\left(\frac{3}{2}, \frac{2}{2}\right)$

$= \sqrt{16 + 81}$

$= \sqrt{97}$    $\left(1\frac{1}{2}, 1\right)$

$\approx 9.85$

**37.** $d = \sqrt{(-4-0)^2 + (4-1)^2}$    $\left(\frac{(1+4)}{2}, \frac{[0+(-4)]}{2}\right)$

$= \sqrt{(-4)^2 + 3^2}$

$= \sqrt{16 + 9}$    $\left(\frac{5}{2}, -\frac{4}{2}\right)$

$= \sqrt{25}$

$= 5$    $\left(2\frac{1}{2}, -2\right)$

**38.** $d = \sqrt{(-2-0)^2 + (3-0)^2}$    $\left(\frac{0+3}{2}, \frac{0+-2}{2}\right)$

$= \sqrt{(-2)^2 + 3^2}$

$= \sqrt{4 + 9}$    $\left(\frac{3}{2}, -\frac{2}{2}\right)$

$= \sqrt{13}$

$\approx 3.61$    $\left(1\frac{1}{2}, -1\right)$

**39.** $d = \sqrt{(-6+6)^2 + (-1-7)^2}$    $\left(\frac{[7+(-1)]}{2}, \frac{(-6+-6)}{2}\right)$

$= \sqrt{0^2 + (-8)^2}$

$= \sqrt{0 + 64}$    $\left(\frac{6}{2}, -\frac{12}{2}\right)$

$= \sqrt{64}$

$= 8$    $(3, -6)$

**40.** $d = \sqrt{(-4-2)^2 + (5-5)^2}$    $\left(\frac{5+5}{2}, \frac{2+(-4)}{2}\right)$

$= \sqrt{(-6)^2 + 0^2}$

$= \sqrt{36 + 0}$    $\left(\frac{10}{2}, -\frac{2}{2}\right)$

$= \sqrt{36}$

$= 6$    $(5, -1)$

**41.** $d = \sqrt{(2+7)^2 + (-4-12)^2}$    $\left(\frac{[12+(-4)]}{2}, \frac{(-7+2)}{2}\right)$

$= \sqrt{9^2 + (-16)^2}$

$= \sqrt{81 + 256}$    $\left(\frac{8}{2}, -\frac{5}{2}\right)$

$= \sqrt{337}$

$\approx 18.36$    $\left(4, -2\frac{1}{2}\right)$

**42.** $d = \sqrt{(-9+5)^2 + (-8+4)^2}$

$= \sqrt{(-4)^2 + (-4)^2}$

$= \sqrt{16 + 16}$

$= \sqrt{32}$

$\approx 5.66$

$\left(\frac{[-4+(-8)]}{2}, \frac{[-5+(-9)]}{2}\right)$

$\left(-\frac{12}{2}, -\frac{14}{2}\right)$

$(-6, -7)$

**43.** $\sin D = \frac{8}{10} = 0.8$      $\sin E = \frac{6}{10} = 0.6$

$\cos D = \frac{6}{10} = 0.6$      $\cos E = \frac{8}{10} = 0.8$

$\tan D = \frac{8}{6} = 1.\overline{3}$      $\tan E = \frac{6}{8} = 0.75$

# Extra Practice *continued*

**44.** $\sin D = \frac{15}{17} \approx 0.88$  $\quad \sin E = \frac{8}{17} \approx 0.47$

$\quad\cos D = \frac{8}{17} \approx 0.47$  $\quad \cos E = \frac{15}{17} \approx 0.88$

$\quad\tan D = \frac{15}{8} = 1.875$  $\quad \tan E = \frac{8}{15} = 0.5\overline{3}$

**45.** $\sin D = \frac{3}{3\sqrt{2}} = \frac{\sqrt{2}}{2} \approx 0.707$

$\quad\sin E = \frac{3}{3\sqrt{2}} = \frac{\sqrt{2}}{2} \approx 0.707$

$\quad\cos D = \frac{3}{3\sqrt{2}} = \frac{\sqrt{2}}{2} \approx 0.707$

$\quad\cos E = \frac{3}{3\sqrt{2}} = \frac{\sqrt{2}}{2} \approx 0.707$

$\quad\tan D = \frac{3}{3} = 1$  $\qquad \tan E = \frac{3}{3} = 1$

**46.** $\sin 45° = \frac{7}{t}$  $\qquad \cos 45° = \frac{r}{9.9}$

$\quad 0.707 = \frac{7}{t}$  $\qquad 0.707 = \frac{r}{9.9}$

$\qquad 9.9 = t$  $\qquad\qquad 7 = r$

**47.** $\sin 20° = \frac{b}{27}$  $\qquad \cos 20° = \frac{a}{27}$

$\quad 0.3420 = \frac{b}{27}$  $\qquad 0.9397 = \frac{a}{27}$

$\qquad 9.23 = b$  $\qquad\qquad 25.37 = a$

**48.** $\cos 30° = \frac{10}{w}$  $\qquad \tan 30° = \frac{v}{10}$

$\quad 0.866 = \frac{10}{w}$  $\qquad 0.5774 = \frac{v}{10}$

$\qquad 11.55 = w$  $\qquad\qquad v = 5.77$

**49.** Any nonzero numbers for $b$ and $c$ and any nonzero number for $a$ except 1

**Algebra 1**
Extra Practice  Worked-out Solution Key